EDEXCEL
GCSE MATHS

HIGHER

Marguerite Appleton, David Bowles, Dave Capewell, Geoff Fowler, Derek Huby, Jayne Kranat, Steve Lomax, Peter Mullarkey, James Nicholson, Matt Nixon and Katherine Pate

Powered by  **MyMaths**.co.uk

OXFORD
UNIVERSITY PRESS

OXFORD
UNIVERSITY PRESS

Great Clarendon Street, Oxford, OX2 6DP, United Kingdom

Oxford University Press is a department of the University of Oxford. It furthers the University's objective of excellence in research, scholarship, and education by publishing worldwide. Oxford is a registered trade mark of Oxford University Press in the UK and in certain other countries

British Library Cataloguing in Publication Data
Data available

978-0-19-835151-1

10 9 8 7 6

Paper used in the production of this book is a natural, recyclable product made from wood grown in sustainable forests. The manufacturing process conforms to the environmental regulations of the country of origin.

Printed in China by Golden Cup

Acknowledgements

The publisher would like to thank David Bowles, John Guilfoyle and Katie Wood for their contributions to this book.

Although we have made every effort to trace and contact all copyright holders before publication this has not been possible in all cases. If notified, the publisher will rectify any errors or omissions at the earliest opportunity.

Approval message from Edexcel

In order to ensure that this resource offers high-quality support for the associated Pearson qualification, it has been through a review process by the awarding body. This process confirms that; this resource fully covers the teaching and learning content of the specification or part of a specification at which it is aimed. It also confirms that it demonstrates an appropriate balance between the development of subject skills, knowledge and understanding, in addition to preparation for assessment.

Endorsement does not cover any guidance on assessment activities or processes (e.g. practice questions or advice on how to answer assessment questions), included in the resource nor does it prescribe any particular approach to the teaching or delivery of a related course. While the publishers have made every attempt to ensure that advice on the qualification and its assessment is accurate, the official specification and associated assessment guidance materials are the only authoritative source of information and should always be referred to for definitive guidance.

Pearson examiners have not contributed to any sections in this resource relevant to examination papers for which they have responsibility. Examiners will not use endorsed resources as a source of material for any assessment set by Pearson.

Endorsement of a resource does not mean that the resource is required to achieve this Pearson qualification, nor does it mean that it is the only suitable material available to support the qualification, and any resource lists produced by the awarding body shall include this and other appropriate resources.

Contents

1   Calculations 1
Introduction 2
Place value and rounding 4
Adding and subtracting 8
Multiplying and dividing 12
Summary and review 16
Assessment 1 18

2   Expressions
Introduction 20
Simplifying expressions 22
Indices 26
Expanding and factorising 1 30
Algebraic fractions 34
Summary and review 38
Assessment 2 40

3   Angles and polygons
Introduction 42
Angles and lines 44
Triangles and quadrilaterals 48
Congruence and similarity 52
Angles in polygons 56
Summary and review 60
Assessment 3 62

4   Handling data 1
Introduction 64
Sampling 66
Organising data 70
Representing data 1 74
Averages and spread 1 78
Summary and review 82
Assessment 4 84

5   Fractions, decimals and percentages
Introduction 86
Fractions and percentages 88
Calculations with fractions 92
Fractions, decimals and percentages 96
Summary and review 100
Assessment 5 102
Lifeskills 1: The business plan 104

6   Formulae and functions
Introduction 106
Formulae 108
Functions 112
Equivalences in algebra 116
Expanding and factorising 2 120

Summary and review 124
Assessment 6 126
Revision exercise 1 128

7   Working in 2D
Introduction 130
Measuring lengths and angles 132
Area of a 2D shape 136
Transformations 1 140
Transformations 2 144
Summary and review 148
Assessment 7 150

8   Probability
Introduction 152
Probability experiments 154
Theoretical probability 158
Mutually exclusive events 162
Summary and review 166
Assessment 8 168

9   Measures and accuracy
Introduction 170
Estimation and approximation 172
Calculator methods 176
Measures and accuracy 180
Summary and review 184
Assessment 9 186

10 Equations and inequalities
Introduction 188
Solving linear equations 190
Quadratic equations 194
Simultaneous equations 198
Approximate solutions 202
Inequalities 206
Summary and review 210
Assessment 10 212
Lifeskills 2: Starting the business 214

11 Circles and constructions
Introduction 216
Circles 1 218
Circles 2 222
Circle theorems 226
Constructions and loci 232
Summary and review 236
Assessment 11 238

12 Ratio and proportion

Introduction 240
Proportion 242
Ratio 246
Percentage change 250
Summary and review 254
Assessment 12 256
Revision exercise 2 258

13 Factors, powers and roots

Introduction 260
Factors and multiples 262
Powers and roots 266
Surds 270
Summary and review 274
Assessment 13 276

14 Graphs 1

Introduction 278
Equation of a straight line 280
Linear and quadratic functions 284
Properties of quadratic functions 288
Kinematic graphs 292
Summary and review 296
Assessment 14 298

15 Working in 3D

Introduction 300
3D shapes 302
Volume of a prism 306
Volume and surface area 310
Summary and review 316
Assessment 15 318
Lifeskills 3: Getting ready 320

16 Handling data 2

Introduction 322
Frequency diagrams 324
Averages and spread 2 328
Box plots and cumulative
      frequency graphs 332
Scatter graphs and correlation 336
Time series 340
Summary and review 344
Assessment 16 346

17 Calculations 2

Introduction 348
Calculating with roots and indices 350
Exact calculations 354
Standard form 358
Summary and review 362
Assessment 17 364

18 Graphs 2

Introduction 366
Cubic and reciprocal functions 368
Exponential and trigonometric functions 372
Real-life graphs 376
Gradients and areas under graphs 380
Equation of a circle 384
Summary and review 388
Assessment 18 390
Revision exercise 3 392

19 Pythagoras and trigonometry

Introduction 394
Pythagoras' theorem 396
Trigonometry 1 400
Trigonometry 2 404
Pythagoras and trigonometry problems 408
Vectors 412
Summary and review 416
Assessment 19 418

20 Combined events

Introduction 420
Sets 422
Possibility spaces 426
Tree diagrams 430
Conditional probability 434
Summary and review 438
Assessment 20 440
Lifeskills 4: The launch party 442

21 Sequences

Introduction 444
Linear sequences 446
Quadratic sequences 450
Special sequences 454
Summary and review 458
Assessment 21 460

22 Units and proportionality

Introduction 462
Compound units 464
Converting between units 468
Direct and inverse proportion 472
Rates of change 476
Growth and decay 480
Summary and review 484
Assessment 22 486
Revision exercise 4 488

Formulae 490
Key phrases and terms 491
Answers 493
Index 550

About this book

This book has been specially created for the new Edexcel GCSE Mathematics 9-1.

It has been written by an experienced team of teachers, consultants and examiners and is designed to help you obtain the best possible grade in your maths GCSE.

As well as mathematical fluency, Assessment Objective 1 (AO1), the new course places an increased emphasis on your ability to reason, AO2, and your ability to apply mathematical knowledge to problem solving, AO3. This change of emphasis is built into the way topics are covered in this book.

In each chapter the lesson are organised in pairs. The first lesson is focussed on helping you to master the basic skills required (AO1) whilst the second lesson applies these skills in questions that develop your reasoning and problem solving abilities (AO2 & 3).

Throughout the book four-digit MyMaths codes are provided allowing you to link directly, using the search bar, to related lessons on the MyMaths website: so you can see the topic from a different perspective, work independently and revise.

At the end of a chapter you will find a summary of what you should have learnt together with a review section that allows you to test your fluency with the basic skills (AO1). Depending on how well you do a *What next?* box provides suggestions on how you could improve even further. This includes links to InvisiPen worked solution videos contained on the accompanying online Kerboodle. Finally there is an Assessment section which allows you to practise exam-style questions (AO1 – 3).

At the end of the book you will find a guide to understanding exam questions and a full set of answers to all the exercises.

The GCSE maths specification identifies three types of content at higher level. A coloured band in the top-right corner indicates what type of content is included in a lesson.

	Standard	All students should develop confidence and competence with this content.
	Underlined	All students will be assessed on this content; more highly attaining students should develop confidence and competence with this content.
	Bold	Only the more highly attaining students will be assessed on this content. The highest attaining students should develop confidence and competence with this content.

We wish you well with your studies and hope that you enjoy this course and achieve exam success.

1 Calculations 1

Introduction

When you go shopping in a supermarket, you are presented with hundreds of products, often looking similar, as well as lots of different offers. You need to be able to do arithmetic in your head to ensure that you are keeping within your budget and also that you choose the best value offer.

What's the point?

Being able to add, subtract, multiply and divide doesn't just mean that you're good at maths at school – it means that you can confidently look after your own finances in the real world.

Objectives

By the end of this chapter you will have learned how to...

- Order positive and negative integers and decimals.
- Round numbers to a given number of decimal places or significant figures.
- Use mental and written methods to add, subtract, multiply and divide with positive and negative integers and decimals.
- Use BIDMAS to complete calculations in the correct order.

Check in

1 Write in words the value of the digit 4 in each of these numbers.

    a 4506   **b** 23 409   **c** 200.45   **d** 13.054

2 **a** Sketch a number line showing values from -5 to $+5$ and mark these numbers on it.

        i -3 **ii** $+4$ **iii** -2.5 **iv** 0

    b Write this set of directed numbers in ascending order.

        $+5, -2.4, -3, +6, 0, +1.5, -1.8$

Chapter investigation

Ternary uses a base-three system and the digits 0, 1 and 2 to write numbers.

In ternary, you use units, threes, nines, twenty-sevens, eighty-ones … to record place value.

100 = 1 eighty-one + 0 twenty-sevens + 2 nines + 0 threes + 1 unit
100 is 10 201 in ternary

Investigate these ternary calculations.

$1 + 1 = 2$	$1 + 2 = 10$	$2 + 2 = 11$
$1 \times 1 = 1$	$1 \times 2 = 2$	$2 \times 2 = 11$
$10 \times 12 = 120$	$11 \times 11 = 121$	$12 \times 12 = 221$
$101 \times 12 = 1212$	$202 \times 21 = 12 012$	$121 \times 21 = 11 011$

Show that the rules for long multiplication and division apply to ternary numbers.

1.1 Place value and rounding

- < means less than
- > means greater than
- ≤ means less than or equal to
- ≥ means greater than or equal to

Place the correct symbol $<$ $>$ or $=$ between the numbers in each pair.

a 5.07　5.7　**b** 397　379　**c** -10　5　**d** -19　-24　**e** $\dfrac{3}{2}$　1.5

a $5.07 < 5.7$　**b** $397 > 379$　**c** $-10 < 5$　**d** $-19 > -24$　**e** $\dfrac{3}{2} = 1.5$

- Numbers round up if the next digit is a 5 or more.
- Numbers round down if the next digit is a 4 or less.

Round 72 456.0374 to the nearest

a 10　**b** 100　**c** 1000　**d** tenth　**e** hundredth　**f** thousandth.

a 72 460　**b** 72 500　**c** 72 000　**d** 72 456.0　**e** 72 456.04　**f** 72 456.037

- When rounding to **significant figures**, count from the first non-zero digit.

dp and **sf** mean 'decimal places' and 'significant figures'.

a Round these numbers to 2 dp.

　i 34.567　**ii** 3.887 126　**iii** 215.587 54

b Round these numbers to 2 sf.

　i 39.54　**ii** 217　**iii** 0.000 455　**iv** 12 019　**v** 25.505

a **i** 34.57　**ii** 3.89　**iii** 215.59

b **i** 40　**ii** 220　**iii** 0.000 46　**iv** 12 000　**v** 26

- Multiplying or dividing a number by a power of 10 changes the place value of each digit.

Multiplying a number by 10 moves the digits one place to the left.
Dividing by 100 moves the digits two places to the right.

Work out　**a** $3.72 \div 100$　**b** $0.0349 \times 10\,000$　**c** $17.3 \div 1000$

a Move the digits two places right.

$3.72 \div 100 = 0.0372$

b Move the digits four places left.

$0.0349 \times 10\,000 = 349$

c Move the digits three places right.

$17.3 \div 1000 = 0.0173$

Exercise 1.1S

1 Write these numbers in words.

    a 1307   **b** 29006

    c 300000   **d** 605030

2 Write these numbers in figures.

    a Eight thousand and forty-three

    b Seventy million

    c Two hundred thousand and fifty-one

    d Two thousand and ten

3 Write these sets of numbers in ascending (increasing) order.

    a 0.3, 3.1, 1.3, 2, 1, 0.1

    b 607, 77.2, 27.6, 7.06, 6.07

4 Write these sets of numbers in descending (decreasing) order.

    a 6008, 682.8, 862.6, 6000.8, 8000.6

    b 47.9, 94.7, 49.7, 79.4, 74.9, 97.4

5 Multiply these numbers by 10.

    a 16.7   **b** 24.8   **c** 0.716

    d 1.095   **e** 243   **f** 281.3

6 Divide these numbers by 10.

    a 214   **b** 67.3   **c** 4106

    d 200.7   **e** 6.025   **f** 86

7 Decide which number in each pair is bigger. Explain your answers.

    a 4.52 and 4.05   **b** 5.5 and 5.05

    c 16.8 and 16.75   **d** 16.8 and 16.15

8 Decide whether each statement is true or false.

    a $4.1 < 4$ **b** $6.33 < 6.333$

    c $0.23 \leqslant 0.24$ **d** $-2.3 \geqslant -2.4$

    e $5.31 < 5.31$ **f** $5.31 \leqslant 5.31$

9 Write these sets of numbers in ascending order.

    a 7.83, 7.3, 7.8, 7.08, 7.03, 7.38

    b 4.2, 8.24, 8.4, 4.18, 2.18, 2.4

10 Write these sets of numbers in descending order.

    a 16.7, 18.16, 16.18, 17.16, 18.7, 17.6

    b 1.06, 13.145, 1.1, 2.38, 13.2, 2.5

11 Round these numbers to the nearest
i 10   **ii** 100.

    a 3048   **b** 1763   **c** 294

    d 51   **e** 43   **f** 743

12 Round these numbers to the nearest 1000.

    a 2964   **b** 1453   **c** 17

    d 24598   **e** 16344   **f** 167733

13 Round these numbers to
i 1 decimal place   **ii** 2 decimal places.

    a 39.114   **b** 7.068   **c** 5.915

    d 512.715   **e** 4.259   **f** 12.007

    g 0.833   **h** 26.8813

14 Round these numbers to the nearest
i tenth   **ii** hundredth   **iii** thousandth.

    a 0.07   **b** 15.9184   **c** 127.9984

    d 887.172   **e** 55.14455   **f** 0.00749

15 Round these whole numbers to two significant figures.

    a 483   **b** 1206   **c** 488

    d 13562   **e** 533   **f** 14511

16 Round these numbers to two significant figures.

    a 0.355   **b** 0.421   **c** 0.0566

    d 0.004673   **e** 1.357   **f** 0.000004152

17 Round these numbers to one significant figure.

    a 157   **b** 2488   **c** 4.66

    d 13.77   **e** 0.000453   **f** 121450

18 Round each number to the accuracy given in brackets.

    a 9.732 (3 sf)   **b** 0.36218 (2 dp)

    c 147.49 (1 dp)   **d** 28.613 (2 sf)

    e 0.5252 (2 sf)   **f** 4.1983 (2 dp)

    g 1245.4 (3 dp)   **h** 0.00425 (3 dp)

    i 273.6 (2 sf)   **j** 459.97314 (1 dp)

19 Calculate

    a 13.06×100 **b** $208.5 \div 100$

    c 1.085×1000 **d** $2487 \div 1000$

    e $0.008 \div 10$ **f** 0.00619×1000

    g $45.13 \div 1000$ **h** 0.000045×100

1001, 1005, 1013, 1072   SEARCH

1.1 Place value and rounding

- Numbers round up if the next digit is a 5 or more and round down if the next digit is a 4 or less.
- Multiplying or dividing a number by a power of 10 changes the place value of each digit.

< less than
≤ less than or equal to
> greater than
≥ greater than or equal to
= equal to
≠ not equal to

HOW TO

① Take time to understand the situation in the question.

② Use your knowledge of place value or rounding.

③ Write the answer, include any units.

EXAMPLE

These are the top results from the Men's 400 m hurdles final at the London 2012 Olympics. The times are given in seconds.

A commentator says: "There is less than a quarter of a second between the gold and silver medallists."

Is the commentator correct?

Culson	48.10
Greene	48.24
Sanchez	47.63
Taylor	48.25
Tinsley	47.91

①② Find the two fastest (shortest) times.

Gold medal: Sanchez   47.63
Silver medal: Tinsley   47.91

③ Find the difference between the times.
A quarter of a second is 0.25 seconds.
Write a conclusion.

47.91 – 47.63 = 0.28 seconds
0.28 > 0.25 (quarter of a second)
The commentator is wrong.

EXAMPLE

Use $14 \times 35 = 490$ to help solve this problem, without using a calculator.

Maria buys 35 bottles of water. Each bottle costs £1.40.

How much does Maria spend?

① Compare the cost of the water with the calculation you have been given.
Cost of water = 1.40 × 35

② 14 = 1.40 × 10. The answer in the calculation, 490, is 10 times too big.

③ Divide 490 by 10. Give your answer in pounds.
490 ÷ 10 = 49
Maria spends £49.

EXAMPLE

A shelf is 2 m long. How many 12 cm files will fit on the shelf?

① Convert 2 m to cm and divide by 12.
2 m = 200 cm   200 ÷ 12 = 16.6666...

②③ You round down here as you will not be able to fit in a seventeenth file.
16 files

Number   Calculations 1

Exercise 1.1A

1 These are the top results from the Women's long jump final at the London 2012 Olympics. The distances are given in metres.

DeLoach	6.89
Kolchanova	6.76
Nazarova	6.77
Sokolova	7.07
Radevica	6.88
Reese	7.12

a Who won the gold, silver and bronze medals?

b Radevica's personal best is 6.92 m. If she had jumped 6.92 m at the Olympics, would she have won a medal?

2 Here are the results of the Men's 100 m backstroke final. The times are in seconds.

Cheng	53.77
Grevers	52.16
Irie	52.97
Lacourt	53.08
Meeuw	53.48
Stoeckel	53.55
Tancock	53.35
Thoman	52.92

Which two swimmers had the closest times? Explain your answer.

3 You are given that $\boxed{15 \times 48 = 720}$

Use this to help solve these problems, without using a calculator.

a Liza's bank will give her a mortgage of four times her annual salary.
Liza's monthly salary is £1500.
How much deposit will Liza need to buy a house worth £100 000?

$\boxed{\text{Cost of the house} = \text{mortgage} + \text{deposit}}$

b Malik's kitchen has an area of 72 m². Each pack of tiles covers 1.5 m². How many packs of tiles will Malik need to cover the floor?

4 **a** A school is planning a coach trip. There are 80 people on the trip. Each coach can carry 24 people. How many coaches are needed?

b A lorry can carry a maximum of 28 tonnes. A crate weighs 800 kg. How many crates can the lorry carry?

5 Andrea chooses two numbers from the list.

44.37	44.44	44.48	44.53
44.55	44.63	44.67	44.71

When she rounds the two numbers to 1 decimal place they are equal.

When she rounds the two numbers to 2 significant figures, they are not equal.

Find Andrea's numbers.

*6 Jessica has five symbol cards

<	≥	=	≤	>

and three pairs of numbers.

3.118	3.112
4.5	$\frac{9}{2}$
3.004	2.9961

Jessica scores one point for each card that she places between a pair of numbers.

Jessica says that she can score seven points. Explain why she is correct.

*7 In our ordinary number system we use the digits 0 to 9 to write numbers. We use units, tens, hundreds, ... to record place value.

Binary uses the digits 0 and 1 to write numbers. Binary uses units, twos, fours, eights, sixteens, ... to record place value.

Eleven is 1 eight + 0 four + 1 two + 1 unit
Eleven is 1011 in binary.

a Write 15 as a binary number.

b Write 11010 as an ordinary number.

c Explain why 1011 + 1111 = 11 010 in binary. Can you create a set of rules for adding and subtracting in binary?

Q 1001, 1005, 1013, 1072　SEARCH

1.2 Adding and subtracting

There are two rules for adding and subtracting negative numbers.

- Adding a **negative** number counts as subtraction.
- Subtracting a negative number counts as addition.

EXAMPLE

Calculate   **a** $-5 + -6$ **b** $+4 - -2$

a $-5 + -6 = -5 - 6 = -11$ **b** $+4 - -2 = 4 + 2 = 6$

You can use mental methods to add and subtract whole numbers and decimals.

- Use **partitioning** to split the numbers you are adding or subtracting into parts.
- Use **compensation** when the number you are adding or subtracting is nearly a whole number, a multiple of 10 or a multiple of 100.

EXAMPLE

Calculate   **a** $19.5 - 7.2$ **b** $5.8 + 4.8$

a $19.5 - 7.2 = 19.5 - 7 - 0.2$ Split the smaller number into parts.
           $= 12.5 - 0.2$ Subtract the units from the highest number.
           $= 12.3$ Subtract the tenths.

b $5.8 + 4.8 = 5.8 + 5 - 0.2$ Rewrite **add 4.8** as **add 5 − 0.2**.
           $= 10.8 - 0.2$ Add the 5 to the highest number.
           $= 10.6$ Subtract 0.2.

Part **a** uses partitioning and part **b** uses compensation.

To use a written method, set out the calculation in columns and line up the decimal points.

- When adding, you may need to carry digits into the next column.
- When subtracting, you may need to borrow from the next column.
- Always estimate the answer before you start.

EXAMPLE

Calculate these using a written method.

   a $102.773 + 28.47$ **b** $26.44 - 1.105$

a   Initial estimate: $100 + 30 = 130$ **b**   Initial estimate: $30 - 1 = 29$

```
   1 0 2 . 7 7 3              ³ ¹⁰
 +   2 8 . 4 7 0           2 6 . 4 4̸ 0
 ─────────────           −   1 . 1 0 5
   1 3 1 . 2 4 3          ──────────────
                           2 5 . 3 3 5
```

If the numbers have different numbers of decimal digits, add extra zeros to the number with fewer digits.

Exercise 1.2S

1 Calculate

   a $+8 - -14$ **b** $-1 + -11$

   c $-9 - -7$ **d** $+3 + -17$

   e $+8 - -4$ **f** $+13 + -1$

   g $+48 - +29$ **h** $-19 + +4$

   i $+34 + -23$ **j** $-104 + +43$

   k $+208 - -136$ **l** $+347 + -298$

2 Calculate

   a $-4.5 + -6.3$ **b** $-2.8 - -3.5$

   c $+5.6 - -7.9$ **d** $-9.4 + +8.7$

   e $-26.5 + -11.7$ **f** $+45.9 - -66.8$

3 Work these out in your head and write the answers.

   a $4.7 + 5.3$ **b** $4.7 + 5.4$

   c $3.6 + 6.7$ **d** $6.8 + 4.3$

   e $7.5 + 8.9$ **f** $2.7 + 4.8$

4 Work these out in your head and write the answers.

   a $3.55 + 4.22$ **b** $2.13 + 3.12$

   c $3.18 + 0.42$ **d** $3.72 + 0.18$

   e $1.42 + 0.71$ **f** $8.39 + 4.65$

5 Work these out in your head and write the answers.

   a $6.6 - 4.1$ **b** $7.5 - 5.4$

   c $8.1 - 5.9$ **d** $7.2 - 3.3$

   e $3.1 - 1.7$ **f** $7.3 - 6.6$

6 Work these out in your head and write the answers.

   a $2.16 - 1.42$ **b** $1.51 - 0.46$

   c $6.39 - 4.88$ **d** $15.46 - 8.32$

   e $5.17 - 4.09$ **f** $4.29 - 3.65$

7 Use a written method to calculate

   a $23.45 + 12.51$ **b** $13.44 + 21.17$

   c $77.33 + 19.02$ **d** $65.47 + 38.95$

   e $8.152 + 6.779$ **f** $5.426 + 4.975$

   g $11.625 + 14.586$ **h** $25.341 + 38.495$

8 Use a written method to calculate

   a $24.72 - 14.04$ **b** $1.52 - 1.09$

   c $6.149 - 2.052$ **d** $16.64 - 15.88$

   e $5.23 - 3.11$ **f** $17.45 - 13.26$

   g $6.41 - 4.37$ **h** $23.6 - 17.9$

9 Use a written method to calculate

   a $1.09 + 1.54$ **b** $0.09 + 0.36$

   c $14.52 + 9.8$ **d** $13.92 + 0.8$

10 Use a written method to calculate

   a $4.5 - 0.53$ **b** $3.085 - 2.99$

   c $16.3 - 3.86$ **d** $112.14 - 53.8$

11 Use a written method to calculate

   a $11.1 - 8.29$ **b** $2.09 - 1.333$

   c $102.8 - 14.79$ **d** $978 + 148.72$

12 Work out these calculations using an appropriate method (written or mental). Show your method clearly.

   a $5.8 - 3.2$ **b** $16.73 - 8.87$

   c $9.6 - 3.7$ **d** $109.54 - 17$

   e $2.37 - 1.4$ **f** $26.25 - 1.98$

13 Work out these calculations using a standard written method. Remember to write an estimate before you do the calculation.

   a $21.864 - 7.968$ **b** $104.87 - 85.42$

   c $417.48 - 57.69$ **d** $24.503 - 16.82$

   e $19.21 - 18.884$ **f** $102.01 - 90.59$

14 Use a calculator to check your answers to question **13**.

15 Work out these calculations using an appropriate method, and show your working clearly.

   a $8.6 - 4.5$ **b** $26.4 + 13.8$

   c $18 - 6.712$ **d** $15.808 - 9.84$

   e $4.008 - 3.116$ **f** $4.109 - 3.64$

16 Work out these calculations using an appropriate method, and show your working clearly.

   a $23.11 + 31.24 + 45.53$

   b $14.58 + 13.35 + 1.056$

   c $26.87 + 13.11 - 25.44$

   d $4.162 + 5.516 - 1.0023$

Q 1007, 1068   SEARCH

1.2 Adding and subtracting

- You can use mental or written methods to add and subtract whole numbers and decimals.
- When you use a written method for adding or subtracting decimals, you should estimate first.
- When you do decimal addition or subtraction, you must take care with the position of the decimal points.

Line up the decimal points, and add or subtract the columns from right to left.

```
  135.23
+  27.8
  163.03
    1 1
```

```
   2 13  1 5 10
   3̶4̶.5̶6̶
 -18.729
  15.831
```

HOW TO

1. Take time to understand the situation in the question. Estimate the answer when working with decimals.
2. Add or subtract using a mental or written method. Check that your answer agrees with your estimate.

You may need to carry digits into the next column.

You may need to borrow from the next column.

EXAMPLE

Hera the baby gorilla weighs 21.62 kg, and Horace her brother weighs 46.34 kg. The total weight of Henna the mother gorilla and her two children is 254.66 kg.

How much does Henna weigh?

1. Estimate the answer first. Round to the nearest 10.

   20 + 50 = 70
   250 – 70 = 180

2. Set out the calculation in columns, making sure you line up the decimal points.

```
  21.62
+46.34
  67.96
```

```
  2 14 8 13 4 .66
 -  67.96
  186.70
```

Borrow digits from the next column. Try this subtraction for yourself!

Henna weighs 186.7 kg.

EXAMPLE

One of these calculations is correct. Decide which calculation has been carried out correctly without working out the answers. Explain what went wrong in the incorrect answers.

a
```
 1 0 1 9 1
 1̶1̶.0̶3̶
 -2.55
  8.48
```

b
```
 38.53
+2.474
 6.327
   1 1
```

c
```
 100.773
 -28.782
 128.011
```

1. Estimate the answer to each calculation to decide which is correct.
   a 11 – 3 = 8
   b 39 + 2 = 41
   c 100 – 30 = 70
2. Explain what went wrong.

   This one is correct.

   Here the decimal points, and therefore all of the columns, were not properly aligned.

   Here the smaller digit in each column has been subtracted from the bigger one.

Exercise 1.2A

1 Use a mental method of calculation to solve each of these problems.

a Charlie has to travel 435 km. After 2 hours he has travelled 187 km. How much further does he have to travel?

b In a test, Alex scores 93 marks and Sophie scores 75 marks. Alex, Sophie and Louise score 265 marks altogether. How many marks did Louise score?

c A recycling box is full of things to be recycled. The empty box weighs 1.073 kg.

Bottles	12.45 kg
Cans	1.675 kg
Paper	8.7 kg
Plastic objects	? kg

The total weight of the box and all the objects to be recycled is exactly 25 kg. What is the weight of the plastic objects?

2 Copy and complete this addition pyramid.

Each number is the sum of the two numbers beneath it.

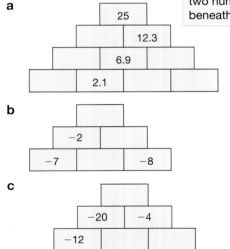

a
25
12.3
6.9
2.1

b
−2
−7   −8

c
−20   −4
−12
−5   5

3 Amy makes a pair of two-digit numbers using the digits 3, 4, 7 and 8. One number is even and the other is odd.

a List the possible pairs of numbers.

b Amy adds her numbers together. What is the largest total that Amy can make?

4 Find the balance in these bank accounts after the transactions shown.

a Opening balance £133.45. Deposits of £45.55 and £63.99, followed by withdrawals of £17.50 and £220.

A negative number represents an overdraft.

b Opening balance − £459.77. Deposit of £650, followed by a withdrawal of £17.85.

5 Each calculation is incorrect. Explain the mistake in each answer and find the correct answer.

a
```
  3 6 . 4 1
+ 6 4 . 5 6
  9 0 . 9 7
```

b
```
  1 4 6 . 5
+  9 8 . 6
  2 3 5 . 1
  1   1   1
```

c
```
  3 8 5 . ⁸9 ¹5
− 2 2 4 . 6 7
  1 6 9 . 2̶ 8
```

d
```
  ⁰4̶ ¹²3̶ ¹1 . ¹0 6
−       4 3 . 2 4
        8 8 . 8 2
```

6 Fit the digits 1, 2, 3, 4, 5, 6, 7, 8 into the eight spaces to make this subtraction correct.

```
  □ □ □
− □ □ □
    □ □
```

7 a Find a pair of numbers in this table where their total is twice their difference.

1026	432
724	342
1448	522

b Can you find a pair where the total is three times the difference?

c Can you find four numbers where one pair adds up to twice the total of the other pair?

8 a List the three-digit numbers that can be made from the digits 2, 4 and 6.

b A three-digit number is made from the digits 2, 4 and 6. Another three-digit number is made from the digits 3, 5 and 7. List the six possible pairs of numbers that add together to make 999.

c The difference between the numbers is 75. Write down the two numbers.

Q 1007, 1068   SEARCH

1.3 Multiplying and dividing

● Multiplying or dividing a positive number by a negative number gives a negative number.
● Multiplying or dividing a negative number by a negative number gives a positive number.

Multiplying by **numbers < 1** makes positive numbers smaller	Multiplying by **numbers > 1** makes positive numbers bigger
Dividing by **numbers < 1** makes positive numbers bigger	Dividing by **numbers > 1** makes positive numbers smaller

0   1

Multiplying doesn't always make numbers bigger and **dividing** doesn't always make them smaller.

● For standard methods of multiplication and division, work with the significant digits from the numbers in the question.
● Use an **estimate** to adjust the **place value** correctly.

EXAMPLE

Calculate

a 18.5×7.9 **b** $47.52 \div 1.8$

a   Estimate $20 \times 8 = 160$

```
      1 8 5
    ×   7 9
    1 6₇6₄5
  1 2₅9₃5 0
  1 4₁6₁1 5
```

Use your estimate to adjust the place values.

$18.5 \times 7.9 = 146.15$

b   Estimate $50 \div 2 = 25$

```
          2 6 4
    18)4 7 5 2
        3 6
        1 1 5
        1 0 8
            7 2
            7 2
             0
```

So $47.52 \div 1.8 = 26.4$

● The order in which **operations** are carried out is **BIDMAS**
(**B**rackets, **I**ndices or powers, **D**ivision, **M**ultiplication, **A**ddition, **S**ubtraction)

Always write the calculation a line at a time, so that you can see each operation clearly.

EXAMPLE

Evaluate

a $4 + 3 \times 2$ **b** $5 + 3^2$ **c** $\sqrt{(5 + 4 \times 11)}$

a $4 + 3 \times 2 = 4 + 6 = 10$

b $5 + 3^2 = 5 + 9 = 14$

c $\sqrt{(5 + 4 \times 11)} = \sqrt{(5 + 44)} = \sqrt{49} = 7$

Using brackets in part **c** shows that **all** the values are contained in the square root.

Exercise 1.3S

1 Calculate

a	$+5 \times -5$	**b**	$+4 \times -8$
c	$-8 \times +9$	**d**	$-4 \times +5$
e	-3×-10	**f**	-7×-7
g	$+8 \times +2$	**h**	$+5 \times -4$
i	$-2 \times +9$	**j**	-13×-2
k	$-7 \times +6$	**l**	$+12 \times -4$

2 Calculate

a	$-18 \div +9$	**b**	$-20 \div +4$
c	$-30 \div -6$	**d**	$-12 \div -3$
e	$-66 \div +3$	**f**	$+47 \div -47$
g	$-80 \div -2$	**h**	$+24 \div +6$
i	$-45 \div -9$	**j**	$-51 \div +3$
k	$+57 \div -19$	**l**	$-81 \div -3$

3 Use a mental method to work out

a	14×7	**b**	19×8
c	21×13	**d**	17×19
e	11×28	**f**	21×29

4 Now use a written method to work out the answers for question **3**. Check that you get the same answers with both methods.

5 Write a division that is equivalent to each of these multiplications. The first one has been done for you.

a	$4 \times 0.5 = 4 \div 2$	**b**	6×0.2
c	12×0.2	**d**	2×0.001

6 Write a multiplication that is equivalent to each of these divisions. The first one has been done for you.

a	$4 \div 0.5 = 4 \times 2$	**b**	$6 \div 0.25$
c	$16 \div 0.01$	**d**	$15 \div 0.05$

7 Use a mental method to work out these calculations. Start each one with an estimate, and show your method.

a	31×0.3	**b**	$49 \div 0.07$
c	$3.66 \div 0.3$	**d**	$4.24 \div 0.4$
e	$13.9 \div 0.03$	**f**	$3.9 \div 0.03$
g	$171 \div 0.3$	**h**	5.2×0.125

8 Use a written method to work out

a	4.7×5.3	**b**	1.53×2.8
c	21.6×4.9	**d**	33.65×3.89
e	21.58×1.99	**f**	42.77×8.64

9 Use a written method to work out

a	$34.83 \div 9$	**b**	$5.425 \div 7$
c	$7.328 \div 8$	**d**	$451.8 \div 60$
e	$54.39 \div 3$	**f**	$58.65 \div 17$
g	$66.4 \div 16$	**h**	$185.76 \div 24$
i	$7.752 \div 1.9$	**j**	$3.055 \div 1.3$

10 Use a written method to work these out, giving your answers correct to 2 dp.

a	$14.73 \div 2.8$	**b**	$51.99 \div 1.8$
c	$193.8 \div 0.14$	**d**	$1013 \div 5.77$
e	$23.78 \div 0.83$	**f**	$65.79 \div 0.59$

11 Use a calculator to check your answers to questions **8–10**.

12 Given that $43 \times 67 = 2881$, find

a	4.3×6.7	**b**	430×0.067
c	$2881 \div 670$	**d**	$28.81 \div 430$
e	$2.881 \div (0.43 \times 0.67)$		

13 Find the values of these expressions.

a	$(5^2 + 3) \times 7$	**b**	$(9 - 7)^2$
c	$(5 - 3) \times (4^2 - 7)$	**d**	$(5^2 - 8)^2$

14 Calculate the values of these expressions.

a	$(4 + 7)^2$	**b**	$(6 + 7) \times 9 \div 3$
c	$\dfrac{6 \times (5^2 - 13)}{4}$	**d**	$\sqrt{100 - 2 \times 6^2}$

15 Solve each of these calculations.

a $(15.7 + 1.3) \times (8.7 + 1.3)$

b $\dfrac{7^2}{(2.3 \times 4)^2}$ **c** $\dfrac{(7 + 5)^2}{(25 + 7 \times 8)}$

> Decide whether to use a mental, written or calculator method. Where appropriate give your answer to 2 decimal places.

16 Find the values of these expressions.

a $\dfrac{28}{4} + \sqrt{100 - (9^2 + 5 \times 2)}$

b $\sqrt{28 + 4^2 - (10 - 8)} + 4 \times 3$

Q 1011, 1068, 1916, 1917   SEARCH

1.3 Multiplying and dividing

- You can make an **estimate** before starting a written calculation. You can use it to check your answer or to adjust place value.
- In a multi-stage calculation, you need to remember the **order of operations, BIDMAS**.

BIDMAS

Brackets
Indices (or powers)
Division or
Multiplication
Addition or
Subtraction

HOW TO

① Take time to understand the situation in the question.
Estimate the answer when working with decimals.

② Multiply or divide using a mental or written method.
Remember to use BIDMAS.
Check that your answer agrees with your estimate.

EXAMPLE

A firework display costs £28 800.
The display lasts for 15 minutes.

On average eight fireworks are set off every second.
What is the average cost of one firework?

① You will need to divide the cost by the total number of fireworks.

② Find the number of fireworks set off in the display.

15 minutes = 15 × 60 seconds = 900 seconds $15 \times 6 \times 10 = 90 \times 10$

Total fireworks = 900 × 8 = 7200 $9 \times 8 \times 100 = 72 \times 100$

Average cost = 28800 ÷ 7200 = 4 $288 \div 72 = 4$

The average cost is £4 per firework.

EXAMPLE

Adam explained how he would calculate $\dfrac{5 + \sqrt{9}}{4}$

What is the problem here?

Find the correct answer.

There is a root, an addition and a division. I'll do the root first, then divide, and then add.

① Explain how Adam applied BIDMAS incorrectly.

Even though there are no brackets in this expression, the whole of the 'top line' is divided by 4, so you need to find the square root, then add, and then divide.

② The expression could be written as $(5 + \sqrt{9}) \div 4$.

$(5 + \sqrt{9}) \div 4 = (5 + 3) \div 4$

$= 8 \div 4$

$= 2$

Exercise 1.3A

1 Charlie is paid £4.46 per hour for her part time job.

   a She works for 17 hours. How much is she paid?

   b One weekend she earns £57.98. She works two hours longer on Saturday than she does on Sunday. How long does she work for each day?

   c How long does she have to work to earn more than £100?

2 Skye buys 18 packets of rice. Each packet of rice costs £1.17. Skye also buys 12 packets of pasta. She spends £37.50 in total. How much does one packet of pasta cost?

3 Nancy says, 'I am working out $36 \div (2 + 3)$. I'll get the same result if I do $36 \div 2$ and $36 \div 3$, and then add the answers together.' Explain why Nancy is wrong.

4 Insert brackets where necessary to make each of the calculations correct.

   a $5 \times 2 + 1 = 15$

   b $5 \times 3 - 1 \times 4 = 40$

   c $20 + 8 \div 2 - 7 = 17$

   d $2 + 3^2 \times 4 + 3 = 65$

   e $2 \times 6^2 \div 3 + 9 = 33$

   f $4 \times 5 + 5 \times 6 = 150$

5 Fill in the missing numbers to make the calculations complete.

   a

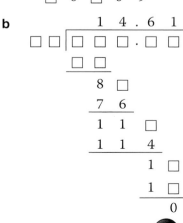

6 There are 18 questions in a quiz.

A correct answer scores 4 points.
An incorrect answer scores −3 points.

A question not answered scores nothing. It is possible to have a negative total.

   a What are the maximum and minimum points that you could score on the quiz?

   b Write down three different ways in which a student could have a total score of 12 points.

7 Steve's car manual says his car does 42 mpg. Mpg is miles per gallon. A gallon is 4.55 litres.
A litre of diesel costs £1.38. Steve drives from Leeds to Edinburgh, a distance of 215 miles. How much should it cost for diesel?

8 Explain whether each statement is true or false, giving examples to support your argument.

   a When you divide a number by 2, the answer is always smaller than the number you started with.

   b Dividing by $\frac{1}{2}$ is always the same as doubling a number.

   c When you multiply by 5, the answer is always bigger than the number you started with.

   d When you multiply by 10, the answer will always be different from the original number.

9 **a** Explain the effect of repeatedly multiplying a number by +0.9.

   b Explain the effect of repeatedly multiplying a number by −0.9.

10 Think of a number between 100 and 999. Repeat it to make it a 6-digit number. Enter this number on your calculator. Now divide it by 13. And then by 11. And then by 7.
Do you end up with what you first thought of?
Does it work for any 3-digit number?

> Hint: Multiply your number by 1001. What do you notice?

Summary

Checkout

You should now be able to...

	Test it Questions
✔ Order positive and negative integers and decimals.	1, 2
✔ Round numbers to a given number of decimal places or significant figures.	3
✔ Use mental and written methods to add, subtract, multiply and divide with positive and negative integers and decimals.	4 – 7
✔ Use BIDMAS to complete calculations in the correct order.	8, 9

Language	Meaning	Example
Place value	The value of a digit according to its position in a number.	123.4 2 means 2 tens = 20 4 means 4 tenths = $\frac{4}{10}$

Rounding	Making a number easier to work with by giving less digits.	103.67 = 103.7 (1 dp)
		      = 100 (1 sf)
		
		 0.0055 = 0.0 (1 dp)
		      = 0.006 (1 sf)

Decimal places	The number of digits after the decimal point.	
Significant figures	The number of digits after the first non-zero digit.	
Directed number	A positive or negative number.	... −4 −3 −2 −1 0 1 2 3 ....
Negative	A number that is less than zero	3 − 6 = −3
Estimate	An approximate calculation or a judgement of a quantity.	Estimate 68.89 × 21.1 ≈ 70 × 20 = 1400 Exact = 1453.579

Partitioning	Splitting a larger number into smaller numbers which add up to the original number.	85 + 25.6 = 85 + (15 + 10.6)
		      = 100 + 10.6
		      = 110.6

Compensation	Replacing a number by a simpler approximate value and a correction.	158 − 18.9 = 158 − (20 − 1.1)
		      = (158 − 20) + 1.1
		      = 139.1

Operations	Rules for processing numbers.	Addition, subtraction, multiplication and division.
Order of operations	The order in which operations have to be carried out to give the correct answer to a calculation.	2 + 4 × 3 − 1 = 2 + 12 − 1 = 13 (2 + 4) × 3 − 1 = 6 × 3 − 1 = 18 − 1 = 17 (2 + 4) × (3 − 1) = 6 × 2 = 12 2 + 4 × (3 − 1) = 2 + 4 × 2 = 2 + 8 = 10
BIDMAS	An acronym for the correct order of operations: **B**rackets, **I**ndices (or powers), **D**ivision or **M**ultiplication, **A**ddition or **S**ubtraction.	

Review

1 Copy the numbers and write $>$ or $<$ between them to show which is larger.

   a   24.3 $\square$ 24.5

   b -0.5 $\square$ -0.9

   c   0.5 $\square$ 0.06

   d   1.456 $\square$ 1.46

2 Are the following statements true or false? In cases where the statement is false write a true statement relating the two expressions.

   a $0.85 \div 10 > 85 \div 1000$

   b $0.85 \div 10 \geqslant 85 \div 1000$

   c $1 \leqslant \dfrac{7}{5} \leqslant 1\dfrac{1}{2}$

   d $3^2 = 3 \times 2$

   e $9^2 \neq 9 + 9$

3 Round

   a   45.892   to 1 decimal place

   b   0.0752   to 2 decimal places

   c   0.0854   to 2 significant figures

   d   1521   to 1 significant figure

   e   78 025   to 3 significant figures

   f   3.529   to 2 significant figures.

4 Work out the value of these expressions.

   a 83×100 **b** 2.59×10

   c 0.31×1000 **d** 0.05×100

   e $764 \div 10$ **f** $5490 \div 1000$

   g $8.5 \div 100$ **h** $0.08 \div 10$

5 Calculate these using written or mental methods.

   a $627 + 3215$ **b** $27.3 + 164.7$

   c $302.8 + 6.52$ **d** $0.34 + 52.713$

   e $8124 - 398$ **f** $104.1 - 69.5$

   g $1589.4 - 672$ **h** $18.31 - 8.4$

6 Calculate these using written or mental methods.

   a 18×30 **b** 3.5×18

   c $3.36 \div 6$ **d** $650 \div 1.3$

   e 8.31×6.2 **f** 1.083×2.45

   g $15.2 \div 8$ **h** $406 \div 1.4$

7 Work out these calculations involving negative numbers.

   a $17 - 28$ **b** $-189 + -52$

   c $9.3 - -3.3$ **d** $-0.62 + 0.19$

   e 14×-3 **f** -1.1×-5

   g $170 \div -10$ **h** $-6.5 \div -1.3$

8 Evaluate these without using a calculator.

   a $19 - 3 \times 6$ **b** $35 + 25 \div 5$

   c $8 \times (17 - 3)$ **d** 7×4^2

   e $3.2 + \sqrt{49}$ **f** $\sqrt{6} + 5 \times 6 \div 0.2$

9 Use your calculator to work these out.

   a $12.78 + 6.41 \times 8.32 - 9.3$

   b $\dfrac{3 - \sqrt{(-3)^2 - 4 \times (4) \times (-1)}}{2}$

What next?

Score		
0 – 4		Your knowledge of this topic is still developing. To improve look at MyMaths: 1001, 1005, 1007, 1011, 1013, 1068, 1072, 1916, 1917
5 – 8		You are gaining a secure knowledge of this topic. To improve look at InvisiPens: 01Sa – q
9		You have mastered these skills. Well done you are ready to progress! To develop your exam technique looks at InvisiPens: 01Aa – f

Assessment 1

1 Carli tries to order each set of numbers from smallest to largest.
One pair of numbers in each list is in the wrong order.
Say where Carli has made a mistake and put each list of numbers in order,
starting with the smallest

    a   0.8 1.9 3.3 44 303 57.6   [3]

    b -0.07 -2.19 30 43.56 188.0 194.7   [3]

2 Carli then tries to put these numbers in ascending order:
$42 \div 100, 0.3 \times 10, 4236 \div 1000, 516 \div 10, 42 \times 100, 216 \times 1000$
Has she ordered the numbers correctly? Give your reasons.   [4]

3 The world's tallest man, Robert Wadlow, was 271.78 cm (2 dp) tall.
The world's tallest woman, Yao Defen, is 233.34 cm (2 dp) tall.

    a   A challenger to the world's tallest man record measured his height as 271.8 cm
to one decimal place.
Has the challenger definitely beaten the world record? Give your reasons.   [1]

    b   A challenger to the world's tallest woman record measured her height as 233.341 cm.
Has the challenger definitely beaten the world record? Give your reasons.   [1]

4 Dave is 36 and Jane is 44. Jane says that she and Dave are the same age to one significant figure.

    a   Is Jane correct?   [1]

    b   Will Dave and Jane be the same age to one significant figure in one years' time?
Give your reasons.   [2]

    c   How old will Dave and Jane be the next time their ages are the same to
one significant figure?   [2]

5 As the Earth spins on its axis, everything on the
Earth's surface moves with it.
The distance travelled in one day due to the Earth's rotation
is $3.142x$, where x is the diameter of the circular path.
Abena lives on the equator and Edward lives in the UK.
$x = 12756$ for Abena and $x = 8134$ for Edward.

Edward's path

Abena's path

    a   Write both values of x to two significant figures.   [2]

    b   Use your answers to part **a** to estimate how much
further Abena travels than Edward in one day.   [3]

    c   Explain how using values of x correct to one significant figure would affect
the estimate in part **b**.   [3]

6 Jasmine's bike has wheels of circumference (i.e. perimeter) 2.5 m.
When Jasmine cycles to school, the wheels go round 850 times.
How far does Jasmine cycle to school?   [2]

7 Work out the following without using a calculator. Include units in your answer.

    a   A bag of sweets weighing 113 g includes wrappings of 0.5 g.
Each sweet weighs 4.5 g.
How many sweets are in the bag?   [2]

    b   A football stadium has 135 400 m² of seating for its fans. Each fan is allowed 5.4 m² of space.
How many fans, to the nearest 1000, is the stadium capable of holding?   [2]

8 There are 10 questions in a quiz.
A **correct** answer scores 3 points. A **wrong** answer **loses 2 points**.
Any question not answered loses 1 point. A negative total is possible.

    a Write down the maximum and minimum points any player can score. [2]

    b Jenny answers 8 of the ten questions. 5 are correct. How many points does Jenny score? [3]

    c Describe 3 different ways of scoring −10 points. [3]

9 To find your BMI (Body Mass Index) you use this process

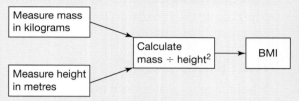

Name	Mass (kg)	Height (m)	BMI
Sue	57.5	1.7	
Clive	105.8	1.95	
Ben		1.77	19.7
Henry	71.3		21.3

    Copy and complete the table. Give your answers to 1 decimal place. [6]

10 A supermarket stocks packets of the new breakfast cereal Maltibix.
Each packet of Maltibix holds 650 g inside a cardboard box weighing 68 g.
50 boxes, each holding 36 of these packets are delivered to the supermarket.
Does the mass of the delivery exceed 1000 kg? [4]

11 In the number grids shown, the number in each cell is the sum of the two cells above it.
Copy and complete the grids shown.

    a 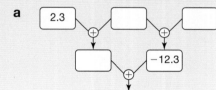   **b**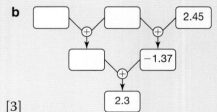

    [3]   [3]

12 A magic square is a square grid of numbers where each number is **different**. The sum of the numbers in each row, each column and each diagonal is the same.
Fill in the missing values in the magic square. [4]

13 Amanda's garden is rectangular and measures 12.5 m by 9.2 m.
The garden is to be sown with grass seed.
The gardener needs 25 g of grass seed for each square metre of ground.
Grass seed costs £5.35 per kg. The gardener estimates that he will need £20 to buy the grass seed.
Amanda disagrees with the gardener's estimate.
She gives the gardener £15 to buy the grass seeds.
Which estimate do you agree with? Explain your answer. [4]

14 The *reciprocal* of a number n is $1 \div n$.

    a What is the only number which is the same as its reciprocal? [1]

    b What is the only number which has no reciprocal? Explain your answer. [1]

    c Lewis says that every positive number is greater than its reciprocal.
Find an example that disproves Lewis' claim. [1]

15 Write one pair of brackets in each calculation to make the answer correct.

    a $3 + 4 \times 5 = 35$ [1]   **b** $4 \div 3 + 5 = 6\frac{1}{3}$ [1]   **c** $5(2^3 + 0.4) \div 4 - 3 \times -1 = 6$ [2]

2 Expressions

Introduction

The real world is messy, complicated and always in motion. Algebra is a vital part of maths because it attempts to describe aspects of the world, such as fluid flow or the forces acting on a suspension bridge. Through equations and formulae, algebra provides a mathematical model to describe a real-world situation, from which understanding can be gleaned and predictions made. You can only do this if you make assumptions that simplify the situation. Although simplifying the situation means that the model is only an approximation to the real world, it helps us to understand the forces that lie behind it.

What's the point?

Without algebra, you would not be able to work with large mechanical forces. There would be no skyscrapers or suspension bridges. You would also not be able to understand electronics, so there would be no tablets or mobile phones.

Objectives

By the end of this chapter, you will have learned how to …

- Use algebraic notation and simplify expressions by collecting like terms.
- Substitute numbers into formulae and expressions.
- Use the laws of indices.
- Multiply a single term over a bracket.
- Take out common factors in an expression.
- Simplify algebraic fractions and carry out arithmetic operations with algebraic fractions.

Check in

1   Work out these multiplications and divisions mentally.

    a 15×3 **b** 4×13 **c** $(-2) \times 13$ **d** 14×14

    e $56 \div 8$ **f** $91 \div 7$ **g** $150 \div (-3)$ **h** $1200 \div 40$

2   Explain why the answer to each of these questions is 15.

    a $9 + 3 \times 2$ **b** $24 \div 3 + 7$ **c** $(3 + 2) \times 3$ **d** $3^3 - 4 \times 3$

3   Find the highest common factor of these number pairs.

    a 6 and 9   **b** 8 and 12   **c** 20 and 30   **d** 12 and 18

    e 24 and 52   **f** 50 and 75   **g** 99 and 132   **h** 7 and 14

Chapter investigation

Think of a number between 1 and 10.

- Double it.
- Add 4.
- Halve your answer.
- Take away the number you first thought of.

What do you notice? Investigate why this is the case.

Can you invent different instructions that give similar results?

2.1 Simplifying expressions

- An **expression** is a collection of letters and numbers with no = sign, for example $3x + 1$
- An **equation** contains an = sign and an unknown letter to be solved, for example $3x + 1 = 10$
- A **formula** is a relationship between two or more letters and it contains an = sign, for example $P = IV$

- You can substitute values into expressions and formulae.

EXAMPLE

Work out the value of each expression when $a = 3$ and $b = -2$.

a $\dfrac{4a}{6}$

b $\dfrac{a - b}{5}$

a $\dfrac{4a}{6} = \dfrac{4 \times a}{6}$

$= \dfrac{4 \times 3}{6} = \dfrac{12}{6} = 2$

b $\dfrac{a - b}{5} = \dfrac{3 - -2}{5}$

$= \dfrac{3 + 2}{5} = \dfrac{5}{5} = 1$

There are conventions for writing expressions in algebra.

- Do not include the multiplication sign. $3 \times p \rightarrow 3p$
- Write divisions as fractions. $3 \div p \rightarrow \dfrac{3}{p}$
- Write numbers first in products. $p \times 3 \rightarrow 3p$
- Write letters in products in alphabetical order. $4 \times q \times r \times p \rightarrow 4pqr$

- To simplify an expression, you collect like terms together.
  $2a$ and $6a$ are like terms, $2a$ and $4b$ are not like terms, a and a^2 are not like terms.

You can add, subtract, multiply or divide algebraic terms.

$3n + 5n + 8n = 16n$ $4p - p = 3p$ $2 \times 6p = 12p$ $\dfrac{8r}{4} = 2r$

- To simplify an expression, you follow the same order of operations as in arithmetic.
  Brackets → **I**ndices → **D**ivision or **M**ultiplication → **A**ddition or **S**ubtraction

EXAMPLE

Simplify

a $4n + 2 \times 5n$ **b** $3r \times 2s$ **c** $4t^2 - 3 \times t^2 + t$

a $4n + 2 \times 5n = 4n + (2 \times 5n) = 4n + 10n = 14n$

b $3r \times 2s = 3 \times r \times 2 \times s$

$= 3 \times 2 \times r \times s$ Rearrange: numbers first, then letters.

$= 6rs$

c $4t^2 - 3 \times t^2 + t = 4t^2 - (3 \times t^2) + t$

$= 4t^2 - 3t^2 + t$ Collect like terms.

$= t^2 + t$

You can use the acronym BIDMAS to remember the order.

Algebra Expressions

Exercise 2.1S

1 Write these expressions using the rules of algebra.

  a $5 \times w$ **b** $6 \div k$

  c $y \times y$ **d** $ab6$

  e $k \times k \times 8 \times k$ **f** $3k \times 4k^2$

2 Evaluate these expressions, given that $x = 6$.

  a $3x + 2$ **b** $10 - x$

  c $\dfrac{10x - 16}{2}$ **d** $3x^2$

3 Work out the value of these when $a = 3$, $b = 5$, $c = 4$ and $d = 6$.

  a a^2 **b** $2a^2$ **c** b^2

  d $2b^2$ **e** c^2 **f** $2c^2$

  g d^2 **h** $2d^2$ **i** $abcd$

4 If $x = -3$, $y = 2$ and $z = 4$, work out the value of these expressions.

  a $x + y$ **b** $y^2 - 5$

  c $x^2 + 2$ **d** $2y + z$

  e $3y + 2x$ **f** $z^2 - 2y$

  g $3z + 2x$ **h** $3x + 2y - z$

5 Calculate the value of each expression when $r = 2$, $s = 4$ and $t = -3$.

  a $\dfrac{s}{2}$ **b** $\dfrac{6r}{3}$ **c** $\dfrac{t}{3}$

  d $\dfrac{s + r}{3}$ **e** $\dfrac{t - 5}{2}$ **f** $\dfrac{s \times r}{3}$

  g $\dfrac{3r}{t}$ **h** $\dfrac{st}{r}$ **i** $\dfrac{r + s}{t}$

6 Simplify each expression.

  a $8n + 3n + 4n$ **b** $3m + 2m + 7m$

  c $8p + 6p + 3p$ **d** $5q + 7q + 6q$

  e $12x - 7x - 4x$ **f** $8w - w - 3w$

  g $4a + 6a - 3a$ **h** $12b - 3b - 4b$

  i $3j - 4j + 2j$ **j** $k - 5k + 6k$

7 Simplify these divisions. The first one has been done for you.

  a $d \div 4 = \dfrac{d}{4}$ **b** $x \div 3$

  c $y \div 7$ **d** $t \div 9$

  e $2a \div 3$ **f** $3n \div 4$

  g $5p \div 7$ **h** $2v \div 4$

8 Simplify these expressions.

  a $2p + 5q + 3p + q$

  b $6x + 2y + 3x + 5y$

  c $4m + 2n - 2m + 6n$

  d $5x + 3y - 4x + 2y$

  e $7r - 4s + r - 2s$

  f $2f - 3g + 5g - 6f$

  g $3a + 2b + 5c - a + 4b$

  h $7u - 5v + 3w + 3v - 2u$

  i $5x - 3y - 2x + 4z - y + z$

  j $4r + 6s - 3t + 2r + 5t - s$

9 Simplify these expressions.

  a $3a + 4b + 8a + 2b$

  b $3t + 9 - t + 17$

  c $3x - 4y - 2x - 8y$

  d $9p + p^2 + 5p$

  e $10xy + 10yx$

  f $6ab + 2ba - ba$

10 Simplify these expressions.

  a $4m \times 7n$ **b** $6m \times 2m$

  c $\dfrac{20p}{2}$ **d** $\dfrac{14a}{7a}$

  e $2a \times 3b \times 4c$ **f** $k \times 2k \times 3k$

  g $\dfrac{20ab}{5a}$ **h** $\dfrac{45c^2}{5c}$

11 Simplify these expressions.

  a $3r + 3 \times 2r$ **b** $2m^2 + 2m \times m$

  c $6x \div 2 + x$ **d** $2t \times 4v$

  e $5m \times 2n$ **f** $3x \times 2y^2$

  g $x^2 + x^2 + x$ **h** $3 \times 3w - 2 \times 4$

  i $z \times z^2 + 3z + 1$

12 Simplify each expression.

  a $3k^2 - 2 \times k^2 + k$

  b $4m + 6m + 2 + m^2$

  c $6t - (4 \times -t) + 5$

13 Copy this grid, replacing each expression with its simplified form (where possible).

$3a + 7b - 5a + 2b$	$3a \times 4a$	$\dfrac{20b}{5}$
$\dfrac{16ab^2}{8b}$	$2p + 7p^2 + 5p^3 + 8p$	$11abc + 2cab$
$5m - 4$	$3m \times 4m \times 5m$	$\dfrac{4a}{2a^3}$

Q 1178, 1179, 1186   SEARCH

2.1 Simplifying expressions

- You can substitute values into expressions and formulae.
- You can add, subtract, multiply or divide algebraic terms.
- To simplify an expression, you follow the same order of operations as in arithmetic.

> You can use the acronym BIDMAS to remember the order.

HOW TO

To write expressions and equations in algebra

① Give every '**unknown**' a letter and write it down. Translate the words into letters and symbols.

② Simplify the expression by collecting like terms. You may need to substitute values into your expression.

③ Interpret your answer in the context of the question.

EXAMPLE

Paul, Rashid and Sara count the number of books in their school bags.
Paul has p books, Rashid has r books and Sara has s books.
These two equations are true: $r = 2p$ and $s = r + 2$

a Paul says that he has half the number of books as Rashid. Is he correct?

b Rashid says that he has two more books than Sarah. Is he correct?

c Show that $p + r + s = 12$ when $p = 2$.

a Paul has half the number of books as Rashid. This is the same as saying Rashid has twice as many books as Paul.
  ① Translate into letters and symbols.
  $r = 2p$. This is true, so Paul is correct.

b Rashid has two more books than Sarah.
  ① Translate into letters and symbols.
  $r = s + 2$ this is the same as saying
  $r - 2 = s$
  $s = r - 2$ is not true, so Rashid is wrong.

c ② Substitute $p = 2$ to find r.
  $r = 2p = 2 \times 2 = 4$
  Substitute $r = 4$ to find s.
  $s = r + 2 = 4 + 2 = 6$
  Substitute $p = 2$, $r = 4$ and $s = 6$.
  ③ $p + r + s = 2 + 4 + 6 = 12$

EXAMPLE

Find an expression for the area of the shape.

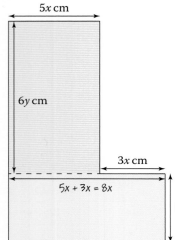

5x cm

6y cm

3x cm

5x + 3x = 8x

2y cm

① Find the area of each part of the composite shape.
  Area = length × width
  Area of red rectangle = $5x \times 6y = 30xy$
  The length of the blue rectangle is $5x + 3x = 8x$
  Area of blue rectangle = $8x \times 2y = 16xy$
② Collect the terms together to find the total area.
③ Total area = $30xy + 16xy = 46xy$ cm²
  Include units in your final answer.

Exercise 2.1A

1 Three students tried to simplify $3m + 5$.
Which of them did it correctly?

Sara
$$3m + 5 = 8m$$

Caitlin
$$3m + 5 = 15m$$

Abdul
$$3m + 5 = 3m + 5$$

2 Gemma and Paul evaluated $2x^2$ when $x = 6$.
Who was right? Explain why.

Gemma
when $x = 6$, $2x^2 = 144$

Paul
when $x = 6$, $2x^2 = 72$

3 Audrey, Billie and Cerys count the amount of money they each have.
Audrey has £a, Billie has £b and Cerys has £c.
These two equations are true:
$$b = 3a$$
$$c = a + b$$

a Audrey says that she has three times as much money as Billie.
Cerys says that she has four times as much money as Audrey.
Are they both correct?
Explain your answer fully.

b Audrey has £5. How much money do the three friends have in total?

4 Simplify the expressions in the grid and find the 'odd one out' for each row.

$3p + 2q + p + 5q$	$6p + 3q - 2p + 4q$	$5p - 3q - p + 5q$
$2m \times 3n$	$2 \times n \times m \times 5$	$6mn$
$\dfrac{24cd}{12c}$	$\dfrac{2d^2}{d^2}$	$\dfrac{2d^2}{d}$
$2n - 8$	$3m + 2n - m - 2m$	$3n - 2 - 6 - n$

5 Rearrange each set of cards to make a correct statement.

a

$(x - 3y)$ $+$ $(2x + 3y)$ $=$

$(3x + 2y)$ $-$ $(2x - 4y)$

b

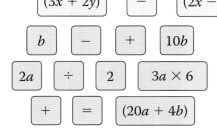

6 a Write a simplified expression for the

  i perimeter

  ii area of this rectangle.

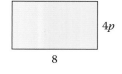
4p
8

b Two of the rectangles are joined to make a composite shape.

  i Write an expression for the area of the composite shape.

  ii The perimeter of the composite shape is $32 + 8p$.

    Draw a possibility for the composite shape.

7 What are the measurements of a rectangle with perimeter $6x + 4y$ and area $6xy$?

8 Find the area of the shape.

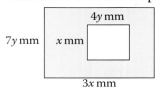

$7y$ mm x mm $4y$ mm $3x$ mm

9

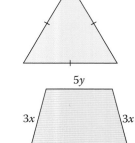

$2y + 3x$ $3y + x$ $5y + 2x$ $4y + x$

$5y$ $3x$ $3x$ $5y + x$

a The triangle and the parallelogram have the same perimeter.
Find the side length of each side of the triangle.

b Does the rectangle or the trapezium have the greatest perimeter?
Explain how you decide.

10 The area of the trapezium is $12p^2$. Find an expression for the height.

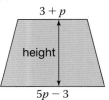

$3 + p$
height
$5p - 3$

   1178, 1179, 1186   SEARCH

2.2 Indices

- You can use **index** notation to write repeated multiplication.

Indices is the plural of index.

$5 \times 5 \times 5 \times 5 = 5^4 \qquad m \times m \times m \times m = m^4$

5 is the **base**, 4 is the index. You say 'm to the **power** of 4'.

You can simplify expressions with **indices** and numbers.

EXAMPLE

Simplify

a $3 \times p \times p \times p \times q \times q$

b $2 \times s \times s \times 3 \times t \times t \times t$

a $3 \times p \times p \times p \times q \times q$
$= 3 \times p^3 \times q^2$
$= 3p^3q^2$

b $2 \times s \times s \times 3 \times t \times t \times t$
$= 2 \times s^2 \times 3 \times t^3$
$= 2 \times 3 \times s^2 \times t^3$
$= 6s^2t^3$

You can simplify expressions with powers of the same base.

$n^2 \times n^2$

$= n \times n \times n \times n = n^4$

$t^5 \div t^2 = \dfrac{t^5}{t^2} = \dfrac{{}^1\!t \times {}^1\!t \times t \times t \times t}{{}^1\!t \times {}^1\!t} = t^3$

$(v^2)^3 = v^2 \times v^2 \times v^2 = v^{2 \times 3} = v^6$

- To multiply powers of the same base, add the indices.

$x^a \times x^b = x^{(a + b)}$

- To divide powers of the same base, subtract the indices.

$x^a \div x^b = x^{(a - b)}$

- When finding the 'power of a power', multiply the indices.

$(x^a)^b = x^{a \times b}$

You can apply the index laws to positive and negative indices.

- When terms have numerical **coefficients**, deal with these first.

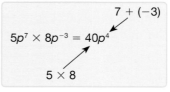

$7 + (-3)$

$5p^7 \times 8p^{-3} = 40p^4$

5×8

EXAMPLE

Simplify each of these using the index laws.

a $\dfrac{(k^3 \times k^2)^7}{k}$

b $(5p^2)^3 \times 2p^{-7}$

a $\dfrac{(k^3 \times k^2)^7}{k} = \dfrac{(k^5)^7}{k}$
$= \dfrac{k^{35}}{k}$
$= k^{34}$

$3 + 2 = 5$
$5 \times 7 = 35$

Remember k is really k^1.

$35 - 1 = 34$

b $(5p^2)^3 \times 2p^{-7}$
$= 125p^6 \times 2p^{-7}$
$= 250p^{-1}$

$2 \times 3 = 6$
$125 \times 2 = 250$
$6 + -7 = -1$

EXAMPLE

Expand $(2p^2q)^3$

$2^3 = 8, (p^2)^3 = p^6, (q^1)^3 = q^3.$
$(2p^2q)^3 = 8p^6q^3$

Everything inside the bracket is cubed.

Exercise 2.2S

1 Write these expressions in index form.

   a $y \times y \times y \times y$

   b $m \times m \times m \times m \times m \times m$

   c $6 \times v \times v \times w \times w \times w$

   d $2 \times r \times r \times r \times r \times s$

   e $2 \times m \times m \times 3 \times n$

   f $4 \times y \times y \times y \times 2 \times z \times z$

2 Simplify

   a $3m^2 \times 2$ **b** $3 \times 4p^3$

   c $2x \times 3y^2$ **d** $5r^2 \times 2s^2$

3 Kyle thinks that $a^5 \times a^2 = a^{10}$.
   Do you agree with Kyle? Give your reasons.

4 Simplify

   a $n^2 \times n^3$ **b** $s^3 \times s^4$

   c $p^3 \times p$ **d** $t \times t^3$

5 Write each of these as a single power in the form x^n.

   a $x^2 \times x^2 \times x^3$ **b** $x \times x^5 \times x^2$

   c $x^3 \times x^2 \times x^4$ **d** $x^5 \times x \times x$

6 Write each of these as a single power in the form r^n.

   a $r^4 \div r^2$ **b** $r^5 \div r^4$

   c $r^7 \div r^2$ **d** $r^8 \div r^5$

7 Simplify

   a $\dfrac{m^6}{m^2}$ **b** $\dfrac{x^4}{x^3}$ **c** $\dfrac{t^7}{t^5}$ **d** $\dfrac{y^4}{y}$

8 Simplify

   a $\dfrac{x^2 \times x^3}{x^4}$ **b** $\dfrac{m^3 \times m}{m^2}$

   c $\dfrac{s^2 \times s^4}{s^3}$ **d** $\dfrac{v \times v^3 \times v^3}{v^4}$

   e $\dfrac{q^2 \times q^3 \times q^2}{q^4}$ **f** $\dfrac{t^3 \times t \times t^2}{t^2}$

   g $\dfrac{p^4 \times p^2 \times p^2}{p^7}$ **h** $\dfrac{y^2 \times y^4 \times y}{y^3 \times y^2}$

9 Simplify these expressions, giving your answer in index form.

   a $x^{10} \times x \times x^4$ **b** $x^4 \times x^{11} \div x^3$

   c $x^5 \times x^7 \times y^3 \div y$ **d** $x^6 \times y^3 \times x^9 \times y^2$

   e $\dfrac{y^4 \times x^8}{x^0 \times y}$ **f** $\dfrac{x^3 \times y^7 \times z^4 \div y^2}{z^3 \times x \times y^3}$

10 Tracey thinks that $4y^5 \times 2y^2 = 6y^7$ because *'the index rules say that you add the powers when two terms are multiplying each other'.* Do you agree with Tracey? Give reasons for your answer.

11 Simplify these expressions.

   a $3x^5 \times x^2$ **b** $5y^2 \times y^5$

   c $4b^2 \times 3b^6$ **d** $2p^4 \times 5p^7$

   e $5h^5 \times 6h^6$ **f** $4s^3 \times 3t^4$

12 Andy thinks that $12p^{12} \div 3p^4 = 9p^8$ because *'the index rules say that you subtract the powers when two terms are dividing each other'.* Do you agree with Andy?

   Give reasons for your answer.

13 Simplify these expressions.

   a $10y^6 \div 5y^2$ **b** $6a^9 \div 3a^3$

   c $20k^7 \div 4k^3$ **d** $18p^8 \div 6p^3$

   e $35x^{10} \div 7x^4$ **f** $4x^8 \div 8y^4$

14 Simplify these expressions.

   a $(a^3)^2$ **b** $(y^2)^6$ **c** $(k^3)^5$

   d $(p^7)^8$ **e** $(a^3)^7$ **f** $(a^3)^7$

15 Simplify these expressions.

   a $(2a^3)^2$ **b** $(3y^2)^6$ **c** $(5k^3)^2$

   d $(6p^7)^3$ **e** $(2a^3)^7$ **f** $(4a^4)^4$

16 Simplify these expressions.

   a $y^{-5} \times y^7$ **b** $x^2 \times x^{-4}$

   c $a^{-1} \times a^{-5}$ **d** $h^{-2} \div h^4$

   e $\dfrac{p^3}{p^{-1}}$ **f** $\dfrac{p^{-4}}{p^{-3}}$

17 Simplify

   a $g^8 \times g^{-5}$ **b** $\dfrac{h^{-2}}{h^4}$

   c $(b^{-4})^3$ **d** $j^{-4} \times j^{-2}$

   e $(t^{-5})^{-2}$ **f** $n^{-8} \div n^{-6}$

18 Simplify fully

   a $(2p^8)^2$ **b** $10r^3 \times 6r^{-4}$

   c $(3h^{-3})^3$ **d** $9b^3 \div 3b^{-5}$

   e $(3m^3 \times 2m^{-7})^2 \div 18m$

   f $18(f^{-4})^4 \div 9f^{-16}$

1033, 1045, 1301, 1951   SEARCH

2.2 Indices

RECAP

- An **index** is a power. The **base** is the number which is raised to this power.
- You can simplify expressions with the same base using the three index laws.
  - When multiplying, add the indices.
  - When dividing, subtract the indices.
  - To find the 'power of a power', multiply the indices.

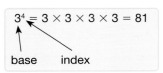

$3^4 = 3 \times 3 \times 3 \times 3 = 81$

base   index

$x^2 \times x^5 = x^7$

$y^8 \div y^4 = y^4$

$(z^3)^2 = z^6$

HOW TO

① Read the question carefully. You may need to apply your knowledge of other topics.

② Use the index laws to calculate or simplify. Deal with numbers first, then powers.

③ Answer the question. Give an explanation if the question asks for one.

EXAMPLE

A rectangle has area $28x^2y^6$.

The width of the rectangle is $4xy^3$.

Find the perimeter of the rectangle.

> The perimeter is the distance around the edge of the rectangle.

① To find the perimeter, you need to find the length first.

Area = length × width

Length = area ÷ width

② Length $= \dfrac{28x^2y^6}{4xy^3} = \dfrac{28}{4} \times \dfrac{x^2}{x} \times \dfrac{y^6}{y^3} = 7xy^3$

③ Find the distance around the rectangle.

Perimeter $= 4xy^3 + 7xy^3 + 4xy^3 + 7xy^3 = 22xy^3$

EXAMPLE

The three terms along each diagonal must multiply to give the same expression.

$6x$		$9x^2$
	$2x$	
$4x^{-2}$		

Complete the grid.

① Find the product of the diagonal

② $4x^{-2} \times 2x \times 9x^2 = 72x^1$ $4 \times 2 \times 9 = 72$

                  $= 72x$ $-2 + 1 + 2 = 1$

The diagonals multiply to give $72x$.

$\dfrac{72x}{6x \times 2x} = \dfrac{72x}{12x^2}$ Simplify the denominator first.

         $= 6x^{-1}$ $72 \div 12 = 6$ $x \div x^2 = x^{-1}$

③ The missing entry is $6x^{-1}$.

Exercise 2.2A

1 Find an expression for the area of each shape.

a

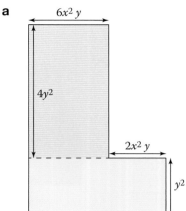

b

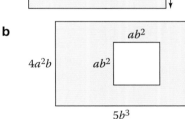

2 A rectangle has area $20p^4q^2$.
The width of the rectangle is $2p^2q$.
Find the perimeter of the rectangle.

3 A square has area $16a^2b^6$.
Find the length of the sides of the square.

4 Rearrange each set of cards to make a correct statement.

a

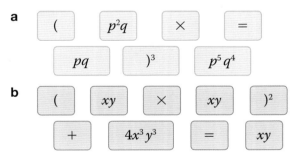

b

5 Jenna writes the first four terms of a geometric sequence.

$$xy, \quad 2x^3y, \quad 4x^5y, \quad 6x^7y \dots$$

Which term is wrong?

Explain your answer.

6 Simplify $((x^4)^2)^5$

7 Is this statement true or false?
$$(xy^2)^3 = (xy^3)^2$$
Explain your answer.

8

> $0 = 1 - 1$
>
> You can use this fact to show that $x^0 = 1$ for any value of x.
> $$x^0 = x^{1-1}$$
> $$= x^1 \div x^1$$
> $$= x \div x$$
> $$= 1$$

$-1 = 0 - 1$

Use this fact to show that $x^{-1} = \frac{1}{x}$.

9 $(\sqrt{x^2}) = x$

Use this fact to show that $\sqrt{x} = x^{\frac{1}{2}}$.

Hint: What happens when you square $x^{\frac{1}{2}}$?

10 Write a simplified expression for the area of this triangle.

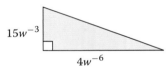

11 Show that the expression $(4p^4)^3 \div (8p^7)^2$ simplifies to $\frac{1}{p^2}$.

12 True or false? $3^x \times 3^y$ simplifies to give 9^{x+y}.
Explain your answer.

13 Which expression is the odd one out?
Explain your answers.

$$5t^2 \times 10t^{-4} \div (5t^{+3})^2 \qquad \frac{2}{t^6} \qquad \left(\frac{4t^2}{16t}\right) \times 8t^{-7}$$

14 Find the value of x in this equation.
$$(2^2)^x \times 2^{3x} = 32$$

15 a If $u = 3^x$, show that $9^x + 3^{x+1}$ can be written as $u^2 + 3u$.

b Write an expression in terms of x for $u^3 + 9u$.

c Write an expression in terms of x for $u^2 - \frac{1}{u}$.

d Write $81^x - 9^{x-1}$ in terms of u.

2.3 Expanding and factorising 1

You can use **brackets** in algebraic expressions.

- You can multiply out brackets.
  - You multiply each term inside the bracket by the term outside.

$$2(x + 4) = 2 \times x + 2 \times 4 = 2x + 8$$

Expand means 'multiply out'.

EXAMPLE

Expand these brackets.

a $3p(2p + 7 - q)$

b $-4m(m - 2)$

a $3p\,(2p + 7 - q) = 6p^2 + 21p - 3pq$

b $-4m(m - 2) = -4m^2 + 8m$

EXAMPLE

Expand and simplify $3(t - 2) + 5(2 + t)$.

$$3(t - 2) + 5(2 + t) = 3t - 6 + 10 + 5t$$
$$= 3t + 5t - 6 + 10$$
$$= 8t + 4 \quad \text{Collect like terms.}$$

To factorise completely use the highest common factor.

- To factorise an expression, look for a **common factor** which divides into all the terms.

$$3x + 9$$
$$\div 3) \quad) \div 3$$
$$3(x + 3)$$

Write the common factor outside the bracket.

$$a^2 - a$$
$$\div a) \quad \div a)$$
$$a(a - 1)$$

Sometimes the common factor is a letter.

EXAMPLE

Factorise fully

a $6x + 9$

b $12pq - 4pw$

c $5x + 10x^2 - 25xy$

a The HCF of $6x$ and 9 is 3. $\quad 6x + 9 = 3(2x + 3)$

b HCF of $12pq$ and $4pw$ is $4p$. $\quad 12pq - 4pw = 4p(3q - w)$

c The HCF of $5x$, $10x^2$ and $25xy$ is $5x$.

$$5x + 10x^2 - 25xy = 5x(1 + 2x - 5y)$$

EXAMPLE

a Factorise $y^2 - 3y$

b Factorise $(p + q)^2 - 2(p + q)$

a $y^2 - 3y = y(y - 3)$

b Each part in **b** has $(p + q)$ in common

$$(p + q)^2 - 2(p + q)$$
$$= (p + q)((p + q) - 2)$$
$$= (p + q)(p + q - 2)$$

Algebra Expressions

Exercise 2.3S

1 Expand these brackets.

    a $4(n + 5)$ **b** $6(b - 7)$

    c $a(a + 3)$ **d** $a(b - c)$

    e $4(2x + 3y - 4z)$ **f** $2h(h + 9)$

2 Expand these brackets.

    a $-3(k + 9)$ **b** $-2(h - 5)$

    c $-(w - 4)$ **d** $-(t - p)$

    e $-k(k + 7)$ **f** $-9(2m - k + 4)$

    g $-(x^2 - x - 8)$ **h** $-2(x^2 + 3)$

    i $-3(1 - x)$ | Be careful with negatives. |

3 Find all the common factors of

    a $2x$ and 6

    b $4y$ and 12

    c 10 and $20j$

    d 6 and $12p$

    e 9 and $6q$

    f $6t$ and 4

    g $4x$ and 10

    h $24t$ and 8

> Hint for **3a**: 2 and x are factors of $2x$. 1, 2, 3 and 6 are factors of 6. 2 is a common factor of $2x$ and 6.

4 Find the highest common factor of

    a $3x$ and 9 **b** $12r$ and 10

    c $6m$ and 8 **d** 4 and $4z$

5 Find the highest common factor of

    a y^2 and y **b** $4s^2$ and s

    c $7m$ and m^3 **d** $2y^2$ and $2y$

6 Factorise these expressions.

    a $2x + 10$ **b** $3y + 15$

    c $8p - 4$ **d** $6 + 3m$

    e $5n + 5$ **f** $12 - 6t$

    g $14 + 4k$ **h** $9z - 3$

7 Factorise each of these fully by removing common factors.

    a $2x + 4$ **b** $3y - 6$

    c $12p + 36q$ **d** $25w - 5$

    e $6xy + xw$ **f** $ab - 2bc$

    g $pqr + qrt - qsw$ **h** $5xy - x$

7   **i** $2xy + 6x$ **j** $4ab + 6a^2$

    k $25p^2 - 10p$ **l** $7x + 14xy$

    m $2ac + 4a^2 - 8a$

    n $15mn - 5m + 10m^3$

    o $6p^4 - 12p$

8 Expand and simplify each of these expressions.

    a $3(p + 3) + 2p$ **b** $2(m + 4) + 5m$

    c $4(x + 1) - 2x$ **d** $2(5 + k) + 3k$

    e $4(2t + 3) + t - 2$ **f** $3(2r + 1) - 2r + 4$

9 Expand and simplify these expressions.

    a $3(c + 2) + 7(c + 8)$

    b $4(2x + 8) + 5(3x + 7)$

    c $x(x + 8) + x(x + 2)$

    d $5t(3t + 6) + 2t(t + 1)$

    e $3(x - 7) + 4(x - 6)$

    f $5(2 - x) + 7(x - 3)$

    g $4(m - 6) - 2(m + 1)$

    h $3(g - 3) - 7(2g - 6)$

    i $2(p + 5) - (p - 4)$

    j $(q - 4) - (3 - q)$

10 Factorise these expressions

    a $10(x + y) + 13(x + y)$

    b $(a - b)^2 + 5(a - b)$

    c $6(q + r) - (q + r)^3$

    d $(pt - w) + 6(pt - w)$

11 Expand and simplify

    $2x(x + 7) + x(9 - x) - 3x(2x - 7)$

12 Expand and simplify

    a $3(5x + 9)$

    b $2p(4p - 8)$

    c $3m(5 - 2m)$

    d $3(2y + 9) + 5(3y - 2)$

    e $5x(2x + 2y - 9)$

    f $4(t + 9) - 3(2t - 7)$

    g $(7h + 9) - (3h - 7)$

    h $x(3x^2 + x^3)$

2.3 Expanding and factorising 1

- Expand means multiply each term inside the bracket by the term outside.
- Factorising is the 'opposite' of expanding brackets.

$$2(x + 4) \overset{\text{expand}}{\underset{\text{factorise}}{=}} 2x + 8$$

① Read the question carefully. Give any unknown values a letter.

② Expand the brackets   …or…   find the HCF of all of the terms and collect like terms   and factorise.

③ Answer the question fully.

Look for a **common factor** for all the terms.

EXAMPLE

Fill in the missing values.
$3(2x + 11) + \square(2x - 5) = 8(2x + \square)$

① Label the empty boxes so that you can tell them apart.
$3(2x + 11) + \boxed{*}(2x - 5) = 8(2x + \boxed{\cdot})$

② Expand the brackets on both sides.
$6x + 33 + \boxed{*} \times 2x - 5 \times \boxed{*} = 16x + 8 \times \boxed{\cdot}$

Look at the x terms. $6x + \boxed{*} \times 2x = 16x$
$6 + 2 \times \boxed{*} = 16$
$2 \times \boxed{*} = 10 \text{ so } \boxed{*} = 5$

Now look at the constant terms. $33 - 5 \times \boxed{*} = 8 \times \boxed{\cdot}$
Substitute $\boxed{*} = 5$ $33 - 5 \times 5 = 8 \times \boxed{\cdot}$
$33 - 25 = 8 \times \boxed{\cdot}$
$8 = 8 \times \boxed{\cdot} \text{ so } \boxed{\cdot} = 1$

③ $3(2x + 11) + \boxed{5}(2x - 5) = 8(2x + \boxed{1})$

EXAMPLE

A rectangle of width x has length 1 cm more than its width. Its area is 200 cm².
Show that $x^2 + x = 200$. Explain why x must be between 13 cm and 14 cm.

① The length is $(x + 1)$ cm
Area of rectangle = length × width
$200 = x(x + 1)$

② $200 = x^2 + x$

Sketch a diagram to help:

③ Substitute $x = 13$ and $x = 14$. $13^2 + 13 = 169 + 13 = 182$ smaller than 200
x is between 13 cm and 14 cm. $14^2 + 14 = 196 + 14 = 210$ larger than 200

EXAMPLE

A cuboid has volume $48x^2 + 16xy$ cm³.

Write down a possibility for the dimensions of the cuboid.

② Factorise the expression. The HCF of $48x^2$ and $16xy$ is $16x$.
$48x^2 + 16xy = 16x(3x + y)$

③ Volume of a cuboid = length × width × height
$16x(3x + y) = 16 \times x \times (3x + y)$
Possible dimensions are 16 cm × x cm × $(3x + y)$ cm
There are many other possibilities including 8 cm × $2x$ cm × $(3x + y)$ cm,
4 cm × $4x$ cm × $(3x + y)$ cm

Algebra Expressions

Exercise 2.3A

1 All three students have completed their factorisations incorrectly.
Explain what they have done wrong.

Clare	Ben	Vicky
$5x + 10xy$ $= 5x(0 + 2y)$	$6pq + 3p$ $= 3(2pq + 1)$	$21p + 14pq$ $= 7p(14 + 7q)$

2 Factorise fully

   a $ax + bx + ay + by$

   b $cd + bd + cm + bm$

   c $a^2 + ab + 2a + 2b$

   d $cd + ce - me - md$

3 Fill in the missing values.

   a $3(x + \square) = 3x + 12$

   b $2(x + 4) + \square = 2x + 11$

   c $\square(2x - 1) = 8x - \square$

   d $6(x - \square) + \square(2x + 1) = 16x - 19$

   e $2(x + 1) + \square(3x - 1) = 20x - \square$

   f $\square(4x + 3) + 2(x - 1) = 7(2x + \square)$

4 Write a factorised expression for

   a the perimeter of this rectangle

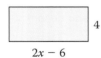

       4

     $2x - 6$

   b the perimeter of a square with sides $5b + 10$.

5 **a** Using brackets, write a formula for the area of this rectangle.

       3

     $2x - 1$

   b Expand the brackets.

   c The area of the rectangle is $15\,\text{cm}^2$.
Show that $6x - 18 = 0$.

6 An expression expands to give $24x + 16$.

   a What could the expression have been if it involved one pair of brackets?

   b What could the expression have been if it involved adding two single brackets?

7 Write an expression involving brackets for the area of this trapezium.

Expand the brackets and simplify your expression.

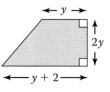

8 Use factorisation to help you to evaluate these, without a calculator.

   a $2 \times 1.86 + 2 \times 1.14$

   b $3 \times 5.87 - 3 \times 0.37$

   c $5.86^2 + 5.86 \times 4.14$

   d $3.32 \times 6.68 + 3.32^2$

9 Show that the shaded area of this rectangle is $2(4x + 5)$.

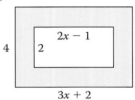

10 A cuboid has volume $20xy + 50x^2y\,\text{cm}^3$.

Write down three possibilities for the dimensions of the cuboid.

11 A cuboid has volume $16x^2y - 54x^2\,\text{cm}^3$.

   a Write down three possibilities for the dimensions of the cuboid.

   b Explain why y must be greater than $3.375\,\text{cm}$.

12 A trapezium has area $9pq + 7p^2q$.
Find the length of the missing side.
Give your answer in terms of p and q.

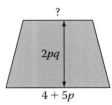

13 Find the missing term in this geometric sequence.

?
$6x^2 - 12x$
$12x^3 - 24x^2$
?
$48x^5 - 96x^4$

In a geometric sequence, the next term is found by multiplying the previous term by a number.

Q 1155, 1247   SEARCH

2.4 Algebraic fractions

To convert to an equivalent fraction, multiply numerator and denominator by the same expression.

● You can add (or subtract) fractions by converting them to equivalent fractions with a common **denominator**, and then adding (or subtracting) the **numerators**.

EXAMPLE

Simplify these expressions.

a $\dfrac{x + 4}{6} - \dfrac{2x - 1}{5}$

b $\dfrac{5}{x + 2} + \dfrac{3}{x - 4}$

a $\dfrac{x + 4}{6} - \dfrac{2x - 1}{5} = \dfrac{5(x + 4)}{30} - \dfrac{6(2x - 1)}{30}$

$= \dfrac{5x + 20}{30} - \dfrac{(12x - 6)}{30}$

$= \dfrac{5x + 20 - 12x + 6}{30}$

$= \dfrac{26 - 7x}{30}$

b $\dfrac{5}{x + 2} + \dfrac{3}{x - 4}$

$= \dfrac{5(x - 4)}{(x + 2)(x - 4)} + \dfrac{3(x + 2)}{(x + 2)(x - 4)}$

$= \dfrac{5x - 20 + 3x + 6}{(x + 2)(x - 4)}$

$= \dfrac{8x - 14}{(x + 2)(x - 4)} = \dfrac{2(4x - 7)}{(x + 2)(x - 4)}$

Factorise the numerator and denominator – sometimes you may be able to cancel further.

● You can **cancel** common **factors** in algebraic expressions.

EXAMPLE

Cancel these fractions fully.

a $\dfrac{5p^3}{10p}$

b $\dfrac{3x + 6}{3}$

c $\dfrac{3 + p}{3}$

d $\dfrac{x - 3}{x^2 - 3x}$

a $\dfrac{5p^3}{10p} = \dfrac{p^2}{2} = \dfrac{1}{2}p^2$ — Divide numerator and denominator by the common factor $5p$.

b $\dfrac{3x + 6}{3} = \dfrac{3(x + 2)}{3} = x + 2$ — Factorise first.

c $\dfrac{3 + p}{3}$ — You cannot cancel the expression because 3 and $3 + p$ have no common factors.

d $\dfrac{x - 3}{x^2 - 3x} = \dfrac{x - 3}{x(x - 3)} = \dfrac{1}{x}$ — Factorise the denominator and then cancel.

● You can cancel common factors before multiplying or dividing.

EXAMPLE

a $\dfrac{x}{3x + 6} \times \dfrac{x + 2}{x^2}$

b $\dfrac{3}{x} \div \dfrac{3x - 6}{x^3}$

Cancel the common factors before multiplying.

a $\dfrac{x}{3x + 6} \times \dfrac{x + 2}{x^2} = \dfrac{x}{3(x + 2)} \times \dfrac{x + 2}{x^2_x}$

$= \dfrac{1}{3} \times \dfrac{1}{x}$

$= \dfrac{1}{3x}$

b $\dfrac{3}{x} \div \dfrac{3x - 6}{x^3} = \dfrac{3}{x} \times \dfrac{x^3}{3x - 6}$

$= \dfrac{3}{x} \times \dfrac{x^{3x^2}}{3(x - 2)}$

$= \dfrac{x^2}{x - 2}$

Exercise 2.4S

1 Cancel down each of these fractions into its simplest form.

a $\dfrac{4}{12}$ **b** $\dfrac{21}{28}$ **c** $\dfrac{24}{40}$

d $\dfrac{28}{63}$ **e** $\dfrac{45}{72}$ **f** $\dfrac{42}{126}$

2 Add these fractions.

a $\dfrac{1}{2} + \dfrac{1}{2}$ **b** $\dfrac{1}{2} + \dfrac{1}{4}$ **c** $\dfrac{1}{2} + \dfrac{1}{3}$

3 Subtract these fractions.

a $\dfrac{2}{3} - \dfrac{1}{3}$ **b** $\dfrac{1}{2} - \dfrac{1}{6}$ **c** $\dfrac{1}{3} - \dfrac{1}{4}$

4 Do these multiplications and divisions.

a $\dfrac{3}{8} \times 4$ **b** $2 \times \dfrac{2}{5}$ **c** $\dfrac{7}{8} \div 2$

d $\dfrac{15}{16} \div 3$ **e** $12 \times \dfrac{3}{8}$ **f** $\dfrac{8}{11} \div 6$

5 Calculate

a $\dfrac{2}{3} \times \dfrac{1}{3}$ **b** $\dfrac{5}{8} \div \dfrac{1}{4}$ **c** $\dfrac{5}{9} \times \dfrac{1}{5}$

d $\dfrac{9}{20} \div \dfrac{1}{5}$ **e** $\dfrac{4}{15} \times \dfrac{5}{8}$ **f** $\dfrac{12}{21} \div \dfrac{6}{7}$

6 Cancel these fractions fully.

a $\dfrac{15w}{5}$ **b** $\dfrac{3b}{9}$ **c** $\dfrac{10c^2}{5c}$

d $\dfrac{12bd}{3d^2}$ **e** $\dfrac{100(bd)^2}{25b}$ **f** $\dfrac{60x^2y^4}{40xy}$

g $\dfrac{2x + 6}{2}$ **h** $\dfrac{x^2 + x}{x}$ **i** $\dfrac{5y - 10}{15}$

7 David is simplifying $\dfrac{5x + 35}{x + 4}$

$$\dfrac{5x + 35}{x + 4} = \dfrac{5 + 35}{4} = \dfrac{40}{4} = 10$$

Do you agree with David's solution?

Explain your answer.

8 Simplify these fractions.

a $\dfrac{x^2 + 2x}{x + 2}$ **b** $\dfrac{p^2 - 3p}{p - 3}$ **c** $\dfrac{y - 5}{y^2 - 5y}$

d $\dfrac{6y^2}{y^3 - y^2}$ **e** $\dfrac{3x^2 + 6x}{x^2 + 2x}$ **f** $\dfrac{3x^3 + 4x}{x^2 + 2}$

9 Simplify these fractions.

a $\dfrac{(y + 2)^2}{y + 2}$ **b** $\dfrac{(x - 4)^2}{x - 4}$ **c** $\dfrac{x + 3}{(x + 3)^2}$

d $\dfrac{(p - 1)^3}{p - 1}$ **e** $\dfrac{(y + 4)^3}{(y + 4)^2}$ **f** $\dfrac{b - 2}{(b - 2)^4}$

10 Simplify these expressions.

a $\dfrac{3p}{5} + \dfrac{p}{5}$ **b** $\dfrac{y}{7} + \dfrac{3y}{7}$ **c** $\dfrac{1}{3p} + \dfrac{8}{3p}$

d $\dfrac{5y}{4} + \dfrac{y}{8}$ **e** $\dfrac{2p}{5} - \dfrac{p}{3}$ **f** $\dfrac{6}{x} - \dfrac{7}{y}$

11 Sort these expressions into equivalent pairs. Which is the odd one out? Create its pair.

A $\dfrac{5x}{12} - \dfrac{3x}{12}$ **B** $\dfrac{x}{6} + \dfrac{x}{4}$

C $\dfrac{2}{3}x - \dfrac{1}{3}x$ **D** $\dfrac{x}{3} - \dfrac{x}{4}$ **E** $\dfrac{x}{6}$

F $\dfrac{x}{12}$ **G** $\dfrac{4x^2}{12x}$

12 By cancelling where possible, simplify these multiplication and division calculations.

a $\dfrac{4p}{3} \times \dfrac{9}{4p}$ **b** $\dfrac{6ab}{7} \times \dfrac{2}{b}$ **c** $\dfrac{4m^2}{8} \times \dfrac{2n}{5m^3}$

d $\dfrac{3}{g} \div \dfrac{g}{5}$ **e** $\dfrac{4w}{3} \div \dfrac{w}{2}$ **f** $\dfrac{2f^2}{p^3} \times \dfrac{p}{4f}$

13 Simplify these expressions.

a $\dfrac{5}{x + 2} \times \dfrac{2}{x}$ **b** $\dfrac{3}{x} \times \dfrac{x}{x - 1}$

c $\dfrac{3}{x + 2} \times \dfrac{3x + 6}{2x + 3}$ **d** $\dfrac{x}{x + 2} \times \dfrac{x^2 + 2x}{x^2}$

e $\dfrac{x}{x + 2} \div \dfrac{x}{2}$ **f** $\dfrac{5y - 10}{15} \div \dfrac{y - 2}{3y}$

14 Write each expression as a single fraction.

a $\dfrac{x + 2}{5} + \dfrac{2x - 1}{4}$ **b** $\dfrac{3x - 2}{7} + \dfrac{5 - 3x}{11}$

c $\dfrac{2y - 5}{3} - \dfrac{3y - 8}{5}$

d $\dfrac{3(p - 2)}{5} - \dfrac{2(7 - 2p)}{7}$

e $\dfrac{2}{x - 7} + \dfrac{3}{x + 4}$ **f** $\dfrac{5}{x - 2} + \dfrac{3}{x + 3}$

g $\dfrac{3}{y - 2} - \dfrac{4}{y + 1}$ **h** $\dfrac{2}{p + 3} + \dfrac{5}{p - 1}$

i $\dfrac{3}{w} + \dfrac{9}{w - 8}$ **j** $\dfrac{4}{x - 2} + \dfrac{5}{(x - 2)^2}$

Q 1149, 1151, 1164   SEARCH

2.4 Algebraic fractions

- Add and subtract algebraic fractions by finding a common denominator.
- Simplify algebraic fractions by cancelling common factors.

HOW TO

To simplify expressions that use algebraic fractions

① Apply the same techniques you use when calculating with numerical fractions to algebraic fractions.

② Use your knowledge of collecting like terms and factorising to simplify the fraction.

EXAMPLE

In a multiplication pyramid, you multiply the two numbers directly below to get the number above.

Complete this multiplication pyramid.

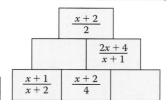

① The numbers below multiply to give the number above.

$$\frac{x+1}{x+2} \times \frac{x+2}{4} = \frac{x+1}{4}$$ ② Multiply the numerators. Then multiply the denominators.

① Divide the top number by one of the numbers below to find the missing entry.

$$\frac{2x+4}{x+1} \div \frac{x+2}{4} = \frac{2x+4}{x+1} \times \frac{4}{x+2}$$ ② Invert the divisor and then multiply.

$$= \frac{2(x+2)}{x+1} \times \frac{4}{x+2}$$ Factorise the numerator and cancel common factors.

$$= \frac{8}{x+1}$$

EXAMPLE

A linear sequence has first term $\frac{2}{x+1}$ and second term $\frac{3}{x+2}$.

Lucia says that the next term is $\frac{4}{x+3}$. Is she correct?

2, 7, 12, 17, 22 .... is an example of a linear sequence. The next term is always five more than the previous term.

If the sequence is linear then the difference between consecutive terms is the same.

$$\frac{2}{x+1} \qquad \frac{3}{x+2} \qquad \frac{4}{x+3}$$

① Find the difference between the second and first term.

$$\frac{3}{x+2} - \frac{2}{x+1} = \frac{3(x+1) - 2(x+2)}{(x+2)(x+1)} = \frac{3x+3-2x-4}{(x+2)(x+1)} = \frac{x-1}{(x+2)(x+1)}$$

② Expand the numerators and collect like terms. Don't expand the denominator. Find the difference between the third and second term.

$$\frac{4}{x+3} - \frac{3}{x+2} = \frac{4(x+2) - 3(x+3)}{(x+3)(x+2)} = \frac{4x+8-3x-9}{(x+3)(x+2)} = \frac{x-1}{(x+3)(x+2)}$$

The common difference is not the same. Lucia is wrong.

Exercise 2.4A

1 The three terms along each diagonal must add to give the same expression.

Complete the grid.

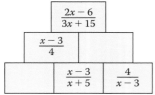

2 The three terms along each side multiply to give $12x$.

Complete the grid.

3 Complete this multiplication pyramid.

$$\begin{array}{c} \boxed{\dfrac{2x-6}{3x+15}} \\ \boxed{\dfrac{x-3}{4}} \quad \boxed{} \\ \boxed{} \quad \boxed{\dfrac{x-3}{x+5}} \quad \boxed{\dfrac{4}{x-3}} \end{array}$$

4 A linear sequence has first term $\dfrac{x+4}{3}$ and second term $\dfrac{x+5}{4}$.

Find the next term in the sequence.

5 Show that $\dfrac{1}{(x+1)(x+2)}$, $\dfrac{1}{(x^2+2x)}$ and $\dfrac{1}{x^2+x}$ are the first three terms of a linear sequence.

6 Amelia has tried to calculate

$\dfrac{2}{x-3} - \dfrac{3}{x+5}$ and $\dfrac{6}{x+4} + \dfrac{5}{x-2}$.

Both of her answers are wrong.
Correct Amelia's mistakes and find the correct answers.

$$\dfrac{2}{x-3} - \dfrac{3}{x+5} = \dfrac{2(x+5) - 3(x-3)}{(x-3)(x+5)}$$

$$= \dfrac{2x + 10 - 3x - 9}{(x-3)(x+5)}$$

$$= \dfrac{1-x}{(x-3)(x+5)}$$

$$\dfrac{6}{x+4} + \dfrac{5}{x-2} = \dfrac{6x - 2 + 5x + 20}{(x+4)(x-2)}$$

$$= \dfrac{11x + 18}{(x+4)(x-2)}$$

7 Here is a rectangle.

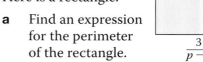

a Find an expression for the perimeter of the rectangle.

b Explain why p must be greater than -1.

c Explain why p cannot equal 2.

8 $\dfrac{6}{2x+1} + \dfrac{1}{7-x}$

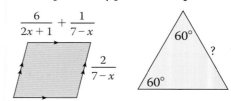

The perimeter of the parallelogram is equal to the perimeter of the triangle.

Show that the unknown side of the triangle has length $\dfrac{30}{(2x+1)(7-x)}$.

9 The large rectangle is an enlargement of the smaller rectangle. Find the length of the missing side.

10 Nina starts with a rectangle.

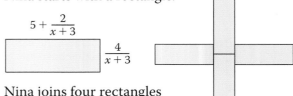

Nina joins four rectangles together to make this composite shape. Show that the perimeter of Nina's shape is $\dfrac{8(5x+18)}{x+3}$.

11 Find the mean of these three numbers.

$$\dfrac{x-1}{12} \qquad \dfrac{x+2}{6} \qquad \dfrac{x}{4}$$

12 a Show that

$$\dfrac{3x+6}{x-1} \times \dfrac{2x-2}{5x+20} \times \dfrac{5x-15}{x+2} \times \dfrac{x+4}{6x-18} = 1$$

b Fill in the blanks to make the equation correct.

$$\dfrac{\square}{6x-3} \times \dfrac{6-3x}{2x+8} \times \dfrac{3x+15}{\square} \div \dfrac{6x+30}{2x-1} = 1$$

Q 1149, 1151, 1164   SEARCH

Summary

Checkout

You should now be able to...

Test it
Questions

✓	Use algebraic notation and simplify expressions by collecting like terms.	1 – 3
✓	Substitute numbers into formulae and expressions.	4, 5
✓	Use the laws of indices.	6, 7
✓	Multiply a single term over a bracket.	8
✓	Take out common factors in an expression.	9
✓	Simplify algebraic fractions and carry out arithmetic operations with algebraic fractions.	10, 11

Language   Meaning   Example

Language	Meaning	Example
Expression	A meaningful collection of letters, numbers and operations.	$5x - 2$ Terms $\qquad$ $5x$ and -2
Term	One of the component parts in an expression. Terms are linked with addition or subtraction signs.	Variable $\qquad x$ Coefficient of x $\quad$ 6
Variable	An unknown quantity represented by a letter.	
Coefficient	A number in front of a letter that shows how many of that letter are required.	
Substituting	Replacing a letter with a numerical value.	If $x = 2$, $5x - 2 = 5 \times 2 - 2 = 8$
Like terms	Terms that contain exactly the same combination of variables. It is usual to collect like terms.	$5x^2 + 3x - 7x = 5x^2 - 4x$
Index/Indices **Base** **Power**	In index notation, the index or power shows how many times the base has to be multiplied by itself. The plural of index is indices.	Power or index $5^3 = 5 \times 5 \times 5$ Base
Index laws	A set of rules for calculating with numbers written in index notation.	$a^m \times a^n = a^{m+n}$ $\qquad 3^2 \times 3^5 = 3^7$ $a^m \div a^n = a^{m-n}$ $\qquad 5^6 \div 5^2 = 5^4$ $(a^m)^n = a^{mn}$ $\qquad (2^3)^4 = 2^{12}$
Expand	Multiply out brackets and collect like terms.	$5x(5 + x) + 10(x - 3)$ $= 10x + 2x^2 + 10x - 30$ $= 2x^2 + 20x - 30$
Factorise	Rewrite an expression using brackets by taking out the highest common factor.	$10xy^2 + 5x^2y = 5xy \times 2y + 5xy \times x$ $= 5xy(2y + x)$

Review

1   In the expression
$$7x^2 + 3x - 10$$

   a   how many terms are there
   b   what is the largest coefficient
   c   how many variables are there?

2   Simplify these expressions.
   a $y \times 13$ b $y \times 7 \times x$
   c $x + x + x$ d $y \times y \times y$
   e $2 \times x \div 4$ f $4yx + yx$

3   Simplify these expressions.
   a $4a - 3b + 2a$ b $7 - 5a - 5 + 7b$
   c $4a + 4a^2$ d $3a^2b + 2a^2b - 5ab^2$
   e $6ab + 3b - 9$ f $8a^2 \div 4 + 2a - 3a^2$

4   Evaluate these expressions when $x = 2$ and $y = -7$.
   a $5x$ b $3y$
   c $-2x$ d $-4y$
   e $3x^2$ f xy
   g $\dfrac{4x}{6}$ h $\dfrac{-21}{y}$

5   $v^2 = u^2 + 2as$
   a   Find v when $u = 8$, $a = 3$ and $s = 6$.
   b   Find s when $v = 12$, $u = 9$ and $a = 9$.

6   Simplify these expressions.
   a $a^3 \times a^4$ b $b^8 \div b^3$
   c $(c^3)^2$ d $4d^3 \times 5d^6$
   e $8e^9 \div 2e^3$ f $18f^6 \times 2f^2 \div 6f^8$

7   Simplify these expressions.
   a $x^6 \times x^4 \times x^2 \div x^7$
   b $\dfrac{y^6 \times y^3 \times y^3}{y^5 \times y^5}$
   c $(2z^2)^3$ d $(27x^2y)^0$
   e $14u^7 \times 2u^{-3}$ f $10p^3 \div 2p^{-5}$
   g $(3r^{-2})^3$ h $(2s^2t^{-3})^{-2}$

8   Expand the brackets in these expressions.
   a $5(2a + 3)$ b $3(6b - 3c)$
   c $-4d(8d - 2c)$ d $y(y + 3) - 2y(y + 1)$

9   Factorise fully these expressions.
   a $5x^2 + 10x$ b $21ab^2 - 14a$
   c $30p^2 + 15pq^2 - 45pq^3$

10   Simplify these algebraic fractions.
   a $\dfrac{2x^2 + 4x}{8x}$ b $\dfrac{x^2 + x}{(x + 1)^2}$

11   Simplify these expressions involving algebraic fractions.
   a $\dfrac{1}{a} + \dfrac{2}{a}$ b $\dfrac{2b}{5} - \dfrac{3b}{8}$
   c $\dfrac{5}{6c} + \dfrac{2}{3c}$ d $\dfrac{3}{4d} - \dfrac{1}{2d^2}$
   e $\dfrac{3}{d} + \dfrac{5}{f}$ f $\dfrac{3a}{b} \times \dfrac{5b}{3a^2}$
   g $\dfrac{3a^2}{b} \div \dfrac{9a}{b^2}$ h $\dfrac{a}{a+1} - \dfrac{a}{a+2}$

What next?

Score		
	0 – 4	Your knowledge of this topic is still developing. To improve look at MyMaths: 1033, 1045, 1064, 1149, 1151, 1155, 1164, 1178, 1179, 1186, 1247, 1301
	5 – 9	You are gaining a secure knowledge of this topic. To improve look at InvisiPens: 02Sa – g
	10 – 11	You have mastered these skills. Well done you are ready to progress! To develop your exam technique looks at InvisiPens: 02Aa – f

Assessment 2

1 The formula for the curved surface area of a cone is $A = \pi r l$, where r is the radius of the base and l is the slant height.

    a Find A for a cone with base radius 5 cm and slant height 10 cm. [2]

    b Find r when $A = 45\,\text{m}^2$ and $l = 4\,\text{m}$. [3]

    c Find l when $A = 126.4$ inches^2 and $r = 12.3$ inches. [3]

2 The area of a trapezium is given by the formula $A = \dfrac{(a + b) \times h}{2}$ where a and b are the parallel sides and h is the height.

    Find the area of the trapezium where $a = 2z^2$, $b = 3z^2$ and $h = 4z$. [3]

3 George divides the square *STUV* into four rectangles *SADC*, *ATED*, *DEUB* and *CDBV*.

    a George says that in simplest form

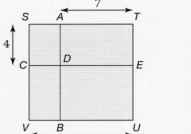

        Area of *SADC* = $4y$
        Area of *ATED* = 28
        Area of *DEUB* = $7y$
        Area of *CDBV* = y^2

        Give a geometrical reason why George's areas can not be correct. [4]

    b Give the correct areas of the four rectangles. [9]

4 You can measure the height of a cliff, H m, by dropping a stone from the top and timing it until it hits the sea. The formula is $H = 5t^2$ where t is the time in seconds.

    a Use this formula to find the height of the cliff if you hear the splash after

        i 4 seconds [2]   **ii** 7 seconds. [1]

    b A cliff is 320 m high.
        How long will it take to hear the splash if you drop a stone from the top? [3]

5 Charlie tried to factorise the expressions below.

For each expression, decide if Charlie

- factorised the expression completely
- factorised the expression, but not completely
- factorised the expression incorrectly.

In the last two cases, give the correct answer.

    a $vwx + xyz - vxz = x(vw + yz - vz)$ [1]

    b $a^2bc^3 - ab^4c^2 + a^5b^3c^4 = ab(ac^3 - b^3c^2 + a^4b^2c^4)$ [3]

    c $12p^3q^2r^9 - 18p^3q^5r^5 - 30p^5q^7r^4 = p^3q^2r^4(12r^5 - 18q^3r - 30p^2q^3r)$ [3]

    d $14pq + 21q^2 - 56pq^2 = 7q(2p + 3q - 8pq)$ [2]

    e $g^3 + g^2 - g = g(g^2 + g + 1)$ [3]

    f $g^3 + g^2 - g + 1 = g(g^2 + g - 1)$ [2]

    g $4(p - q) - 6(p - q)^2 = (p - q)(4 - 6(p - q))$ [2]

    h $3(y + 2z)^2 + 9(y + 2z)^3 = 3(y + 2z)^2(3y + 6z)$ [3]

    i $4(x^2 - 3x + 2) - 6(x^2 - 3x + 2) = -2(x^2 - 3x + 2)$ [3]

6 Romeo buys Juliet a gift. It is a cuboid with a square base of side y cm and height 20 cm.

    a Calculate, in terms of y, the total surface area of the cuboid in its simplest form. [4]

    b The cuboid has volume 200 cm³. Romeo says that the *exact* value of y is 10.

       Is he correct? Give your reasons. [4]

7 Wanda is W years old.

    a Wanda has twin brothers who are 5 years less than twice Wanda's age.

       i Find an expression for the total of the ages of the three children, giving your answer in its simplest form. [3]

       ii Factorise your answer to part **i**. [1]

    b Wanda's dad's age is three years more than four times Wanda's age. Her mum is 2 years younger than her dad.

       i Find an expression for the sum of her parents' ages, giving your answer in its simplest form. [4]

       ii Factorise your answer to part **i**. [1]

8 An L-shaped room has width w, and a floor area of $w(3w - 1)$.

Greta draws a plan of the room.

Find the values of x and w. [9]

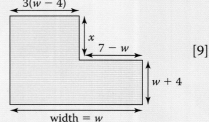

9 A pyramid has a rectangular base with width $2p$ and length $5p$. Two of the triangular sides have area $2p(3p + 2)$ each and the other two have area $5p(p - 3)$ each.

    a Find a simplified expression for the total surface area of the pyramid. [4]

    b Factorise your answer to part **a**. [1]

10 Simplify these fractions.

    a $\dfrac{55}{99}$ [1]   **b** $\dfrac{55}{100}$ [1]   **c** $\dfrac{26}{36}$ [1]   **d** $\dfrac{4q}{10}$ [1]

    e $\dfrac{15x^2}{3x}$ [1]   **f** $\dfrac{8pq}{16q^2}$ [1]   **g** $\dfrac{(3ab)^4}{6a^2b^4}$ [2]   **h** $\dfrac{2z-8}{12}$ [2]

11 Write each of the following expressions as a single fraction in its simplest form.

    a $\dfrac{z - 3}{6} + \dfrac{2z + 1}{8}$ [5]   **b** $\dfrac{7 - 2y}{5} - \dfrac{6y - 2}{7}$ [4]

    c $\dfrac{4}{x + 1} + \dfrac{3}{x + 2}$ [4]   **d** $\dfrac{7}{2p + 3} - \dfrac{4}{3p - 2}$ [4]

    e $\dfrac{2a}{a - 2} + \dfrac{3a}{2a + 9}$ [4]   **f** $\dfrac{3}{(2x + 1)^2} - \dfrac{4}{(2x + 1)^3}$ [3]

12 a Find an expression for the perimeter of this triangle. Write your answer in its simplest form. [7]

    b The base is 5 metres long. Find the value of w. [2]

    c Use this value of w to find the lengths of the other two sides. [4]

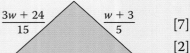

3 Angles and polygons

Introduction

Tiling is a fascinating topic that is highly mathematical, involving angles and shapes. There are some wonderful tiling patterns to be found in architecture, particularly in Islamic palaces and mosques.

The tiling shown in this picture is from the Alhambra Palace in Granada, Spain.

What's the point?

An understanding of angles and shapes allows us to create beautiful things.

Objectives

By the end of this chapter, you will have learned how to …

- Use angle facts including at a point, on a line, at an intersection and for parallel lines.
- Use bearings to specify directions.
- Identify types of triangle and quadrilateral and use their properties.
- Identify congruent shapes and use congruence to prove geometric results.
- Identify similar shapes and use similarity to find lengths and areas.
- Calculate the properties of polygons including interior and exterior angles for regular polygons.

Check in

1   Draw a grid with axes from −3 to +5.
    Draw these points on the grid.

    a (1, 3)   **b** (3, −2)   **c** (4, 5)   **d** (5, 2)

    e (−3, 3)   **f** (0, 4)   **g** (−1, 0)   **h** (−3, −2)

2   The diagrams show angles on a straight line or at a point.
    Work out the missing angles.

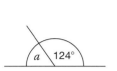 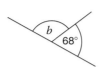

3   Work out the missing angle in these shapes.

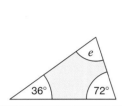 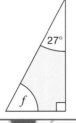 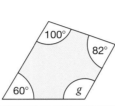 

Chapter investigation

A quadrilateral has been drawn on a 3 × 3 square dotty grid.

How many different quadrilaterals can you find?

What if the grid were extended to 4 × 4?

3.1 Angles and lines

- An **angle** is a measure of turn. You measure angles in **degrees**.

An **acute** angle is less than 90°.

A **right angle** is exactly 90°.

An **obtuse** angle is between 90° and 180°.

A **reflex** angle is between 180° and 360°.

- You should know these facts:

There are 360° at a point.

There are 180° on a straight line.

Vertically opposite angles are equal.

EXAMPLE

Calculate the values of p, q and r.

a

b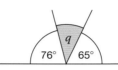

c

a $360° - 260° = 100°$

$p = 100°$

Angles at a point add to 360°.

b $76° + 65° = 141°$

$180° - 141° = 39°$

$q = 39°$

Angles on a straight line add to 180°.

c $r = 85°$

Vertically opposite angles are equal.

- **Parallel** lines are always the same distance apart.

Parallel lines are shown by sets of arrows.

- **Perpendicular** lines meet at a right angle.

Parallel lines never **intersect** (cross) each other.

When a line crosses **parallel** lines, eight angles are formed.

- **Alternate angles** are equal.

- **Corresponding angles** are equal.

- **Interior angles** add up to 180°.

Angles that add up to 180° are said to be supplementary.

EXAMPLE

Find the angles marked by letters. State whether each answer is acute, obtuse or reflex

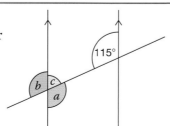

$a = 115°$ Alternate angles are equal.

$b = 115°$ Corresponding angles are equal.

$c = 180° - 115° = 65°$

Angles on a straight line add to 180°.

115° is an obtuse angle.

65° is an acute angle.

Exercise 3.1S

1 Choose one of these to describe each angle.

| acute | right angle | obtuse | reflex |

a
b
c
d

2 Choose one of these to describe each angle.

| acute | right angle | obtuse | reflex |

a 90°   **b** 40°   **c** 140°
d 200°   **e** 270°   **f** 36°
g 137°   **h** 248°   **i** 302°
j 33°   **k** 96°   **l** 239°

3 a Draw two lines that are parallel. Label them with >.

    b Draw two lines that are perpendicular. Label them with ∟.

4 Give the values in degrees of the coloured angles.

a   **b**

5 Calculate the size of the angle marked by a letter in each diagram.
Give a reason for each answer.
The diagrams are not accurately drawn.

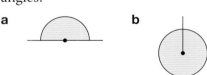

a 320° (*a*)   **b** 120° (*b*)

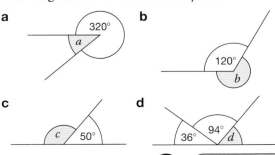

c *c* 50°   **d** 36° 94° *d*

6 Find the value of each angle marked with a letter.
Give a reason for each answer.

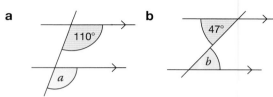

a 110° *a*   **b** 47° *b*

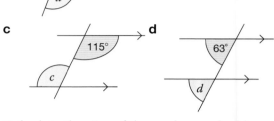

c 115° *c*   **d** 63° *d*

7 Calculate the size of the angles marked by letters in each diagram.
Give a reason for each answer.
The diagrams are not accurately drawn.

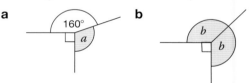

a 160° *a*   **b** *b* *b*

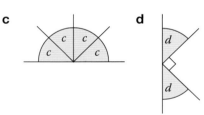

c *c* *c* *c* *c*   **d** *d* *d*

8 Calculate the sizes of the angles marked by letters in each diagram.
Give a reason for each of your answers.

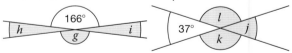

h 166° *i* *g*   37° *l* *j* *k*

These diagrams are not drawn to scale.

9 Find the value of each angle marked with a letter. Give a reason for each answer.

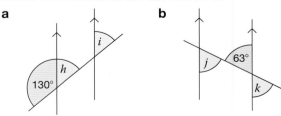

a *i* *h* 130°   **b** *j* 63° *k*

3.1 Angles and lines

RECAP

- Angles at a point add up to 360°.
- Angles at a point on a straight line add up to 180°.
- Vertically opposite angles are equal.
- When two lines are parallel, alternate angles are equal and corresponding angles are equal.

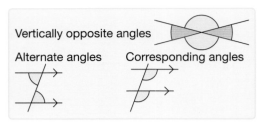

Vertically opposite angles

Alternate angles   Corresponding angles

North, east, south or west are often not enough to give an accurate direction.

- A **bearing** is an angle measured clockwise from north.

You use a 360° **scale** or a **bearing** to give a direction accurately.

- To give a bearing accurately you measure from north, measure clockwise and use three figures.

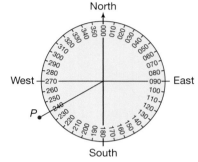

The bearing of *P* is 240°.

HOW TO

① You may need to draw a sketch and label the angles that you know.

② Look for parallel lines or places where angles meet at a point.

③ Write down each rule that you use during each stage of your calculation.

```
000° = North
090° = East
180° = South
270° = West
```

EXAMPLE

Find the unknown angle in this diagram. Give reasons for your answer.

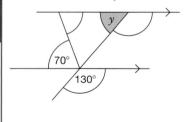

①

② Corresponding angles

$y = 180° - 130°$

$= 50°$

③ Angles on straight line add to 180°.

EXAMPLE

The bearing of *G* from *B* is 028°.

Find the bearing of *B* from *G*.

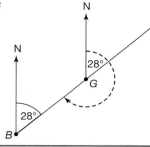

① Draw a sketch.

② Corresponding angle

Bearing of *B* from *G* is the angle at *G* measured clockwise from north to *B*.

③ Bearing of *B* from *G* = 28° + 180° = 208°

Exercise 3.1A

1 This diagram is wrong. Explain why.

2 Find the missing angles in each diagram. Write down which angle fact you are using each time.

a 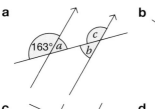   **b**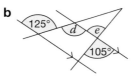

c 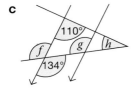   **d**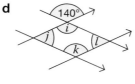

3 These diagrams have not been drawn accurately.
Find the bearing of *X* from *Y* in each case.

a 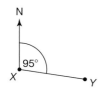   **b**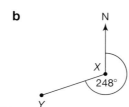

Draw a sketch to help you.

4 Find these bearings.

a The bearing of *A* from *B* is 104°.
Work out the bearing of *B* from *A*.

b The bearing of *E* from *F* is 083°.
Work out the bearing of *F* from *E*.

c The bearing of *J* from *K* is 297°.
Work out the bearing of *K* from *J*.

5 A plane takes off from a runway in a north-west direction.
It then turns through an angle of 75° to its right.
A helicopter, flying on a bearing of 232°, needs to turn to fly in the same direction.
What turn must the helicopter make?

6 Jennifer draws this diagram.

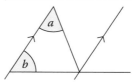

Use Jennifer's diagram to prove these statements.

a The exterior angle of a triangle is equal to the sum of the interior angles of the other two vertices.

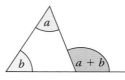

An exterior angle is formed by extending a side.

b The interior angles of a triangle add up to 180°.

$a + b + c = 180°$

7 Prove that the opposite angles in a parallelogram are equal.

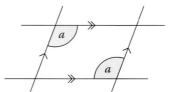

8 Work out the missing angles.

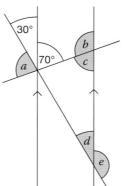

Q 1082, 1086, 1109   SEARCH

3.2 Triangles and quadrilaterals

There are 180° on a straight line.

You can draw any triangle ... tear off the corners ... and put them together to make a straight line.

● The angles in a **triangle** add to 180°.

> For an equilateral triangle each angle is 60° as 180° ÷ 3 = 60°

● You should know these names for special triangles:

Right-angled	Equilateral	Isosceles	Scalene
		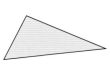	
One 90° angle marked ∟	3 equal angles 3 equal sides	2 equal angles 2 equal sides	No equal angles and no equal sides

> Lines with the same mark are equal in length.

● A **quadrilateral** is a 2D shape with four sides and four angles.

Square	Rectangle	Rhombus	Parallelogram
			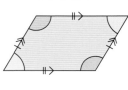

Trapezium	Isosceles trapezium	Kite
	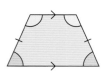	

> The equal angles are coloured the same.

● The angles in a quadrilateral add to 360°.

You can draw a diagonal in a quadrilateral to form two triangles.

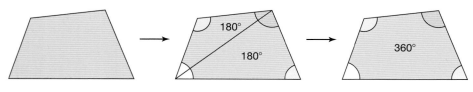

> The angles in each triangle add to 180°. 2 × 180° = 360°

Exercise 3.2S

1 Give the mathematical name of each coloured shape in the regular hexagon.

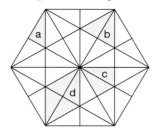

2 Calculate the size of the unknown angle in each diagram.
The diagrams are not drawn to scale.

a **b**

 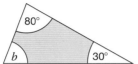

c **d**

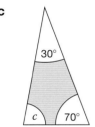 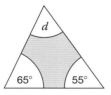

e **f**

 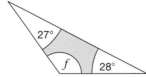

3 List any triangles in question **2** that are

a right-angled

b isosceles.

4 State the type of each triangle, if the sides of the triangle are

a 8 cm, 8 cm, 8 cm

b 6 cm, 7 cm, 8 cm

c 3 cm, 5 cm, 5 cm

d 25 mm, 10 mm, 2.5 cm

5 Calculate the third angle of the triangle and state the type of each of these triangles.

a 30°, 60° **b** 70°, 40°

c 60°, 60° **d** 35°, 65°

e 45°, 45° **f** 20°, 30°

6 State the value of each unknown angle.

a **b**

  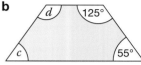

Parallelogram Isosceles trapezium

c **d**

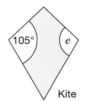 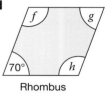

Kite Rhombus

7 Find the unknown angle in each quadrilateral and state the type of quadrilateral.

a **b**

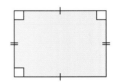 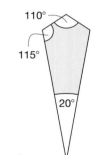

c **d**

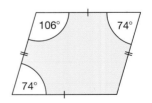 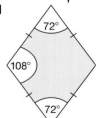

e **f**

3.2 Triangles and quadrilaterals

- A triangle has three sides and its angles add to 180°.
- A quadrilateral has four sides and its angles add to 360°.

▲ Osaka Castle in Japan.

HOW TO

① Sketch a diagram (unless one is given).
Mark (or look for) known angles and equal or parallel sides.

② Use the properties of triangles and quadrilaterals.
Look for parallel lines or places where angles meet at a point.

③ Write down each rule that you use during each stage of your calculation.

EXAMPLE

Calculate the values of x and y.

Give reasons for your answers.

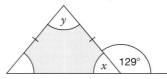

① As the triangle is isosceles, two of the angles are equal.

② $180° - 129° = 51°$

$x = 51°$

③ Angles on a straight line add to 180°.

② $51° + 51° = 102°$

$180° - 102° = 78°$

$y = 78°$

③ Angles in a triangle add to 180°.

EXAMPLE

Calculate the values of y and z.
Give a reason for each of
your answers.

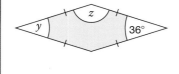

① The quadrilateral is a rhombus.

② $y = 36°$

③ Opposite angles of a rhombus are equal.

② $36° + 36° = 72°$

$360° - 72° = 288°$

$288° ÷ 2 = 144°$

$z = 144°$

③ Opposite angles of a rhombus are equal and angles in a
quadrilateral add to 360°.

EXAMPLE

Are these statements true or false?
Give reasons for your answer.

a All squares are rhombuses.

①

b Parallelograms are rectangles.

①

a ② A **square** is a special type of rhombus with all
angles equal.

③ True: all properties of a rhombus are also
properties of a square.

b ② A **rectangle** is a special type of parallelogram with
all angles equal.

③ False: a parallelogram does not have all its angles
equal.

Exercise 3.2A

1 Choose three of these angles that could be put together to make the angles in a triangle.

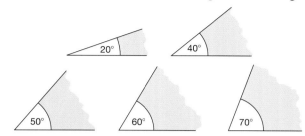

2 Calculate the size of the unknown angles in each diagram.
The diagrams are not drawn to scale.

a   **b**

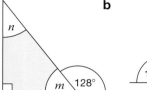

c   **d**

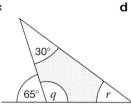

3 The points A $(-1, -2)$ and B $(3, -2)$ are shown. Give the coordinates of a point C, so that triangle ABC

  a is isosceles

  b is right-angled but scalene

  c is right-angled and isosceles

  d is scalene

  e is equilateral (only an approximate value of y is needed)

  f has an area of $4\,cm^2$.

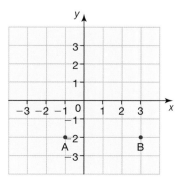

4 Calculate the value of x for each quadrilateral.

a   **b**

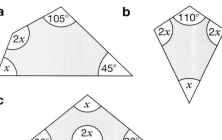

c

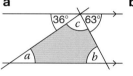

5 Find the value of each angle marked with a letter. Give a reason.

a   **b**

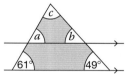

c

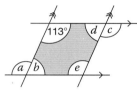

6 Find the values of the angles marked by letters. Give reasons for each step in your answer.

a   **b**

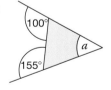
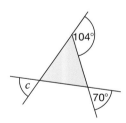

7 Are these statements true or false? Give reasons for your answers.

  a All squares are rectangles.

  b All kites are rhombuses.

  c All rhombuses are rectangles.

8 Prove that the pink triangle is isosceles.

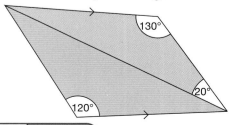

Q 1080, 1102, 1130, 1141 SEARCH

3.3 Congruence and similarity

● **Congruent** shapes are exactly the same shape and size.

Congruent shapes fit exactly on top of each other.

In congruent shapes

● corresponding angles are equal

● corresponding sides are equal.

Two triangles are congruent if they satisfy one of four sets of conditions:

SSS: three sides the same

SAS: two sides and the included angle the same

You should learn the four conditions for congruency.

ASA: two angles and the included side the same

RHS: right-angled triangles with **hypotenuse** and one other side the same.

EXAMPLE

This is triangle *A*.

Are triangles *B*, *C* or *D* congruent to triangle *A*?

A — 8 cm, 40°, 20°, 120°

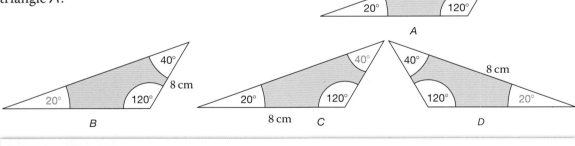

B — 40°, 8 cm, 20°, 120°

C — 40°, 20°, 120°, 8 cm

D — 40°, 120°, 8 cm, 20°

Fill in the missing information.

No: 8 cm in the wrong place.   No: 8 cm in the wrong place.   Yes, congruent (SAS).

In an **enlargement**, the object and the image are **similar**:

● the angles stay the same

● the lengths increase in proportion.

You use corresponding lengths to find the **scale factor**.

● Scale factor = $\dfrac{\text{length of image}}{\text{length of object}}$

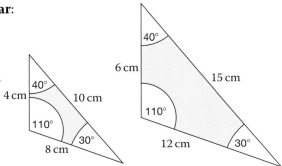

4 cm, 40°, 10 cm, 110°, 8 cm, 30°

6 cm, 40°, 15 cm, 110°, 12 cm, 30°

Exercise 3.3S

1 Explain whether or not these pairs of triangles are congruent.

a
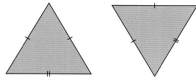

b
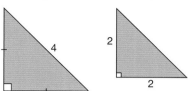

c
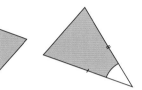

2 Which of these triangles are congruent to triangle *A*?

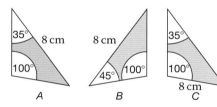

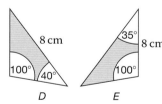

3 In each question, the two triangles are similar.
Find the value of the unknown angles.

| Angles in a triangle add to 180°. |

a
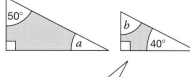

b
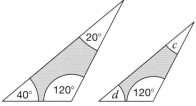

4 Which of these rectangles are similar to the green rectangle?

For the ones that are similar, give the scale factor of the enlargement.

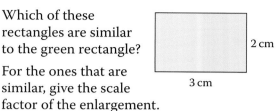

2 cm
3 cm

a 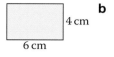 4 cm / 6 cm   **b** 10 cm / 15 cm   **c** 4 cm / 5 cm

d 3 cm / 6 cm   **e** 8 cm / 12 cm   **f** 2 cm / 2 cm

5 In each question, the two triangles are similar. Calculate the scale factor of the enlargement and the unknown length.

a
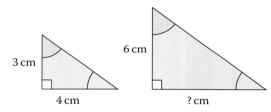
3 cm, 4 cm, 6 cm, ? cm

b
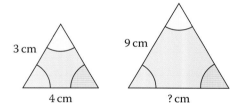
3 cm, 4 cm, 9 cm, ? cm

6 The two trapezia are similar. Find the lengths *a* and *b*.

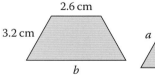
2.6 cm, 3.2 cm, *b*

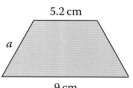
5.2 cm, *a*, 9 cm

7 The two quadrilaterals are similar. Find the lengths *c* and *d*.

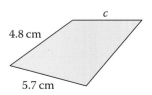
c, 4.8 cm, 5.7 cm

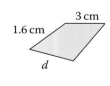
3 cm, 1.6 cm, *d*

Q 1119, 1148   SEARCH

3.3 Congruence and similarity

- In congruent shapes, corresponding lengths are equal and corresponding angles are equal.
- In similar shapes, corresponding angles are equal and the lengths increase in proportion.

HOW TO

1. Sketch a diagram (unless one is given).
   Mark (or look for) known angles and equal or parallel sides.

2. Look for congruent or similar shapes by comparing (or finding) angles and sides.

3. Apply your knowledge of congruence or similarity to find the answer.

▲ A self-similar object is similar to a smaller part of itself. The Romanesco broccoli is made up of smaller florets that mimic the shape of the whole broccoli.

EXAMPLE

Explain whether or not these triangles are congruent.

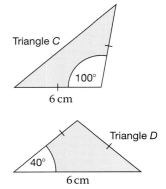

Triangle C

100°

6 cm

Triangle D

40°

6 cm

1. Both triangles are isosceles.

   Both triangles have angles 40°, 40°, 100°.

2. Congruent triangles satisfy SSS.

   If both triangles have sides 6 cm, 6 cm, 6 cm then the triangles will be equilateral.

   In an equilateral triangle, angles are 60°, 60°, 60°.

   The triangles are not equilateral, so they don't satisfy SSS.

3. They have equal angles, but the side lengths are *not* equal.

   So the two triangles are not congruent.

EXAMPLE

a Show that triangle *ABE* is similar to triangle *ACD*.

b Calculate the value of *x*.

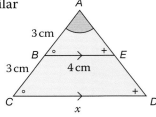

A

3 cm

B   E

3 cm   4 cm

C

x

D

Similar shapes have the same angles but are different sizes.

1. Draw a sketch of each triangle.

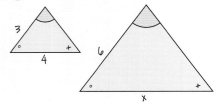

3

4

6

x

a 2. Compare the angles.

Angle B = angle C   (corresponding angles are equal)

Angle E = angle D   (corresponding angles are equal)

Angle A is common to both triangles.

3. So △ABE and △ACD are similar.

b 2. 3 cm and 6 cm are corresponding sides.

The scale factor is 6 ÷ 3 = 2

3. x = 4 cm × 2 = 8 cm

Exercise 3.3A

1 Explain why triangles *PQR* and *WXY* are congruent.

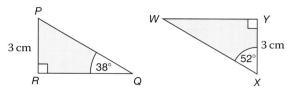

2 *ABCD* is a rectangle.

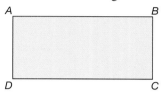

Prove that triangles *ABD* and *CDB* are congruent.

3 *KLMN* is a kite.

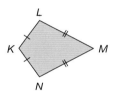

a Explain why triangles *KLN* and *MLN* are not congruent.

b Explain why triangles *KLM* and *KNM* are congruent.

4 Calculate the value of each unknown length.

a

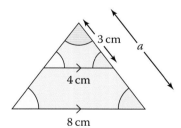

b

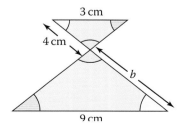

5 In the diagram *PQ* is parallel to *TR*. Work out the lengths *RT*, *QR* and *QS*.

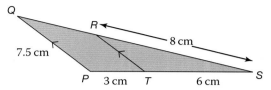

6 In the diagram *WZ* is parallel to *XY*.

a Work out the lengths *XY* and *VY*.

b Work out the perimeter of the trapezium *WXYZ*.

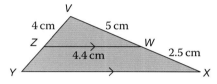

7 *KN* is parallel to *LM*.

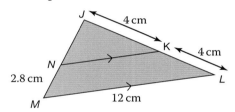

Work out the lengths *KN* and *JN*.

8 *EFGH* is a parallelogram. The diagonals *EG* and *FH* meet at the point *M*.

a Prove that triangles *MEF* and *MGH* are congruent.

b Hence prove that *M* is the midpoint of *EG*.

9 *RSTU* is a rhombus. *O* is the point where the diagonals cross.

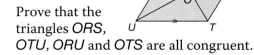

a Prove that the triangles *ORS*, *OTU*, *ORU* and *OTS* are all congruent.

b Hence prove that diagonals *RT* and *SU* cross at right angles.

3.4 Angles in polygons

A **polygon** is a 2D shape with three or more straight sides.

- A **regular** shape has equal sides and equal angles.

A regular hexagon has six equal sides and six equal angles.

Sides	Name	Sides	Name
3	triangle	7	heptagon
4	quadrilateral	8	octagon
5	pentagon	9	nonagon
6	hexagon	10	decagon

- You should know the names of the polygons in the table.

A line of symmetry divides the shape into two identical halves, each of which is the mirror image of the other.

- A **regular polygon** with *n* sides has *n* lines of symmetry.

The **order of rotational symmetry** is the number of times a shape looks exactly like itself in a complete turn.

- A **regular polygon** with *n* sides has rotational symmetry of order *n*.

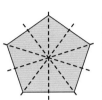

A regular pentagon has 5 lines of symmetry and rotational symmetry of order 5.

The **interior angles** are inside the polygon.

The **exterior angles** are made by extending each side in the same direction. Exterior angles are outside the polygon.

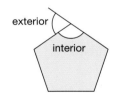

- The exterior angles of any polygon add to 360°.

- Interior angle + exterior angle = 180° (angles on a straight line add to 180°).

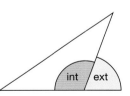

You can divide any polygon into triangles by drawing diagonals from a **vertex** (corner). The number of triangles is always two less than the number of sides.

- The sum of the interior angles of any polygon = (number of sides − 2) × 180°

EXAMPLE

Calculate the sum of the interior angles for a pentagon.

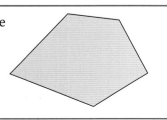

Draw in the two diagonals. Three triangles formed:

3 × 180° = 540°
Sum of interior angles = 540°

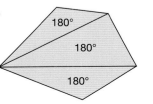

Exercise 3.4S

1 State the number of lines of symmetry for each of these regular polygons.

a

b

c

d

e

f

2 State the order of rotational symmetry for each polygon in question **1**.

3 **a** Calculate the value of one interior angle of an equilateral triangle.

Two equilateral triangles are placed together to form a rhombus.

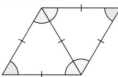

b Calculate the value of each interior angle of this rhombus.

c Calculate the sum of the interior angles of a rhombus.

4 Copy and complete this table.

Regular polygon	No. of sides	Exterior angle	Interior angle
Triangle	3		
Quadrilateral	4		
Pentagon	5		
Hexagon	6		
Heptagon	7		
Octagon	8		
Nonagon	9		
Decagon	10		

5 A regular polygon has 15 sides.

   a Use the angle sum of this polygon to work out the size of an interior angle.

   b Use the sum of the exterior angles to check your answer to part **a**.

6 A regular polygon has 18 sides.

   a Calculate the size of an interior angle of this polygon.

   b Use another method to check your answer to part **a**.

7 Explain how to find the sum of all the interior angles in an octagon.

8 Find the angle x.

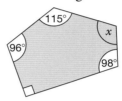

9 Two of the angles of a hexagon are right angles and three angles are equal to 132°. Find the other angle.

10 A dodecagon is a polygon with 12 sides. Eleven of the angles of a dodecagon are equal to 154°.
Calculate the other angle.

*11 The diagram shows a regular nonagon divided into congruent triangles.

   a Find the angles marked x and y.

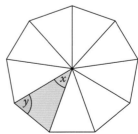

   b Use the value of y to check your answers in question **4** for a regular nonagon.

Q 1100, 1320   SEARCH

3.4 Angles in polygons

- The exterior angles of any polygon add to 360°.
- Interior angle + exterior angle = 180°
- The sum of the interior angles of any polygon = (number of sides − 2) × 180°

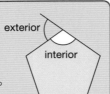

▲ Stained glass windows often contain elaborate patterns using simple polygons.

HOW TO

① Draw a sketch (if needed).

② Decide whether to use the sum of the interior or exterior angles.

③ Find the answer.

EXAMPLE

Calculate the value of x in this regular hexagon.

Four triangles

$4 × 180° = 720°$

② Sum of interior angles = 720°

There are six interior angles, so:

③ One interior angle, $x = 720° ÷ 6 = 120°$

①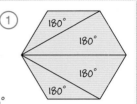

EXAMPLE

An irregular hexagon has angles 108°, 92°, 120°, 134°, 115° and x. Find the size of angle x.

Any hexagon has six sides and divides into four triangles.

② Sum of interior angles is $4 × 180° = 720°$

Sum of the five given angles is

$108° + 92° + 120° + 134° + 115° = 569°$

③ $x = 720° − 569° = 151°$

①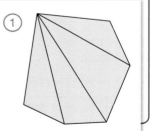

EXAMPLE

A regular polygon has interior angles of 156°.

How many sides does the polygon have?

② Find the exterior angle.

Exterior angle = $180° − 156° = 24°$ Angles on a straight line.

③ The sum of the exterior angles is 360°,

so divide to find out how many there are.

Number of exterior angles = $360° ÷ 24° = 15$

The polygon has 15 sides.   The number of sides = number of angles.

①

Geometry Angles and polygons

Exercise 3.4A

1 The interior angle of a regular polygon is 162°.

    a Calculate the value of an exterior angle.

    b State the sum of the exterior angles of the polygon.

    c Calculate the number of exterior angles in the polygon.

    d State the number of sides of the regular polygon.

2 A regular polygon has 15 sides.

    a Calculate the value of an exterior angle.

    b Calculate the value of an interior angle.

3 Calculate the size of the angles marked with letters in these polygons.

    a

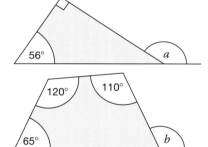

    b

4 An interior angle of a regular polygon is three times the exterior angle.

    a Calculate the value of each exterior angle.

    b Calculate the value of each interior angle.

    c Give the name of the regular polygon.

5 Find the missing angles in these irregular polygons.

    a   **b**

6 Find the size of smallest angle of this pentagon.

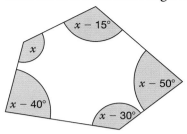

7 How many sides does a regular polygon have if

    a it has exterior angle 30°?

    b it has interior angle 135°?

8 A tessellation is a tiling pattern with no gaps or overlaps. Explain why a regular hexagon tessellates.

9 Does a regular pentagon tessellate? Explain your answer.

*10 The diagram shows a regular hexagon attached to a square.

    a Calculate angle z.

    b Name one or more regular polygons that would fit exactly into this space.

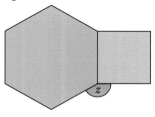

11 A regular polygon can be made by fitting together squares and equilateral triangles. Find angle x.

12 *ABCDEF* is a regular hexagon.

    Show that angle *ACD* is a right angle.

Summary

Checkout

You should now be able to...

Test it

Questions

	Test it Questions
✔ Use angle facts including at a point, on a line, at an intersection and for parallel lines.	1, 2
✔ Use bearings to specify directions.	3
✔ Identify types of triangle and quadrilateral and use their properties.	4 – 7
✔ Identify congruent shapes and use congruence to prove geometric results.	8
✔ Identify similar shapes and use similarity to find lengths and areas.	9
✔ Calculate the properties of polygons including interior and exterior angles for regular polygons.	10

Language   Meaning   Example

Language	Meaning	Example
Acute angle	$0 <$ acute angle $< 90°$	obtuse · right angle · reflex · acute · point
Right angle	right angle $= 90°$	
Obtuse angle	$90° <$ obtuse angle $\leqslant 180°$	
Reflex angle	$180° <$ reflex angle $\leqslant 360°$	
Alternate and Corresponding angles	When a line crosses a pair of parallel lines, **alternate angles** lie on opposite sides of the crossing line and opposite sides of the parallel lines. **Corresponding angles** lie on the same side of the crossing line and the same side of the parallel lines.	Alternate angles Corresponding angles
Three-figure bearing	A direction defined by a three-figure angle measured clockwise from north.	North-east is 045° North-west is 315°
Congruent	Exactly the same shape and size.	A B C
Similar	The same shape but different in size.	4 cm · 8 cm · 5 cm
Scale factor	The ratio of corresponding lengths in two similar shapes.	5 cm · 10 cm · 4 cm A and B are similar; the scale factor is 2. A and C are congruent.
Polygon	A 2D shape with straight edges.	Triangle, square, hexagon.
Quadrilateral	A polygon with four sides.	Square, rectangle, rhombus, parallelogram, trapezium, kite.
Interior angle	The angle between two adjacent sides inside a polygon.	← exterior angle
Exterior angle	The angle between one side of a polygon and the next side extended.	interior angle

Review

1 Calculate the size of angles *a*, *b*, and *c*.

2 Calculate the size of angles *a*, *b* and *c*. Give a reason for each answer.

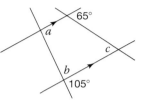

3 Find these three-figure bearings.

   a The bearing of A from B is 063°. Work out the bearing of B from A.

   b The bearing of C from D is 141°. Work out the bearing of D from C.

   c The bearing of E from F is 205°. Work out the bearing of F from E.

4 Calculate the size of angle *a* in this triangle.

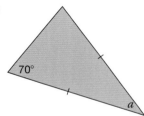

5 Calculate the size of angle *a* in this kite.

6 Draw coordinate axes with *x* and *y* from 0 to 6. Now plot these points: A (1, 1), B (3, 3), C (5, 1).

Join up the dots to form a triangle, what type of triangle is this?

7 Which quadrilateral is being described below?

   a One pair of parallel sides and no equal angles.

   b Two pairs of equal angles, two pairs of parallel sides and all sides equal.

8 Are these two triangles congruent?

Give a reason for your answer.

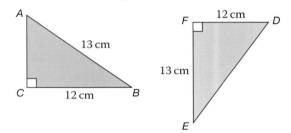

9 These hexagons are similar. Find *x*.

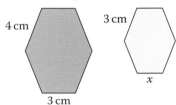

10 a A heptagon has seven sides. What do the interior angles of a heptagon add up to?

   b Prove that the sum of the exterior angles of a regular pentagon is 360°.

What next?

	Score		
	0 – 4		Your knowledge of this topic is still developing. To improve look at MyMaths: 1080, 1082, 1086, 1100, 1102, 1109, 1119, 1130, 1141, 1148, 1320
	5 – 9		You are gaining a secure knowledge of this topic. To improve look at InvisiPens: 03Sa – k
	10		You have mastered these skills. Well done you are ready to progress! To develop your exam technique looks at InvisiPens: 03Aa – g

Assessment 3

1 **a** What is the angle between the hour hand and the minute hand
of a clock at five minutes past seven? [6]

 b Susan claims that at some point between 1:05 pm and 1:06 pm the hour and
minute hands of a clock are pointing in the same direction. Is she correct?
Explain your answer. [8]

2 The cruise ship 'Black Watch' sails on a bearing of 135°. To avoid a storm it
changes course to a bearing of 032°. What angle has it turned through? [2]

3 Three villages, East Anywhere, Middle Anywhere and West
Anywhere, are the vertices of the triangle shown. The
bearing of East Anywhere from West Anywhere is 075°.

 Write down the bearing of

 a Middle Anywhere from West Anywhere [1]

 b West Anywhere from East Anywhere [2]

 c Middle Anywhere from East Anywhere [2]

 d West Anywhere from Middle Anywhere. [3]

4 Rafa and Sunita were standing on top a hill. Rafa was facing due north and Sunita was
facing due south. Explain why they could see each other. [1]

5 Sienna says that the missing angles have these values:

$a = 46°$	$b = 46°$	$c = 139°$	$d = 41°$
$e = 92°$	$f = 88°$	$g = 128°$	$h = 101°$

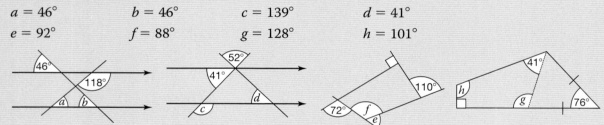

 Decide if her value for each angle is correct or incorrect.
For the incorrect angles, give the correct value. Give reasons for your answers. [20]

6 Manuel made the following statements about triangles. Write down if each of his
statements is true or false. If the statement is true draw an example. If the statement is
false explain why it is false.

 a Some triangles have two obtuse angles. [2]

 b Some triangles have one obtuse angle and two acute angles. [2]

 c Some triangles have one right angle, one acute angle and one obtuse angle. [2]

 d Some triangles have two acute angles and one right angle. [2]

7 **a** Natasha draws a pentagon with angles of 125°, 155° and 74°. She wants to draw the other
2 angles so that they are are equal. What size should she draw these 2 angles? [5]

 b Shivani draws an octagon that has 5 angles each of 114°. She wants to draw the
other 3 angles so that they are equal. What size should she draw these 3 angles? [5]

 c Tyler draws a hexagon with 3 angles of 137° each. He wants to draw the remaining 3 angles
so that they have the ratio $w:w + 120:w - 30$. What value should he use for w? [7]

 d Dean draws a regular polygon containing 60 sides.
What size should he use for the interior angle? [2]

8 Rose Bloom has designed a patio for her garden in the shape of a regular pentagon of side 8 m.

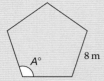

a Calculate angle *A*. [2]

b Rose borders the patio with square paving slabs of side length 2 m, as shown.

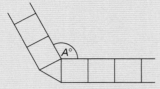

She cuts each triangular corner slab from one square slab.
Calculate the number of slabs needed. [2]

c Using the value of *A* that you found in part **a**, calculate
the three angles in the triangular corner slab. [4]

9 In a photo a statue's height is 19.2 cm and its legs measure 8 cm.
The statue is actually 288 cm tall.
How long are its legs? [4]

10 a Show that triangles *PQR* and *PTS* are similar. [2]

b Show that trapezia *ABCD* and *ABFE* are not similar. [4]

c Find length *RQ*. [2]

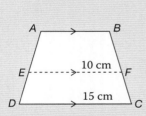

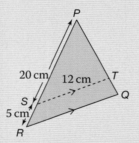

11

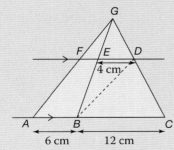

Triangle *BCG* has area 36 cm².

a Find length *FE*. [2]

b Find the area of triangle *ABG*. [5]

c Find the area of triangle *EFG*. [4]

d Show that ∠*DBC* = ∠*FAB*. [2]

4 Handling data 1

Introduction

In the modern world, there is an ever-increasing volume of data being continually collected, analysed, interpreted and stored. It was estimated that in 2007, there was 295 exabytes (or 295 billion gigabytes) of data being stored around the world. It has increased significantly since then. To put this into perspective, if all that data were recorded in books, it would cover the area of China in 13 layers of books. With all this information being generated, it is important that we have the mathematical techniques to cope with it. Statistics is the branch of maths that deals with the handling of data.

What's the point?

Availability of data helps us to understand the world around. Whether it is scientists looking for trends in global warming or consumers looking at cost comparison data to help inform purchasing decisions.

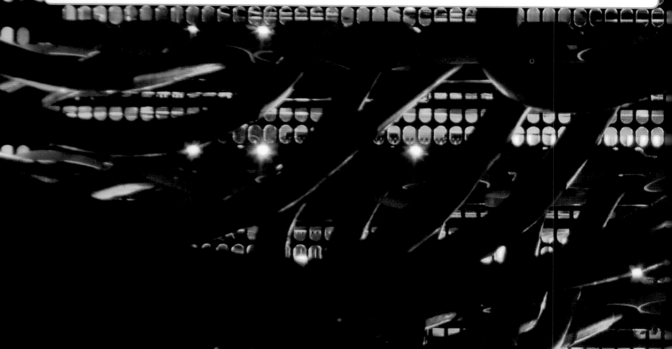

Objectives

By the end of this chapter, you will have learned how to ...

- Identify when a sample may be biased.
- Construct and interpret frequency tables, bar charts and pie charts.
- Calculate the mean, median and mode of a data set.
- Calculate the range and inter-quartile range of a data set.
- Use averages and measures of spread to compare data sets.

Check in

1 Find

a $\frac{1}{2}$ of 45   **b** $\frac{1}{2}$ of $(41 + 1)$ **c** $\frac{1}{4}$ of 24   **d** $\frac{3}{4}$ of $(27 + 1)$

2 List these two sets of numbers in ascending order.

a 45, 66, 89, 101, 98, 55, 112, 90, 29, 65

b 11.3, 8.9, 7.88, 9.0, 8.78, 8.95, 11.25, 10.9, 8.87, 9.8.

Chapter investigation

What is the average age of students in your class?

4.1 Sampling

To find out information about a **population**, you can collect **data** using a **survey**.

It may be time-consuming, too costly, or too impractical to collect data from everyone. In these cases you should take a representative **sample**.

In a statistical survey, the larger the **sample**, the more reliable the results.

Your sample should represent the whole population you are studying.

- You need to plan your survey to minimise **bias**. An unrepresentative sample will distort your results.

> A population is the group of people or items that you want to find out information about.

- In a **random sample**
  - Each member of the population has the same chance of being included.
  - The person choosing the sample has no control over who is included.

> The most reliable **data** comes from a census – a survey of the whole population.

To select a random sample:
- Assign a number to every member of the population.
- Use a calculator to generate random numbers.
- Choose the members with these numbers.

Organisations carry out statistical surveys to collect data that help them plan for the future.

You can use a questionnaire and collect data in a survey.

You have to choose suitable questions for a questionnaire.

> 'Picking names from a hat' is another method of random sampling.

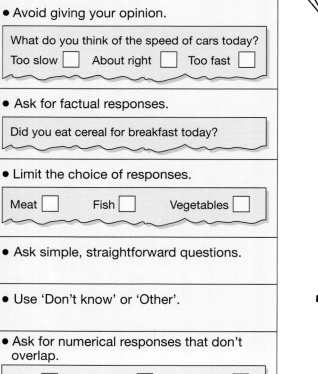

Never ask a leading question, such as: The speed of cars today, is very bad, isn't it?	• Avoid giving your opinion. What do you think of the speed of cars today? Too slow ☐ About right ☐ Too fast ☐
Never ask a vague question, such as: Do you eat cereal?	• Ask for factual responses. Did you eat cereal for breakfast today?
Never ask a question that gives too many responses, such as: What do you like to eat?	• Limit the choice of responses. Meat ☐ Fish ☐ Vegetables ☐
Never ask a complicated, wordy question.	• Ask simple, straightforward questions.
Don't forget to allow for all possible responses.	• Use 'Don't know' or 'Other'.
Never use 'occasionally', 'regularly' or 'sometimes' as they mean different things to different people.	• Ask for numerical responses that don't overlap. 1–10 ☐ 11–20 ☐ 21–30 ☐

Exercise 4.1S

1 Megan made this table to show the advantages and disadvantages of using a sample or carrying out a census of the whole population.

Census	**Advantages**
	• Everyone is represented
	Disadvantages
	• Times a long time
	• Expensive
	• Produces a lot of data
Sample	**Advantages**
	• Quick to do
	• Cheap to carry out
	Disadvantages
	• Could be biased depending on the sample chosen

Copy the table. Add these statements to the correct region of Megan's table.

- Difficult to make sure that the whole population is included
- Less data to analyse
- Unbiased
- Not everyone is represented

2 These questions appeared on a questionnaire.

Write one criticism of each question.

a What is your favourite sweet?

b How often do you use a computer?

c How much pocket money/allowance are you given?

d Did you have a shower this morning?

e These buses are always late. What do you think of the bus service?

f How much do you spend on clothes?

g How tall are you?

h Do you like shopping?

i Where do you live?

j How many DVDs do you own?

loads ☐   a lot ☐   many ☐

3 Katy is doing a survey to find out how often people go to the cinema and how much they spend.

She writes this question.

How many times a month do you go to the cinema?

a What is wrong with this question?

b Write an improved question to find out how often people visit the cinema.

c Write a question to find out how much people spend when they go to the cinema.

4 Sally put this question in a questionnaire.

Do you agree that tennis is the best sport?

a **i** What is wrong with this question?

ii Write a better question to find out the favourite sport.
Include some response boxes.

Sally also wants to find out how often people play sport.

b Design a question for Sally to use. Include some response boxes.

5 Katy is doing a survey to find out how often people go to the cinema and how much they spend. She stands outside a cinema and asks people as they go in.
Write a reason why this sample could be biased.

6 Sally wants to find out how often people play sport. Sally belongs to an athletics club. She asked members in her athletics club. How could this sample be biased?

7 The school cook wants to know the favourite baked potato fillings.
She plans to carry out a random sample of Year 11 students.

a Say why the results could be biased.

b There are 1000 students in the school. Describe how the cook could select a sample of 50 students, using random sampling.

4.1 Sampling

- You can use a questionnaire to collect data in a survey.
- A good sample should represent the whole population. It must not be biased.
- In a random sample, each person or item has the same chance of being included.

When a population is made up of different groups you can reduce **bias** by ensuring each group is represented in your sample.

12.1
- In a **stratified sample**, each group (or strata) is represented in the same proportion as in the whole population.

HOW TO
(1) Work out the sample size as a fraction of the whole population.
(2) Multiply the fraction by the number in each group to find the number in the sample.
(3) Round up or down where necessary.

EXAMPLE

A sports centre has 600 school-aged members.
Work out the number from each group
needed for a stratified sample of 90 members.

Age	5–15	16–18
Girls	140	96
Boys	205	159

(1) Work out the sample as a fraction of the population.

Total population = 140 + 96 + 205 + 159 = 600

Proportion of sample size to population = $\frac{90}{600} = \frac{3}{20}$

(2) Multiply the fraction by the number in each group to find the number in the sample.

Girls 5–15 $\frac{3}{20} \times 140 = 21$

(3) 30.75 boys is not possible so round to 31.

Boys 5–15 $\frac{3}{20} \times 205 = 30.75 \approx 31$

Girls 16–18 $\frac{3}{20} \times 96 = 14.4 \approx 14$

Boys 16–18 $\frac{3}{20} \times 159 = 23.85 \approx 24$

Check $21 + 31 + 14 + 24 = 90$

EXAMPLE

A sixth form college has 400 students aged 16 to 18.
A sample is chosen, stratified by age and gender, of 50 of the 400 students.
The sample includes six 16-year-old female students.
Find the least possible total number of 16-year-old female students.

(1) Work out the sample as a fraction of the population.

Sample size as a fraction of population = $\frac{50}{400} = \frac{1}{8}$

(2) n is the number of 16-year-old females in the college.
There are 6 students in the sample, and 5.5 is the lowest number that would round to 6.

(3) $\frac{1}{8} \times n \geq 5.5$

$n \geq 5.5 \times 8$

$n \geq 44$

There are at least 44 16-year-old females in the college.

Exercise 4.1A

1 Merlin wants to find out how far people would travel to see their favourite band perform.

He goes into town one Saturday morning and asks anyone listening to music on a MP3 player.

> How far would you travel to see your favourite band?
>
> less than 1 mile ☐   5–10 miles ☐   any distance ☐

Make two criticisms of Merlin's questionnaire.

2 James wants to know which flavour crisps he should stock in the school tuck shop. He asks five students in year 11 these questions.

> Do you prefer plain or ketchup flavoured crisps?
>
> How many times have you visited the tuck shop?
>
> Once ☐   Lots of times ☐

Make four criticism of James' questionnaire.

3 Which of the following statements are true? Give reasons for your answer.

  a A sample will always give the same results if the data is collected using a unbiased method.

  b A sample uses half the population.

4 Energising! make batteries. The company wants to investigate how long their batteries last.

Should Energising! use a census or a sample? Give reasons for your answer.

5 At an infant school there are 95 Reception students, 105 Year 1 and 100 Year 2.

A sample of 60 students stratified by year group is chosen.

How many students from each year will be chosen for the sample?

6 A company has a database of 4000 customers, of which 2750 are women and the rest men. The company wants to survey 800 customers. Describe how to select a sample stratified by gender.

7 A tennis club wants to find out what facilities to offer.

The club's membership is

	18–30	31–50	over 50
Male	100	97	83
Female	140	133	147

Describe how to select a stratified sample by age and by gender.

8 There are 600 students in Years 6, 7 and 8 at a middle school. This incomplete table shows information about the students.

	Boys	Girls
Year 6	96	81
Year 7	87	102
Year 8		

A sample is chosen, stratified by both age and gender, of 60 of the 600 students.

  a Calculate the number of Year 6 boys and Year 6 girls to be sampled.

In the sample there are nine Year 8 boys.

  b Work out the least possible number of Year 8 boys in the middle school.

9 Shona carries out a survey of favourite sports stars at a gym club which has 178 girls and 42 boys as members. She selects a random sample of 50. In her sample there are 20 girls and 30 boys. Say why this sample could be biased.

10 Of 200 people on an activity holiday, 30 are adults, 85 are boys and 85 girls. A representative sample is to be chosen. Say why it would be difficult to choose a stratified sample of 60.

4.2 Organising data

- You can use a data collection sheet to collect **data** from a questionnaire or experiment.
- You can use a **two-way table** to collate the two sets of results.

Two questions on a questionnaire are:

'Are you male or female?'   and   'How old are you?'

Design a two-way table to collect this data.

	Under 10	10–19	20–29	30–40	40+	Total
Male						
Female						
Total						

- You can use a **stem-and-leaf diagram** to display numerical data.

A stem-and-leaf diagram shows

- the shape of the distribution
- each individual value of the data.

This stem-and-leaf diagram is **ordered**, as the data is in numerical order.

13	5
12	0 6 9
11	2 2 6 8
10	3 7

This means 135.

This means 120.

stem   leaf

Key: 11 | 2 means 112   Always give a key.

The number of televisions in each house in Fern's street is shown in the frequency table.

a   Calculate the number of houses in Fern's street.

b   Calculate the total number of televisions in Fern's street.

No. of TVs	No. of houses
0	1
1	5
2	12
3	9
4	1

① Complete the table to show the total number of televisions.

a   ② Total number of houses = 1 + 5 + 12 + 9 + 1 = 28

b   ③ Multiply the number of TVs by the number of houses for each row in the table. Find the total.

Total number of televisions = 0 + 5 + 24 + 27 + 4 = 60

No. of TVs × no. of houses
0 × 1 = 0
1 × 5 = 5
2 × 12 = 24
3 × 9 = 27
4 × 1 = 4

Exercise 4.2S

1 Katy is doing a survey to find out how often people go to the cinema and how much they spend.
Design a suitable data collection sheet in the form of a two-way table that she could use.

2 Sally wants to find out how often people play sport. She wants to divide the results into those from males and those from females.
Design a suitable data collection sheet in the form of a two-way table that she could use.

3 James wants to know which flavour crisps he should stock in the school tuck shop.
He also wants to know which year groups prefer which flavours.
Design a suitable data collection sheet in the form of a two-way table that he could use.

4 One hundred mobile phone users are surveyed about colour of their phone (either black or silver) and for their payment method (either pay as you go or contract).

  a Devise a two-way table that would show this information.

  b Choose four suitable numbers for your table.

5 A car salesperson sells vehicles that are either saloons or hatchbacks and are bought part-exchange or cash.
Devise a suitable two-way table to summarise their sales.

6 There are two cinemas in a Cinecomplex. The number of people in each cinema is shown in the two-way table.

	Cinema	
	1	**2**
Adult	31	47
Child	12	8

Calculate the number of

  a adults in the Cinecomplex

  b children in the Cinecomplex

  c people in Cinema 1

  d people in Cinema 2

  e people in the whole Cinecomplex.

7 Sophie did a survey to find the number of DVDs owned by the students in her class. The results are shown in the frequency table.

Number of DVDs	Number of students
0	1
1	8
2	6
3	8
4	2
5	3

  a Calculate the number of students in Sophie's class.

  b Calculate the total number of DVDs owned by the whole class.

8 The times taken, in seconds, for 24 athletes to run 400 metres are given.

44.3 44.4 43.3 43.2 44.0 45.2

45.0 44.5 45.6 43.9 46.5 46.3

46.0 46.5 44.7 46.9 44.1 43.8

45.0 46.9 43.0 46.1 45.1 43.8

Draw an ordered stem-and-leaf diagram, using stems of 43, 44, 45 and 46.

Use the key: 43 | 3 means 43.3.

9 The weights, in kilograms, of 30 students are shown.

48 47 48 53 61 70 45 56 57 60

42 46 44 55 63 65 49 50 55 65

70 53 64 61 46 47 56 40 41 54

Draw an ordered stem-and-leaf diagram. Remember to give the key.

10 The attempted heights, in centimetres, during a high jump event are shown.

214 204 225 230 244

210 207 209 240 232

230 216 242 233 238

206 217 236 216 211

208 230 209 237 241

Draw an ordered stem-and-leaf diagram.

Q 1193, 1214, 1215   SEARCH

4.2 Organising data

- You can organise data in a frequency table.
- A two-way table links two types of information.
- You can also use a stem-and-leaf diagram to show numerical data.

Frequency tables and two-way tables make it easy to spot trends within the data set. They are also useful for organising large data sets and non-numerical data.

1. Construct or complete the table or diagram.
2. Interpret or describe the trend of the data.
3. Work out information from the data.

Stem-and-leaf diagrams are useful because they display all the raw data.

It can be hard to spot trends from stem-and-leaf diagrams for large data sets because there is so much information!

EXAMPLE

A class of 30 students study either History or Geography.

There are 13 girls in the class, with 8 girls and 9 boys studying Geography.

How many boys study History?

1. Draw a two-way table to show the information that you know.
2. Use the totals to fill in the other entries.

	History	Geography	Total
Boys		9	30 − 13 = 17
Girls	13 − 8 = 5	8	13
Total		9 + 8 = 17	30

↓

	History	Geography	Total
Boys	17 − 9 = 8	9	17
Girls	5	8	13
Total	30 − 17 = 13	17	30

3. Read the answer from the table.

8 boys study History.

EXAMPLE

Draw a back-to-back stem-and-leaf diagram to show the test results of a group of girls and a group of boys.

Girls 52   34   58   46   41   57   47   35   49   47   54

Boys 61   43   47   56   59   39   58   69   52   46   54

Compare the performances of the boys and the girls.

1. Construct the back-to-back stem-and-leaf diagram, making sure the leaves are in order.

```
        Girls            Boys
       5 4 | 3 | 9
   9 7 7 6 1 | 4 | 3 6 7
     8 7 4 2 | 5 | 2 4 6 8 9
             | 6 | 1 9
```

Key: 1 | 4 | 3
stands for 41% for girls, 43% for boys

2. Interpret the trends of the data.

The girls' results are all lower than 60, and mostly in the 40s and 50s. The boys' results are more spread out. Most of the boys have marks in the 50s and 60s.

3. Compare the performances.

The boys have generally performed better than the girls.

Exercise 4.2A

1 A group of 40 students study either French or Spanish.
There are 17 boys in the group, with 15 girls and 10 boys studying French.
How many girls study Spanish?

2 The table shows the number of students in a school who take part in the Duke of Edinburgh scheme.

There are

- seven year 8 boys
- five Year 10 girls
- equal numbers of boys and girls in year 9.

Year 8	13
Year 9	12
Year 10	11

Emily says that there are the same number of boys and girls taking part.
Do you agree with Emily?

> For each of the data sets **3–4**.
> **a** Draw a back-to-back stem-and-leaf diagram
> **b** Make comparisons between the data.

3 Height to the nearest cm of a sample of men and a sample of women.

Men	176	183	184	172	168
	175	183	159	169	174
	160	180	178	167	182
	188	171	178	158	165
Women	157	148	151	167	174
	165	169	158	153	155
	161	158	155	172	156
	162	166	149	178	154

4 Reaction times to the nearest tenth of a second of a sample of boys and a sample of girls.

Boys	4.2	5.7	3.2	3.8	6.4	3.8
	6.1	5.9	5.3	5.6	3.6	4.4
	5.2	3.2	3.8	5.8	4.7	4.5
	6.2	6.8	7.1	6.6	7.2	5.9
Girls	4.4	4.6	5.2	4.3	6.7	7.2
	8.0	4.0	8.2	7.7	6.3	7.6
	4.8	5.9	5.2	7.4	7.3	6.2
	6.5	5.6	6.6	6.3	5.5	4.8

5 In a traffic survey, the colour and speed of 100 cars are recorded. The results are summarised in the two-way table.

	Not speeding	Over the speed limit
Red	5	55
Not red	10	30

Maria claims that drivers of red cars tend to break speed limits. Use the two-way table to decide whether you agree with Maria.

6 Kyra and Ravi collect data on the number of minutes per day students in their year play computer games.

Kyra collects this data from a sample of 14 students.

40	26	75	55	33	39	28
66	67	71	64	37	52	47

a Kyra starts to make a frequency table to display the data.

Minutes	Frequency
26	1
27	0
28	1
29	0

Give two advantages of using a stem-and-leaf diagram instead of Kyra's table.

b Ravi suggest that Kyra uses this grouped frequency table instead.

Minutes	Frequency
20 to 29	2
30 to 39	3
40 to 49	2
50 to 59	2
60 to 69	3
70 to 79	2

Give one advantage and one disadvantage of using a grouped frequency table to display the data.

c Ravi takes a census all of all 152 students in the year. Give one disadvantage of using a stem-and-leaf diagram to display the data.

4.3 Representing data 1

You can use a **bar chart** to display data.

Bar charts give a visual picture of the size of each category.

- A bar chart shows
  - how each category compares with the others
  - all the data, but in categories.

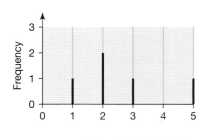

The bars can be horizontal or vertical.

Lines are drawn instead of bars.

- **Bar-line charts** are a good way to display (discrete) numerical data.

EXAMPLE

A class are asked to name one favourite pet. The results are shown in the table.

Pet	Dog	Cat	Guinea pig	Other
Number of students	16	9	12	3

Draw a bar chart to show this information.

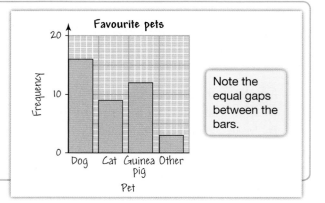

Note the equal gaps between the bars.

You can use a **pie chart** to display data.

Pie charts use a circle to give a quick visual picture of all the data.

The size of each **angle** shows the size of each **category**.

12.1

- A pie chart shows the **proportion** or fraction of each category compared to the whole circle.

EXAMPLE

60 vehicles are shown on the pie chart.

Calculate the numbers of cars, vans, buses and lorries.

Vehicles parked in the High St.

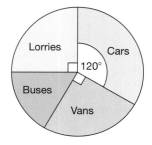

60 vehicles represents 360°.

1 vehicle represents   360° ÷ 60 = 6°

The angle for buses is 360° − (90° + 90° + 120°) = 60°

Number of cars   = 120 ÷ 6 = 20

Number of vans   = 90 ÷ 6 = 15

Number of buses   = 60 ÷ 6 = 10

Number of lorries = 90 ÷ 6 = 15

Total number of vehicles =   60

Use a protractor to measure the angles in the pie chart.

Exercise 4.3S

1 The frequency table shows the results of a 'best live band' survey.
Draw a bar chart to show this information.

U2	The White Stripes	Kings of Leon	Arcade Fire	Radiohead
500	350	300	250	150

2 The number of concerts held at various venues is given.

Venue	Number of concerts
NEC	10
Arena	15
MEN	20
NIA	18
Wembley	9

Draw a bar chart to show the information.

3 The number of Bank Holidays in different countries is shown in the table.
Draw a bar chart to show this information.

Country	Number of Bank Holidays
UK	8
Italy	16
Iceland	15
Spain	14

4 The cost to fly to certain resorts is given in the frequency table.
Draw a bar chart to show this information.

Resort	Cost (£)
Lisbon	90
Crete	100
Malta	80
Menorca	70
Cyprus	110

5 Seven boys and five girls attend an after school homework club.

a Calculate the angle one student represents in a pie chart.

b Calculate the angles to represent boys and girls.

c Draw a pie chart to show the information.

6 The weather record for 60 days is shown in the frequency table. This gives the predominant weather for that particular day.

Weather	Number of days
Sunny	15
Cloudy	18
Rainy	14
Snowy	3
Windy	10

a Calculate the angle that one day represents in a pie chart.

b Calculate the angle of each category in the pie chart.

c Draw a pie chart to show the data.

7 A school fete is open from 10 am to 4 pm. A teacher has offered to help.
She spends these times on each stall.

Stall	Time
Bat the Rat	30 mins
Hook a Duck	25 mins
Smash a Plate	35 mins
Roll a Coin	80 mins
Tombola	70 mins
Break 1	60 mins
Break 2	60 mins

Draw a pie chart to show this information.

8 A car dealer sells 18 cars in one week of three different types: diesel, petrol and electric.

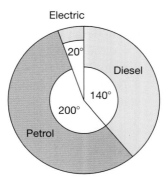

a Calculate the angle that represents one car.

b Calculate the number of cars sold that are

i diesel   **ii** petrol

iii electric.

1205, 1206, 1207 SEARCH

4.3 Representing data 1

RECAP

RECAP

- You can use a bar chart or a pie chart to display data.
- A bar chart uses a bar to show the size of each category.
- A pie chart uses a circle to give a visual picture of all the data. The sectors of the circle represent the size of each category. The size of the sector angle is proportional to the frequency.

HOW TO

1. Read information from the diagram. Make sure that you read the scale or key carefully.
2. Use the information from the diagram to answer the question.

EXAMPLE

A class are asked to name one favourite pet. The results are shown in the table.

Pet	Dog	Cat	Guinea pig	Other
Number of students	16	9	12	3

Jess uses her computer to create this graph.

Make two criticisms of the graph.

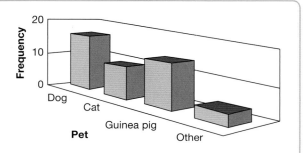

1. Compare the graph with the frequency table and graph in 4.3 Skills. Look at the scale of the graph.

   The scale chosen makes it is difficult to read information from the graph.

2. Bar charts should let you compare the different categories.

   The perspective makes it difficult to compare the heights of the bars.

EXAMPLE

Janelle asks people in Easton and Westville to name their favorite fruit. The results are shown in the pie chart.

Janelle says that there *are more people* in Easton whose favorite fruit is banana than there are in Westville. Explain why Janelle is wrong and rewrite her statement so that it is true.

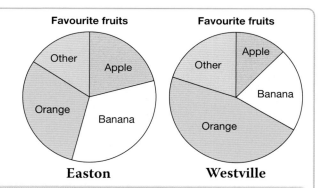

1. Pie charts don't show the frequency of each category.
2. Janelle doesn't know how many people in Westville are represented in the pie chart. Although the sector is larger for Easton, it could represent fewer people.

   Pie charts show the proportion of each category.

   There is a higher proportion of people in Easton whose favorite fruit is banana than there are in Westville.

Exercise 4.3A

1 The number of students in different year groups is given.
Marnie draws this bar chart for her data.

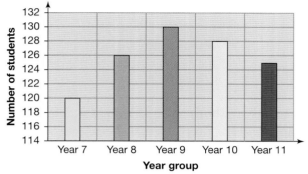

Give one reason why Marnie's graph is misleading.

2 Josh records the number of jars of sauce sold in his deli.

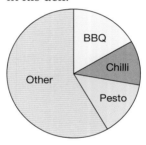

He draws a pie chart to show the information. Josh wants to compare the sales of each flavour he sells. Give three disadvantages of Josh's chart.

3 The number of drink sold during lunchtime in a shop are shown in the table.

Drink	No. sold	Drink	No. sold
Apple	6	Lemonade	3
Blackcurrant	5	Mocha	2
Cappuccino	8	Orange	7
Cola	12	Pineapple	4
Espresso	3	Tea	9
Latte	5	Water	9

a Give one disadvantage of using a pie chart to display the data.

b Give one advantage of using a bar chart to display the data.

4 The pie charts show the number of Year 11 students studying languages at two nearby schools.

Ashford School

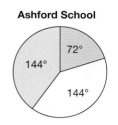

Benton School

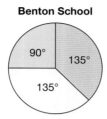

☐ French ☐ Spanish ☐ German

a Jermaine says twice as many students at Ashford school study German compared to those studying Spanish. Do you agree with Jermaine?

b Lydia says it must be the case that more students at Ashford schools study French than at Benton school. Do you agree with Lydia?

5 The bar chart shows the number of letters delivered in one week to No. 10 and No. 12.

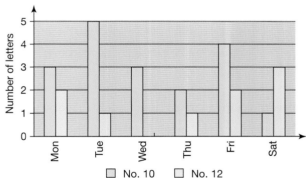

☐ No. 10 ☐ No. 12

a Draw an accurate pie chart to show the number of letters delivered each day to No. 10.

b Draw an accurate pie chart for No. 12.

c Mark wants to compare the proportion of letters received on each day for the two houses.
Should he use the bar chart or the pie charts? Give your reasons.

d Peter want to compare the number of letters received each day for the two houses.
Should he use the bar chart or the pie charts? Give your reasons.

 🔍 1205, 1206, 1207   SEARCH

4.4 Averages and spread 1

- The **mean** of a set of data is the total of all the values divided by the number of values.
- The **mode** is the value that occurs most often.
- The **median** is the middle value when the data is arranged in order.
- The **range** is the highest value – lowest value.

Outliers are values that lie outside most of the other values of a set of data.

In this data set

1, 1, 2, 2, 3, 4, 4, 4, 16

16 is an outlier.

The mean and range are both affected by outliers.

EXAMPLE

Ten people took part in a golf competition. Their scores are shown in the frequency table.

Calculate the mean, mode and median of the scores.

You can calculate directly from the frequency table.

Score	Frequency	Score × Frequency
67	1	67
68	4	272
69	3	207
70	1	70
71	1	71
	10	687

68 + 68 + 68 + 68 or 68 × 4.

The total of all the scores of the 10 golfers.

Mean = 687 ÷ 10 = 68.7

Median = (5th value + 6th value) ÷ 2 = (68 + 69) ÷ 2 = 68.5

There are 10 values so the median is the mean of the 5th and 6th value.

Mode = 68   The mode is 68, not 4.

Spread is a measure of how widely dispersed the data are.

The range and the **interquartile range** (IQR) area measures of spread.

If there are one or more extreme values the IQR is a better measure of spread than the range.

- IQR = upper quartile – lower quartile

Lower quartile
$= \frac{1}{4} (n + 1)$th value.

Upper quartile
$= \frac{3}{4} (n + 1)$th value.

EXAMPLE

Louise collected data on the number of times her friends went swimming in one month.

4   7   22   1   6   2   1   5   6   6   4

Work out the interquartile range.

The outlier (22) does not affect the IQR.

Put the data in order.   1   1   ②   4   4   5   6   6   ⑥   7   22
There are 11 pieces of data.

Lower quartile $= \left(\frac{11 + 1}{4}\right)$th value   Upper quartile $= \left(\frac{3(11 + 1)}{4}\right)$th value

= 3rd value = 2   = 9th value = 6

IQR = 6 – 2 = 4

Exercise 4.4S

1 Calculate the mean for each set of numbers.

    a 4, 9, 7, 12   **b** 8, 11, 8

    c 3, 2, 2, 3, 0   **d** 1, 9, 8, 6

    e 2, 2, 4, 1, 0, 3 **f** 23, 22, 25, 26

    g 17, 19, 19, 20, 18, 17, 16

    h 103, 104, 105

    i 14, 10, 24, 12

    j 4, 6, 7, 6, 4, 4, 3, 7, 3, 6

2 **a** Find the median of each set of numbers.

        i 7, 8, 8, 5, 4, 3, 3

        ii 11, 12, 10, 9, 9, 10, 26

        iii 38, 35, 30, 37, 34, 32, 36

        iv 101, 98, 103, 97, 99, 97, 95

        v 3, 2, 0, 0, 1, 2, 1, 2, 3

    b Which data sets contain outliers? Have the outliers affected the median?

3 Calculate the mode and range of each set of numbers.

    a −6, 0, 1, 1, 1, 1, 2, 2, 2, 3, 3, 4, 4

    b 5, 5, 6, 6, 6, 7, 7, 7, 7, 8, 8, 8, 8, 8

    c 10, 11, 11, 11, 12, 12, 13, 14

    d 21, 22, 23, 24, 24, 25, 25, 25, 36

    e 8, 8, 8, 9, 9, 10, 11

    f 4, 3, 5, 5, 6, 6, 4, 3, 4, 5, 6, 5, 3

Which data sets contain outliers? Have the outliers affected the mode or range.

4 The numbers of flowers on eight rose plants are shown in the frequency table.

Number of flowers	Tally	Frequency
3	||||	4
4	||	2
5	||	2

    a List the eight numbers in order of size, smallest first. Calculate the mean, mode, median and range of the eight numbers.

    b Use the table to calculate the mean, mode and median. Check your answers against your answer for part **a**.

5 The number of days that 25 students were present at school in a week are shown in the frequency table.

Number of days	Tally	Frequency
0		0
1	||||	4
2	Ⱶ卌 |	6
3	||	2
4	卌	5
5	卌 |||	8

    a List the 25 numbers in order of size, smallest first. Calculate the mean, mode, median and range of the 25 numbers.

    b Use the table to calculate the mean, mode and median. Check your answers against your answer for part **a**.

6 For these sets of numbers work out the

    i range   **ii** mode

    iii mean   **iv** median

    v interquartile range.

    a 5, 9, 7, 8, 2, 3, 6, 6, 7, 6, 5

    b 45, 63, 72, 63, 63, 24, 54, 73, 99, 65, 63, 72, 39, 44, 63

    c 97, 95, 96, 98, 92, 95, 96, 97, 99, 91, 96

    d 13, 76, 22, 54, 37, 22, 21, 19, 59, 37, 84

    e 89, 87, 64, 88, 82, 88, 85, 83, 81, 89, 90

    f 53, 74, 29, 32, 67, 53, 99, 62, 34, 28, 27, 27, 27, 64, 27

    g 101, 106, 108, 102, 108, 105, 106, 109, 103, 105, 107, 104, 104, 105, 105

7 **a** Subtract 100 from each of the numbers in question **6g** and write down the set of numbers you get.

    b Work out the

        i range   **ii** mode

        iii mean   **iv** median

        v interquartile range.

    c Compare your answers for the measures of spread in part **b i** and **v** and **6g i** and **v**.

    What do you notice?

🔍 1192, 1202, 1254   SEARCH

4.4 Averages and spread 1

RECAP

- To calculate the mean of a set of data, add all the values and divide by the number of values.
- To find the median, arrange the data in order and choose the middle value.
- To find the mode, choose the value that occurs most often.
- To find the range, subtract the smallest value from the largest value.
- To find the interquartile range, subtract the lower quartile from the upper quartile.

> The mean, median and mode are averages. The range and interquartile range are measures of spread.

HOW TO

To compare data sets

① Calculate the mean, median, mode, range or interquartile range for each set of data.

② To compare the averages, look at the mean, median or mode.

③ To compare the spread, look at the range or interquartile range.

EXAMPLE

A team of 7 girls and a team of 8 boys did a sponsored run for charity.
The distances the girls and boys ran are shown.

Girls

Distance (km)	Frequency
1	3
2	2
3	1
4	1
5	0

Boys

3, 5, 5, 3, 4, 4, 5, 5
all distances in kilometres

By calculating the mean, median and range, compare each set of data.

① Calculate the mean, median and range of each set of data.

Girls

$$Mean = (3 + 4 + 3 + 4 + 0) \div 7 = 14 \div 7 = 2$$

Median = 2   (the 4th distance is the middle value)

Range = 4 − 1 = 3   (highest value − lowest value)

Boys

Mean = 34 ÷ 8 = 4.25

Median = (4 + 5) ÷ 2 = 9 ÷ 2 = 4.5   (the mean of the 4th and 5th distances)

Range = 5 − 3 = 2   (highest value − lowest value)

② Compare the mean and median.

The mean and median show the boys ran further on average than the girls.

③ Compare the range.

The range shows that the girls' distances were more spread out than the boys' distances.

Exercise 4.4A

1 The number of bottles of milk delivered to two houses is shown in the table.

	Sat	Sun	Mon	Tues	Wed	Thur	Fri
Number 45	2	0	1	1	1	1	1
Number 47	4	0	2	2	2	2	2

   a Calculate the range for Number 45 and Number 47.

   b Use your answers for the range to compare the number of bottles of milk delivered to each house.

2 The number of cars at each house on Ullswater Drive are

   2 4 1 0 1 2 1 2 3 2

   a Copy and complete the frequency table.

Number of cars	Tally	Number of houses
0		
1		
2		
3		
4		

   b Calculate the mean, mode and median number of cars for Ullswater Drive. The mean, mode and median number of cars at each house on Ambleside Close are

Mean	Mode	Median
0.7	0	1

   c Use the mean, mode and median to compare the number of cars on Ullswater Drive and Ambleside Close.

3 The range of these numbers is 17.

   Find two possible values for the unknown number.

4 Monica has five numbered cards. One of the cards is numbered −2.6 Monica's cards have

   ● range = 7.2

   ● median = 3.5

   ● mode = 4.1

   Write down the five numbers on Monica's cards.

5 Write down three sets of seven numbers that have a median 6, range 14 and interquartile range 5.

6 A set of five numbers satisfies these criteria:

   Range = 10   Median = 8   Mode = 6

   Explain why the mean of the numbers must be between 8.8 and 10.4.

7 The bar chart shows the test results for a class of 19 students.

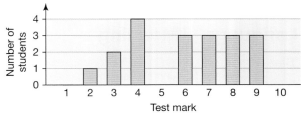

   a Calculate the mean, mode, median, range and interquartile range.

   b If a new student joined the class and got a mark of 10, how would this affect your answer to part **a**?

8 There are nine passengers on a bus. The mean age of the passengers is 44 years old. Jasmine gets on the bus. Jasmine is 14 years old. Find the new mean age of the passengers on the bus.

9 Reuben counted the raisins in 21 boxes. The mean number of raisins per box was 14.095 (3 dp). Reuben records the information for 20 of the boxes in the table.

Number of raisins	Number of boxes
13	5
14	8
15	7

   Find number of raisins in the last box.

10 The mean mark in a statistics test for a class was 84%. There are 32 students in the class, 12 of whom are girls. The mean mark in the test for the girls was 93%. Work out the mean mark in the statistics test for the boys.

Q 1192, 1202, 1254   SEARCH

Summary

Checkout
You should now be able to...

	Test it Questions
✔ Identify when a sample may be biased.	1
✔ Construct and interpret frequency tables, bar charts and pie charts.	2
✔ Calculate the mean, median and mode of a data set.	2, 4
✔ Calculate the range and inter-quartile range of a data set.	2, 4
✔ Use averages and measures of spread to compare data sets.	4

Language   Meaning   Example

Language	Meaning	Example
Bias	An in-built error caused by choosing an unrepresentative sample so that that some outcomes are more likely than others.	Choosing fifty students from the same year group will not give an accurate representation of the homework set in other year groups.
Two-way table	A table that links information about two different categories.	(see table below)
Bar-chart	A way of displaying data where the height of each bar represents the frequency.	See lesson 4.3
Bar-line chart	A way of displaying data where the length of each line represents the frequency.	See lesson 4.3
Pie chart	A circular chart divided into sectors. Each sector represents one category. The size of the angle of each sector is proportional to the frequency.	See lesson 4.3
Mean	An average found by adding all the values together and dividing by the number of values.	(see example below)
Mode	The value that occurs most often	
Median	The middle value when the data is arranged in order of size. If there is an even number of data, the median is the mean of the middle two values	
Range	The difference between the largest value and the smallest value.	
Lower quartile	The middle value or median of the lower half of the set of data.	
Upper quartile	The middle value or median of the upper half of the set of data.	
Interquartile range (IQR)	The difference between the upper quartile and the lower quartile.	

Two-way table example:

	Own a pet	Don't own a pet
Year 10	73	67
Year 11	59	81

Mean / Mode / Median example:

Data: 1, 1, 2, 2, 4, 5, 7, 8, 8, 8, 9

Mean = (1 + 1 + 2 + 2 + 4 + 6 + 6 + 8 + 8 + 8 + 9) ÷ 11

= 55 ÷ 11

= 5

Mode = 8

Median = 6

Range = 9 − 1 = 8

Lower Quartile = $\frac{11 + 1}{4}$ = 3rd value

= 2

Higher Quartile = $\frac{3(11 + 1)}{4}$ = 9th value

= 8

Interquartile range = 8 − 2

= 6

Statistics   Handling data 1

Review

1 Rufus wants to know the average age of people who watch football on television. He visits a pub which is showing a football match and surveys those in the pub to estimate the average age.
What is wrong with this method of sampling?

2 A class of 36 students choose to study either French or Spanish, but not both. There are 15 boys in the class. 7 boys study French. 19 students study Spanish. How many girls study Spanish?

3 The table shows the amount of money spent by a County Council in one year on different services.

Service	Expenditure (£1000s)
Children's Services	120
Corporate/Finance	60
Adult Social Care	210
Transport	45
Residents' Services	105

a Construct a pie chart to show this information.

A second council spends the same proportion on each service but has a larger budget.

The second council spends £252 000 on Adult Social Care.

b What is the total service budget of the second council?

c Draw a bar chart to show the expenditure of the second council.

4 Molly counts the number of seconds people can hold their breath for.

25, 32, 39, 41, 17, 23, 29, 37, 35, 40, 72, 39, 31, 39, 42

a Calculate the
    i   mean   ii   range of the data.

b Calculate the
    i   median   ii   upper quartile
    iii lower quartile of this data.

c What is the interquartile range?

5 The ages of members of a choir are given below.

Women: 31, 39, 27, 56, 58, 60, 28, 34, 37, 31, 43

Men: 52, 61, 35, 47, 48, 25, 30

a For the men's ages find
    i   the median
    ii the interquartile range.

b For the women's ages find
    i   the median
    ii the interquartile range.

c Compare the two sets of data.

What next?

	Score			
	0 – 1		Your knowledge of this topic is still developing. To improve look at MyMaths: 1192, 1193, 1202, 1205, 1206, 1207, 1214, 1248, 1254	
	2 – 3		You are gaining a secure knowledge of this topic. To improve look at InvisiPens: 04Sa – n	
	4		You have mastered these skills. Well done you are ready to progress! To develop your exam technique look at InvisiPens: 04Aa – i	

Assessment 4

1 Five questions for investigation are listed below. Answer the following for each investigation.

   i   Would you investigate the whole population or take a sample? [4]

   ii   How you would collect the data needed? [4]

   iii How would you avoid bias? [4]

      a   What is the most common make of car amongst teachers in your school?

      b   How tall are the children in your class?

      c   Are winters in the UK getting warmer?

      d   Can boys estimate lengths better than girls?

2 The populations of 3 villages are: 410   260   320
A local historian is conducting a survey. She selects 40 people from each village.
Explain why this is not a fair sample. [1]

3 A manager recorded the time that staff members spent on personal emails during working hours.

Time (nearest min.)	5	6	7	8	9	10	11	12
Frequency	75	52	47	31	22	18	12	6

   a   Draw a bar chart to represent this data [4]

   b   Draw a pie chart to represent this data [6]

   c   Find the

      i   mode   [1]   **ii**   median   [2]

      of the time spent on personal emails.

   d   Estimate the mean time spent on personal emails, giving your answer to the nearest second. [5]

4 Display the following information in a two-way table:

In a group of 35 business executives, 15 wear glasses and 5 are left-handed.
4 executives are left-handed **but don't** wear glasses. [5]

5 Harry Bows, a sweetshop owner, counts the coins in his till.

This data is shown in the bar chart:

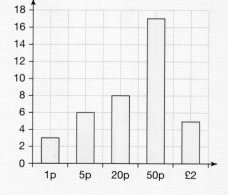

   a   Harry adds 12 £1 coins, 14 ten pence coins and 7 two pence coins to the till. Add this data to the bar chart. [3]

   b   Which coin is found most often in Harry's till? [1]

   c   What is the value of the median coin in Harry's till? [3]

   d   Find the mean value of the coins in Harry's till. [5]

6 These two pie charts show how 1200 hospital staff members get to work in summer.

a Which is the modal form of travel? [1]

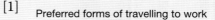
Preferred forms of travelling to work

b Which is the least favourite form
of travel? [1]

c What percentage of staff take the
bus to work? [2]

d How many staff drive to work? [3]

e How many staff cycle to work? [4]

☐ Bus = 72°
☐ Walk = 162°
☐ Cycle
☐ Car = 36°

7 A lorry driver recorded the distances driven in 10 journeys (in miles):

56   113   88   67   163   90   88   109   135   121

a Calculate the
i mean   [3]   **ii** median   [2]   **iii** mode   [1]

b Does he average 100 miles per day? Explain your answer. [2]

c Calculate:

i the range   [1]   **ii** the interquartile range. [3]

d He realises that one distance recorded as 88 should have been 98.
Without performing any additional calculations, decide what effect this will have on the
mean, median, mode, range and interquartile range of the distances. Give reasons for your
answers. [5]

8 The table shows information about the number of children and pets in different families.

		Number of pets					
		0	**1**	**2**	**3**	**4**	**5**
	1	3	5	4	1	0	1
Number of	**2**	4	2	2	0	0	0
children in family	**3**	1	0	2	1	0	0
	4	0	1	0	0	0	0
	5	2	1	0	0	0	0

a How many families have 1 pet? [1]

b How many 2 children families have 1 pet? [1]

c How many families are there where the number of pets is the same as the
number of children in the family? [1]

d What is the modal number of pets? [1]

e Calculate the mean number of pets per family. [5]

5 Fractions, decimals and percentages

Introduction

Many food and drink products that you buy come with nutrition information clearly displayed on the labelling. This is usually in the form of percentages, for example 'saturated fat 5%, total carbohydrate 12%, calcium 9%, etc'. Product manufacturers are expected by law to display this information, and the labels often tend to have a similar format.

What's the point?

Awareness of what you are eating and drinking is important in achieving a healthy, balanced diet and percentages allow you to monitor this.

Fat Total	1g
Fat Saturated	2g
Fat Trans	2g
Carbohydrate	0g
Sugar	23g
Dietary Fibre	15g
Sodium	0g
	45mg

Recommended Daily Allowances

| Vitamin A | 0% | Vitamin C | |
| Calcium | 15% | Iron | |

	per Serving	Intake (per Serving)	
Energy	607 kJ (145Cal)	7%	243 kJ (58Cal)
Protein	2.5g	5%	1.0g
Fat, Total†	0.6g	0.9%	0.2g
- saturated	0.2g	0.9%	0.1g
Carbohydrate	31.2g	10%	12.5g
- sugars	8.0g	9%	3.2g
Sodium	815mg	36%	325mg

*Percentage Daily Intakes are based on an average adult diet of 8700kJ.

Ingredients: Sugar, non dairy
nut oil

Objectives

By the end of this chapter, you will have learned how to ...

- Find fractions and percentages of amounts.
- Add, subtract, multiply and divide with fractions and mixed numbers.
- Convert between fractions, decimals (including recurring decimals) and percentages.
- Order fractions, decimals and percentages.

ving size: 26g

per 26g serve

Check in

1 Shade five copies of this diagram to represent each of these fractions.

a $\frac{1}{2}$ **b** $\frac{2}{3}$ **c** $\frac{3}{4}$ **d** $\frac{5}{6}$ **e** $\frac{7}{12}$

2 Cancel each of these fractions down to their simplest terms.

a $\frac{2}{4}$ **b** $\frac{15}{20}$ **c** $\frac{8}{10}$ **d** $\frac{95}{100}$ **e** $\frac{6}{8}$

3 Write notes to show how you would find a mental estimate for each of these calculations.

a 19.2×28.9 **b** $355.72 \div 58.91$ **c** $1206 - 816$ **d** $6987 + 6039$

according to direct

approximately

Average Quantity Per

Average Quantity Per Serving

Average Quantity Per 100ml

146 kJ   86 kJ

1.9 g   1.1 g

0.6 g   0.4 g

0.1 g   0.1 g

Total   3.2 g

urated

rbohydrate   5.5 g   2.4 g

ugars   4.0 g

odium   620 mg

PROTEIN

FAT

Total	6.3g	7.4g
Saturated	1.4g	1.7g
Trans	Less than 0.1g	
Polyunsaturated	2.4g	2.8g
Omega-3	0.5g	0.6g
EPA	210mg	250mg
DHA	290mg	350mg
Monounsaturated	2.5g	2.9
Total	0.3g	0.4
Sugars	0.2g	0.
	391mg	460

CARBOHYDRATE

SODIUM   365 mg

Nutrition Facts

Serving Size 1 Rounded Scoop (32g)
Servings Per Container 73

Amount Per Serving

Calories 132   Calories from Fat 10

% Daily Value*

Total Fat 1g   **2%**

     3%

     0%

     3%

     2%

Sugars 3g

Protein 24g

Chapter investigation

$\pi = 3.14159265358979323846$ (20 dp)

Find an approximation to π in the form $3 + \frac{1}{a}$

Find another approximation to π in the form $3 + \cfrac{1}{a + \frac{1}{b}}$

Which approximation is more accurate?

5.1 Fractions and percentages

● You find fractions of a quantity by multiplying.

For example, two thirds of $5 = \frac{2}{3} \times 5 = \frac{2 \times 5}{3} = \frac{10}{3} = 3\frac{1}{3}$

> Notice that you cannot give $3\frac{1}{3}$ as an **exact** answer in decimals because $\frac{1}{3}$ is a recurring decimal.

● When the quantity and the denominator of the fraction have a **common factor**, you can **cancel** this factor before multiplying.

For example, $\frac{2}{\underset{1}{3}} \times 24^8 = \frac{2}{1} \times 8 = 16$

EXAMPLE

Calculate **a** $\frac{3}{4}$ of 28   **b** $\frac{5}{8}$ of 6   **c** $\frac{4}{9}$ of 12   **d** $\frac{5}{9}$ of 25

a $\frac{3}{\underset{1}{4}} \times 28^7 = 3 \times 7 = 21$

b $\frac{5}{\underset{4}{8}} \times 6^3 = \frac{5 \times 3}{4} = \frac{15}{4} = 3\frac{3}{4}$

c $\frac{4}{\underset{3}{9}} \times 12^4 = \frac{4 \times 4}{3} = \frac{16}{3} = 5\frac{1}{3}$

d $\frac{5}{9} \times 25 = \frac{125}{9} = 13\frac{8}{9}$

● You can use mental methods to find percentages of amounts.

To find 50%, divide by 2.   To find 25%, divide by 4.
To find 10%, divide by 10.   To find 1%, divide by 100.

> A percentage tells you how many parts per 100 there are. 78% means 78 out of 100.

● You can find a percentage of an amount by multiplying the quantity by an equivalent decimal or fraction.

To find 47% of an amount, multiply the amount by $\frac{47}{100}$ or 0.47

EXAMPLE

Calculate   **a** 45% of 60 cm   **b** 34% of 85 kg   **c** 16% of £25

a 50% of 60 = 30   50% is a half.
   5% of 60 = 3
   45% of 60 cm = 30 cm − 3 cm = 27 cm

b 34% = $\frac{34}{100}$ = 0.34   0.34 is the **decimal equivalent** of 34%.
   34% of 85 kg = 0.34 × 85
        = 28.9 kg   By calculator.
        = 29 kg to 2 sf.

c 16% of £25 = $\frac{\overset{4}{16}}{\underset{1}{100}} \times 25^1$ Cancel the quantity (25) and the denominator (100). Then cancel 16 and 4.
   = £4

Exercise 5.1S

1 Calculate these. Give your answers as proper fractions or mixed numbers.

    a $\frac{1}{2}$ of 7   **b** $\frac{1}{5}$ of 8   **c** $\frac{1}{3}$ of 10

    d $\frac{1}{3}$ of 2   **e** $\frac{2}{5}$ of 6

2 Work these out. Show how common factors can be cancelled in each.

    a $\frac{1}{2}$ of 20   **b** $\frac{1}{4}$ of 84   **c** $\frac{1}{3}$ of 36

    d $\frac{1}{5}$ of 65   **e** $\frac{2}{3}$ of 33

You should show all of your working for questions **3**, **4** and **5**. In particular, show how the calculations can be simplified by cancelling common factors.

3 A reel holds 60 m of wire when new. $\frac{2}{5}$ of the wire has been used.

    a What length of wire has been used?

    b What length of wire is left on the reel?

4 Calculate the amount of liquid in these containers.

    a A 40 litre barrel that is $\frac{3}{8}$ full.

    b A 240 cl jar that is $\frac{3}{4}$ full.

    c A 120 cl glass that is $\frac{2}{5}$ full.

    d A 750 ml bottle that is $\frac{2}{3}$ empty.

5 Calculate

    a $\frac{5}{8}$ of 48 m   **b** $\frac{2}{9}$ of 36 km

    c $\frac{4}{7}$ of 28 mm   **d** $\frac{3}{4}$ of 120 m

The answers to questions **6** and **7** are not whole numbers.
You should show your working as before, and give your answers as proper fractions or mixed numbers.

6 Calculate

    a $\frac{1}{6}$ of 10   **b** $\frac{1}{4}$ of 22   **c** $\frac{3}{10}$ of 15

    d $\frac{1}{12}$ of 8   **e** $\frac{4}{9}$ of 21

7 Calculate these lengths.

    a $\frac{1}{9}$ of 24 miles   **b** $\frac{5}{6}$ of 40 miles

    c $\frac{5}{18}$ of 45 miles   **d** $\frac{3}{20}$ of 25 miles

8 Calculate these and convert your answers to (approximate) decimal numbers.

    a $\frac{5}{12}$ of 16 m   **b** $\frac{4}{9}$ of 12 mm

    c $\frac{3}{22}$ of 64 cm   **d** $\frac{3}{14}$ of 104 km

9 Calculate mentally.

    a 25% of 42   **b** 90% of 140

    c 20% of 1200   **d** 60% of 500

    e 30% of 440   **f** 11% of 900

10 Calculate these, using a mental method wherever possible.

    a 30% of £750   **b** 55% of 1800 m

    c 90% of 2800 kg   **d** 60% of €240

11 Use an appropriate method to work these out. Do not use a calculator.

    a 9% of 1500   **b** 13% of 700

    c 31% of 2400   **d** 36% of 50

    e 43% of 900   **f** 6% of 3200

12 Calculate these, using an appropriate mental or written method. Show all your working and do not use a calculator.

    a 23% of 4800 mm   **b** 61% of 3200 kg

    c 39% of €3700   **d** 17% of £2900

13 Convert to decimals.

    a 50%   **b** 60%   **c** 25%

    d 51%   **e** 64%   **f** 22%

    g 15%   **h** 70%   **i** 7%

    j 8.5%   **k** 0.15%   **l** 0.01%

14 Calculate these, using an appropriate method. Use a calculator where necessary.

    a 15% of 38   **b** 25% of 800

    c 27% of 59   **d** 96% of 104

    e 41% of 41   **f** 80% of 25

15 Calculate these, rounding your answers to the nearest penny.

    a 16% of £24   **b** 63% of £85

    c 93% of £15   **d** 42% of £405

    e 88% of £32   **f** 6% of £265

5.1 Fractions and percentages

- You find a fraction of an amount by multiplying the fraction and the quantity.
- You can find a percentage of an amount by multiplying the quantity by an equivalent decimal or fraction.

▲ The term percent is derived from the Latin per centum, meaning "by the hundred".

HOW TO

① Read the problem and decide what fraction or percentage of the amount you need.

② Calculate the answer mentally or with a calculator.

③ Answer the question and include any units.

EXAMPLE

Jane earns £320 each week. She spends $\frac{3}{5}$ of the money and deposits the rest into a savings account.
How much money will Jane save each year?

① Jane will save $\frac{2}{5}$ of the money. $\quad 1 - \frac{3}{5} = \frac{2}{5}$

② Find $\frac{2}{5}$ of £320. $\quad \frac{2}{5} \times 320 = 128$

③ There are 52 weeks in a year. $\quad 128 \times 52 = 6656$

Jane saves £6656 each year.

EXAMPLE

Which deal gives most money off?

Was £255.
Now ⅓ off!

Was £238.
Now 35% off!

① Use a calculator to work out the discount on each deal. Round to the nearest penny.

② Multiply the amount by the fraction.

$\frac{1}{3}$ of £255 = £255 $\times \frac{1}{3}$ = £85 $\quad$ or you could divide by 3.

35% = 0.35. Multiply the amount by the decimal.

35% of £238 = £238 $\times$ 0.35 = £83.30

Write in the answer in pounds and pence.

③ The first deal gives more money off.

EXAMPLE

Tom is given money for his birthday.
He spends $\frac{1}{3}$ of it on Monday.
On Tuesday he spends $\frac{3}{4}$ of the remainder.
What fraction of the total amount does he spend on Tuesday?

① There is no amount in the question.
Let m be the total amount of money.

② Tom has $\frac{2}{3}$ of the money remaining on Tuesday.

Remaining money = $\frac{2}{3}$ of m = $\frac{2}{3} \times m$

Tom spends $\frac{3}{4}$ of the remainder.

$\frac{3}{4}$ of $\frac{2}{3} \times m = \frac{3}{4} \times \frac{2}{3} \times m = \left(\frac{3}{4} \times \frac{2}{3} \right)$ of m

③ Multiply the fractions together to find the fraction of the total.

$\frac{3}{4} \times \frac{2}{3} = \frac{3 \times 2}{4 \times 3} = \frac{6}{12} = \frac{1}{2}$

Tom spends half of the money on Tuesday.

Exercise 5.1A

1 Rearrange these cards to make three correct statements.

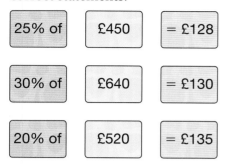

25% of	£450	= £128
30% of	£640	= £130
20% of	£520	= £135

2 Calculate these fractions of amounts without using a calculator.

a A jacket normally costs £130. In a sale the jacket is priced at $\frac{4}{5}$ of its normal selling price. What is the new price of the jacket?

b Hector rents out a holiday flat. His flat is available for 45 weeks of the year. He rents the flat out to tourists for $\frac{7}{9}$ of the time it is available. For how many weeks is Hector's flat occupied by tourists?

3 An empty swimming pool is to be filled with water. It takes 12 hours to fill the pool and the full pool contains 98 m³ of water. How much water will the pool contain after 5 hours?
Show your working.

4 Mrs Jones has a conservatory built, which costs £12 000.
She pays an initial deposit of 15%. The remainder is to be paid in 24 equal monthly instalments. How much is each of these instalments?

5 A restaurant adds a 12% service charge to the bill. What will be the total cost for a meal that is £65.80 before the service charge is added?

6 A school has 1248 pupils, and 48% of them are girls. How many boys are there in the school? Show your working.

7 Julia earns a salary of £47 800 per year. She is awarded a 2.7% pay rise. Calculate her new salary.

8 Rajin earns £18 500 per year; the first £6550 is untaxed; the remainder is taxed at 20%. How much tax does he pay per month?

9 To obtain a driving licence, you must pass both the theory and the practical test.
You cannot take the practical test if you fail the theory test.
200 people applied for a driving licence.
80% of them passed the theory test.
Three out of four people who took the practical test passed.
How many people obtained their driving licence?

10 Alfie sees the same coat in two different shops.

| **Top Bloke** | **Merton** |
| £85 with 15% off | £95 with $\frac{1}{4}$ off |

Where should Alfie buy the coat?

11 Fill in the gaps.

a 45% of £48 = □% of £45

b 2% of 400 m = $\frac{1}{5}$ of □ m

c 25% of 8 kg = $\frac{1}{\square}$ of 4 kg

12 The diagram shows three identical circles A, B and C.
$\frac{2}{3}$ of shape A is shaded.
60% of shape C is shaded.

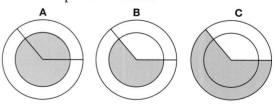

What fraction of shape B is shaded?

13 At a gym, 70% of the members are women.
50% of the women use the swimming pool.
80% of the men use the swimming pool.
What percentage of the members use the swimming pool?

1030, 1031, 1046   SEARCH

5.2 Calculations with fractions

You can only add and subtract fractions if they have common denominators.

$$\frac{2}{8} + \frac{5}{8} = \frac{7}{8}$$ Add the numerators.

- To add or subtract fractions with different denominators, change them to equivalent fractions with the same denominator.

EXAMPLE

Work out these calculations.

a $\frac{11}{12} - \frac{1}{3}$ **b** $\frac{3}{4} + \frac{2}{5}$ **c** $\frac{4}{5} - \frac{3}{10}$

If your answer is an **improper fraction**, change it to a **mixed number**.

Cancel any common factors in the numerator and denominator.

$\frac{11}{12} - \frac{1}{3} = \frac{11}{12} - \frac{4}{12} = \frac{7}{12}$ The **lowest common denominator** is the lowest common multiple (LCM) of 12 and 3, which is 12.

$\frac{3}{4} + \frac{2}{5} = \frac{15}{20} + \frac{8}{20} = \frac{23}{20} = 1\frac{3}{20}$ $\frac{23}{20}$ is an improper fraction – its numerator is larger than its denominator. $1\frac{3}{20}$ is a mixed number – a whole number and a fraction.

$\frac{4}{5} - \frac{3}{10} = \frac{8}{10} - \frac{3}{10} = \frac{5}{10} = \frac{1}{2}$ 5 and 10 have common factor 5.

- To multiply fractions, multiply the numerators and then the denominators, and cancel any common factors.
- To divide by a fraction, multiply by its reciprocal. The reciprocal of a fraction is the original fraction 'turned upside down'.

The reciprocal of $\frac{3}{5}$ is $\frac{5}{3}$.

EXAMPLE

Work out each of these calculations.

a $\frac{3}{5} + \frac{11}{16}$ **b** $\frac{2}{7} - \frac{1}{4}$ **c** $\frac{3}{5} \div 8$ **d** $\frac{4}{9} \div \frac{1}{3}$ **e** $\frac{3}{8} \times \frac{5}{9}$ **f** $2\frac{4}{5} \div 3\frac{1}{2}$

a $\frac{3}{5} + \frac{11}{16} = \frac{48}{80} + \frac{55}{80} = \frac{103}{80} = 1\frac{23}{80}$

The LCM of 5 and 16 is 80.

b $\frac{2}{7} - \frac{1}{4} = \frac{8}{28} - \frac{7}{28} = \frac{1}{28}$

The LCM of 7 and 4 is 28.

c $\frac{3}{5} \div 8 = \frac{3}{5} \times \frac{1}{8} = \frac{3 \times 1}{5 \times 8} = \frac{3}{40}$

The reciprocal of 8 is $\frac{1}{8}$.

d $\frac{4}{9} \div \frac{1}{3} = \frac{4}{9} \times \frac{3}{1} = \frac{12}{9} = \frac{4}{3} = 1\frac{1}{3}$

The reciprocal of $\frac{1}{3}$ is $\frac{3}{1}$ (which is 3).

e $\frac{\cancel{3}}{8} \times \frac{5}{\cancel{9}_3} = \frac{1}{8} \times \frac{5}{3} = \frac{5}{24}$

Notice how you can cancel common factors before multiplying.

f $2\frac{4}{5} \div 3\frac{1}{2} = \frac{14}{5} \div \frac{7}{2} = \frac{\cancel{14}^2}{5} \times \frac{2}{\cancel{7}_1} = \frac{4}{5}$

Change mixed numbers to improper fractions first.

Cancelling the fractions before multiplying gives the same answer as simplifying after multiplying.

Exercise 5.2S

1 Calculate

a $\frac{1}{5} + \frac{2}{5}$ **b** $\frac{1}{4} + \frac{3}{4}$

c $\frac{5}{6} - \frac{1}{6}$ **d** $\frac{9}{10} - \frac{3}{10}$

2 Convert to the equivalent fraction.

a $\frac{2}{3} = \frac{}{30}$ **b** $\frac{3}{7} = \frac{}{42}$

c $\frac{7}{9} = \frac{}{45}$ **d** $\frac{5}{8} = \frac{}{40}$

3 Calculate

a $\frac{1}{5} + \frac{1}{10}$ **b** $\frac{2}{3} + \frac{1}{6}$

c $\frac{2}{5} + \frac{3}{20}$ **d** $\frac{1}{8} + \frac{1}{4}$

4 Convert to a mixed number.

a $\frac{5}{4}$ **b** $\frac{9}{5}$ **c** $\frac{13}{8}$ **d** $\frac{17}{4}$

5 Convert to an improper fraction.

a $1\frac{3}{4}$ **b** $1\frac{7}{16}$ **c** $1\frac{5}{9}$ **d** $2\frac{4}{7}$

6 Do these calculations with mixed numbers.

a $1\frac{1}{2} + \frac{1}{4}$ **b** $2\frac{1}{3} - \frac{2}{3}$

c $1\frac{1}{5} + 2\frac{1}{10}$ **d** $3\frac{1}{4} - \frac{1}{8}$

e $2\frac{3}{5} + 1\frac{1}{3}$ **f** $2\frac{3}{4} + 1\frac{5}{6}$

g $4\frac{3}{7} + 3\frac{1}{2}$ **h** $5\frac{4}{9} + 2\frac{3}{7}$

i $3\frac{3}{5} - 2\frac{1}{4}$ **j** $2\frac{1}{2} - 1\frac{3}{4}$

k $3\frac{3}{4} - 1\frac{4}{5}$ **l** $7\frac{3}{7} - 2\frac{1}{2}$

7 Calculate these, giving your answers in their simplest form.

a $\frac{3}{4} \times \frac{1}{5}$ **b** $\frac{2}{3} \times \frac{2}{9}$

c $\frac{2}{7} \times \frac{1}{4}$ **d** $\frac{5}{16} \times \frac{4}{5}$

e $\frac{5}{9} \times \frac{4}{7}$ **f** $\frac{7}{8} \times \frac{2}{21}$

g $\frac{4}{5} \times \frac{3}{13}$ **h** $\frac{8}{35} \times \frac{7}{24}$

8 Calculate these, giving your answers as fractions in their simplest terms.

a $3 \div 4$ **b** $6 \div 8$

c $4 \div 5$ **d** $8 \div 10$

e $3 \div 7$ **f** $9 \div 5$

g $24 \div 7$ **h** $22 \div 8$

> For example,
> $13 \div 4 = 13 \times \frac{1}{4}$
> $= \frac{13}{4} = 3\frac{1}{4}$.

9 Calculate

a $\frac{5}{8} \div 4$ **b** $\frac{3}{4} \div 6$ **c** $\frac{2}{3} \div 7$

d $\frac{3}{16} \div 9$ **e** $4 \div \frac{1}{5}$ **f** $6 \div \frac{2}{3}$

g $5 \div \frac{2}{5}$ **h** $11 \div \frac{3}{7}$

10 Calculate

a $\frac{1}{8} \div \frac{3}{8}$ **b** $\frac{1}{5} \div \frac{4}{5}$ **c** $\frac{1}{14} \div \frac{3}{7}$

d $\frac{1}{10} \div \frac{2}{5}$ **e** $\frac{2}{3} \div \frac{3}{4}$ **f** $\frac{5}{8} \div \frac{7}{9}$

g $\frac{1}{12} \div \frac{3}{8}$ **h** $\frac{2}{7} \div \frac{3}{4}$

11 Calculate

a $1\frac{1}{2} \times \frac{3}{4}$ **b** $2\frac{3}{4} \times \frac{2}{5}$

c $1\frac{1}{5} \times 2\frac{1}{2}$ **d** $1\frac{2}{3} \times 1\frac{4}{5}$

e $2\frac{7}{8} \div \frac{3}{5}$ **f** $4\frac{1}{4} \div 3\frac{1}{2}$

g $5\frac{3}{8} \div 2\frac{3}{4}$ **h** $9\frac{1}{3} \div 2\frac{1}{4}$

12 Calculate

a $3\frac{7}{8} + 2\frac{1}{4}$ **b** $3\frac{7}{8} - 3\frac{1}{4}$

c $5\frac{1}{2} \times 1\frac{7}{8}$ **d** $2\frac{1}{2} \div 3\frac{3}{4}$

13 Calculate

a $\dfrac{\frac{3}{4} + \frac{1}{2}}{\frac{5}{8} - \frac{1}{4}}$ **b** $\left(\frac{3}{4} - \frac{1}{8}\right)\left(\frac{3}{4} + \frac{1}{8}\right)$

c $\dfrac{1\frac{1}{2} - \frac{7}{8}}{2\frac{3}{4} - 1\frac{1}{6}}$ **d** $\left(2\frac{3}{5} - 1\frac{3}{4}\right)\left(1\frac{7}{8} + 3\frac{1}{4}\right)$

14 Check your answers to questions **7–13** using a calculator.

Q 1017, 1040, 1047, 1074 SEARCH

5.2 Calculations with fractions

- To add or subtract fractions, convert to equivalent fractions with the same denominator, and then add or subtract the numerators.
- To multiply fractions, multiply the numerators and then the denominators, and cancel any common factors.
- Dividing by a fraction is the same as multiplying by its reciprocal.

Convert mixed numbers to improper fractions before calculating.

HOW TO

(1) Read the problem and decide what calculation you must do.

(2) Calculate using your knowledge of fractions. Simplify fully and include any units.

EXAMPLE

Jodi wrote

$$\frac{2}{3} + \frac{3}{4} = \frac{2+3}{3+4} = \frac{5}{7} \text{ ✗}$$

Without calculating the correct answer, explain how you know that Jodi's answer is incorrect.

(1) Explain how you know that the answer is incorrect.

The answer $\frac{5}{7}$ cannot be correct. Both $\frac{2}{3}$ and $\frac{3}{4}$ are bigger than $\frac{1}{2}$, so the answer must be greater than 1.

(2) Explain the mistake that Jodi made in her calculation.

Jodi added the numerators and denominators instead of finding the common denominator first.

EXAMPLE

Paul cuts 75 cm from a piece of rope $2\frac{2}{3}$ m long. Find the exact length of the remaining rope.

(1) Subtract 75 cm from $2\frac{2}{3}$ m. First, change 75 cm to metres.

$$75 \text{ cm} = 0.75 \text{ m} = \frac{3}{4} \text{ m}$$

(2) $$2\frac{2}{3} - \frac{3}{4} = \frac{8}{3} - \frac{3}{4}$$

Convert $2\frac{2}{3}$ into an improper fraction.

$$= \frac{32}{12} - \frac{9}{12}$$

Find the lowest common denominator.

$$= \frac{23}{12}$$

$$= 1\frac{11}{12} \text{ m}$$

Convert the answer to a mixed number and include the units.

EXAMPLE

Write in ascending order $\frac{7}{8} \quad \frac{5}{6} \quad \frac{3}{4}$

Ascending – going up

Descending – going down

(1) Convert the fractions to fractions with a common denominator.

LCM of 8, 6 and 4 = 24

$$\frac{7}{8} = \frac{21}{24} \qquad \frac{5}{6} = \frac{20}{24} \qquad \frac{3}{4} = \frac{18}{24}$$

(2) Order the fractions by comparing the numerators.

$$\frac{18}{24} < \frac{20}{24} < \frac{21}{24}$$

In ascending order the fractions are: $\frac{3}{4}, \frac{5}{6}, \frac{7}{8}$

Exercise 5.2A

1 Write each set of fractions in ascending order. Show your working.

a $\dfrac{3}{7}, \dfrac{3}{8}$ and $\dfrac{5}{14}$ **b** $\dfrac{2}{3}, \dfrac{5}{6}$ and $\dfrac{2}{7}$

2 Write each set of fractions in descending order. Show your working.

a $\dfrac{2}{5}, \dfrac{3}{8}, \dfrac{3}{4}$ and $\dfrac{17}{40}$ **b** $\dfrac{5}{6}, \dfrac{11}{24}, \dfrac{7}{12}$ and $\dfrac{5}{8}$

3 $\dfrac{3}{8}$ of the students in a year group study history. $\dfrac{1}{3}$ of the students study geography. The remaining students in the year study computing. What fraction of the students in the year study computing?

4 A furniture manufacturer makes sets of storage boxes that nest inside one another. The length of each box is $1\frac{1}{3}$ times the width.

a If the width of the largest box in the set is $4\frac{1}{2}$ inches, what is the length?

b If the length of the smallest box is $4\frac{1}{6}$ inches, what is the width?

5 Choose from this list

$\dfrac{1}{2}$ $\dfrac{3}{8}$ $\dfrac{2}{5}$ $\dfrac{4}{7}$ $\dfrac{1}{10}$

a two fractions that have the lowest product

b three fractions that add up to 1.

6 Copy the equation grid and fill in the empty squares to make all of the equations correct both horizontally and vertically.

$\dfrac{1}{2}$	$+$	$\dfrac{1}{6}$	$=$	
$\times$		$+$		$\bigstar$
	$\times$		$=$	
$=$		$=$		$=$
$\dfrac{3}{20}$	$\times$		$=$	$\dfrac{1}{10}$

Enter each fraction in its simplest form. Which of the four operations ($+$, $-$, $\times$ or $\div$) must replace the '$\bigstar$'?

7 The diagram shows three identical pentagons A, B and C.

$\dfrac{5}{7}$ of shape A is shaded.

$\dfrac{4}{5}$ of shape C is shaded.

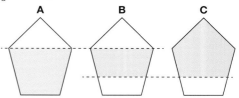

What fraction of shape B is shaded?

8 Which of $\dfrac{4}{9}$ or $\dfrac{6}{11}$ is closer in value to $\dfrac{1}{2}$? Explain your answer.

9 A unit fraction has 1 as a numerator and an integer as a denominator.

$\dfrac{1}{2}$ and $\dfrac{1}{3}$ are both unit fractions.

Find three unit fractions that add up to $\dfrac{1}{2}$.

10 Jennie says that if $a > b$, then $\dfrac{1}{a} > \dfrac{1}{b}$.

a Find an example that disproves Jennie's statement.

b Can you find a pair of values for which Jennie's statement is true?

11 $\dfrac{3}{x} \times \dfrac{5}{6} \div \dfrac{y}{2} = 1$

Prove that $xy = 5$.

12 This process can be used to find an approximation to $\sqrt{3}$.

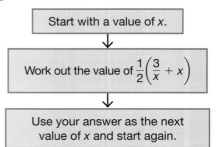

a Use the process twice to find an approximation to $\sqrt{3}$. Start with the value $x = 3$.

b What happens when you square your answer to part **a**? Comment on the accuracy of your approximation to $\sqrt{3}$.

Q 1017, 1040, 1047, 1074   SEARCH

5.3 Fractions, decimals and percentages

- To convert a **terminating** decimal to a fraction, write the decimal as a fraction with **denominator** 10, 100, 1000, ... according to the number of decimal places.
- To convert a percentage to a fraction, divide the percentage by 100.

EXAMPLE

Convert these to fractions. **a** 0.306 **b** 45% **c** 32.5% **d** 0.52

3 decimal places so denominator is 1000.

a $0.306 = \dfrac{306}{1000} = \dfrac{153}{500}$ **b** $45\% = \dfrac{45}{100} = \dfrac{9}{20}$

c $32.5\% = \dfrac{32.5}{100} = \dfrac{65}{200} = \dfrac{13}{40}$ **d** $0.52 = \dfrac{52}{100} = \dfrac{26}{50} = \dfrac{13}{25}$

- To convert a fraction to a decimal divide the numerator by the denominator.

EXAMPLE

Write these fractions as decimals.

a $\dfrac{5}{8}$ **b** $\dfrac{1}{7}$

a $\dfrac{5}{8} = 5 \div 8 = 8\overline{)5.000}^{0.625}$

b $\dfrac{1}{7} = 1 \div 7 = 7\overline{)1.000000000...}^{0.142857142...} = 0.\dot{1}4285\dot{7}$

0.625 is a **terminating** decimal.

$0.\dot{1}42857$ is a recurring decimal. The dots show the recurring group of digits.

To decide if a fraction will be a terminating or a recurring decimal, write it in its simplest form and look at the denominator.

- If the only factors of the denominator are combinations of 2 and 5 then the fraction will be a terminating decimal.
- If the denominator has any factors other than 2 and/or 5 then the fraction will be a recurring decimal.

EXAMPLE

Say whether these fractions are terminating or recurring decimals. **a** $\dfrac{4}{30}$ **b** $\dfrac{7}{16}$

a $\dfrac{4}{30} = \dfrac{2}{15}$
denominator $= 15 = 3 \times 5 \longrightarrow \dfrac{4}{30}$ is a recurring decimal

b denominator $= 16 = 2^4 \longrightarrow \dfrac{7}{16}$ is a terminating decimal

- To convert a fraction to a percentage, write it as a decimal and then multiply the decimal by 100%.

$100\% = \dfrac{100}{100} = 1$ so you are multiplying by 1 which does not change the value of the decimal.

EXAMPLE

Write these fractions as percentages. **a** $\dfrac{5}{8}$ **b** $\dfrac{1}{7}$

a $\dfrac{5}{8} = 0.625 = 0.625 \times 100\% = 62.5\%$

b $\dfrac{1}{7} = 0.\dot{1}4285\dot{7} = 0.\dot{1}4285\dot{7} \times 100\% = 14.\dot{2}8571\dot{4}\% = 14.3\%$ to 1 dp

Number Fractions, decimals and percentages

Exercise 5.3S

1 Write these fractions as decimals.

a $\frac{1}{2}$ b $\frac{3}{4}$ c $\frac{2}{5}$

d $\frac{1}{10}$ e $\frac{1}{5}$ f $\frac{1}{4}$

2 Convert each of the decimals from question **1** to a percentage.

3 Convert these percentages to decimals.

a 43%   b 86%   c 94%

d 45.5%   e 3.75%   f 105%

4 Convert these decimals to fractions, using a mental method.

a 0.5   b 0.25   c 0.2

d 0.125   e 0.75   f 0.9

5 Convert each fraction to a percentage, using a written method. Show your working.

a $\frac{5}{8}$ b $\frac{4}{5}$ c $\frac{7}{8}$

d $\frac{3}{5}$ e $\frac{3}{8}$ f $\frac{1}{8}$

6 Use a calculator to check your answers to question **5**.

7 Convert these decimals to percentages.

a $0.5\dot{5}$ b $0.3\dot{4}$ c $0.7\dot{5}$

d $0.51\dot{2}\dot{8}$ e $0.43\dot{7}$ f 0.838

g $0.83\dot{8}$ h $1.0\dot{5}$

8 Convert these percentages to decimals.

a 22%   b 18.5%   c $55.5\dot{5}$%

d $35.\dot{5}$%   e $6.5\dot{6}$%   f $61.4\dot{9}$%

g $54.4\dot{6}$%   h $152.\dot{2}$%

9 Convert these fractions to percentages. You should be able to do these without a calculator.

a $\frac{4}{5}$ b $\frac{3}{20}$ c $\frac{7}{8}$ d $\frac{3}{4}$

e $\frac{7}{25}$ f $\frac{3}{16}$ g $\frac{9}{20}$ h $\frac{7}{50}$

10 Convert these decimals to fractions.

a 0.51   b 0.43   c 0.413

d 0.719   e 0.91   f 0.871

11 Convert these percentages to fractions.

a 49%   b 53%   c 73%

d 81%   e 37%   f 19%

12 Use a calculator to convert these fractions to decimals.

a $\frac{1}{16}$ b $\frac{7}{25}$ c $\frac{7}{125}$

d $\frac{3}{40}$ e $\frac{7}{16}$ f $\frac{1}{32}$

13 Convert the decimal answers from question **12** to percentages.

14 Convert these decimals to fractions. Give your answers in their simplest forms.

a 0.32   b 0.55   c 0.44

d 0.155   e 0.64   f 0.265

15 Convert these percentages to fractions. Give your answers in their simplest forms.

a 55%   b 62%   c 84%

d 65%   e 72%   f 18.5%

16 Which of these fractions will make recurring decimals? Explain your answer.

$\frac{22}{25}$ $\frac{17}{20}$ $\frac{8}{11}$ $\frac{2}{5}$

17 Write each of these recurring decimals using 'dot' notation.

a 0.111…   b 0.555…

c 0.755 55…   d 0.346 346 346…

e 0.765 656 5…   f 0.012 612 6126…

18 Use a written method to convert each fraction to a decimal. Use the 'dot' notation to represent recurring decimals.

a $\frac{1}{3}$ b $\frac{1}{6}$ c $\frac{2}{3}$

d $\frac{1}{7}$ e $\frac{1}{9}$ f $\frac{5}{6}$

19 Convert your decimal answers from question **18** to percentages.

20 Use a calculator to check your answers to questions **18** and **19**.

21 Use an appropriate method to convert these fractions to decimals.

a $\frac{3}{7}$ b $\frac{3}{16}$ c $\frac{17}{80}$

d $\frac{5}{9}$ e $\frac{4}{25}$ f $\frac{5}{7}$

Q 1015, 1016, 1063, 1066   SEARCH

5.3 Fractions, decimals and percentages

- You can convert between fractions, decimals and percentages.
- Recurring decimals contain digits that repeat over and over again.
- A terminating decimal has a finite number of digits.
- If the only factors of the denominator are combinations of 2 and/or 5, then a fraction can be represented by a terminating decimal.

▲ The line separating the numerator and denominator is called the vinculum. The notation was developed by the mathematician Al-Hassar who lived in Morocco during the 12th century.

HOW TO

① Read the problem and decide how to use the fact that fractions, decimals and percentages are equivalent.

② You may need to use your knowledge from other topics to solve complex problems.

EXAMPLE

Write these numbers in order of size. Start with the smallest number.

0.8 70% $\dfrac{7}{8}$ $\dfrac{3}{4}$

① Order fractions, decimals and percentages by converting them into decimals.

Rewrite $\dfrac{7}{8}$ as an equivalent fraction with a denominator of 1000.

$$70\% = \frac{70}{100} = 0.7 \qquad \frac{7}{8} = \frac{35}{40} = \frac{175}{200} = \frac{875}{1000} = 0.875 \qquad \frac{3}{4} = 0.75$$

Place the decimals in order 0.7 0.75 0.8 0.875

                 70% $\dfrac{3}{4}$ 0.8 $\dfrac{7}{8}$

EXAMPLE

Find $0.5\dot{4} + 0.\dot{4}\dot{5}$.

① Change the recurring decimals to fractions, then find the sum.

$$x = 0.5\dot{4} \qquad\qquad y = 0.\dot{4}\dot{5}$$

② Use your knowledge of place value and algebra to eliminate the recurring digits.

$$100x = 54.5\dot{4} \qquad\qquad 100y = 45.\dot{4}\dot{5}$$

Multiply by 100 so that the decimal part matches the original decimal part.

$$-x = 0.5\dot{4} \qquad\qquad -y = 0.\dot{4}\dot{5}$$

Subtract one from the other.

$$99x = 54 \qquad\qquad 99y = 45$$

$$x = \frac{54}{99} = \frac{6}{11} \qquad\qquad y = \frac{45}{99} = \frac{5}{11}$$

$$0.5\dot{4} + 0.\dot{4}\dot{5} = \frac{6}{11} + \frac{5}{11} = \frac{11}{11} = 1$$

EXAMPLE

Prove that $0.4\dot{9} = \dfrac{1}{2}$

① Change the recurring decimal to a fraction.

$$x = 0.4\dot{9}$$

② Use your knowledge of place value and algebra to eliminate the recurring digits.

$$10x = 4.\dot{9}$$

Multiply x by 10 so that the decimal part contains the recurring digit.

$$100x = 49.\dot{9}$$

Multiply x by 100 so that the decimal part matches $10x$.

$$90x = 45$$

$$x = \frac{45}{90} = \frac{1}{2}$$

Subtract one from the other.

Exercise 5.3A

1 Findlay is told that he must get 80% on his homework for a grade 7, 70% for a grade 6, 60% for a grade 5 and 50% for a grade 4.

What grades does he get for

a 36 out of 50   **b** 17 out of 20

c 15 out of 25   **d** 12 out of 15?

2 Jessica sees bargain signs in two bookshops. She wants to buy two books, each costing £6.90.

H W Smith	**Stonewaters**
Book sale! 25% off	Buy one get one half price!

a Her friend Shannon says it doesn't matter which shop she goes to. Is she correct?

b Chloe wants to buy two books, one costing £5.60, the other £4.68. Stonewaters says the half price offer applies to the cheaper book. Which shop should she go to?

3 Rewrite these sets of numbers in ascending order. Show your working.

a 33.3%, 0.33, 33, $33\frac{1}{3}\%$

b 0.45, 44.5%, 0.454, $0.\dot{4}$

c $0.2\dot{3}$, 0.232, 22.3%, 23.22%, 0.233

d $\frac{2}{3}$, 0.66, $0.6\dot{5}$, 66.6%, 0.6666

e $\frac{1}{7}$, 14%, 0.142, $\frac{51}{350}$, $14.\dot{1}\%$

f 86%, $\frac{5}{6}$, $0.8\dot{6}$, 0.866, $\frac{6}{7}$

4 All fractions can be turned into a decimal by dividing the numerator by the denominator. Some produce recurring decimals.

For example $\frac{1}{3} = 1 \div 3 = 0.333\,333\,333 \ldots$

a Convert each of the fractions less than 1 with a denominator of 7 into a decimal using your calculator. Write down all the decimal places in your answer.

For example $\frac{1}{7} = 1 \div 7 = 0.142\,857\,142 \ldots$

$$\frac{2}{7} = 2 \div 7 = \ldots$$

b Write what you notice about each of your answers.

5 Shula says, 'I used my calculator to change $\frac{1}{13}$ to a decimal, and I got the answer $0.076\,923\,08$. There is no repeating pattern, so the decimal does not recur.' Explain why Shula is wrong.

6 Jodie says that the recurring decimal $0.\dot{9}$ is a little smaller than 1. Abby says that $0.\dot{9}$ is equal to 1. Who is correct? Explain your reasoning.

7 Convert each of these recurring decimals to a fraction in its simplest form. Show your working.

a $0.1\dot{1}$ **b** $0.2\dot{2}$

c $0.\dot{1}\dot{5}$ **d** $0.\dot{1}2\dot{5}$

e $0.\dot{2}1\dot{6}$ **f** $0.2\dot{1}$

g $0.7\dot{2}$ **h** $0.8\dot{2}\dot{7}$

i $0.6\dot{3}2\dot{1}$ **j** $0.81\dot{7}\dot{5}$

8 a Which one of these is a recurring decimal?

$$\frac{18}{25} \quad \frac{19}{20} \quad \frac{8}{11} \quad \frac{9}{18}$$

b Write $\frac{7}{9}$ as a recurring decimal.

c You are told that $\frac{1}{54} = 0.0\dot{1}8\dot{5}$. Write $\frac{4}{54}$ as a recurring decimal.

9 a Prove that $0.\dot{5}\dot{7} = \frac{19}{33}$.

b Hence, or otherwise, write the decimal number $0.3\dot{5}\dot{7}$ as a fraction.

10 Which is closer to 0.5: $0.\dot{3}\dot{6}$ or $0.\dot{6}\dot{3}$? Justify your answer.

11 Work out $\frac{22}{7}$ as a decimal.

Now work out

$$4 - \frac{4}{3} + \frac{4}{5} - \frac{4}{7} + \frac{4}{9} - \frac{4}{11} + \ldots$$

Continue the sum following the same pattern with the signs and the denominators.

What special number do you find?

12 Find a fraction equal to the recurring decimal $0.\dot{0}12\,345\,678\,\dot{9}$, giving your answer in its simplest form. Show your working.

1015, 1016, 1063, 1066   SEARCH

Summary

Checkout

You should now be able to...

	Test it Questions
✔ Find fractions and percentages of amounts.	**1, 2**
✔ Add, subtract, multiply and divide with fractions and mixed numbers.	**3 – 6**
✔ Convert between fractions, decimals (including recurring decimals) and percentages.	**7 – 11**
✔ Order fractions, decimals and percentages.	**12**

Language	Meaning	Example
Fraction	A number of equal parts of the whole.	$\frac{3}{4}$
Denominator	The number of equal parts in the whole.	Denominator 4
Numerator	The number of equal parts in the fraction.	Numerator 3
Common factor	A factor that is shared by two or more numbers or terms.	$15 = 3 \times 5 \quad 35 = 5 \times 7$ 5 is a common factor of 15 and 35
Cancel	Common factors in the numerator and denominator of a fraction can be cancelled.	$\frac{15}{35} = \frac{3}{7}$
Improper fraction	A fraction with a larger numerator than denominator.	$\frac{4}{3} \quad 4 > 3$
Mixed number	A number made up of two parts: a whole number followed by a proper fraction.	$1\frac{1}{3}$
Percentage	A fraction with a denominator of 100.	$75\% = \frac{75}{100} = \frac{3}{4}$
Decimal	A fraction written as a sum of powers of a tenth.	$0.125 = \frac{1}{10} + \frac{2}{100} + \frac{5}{1000} = \frac{1}{8}$
Terminating	A decimal with a finite number of digits.	$0.125 = \frac{1}{8}$
Recurring	A decimal with a repeating pattern that goes on forever.	$0.\dot{8}\dot{1} = 0.818181... = \frac{9}{11}$
Reciprocal	The reciprocal of a number is what you multiply it by to get 1.	Reciprocal of 5 is $\frac{1}{5}$ Reciprocal of $\frac{2}{3}$ is $\frac{3}{2}$

Review

1 Calculate the following fractions of amounts.

   a $\frac{1}{7}$ of 28   **b** $\frac{3}{8}$ of 40

   c $\frac{4}{9}$ of 27   **d** $1\frac{1}{5}$ of 20

2 Calculate the following percentages of amounts.

   a 35% of 60   **b** 85% of 12

   c 2.5% of 40   **d** 20% of 25

3 Convert these mixed numbers to improper fractions.

   a $2\frac{6}{7}$ **b** $1\frac{5}{11}$

4 Convert these improper fractions to mixed numbers.

   a $\frac{20}{9}$ **b** $\frac{50}{7}$

5 Write down the reciprocal of these numbers.

   a 5   **b** $\frac{1}{3}$

   c $\frac{2}{7}$ **d** $1\frac{2}{5}$

6 Calculate these expressions and write your answer in its simplest form.

   a $\frac{3}{5} \times \frac{1}{6}$ **b** $9 \times \frac{3}{8}$

   c $\frac{2}{9} \div \frac{1}{3}$ **d** $12 \div \frac{2}{3}$

   e $\frac{1}{12} + \frac{5}{12}$ **f** $\frac{5}{6} - \frac{1}{3}$

   g $\frac{2}{11} + \frac{1}{3}$ **h** $2\frac{3}{4} - 1\frac{2}{7}$

7 Convert these numbers to decimals.

   a $\frac{4}{5}$ **b** $\frac{17}{20}$

   c $\frac{3}{100}$ **d** $\frac{15}{40}$

   e 22%   **f** 0.4%

   g 200%   **h** 5.5%

8 Convert these numbers to fractions in their simplest form.

   a 0.7   **b** 0.214

   c 0.36   **d** 0.01

   e 12%   **f** 150%

   g 44.4%   **h** 0.2%

9 Convert these numbers to percentages.

   a $\frac{3}{5}$ **b** $\frac{13}{20}$

   c $\frac{7}{1000}$ **d** $\frac{18}{40}$

   e 0.35   **f** 0.8

   g 0.09   **h** 1.8

10 Write these fractions as recurring decimals.

   a $\frac{2}{9}$ **b** $\frac{6}{7}$

11 Write these recurring decimals as fractions in their simplest form. Show your method.

   a $0.\dot{8}$ **b** $0.2\dot{3}$

12 Write these numbers in ascending order

   a $\frac{9}{16}$ $\frac{5}{8}$ $\frac{2}{3}$ **b** 22.2% $0.\dot{2}$ $\frac{1}{5}$

What next?

Assessment 5

1 The cost of buying a certain DVD in 2012 was £20. This figure was made up as shown in the table. Do not use a calculator for this question.

Labour	$\frac{7}{25}$
Materials	$\frac{1}{10}$
Advertising	$\frac{2}{5}$
Profit	P

  a Find the value of P. [3]

  b Calculate the actual cost of each of the four parts. [5]

By 2014 the costs had altered. Labour costs had decreased by $\frac{1}{10}$ and material costs had decreased by $\frac{3}{40}$.

The profit had increased by $\frac{2}{11}$ and advertising had not changed.

  c Calculate the new price of the DVD. [5]

2 Ben asked some friends which e-mail provider they use.

Provider	Number of boys	Number of girls
Space	5	0
Cyber	1	2
Chat-chat	3	6
UK Telecom	8	8
Blue sky	8	4
	Total 25	Total 20

  a Which provider do 20% of the girls use? [1]

  b Which provider do 20% of the boys use? [1]

  c Which provider do 20% of the total number of pupils use? [1]

  d Ben said: 'In my survey, UK Telecom was equally popular with both boys and girls.' Was Ben correct or incorrect? Give your reasons. [2]

3 **a** What percentage of the word MISSISSIPPI is made up by the letter S? [2]

  b Old Macdonald had a farm. He had 21 pigs, 8 lambs, 150 sheep, 8 calves, 27 cows, 35 chickens and 1 bull.
What percentage of his livestock were chickens? [2]

4 In the $4 \times 100\,\text{m}$ relay, the first 3 runners took, respectively, 25%, $\frac{7}{25}$ and 0.35 of the total time.

  a What proportion of the time was taken by the 4th member?
Give your answer as a decimal. [2]

  b Which team member ran the fastest leg? [1]

  c Which team member ran the slowest leg? [1]

5 S. Crumpy has an orchard. The orchard contains $3\frac{3}{4}$ hectares of apple trees. Today he needs to treat $\frac{2}{5}$ of the area for disease prevention. What area does he need to treat? Do not use a calculator. [3]

6 In a football match the goalkeeper kicked the ball from the goal line for $\frac{5}{9}$ of the length of the pitch and a player then kicked it a further $\frac{7}{20}$. If the length of the pitch is 90 yards, how many yards further is the opposite goal line? Do not use a calculator. [4]

7 a In her garden, Lizzie planted flowers in 27% of the total area. She covered 145.8 m² of the garden with flowers. What is the total area of her garden? [2]

Patio	Lawn	Flowers

b 36% of the remainder of her garden was taken up by the patio. Find the area of the lawn, to the nearest m². [4]

8 At Topmarks College, $\frac{8}{11}$ of the students are girls. Of these girls $\frac{3}{4}$ are brunette and of these brunettes $\frac{5}{9}$ wear earrings. What fraction of the school students are brunette girls who wear earrings? Do not use a calculator. [2]

9 'FALSEPRINT' film laboratories sell prints in sizes 12.5 cm by 7.5 cm and 15 cm by 10 cm. Their adverts say that their 15 cm by 10 cm prints are more than 50% bigger than the 12.5 cm by 7.5 cm size. Are they correct? [3]

10 Sunil said, '$\frac{1}{7}$ is 14% to the nearest per cent'.

Sirendra then said, 'so $\frac{2}{7}$ must be 28% to the nearest per cent'.

Explain why Sirenda is wrong. [2]

11 Jordan asked some girls in her college if they read "Hiya" magazine and wrote the following results.

YES	200 PEOPLE = 66%
NO	102 PEOPLE = 34%

Explain why the values in the table **must** be rounded. [2]

12 Write these numbers in ascending order. [5]

$0.34 \qquad \frac{3}{8} \qquad 33\frac{1}{3}\% \qquad \frac{5}{14} \qquad 0.334 \qquad 33.3\%$

13 Anusha says that $0.\dot{9} = 1$. Show that she is correct. [2]

14 a Convert the following fractions to decimals. Do not use a calculator.

i $\frac{2}{9}$ [2]   **ii** $\frac{13}{25}$ [2]   **iii** $\frac{11}{12}$ [2]

iv $\frac{7}{20}$ [2]   **v** $\frac{7}{11}$ [2]   **vi** $\frac{11}{16}$ [2]

b State a rule for determining how to identify a fraction which can be written as a terminating decimal. [1]

c Without calculating the values, state which of these fractions will have a recurring decimal expansion.

$\frac{23}{25}, \frac{7}{8}, \frac{14}{30}, \frac{19}{99}, \frac{425}{612}, \frac{10}{512}$ [3]

15 Convert the following recurring decimals to fractions. Give each fraction in its simplest form.

a $0.\dot{1}\dot{8}$ [4]   **b** $0.2\dot{6}$ [4]   **c** $0.\dot{1}0\dot{2}$ [4]

d $0.\dot{4}16\dot{5}$ [4]   **e** $0.2\dot{2}\dot{7}$ [4]   **f** $0.35\dot{1}\dot{4}$ [4]

Life skills 1: The business plan

Four friends – Abigail, Mike, Juliet and Raheem – are planning to open a new restaurant in their home town of Newton-Maxwell. They have a lot to think about and organise!

They start by creating a business plan. This plan needs to include: market research to understand their potential customers, what their costs and revenues are expected to be, how big a loan they could afford to borrow, and how any profits should be shared.

Task 1 – Market research

The friends decide to investigate how much people are prepared to pay for a three-course meal. They carry out a small pilot survey. This involves stopping people in the street and asking them a few questions.

a   Draw a comparative bar chart to show the ages of the women and men interviewed. Describe what this shows.

b   Abigail and Mike think that men will be prepared to pay more for a good meal than women. Do the results back up this theory? Calculate averages to justify your conclusions.

Pilot survey results (15 men and 15 women)

M 24, £33	F 20, £23	F 22, £25
M 37, £36	M 62, £33	F 47, £36
M 47, £35	M 42, £32	F 19, £16
F 52, £32	M 31, £22	M 66, £25
M 26, £24	M 55, £40	F 38, £35
F 18, £20	M 39, £35	M 40, £30
M 21, £21	F 23, £21	M 20, £30
F 58, £40	F 35, £32	F 32, £28
F 22, £30	F 61, £37	F 28, £20
M 23, £27	M 51, £27	F 44, £34

Key   Gender (M/F) Age (years), amount prepared to spend

Task 2 – Projected revenue

The friends are estimating the revenue (money coming in) for their restaurant in the first year.

To do this they make some assumptions.

a   Use their assumptions to estimate the revenue for the first year, after the VAT paid by customers has been taken away.

b   Write down a formula for the profit (money left after costs), P, in terms of the other variables listed to the right.

c   In one year $G = £40\,000$, $S = £80\,000$ and $C = £50\,000$. Find P for the value of R you found in in part **a**.

As revenue increases, they want to be able to give staff pay rises.

d   Make S the subject of the formula you found in part **b**. Find the new value of S for $R = £250\,000$ and $G = £50\,000$. Take C and P as having the same values as in part **c**.

The cost of a meal includes Value Added Tax (VAT) charged at 20%.

Assumptions

- The mean amount paid for a meal equals the mean that people in the pilot survey are prepared to pay.
- The restaurant is open 364 days a year.
- The mean number of meals sold a day is 25.

Variables

G = cost of food bought by restaurant

S = salary costs C = other costs

R = revenue P = profit

Age band	Persons
16–24	67 400
25–49	166 100
50–64	64 100
65 and over	58 000

▲ Age distribution of the population in Newton-Maxwell.

Ownership shares

Abigail $\frac{2}{5}$
Raheem $\frac{1}{4}$

Mike and Juliet both own an equal share of the remainder.

Task 3 – The survey

Following from the pilot survey, the friends decide to do a much larger survey based on a sample of 200 people from Newton-Maxwell stratified by age. The table shows the population of the town by age.

How many people from each age group should they include in their sample?

Task 4 – Shares in the business

The friends invested different capital (initial amounts of money) into the business. Based on this, they each own shares that determine the fraction of profit they are entitled to.

a What fraction of the business do Mike and Juliet each have?

b Draw a pie chart to show how much of the business each person owns.

c How much profit would each owner get from a yearly overall profit of £50 000?

Five year repayment formula

C = amount borrowed A = amount repaid each year

i = annual interest rate (AIR), expressed as a decimal

$$C = \frac{A}{1+i} + \frac{A}{(1+i)^2} + \frac{A}{(1+i)^3} + \frac{A}{(1+i)^4} + \frac{A}{(1+i)^5}$$

Task 5 – The business loan

The friends decide to take out a business loan in order to equip the restaurant.

a If the maximum they can repay each year is £5000, how much can they borrow at an AIR of

  i 6%   **ii** 8%?

b If they borrow £30 000, how much will each yearly repayment be at an AIR of

  i 6%   **ii** 8%?

c If they borrow £30 000 but cannot afford to repay more than £8000 per annum, what is the maximum interest rate they can borrow at?

d Mike complains that the repayment formula is too long.

Abigail tells him that it can also be written as

Raheem says that the formula can be simplified further.

$$C = \frac{A}{1+i}\left(\frac{1 - (\frac{1}{1+i})^5}{1 - (\frac{1}{1+i})}\right)$$

$$C = \frac{A}{i}\left(1 - \left(\frac{1}{1+i}\right)^5\right)$$

Starting from Abigail's formula, show that Raheem is correct.

e Mike then complains he can't use it to find the amount to repay. Show that he can, by re-arranging the formula in part **d** to make A the subject.

6 Formulae and functions

Introduction

Nurses often use mathematical formulae when they are administering drugs, for example, converting from one unit to another, calculating the amount of a drug based on somebody's weight, or working out concentrations from solutions.

Working with formulae is a topic within algebra and is a good example of the practical use of mathematics.

What's the point?

The ability to apply a mathematical formula accurately when calculating a patient's dose of a particular medicine is vitally important.

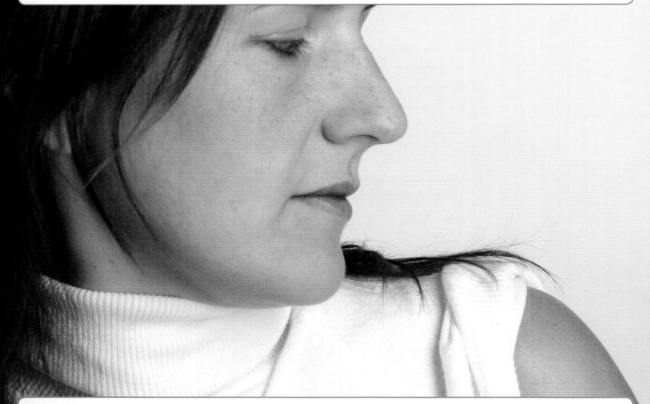

Objectives

By the end of this chapter, you will have learned how to …

- Substitute values into formulae and rearrange formulae to change their subject.
- Write an equation to represent a function, and find inputs and outputs. Find the inverse of a function and construct and use composite functions.
- Use the terms expression, equation, formula, identity, inequality, term and factor.
- Construct proofs of simple statements using algebra.
- Expand brackets to get a quadratic expression and factorise quadratics into brackets.

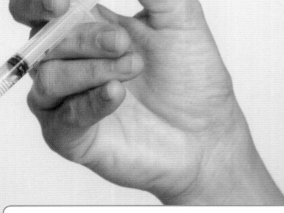

Chapter investigation

This grid of numbers uses each of the numbers from 1 to 9.

4	9	2
3	5	7
8	1	6

Every row, column and diagonal adds up to 15. It is called a magic square.

Create your own magic square.

6.1   Formulae

● The **subject** of a formula is the variable before the equals sign.
  For example, in $A = \pi r^2$, A is the subject of the formula.

Formulae can be **derived**.

EXAMPLE

The cost of a taxi journey, £C, is
£2 per mile plus £3 standard charge.

a Derive a formula for the cost, £C,
of a taxi journey in terms of the
number of miles, m, travelled.

b Use the formula to calculate the
cost of a taxi journey of 14.5 miles.

a $C = 3 + 2m$ A 1-mile journey costs £3 + 1 × £2
                        A 2-mile journey costs £3 + 2 × £2
                        A journey costs £3 plus the number
                        of miles multiplied by 2.

b $C = 3 + 2 \times 14.5 = 3 + 29 = £32$

● You can **rearrange** a formula in order to change its subject.

EXAMPLE

a Rearrange the formula $A = \pi r^2$ to make r the subject.

b Make l the subject of $T = 2\pi\sqrt{\dfrac{l}{g}}$.

a $A = \pi r^2$ Divide both sides by π.

$\dfrac{A}{\pi} = r^2$ Square root both sides.

$\sqrt{\dfrac{A}{\pi}} = r$ r is now the subject.

$r = \sqrt{\dfrac{A}{\pi}}$ Put the subject on
the left-hand side.

b $T = 2\pi\sqrt{\dfrac{l}{g}}$ Divide both sides by 2π.

$\dfrac{T}{2\pi} = \sqrt{\dfrac{l}{g}}$ Square both sides.

$\dfrac{T^2}{4\pi^2} = \dfrac{l}{g}$ Multiply both sides by g.

$\dfrac{T^2 g}{4\pi^2} = l$

● You can **rearrange** formulae using inverse operations.

● When the **subject** appears twice, you can rearrange the
formula by collecting like terms and then factorising.

To make x the subject of $ax + b = cx + d$

$ax + b = cx + d \longrightarrow ax - cx = d - b \longrightarrow x(a - c) = d - b \longrightarrow x = \dfrac{d - b}{a - c}$

                  Collect x-terms   **Factorise**   Divide by
                  on one side $a - c$

EXAMPLE

Rearrange this formula
to make R the subject.

$\dfrac{2p + 10}{50} = \dfrac{R}{15 + R}$

$(2p + 10)(15 + R) = 50R$ Cross-multiply to remove the fractions.

$30p + 150 + 2pR + 10R = 50R$ Expand the brackets.

$30p + 150 + 2pR = 40R$

$30p + 150 = 40R - 2pR$ Collect all R terms on one side.

$30p + 150 = R(40 - 2p)$ Factorise.

$R = \dfrac{30p + 150}{40 - 2p} \Rightarrow R = \dfrac{15p + 75}{20 - p}$ Divide and simplify.

Algebra   Formulae and functions

Exercise 6.1S

1 If $T = 4a - b^2$ calculate T when

   a $a = 4$ and $b = 2$

   b $a = 5$ and $b = 3$

   c $a = -5$ and $b = 2$

   d $a = 4$ and $b = -2$

2 If $a = 4$, $b = 3$ and $c = 2$ work out the value of each expression.

   a $2(a + b) - c$ **b** $c + \dfrac{b}{a}$

   c ca^b **d** $(ab)^c$

3 An electrician charges £35 for each job plus £20 per hour.

   a   Write a formula for the electrician's charge in pounds.

   b   Use your formula to find the charge of a job that takes 3 hours.

4 The cost of a taxi is £2 for a callout 160p for each mile.

   a   Write a formula for the cost of a taxi in pounds.

   b   Work out the cost for a journey of

      i   5 miles   **ii**   15 miles.

5 In Spain a hire car costs €75 plus €35 a day.

   a   Write a formula for the cost of hiring a car in euros.

   b   How much does it cost to hire a car for 7 days?

   c   Louise paid €495 to hire a car. How many days did she hire it for?

6 Make m the subject of each formula.

   a $y = mx + c$ **b** $t = \dfrac{m - k}{w}$

   c $m^2 + kt = p$ **d** $\sqrt[3]{m} - k = l$

7 Make x the subject of each formula.

   a $y = \dfrac{1}{2}x + kw$ **b** $m = \dfrac{1}{4}(ax - t^2)$

   c $2y = \sqrt{x}$ **d** $y = \sqrt{k - lx}$

   e $k = \dfrac{t - a\sqrt{x}}{h}$ **f** $m = \dfrac{p}{x} - t$

   g $w - \dfrac{p}{x} = c$ **h** $\dfrac{y}{ax} - b = j$

8 A formula to change from degrees Celsius to degrees Fahrenheit is

$$F = \frac{9(C + 40)}{5} - 40$$

   a   Use this to change $30\,°C$ into $°F$.

   b   Rearrange to make C the subject of the formula.

   c   Use your new formula to find the Celsius equivalent of $-32\,°F$.

9 The formula $T = 2\pi \sqrt{\dfrac{L}{g}}$ is used to find the time, T, that a pendulum of length L takes to swing freely under gravity, g.

   a   Rearrange to make g the subject.

   b   Find a value of g (to 2 sf), given that a pendulum of length 0.4 m takes 1.27 seconds to complete its swing.

10 Show that each formula can be rearranged into the given form.

   a $\dfrac{p(c - qt)}{m - x^2} = wr$

       into $x = \sqrt{\dfrac{mwr + pqt - pc}{wr}}$

   b $\dfrac{1}{a} + \dfrac{1}{b} = \dfrac{1}{c}$ into $b = \dfrac{ac}{a - c}$

   c $\sqrt[4]{t - qx} = 2p$ into $x = \dfrac{t - 16p^4}{q}$

11 Rearrange the lens formula $\dfrac{1}{f} = \dfrac{1}{u} + \dfrac{1}{v}$

   a   to make f the subject

   b   to make v the subject.

12 Rearrange each formula to make w the subject.

   a $pw + t = qw - r$

   b $a - cw = k - lw$

   c $p(w - y) = q(t - w)$

   d $r - w = t(w - 1)$ **e** $w + g = \dfrac{w + c}{r}$

   f $\dfrac{wx - t}{r} = 5 - w$ **g** $\dfrac{w + t}{w - 5} = k$

   h $\dfrac{w + p}{q - w} = \dfrac{3}{4}$ **i** $\sqrt{\dfrac{w - t}{w + q}} = 5$

Q 1170, 1171, 1186   SEARCH

6.1 Formulae

- A formula is a rule linking two or more variables.
- The subject of the formula appears before the equals sign.
- You can rearrange the formula to change the subject.

You rearrange formulae by using inverse operations.

HOW TO

① Define what letters you will use and write the formula using letters.

② Use your knowledge of substitution or changing the subject of a formula.

③ Give your answer in the context of the question.

EXAMPLE

A plumber charges £25 for a callout and £30 per hour of work.

Write a formula for the number of hours worked.

Use your formula to work out how many hours it would take for the plumber to earn £475.

① It's easier to find a formula for the total charge first.

Charge in pounds = 25 + 30 × number of hours of work

$$C = 25 + 30h$$

where C = charge in pounds, h = number of hours of work.

② Change the subject of the formula to make h the subject.

$$C - 25 = 30h$$

$$\frac{C - 25}{30} = h$$

Substitute C = £475 into the formula.

$$h = \frac{475 - 25}{30} = \frac{450}{30} = 15$$

③ The plumber would need to work for 15 hours.

EXAMPLE

15.3

The formula for the volume of a sphere is $V = \frac{4}{3}\pi r^3$ where r is the radius of the sphere.

a The radius of a rubber ball is 2.3 cm. Find the volume of rubber in the ball.

b Rearrange the formula to find an expression for r.

a $r = 2.3 \Rightarrow V = \frac{4}{3}\pi r^3 \Rightarrow V = \frac{4}{3} \times \pi \times 2.3^3 \Rightarrow V = 50.965...$

The volume of rubber is 51.0 cm³ (1 dp)   Include the units in the final statement.

b $V = \frac{4\pi r^3}{3}$

① The order of operations is $r \rightarrow r^3 \rightarrow \times 4\pi \rightarrow \div 3$.

$3V = 4\pi r^3$

② Multiply both sides by 3.

$\frac{3V}{4\pi} = r^3$

Divide both sides by 4π.
Cube root both sides.

You can treat r^3 as a single term. That's because powers have priority over multiplication.

$r = \sqrt[3]{\dfrac{3V}{4\pi}}$

③

Exercise 6.1A

1 For each shape, write your own formula to represent

    i the perimeter, P **ii** the area, A.

    a   **b**

 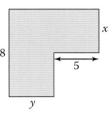

2 Write your own formula to represent these quantities.

    a The total cost, P, of buying d dollars if the exchange rate is $\$1 = £0.64$ and there is a fixed commission fee of £12.

    b The time in minutes, T, to complete my homework if I take 10 minutes to get my books organised and then 35 minutes to do each piece, p.

    c The total phone bill, B, if you send n texts at 10p each, make m minutes of phone calls at 30p per minute and pay a line rental of £5.60.

3 James and Sebastian are rearranging the formula $C = a(x - b)$ in order to make x the subject. They both come up with solutions that look different but are, in fact, correct. Can you explain why?

James' solution	Sebastian's solution
$\dfrac{C}{a} + b = x$	$\dfrac{C + ab}{a} = x$

4 Richard has made a mistake with his rearranging whilst trying to make p the subject of this formula. Copy his working and explain where he has gone wrong.

$$k = mp^3$$
$$\sqrt[3]{k} = mp$$
$$\frac{\sqrt[3]{k}}{m} = p$$

5 These are the stages in changing the subject of the formula $c = \dfrac{8(D + k)}{ab}$.
Put them in order.

$D + k = \dfrac{1}{8}abc$	$\dfrac{8(D + k)}{ab} = c$
$D = \dfrac{1}{8}abc - k$	$8(D + k) = abc$

6 These three formulae are equivalent – true or false? Explain your answer.

$x = \dfrac{t}{p} - wx$	$x = \dfrac{t - wpx}{p}$

$$\frac{w - px}{-p} = x$$

7 There are three different ways to find the area of a triangle. The Greek mathematician Heron discovered that if all three sides (a, b and c) are known then a formula for the area is

$$A = \sqrt{s(s - a)(s - b)(s - c)}$$

where s is the semiperimeter $s = \dfrac{a + b + c}{2}$

    a Find the area of these triangles.

    i   4 cm, 6 cm, 8 cm

    ii   15 cm, 7 cm, 10 cm

    iii   4 cm, 6 cm, 5 cm

    iv   11 cm, 12 cm, 14 cm

    *b Rearrange the formula to make c the subject.

8 The volume of a torus ('doughnut') with inner radius r and outer radius R is given by the formula

$$V = \frac{\pi^2(R + r)(R - r)^2}{4}$$

    a Find V when $R = 10\,\text{cm}$ and $r = 4\,\text{cm}$.

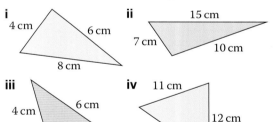

    *b If $V = 352\pi^2\,\text{mm}^3$ and the difference between the inner and outer radius is 8 mm, find the inner and outer radius.

1170, 1171, 1186   SEARCH

6.2 Functions

- A **function** is a relation between a set of **inputs**, the '**domain**', and a set of **outputs**, the '**range**', such that each input is related to an output.

For a function f, input x gives output $f(x)$.

EXAMPLE

Find the functions shown in these **mapping diagrams**.

a
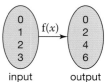

input   output

b
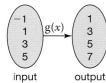

input   output

c
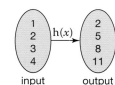

input   output

   a $f(x) = 2x$

The rule is 'double the input'

   b $g(x) = x + 2$

The rule is 'add 2 to the input'

   c $h(x) = 3x - 1$

The rule is 'multiply the input by 3 and subtract 1'

EXAMPLE

The function
$f(x) = 2x + 1$

a   Find $f(3)$

b   Find $f(-2)$

c   Solve $f(x) = 0$

a $f(3) = 2 \times 3 + 1 = 7$ $f(3)$ means the input to the function $f(x)$ is $x = 3$.

b $f(-2) = 2 \times -2 + 1 = -3$ $f(-2)$ means the input to the function $f(x)$ is $x = -2$.

c $2x + 1 = 0$ $f(x) = 0$ means the output to the function $f(x)$ is 0.

$$2x = -1$$
$$x = -\frac{1}{2}$$

The inverse of a function $f(x)$ is written $f^{-1}(x)$.

EXAMPLE

The function
$g(x) = 4x - 3$.

Find $g^{-1}(x)$.

$$y = 4x - 3$$
$$y + 3 = 4x$$
$$\frac{y + 3}{4} = x$$
$$g^{-1}(x) = \frac{x + 3}{4}$$

If a function maps the input, x, to an output, y, then the inverse function maps the output, y, to the input, x.

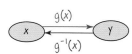

Functions can also be combined to form **composite functions** $fg(x) = f(g(x))$.

EXAMPLE

The function $f(x) = x^2$ and $g(x) = 2x + 3$.

Find   **a** $ff(3)$ **b** $fg(2)$ **c** $gf(x)$.   **d**   Solve $fg(x) = 2gf(x)$.

a $f(3) = 3^2 = 9$
    $ff(3) = f(9) = 9^2 = 81$

b $g(2) = 2 \times 2 + 3 = 7$
    $fg(2) = f(7) = 7^2 = 49$

c $f(x) = x^2$
    $gf(3) = g(x^2) = 2 \times x^2 + 3 = 2x^2 + 3$

d $fg(x) = 2gf(x)$
    $(2x + 3)^2 = 2(2x^2 + 3)$
    $4x^2 + 12x + 9 = 4x^2 + 6$
    $12x = -3$
    $x = -\frac{1}{4}$

The order is important.
$fg(x) = f(g(x))$
$\neq gf(x) = g(f(x))$.

Algebra   Formulae and functions

Exercise 6.2S

1 Find the missing functions.

a
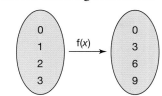

b
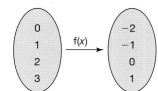

c
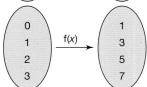

d
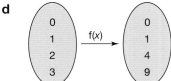

e
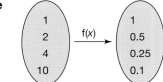

f
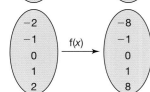

2 The function $f(x) = 2x + 1$. Find

  a $f(2)$ **b** $f(5)$

  c $f(-3)$ **d** $f(0)$

3 The function $g(x) = 2x - 4$. Solve

  a $g(x) = 0$ **b** $g(x) = 10$

  c $g(x) = -6$ **d** $g(x) = -4$

4 Complete the mapping diagram for the function $f(x) = 5x - 2$.

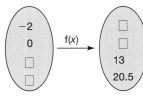

5 The function $h(x) = \frac{1}{x + 2}$. Evaluate

  a $h(2)$ **b** $h(0)$

  c $h(0.5)$ **d** $h(-2)$

6 Find the inverse of these functions.

  a $f(x) = x + 4$ **b** $g(x) = x - 3$

  c $h(x) = \frac{x}{2}$ **d** $f(x) = 5x$

  e $g(x) = 2x - 3$ **f** $h(x) = 7x + 2$

  g $f(x) = \frac{x}{3} - 4$ **h** $f(x) = \frac{x + 5}{2}$

7 Find the inverse of these functions.

  a $f(x) = \frac{1}{x}$ **b** $g(x) = 2 - x$

  Comment on your answers.

8 The functions $f(x)$ and $g(x)$ are defined as $f(x) = 2x + 3$ and $g(x) = 4x$. Find

  a $ff(2)$ **b** $gg(5)$ **c** $fg(3)$

  d $gf(3)$ **e** $fgf(1)$ **f** $gfg(1)$

  g $fg(x)$ **h** $gf(x)$

9 The functions $f(x)$ and $g(x)$ are defined as $f(x) = x^2$ and $g(x) = 3x + 1$.

  a Find $fg(x)$.   **b** Find $gf(x)$.

  c Solve $fg(x) = gf(x)$.

10 The function $f(x)$ and $g(x)$ are defined as $f(x) = 2x + 1$ and $g(x) = x^2 + 1$.

  Find

  a $f(0)$ **b** $g(3)$

  c $g(0)$ **d** $f(1)$

  e $fg(x)$ **f** $gf(x)$

  g Solve $2fg(x) = gf(x)$.

  Use your answers to parts **a**–**d** to check your answer.

11 The function $f(x) = 4x + 5$. Prove $ff^{-1}(x) = x$.

*12 Using the functions and inverse functions from question 7, explain why the graph of $f^{-1}(x)$ is a reflection of the graph of $f(x)$ in the line $y = x$.

13 By considering the function $f(x) = x^2$, investigate why it is important to 'restrict the domain' for inverse functions.

1159, 1940, 1941   SEARCH

6.2 Functions

14.1

RECAP

- A function is a relation between a set of inputs, the 'domain', and a set of outputs, the 'range', such that each input is related to *one* output.
- For a function f, input x gives output $f(x)$.
- The inverse of a function $f(x)$ is written $f^{-1}(x)$.
- Functions can also be combined to form composite functions $fg(x) = f(g(x))$.

HOW TO

Here is a general strategy for representing a function given in words.
1. Write the function as a mapping diagram.
2. Write the function as a formula.
3. Plot the inputs and outputs on a graph.

EXAMPLE

The **mapping diagram** shows information about the function 'multiply by 2 and then add 1'.

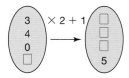

a   Use the function to find the missing numbers.
b   Create a formula that describes the function.
c   Write the function using function notation.
d   Plot the inputs and outputs on a graph.

a   1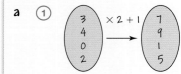

A mapping diagram is a way of showing how inputs and outputs of a function are connected.

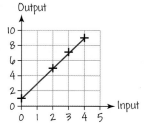

b   2 $y = 2x + 1$ where x = input and y = output.

c $f(x) = 2x + 1$ It is also true that $y = f(x)$.

d   3   Plot points with coordinates (input, output).
The points could be joined with a straight line.

EXAMPLE

Use the information given to find g where g is a linear function.
$g(x) = mx + c$ for some numbers m and c.
$fg(x) = 3 - 10x$ and $f(x) = 5x - 2$

Remember $fg(x)$ means apply g then f.

Let $g(x) = mx + c$

$fg(x) = 5(mx + c) - 2$

$= 5mx + 5c - 2$

$= 3 - 10x$ Given in the question.

$5m = -10 \implies m = -2$ Equate the coefficient of x

$5c - 2 = 3$ and the constant term.

$5c = 5 \implies c = 1$

$g(x) = -2x + 1$ or $g(x) = 1 - 2x$

Algebra   Formulae and functions

Exercise 6.2A

1 **i** Copy and complete each mapping diagram.

ii Write a formula to describe the function.

iii Write the formula using function notation.

iv Plot the inputs and outputs on a graph.

a 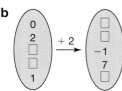 **b**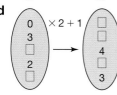

c **d**

> For questions **2** and **3** you can use graphing software.

2 **i** Find the inverse of each function in **1**. Write your solution using function notation.

ii Plot the graph of each function, and its inverse, on the same grid. Use equal scales on the x- and y-axes.

iii Add the line $y = x$ to each of your graphs. What do you notice? Suggest a reason for your observations.

3 Plot the graph of each of these functions and their inverses.

a $y = 5 - x$ **b** $y = \dfrac{2}{x}$

c $y = -\dfrac{10}{x}$ **d** $y = -x$

Comment on your answers.

4 Are these the graphs of functions? Give your reasons.

a **b**

5 The graph shows the inverse of a function.

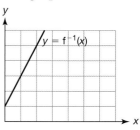

a Copy the diagram and add the graph of $y = f(x)$.

b Write a formula for $f(x)$.

6 The graph shows the inverse of a function $f^{-1}(x)$.

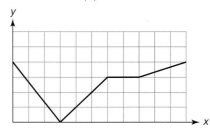

On a copy of the graph add the function $y = f(x)$.

Comment on your answer.

7 f is a linear function, $f(x) = mx + c$.

a If f is its own inverse, $f^{-1}(x) = f(x)$, find $f(x)$. > There are two solutions.

b If $ff(x) = f(x)$ find $f(x)$.

8 The 'index functions' are defined by $f_n(x) = x^n$, where n is a number.

a Show that $f_n f_m(x) = f_{nm}(x)$.

b Hence or otherwise prove that

i $f_n f_m(x) = f_m f_n(x)$

ii $f_n^{-1}(x) = f_{\frac{1}{n}}(x)$

9 Use the information given to find g. In each case g is a linear function. $g(x) = mx + c$, for some numbers m and c.

a $f(x) = 2x + 3$ $fg(x) = 6x + 7$

b $f(x) = 5 - 8x$ $fg(x) = 21 - 4x$

c $f(x) = \dfrac{7 - 2x}{5}$ $fg(x) = \dfrac{15 - 6x}{5}$

*d $f(x) = 6x$ $(fg)^{-1}(x) = \dfrac{x + 30}{18}$

Q 1159, 1940, 1941   SEARCH

6.3 Equivalences in algebra

You need to know some key vocabulary used in algebra.

- An **expression** is made up of algebraic **terms**. It has no equals sign.

  $2x + 3b$ and $2(l + w)$ are expressions.

- A **formula** is a rule linking two or more variables.

  $P = 2l + 2w$ is a formula for the perimeter of a rectangle.

- An **equation** is only true for particular values of a variable. You can solve the equation to find the solution.

  $x + 4 = 10$ is an equation. Its solution is $x = 6$.

- An **identity** is true for any values of the variables. The sign $\equiv$ means identically equal to.

  $a + a + a + a \equiv 4a$ is an identity.

- A **function** links two variables. When you know one, you can work out the other.

  $x \rightarrow 3x + 2$ or $y = 3x + 2$ is a function.

EXAMPLE

Decide if each of these statements is an identity, an equation or a formula.

a $x^3 - 2x = x(x^2 - 2)$ **b** $5x - 3 = 2x + 3$ **c** $A = bh$

> If you know the values of b and h, you can find A from the formula or b given A and h or h given b and A.

a Expand the right-hand side: $x(x^2 - 2) = x^3 - 2x$
$x^3 - 2x$ = left-hand side so the statement is an identity.

b $5x - 3 = 2x + 3$

$3x - 3 = 3$ This statement is only true for one value of x.

$3x = 6$

$x = 2$

$5x - 3 = 2x + 3$ is an equation.

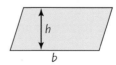

c $A = bh$ is a formula showing the relationship between the length of a pair of parallel sides, the distance between them and the area of a parallelogram.

EXAMPLE

Write an identity for these expressions.

a $(a + b)^2$

b $(x + 1)(x + 2)(x + 3)$

a $(a + b)^2 \equiv (a + b)(a + b)$
$\equiv a^2 + ab + ab + b^2$
$\equiv a^2 + 2ab + b^2$

b $(x + 1)(x + 2) \equiv x^2 + 3x + 2$
$(x^2 + 3x + 2)(x + 3)$
$\equiv x^3 + 3x^2 + 2x + 3x^2 + 9x + 6$
$\equiv x^3 + 6x^2 + 11x + 6$

Exercise 6.3S

1 Multiply out the brackets and write each of these as an identity.
The first one has been done for you.

  a $7(x + 4) \equiv 7 \times x + 7 \times 4 \equiv 7x + 28$

  b $3(x - 2)$

  c $2(3 + x)$

  d $5(2 - x)$

2 Copy and complete these identities by factorising.

  a $8m + 4 \equiv 4 (\quad)$

  b $12n - 9 \equiv \square (\quad)$

  c $15p + 55 \equiv \square (\quad)$

  d $q^2 + 2q \equiv q (\quad)$

  e $16r - 28 \equiv \square (\quad)$

  f $4pq - 10q \equiv \square (\quad)$

3 Copy the table.

Expressions	Equations	Functions	Formulae	Identities

Write these under the correct heading in your table.

  a $a + bc = bc + a$ **b** $y = 3x + 2$

  c $a + bc = d$ **d** $3a + 5 = -4$

  e $4xy + 3x - z$ **f** $E = mc^2$

  g $s = ut$ **h** $x - 1 = y$

4 Copy these statements and say whether they are identities, equations or formulae.

a $c = 2\pi r$	**b** $3x(x + 1) = 3x^2 + 3x$	**c** $3x + 1 = 10$
d $y \times y = y^2$	**e** $2x + 5 = 3 - 7x$	**f** $A = \frac{1}{2}(a + b)h$
g $a^2 + b^2 = c^2$	**h** $20 - x = -(x - 20)$	**i** $2x^2 = 50$

5 Write an example of

  a an expression   **b** an equation

  c a formula   **d** an inequality

  e an identity.

6 Are these identities true or false? Explain your answer.

  a $4(a + 2) \equiv 4a + 2$

  b $3(x + 2) \equiv 3x + 6$

  c $5(y - 2) \equiv 5y - 10$

  d $y(y + 3) \equiv 2y + 3$

  e $x(x - 4) \equiv x^2 - 4x$

7 Prove that these are all identities.

  a $5a + 10 \equiv 5(a + 2)$

  b $3x + 12 \equiv 3(x + 4)$

  c $5y - 15 \equiv 5(y - 3)$

  d $y^2 + 3y \equiv y(y + 3)$

  e $x^2 - 4 \equiv (x + 2)(x - 2)$

8 Prove that these are identities.

  a $4(a + 2) + 2(a + 1) \equiv 6a + 10$

  b $3(x + 2) + 4(x - 1) \equiv 7x + 2$

  c $5(y - 2) + 3(y - 3) \equiv 8y - 19$

  d $y(y + 3) + 2(y + 3) \equiv y^2 + 5y + 6$

  e $x(x - 4) + x(x + 2) \equiv 2x^2 - 2x$

9 Find the values of a and b such that

  a $2(x + 2) + 5(x + 1) \equiv ax + b$

  b $3(x - 2) + 4(x + 3) \equiv ax + b$

  c $5(y + a) + 3(y - b) \equiv 8y - 19$

  d $y(y + a) + 2(y + b) \equiv y^2 + 3y + 4$

  e $x(x - 4) + 2x(x - 3) \equiv ax^2 - bx$

  f $4(y - 3) - 2(5 - y) \equiv ay + b$

10 Classify these expressions as equations, identities or formulae.

  a $V = \frac{4}{3}\pi r^3$ **b** $x + yz = yz + x$

  c $3f = 4(f - 2) + 8$ **d** $a + bc = d$

  e $x - 1 = y$ **f** $x^n x^m = x^{n+m}$

  g $20 - p = -(p - 20)$

  h $a^2 + b^2 = c^2$ **i** $s = ut + \frac{1}{2}at^2$

  j $3(x - 2)^2 = 27$

  k $4a + 2 (a - 6) = 6(a - 2)$

6.3 Equivalences in algebra

- Letters can either represent variables or unknowns.
- An expression is a meaningful collection of mathematical symbols.
- An equation is an expression that uses one '=' symbol and at least one unknown.
- A formula is an equation that describes the connection between two or more variables.
- An identity is an equation that is true for every possible value of the unknown.
- An inequality is a statement about two expressions that are not equal.

EXAMPLE

Look at the diagram

a State an inequality that can be derived from the shape.

b Create an identity by considering the area of the shape.

c The area is 144 square units. State an equation that could be solved to find the value of x.

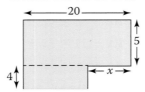

a $x < 20$ part < whole

b Calculate the area in two ways as a sum or as a difference.

Area $= 20 \times 5 + 4 \times (20 - x)$
$\equiv 20 \times (5 + 4) - 4 \times x$
$100 + 4(20 - x) \equiv 180 - 4x$

c $180 - 4x = 144 \implies 45 - x = 36$
$\implies x = 45 - 36 = 9$

p.136

HOW TO

To prove a statement is true

① Use algebra to create statements about the information in the problem.

② Use the rules of algebra to manipulate your expression.

③ Make a final conclusion using correct symbols.

Establishing an identity is a form of proof.

- To prove a statement is true, you need to show that it works for *all* cases.
- To prove a statement is false, you need to find *one* **counter-example**.

EXAMPLE

Prove that the sum of two **consecutive** odd numbers is an even number.

An even number is a multiple of 2: $2n$.

An odd number is one more than an even number: $2n + 1$.

Consecutive odd numbers:
$2n + 1$, $(2n + 2)$ $2n + 3$.
$(2n + 1) + (2n + 3) = 4n + 4$
$= 4(n + 1)$
$= 2 \times 2(n + 1)$

The sum is even because it is a multiple of 2.

EXAMPLE

For every integer value of n, $n^2 + n + 41$ is prime. Show that this statement is false.

Try different values of n until you find one that does not work.

$n = 1$ $n^2 + n + 41 = 1^2 + 1 + 41 = 43$ prime ✓
$n = 2$ $= 2^2 + 2 + 41 = 47$ prime ✓
$n = 3$ $= 3^2 + 3 + 41 = 53$ prime ✓
Try $n = 41$ 41 will appear in each term.
$n^2 + n + 41 = 41^2 + 41 + 41 = 1763$
$= 41(41 + 1 + 1)$
$= 41 \times 43$ – <u>not</u> prime ✗

The statement is false because it does not work for $n = 41$. (or for $n = 40$)

Algebra   Formulae and functions

Exercise 6.3A

1 For each of the following diagrams

   i create an identity

   ii write an equation.

a 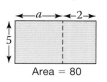   **b**

Area = 80

Area = 105

c 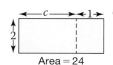   **d**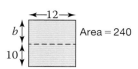

Area = 24

Area = 240

2 For each of these diagrams

   i state an inequality that can be derived from the shape

   ii create an identity by considering how to find the area

   iii use your answers to parts **i** and **ii** to find a possible value for the area

   iv write an equation for your value. Swap with a partner and solve each other's equations.

a   **b**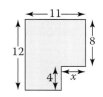

3 Create an identity using each of these diagrams.

a 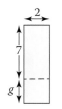   **b**

c   **d**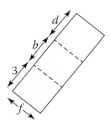

3 **e**   **f**

4 Decide whether each of the following statements is always true, sometimes true or never true

   a An equation is a formula.

   b A formula is an equation.

   c A formula is a function.

   d A function is a formula.

   e An inequality is an equation.

   f An inequality is a function.

   g An equation uses the letter 'x'.

   h A formula is written using algebra.

5 The sum of five consecutive integers is exactly divisible by 5.

   a Check this statement is true for three examples.

   b Prove that the statement is always true.

> Let the numbers be
> $n, n + 1, n + 2, n + 3$ and $n + 4$.

6 Prove that each of these statements is false.

   a Adding 7 to an integer gives an odd numbers.

   b The sum of two primes is always odd.

   c $x^2 > x$ for any number x.

   d The product of two consecutive integers is always odd.

7 Prove each of these statements.

   a The square of any even number is a multiple of 4.

   b If you multiply two consecutive integers and subtract the smaller number you always get a square number.

   c The product of an odd number and an even number is an even number.

   d The sum of two odd numbers is even.

🔍 1150, 1247, 1938, 1942   SEARCH

6.4 Expanding and factorising 2

You can **expand** a double bracket in algebra by multiplying pairs of terms.

Each term in the first bracket multiplies each term in the second bracket:

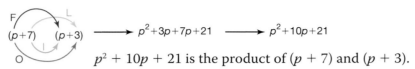

F...	Firsts
O ...	Outers
I ...	Inners
L ...	Lasts

$(p+7)$ $(p+3)$ $\longrightarrow$ $p^2+3p+7p+21$ $\longrightarrow$ $p^2+10p+21$

$p^2 + 10p + 21$ is the product of $(p + 7)$ and $(p + 3)$.

● The two numbers in the brackets **multiply** to give the number at the end and **add** to give the number of xs.

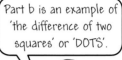

Part b is an example of 'the difference of two squares' or 'DOTS'.

EXAMPLE

Factorise

a $x^2 + 8x + 15$

b $x^2 - 16$

a Look for two numbers that multiply to give $+15$ and add to give $+8$.
These are $+3$ and $+5$.

$x^2 + 8x + 15 = (x + 3)(x + 5)$

b Look for two numbers that multiply to give -16 and add to give 0.
These are $+4$ and -4.

$x^2 - 16 = x^2 + 0x - 16 = (x + 4)(x - 4)$

● You simplify algebraic fractions by factorising and cancelling common factors.

$$\frac{x^2 + 5x + 6}{x + 2} = \frac{(x + 2)(x + 3)}{(x + 2)} = \frac{\cancel{(x + 2)}(x + 3)}{\cancel{(x + 2)}} = x + 3$$

$x + 2$ is a common factor of the **numerator** and **denominator**.

● You can factorise a **quadratic** where the coefficient of x^2 is not 1.

● Multiply the **coefficient** of x^2 and the constant.

$2x^2 + 11x + 12 \rightarrow 2 \times 12 = 24$

This is a suggested method but you may find a method that works better for you.

● Find two numbers that multiply to give this value and add to give the coefficient of x.

Find two numbers that multiply to give $+24$ and add to give $+11 \rightarrow +3, +8$

● Write the quadratic with the x-term split into two x-terms, using these numbers.

$2x^2 + 3x + 8x + 12$

● Factorise the pairs of terms.

$x(2x + 3) + 4(2x + 3)$

● Factorise again, taking the bracket as the common factor.

$(2x + 3)(x + 4)$

● Check by expanding.

$(2x + 3)(x + 4) = 2x^2 + 8x + 3x + 12$
$= 2x^2 + 11x + 12$ ✓

Algebra Formulae and functions

Exercise 6.4S

1 Expand and simplify these expressions involving double brackets.

a $(x + 2)(x + 3)$ **b** $(p + 5)(p + 6)$

c $(w + 1)(w + 4)$ **d** $(c + 5)^2$

e $(x + 4)(x - 2)$ **f** $(y - 2)(y + 7)$

g $(t + 6)(t - 2)$ **h** $(x - 2)(x - 5)$

i $(y - 4)(y - 10)$ **j** $(w - 1)(w - 2)$

k $(p - 5)^2$ **l** $(q - 12)^2$

2 Expand and simplify

a $(2x + 1)(3x + 7)$ **b** $(5p + 2)(2p + 3)$

c $(3y + 4)(2y + 1)$ **d** $(2y + 6)^2$

e $(5t - 4)(2t + 4)$ **f** $(5w - 1)(3w + 9)$

g $(2x + 2y)(3x - 3y)$ **h** $(3m - 4)^2$

i $(2p + 5q)(3p - 8q)$ **j** $(2m - 3n)^2$

3 Factorise each of these using double brackets.

a $x^2 + 7x + 10$ **b** $x^2 + 8x + 15$

c $x^2 + 8x + 12$ **d** $x^2 + 12x + 35$

e $x^2 - 3x - 10$ **f** $x^2 - 2x - 35$

g $x^2 - 8x + 15$ **h** $x^2 - x - 20$

i $x^2 - 8x - 240$ **j** $x^2 + 3x - 108$

k $x^2 - 25$ **l** $x^2 - 6 - x$

4 Expand and simplify

a $(x + 1)(x - 1)$ **b** $(5x - 1)(5x + 1)$

c $(2x + 3)(2x - 3)$ **d** $(x + y)(x - y)$

5 Factorise these expressions.

a $x^2 - 100$ **b** $y^2 - 16$

c $m^2 - 144$ **d** $p^2 - 64$

e $x^2 - \dfrac{1}{4}$ **f** $k^2 - \dfrac{25}{36}$

g $w^2 - 2500$ **h** $49 - b^2$

i $4x^2 - 25$ **j** $9y^2 - 121$

k $16m^2 - \dfrac{1}{4}$ **l** $400p^2 - 169$

m $x^2 - y^2$ **n** $4a^2 - 25b^2$

o $9w^2 - 100v^2$ **p** $25c^2 - \dfrac{1}{4}d^2$

q $x^3 - 16x$ **r** $50y - 2y^3$

s $\left(\dfrac{16}{49}\right)x^2 - \left(\dfrac{64}{81}\right)y^2$

Remember DOTS.

6 Factorise these algebraic expressions.

a $6x^2 - 15xy + 9y^2$ **b** $16a^2 - 9b^2$

c $x^2 - 11x + 28$ **d** $2x^2 + 11x - 21$

e $x^3 + 3x^2 - 18x$ **f** $5ab + 10(ab)^2$

g $10 - 3x - x^2$ **h** $10 - 10x^2$

i $2y + y^2 - 63$ **j** $2x^3 - 132x$

k $6x^2 + 6 - 13x$ **l** $x^4 - y^4$

7 Factorise fully

a $2x^2 + 5x + 3$ **b** $3x^2 + 8x + 4$

c $2x^2 + 7x + 5$ **d** $2x^2 + 11x + 12$

e $3x^2 + 7x + 2$ **f** $2x^2 + 7x + 3$

g $2x^2 + x - 21$ **h** $3x^2 - 5x - 2$

i $4x^2 - 23x + 15$ **j** $6x^2 - 19x + 3$

k $12x^2 + 23x + 10$ **l** $8x^2 - 10x - 3$

m $6x^2 - 27x + 30$ **n** $4x^2 - 9$

8 If a quadratic expression has a negative x^2 term, you can factorise it by first taking out '-1'. Factorise these expressions.

a $3 - 7y - 6y^2$

b $10p + 3 - 8p^2$

c $11y - 3y^2 - 10$

d $27m - 6m^2 - 30$

e $5xy + 2y - 3x^2y$

> $21 - x - 2x^2$
> $= -(-21 + x + 2x^2)$
> $= -(2x^2 + x - 21)$
> $= -(x - 3)(2x + 7)$
> $= (3 - x)(2x + 7)$

9 Cancel these fractions fully.

a $\dfrac{x^2 + 5x + 6}{x + 3}$ **b** $\dfrac{x^2 - 3x - 28}{x + 4}$

c $\dfrac{x - 5}{x^2 - 12x + 35}$ **d** $\dfrac{x^2 - 4}{x + 2}$

e $\dfrac{4y^2 - 25}{2y + 5}$ **f** $\dfrac{x - 9}{x^2 - 81}$

g $\dfrac{2x^2 - 7x + 5}{x - 1}$ **h** $\dfrac{3x^2 + 10x + 8}{x^2 - 4}$

10 Simplify these calculations.

a $\dfrac{(x^2 + 11x + 28)}{5} \times \dfrac{15}{(x + 4)}$

b $\dfrac{x^2 - 11x + 18}{12} \div \dfrac{x^2 + 17x + 18}{24}$

c $\dfrac{2x^2 - 7x - 15}{x^2 - 36} \times \dfrac{2x + 12}{2x^3 + 3x^2}$

Q 1150, 1151, 1156, 1157   SEARCH

6.4 Expanding and factorising 2

RECAP

- To expand double brackets, you multiply each term in the second bracket by each term in the first bracket.

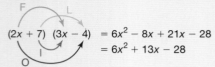

$$(2x + 7)(3x - 4) = 6x^2 - 8x + 21x - 28$$
$$= 6x^2 + 13x - 28$$

F ...	Firsts
O ...	Outers
I ...	Inners
L ...	Lasts

- To factorise into double brackets, look for two numbers that add to give the coefficient of x and multiply to give the constant.

EXPAND

$$(x + 4)(x + 7) \qquad x^2 + 11x + 28$$
$$4 + 7 \quad 4 \times 7$$

FACTORISE

> Quadratic expressions often factorise into double brackets.

HOW TO

① Read the question carefully. Give any unknown values a letter.
Decide whether to

expand ...or... factorise into double brackets.

② Collect like terms ...or... check your answer by expanding.

③ Use your expansion or factorisation to answer the question.

EXAMPLE

Marina arranges a group of tiles to make a rectangle.
The rectangle is $2x + 5$ tiles wide and $2x - 1$ tiles long.
Show that adding 9 more tiles will let Marina make a square.

① Sketch a diagram:

2x+5

2x-1

Area of rectangle = length × width

$$A = (2x + 5)(2x - 1)$$
$$= 4x^2 - 2x + 10x - 5 \quad ② \text{ Expand and collect like terms.}$$
$$A = 4x^2 + 8x - 5$$

③ Adding 9 tiles: $4x^2 + 8x - 5 + 9 = 4x^2 + 8x + 4$
$$= (2x + 2)(2x + 2) \qquad \text{Factorise into double brackets.}$$
$$= (2x + 2)^2$$

Adding 9 tiles will make a square.

EXAMPLE

a Expand $(x + 6)(x - 6)$.

b Hence calculate 106×94 without using a calculator.

a ① Expand using FOIL. $(x + 6)(x - 6) = x^2 - 6x + 6x - 36$
② Collect like terms. $= x^2 - 36$

b $106 \times 94 = (100 + 6)(100 - 6)$
$$= 100^2 - 36 \qquad ③ \text{ Use } (x + 6)(x - 6) = x^2 - 36.$$
$$= 10\,000 - 36$$
$$= 9964$$

Algebra Formulae and functions

Exercise 6.4A

1 Write an expression for the areas of this rectangle and square.

a
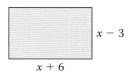
$x - 3$
$x + 6$

b
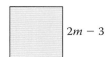
$2m - 3$

2 a Write an expression for the perimeter of this triangle.

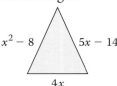
$x^2 - 8$ $5x - 14$
$4x$

b Factorise your expression.

c Explain why x cannot equal 2.

3 a Expand $(a + b)^2$.

b Hence, or otherwise, calculate $1.32^2 + 2 \times 1.32 \times 2.68 + 2.68^2$

c Write another calculation that you could work out using this expansion.

4 Factorise $2.3^2 + 2 \times 2.3 \times 1.7 + 1.7^2$ and use this to show that the calculation results in 16.

5 Copy and fill in the missing values.

a $(x + \square)(x + 6) = x^2 + 9x + \square$

b $(x + 4)(x - \square) = x^2 + \square x - 8$

c $(x + 1)(x + 2) + (x + 4)(x - \square)$
$= \square x^2 + 3x - \square$

d $(3x - \square)(\square x + 3) = 6x^2 + \square x - 6$

6 Find the mean of these three quadratic expressions.

$(x - 9)^2$ $2(x + 3)^2$ $3(x + 1)^2$

Give your answer in the form $a(x^2 + b)$, where a and b are whole numbers.

7 In a multiplication pyramid, you multiply the two numbers directly below to get the number above.

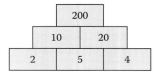
	200	
10		20
2	5	4

Complete this multiplication pyramid.

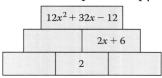
$12x^2 + 32x - 12$
$2x + 6$
2

8 Show that the mean average of these expressions is $(2x + 3)(x + 5)$.

$6x^2 + 10x + 23$ $2x^2 + 20x - 11$

$4x^2 + 13x + 21$ $9x - 4x^2 + 27$

9 Without using a calculator, find the missing side in this right-angled triangle in surd form.

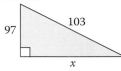
97 103
x

10 Show that the shaded area in the diagram is 6160 cm².

Do not use a calculator.

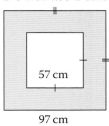
57 cm
97 cm

11 The product of these three expressions is 8. Use this information to find the value of y.

$\dfrac{y^2 + 10y + 21}{2y + 8}$ $\dfrac{y^2 - 16}{15}$ $\dfrac{60}{y^2 - y - 12}$

19.1

Q 1150, 1151, 1156, 1157   SEARCH

Summary

Checkout

You should now be able to...

	Test it Questions
✓ Substitute values into formulae and rearrange formulae to change their subject.	1, 2
✓ Write an equation to represent a function, and find inputs and outputs. Find the inverse of a function and construct and use composite functions.	3
✓ Use the terms expression, equation, formula, identity, inequality, term and factor.	4, 5
✓ Construct proofs of simple statements using algebra.	6, 7
✓ Expand brackets to get a quadratic expression and factorise quadratics into brackets.	8 – 11

Language	Meaning	Example
Equation	An algebraic expression containing an $=$ sign and at least one unknown. Only true for specific values of the unknown.	$x^2 - 5x + 6 = 0$ Only true for $x = 2$ or 3.
Formula	An equation linking two or more variables.	$V = IR$
Subject	The variable before the $=$ sign in a formula.	V is the subject.
Rearrange	Rewrite a formula with a different variable as the subject.	Rearranged to make I the subject. $I = \dfrac{V}{R}$
Function	A rule that links each input value with *one* output value.	
Domain	The set of input values for the function.	
Range	The set of output values for the function.	
Composite function	A two-step function. The output of the first step is used as the input for the second step.	$f(x) = 2x \qquad g(x) = x + 1$ $fg(x) = 2x + 2 \qquad gf(x) = 2x + 1$
Identity	An equation that is true for every possible value for the variables.	$\dfrac{a}{4} \equiv 0.25 \times a$
Proof	A series of logical statements that show that if certain facts are true then something else must always be true.	The difference of the squares of two consecutive integers is always odd. $(n + 1)^2 - n^2 = n^2 + 2n + 1 - n^2$ $= 2n + 1$
Counter-example	An example which shows that a statement can be false, that is, not always true.	Statement: All primes are odd. Counter-example: 2 is a prime.
Expand	Remove the brackets in an expression by multiplying.	$(3x + 1)(2x - 3) = 6x^2 - 9x + 2x - 3$ $= 6x^2 - 7x - 3$
Factorise	Write an expression as a product of terms.	$6x^2 - 7x - 3 = (3x + 1)(2x - 3)$
Quadratic	An expression of the form $ax^2 + bx + c$ where a, b and c are numbers and $a \neq 0$.	$6x^2 - 7x - 3$ $a = 6, b = -7, c = -3$

Review

1 Use the formula $v = u + at$ to calculate

  a v if $u = 5$, $a = -2$ and $t = 8$

  b u if $v = 60$, $a = 4$ and $t = 10$

  c a if $v = 36$, $u = 0$ and $t = 4$

  d t if $v = 20$, $u = 50$ and $a = -6$

2 Rearrange each formula to make X the subject.

  a $3 + 2X = A$ **b** $AX - B = 3C$

  c $\dfrac{3X + Y}{4Z} = 5$ **d** $\sqrt{X + 4} = 2Y$

  e $X^2 - 2K = L^2$

3 The functions f and g are defined as $f(x) = 5x$ and $g(x) = x + 3$.

  a Write down the value of

    i $f(7)$ **ii** $g(-5)$.

  b Write down the inverse of

    i $f(x)$ **ii** $g(x)$.

  c Write down the composite functions

    i $gf(x)$ **ii** $fg(x)$

4

$5x^2 + 3$	$v^2 = u^2 + 2as$
$7a + 5 = 19$	$3(x + 2) \equiv 3x + 6$
	$2y + 9 > 37$

From the box chose an example of

  a a formula   **b** an identity

  c an expression **d** an equation.

5 Write an inequality for each of these statements.

  a x is greater than -2.

  b y is less than or equal to 0.

6 Use algebra to show that this identity is true.

$$5(2x + 3) + 2(4 - 5x) \equiv 23$$

7 Prove that the sum of any three consecutive integers is a multiple of three.

8 Expand and simplify these expressions.

  a $(x + 9)(x + 2)$

  b $x(x - 7)(x - 6)$

  c $(3x - 2)(x + 11)$

  d $(4x + 2)(3x - 1)$

9 Factorise these quadratic expressions.

  a $x^2 - 81$ **b** $16x^2 - 49$

  c $x^2 - 7x$ **d** $21x^2 + 28x$

10 Factorise these quadratic expressions.

  a $x^2 - 3x - 4$ **b** $x^2 - 7x + 10$

  c $x^2 + 8x - 9$ **d** $2x^2 + 7x + 3$

  e $6x^2 + 7x - 3$ **f** $8x^2 - 16x + 6$

11 Simplify fully these algebraic fractions.

  a $\dfrac{x^2 + 3x + 2}{x + 1}$

  b $\dfrac{x^2 - 16}{x^2 + 8x + 16}$

  c $\dfrac{2x^2 - 3x + 1}{2x^2 + x - 1}$

What next?

Score			
	0 – 4		Your knowledge of this topic is still developing. To improve look at MyMaths: 1150, 1151, 1156, 1157, 1159, 1170, 1171, 1186, 1247
	5 – 9		You are gaining a secure knowledge of this topic. To improve look at InvisiPens: 06Sa – p
	10 – 11		You have mastered these skills. Well done you are ready to progress! To develop your exam technique looks at InvisiPens: 06Aa – f

Assessment 6

1   **a**   Erin and Stina both substitute the values $u = -4$, $v = 12$ and $t = 8$ into the expression $s = (u + v)t$. Erin says that this gives $s = 64$. Stina says that this gives $s = 32$.

Who is correct? Give your reasons. [2]

b   Erin and Stina now both substitute the values $u = 0$, $a = 2$ and $s = 16$ into the expression $v^2 = u^2 + 2as$. Erin says that this gives $v = 64$. Stina says that this gives $v = 8$.

Who is correct? Give your reasons. [3]

2   Lucy rearranged each formula into the given form.

a $a = -n^2\left(x - \dfrac{h}{n^2}\right)$ into $n = \sqrt{\dfrac{h - a}{x}}$ [4]

b $t = \left(\sqrt{\dfrac{1 - e}{1 + e}}\right)x$ into $e = \dfrac{x^2 - t^2}{x^2 + t^2}$ [4]

c $\dfrac{1}{S} = \dfrac{1}{R} + \dfrac{1}{r}$ into $S = \dfrac{r + R}{rR}$ [4]

Show how she rearranged each formula.

3   A mobile phone provider 'TextUnending' charges 15p each minute if you call between the peak times of 09:00 and 20:00 and 12p per minute at other times, called off-peak. Texts cost 14 pence per text at all times.

a   Using the letters C for the cost (in pence), p for the number of peak minutes, q for the number of off-peak minutes and t for the number of text messages sent, write down a formula to work out the cost. [2]

b   Hugo made one call for 3 minutes 20 seconds at 13:15 and another lasting from 20:32 to 21:07 exactly. He also sent 8 text messages. How much did this cost? [3]

c   Cheng made three calls lasting from 9:39 to 10:02, 11:53 to 12:17 and 19:42 to 20:15. She also sent 25 text messages. Work out the total cost. [4]

4   The formula for the length of a skid, S m, for a vehicle travelling at v km/h is $S = \dfrac{v^2}{170}$

a   A car skids while travelling at 100 km/h. How long is the skid? [2]

b   A car and a lorry are travelling head-on towards each other on a narrow road.

They see the danger looming and start to skid at the same instant.

The car is travelling at 70 km/h and the lorry at 52 km/h.

Calculate the minimum distance they are apart if they just stop in time. [3]

c   Find the speed of a vehicle which skidded for 60 m. Give your answer to the nearest km/h. [4]

5   Christopher Columbus is sitting on a cliff ledge above the sea.

When he is x metres above sea level, the horizon is y miles away. y and x are connected by the formula $y = 3.57\sqrt{x}$.

a   How far out to sea can Christopher see when he is 8.5 m above the sea? [2]

b   Calculate Christopher's height above the sea when the distance to the horizon is 8.55 km. [4]

c   A pirate ship sails past the cliff 33 km offshore when Christopher is 85 m above the sea. Can Christopher see the pirate ship? [3]

6 You score a total of x marks in 3 tests. In another x tests you score another 27 marks but your mean score remains the same. How many marks did you score in the first 3 tests? [5]

7 **a** Carlo tried to expand and simplify these expressions. For each expression work out if his answer is correct or not. If not, then give the correct answer.

    i $(p-4)(p-7) = p^2 + 28$ [2]

    ii $(v+9)(v-7) + (4-5v)^2 = 6v^2 - 47$ [5]

  b Carlo then tried to factorise these expressions. For each expression work out if his answer is correct or not. If not, then give the correct answer.

    i $z^2 + 13z + 36 = (z+4)(z+9)$ [2]

    ii $v^2 - 100 = (v-10)^2$ [2]

    iii $30x^2 + 13xy - 3y^2 = 3x(5x + 3y)$ [2]

8 **a** Claire wants to factorise $2.4^2 - 2 \times 2.4 \times 3.6 + 3.6^2$.

    i Show how she can do this. [2]

    ii She then multiplies out the brackets. What total does she get? [2]

  b Show how she can work out the following values without using a calculator.

    i $89^2 - 11^2$ [4]

    ii $6.89^2 - 3.11^2$ [4]

9 Karl says that when $x^2 - 8x + 30$ is written in the form $(x-a)^2 + b$, the values of a and b are 4 and 15 respectively.

Is he correct? If not, what are the correct values of a and b? [5]

10 Selina makes the following statements. For each statement either prove that it is always true or find a counter-example to show that it is false.

  a An odd number times an even number is always odd. [1]

  b Prime numbers have one factor. [1]

  c The sum of two even numbers is always even. [2]

  d Any number squared is more than 0. [1]

  e Two prime numbers multiplied together are always odd. [1]

  f The sum of three consecutive even numbers is always divisible by 6 [2]

  g The square of any number is never a prime number. [2]

11 Work out the rule that turns *each* input in these function machines into its corresponding output.

  a 6, 10, 12 → 25, 41, 49 [2]   **b** 6, 15, 36 → 0, 3, 10 [2]

12 **a** Hassan says that the inverse of $g(x) = 2x + 1$ is $2x - 1$.

Is he correct? If not, what is the correct inverse of $g(x) = 2x - 1$? [3]

  b Hassan now has the functions $f(x) = 3x - 1$ and $g(x) = x^2$.

He evaluates **i** $f(6) = 17$ [2]   and   **ii** $fg(6) = 289$ [2]

In each case, work out if he is correct. If not, then find the correct value for each expression.

Revision 1

1 Jenni gets the bus to work. The fare is £1.75 for a single journey and no return tickets are issued. She works Monday to Wednesday, Friday and Saturday. Explain if Jenni would save money if she bought a weekly ticket costing £15. [4]

2 Ankit buys petrol and collects 'Frequent Traveller Miles' (FTM) vouchers that give him free flights. He gets 5 FTMs for every 30 litres of petrol that he buys. The petrol for Ankit's car costs 136.9 p per litre. He needs 600 FTMs for a free trip to Nice.

  a How much does it cost Ankit to collect enough vouchers? [5]

Ankit's car averages 38 miles per gallon. Take 1 litre as being 0.22 gallons.

  b How many miles does Ankit have to drive to get his free trip? [5]

3 Julie is lining the base and sides of a drawer with paper. The drawer has the dimensions shown.

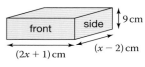

  a Give the following in terms of x. Expand all brackets.

    i The area of the front of the drawer. [2]

    ii The area of a side of the drawer. [2]

    iii The total area of paper Julie uses. [6]

    iv The volume of the drawer. [2]

  b The total area of paper Julie uses is $51x$ cm². Find the numerical volume of the drawer. Show your working. [5]

4 a A hoist is made up of 4 metal girders PQ, QR, QS and RS. The girders are attached to the horizontal ground PST. P, Q and R lie on a straight line. The distances PS and SR are equal, RQS is a right angle and angle QPS is 28°.

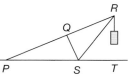

4   **i** Find angles QSR, RST and PSR. Explain your answers. [9]

    ii Explain why $PQ = QR$. [2]

  b Another hoist is made up of 5 girders, AB, BC, CD, DB and AD. AB is parallel to DC and the girders AB, BD and BC are all the same length. Angle BAD is 54°.

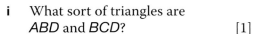

    i What sort of triangles are ABD and BCD? [1]

    ii Find angles ABD and DBC. [8]

    iii Is triangle ABD similar to triangle BCD? Explain your answer. [1]

5 $ABCDEFGH$ is a regular octagon.

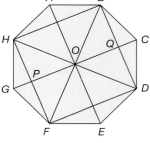

  a Sara makes the following statements. For each statement decide if Sara is correct or not. Fully describe the shape in each case.

    i HBD is an isosceles triangle. [4]

    ii $HBDF$ is a trapezium. [2]

    iii $HBCG$ is a parallelogram. [3]

  b Using the letters in the diagram, name

    i a triangle congruent to HOP [1]

    ii a triangle similar to HOP. [1]

  c Find the value of these angles.

    i ABC [4]   **ii** POF [1]

    iii HOC [1]   **iv** ABH [2]

6 Stars in the night sky are grouped by their surface temperatures, measured in thousands of degrees.

Group	O	B	A	F	G	K	M
Surface Temperature (1000 °C)	40	15	8	6	5	4	2

To illustrate this data draw a

  a bar chart [5]   **b** pie chart. [5]

7 The data shows the number of nights and number of guests staying in a hotel.

No. of nights (x)	1	2	3	4	5	6	7	8	9	10	11	12
No. of guests (f)	4	5	3	6	8	3	9	7	5	4	2	6

a Calculate the

  i mean [4]

  ii median length of stay. [3]

b Find the

  i range [1]

  ii interquartile range. [5]

8 a In an election Joe Trustme, polled 8082 votes and won 36% of all the votes cast. How many votes were cast altogether? [2]

b Ali Imyurwoman polled 7010 votes. What percentage of the total vote did she poll? [2]

c The number of votes cast represented a turnout of 63.7% to the nearest 0.1%. What are the minimum and maximum number of voters in the total electorate? [4]

9 a Solomon sits 6 exam papers and obtains a mean mark of 57.5. He then sits another paper and scores 45. What is his mean score now? [5]

b Sabina sits some exam papers and obtains a mean mark of 32, with 288 marks altogether. She takes another test and her new mean is 35.7. What did she score in the last test? [5]

10 Brianna is diluting sulphuric acid. Flask A contains 80 ml of sulphuric acid and flask B contains 100 ml of water. Brianna transfers 20 ml of water from flask B into flask A.

a What fraction of the liquid in flask A is water? [2]

The sulphuric acid and water in flask A are *thoroughly mixed* together. A 20 ml spoonful is transferred to flask B.

10 b How much

  i sulphuric acid [1]

  ii water does the spoon hold? [1]

c Give the new volumes of sulphuric acid and water in

  i flask A [2]

  ii flask B. [2]

11 Sian says '$10x$ is always bigger than x'. Explain whether she is right. [2]

12 The overtaking distance D, when one vehicle passes another, is given by the formula $D = \dfrac{V(L + 130)}{V - U}$ where U is the speed of the slower vehicle, V the speed of the faster vehicle and L the length of the slower vehicle. V and U are in mph, D and L are in feet.

a Peter, driving at 90 mph, overtakes a van of length 12 ft, travelling at 60 mph. What is the overtaking distance? [2]

b Later, travelling at 110 mph, he passes a car travelling at 70 mph. His overtaking distance is 400 ft. Find the length of the car being overtaken. [5]

c The law in a particular country says that all overtaking must be done in a maximum distance of 850 ft. Use your formula to calculate the minimum speed Peter must be travelling to overtake a coach of length 41 ft travelling at 60 mph. [6]

d A motorway runs parallel to a railway line. A train of 8 coaches, each of length 65 feet including engines, is travelling at 125 mph. Peter drives at 144 mph and decides to 'race' the train. Can he pass the train in under a mile? (1 mile = 5280 ft) [5]

7 Working in 2D

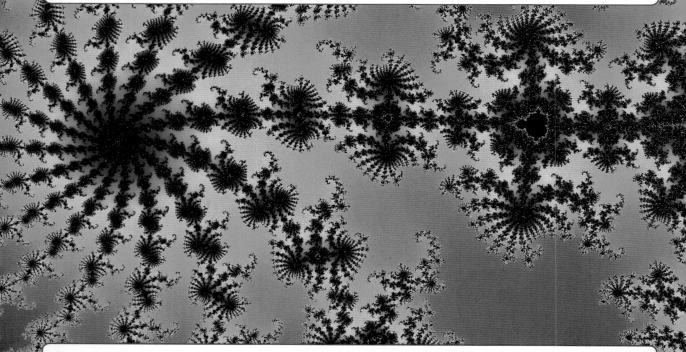

Introduction

Self-similarity is the property in which an entire shape is mathematically similar to a part of itself. This means is that if you 'zoom in' on a small part of the shape, you get an exact replica of the original shape itself. Self-similarity is used in fractal images, like the one you can see here. It has real-world use in describing the structure of coastlines, as well as the natural growth of plants such as ferns and the formation of crystals and snowflakes.

What's the point?

The real world, being mathematically untidy, is never exactly self-similar. However self-similarity provides a very useful model in understanding the complex geometries seen in nature, which can't usually be reduced to simple rectangles and circles.

Objectives

By the end of this chapter, you will have learned how to ...

- Measure line segments and angles accurately.
- Use scale drawings and bearings.
- Calculate the areas of triangles, parallelograms, trapezia and composite shapes.
- Describe and transform shapes using reflections, rotations, translations (described as 2D vectors) and enlargements (including fractional and negative scale factors).
- Identify what changes and what is invariant under a combination of transformations.

Check in

1 These shapes are drawn on a centimetre square grid.

a 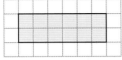 **b** 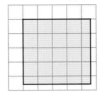 **c**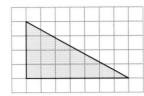

For shapes **a–c**

  i write the lenght of each side in cm.
    Hence find the perimeter in **i** cm, **ii** mm.

  ii calculate the area of the shape in cm².

2 Write the coordinates of the points **A–E**.

3 Write the equations of the lines **a–e**.

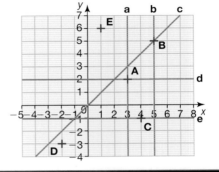

Chapter investigation

Create a snowflake!

Step 1: Draw an equilateral triangle.

Step 2: Draw equilateral triangles on each of the three sides
      (carefully – you'll have to divide each side into three equal parts).

Step 3: You have now got 12 sides.
      Draw equilateral triangles on each of these.

If it's still not snowflaky enough, try once more – but you'll find
it starts getting very fiddly!

Measuring lengths and angles

You measure lengths in **millimetres** (mm), **centimetres** (cm), **metres** (m) or **kilometres** (km).

$10\,\text{mm} = 1\,\text{cm}$ $100\,\text{cm} = 1\,\text{m}$ $1000\,\text{m} = 1\,\text{km}$

A ruler **measures length** in **millimetres** (mm) or **centimetres** (cm).

This line measures 2.5 cm or 25 mm.

This line measures 2.7 cm or 27 mm.

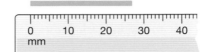

To measure a line, line up the ruler so that the zero mark is at the start of the line.

You can **measure** and draw an **angle** in **degrees** with a **protractor**.
A protractor measures angles up to 180°.
There are 180° in a half turn.

- You can measure a reflex angle by measuring the associated acute or obtuse angle.

A full turn is 360°.

Angles at a point add to 360°.

EXAMPLE

Measure the reflex angle $A\hat{B}C$.

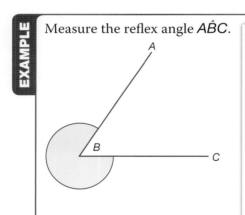

The acute angle $A\hat{B}C = 56°$.
So the reflex angle $A\hat{B}C$ is
$360° - 56° = 304°$

- A **bearing** is an **angle** measured in a clockwise direction from north.

You always write bearings with three digits, for example 070°, 190°, 230°.

To find the bearing of A from B

- Imagine you are standing at B, facing north.
- Turn clockwise until you face A.

The angle you have turned through is the bearing of A from B.

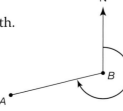

The bearing of A from B is 256°.

Exercise 7.1S

1 Convert these metric measurements of length.

   a 180 cm to mm   **b** 45 mm to cm

   c 350 cm to m   **d** 2000 m to km

   e 3500 m to km   **f** 4500 mm to m

2 Measure the lengths of these lines in

   a centimetres   **b** millimetres.

A ———————————— B

C ———————————————— D

E —————————— F

G ————————————————————— H

I ——————— J

K ————————————————— L

M ——————— N

O ——————— P

Q ————————— R

S ——————— T

3 **a** Draw a line *AB*, so that *AB* = 9 cm.

   b Find the midpoint of *AB* and mark it with a cross.

In questions **4–7**, for each angle state

a the type of angle – acute, right angle, obtuse or reflex

b your estimate in degrees

c the measurement in degrees.

Set out your answers like this:

Question	Type of angle	Estimate	Measurement
4	acute	40°	30°
5			

4    **5**

6   **7**

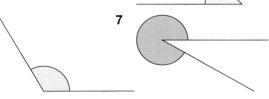

8 Draw and label these angles using a protractor.
State whether each angle is acute, obtuse, reflex or a right angle.

   a 40°   **b** 140°   **c** 90°

   d 36°   **e** 144°   **f** 56°

   g 124°   **h** 38°   **i** 142°

   j 85°   **k** 300°   **l** 200°

   m 320°   **n** 245°   **o** 265°

9 These diagrams are drawn accurately.
Measure the bearing of *T* from *S* in each.

   a
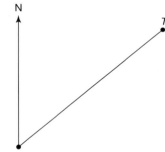

   b
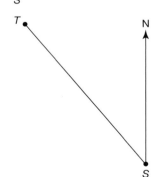

10 *P* and *Q* are points 2 cm apart. Draw diagrams to show the position of points *P* and *Q* where the bearing of *Q* from *P* is

   a 070°   **b** 155°

   c 340°   **d** 260°

Q 1086, 1103, 1117   SEARCH

7.1 Measuring lengths and angles

- A ruler measures lengths in cm and mm.
- You measure and draw angles using a protractor.
- A bearing is an angle measured in a clockwise direction from north. You always write a bearing with three digits.

In **scale drawings**, lines and shapes are **reduced** or **enlarged**.

Corresponding lengths are multiplied by the same **scale factor**. You can write the scale factor as a ratio.

You can write the scale factor 1 cm represents 100 cm as 1 : 100.

real length = 100 × length on the map

map length = real length ÷ 100

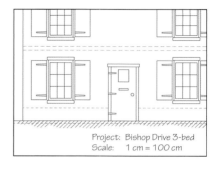

Project: Bishop Drive 3-bed
Scale: 1 cm = 100 cm

HOW TO

1. Choose a scale (unless one is given).
2. Work out or measure the lengths or angles required.
3. Draw a diagram (unless one is given).
4. Give the answers including units.

EXAMPLE

A church, C, is 10 km due west of a school, S.
Joe is 6 km from the school on a bearing of 320°.
He wants to walk directly to the church.

Draw a diagram to show the positions of Joe, the church and the school, and use it to find the bearing Joe should take.

1. Use a scale of 1 cm to 2 km.

2. Represent 10 km by a line 5 cm long.
   Represent 6 km by a line 3 cm long.

   Draw the line JC.
   Draw the north line at J.
   Measure the clockwise angle between the north line and JC.

   Draw the north line at S.

   Measure and draw the 320° bearing from S and, 3 cm from S, mark a point, J, to show Joe's position.

   3 cm

   5 cm

   320°

   Draw and measure lengths and angles carefully or your answer will be inaccurate.

3. Label point S.

   Draw C 5 cm west of S.

4. The bearing Joe needs is 233°.

Exercise 7.1A

1 Measure and write the bearing of

   a   Leeds from Manchester

   b   Sheffield from Leeds

   c   Manchester from Leeds

   d   Manchester from Sheffield

   e   Leeds from Sheffield.

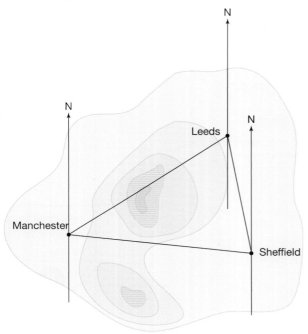

2 Liam sees a lifeboat on a bearing of 122° from Mevagissey. Kim sees the same lifeboat on a bearing of 225° from the Rame Head chapel.

Kim says the lifeboat is nearer to her than to Liam.

Use tracing paper to copy the coastline and work out it Kim is correct.

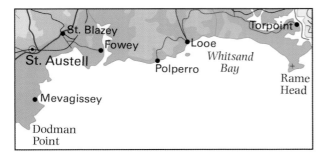

3 Copy the diagram.

The distance from Truro to Falmouth is 14 km.

The bearing of St. Mawes from Falmouth is 080°.

The bearing of St. Mawes from Truro is 170°.

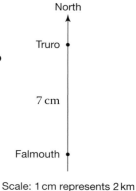

Scale: 1 cm represents 2 km

   a   Mark the position of St. Mawes on your diagram.

   b   Calculate the distance from Falmouth to St. Mawes.

   c   Calculate the distance from Truro to St. Mawes.

4 A youth club (Y) is 4 km due east of a school (S).

Hazel leaves school and walks 5 km on a bearing of 042° to her house (H).

   a   Make a scale drawing to show the positions of Y, S and H.
Use a scale of 1 cm to 1 km.

   b   Hazel walks directly from her house to the youth club.
What bearing does she take?

5 A lighthouse, L, is 6 km on a bearing of 160° from a point H at the harbour.
A boat, B, is 3 km from L on a bearing of 125°.

   a   **i**   Make a scale drawing to show the positions of L, H and B.

      ii   On what bearing should B travel to go directly to H?

      iii   Estimate the distance between B and H.

   b   The boat moves 4 km due west.

      i   Mark on your drawing the new position of B.

      ii   On what bearing should B now travel to go directly to H?

      iii   Estimate the new distance between B and H.

7.2 Area of a 2D shape

- The **perimeter** of a shape is the distance round it.
- The **area** of a shape is the amount of space it covers.

You can use formulae to find the areas of **rectangles** and **triangles**.

Rectangle

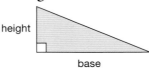

width
length

Triangle

height
base

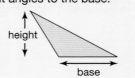

The height of a triangle is always at right angles to the base.

height
base

- Area = length × width

- Area = $\frac{1}{2}$ × base × height

You can find the formula for the **area** of any **parallelogram**.

For this parallelogram ...

cut off one triangle ...

and fit it on the other end ... to make a rectangle.

 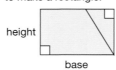

height
base

- Area of parallelogram = **base** × **perpendicular height**

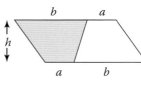

height
base

The height must be perpendicular to the base.

You can find the formula for the area of any **trapezium**.

You can fit two **congruent** (identical) trapeziums together to make a parallelogram.

The base of the parallelogram is $a + b$ and the height is h.

Area of parallelogram = $(a + b) \times h$

Area of trapezium = half area of parallelogram.

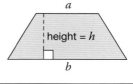

b a

h

a b

- Area of trapezium = $\frac{1}{2} \times (a + b) \times h$

a

height = h

b

The height is the perpendicular distance between the parallel sides.

EXAMPLE

Calculate the area of each shape.

a

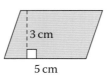

3 cm
5 cm

b

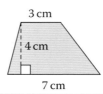

3 cm
4 cm
7 cm

a Area of parallelogram = 5 × 3

= 15 cm²

b Area of trapezium = $\frac{1}{2}$ (3 + 7) × 4

= 5 × 4

= 20 cm²

Exercise 7.2S

1 Calculate the areas of these rectangles.
Remember to give the units of your answers.

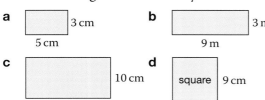

a 5 cm, 3 cm
b 9 m, 3 m
c 20 cm, 10 cm
d square 9 cm

2 Calculate the area of each of these right-angled triangles.
State the units of your answers.

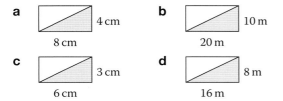

a 8 cm, 4 cm
b 20 m, 10 m
c 6 cm, 3 cm
d 16 m, 8 m

3 Calculate the area of each of these triangles.
State the units of your answers.

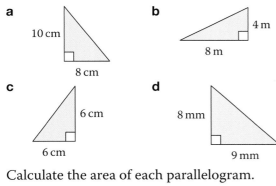

a 10 cm, 8 cm
b 4 m, 8 m
c 6 cm, 6 cm
d 8 mm, 9 mm

4 Calculate the area of each parallelogram.

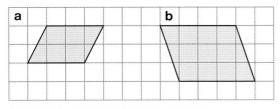

a
b

Give your answers in square units.

5 Calculate the area of each trapezium.

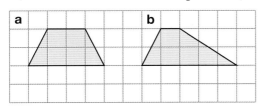

a
b

6 Calculate the area of each parallelogram.
State the units of your answers.

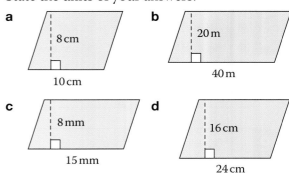

a 8 cm, 10 cm
b 20 m, 40 m
c 8 mm, 15 mm
d 16 cm, 24 cm

7 Calculate the area of each trapezium.
State the units of your answers.

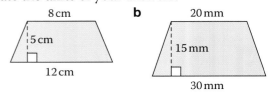

a 8 cm, 5 cm, 12 cm
b 20 mm, 15 mm, 30 mm

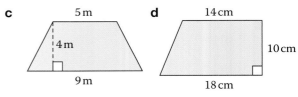

c 5 m, 4 m, 9 m
d 14 cm, 10 cm, 18 cm

8 a Calculate the area of this shape using the formula for the area of a trapezium.

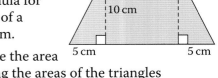

10 cm, 10 cm, 5 cm, 5 cm

b Calculate the area by adding the areas of the triangles and the square.

9 The areas of these shapes are given.
Calculate the unknown lengths.

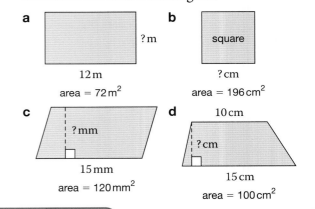

a 12 m, ? m, area = 72 m^2
b square, ? cm, area = 196 cm^2
c ? mm, 15 mm, area = 120 mm^2
d 10 cm, ? cm, 15 cm, area = 100 cm^2

7.2 Area of a 2D shape

- The perimeter of a shape is the distance round it.
- Area of a rectangle = length × width
- Area of a triangle = $\frac{1}{2}$ × base × height
- Area of a parallelogram = base × perpendicular height
- Area of a trapezium = $\frac{1}{2}$ × (a + b) × h

width

length

height

base

height

base

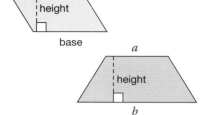

a

height

b

HOW TO

1. Draw a diagram (if needed).
2. Decide which formula (or formulae) to use.
3. Find the answer, including the units.
4. Make sure that you state your conclusion and reasons if the question asks for it.

EXAMPLE

Find the area of this shape.

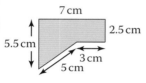

7 cm

2.5 cm

5.5 cm

3 cm

5 cm

1. Split the compound shape into a rectangle and a triangle.
   Area = area of rectangle + area of triangle

2. Area of a rectangle = length × width
   Area of a triangle = $\frac{1}{2}$ × base × height

3. Triangle:
   height = 7 – 3 = 4 cm
   base = 5.5 – 2.5 = 3 cm
   Area = (7 × 2.5) + ($\frac{1}{2}$ × 4 × 3)
   = 17.5 + 6
   = 23.5 cm²

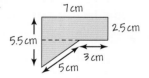

7 cm

2.5 cm

5.5 cm

3 cm

5 cm

height

base

EXAMPLE

A trapezium has area 144 cm². The longer parallel side is 18 cm. The parallel sides are 12 cm apart. Find the length of the other parallel side.

1. Draw a sketch of the diagram.

2. Use the formula for area of a trapezium.
   Area = $\frac{1}{2}$ × (a + b) × h
   144 = $\frac{1}{2}$ × (18 + b) × 12

3. Solve the equation to find the value of b (the other parallel side).
   144 = 6 × (18 + b)
   24 = 18 + b   Divide both sides by 6.
                      Subtract 18 from both sides.
   6 = b

4. The other parallel side has length 6 cm.

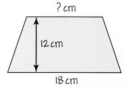

? cm

12 cm

18 cm

▲ The word trapezium originates from the shape made by the ropes and bar of an old-fashioned flying trapeze.

Exercise 7.2A

1 Calculate the perimeter and area of each shape. State the units of your answers.

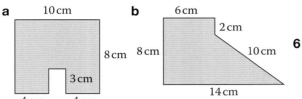

a 10 cm, 8 cm, 3 cm, 4 cm, 4 cm

b 6 cm, 2 cm, 8 cm, 10 cm, 14 cm

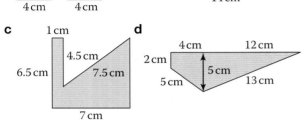

c 1 cm, 4.5 cm, 6.5 cm, 7.5 cm, 7 cm

d 4 cm, 12 cm, 2 cm, 5 cm, 5 cm, 13 cm

2 Pete is making a mobile out of shapes like this. He cuts the shape out of a piece of card that is 30 cm × 20 cm. What is the area of the card left over?

12 cm, 9 cm, 4 cm, 5 cm, 4 cm, 15 cm

3 Caroline has drawn a sandcastle. What is the area of her castle and flag? Start by dividing the shape into parts.

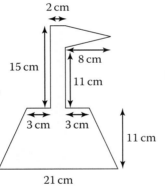

2 cm, 15 cm, 8 cm, 11 cm, 3 cm, 3 cm, 11 cm, 21 cm

4 Use *x* and *y* axes from 0 to 6 on centimetre square paper.

a Plot and join these points to make an arrow.
(0, 3), (3, 6), (5, 6), (3, 4), (6, 4)
(6, 2), (3, 2), (5, 0), (3, 0), (0, 3)

b Find the area of the arrow.

c Use a different method to check your answer to part **b**.

5 A trapezium has area 132 cm². One of the parallel sides is 12 cm. The parallel sides are 4 cm apart. Find the length of the other parallel side.

6 Ryan has two congruent rectangles. The sides of the rectangles are 2 cm and 6 cm.

a Ryan arranges the shape to make this compound shape. Find the perimeter of the shape.

b Ryan wants to arrange the shapes to make a compound shape with perimeter 28 cm. Draw three ways that Ryan can arrange the rectangles.

7 a Draw three different parallelograms that have area 36 cm². Label the base and perpendicular height on your drawings.

b Draw three different triangles that have area 24 cm². Label the width and height on your drawings.

c Draw three different trapeziums that have area 20 cm². Label the two parallel sides and perpendicular height on your drawings.

8 Jeannie has these facts about an allotment.

● The allotment is a rectangle with area 144 m².

● Fencing for the allotment costs £10 per metre.

Jeannie wants to build a fence all around her allotment. She has a budget of £400.

Does Jeannie have enough money to build the fence? Explain how you decide.

9 Find the area of the shaded triangle.

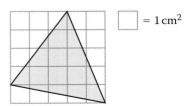

☐ = 1 cm²

7.3 Transformations 1

- You describe a **reflection** using the **mirror line** or reflection line.

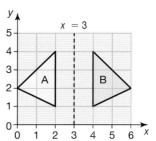

The mirror line is $x = 3$.

Choose a point on the object to find the corresponding point on the image.

Corresponding points are **equidistant** from the mirror line.

To describe a **rotation** you give
- the **centre of rotation** – the point about which it **turns**
- the angle or measure of turn
- the direction of turn – either **clockwise** or **anticlockwise**.

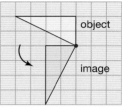

The shape rotates 90° anticlockwise about the dot.

- You can describe a **translation** by specifying the distance moved right or left, and then up or down.

You can write the translation in a column like this: $\begin{pmatrix} \text{right} \\ \text{up} \end{pmatrix}$.

$\begin{pmatrix} 5 \\ 3 \end{pmatrix}$ means 5 right and 3 up. $\begin{pmatrix} -2 \\ -4 \end{pmatrix}$ means 2 left and 4 down.

Left and down are negative directions.

- In a reflection, rotation or translation the object (original shape) and the image (new shape) are **congruent**.

Corresponding angles and lengths are the same in the image and the object.

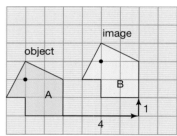

The shape is translated by $\begin{pmatrix} 4 \\ 1 \end{pmatrix}$.

EXAMPLE

Describe the transformation that maps shape A on to

a shape B

b shape C

c shape D.

Maps means changes.

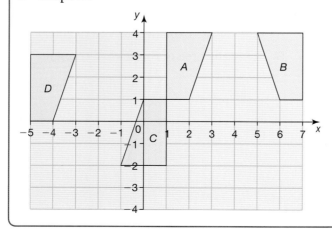

a In a reflection the mirror line bisects the line joining corresponding points on the object and image.

Shape B is a reflection of shape A in the line $x = 4$.

b The vertex (1, 1) does not move during the rotation, so it must be the centre of rotation.

Shape C is a rotation of shape A through 180° about (1, 1).

c Shape D is a translation of shape A by the vector $\begin{pmatrix} -6 \\ -1 \end{pmatrix}$.

Geometry Working in 2D

Exercise 7.3S

1 Match each translation with one of these vectors.

$$\begin{pmatrix} 5 \\ 3 \end{pmatrix} \begin{pmatrix} 2 \\ -4 \end{pmatrix} \begin{pmatrix} 5 \\ -3 \end{pmatrix} \begin{pmatrix} 3 \\ 1 \end{pmatrix} \begin{pmatrix} -4 \\ 2 \end{pmatrix} \begin{pmatrix} -3 \\ 1 \end{pmatrix}$$

a 3 right, 1 up   **b** 5 right, 3 up

c 3 left, 1 up   **d** 5 right, 3 down

e 2 right, 4 down   **f** 4 left, 2 up

2 Copy this diagram.

a Reflect the kite K in the line $x = -1$. Label the image L.

b Reflect the kite K in the line $y = 1$. Label the image M.

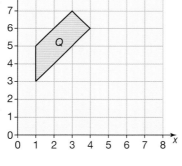

3 Copy this diagram and extend the y-axis to -8.

a Reflect the quadrilateral Q in the x-axis. Label the image R.

b Reflect the quadrilateral Q in the line $y = x$. Label the image S.

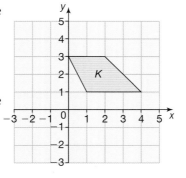

4 Copy this diagram.

a Rotate the kite K through $180°$ about $(1, 1)$. Label the image L.

b Rotate the kite K through $-90°$ about $(2, 0)$. Label the image M.

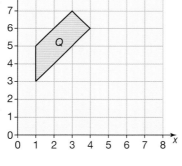

5 Copy this diagram.

a Rotate the triangle A through $90°$ clockwise about $(1, 0)$. Label the image B.

b Rotate the triangle A through $90°$ anticlockwise about $(0, 1)$. Label the image C.

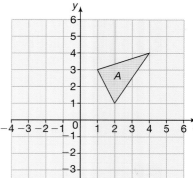

6 Copy this diagram.

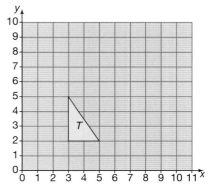

a Translate the triangle T by the vector $\begin{pmatrix} 3 \\ 4 \end{pmatrix}$. Label the image U.

b Translate the triangle T by the vector $\begin{pmatrix} 5 \\ -2 \end{pmatrix}$. Label the image V.

7 Copy this diagram.

a Translate the kite K by the vector $\begin{pmatrix} -4 \\ 3 \end{pmatrix}$. Label the image L.

b Translate the kite K by the vector $\begin{pmatrix} 2 \\ -4 \end{pmatrix}$. Label the image M.

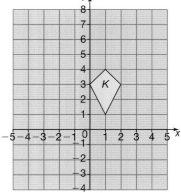

Q 1099, 1113, 1115, 1127   SEARCH

7.3 Transformations 1

RECAP

- Reflections, rotations and translations are all transformations.
- To describe a reflection, you draw or give the equation of the mirror line.
- To describe a rotation, you give the centre and the angle of rotation.
- To describe a translation, you give the distance and direction or the vector.

A reflection flips a shape over. A rotation turns a shape. A translation is a sliding movement.

HOW TO

① Check that the object and image are congruent.
Then decide which type of transformation is involved.

② Give a full description of the transformation.

EXAMPLE

A regular pentagon is divided into five isosceles triangles.
The centre of the pentagon is marked with a dot (•).
The green triangle is rotated about the dot onto the yellow triangle.

a State whether the green and yellow triangles are congruent or similar.

b Calculate the angle and direction of the rotation.

a ① Congruent – same size and same shape.

b ② The five angles at the dot total 360°.
One angle at the dot is 360° ÷ 5 = 72°.

③ Rotation is 72° clockwise about the dot.

EXAMPLE

In this diagram, triangle A undergoes three pairs of transformations.

Triangle A is reflected in the line $x = 0$ (the y-axis) to triangle E.
Then triangle E is reflected in the line $x = -5$ to triangle F.
What single transformation maps triangle A onto triangle F?

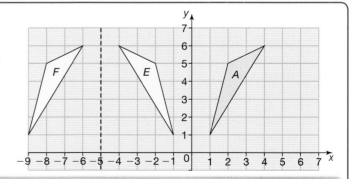

① The combination of reflections is a **translation**.

② A translation by the vector $\begin{pmatrix} -10 \\ 0 \end{pmatrix}$ maps A onto F.

The combination of rotations is equivalent to a single rotation.

The combination of reflections is a translation.

Exercise 7.3A

1 Give the equation of the mirror line for each reflection.

a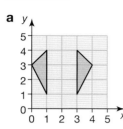

b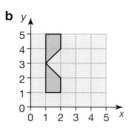

2 Describe fully the transformation that maps

a W to X **b** W to Y **c** W to Z.

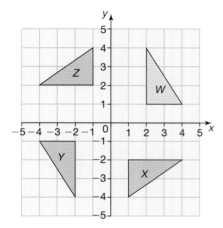

3 A regular octagon is divided into eight isosceles triangles. The centre of the octagon is marked with a dot (•). The green triangle is rotated about the dot onto the yellow triangle. Calculate the angle and direction of the rotation.

4 Copy this diagram.

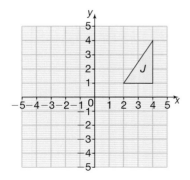

4 **a** Reflect triangle J in the x-axis. Label it K.

b Rotate triangle K 180° about centre $(0, 0)$. Label it L.

c Describe fully the single transformation that takes triangle L to triangle J.

5 Copy this diagram.

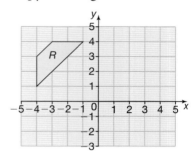

a Translate trapezium R by the vector $\begin{pmatrix} 5 \\ -3 \end{pmatrix}$. Label the image S.

b Translate trapezium S by the vector $\begin{pmatrix} -1 \\ 2 \end{pmatrix}$. Label the image T.

c Describe fully the single transformation that takes trapezium T to trapezium R.

*6 Sue says a rotation of 90° clockwise about $(1, 4)$ maps A onto B. Is Sue correct? Explain your answer.

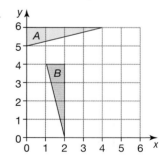

*7 Describe three *different* transformations that map A onto B.

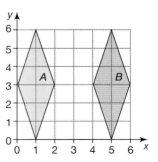

Q 1099, 1113, 1115, 1127 SEARCH

7.4 Transformations 2

In an **enlargement** the angles stay the same and the lengths increase in proportion.

The scale factor of an enlargement is the ratio of corresponding sides.

> ● Scale factor = $\dfrac{\text{length of image}}{\text{length of original}}$

The position of an enlargement is fixed by the **centre of enlargement**.

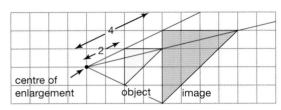

You multiply the distance from the centre to the object by the **scale factor**. This gives the distance to the image along the same extended line.

The scale factor of the enlargement is 2.

The **red lines** start from the centre and pass through corresponding **vertices** of the two shapes.

> ● To describe an enlargement, you give the scale factor and the centre of enlargement.

EXAMPLE

Draw the enlargement of the yellow shape, using scale factor 2 and *P* as the centre of enlargement.

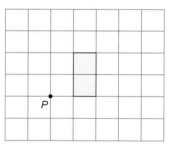

Draw lines from *P* to each vertex.
Multiply the distances from the centre by 2.

2 × 2 = 4   2 × 2 = 4

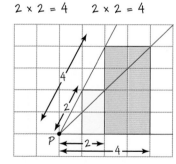

Enlargement with a fractional scale factor reduces the size of the shape.

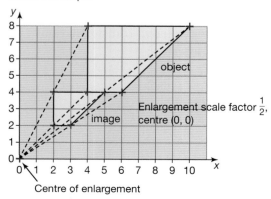

Enlargement scale factor $\frac{1}{2}$, centre (0, 0)

Centre of enlargement

Scale factor $\frac{1}{2}$: all lengths on the image are half the corresponding lengths on the object.

Enlargement with a negative scale factor produces a shape upside down on the opposite side of the centre.

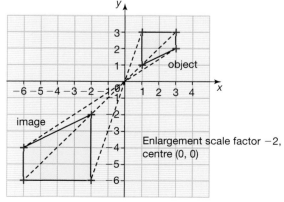

Enlargement scale factor −2, centre (0, 0)

Scale factor −2: all lengths on the image are twice the corresponding lengths on the object; the image is inverted.

Geometry Working in 2D

Exercise 7.4S

1 Copy each diagram on square grid paper. Enlarge each shape by the given scale factor using the given centre of enlargement.

a

b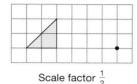

Scale factor 2

Scale factor $\frac{1}{2}$

c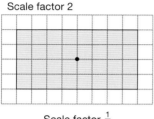

Scale factor $\frac{1}{2}$

2 a Copy this diagram, but extend both axes to 16.

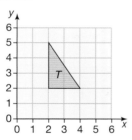

b Enlarge triangle *T* by scale factor 2, centre (0, 0). Label the image *U*.

c Enlarge triangle *T* by scale factor 3, centre (0, 0). Label the image *V*.

3 a Copy this diagram, but extend both axes from −7 to 7.

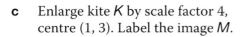

b Enlarge kite *K* by scale factor 2, centre (1, 3). Label the image *L*.

c Enlarge kite *K* by scale factor 4, centre (1, 3). Label the image *M*.

4 a Copy this diagram.

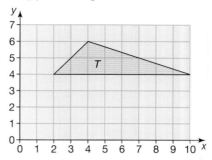

b Enlarge triangle *T* by scale factor $\frac{1}{2}$, centre (0, 0). Label the image *U*.

c Enlarge triangle *T* by scale factor $\frac{1}{2}$, centre (2, 2). Label the image *V*.

5 a Draw a grid with an *x*-axis from −3 to 5 and a *y*-axis from −2 to 10. Plot the points (1, 3) (1, 6) (3, 9) (4, 6). Join them to make quadrilateral *Q*.

b Enlarge quadrilateral *Q* by scale factor $\frac{1}{3}$, centre (4, 3). Label the image *R*.

c Enlarge quadrilateral *Q* by scale factor $\frac{1}{2}$, centre (−2, −2). Label the image *S*.

6 a Draw an *x*-axis from −13 to 5 and a *y*-axis from −7 to 3. Draw triangle *D* with vertices at (−1, 1), (2, 2) and (4, 0).

b Enlarge triangle *D* by scale factor −3, centre (0, 0). Label the image *E*.

c Enlarge triangle *D* by scale factor −$\frac{1}{2}$, centre (0, 0). Label the image *F*.

d How many times longer are the side lengths of triangle *E* than those of triangle *D*?

7 a Draw an *x*-axis from −7 to 6 and a *y*-axis from −9 to 8. Draw triangle *J* with vertices at (−2, 4), (−2, 7) and (4, 4).

b Enlarge triangle *J* by scale factor −$1\frac{1}{2}$, centre (1, 1). Label the image *K*.

c Enlarge triangle *J* by scale factor −$\frac{1}{3}$, centre (1, 1). Label the image *L*.

d How many times smaller are the side lengths of triangle *L* than those of triangle *J*?

 Q 1125   SEARCH

7.4 Transformations 2

RECAP

- In an enlargement, the object and image are similar.
- If the scale factor is less than one (but still postive), then the image is smaller than the object.
- To describe an **enlargement** you give the **scale factor** and the **centre of enlargement**.

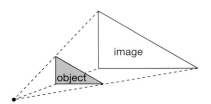

HOW TO

① Draw construction lines to join corresponding vertices on the object and image.

② The construction lines meet at the centre of enlargement.

③ The scale factor of an enlargement is the ratio of corresponding sides. Scale factor = $\dfrac{\text{length of image}}{\text{length of original}}$

To find the centre of enlargement, draw lines between corresponding points.

EXAMPLE

Find the centre of enlargement and calculate the scale factor of the enlargement from *A* to *B*.

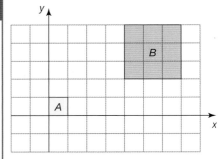

① Draw the red lines to find the centre of enlargement.

② Centre of enlargement is (−2, −1)

③ Scale factor = 3 ÷ 1 = 3

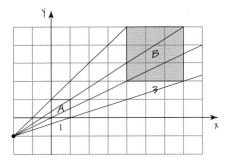

EXAMPLE

Describe the enlargement that maps triangle *X* onto triangle *W*.

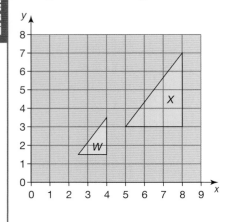

① Draw the red lines to find the centre of enlargement.

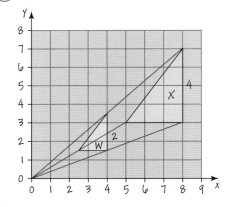

② Centre of enlargement is (0, 0)

③ Scale factor = $2 ÷ 4 = \dfrac{1}{2}$

Exercise 7.4A

1. Copy each diagram on square grid paper. Shape *A* has been enlarged onto shape *B*. Find the centre of enlargement and calculate the scale factor for these enlargements.

   a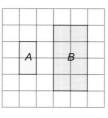

   b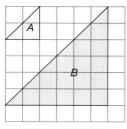

   c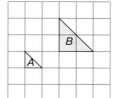

   d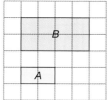

2. Describe fully the single transformation that maps

   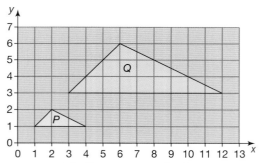

   a   triangle *P* onto triangle *Q*

   b   triangle *Q* onto triangle *P*.

3. Describe fully the single transformation that maps

   a   rectangle *R* onto rectangle *S*

   b   rectangle *S* onto rectangle *R*.

   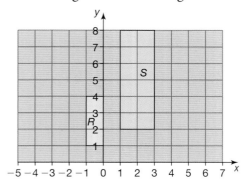

4.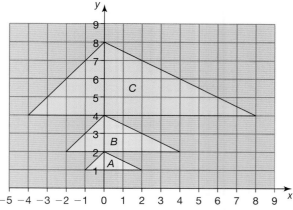

   a   Describe fully the single transformation that maps

      i   triangle *B* onto triangle *A*

      ii   triangle *B* onto triangle *C*

      iii triangle *A* onto triangle *C*.

   b   How many times bigger is the perimeter of triangle *C* than that of triangle *A*?

5. a   Draw *x*- and *y*-axes from −6 to 9. Draw triangle *X* with vertices at (−1, 1), (0, 3) and (3, 2).

   b   Enlarge triangle *X* by scale factor −2, centre (0, 0). Label the image *Y*.

   c   Enlarge triangle *Y* by scale fractor −1½, centre (0, 0). Label the image *Z*.

   d   Describe the enlargement that will transform triangle *X* to triangle *Z*.

   e   Describe the enlargement that will transform triangle *Z* to triangle *X*.

   f   Comment on your answers to **d** and **e**.

6. The table gives the vertices of a kite and its image after a transformation.

Kite	(2, 0)	(3, 2)	(2, 3)	(1, 2)
Image	(6, 0)	(9, 6)	(6, 9)	(3, 6)

   a   What happens to the co-ordinates?

   b   Describe fully the transformation.

Summary

Checkout
You should now be able to...

Test it
Questions

✔ Measure line segments and angles accurately.	1
✔ Use scale drawings and bearings.	1
✔ Calculate the areas of triangles, parallelograms, trapezia and composite shapes.	2
✔ Describe and transform shapes using reflections, rotations, translations (described as 2D vectors) and enlargements (including fractional and negative scale factors).	3 – 5
✔ Identify what changes and what is invariant under a combination of transformations.	6

Language	Meaning	Example
Length	Length is a measure of distance.	Millimetres, centimetres, metres and kilometres are all measures of length. Length can be measured with a ruler.
Angle	The amount that one straight line is turned relative to another that it meets or crosses.	Angles are measured in degrees. One degree is $\frac{1}{360}$ th of a complete turn. Use a protractor to measure an angle.
Area	The amount of space occupied by a 2D shape.	Area = 12 units^2 Perimeter = 14 units
Perimeter	The total distance around the edges that outline a shape.	
Transformation	A geometric mapping that takes the points in an **object** to points in an **image**.	Rotation, reflection, translation, enlargement.
Translation	A transformation in which all the points in the object are moved the same distance and in the same direction.	
Reflection / Mirror line	A transformation that moves points to an equal distance on the opposite side of a minor line.	Mirror line
Rotation / Centre of rotation	A transformation that turns points through a fixed angle whilst keeping their distance from the centre of rotation fixed.	anticlockwise Centre of rotation
Enlargement / Scale factor / Centre of Enlargement	A transformation that moves points a fixed multiple, the scale factor, of their distance from the centre of enlargement.	Scale factor = $\frac{1}{3}$
Invariant	Does not change under a transformation.	A mirror line under reflection.

Review

1 The diagram is drawn to scale where 1 cm = 5 km.

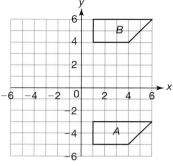

  a What is the bearing of *B* from *A*?

  b What is the bearing of *A* from *B*?

  c What is the length of *AB* in real life?

2 Work out the area of these shapes.

  a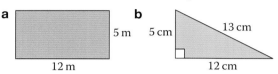

  b

  c

  d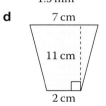

  e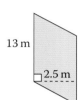

3 Copy this diagram.

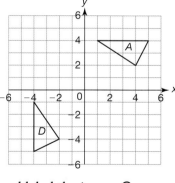

  a Reflect triangle *A* in the line $y = 1$ and label the image *B*.

  b Rotate triangle *A* 90° anti–clockwise about (0, 1) and label the image *C*.

  c Describe the transformation *A* to *D*.

4 Copy this diagram.

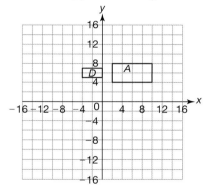

  a Translate shape *A* by the vector $\begin{pmatrix} -5 \\ 2 \end{pmatrix}$ and label the image *C*.

  b Describe the transformation *A* to *B*.

5 Make a copy of this diagram.

  a Enlarge *A* by scale factor 0.25 from centre of enlargement (6, 0) and label the new rectangle *B*.

  b Enlarge *A* by scale factor −1.5 using the origin as the centre of enlargement and label the image *C*.

  c Describe the transformation *A* to *D*.

6 Describe which point(s), if any, are invariant under a

  a rotation   **b** reflection

  c translation   **d** enlargement.

What next?

Score	0 – 2	Your knowledge of this topic is still developing.
		To improve look at MyMaths: 1086, 1099, 1103, 1108, 1113, 1115, 1117, 1125, 1127, 1128, 1129
	3 – 5	You are gaining a secure knowledge of this topic.
		To improve look at InvisiPens: 07Sa – r
	6	You have mastered these skills. Well done you are ready to progress!
		To develop your exam technique looks at InvisiPens: 07Aa – f

Assessment 7

1 a Karl says that to convert from cm to mm you divide by 10. Marta says that you multiply by 10. Who is correct? [1]

b Karl says that to convert from cm to m you divide by 100. Marta says that you multiply by 100. Who is correct? [1]

c Karl says that to convert from km to mm you divide by 1 000 000. Marta says you divide by 10 000. Is either student correct? [1]

2 a Draw a quadrilateral *ABCD* with sides *BC* = 6.5 cm, *AD* = *DC* = 7.5 cm and angles *ADC* = 105° and *DCB* = 60°. [2]

b Measure **i** side *AB* [1]   **ii** ∠*BAD* [1]   **iii** ∠*ABC* [1]

c Give the mathematical name of quadrilateral *ABCD*. [1]

d Use a ruler to find the midpoint of each side. Join the midpoints to form another quadrilateral. What is the mathematical name of this new quadrilateral? [2]

3 Briony makes the following statements about the angles shown below. In each case state whether Briony is right and correct her answer if she is wrong.

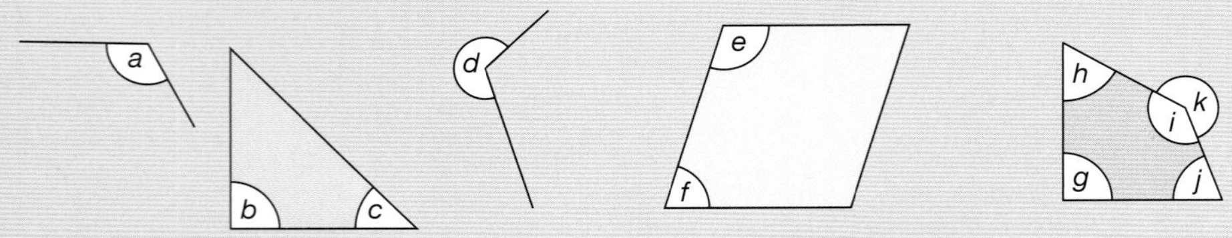

a is obtuse   **b** is acute   **c** is acute   **d** is reflex   **e** is obtuse   **f** is reflex

g is obtuse   **h** is reflex   **i** is obtuse   **j** is obtuse   **k** is acute   [11]

4 a Use a protractor and ruler to construct a regular pentagon with side length 3 cm. [4]

b Find the area of the shape drawn in part **a** by splitting your diagram into appropriate shapes. [7]

5 Three churches, St Mark's, St Spencer's and St Botolph's are the vertices of a triangle. The bearing of St Spencer's from St Mark's is 065°.

a Work out and write down the bearing of

i St Botolph's from St Spencer's [2]

ii St Botolph's from St Mark's [2]

iii St Spencer's from St Botolph's [2]

iv St Mark's from St Botolph's [3]

v St Mark's from St Spencer's [3]

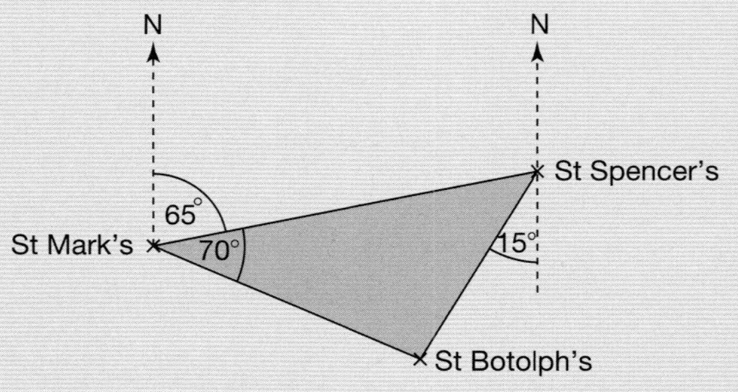

b Look at the 'reverse' bearings of St Mark's from St Spencer's and St Spencer's from St Marks. Repeat this for the other two sets of 'reverse' bearings. Write down a rule connecting a bearing and its 'reverse' bearing. [1]

6 Square patio slabs come in 3 sizes: 1 m × 1 m, 2 m × 2 m and 3 m × 3 m. Farakh wants to build a square patio of side 7 m.

a How many of the 1 m × 1 m slabs would Farakh need? [1]

b Explain whether Farakh can build a patio using

i just 2 m × 2 m slabs [2]   **ii** just 3 m × 3 m slabs. [2]

6 c Can Farakh build a patio using just $2\,m \times 2\,m$ *and* $3\,m \times 3\,m$ slabs?

Draw a diagram to illustrate your answer. [5]

7 One face of a Polo mint is an annulus with an outer diameter of $19\,mm$ and a central hole of diameter $8\,mm$. Calculate the area of this face.

Give your answer

a in terms of π [3] **b** as a decimal correct to 3 sf. [1]

8 A 1-cent coin, a 2-cent coin and a 5-cent coin have the same thickness and have diameters $16\,mm$, $19\,mm$ and $21\,mm$ respectively. These are melted down and recast into another coin with the same thickness.

a Calculate the area of one face of this coin.

Give your answer **i** in terms of π [3] **ii** correct to 3 sf. [1]

b Find the radius of this coin. Give your answer correct to 3 sf. [2]

9 Copy these diagrams and draw enlargements of the shapes with scale factors and centres of enlargement, O, as shown.

Scale Factor 1.5   Scale Factor 2   Scale Factor 3 [9]

10 In a computer game, a rocket takes off from its launching pad and travels 45 miles vertically upwards and 25 miles right.

a Show the path of the rocket on a squared grid. Take 1 square width/length as representing 10 miles. [2]

b Write the vector which represents this translation. [1]

At this position part of the rocket is ejected and the remainder travels a further 15 miles right and 20 miles up.

c Add this new translation to your diagram for part **a**, attaching it to the end of the first translation. [2]

d Write this new translation as a vector. [1]

e Using your diagram, find the *single* translation vector that represents the complete journey. [2]

f How are the vectors in parts **b**, **d** and **e** related? [1]

11 a Jo transforms the triangle, O, into the triangle A. Describe this transformation fully. [3]

b Rotate triangle O 90° clockwise, about the origin. Label the image B. [3]

c Reflect triangle B in the line $y = -x$. Label the image C. [3]

d Fully describe the single transformation that maps O on to C. [2]

8 Probability

Introduction

When did you last look up at the night sky and see a shooting star? It is a rare event, although if you know when and where to look for meteor showers you will greatly increase your chances of seeing one. The world is full of uncertainty, from the unpredictable appearance of shooting stars or earthquakes, to manmade events like the result of a hockey match. Probability is the branch of mathematics that deals with the study of uncertainty and chance.

What's the point?

You apply probability whenever you weigh up everyday risks. For example, should I take an umbrella today? A basic understanding of probability allows you to be more prepared for whatever life throws at you!

Objectives

By the end of this chapter, you will have learned how to …

- Use experimental data to estimate probabilities of future events.
- Calculate theoretical probabilities using the idea of equally likely events.
- Compare theoretical probabilities with experimental probabilities.
- Recognise mutually exclusive events and exhaustive events and know that the probabilities of mutually exclusive exhaustive events sum to 1.

Check in

1 Work out the value of these expressions.

a $1 - \frac{2}{5}$ **b** $1 - \frac{4}{7}$ **c** $1 - \frac{3}{8}$ **d** $1 - \frac{4}{11}$

e $\frac{2}{3} + \frac{1}{6}$ **f** $\frac{1}{5} + \frac{1}{4}$ **g** $\frac{1}{3} + \frac{5}{8}$ **h** $\frac{1}{4} + \frac{3}{8}$

2 Calculate these fractions of an amount.

a $\frac{2}{5} \times 100$ **b** $\frac{1}{9} \times 360$ **c** $\frac{3}{8} \times 56$ **d** $\frac{7}{30} \times 240$

3 Change these fractions to decimals.

a $\frac{7}{10}$ **b** $\frac{3}{4}$ **c** $\frac{3}{8}$ **d** $\frac{2}{5}$

e $\frac{1}{3}$ **f** $\frac{1}{16}$ **g** $\frac{1}{9}$ **h** $\frac{5}{8}$

Chapter investigation

Two people can play an old game called 'Rock, Paper, Scissors'.

In the game, you make a shape with your hand.

- Rock is a closed fist.
- Paper is an open palm with closed fingers.
- And scissors is two fingers held like scissor blades.

The players reveal their 'hand' simultaneously.

Rock beats scissors, paper beats rock, and scissors beats paper.

Is there a best strategy for playing this game?

8.1 Probability experiments

- Probability measures how likely an **event** is to happen.
- All probabilities have a value from 0 (impossible) to 1 (certain).

You can use language such as **likely** and **unlikely** to decribe probabilities.

EXAMPLE

Describe the likelihood of the following events in words.
- **a** You see a head if you toss a fair coin.
- **b** The weather will be sunny for a barbecue in mid-July.

> The event in part **a** has a theoretical value which you can calculate.
>
> For the event in part **b** you can only estimate the probability.

a	Even chance	There are two (equally likely) outcomes.
b	Likely	You could make a better estimate nearer the date using a weather forecast.

- Estimated probability is called the **relative frequency**.
- Relative frequency = $\dfrac{\text{number of favourable trials}}{\text{total number of trials}}$

> You can give probabilities as fractions, decimals, percentages.

EXAMPLE

A drawing pin is thrown up 10 times and lands point in 7 cases.

- **a** Estimate the probability that a drawing pin lands point up

The drawing pin is now thrown another 90 times and lands point up another 67 times.

- **b** Calculate the relative frequency of the drawing pin landing point up using all the observations made so far.

a $P(\text{point up}) = \dfrac{7}{10} = 0.7 = 70\%$

b $P(\text{point up}) = \dfrac{7 + 67}{10 + 90} = \dfrac{74}{100} = 0.74 = 74\%$ There were a total of 74 out of 100.

> 0.74 is a better estimate because it is based on more information.

If you know the probability of an event, you can calculate how many times you **expect** the outcome to happen.

- Expected frequency = number of trials × probability of the event

EXAMPLE

8% of men have red-green colour-blindness. Women are rarely colour-blind.
A group of 65 men and 40 women are attending a meeting.
Estimate how many people in the group suffer from red-green colour-blindness.

65 × 0.08 = 5.2 or about 5 are likely to be colour-blind.

Only the men are likely to be colour-blind.

- If you increase the number of trials, the outcome of an experiment becomes closer to the **theoretical** (expected) outcome.

Exercise 8.1S

1 Use one of the words in the list to describe the probability of each event.

impossible   unlikely   even chance

likely   certain

Give a reason for your answers.

a Rafael and Sebastian have played 12 tennis matches against each other in the last year. Rafael won 6 out of 12 matches.

  i Sebastian wins the next match between the two players.

  ii Sebastian wins the next five matches between the two.

b The average temperature in Edinburgh in February is 4°C.
The average temperature in Sydney, Australia in February is 26°C.

  i The maximum temperature in Edinburgh was under 10°C on February 18th in 1913.

  ii The maximum temperature in Sydney, Australia was at least 10°C on February 29th 1913.

2 Hari works at Quick-Fix garage.
Over the last month, 40 customers have come to the garage with a flat tyre caused by a puncture.
Hari records which tyre was punctured.

Tyre	Frequency
Front left	8
Front right	7
Back left	13
Back right	12

Estimate the probability that the next customer with a flat tyre has a puncture on their

a front left tyre   **b** front right tyre

c back left tyre   **d** back right tyre.

Give your answer as a

i fraction **ii** decimal **iii** percentage.

3 The probability of a drawing pin landing point up when dropped is $\frac{3}{4}$. Estimate how many times it will land point up if it is dropped

a 20 times   **b** 7 times.

4 Rory wants to pick a red ball and can choose bag A or B to pick a ball at random from.

Bag A	Colour	Red	Blue
	Frequency	1	2

Bag B	Colour	Red	Blue
	Frequency	2	8

a Which bag should Rory choose? Give reasons for your answer.

b Rory takes out two balls without replacing. Rory wants to pick two reds. Which bag should he choose? Give reasons for your answer.

5 There are 10 coloured balls in a bag. One ball is taken out and then replaced in the bag. The colours of the ball are shown in the table.

Colour	Red	Green	Blue
Frequency	9	14	27

a How many times was a ball taken out of the bag?

b Estimate the probability of taking out

  i a red ball   **ii** a green ball

  iii a blue ball.

c How many balls of each colour do you think are in the bag?

d How could you improve this guess?

6 Weather records show that approximately 30% of April days in a certain village have some rain. Estimate the number of days the village will have some rain in April next year.

7 A financial advisor claims that, on average, 70% of his recommended shares have increased in value. He has recommended 20 shares in the current year.
How many shares would you expect to increase in value over this year?

8 Throw an ordinary dice until you get a six, counting how many throws you make, including the one on which the six appears.

a Do this twenty times and calculate the relative frequency from your observations for each of the values you have recorded.

b Often when this experiment is done, the outcome with the highest relative frequency is one throw. Give a reason why that happens.

Q 1211, 1264   SEARCH

8.1 Probability experiments

- You can estimate the probability of an event by conducting an experiment.
- Estimated probability is called the **relative frequency**.

  Relative frequency = $\dfrac{\text{number of favourable trials}}{\text{total number of trials}}$

- When you repeat an experiment, you may get a different outcome.
- The expected frequency = number of trials × probability of the event.

HOW TO

1. Think first in terms of words – likely / evens / unlikely
2. Calculate the relative or expected frequency of the event.
3. Answer the question. Keep in mind that the estimated probability becomes more reliable as you increase the number of trials.

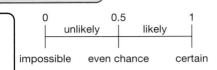

0	0.5	1
	unlikely	likely

impossible   even chance   certain

EXAMPLE

Aleesha takes out a ball out of a bag, records the colour and places the ball back in the bag. She does this 40 times. Aleesha records her results in a table.

a Complete Aleesha's table.

b Which bag **could be** Aleesha's bag? Explain your answer.

Bag 1   Bag 2   Bag 3

2. Use the formula for relative frequency.

Total number of trials = 16 + 4 + 20 = 40

Yellow: 16 ÷ 40 = 0.4

Green: 4 ÷ 40 = 0.1

Blue: 20 ÷ 40 = 0.5

Colour	Yellow	Green	Blue
Frequency	16	4	20
Relative frequency	0.4	0.1	0.5

b

1. Bag 1 has an even chance of yellow and even chance of blue. Green is impossible.
3. **Bag 1** is not her bag because bag 1 has no green balls.
1. Bag 2 has an even chance of yellow, and blue is more likely than green.
3. **Bag 2** could be her bag. The probabilities are close to the relative frequencies from the experiment.
1. Bag 3 yellow is likely, blue and green are unlikely.
3. **Bag 3** could be her bag. Aleesha should do more trials to get more reliable results.

EXAMPLE

Katrina plays a game at a fair. Each try costs £1.
If the spinner lands on the WIN zone you win £2.
If the spinner lands on the SAFE zone you get your £1 back.
Katrina plays the game 10 times.
How much prize money should she expect to win or lose?

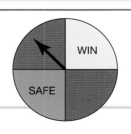

WIN

SAFE

1. There are 10 trials.

2. Probability of winning £2 = $\dfrac{1}{4}$ ➡ 3. Times she wins £2 = $10 \times \dfrac{1}{4} = 2.5$

   Probability of winning £1 back = $\dfrac{1}{4}$ ➡ Times she wins £1 = $10 \times \dfrac{1}{4} = 2.5$

   Expected prize money = $(2.5 \times £2) + (2.5 \times £1) = £5 + £2.50 = £7.50$

   Playing 10 games costs £10. Katrina should expect to lose £2.50.

Probability Probability

Exercise 8.1A

1 Elise spins this spinner 4 times. She says that if the spinner is fair then the spinner will land on each colour once. Is Elise correct?

2 A bag contains 100 coloured balls. Xavier, Yvonne and Zoe each select a ball from the bag, record if the ball is red and replace the ball. The table shows their results

	Number of trials	Number of red balls
Xavier	5	4
Yvonne	20	16
Zoe	100	95

a Xavier says that the probability of choosing a red ball is $\frac{4}{5}$.

Criticise his statement.

b Zoe says that the bag must contain 95 red balls. Is she correct?

c **i** Explain why the most accurate estimate of the relative frequency of choosing a red ball is 0.92.

ii Does the bag contains 92 red balls?

3 Alik spins a spinner and records the result in a table. He adds each column when he lands on a new colour.

Colour	Red	White	Blue
Frequency	10	5	
Relative freq.		0.2	0.4

a Copy and complete Alik's table. What do you notice about the sum of the relative frequencies?

b How many of these spinners could be Alik's?

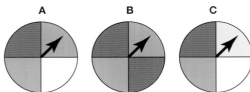

A   B   C

c Draw three possible spinners that could be Alik's spinner. You can divide your spinners into as many sectors as you like.

*4 Keith thinks that fewer people are born on a Saturday or Sunday than on a weekday because any planned births will be scheduled during the week.

He collects information from a hundred friends of different ages about the day their birthday falls on this year and finds that Saturday and Sunday seem to be as common as other days.
Suggest why the information he has collected does not mean that his idea about the days on which people are born is wrong.

*5 A red bag has 5 white and 10 black balls in it, and a blue bag has 5 white and 5 black balls in it. All the balls are identical except for their colour.

a You choose a ball from the red bag at random and note its colour before returning it to the bag. If you do this 30 times, estimate how many times you will see a white ball.

b If you have used the blue bag instead, would you expect to see a white more often, less often or about the same?

*6 The coloured sections on this spinner are all equal. Shawn is playing a game at a school fete which charges 50p a go. She wins a prize if her spinner lands on a blue sector.

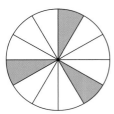

a If Shawn has 24 goes, how often would she be likely to win?

b If the prize is £1, how much profit does the fete make on average per go?

8.2 Theoretical probability

- A trial is an activity or experiement.
- An outcome is the result of a trial.
- An event is one or more outcomes of a trial.

> Rolling a dice is a trial.
>
> The outcomes of rolling a dice are 1, 2, 3, 4, 5, 6.
>
> Rolling an even number is an event.

The outcomes of some experiments are equally likely, for example, throwing a fair coin or rolling a fair dice.

- The **theoretical probability** of an event is

$$\text{probability of event happening} = \frac{\text{number of outcomes in which event happens}}{\text{total number of all possible outcomes}}$$

EXAMPLE

A fair dice is rolled and the number showing on the top is scored. Find the probability of these events.

a 3 is scored.

b A factor of 12 is scored.

c A prime number is scored.

> The probability of an event can be written as P(event).

There are 6 possible outcomes which are equally likely.

a $P(3) = \frac{1}{6}$ There is one 3.

b $P(\text{factor of } 12) = \frac{5}{6}$ Five outcomes are factors of 12: 1, 2, 3, 4 and 6.

c $P(3) = \frac{3}{6} = \frac{1}{2}$ Three outcomes are prime: 2, 3 and 5.

Theoretical probability is based on equally likely outcomes.

- You use experimental probability when the event is unfair or biased or when the theoretical outcome is unknown.

> The closer experimental probability is to theoretical probability the less likely it is that there is bias.

EXAMPLE

Sami threw a drawing pin 10 times and it landed point up 7 times. Cristiano threw a similar drawing pin 43 times and it landed point up 31 times.

a Calculate the relative frequency of the pin landing point up for each person.

b Which of these is the better estimate of the probability?

a Sami $\frac{7}{10} = 0.7$ Cristiano $= \frac{31}{43} = 0.72$ (2 dp)

b Cristiano used the largest sample so he has the better estimate.

- The experimental probability becomes a better estimate for the theoretical probability as you increase the number of trials.

Exercise 8.2S

1 A fair dice is rolled and the number showing on the top is scored.
Find these probabilities

    a P(6)   **b** P(less than 3)

    c P(factor of 10)   **d** P(square number)

2 Are the outcomes in each case equally likely? If they are, give the probability of each outcome. If they are not, explain why.

    a The score showing when a fair die is thrown.

    b Whether a drawing pin lands point up or down when it is thrown in the air.

    c The day of the week a baby is born on.

    d The number of times I have to toss a fair coin until I see a head.

    e What the letter is, if a letter is chosen at random from a book.

3 All the counters in a bag are either green or black. At the start the probability a counter chosen at random from the bag is green is $\frac{1}{4}$. Counters are not replaced when they are taken out.

The first two counters taken out of the bag are green.

What is the least number of black counters that must still be in the bag?

4 A class of 24 students are each given a similar biased dice and asked to record how many times they saw a six when they threw their dice 20 times.

8	9	5	7	6	8	7	8
6	7	4	9	7	7	6	7
8	6	9	5				

    a Give the highest and lowest relative frequencies found by any of the students.

    b Find the best possible estimate of the probability of a six showing on the biased dice from the results given.

5 For the following events give a value between 0 and 1 for the probability of the event described happening.

If you have to estimate the probability then write (est) after your answer.

    a You score a 4 when you throw a fair dice

    b A set of cards has the numbers 1-9 written on one of each of the cards.
You pick a card at random and it is an even number.

    c You score less than 7 when you throw a fair dice.

6 Michelle wonders how often there is a difference of more than 2 when you throw a pair of fair dice. She does an experiment, noting down how often it happens after each block of 10 throws.
Her results are

5	4	5	3	2	4	6
3	3	3	5	4	2	4

    a Give the relative frequencies at the end of each block of 10 throws.

    The first two are $\frac{5}{10} = 0.5$, and $\frac{9}{20} = 0.45$.

    b What is the best estimate Michelle has of the probability of getting a difference of more than 2?

*7 Three friends were investigating how often a biased dice showed a six.
Sergio got 10 sixes from 40 throws.
Annalise got 8 sixes from 40 throws.
Dominika got 21 sixes from 80 throws.

    • Sergio says they should take his as it is the median of the three estimates.

    • Annalise says they should take the mean of the three.

    • Dominika says they should combine all the results and use that relative frequency as the estimate.
Which method is the best?

1211, 1263, 1264   SEARCH

8.2 Theoretical probability

- You use theoretical probability for fair activities, for example rolling a fair dice.
- For unfair or biased activities, or where you cannot predict the outcome, you use experimental probability (relative frequency).
- Relative frequency is the proportion of successful trials in an experiment.

Theoretical probability
$$= \frac{\text{number of favourable outcomes}}{\text{total number of outcomes}}$$
Relative frequency
$$= \frac{\text{number of successful trials}}{\text{total number of trials}}$$

HOW TO

To compare theoretical probability with relative frequency

1. Start by assuming that the activity is fair and calculate the theoretical probability of the event.
2. Use the results of the experiment to calculate the relative frequency.
3. If the relative frequency and theoretical probability are very different, then the activity could be biased.

The more trials you carry out, the more reliable the relative frequency will be.

EXAMPLE

Kaseem rolled a dice 50 times and in 14 of those he scored a 2. Is the dice biased towards 2? Explain your answer.

1. Theoretical probability 2 = $\frac{1}{6}$ = 0.166 ...   If the dice is fair, then every outcome is equally likely.

2. Relative frequency of rolling a 2 = $\frac{14}{50}$ = 0.28   Converting to decimals makes it easier to compare.

3. The relative frequency is higher than the theoretical probability.
The dice appears to be biased towards 2.

EXAMPLE

A fair spinner is divided into blue, red and yellow sectors.
Valerie knows that the blue sector is one quarter of the spinner.
Valerie can't remember the size of angle x, but she knows that it is either 150° or 160°.
She spins the spinner 100 times and records her results in a table.

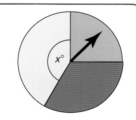

Colour	Blue	Yellow	Red
Frequency	23	42	35

Find the size of angle x. Explain your answer.

1. Calculate the theoretical probabilities for each angle. There are 360° in a circle.

$x = 150°$ P(yellow) = $\frac{150}{360}$ = 0.416.. $x = 160°$ P(yellow) = $\frac{160}{360}$ = 0.444...

Angle for red = 360 - (150 + 90) = 120   Angle for red = 360 - (160 + 90) = 110

If $x = 150°$, P(red) = $\frac{120}{360}$ = 0.333...   P(red) = $\frac{110}{360}$ = 0.305...

2. Use the information in the table to find the relative frequency.

Relative frequency of yellow = $\frac{42}{100}$ = 0.42   Relative frequency of red = $\frac{35}{100}$ = 0.35

3. As the spinner is fair, the relative frequency and theoretical probabilities should be similar.

The relative frequencies are closest to the probabilities for $x = 150°$.

Exercise 8.2A

1 The table below shows the gender and hair colour of 30 students in a class.

	Fair	Dark
Male	7	9
Female	8	6

A student is chosen at random from the class. What is the probability that the student is

a a girl **b** a dark-haired boy.

2 Amber rolled a dice 100 times and in 71 of those rolls her number was a factor of 12.

Is the dice biased towards 5?

Give your reasons.

3 *ABCD* is a square and *E* and *F* are the midpoints of sides *AD* and *BC*.

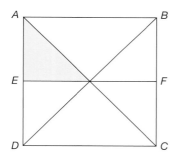

If a point is selected at random in the square what is the probability it is in the shaded area?

4 The square with sides 4 cm shown has four quarter circles shaded.
A point is chosen at random in the square. Calculate the probability the point is in the shaded area.

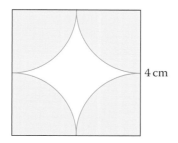

4 cm

5 Bag 1 contains 8 red balls and 12 green balls. Bag 2 contains 5 red balls and 7 green balls. The balls are identical apart from colour.

a You choose a ball from bag 1 at random. What is the probability you choose a green ball?

All the balls are now put into one bag and you choose a ball at random.

b What is the probability you choose a red ball?

6 The coloured sections on the spinner are all equal.

If the spinner lands on a green section, the score is doubled.

If the spinner lands on a blue section, the score is halved.

Shawn is playing a game with her friend Andrea who goes first, with the result shown.

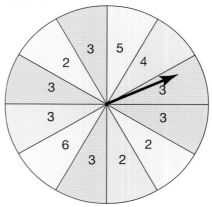

What is the probability that Shawn scores more than Andrea?

7 Iveta has a spinner divided equally into 19 sectors.

The sectors are white, black and red.

There are the same number of black and red sectors.

Iveta spins the spinner 80 times and records her results in a table.

Colour	White	Black	Red
Frequency	17	32	31

Explain why the spinner could be biased.

 Q 1211, 1263, 1264 SEARCH

8.3 Mutually exclusive events

- Two events are **mutually exclusive** if they cannot both happen at the same time.

EXAMPLE

You throw an ordinary dice. Here are three possible events.

A: a prime number   B: an odd number   C: a square number

Which of these pairs of events are mutually exclusive?

a A and B   **b** A and C   **c** B and C

> A: 2, 3 and 5   B: 1, 3 and 5   C: 1 and 4
>
> **a** A and B both contain 3 and 5, so not mutually exclusive
> **b** A and C have nothing in common, so they are mutually exclusive
> **c** B and C both contain 1, so not.

A set of events is exhaustive if they include all possible outcomes.

- If a set of mutually exclusive events are also **exhaustive** then their probabilities sum to 1.

EXAMPLE

The probability that a hockey team wins a match is 0.65. The probability of a draw is 0.2. What is the probability that they lose the match?

> The outcomes win, draw and lose are mutually exclusive.
> $0.65 + 0.2 + P(\text{lose}) = 1$
> $P(\text{lose}) = 1 - (0.65 + 0.2) = 0.15$

- Probability of an event happening = 1 – probability of event not happening p(A) = 1 – p (not A)

EXAMPLE

The probability that you are selected for a random body search by an electronic security monitoring system is 0.06. What is the probability you will not be selected?

> $1 - 0.06 = 0.94$ The probabilities add to 1.

EXAMPLE

The probability that a child in a nursery class has a cat is 0.3.

a Find the probability that the child does not have any pets.
b Find the probability that the child has a hamster.

> **a** Not known, they can have other animals as pets.
> **b** Not known, a child may have either a cat or a hamster, both or neither!

These events may not be mutually exclusive..

Exercise 8.3S

1 For each of the following pairs of events say whether or not they are mutually exclusive.

  a P: a girl has red hair
     Q: she has brown eyes

  b F: a boy gets the same score on 2 dice
     G: his total score is odd

  c M: a girl gets the same score on 3 dice
     N: her total score is odd.

  d X: a girl gets the same score on 3 dice
     Y: her total score is a multiple of 6.

2 A pack of cards contains 60 cards in four colours: black, red, green and blue. There are 15 of each colour.
The black cards carry the numbers 1 to 15.
The red cards are multiples of 2.
The green cards are multiples of 3.
The blue cards are multiples of 6.

The top card is turned over.

For each pair of events, say whether or not they are mutually exclusive.

  a A: the card is black, and
     B: it is an even number

  b C: the card is red, and
     D: it is an odd number

  c E: the card is green, and
     F: it is a factor of 20

  d G: the card is blue, and
     H: it is a prime number

3 All the counters in a bag are either green or black. The probability a counter chosen at random from the bag is green is $\frac{1}{3}$.

  a Find the probability that a counter chosen at random is blue.

  b Find the probability that a counter chosen at random is black.

4 All the counters in a bag are either green, white or black.
The probability a counter chosen at random from the bag is green is $\frac{1}{3}$, and for white it is $\frac{1}{2}$.
What is the probability a counter chosen is

  a black   **b** not white?

5 A fair dice has a triangle, square, circle, rectangle, sphere and a cylinder showing on the six sides. The dice is rolled.
Find the probability that

  a it shows a 3D shape.

  b it does not show a 2D shape with straight edges.

6 Seamons has two packs of 12 cards, each labelled 1 to 12.
He removes all the multiples of 4 from one of the packs, combines the two packs and shuffles them.
He turns as top card.
Find the probability that the card is

  a a multiple of 6

  b an even number

  c not a prime number.

7 A biased dice has the probabilities shown in the table.

Score	1	2	3	4	5	6
Probability	0.1	0.2	x	0.3	0.1	0.1

Find the probability of getting
  a a three

  b an even number

  c not a prime number

*8 A bag contains at least green, blue and white balls. The probability a ball chosen at random is green is $\frac{1}{7}$. Blue is twice as likely to be chosen as green and white is twice as likely to be chosen as blue. Explain why you know there are no other colours in the bag.

*9 A bag contains just green, blue and white balls. Blue is three times as likely to be chosen as green, and white is twice as likely to be chosen as blue.

What is the probability a ball chosen at random from the bag is

  a green   **b** white?

Q 1262, 1263   SEARCH

8.3 Mutually exclusive events

- Events are mutually exclusive if they cannot happen at the same time.
- If the events are also exhaustive, then the probabilities sum to 1.
- $P(A) = 1 - P(\text{not } A)$.

You could list the outcomes in a table.

HOW TO

① Read the question carefully and think about the outcomes possible. It often helps to list them.

② Think what you know about mutually exclusive and exhaustive events to help you calculate probabilities.

EXAMPLE

A red and a blue dice are thrown together and the highest score is recorded.
Find the probability that the score recorded is 5 or less.

① There are 36 possible outcomes, and these can be shown in a table.

	1	2	3	4	5	6
1	1	2	3	4	5	6
2	2	2	3	4	5	6
3	3	3	3	4	5	6
4	4	4	4	4	5	6
5	5	5	5	5	5	6
6	6	6	6	6	6	6

The complementary event is just another way of saying 1 minus the event.

② It is easier to find the **complementary** event

$$(5 \text{ or less}) = 1 - P(6)$$
$$= 1 - \frac{11}{36}$$
$$= \frac{25}{36}$$

EXAMPLE

There are a number of balls in a bag, identical except for their colour.
A ball is taken from the bag at random.
The probability of taking

- green is $\frac{1}{6}$
- blue is $\frac{1}{4}$
- black is $\frac{1}{2}$.

Are any other colours in the bag?

② If the events are exhaustive then their probabilities add up to 1.

$\frac{1}{6} + \frac{1}{4} + \frac{1}{2} = \frac{2}{12} + \frac{3}{12} + \frac{6}{12} = \frac{11}{12}$ There must be at least one other colour.

Exercise 8.3A

1 A red and a blue dice are thrown together and the difference between the scores is recorded.

a Draw a table to show the possible outcomes.

b What is the probability of a difference of
i exactly one **ii** zero **iii** more than 1?

2 There are a number of balls in a bag, identical except for their colour. A ball is taken from the bag at random. Serena says that the probability of taking a green is $\frac{1}{4}$, taking a blue is $\frac{1}{3}$ and taking a black is $\frac{1}{2}$.
Can Serena be correct?
Give reasons for your answer

3 There are a number of balls in a bag, identical except for their colour.
A ball is taken from the bag at random. Arinda says that the probability of taking a green is $\frac{1}{6}$, taking a blue is $\frac{1}{3}$, taking a white is $\frac{1}{8}$, and taking a black is $\frac{3}{8}$.

a Can you tell if Arinda is correct?
Give reasons for your answer

b Can you tell if there are any other colours?
Give reasons for your answer

4 A bag of sweets contains 4 vanilla fudge pieces, 3 white chocolate caramels, 3 chocolate covered fudge pieces, and 2 caramel fudge pieces.
What is the probability that a sweet chosen at random contains

a chocolate **b** fudge **c** caramel?

5 A set of cards with the numbers 1 to 10 is shuffled and a card chosen at random. Here are four possible events.
A a prime number **B** a factor of 36
C an even number **D** an odd number

a List any pairs of mutually exclusive events.

b List any pairs of exhaustive events.

c Explain why A and C are not mutually exclusive.

6 The table below shows the gender and hair colour of the 28 students in a class.

	Fair	Dark	Red
Male	6	9	3
Female	7	6	0

A student is chosen at random from the class, and here are four possible events.
M the student is male
F the student is female
R the student has red hair
D the student has dark hair

Giving a reason, explain whether the following statements are true or false.

a M and F are mutually exclusive

b M and F are exhaustive

c M and D are mutually exclusive

d F and R are mutually exclusive.

7 A red and a blue dice are thrown together and the blue score is subtracted from the red score.

a Draw up a table like the one in the first example to show the possible outcomes.

b What is the probability the score is
i -2 **ii** 5 **iii** less than 5?

8 The probabilities for a biased dice are shown in the table.

Score	1	2	3	4	5	6
Probability	0.1	0.2			0.2	0.1

Getting a 4 is three times as likely as getting a 3.
Find the probability of getting a
a 3 **b** 4 **c** prime number.

*9 A spinner is constructed so that the numbers 1 to 10 are possible, and the probability of seeing a number $k + 1$ is twice as likely as seeing the number k (for $k = 1, 2, ... 9$).
Find the probability of getting a

a 1 **b** 10.

Give your answers as decimals correct to 3 decimal places.

Summary

Checkout

You should now be able to...

Test it

Questions

	Test it Questions
✔ Use experimental data to estimate probabilities and expected frequencies.	1, 2, 7
✔ Use tables to represent the outcomes of probability experiments.	1, 2
✔ Calculate theoretical probabilities and expected frequencies using the idea of equally likely events.	3, 4
✔ Recognise mutually exclusive events and exhaustive events and know that the probabilities of mutually exclusive exhaustive events sum to 1.	3, 5, 6
✔ Compare theoretical probabilities with experimental probabilities.	7

Language	Meaning	Example
Trial	An activity or experiment.	Rolling a single dice.
Outcome	The result of a trial.	Rolling a single dice and getting a 2.
Event	One or more outcomes of a trial.	Getting a prime number when you roll a conventional six-sided dice.
Impossible	It cannot happen. The probability of it happening is 0.	Rolling a conventional six-sided dice and getting the number 7.
Certain	It must happen. The probability of it happening is 1.	Rolling a conventional single dice and getting a number less than 7.
Relative frequency	The **experimental probability** of an outcome after several trials is $$\frac{\text{Number of times the outcome happened}}{\text{Number of times the activity was done}}$$	A drawing pin is thrown 100 times. It lands point down 72 times. The relative frequency of the pin landing point down is $\frac{72}{100} = 0.72$
Expected frequency	How many times you expect the outcome to happen. Number of trials × probability of the event	The expected frequency of getting a 5 when you roll a fair dice 40 times is $\frac{1}{6} \times 40 = 6.666... = 6.\dot{6}$ about 7 times.
Theoretical probability	A value between 0 and 1 describing the likelihood of an event determined by logical consideration of the trial. If all possible outcomes are equally likely this is given by $$\frac{\text{Number of ways the outcome could happened}}{\text{Number of possible outcomes}}$$	There is 1 way of rolling a two on a fair dice. There are 6 different possible outcomes. The theoretical probability of rolling a two is $\frac{1}{6}$.
Bias Biased	All outcomes are not equally likely.	A fair dice has the numbers 1, 1, 2, 3, 4, 5 on its faces. It is more likely that you will roll a 1 than any of the other outcomes. The dice is biased.
Equally likely	All outcomes have the same probability of happening.	When you roll a fair dice all the outcomes {1, 2, 3, 4, 5, 6} are equally likely. They all have a probability of $\frac{1}{6}$.

Review

1 Rosie records the ages of children in a play area.

Age (in years)	Frequency
Younger than 2	6
2 – 4	16
5 – 7	10
8 – 11	7
Older than 11	1

  a What is the relative frequency of children younger than 2 years old?

  b On a different day there are 25 children in the park. Use Rosie's data to estimate the number of children who are in the 2−4 age range.

2 A spinner with five colours on it is repeatedly spun and the results recorded.

Outcome	Relative frequency
Blue	0.2
Red	a
Green	a
Yellow	$3a$
Purple	0.05

  a What is the value of a?

  b Copy and complete the table to show all the relative frequencies.

  c Use these results to estimate how many times the spinner would land on blue if it were spun 300 times.

3 A bag contains 12 blue and 26 green counters. A counter is selected at random. What is the probability the counter is

  a blue   **b** red

  c green or blue?

4 A fair dice is rolled twice. What is the probability that the two numbers add up to

  a 5   **b** 11?

5 The probability that a tennis player wins a point is $\frac{11}{20}$. What is the probability that they lose a point?

6 Archie always chooses either a biscuit, an apple or a banana to eat with a cup of tea. The probability he chooses fruit is $\frac{18}{25}$ and he is exactly twice as likely to pick an apple as a banana.

  Calculate the probability of Archie choosing

  a a biscuit   **b** a banana

  c an apple.

7 An unbiased dice is thrown 60 times and the number 3 occurs 14 times.

  a What is the relative frequency of a 3?

  b What is the theoretical probability of a 3?

  The dice is now thrown an additional 100 times.

  c What would you expect to happen to the relative frequency of a 3?

What next?

<table>
<tr><td rowspan="6">Score</td></tr>
<tr><td>0 – 3</td><td></td><td>Your knowledge of this topic is still developing.
To improve look at MyMaths: 1211, 1262, 1263, 1264</td></tr>
<tr><td>4 – 6</td><td></td><td>You are gaining a secure knowledge of this topic.
To improve look at InvisiPens: 08Sa – e</td></tr>
<tr><td>7</td><td></td><td>You have mastered these skills. Well done you are ready to progress!
To develop your exam technique look at InvisiPens: 08Aa – e</td></tr>
</table>

Assessment 8

1 Amanda says that it is impossible to throw a coin five times and see tails every time.
Is she correct? Give a reason for your answer. [1]

2 A fair coin is thrown and lands on heads twice. What is the probability of a head on
the third throw? Explain your answer. [2]

3 A dice is thrown 100 times and the following
results are obtained.

Score	1	2	3	4	5	6
Frequency	14	17	22	18	15	14

Calculate the relative frequency of scoring an odd number. [3]

4 One in every eight children in a town has asthma. How many children would
you expect to have asthma

   a in the school of 1600 children [1]

   b in Green Lane where 14 children live [1]

   c in Mr Knowles' class of 34 children? [1]

5 The sides of a biased spinner are labelled 1, 2, 3, 4 and 5.
The probability that the spinner will land on
each of the numbers 2, 3 and 4 is given in
the table.

Number	1	2	3	4	5
Probability	x	0.14	0.26	0.22	x

The probability that the spinner lands on either of the numbers 1 or 5 is the same.

   a Valerie says that $x = 0.2$. Explain why Valerie is wrong and find the value of x. [4]

Valerie spins the spinner 200 times.

   b How many times should she expect the spinner to land on 3? [1]

   c How many times should Valerie expect the spinner to land on an odd number? [2]

   d How many times should Valerie expect the spinner to land on a prime number? [2]

6 Bill and Ben buy identical packets of mixed vegetable seeds. They pick out the pea
seeds from their packets. Their results are shown in the tables.

Bill's results

Number of packets	Number of seeds in each packet	Number of pea seeds in each packet
5	30	5, 5, 8, 3, 6

Ben's results

Number of packets	Number of seeds in each packet	Total number of pea seeds
10	30	51

   a Write down an estimate of the probability of selecting a pea seed at random from a
packet using

      i Bill's results   **ii** Ben's results. [4]

   b Whose results are likely to give the better estimate of the probability?
Explain your answer. [2]

7 Jessica buys 20 bottles of juice.
There were four bottles each of apple, blueberry, grape, orange and pineapple.

   a Jessica takes a bottle from the fridge at random. What is the probability that it
is apple juice? [1]

   b Jessica's son drinks all of the orange juice. Jessica takes one bottle at random from the fridge.
What is the probability that it is either grape of pineapple juice? Write your answer
in its simplest form. [2]

8 Harriet buys a box of cat food that contains 30 individual packets.
The probability that a tuna packet is chosen from the box is 0.3.
How many packets of tuna packets are in the box? [2]

9 The probability that the 8:45 train into Manchester Piccadilly station arrives early is 0.18.
The probability that the train arrives on time is 0.31.

   a A frustrated passenger complains that the train is late more often than it is on
time or early. Is the passenger correct? [3]

   b Dave takes 225 journeys on this train during one year.
How many days can he expect to be on time? [3]

10 The table below shows the probabilities of selecting raffle tickets from a drum.
The tickets are coloured red, white or blue and numbered 1, 2, 3 or 4.
A raffle ticket is taken at random from the drum.
Calculate the probability that:

   a it is white and numbered 3 [1]

   b it is numbered 1 [3]

   c it is red [3]

   d it is either blue or numbered 2. [4]

   e What raffle ticket is impossible to draw?
Explain your answer. [1]

		Number			
		1	**2**	**3**	**4**
	Red	$\frac{2}{25}$	$\frac{1}{25}$	$\frac{3}{50}$	$\frac{1}{10}$
Colour	**White**	$\frac{9}{50}$	$\frac{7}{50}$	$\frac{3}{25}$	$\frac{1}{50}$
	Blue	$\frac{3}{50}$	$\frac{4}{25}$	$\frac{1}{25}$	0

11 At a busy road junction, the traffic lights operate as follows
red: 40 seconds   red and amber: 10 seconds   green: 40 seconds   amber: 10 seconds

Work out the probability of each event. Give your answer as a simplified fraction.

   a the lights show green [2]   **b** there is an amber light showing [3]

   c there is not a red light showing [3]   **d** either green or amber lights are showing [3]

12 a Are these pairs of events mutually exclusive?
Give reasons for your answers.

     i Red section and multiples of 3. [2]

     ii Blue section and prime numbers. [2]

     iii White section and multiples of 7. [2]

   b Grace adds 1 to every number on the spinner.
How does this change your answers to part **a**? [6]

13 During December the probability that it is snowing is 0.1. The probability that it is sunny is 0.3.
Janice says that the probability that it is snowing or sunny in December is 0.4.
Is Janice correct? You must give a reason for your answer. [2]

14 Rachel joins a regular pentagon, two squares and a triangle
to create a spinner.
Rachel assumes that the spinner is unbiased.

   a Find the probability of landing on the blue region of
the spinner. [6]

   b How does the assumption that the spinner is unbiased
affect your answer to part **a**? [2]

9 Measures and accuracy

Introduction

During the 1936 Olympic Games held in Berlin in Nazi Germany, the US athlete Jesse Owens broke the 100 m mens' world record, with a time of 10.2 s, a record that remained unbroken for 20 years afterwards. He also took the gold medal in the long jump in the same games and he is widely believed to have been the greatest athlete ever. It is said that Hitler was not impressed by Owens's victories, as he had hoped that the games would be a validation of Aryan supremacy.

The current 100 m world record is held by Usain Bolt of Jamaica, with a time of 9.58 s in 2009. Note the increased accuracy in the reporting of the time in 2009 as compared with 1936.

What's the point?

Being able to rely on accurate decimal measurements is vital in athletics. In a 'photo-finish' with two competitors crossing the finishing line apparently together, 0.01 seconds can be the difference between gold and silver!

Objectives

By the end of this chapter, you will have learned how to ...

- Use approximate values to estimate calculations.
- Use an estimate to check an answer obtained using a calculator.
- Solve problems involving speed and density.
- Look at a value that has been rounded and work out upper and lower bounds for the original value.

Chapter investigation

Most of the quantities you can measure, such as length and mass, are based on the decimal number system; however time is not. During the French Revolution, it was proposed to replace the standard system of hours and minutes with French Revolutionary Time, in which time would be measured using powers of 10.

Investigate French Revolutionary Time and write a short report.

At what times would your school day start and finish using this system?

9.1 Estimation and approximation

● You can use **approximations** to make **estimates**.

Find approximate answers to:

a $12.3 - 8.9$ **b** $76.5 + 184.2$ **c** $20 - 14.53$

a $12.3 - 8.9 \approx 12 - 9 = 3$

b $76.5 + 184.2 \approx 80 + 200 = 280$

c $20 - 14.53 \approx 20 - 15 = 5$

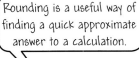

Rounding is a useful way of finding a quick approximate answer to a calculation.

You need to be careful when estimating powers.
For example, 1.3 is quite close to 1, but 1.3^7 is not close to 1^7.

Estimate the values of these calculations.

a $\dfrac{563 + 1.58}{327 - 4.72}$ **b** $\dfrac{3.27 \times 4.49}{1.78^2}$ **c** $\dfrac{\sqrt{2485}}{1.4^3}$ **d** $\dfrac{2.45^3}{2.5 - 2.4}$

Remember the order of operations, BIDMAS.

a Ignore the relatively small amounts added and subtracted.

$\dfrac{563 + 1.58}{327 - 4.72} \approx \dfrac{600}{300} \approx 2$

b 1.78^2 is 'a bit more than 3' so it cancels with 3.27.

$\dfrac{3.27 \times 4.49}{1.78^2} \approx 4.5$

c $1.4^2 \approx 2$ so $1.4^3 \approx 2.8 \approx 3$

$\dfrac{\sqrt{2485}}{1.4^3} \approx \dfrac{\sqrt{2500}}{3} \approx \dfrac{50}{3} \approx 17$

d $2.45 \approx 2.5$ and $2.5^2 \approx 6$,
so $2.5^3 \approx 6 \times 2.5 \approx 15$

$\dfrac{2.45^3}{2.5 - 2.4} \approx \dfrac{15}{0.1} \approx 150$

17.3

● You could use **standard form** to estimate calculations involving very large or very small numbers.

Estimate the value of $\dfrac{4217 \times 0.0625}{23563}$

For standard form, the multiplier must be between 1 and 10.

Writing the calculation in standard form,

$(4.217 \times 10^3) \times (6.25 \times 10^{-2}) \div (2.3563 \times 10^4)$

$\approx (4 \times 10^3) \times (6 \times 10^{-2}) \div (2 \times 10^4)$

$= 12 \times 10^{-3}$

$= 1.2 \times 10^{-2}$

$= 0.012$

You could answer this question without using standard form.

Round to 1 sf and use your knowledge of place value.

$\dfrac{4217 \times 0.0625}{23563} \approx \dfrac{4000 \times 0.06}{20000}$

$= \dfrac{4000 \times 0.06}{20000} = \dfrac{0.24}{20} = 0.012$

You might want to revisit this example when you have covered standard form in lesson **17.3**.

Exercise 9.1S

1 Round each of these numbers to the nearest
   i 3 sf   **ii** 2 sf   **iii** 1 sf.

   a 8.3728   **b** 18.82

   c 35.84   **d** 278.72

   e 1.3949   **f** 3894.79

   g 0.008 372   **h** 2399.9

   i 8.9858   **j** 14.0306

   k 1403.06   **l** 140 306

2 Write a suitable estimate for each of these calculations. In each case clearly show how you estimated your answer.

   a $4.88 + 3.07$ **b** $216 + 339$

   c $0.0049 + 0.003\,02$ **d** $43.89 - 28.83$

   e 3.77×0.85 **f** $44.66 \div 0.89$

   g 3.76×4.22 **h** 17.39×22.98

   i $\dfrac{4.59 \times 7.9}{19.86}$ **j** $54.31 \div 8.8$

3 Write a suitable estimate for each of these calculations. In each case, clearly show how you estimated your answer.

   a 4.98×6.12 **b** $17.89 + 21.91$

   c $\dfrac{5.799 \times 3.1}{8.86}$ **d** $34.8183 - 9.8$

   e $\dfrac{32.91 \times 4.8}{3.1}$

   f $\{9.8^2 + (9.2 - 0.438)\}^2$

4 Explain why approximating the numbers in these calculations to 1 significant figure would *not* be an appropriate method for estimating the results of the calculations.

   a $\dfrac{5.39 + 4.72}{0.53 - 0.46}$ **b** $(2.45 - 0.96)^8$

   c $(1.52 - 1.49)^2$

5 Estimate these square roots mentally, to 1 decimal place.

   a $\sqrt{2}$ **b** $\sqrt{8}$

   c $\sqrt{10}$ **d** $\sqrt{15}$

   e $\sqrt{20}$ **f** $\sqrt{26}$

   g $\sqrt{32}$ **h** $\sqrt{45}$

   i $\sqrt{70}$ **j** $\sqrt{85}$

   Use a calculator to check your estimates.

6 Write a suitable estimate for each of these calculations. In each case clearly show how you estimated your answer.

   a $\dfrac{29.91 \times 38.3}{3.1 \times 3.9}$

   b $\dfrac{16.2 \times 0.48}{0.23 \times 31.88}$

   c $\{4.8^2 + (4.2 - 0.238)\}^2$

   d $\dfrac{63.8 \times 1.7^2}{1.78^2}$

   e $\sqrt{(2.03 \div 0.041)}$

   f $\sqrt{(27.6 \div 0.57)}$

7 Use approximations to estimate the value of each of these calculations. You should show all your working.

   a $\dfrac{317 \times 4.22}{0.197}$ **b** $\dfrac{4.37 \times 689}{0.793}$

   c $\dfrac{4.75 \times 122}{522 \times 0.38}$ **d** $4.8^3 - 8.5^2$

   e $\dfrac{9.32 - 3.85}{0.043 - 0.021}$ **f** $7.73 \times \left(\dfrac{0.17 \times 234}{53.8 - 24.9}\right)$

8 Find approximate values for these calculations. Show your working.

   a $\dfrac{48.75 \times 4.97}{10.13^2}$ **b** $\sqrt{\dfrac{305.3^2}{913}}$

   c $\dfrac{\sqrt{9.67 \times 8.83}}{0.087}$ **d** $\dfrac{6.8^2 + 11.8^2}{\sqrt{47.8 \times 52.1}}$

   e $\dfrac{(23.4 - 18.2)^2}{3.2 + 1.8}$ **f** $\sqrt{\dfrac{2.85 + 5.91}{0.17^2}}$

9 Use approximations to find an estimate for each of these calculations. You could calculate using standard form or using decimals in your calculations. Show your working.

   a $4800 \div 465$ **b** $7326 \div 0.069$

   c $\dfrac{83\,550 \times 0.039}{4378}$ **d** $\dfrac{653 \times 0.415}{0.07 \times 0.38}$

   e $\dfrac{735 + 863}{0.06 \times 0.85}$ **f** $\dfrac{3400 \times 475}{(28.5 + 36.9)^2}$

Q 1005, 1043, 1969   SEARCH

9.1 Estimation and approximation

HOW TO

1. Read the question carefully, and decide which calculation you need to carry out.
2. Estimate using approximations.
3. Round your answer to a suitable degree of accuracy.

EXAMPLE

Monty is planting a new rectangular lawn. His lawn is 4.8 m by 5.1 m. He needs between 45 and 60 grams of seed per square metre of lawn. Seed can be bought in 800 g packets.

  a How many packets of seeds should he buy?   **b** Justify your answer to **a**.

a ① It is better to have too much than too little in this case, so round more up than down.

Area of lawn ≈ 5 × 5 ≈ 25 m²   ② Estimate by rounding to 1 sf.

② Amount of seed ≈ 25 × 60 g ≈ 1500 g   Use the larger amount.

He should buy 2 packets.

b Lowest amount = 4.8 × 5.1 × 45 = 1101 g so one packet is not enough.

Highest 25 × 60 = 1500 g, which is still less than 2 packets (1600 g).

EXAMPLE

Lewis hires a car to travel 118 miles. The hire charge for a small car is £39.98, plus 12 p per mile. A larger car costs £68.92, plus 15 p per mile. Estimate the difference in cost between the two cars.

① Small = 118 × 0.12 + 39.98   Large = 118 × 0.15 + 68.92

② Round all figures a suitable degree of accuracy.

$$≈ 120 × 0.12 + 40$$ $$≈ 120 × 0.15 + 70$$

120 × 0.12 = 14.4   120 × 0.15 = 60 × 2 × 0.15

$$≈ 14 + 40$$ $$≈ 60 × 0.3 + 70$$

$$≈ 54$$ $$≈ 18 + 70$$

③ 88 − 54 = £34 $$≈ 88$$

The large car is about £34 more.

Your answers may not match other people's. It does not necessarily mean they are wrong.

EXAMPLE

A jellybean can be thought of as a cylinder with diameter 1.5 cm and length 2 cm.

Estimate the number of jellybeans that can fit in a 1 litre jar.

① Volume of a jellybean = $\pi r^2 l$ = π × 0.75² × 2

② π ≈ 3, $0.75^2 = \left(\dfrac{3}{4}\right)^2 = \dfrac{9}{16} ≈ 0.5$

Volume of a jellybean ≈ 3 × 0.5 × 2 ≈ 3 cm³

1 litre = 1000 cm³

③ Number of jellybeans ≈ 1000 ÷ 3 ≈ 300

Exercise 9.1A

1 **a** Write a calculation that you can do in your head to estimate the answers to these calculations.

  i $258 + 362$ **ii** $64 \div 27$

  iii 62.7×211.8 **iv** $96.7 - 64.8$

  b Explain carefully whether each of the calculations that you wrote in part **a** will produce an overestimate or an underestimate of the actual result.

  c Use a calculator to find the exact answers, and check your answers to part **b**.

2 Dave is making a garden patio.
The patio is 3.2 m by 4.8 m.
The patio slabs are 0.8 m by 0.5 m.
Estimate the number of patio slabs that Dave needs.

3 Caroline has 490 Mb of space left on her camera's SD card. She is taking photographs which are 6.448 Mb in size.

  a Approximately how many can she take?

  b She can send up to 30 Mb as an attachment to an email. How many photos can she send?

4 Paul's mobile phone gives a summary of his bank account. On 28 February he had about £2000 in his account.

06/03	ATM TOWN	−£100.00
06/03	DC DIRECT DEBIT	−£156.45
06/03	DDEBIT	−£99.30
06/03	CASH TRM MAR 06	−£80.00
04/03	CHQ IN	+£84.37
02/03	ABC & G	+£59.21
01/03	SALARY	+£1758.64
28/02	BALANCE	£2058.63

  a Approximately how much was paid in during the first week of March? (+)

  b Approximately how much was paid out? (−)

  c Approximately how much money did he have in the bank on 6 March?

5 Greta is driving from London to Newcastle, a distance of 278 miles. She estimates she can drive about 60 miles every hour. Her diesel car does 43 miles on one gallon of diesel.

  a Estimate how long the journey takes.

  b She thinks 1 gallon is about 5 litres. Estimate how many litres she thinks she will use.

  c One litre of diesel costs £1.29. Approximately what will the journey from London to Newcastle cost Greta?

  d In fact 1 gallon = 4.54609 litres. Will Greta use more or less diesel than she estimated?

6 A grain of rice can be thought of as a cylinder with length 8 mm and diameter 2 mm.
A 1 kg bag of rice measures
18 cm × 12 cm × 6 cm.

  a Estimate the number of grains of rice in a 1 kg bag.

  b The weight of 1000 grains of rice is approximately 22 grams. Does your estimate in part **a** agree with this statement?

7 The Moon is 3.844×10^5 km from the Earth. The average human walking speed is 5 km per hour. A person in good health can walk for 8 hours a day, for 6 days a week. Estimate how many years it would take to walk to the Moon.

*8 An average adult man weighs 75 kg.
Oxygen, carbon and hydrogen account for 99% of the composition of the human body. Two-thirds of the human body is oxygen, about 20% is carbon and about 10% is hydrogen. The mass of one atom of these elements is shown in the table.

Element	Mass of one atom (kg)
Hydrogen	1.673×10^{-27}
Carbon	1.994×10^{-26}
Oxygen	2.656×10^{-26}

By making a series of estimates, show that there are approximately 7×10^{27} atoms in the human body.

9.2 Calculator methods

You can use a scientific calculator to carry out more complex calculations that involve decimals.

EXAMPLE

Use your calculator to work out this calculation.

$$3.46 + 2.9 \times 4.8$$

Give your answer to 1 decimal place.

Estimate

$3.46 + 2.9 \times 4.8 \approx 3 + 3 \times 5$ (using the order of operations)

$= 3 + 15 = 18$

You type ⟨3⟩⟨.⟩⟨4⟩⟨6⟩⟨+⟩⟨2⟩⟨.⟩⟨9⟩⟨×⟩⟨4⟩⟨.⟩⟨8⟩⟨=⟩

The calculator should display

3.46+2.9×4.8
17.38

So the answer is   17.38 = 17.4   (1 decimal place)

> A scientific calculator has algebraic logic, which means that it understands the **order of operations**.

> In this case the calculator automatically works out the multiplication before the addition.

A scientific calculator has **bracket** keys.

EXAMPLE

a   Use your calculator to work out $(3.9 + 2.2)^2 \times 2.17$.

Write all the figures on your calculator display.

a   Estimate

$(3.9 + 2.2)^2 \times 2.17 \approx (4 + 2)^2 \times 2$ (using the order of operations)

$= 6^2 \times 2$

$= 36 \times 2$

$= 72$

You type ⟨(⟩⟨3⟩⟨.⟩⟨9⟩⟨+⟩⟨2⟩⟨.⟩⟨2⟩⟨)⟩⟨x^2⟩⟨×⟩⟨2⟩⟨.⟩⟨1⟩⟨7⟩⟨=⟩

The calculator should display

(3.9+2.2)2×2.17
80.7457
 = 80.7457

You can use the bracket keys on a scientific calculator to do calculations where the **order of operations** is not obvious.

EXAMPLE

a   Use a calculator to work out the value of

$$\frac{21.42 \times (12.4 - 6.35)}{(63.4 + 18.9) \times 2.83}$$

Write all the figures on the calculator display.

a   Rewrite the calculation as $(21.42 \times (12.4 - 6.35)) \div ((63.4 + 18.9) \times 2.83)$

Type this into the calculator:

(21.42×(12.4−6.35))÷((63.4+18.9)×2.83)	⟶	(21.42×(12.4▪
		0.556401856

So the answer is 0.556 401 856.

Estimate: $\dfrac{20 \times (12 - 6)}{(60 + 20) \times 3} = \dfrac{120}{240} = 0.5$

Exercise 9.2S

1 Use your calculator to work out these. In each case, first write an estimate for your answer.
Write the answer from your calculator to 1 decimal place.

   a $3.4 + 6.2 \times 2.7$

Estimate	$3.4 + 6.2 \times 2.7 \approx 3 + 6 \times 3$
	$= 3 + 18$
	$= \underline{\quad\quad}$
Using the calculator	$3.4 + 6.2 \times 2.7 = \underline{\quad\quad}$

   b $1.98 \times 11.7 - 4.6$

   c $7.8 + 19.3 \div 4.12$

   d $2.09 \times 2.87 + 3.25 \times 1.17$

   e $13.67 \div 1.75 + 3.24$

   f $1.2 + 3.7 \times 0.5$

2 Use your calculator to work out each of these calculations.
Write all the figures on your calculator display.

   a $(2.3 + 5.6) \times 3^2$

   b $2.3^2 \times (12.3 - 6.7)$

   c $(2.8^2 - 2.04) \div 2.79$

   d $7.2 \times (4.3^2 + 7.4)$

   e $11.33 \div (6.2 + 8.3^2)$

   f $(2.5^2 + 1.37) \times 2.5$

3 Calculate each of these, giving your answer to 1 decimal place.

   a $\dfrac{5.4 + 3.8}{4.5 - 2.9}$ **b** $\dfrac{3.8 - 1.67}{4.3 - 2.68}$

   c $\dfrac{12.4 + 5.8}{14.5 - 3.9}$ **d** $\dfrac{13.08 - 2.67}{2.13 + 2.68}$

4 Use a calculator to evaluate these, giving each answer to 2 sf.

   a $3.2 \times (2.8 - 1.05)$

   b $2.8^2 \times (9.4 - 0.083)$

   c $16 \div (5.1^2 \times 7.2)$

   d $(3.8 + 8.9) \times (2.2^2 - 7.6)$

   e $1.8^3 + 4.7^3$

   f $52 \div (4.6 - 1.8^2)$

5 Use your calculator to work out each of these. Write all the figures on your calculator display.

   a $\dfrac{462.3 \times 30.4}{(0.7 + 4.8)^2}$

   b $\dfrac{13.58 \times (18.4 - 9.73)}{(37.2 + 24.6) \times 4.2}$

6 Use your calculator to work out each of these. Write all the figures on your calculator.

   a $\dfrac{165.4 \times 27.4}{(0.72 + 4.32)^2}$

   b $\dfrac{(32.6 + 43.1) \times 2.3^2}{173.7 \times (13.5 - 1.78)}$

   c $\dfrac{24.67 \times (35.3 - 8.29)}{(28.2 + 34.7) \times 3.3}$

   d $\dfrac{1.45^2 \times 3.64 + 2.9}{3.47 - 0.32}$

   e $\dfrac{12.93 \times (33.2 - 8.34)}{(61.3 + 34.5) \times 2.9}$

   f $\dfrac{24.7 - (3.2 + 1.09)^2}{2.78^2 + 12.9 \times 3}$

7 Use your calculator to work out each of these. Write all the figures on your calculator.

   a $\dfrac{317 \times 4.22}{0.197}$ **b** $\dfrac{4.37 \times 689}{0.793}$

   c $\dfrac{4.75 \times 122}{522 \times 0.38}$ **d** $4.8^3 - 8.5^2$

   e $\dfrac{9.32 - 3.85}{0.043 - 0.021}$ **f** $7.73 \times \left(\dfrac{0.17 \times 234}{53.8 - 24.9}\right)$

8 Use your calculator to work out each of these. Write all the figures on your calculator.

   a $\dfrac{48.75 \times 4.97}{10.13^2}$ **b** $\sqrt{\dfrac{305.3^2}{913}}$

   c $\dfrac{\sqrt{9.67 \times 8.83}}{0.087}$ **d** $\dfrac{6.8^2 + 11.8^2}{\sqrt{47.8 \times 52.1}}$

   e $\dfrac{(23.4 - 18.2)^2}{3.2 + 1.8}$ **f** $\sqrt{\dfrac{2.85 + 5.91}{0.17^2}}$

9 How do your answers to questions **7** and **8** compare to the estimates that you made in Exercise 9.1S question **7** and **8**?

Q 1043, 1932, 1933   SEARCH

9.2 Calculator methods

RECAP

- Use your calculator to work out more complex calculations.
- Scientific calculators apply the order of operations.

You can toggle between fractions and decimal answers on your calculator. Try this for yourself using your calculator.

HOW TO

① Take time to understand the situation in the question. Estimate the answer before calculating.

② Use your calculator to work out the answer. Check that the answer agrees with your estimate.

③ Interpret the calculator display in the context of the question.

EXAMPLE

Put brackets in this expression so that its value is 16.26.

$$1.4 + 3.9 \times 2.2 + 4.6$$

① Estimate the answer by rounding to 1 sf.
Put brackets in the expression $1 + 4 \times 2 + 5$ so that the answer is roughly 16.

$(1 + 4) \times 2 + 5 = 5 \times 2 + 5 = 10 + 5 = 15$

② Type $(1.4 + 3.9) \times 2.2 + 4.6$ into your calculator.

$(1.4 + 3.9) \times 2.2 + 4.6$

③ The calculator should display

```
(1.4+3.9)×2.2+4.6
              16.26
```

$= 16.26$ ✓ the correct answer

EXAMPLE

The diagram shows a box in the shape of a cuboid.

Saleem builds boxes of different sizes. He charges £7.89 for each m³ of a box's volume. Work out Saleem's charge for building this box.

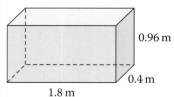

0.96 m

0.4 m

1.8 m

① Make an estimate.

Volume $= 1.8 \times 0.4 \times 0.96 = 2 \times 0.4 \times 1 = 0.8$

So the box should cost about £6.

② Saleem's charge $= 7.89 \times 1.8 \times 0.4 \times 0.96$

$= £5.4535668$

$= £5.45$

③ Interpret the display as pounds and pence.

EXAMPLE

Joseph books 4 tickets online for a show, at £25.90 each. He then finds out he has to pay a £2.90 booking fee per ticket, and a 'single transaction fee' of £2.85.

a How much does he pay altogether? **b** How much does he pay in extra fees?

There are lots of ways of working this out. One using brackets is given here.

a $(25.90 + 2.90) \times 4 + 2.85 = 118.05$

The total cost is £118.05.

b $2.90 \times 4 + 2.85 = 14.45$

He pays £14.45 in extra fees.

Exercise 9.2A

1 Put brackets into each of these expressions to make them correct.

   a $2.4 \times 4.3 + 3.7 = 19.2$

   b $6.8 \times 3.75 - 2.64 = 7.548$

   c $3.7 + 2.9 \div 1.2 = 5.5$

   d $2.3 + 3.4^2 \times 2.7 = 37.422$

   e $5.3 + 3.9 \times 3.2 + 1.6 = 24.02$

   f $3.2 + 6.4 \times 4.3 + 2.5 = 46.72$

2 Put brackets into each of these expressions to make them correct.

   a $3.4 \times 2.3 + 1.6 = 13.26$

   b $3.5 \times 2.3 - 1.04 = 4.41$

   c $2.6 + 6.5 \div 1.3 = 7$

   d $1.4^2 - 1.2 \times 2.3 = 1.748$

   e $2.4^2 \div 1.8 \times 3.2 + 1.6 = 15.36$

   f $3.2 + 5.3 \times 2.4 - 1.2 = 10.2$

3 Work out each of these using your calculator. In each case give your answer to an appropriate degree of accuracy.

   a   Véronique puts carpet in her bedroom. The bedroom is in the shape of a rectangle with a length of 4.23 m and a width of 3.6 m. The carpet costs £6.79 per m².

      i   Calculate the floor area of the bedroom.

      ii   Calculate the cost of the carpet which is required to cover the floor.

   b   Calculate $\frac{1}{3}$ of £200.

4 Elif has 110.5 hours of music downloaded on her computer.
She accidentally deletes 80.15 hours of her music!

How much music does Elif have remaining on her computer hard drive?

Give your answer

   a   in hours and minutes

   b   in minutes.

5 Barry sees a mobile phone offer.

> **Vericheep Fone OFFER**
> Monthly fee £12.99
> FREE – 200 texts every month
> FREE – 200 voice minutes every month
> Extra text messages 3.2p each
> Extra voice minutes 5.5p each

Barry decides to see if the offer is a good idea for him.

His current mobile phone offers him unlimited texts and voice minutes for £22.99 per month.

   a   In February, Barry used 189 texts and 348 voice minutes.
Calculate his bill using the new offer.

   b   In March, Barry used 273 texts and 219 voice minutes.
Calculate his bill using the new offer.

   c   Explain if the new offer is a good idea for Barry.

6 Aliona books 8 tickets online for a show, at £19.85 each. She then finds out she has to pay a £3.25 booking fee per ticket, and a 'single transaction fee' of £2.20.

   a   How much does she pay altogether?

   b   How much does she pay in extra fees?

*7 **a**   The speed of sound depends, amongst other things, on the temperature of the air it travels through. If t stands for temperature in degrees Celsius, then the speed of sound, in feet per second, is

$$\sqrt{(273 + t)} \times 1087 \div 16.52$$

      i   What is the speed at 0 °C?

      ii   What is the speed at 18 °C?

   b   To convert feet per second to miles per hour, you multiply by 15, then divide by 22. Convert your answers to miles per hour.

9.3 Measures and accuracy

- You can measure **length**, **mass** and **capacity** using metric units.

A paper clip weighs 1 gram.

- Length is a measure of distance.

$10\,mm = 1\,cm$ $100\,cm = 1\,m$ $1000\,m = 1\,km$

- Mass is a measure of the amount of matter in an object. Mass is linked to weight.

$1000\,g = 1\,kg$ $1000\,kg = 1$ tonne

- Capacity measures the amount of liquid that a 3D shape holds.

$1000\,ml = 1$ litre $100\,cl = 1$ litre $1000\,cm^3 = 1$ litre

Compound measures describe one quantity in relation to another. These are examples of compound measures.

- **Speed** $= \dfrac{\text{total distance travelled}}{\text{total time taken}}$ Units such as m/s, km/h

- **Density** $= \dfrac{\text{mass}}{\text{volume}}$ Units such as g/cm^3, kg/m^2

Use the triangle to work out which calculation to use.

$M = D \times V$
$D = M \div V$
$V = M \div D$

EXAMPLE

Find the density of a piece of wood with cross-section area $42\,cm^2$, length $12\,cm$ and mass $693\,g$.

Volume = 42 × 12 = 504 cm³
Density = mass ÷ volume
Density = 693 ÷ 504
        = 1.375 g/cm³

- Measurements are not exact. Their accuracy depends on the precision of the measuring instrument and the skill of the person making the measurement.

EXAMPLE

The mass of a meteorite is given as $235.6\,g$. Find the lower and upper bounds of the mass.

For continuous data, the upper bound is *not* a possible value of the data.

The mass is given correct to the nearest 0.1 g.
Lower bound is 235.55 g. Upper bound is 235.65 g.
If the mass of the meteorite is m then
    $235.55 \leq m < 235.65$

$2.3\,cm$ has an **implied accuracy** of 1 dp.

3.50 m has an implied accuracy of 2 dp.

If all measurements are to 1 dp, give your answers to 1 dp.

1.1

Exercise 9.3S

1 Choose one of these metric units to measure each of these items.

millimetre	gram	millilitre	centimetre
kilogram	centilitre	metre	tonne
litre	kilometre		

   a your height

   b amount of tea in a mug

   c your weight

   d length of a suitcase

   e weight of a suitcase

   f distance from Paris to Madrid

   g quantity of drink in a can

   h amount of petrol in a car

   i weight of an elephant

   j weight of an apple.

Write the appropriate abbreviation next to your answers.

2 Convert these measurements to the units shown.

   a 20 mm = ___ cm   **b** 400 cm = ___ m

   c 450 cm = ___ m   **d** 4000 m = ___ km

   e 0.5 cm = ___ mm   **f** 4.5 kg = ___ g

   g 6000 g = ___ kg   **h** 6500 g = ___ kg

   i 2500 kg = ___ t   **j** 3 litres = ___ ml

3 Cheryl takes 4 hours to walk 10 miles. Calculate her average speed, giving the units of your answer.

4 A cyclist travels at 15 mph for $2\frac{1}{2}$ hours. How many miles is the journey?

5 I can usually drive at an average speed of 60 mph on the motorway. How long will a 150-mile journey take?

6 A 420 km journey by car takes 6 hours and uses 30 litres of petrol. Calculate

   a the average speed

   b the petrol consumption in km per litre.

7 An athlete runs 1500 m in 4.5 mins. Calculate the athlete's speed in

   a metres per minute

   b metres per second.

8 A cube of side 2 cm weighs 40 grams.

   a Find the density of the material from which the cube is made, giving your answer in g/cm^3.

   b A cube of side length 2.6 cm is made from the same material. Find the mass of this cube, in grams.

9 A type of emulsion paint has a density of 1.95 kg/litre. Find

   a the mass of 4.85 litres of the paint

   b the number of litres of the paint that would have a mass of 12 kg.

10 Each of these measurements was made correct to one decimal place. Write the upper and lower bounds for each measurement.

   a 5.8 m   **b** 16.5 litres

   c 0.9 kg   **d** 6.3 N

   e 10.1 s   **f** 104.7 cm

   g 16.0 km   **h** 9.3 m/s

11 Find the upper and lower bounds of these measurements, which were made to varying degrees of accuracy.

   a 6.7 m   **b** 7.74 litres

   c 0.813 kg   **d** 6 N

   e 0.001 s   **f** 2.54 cm

   g 1.162 km   **h** 15 m/s

12 Find the maximum and minimum possible total weight of

   a 12 boxes, each of which weighs 14 kg, to the nearest kilogram

   b 8 parcels, each weighing 3.5 kg.

13 Find the upper and lower bounds of these measurements, which are correct to the nearest 5 mm.

   a 35 mm   **b** 40 mm

   c 110 mm   **d** 45 mm

1006, 1067, 1121, 1246, 1968   SEARCH

9.3 Measures and accuracy

RECAP

- You can measure **length**, **mass** and **capacity** using metric units.
- A compound measure, such as speed or density, connects two different measurements.
- The upper bound is the smallest value that is greater than or equal to any possible value of the measurement.
- The lower bound is the greatest value that is smaller than or equal to any possible value of the measurement.

$$\text{Speed} = \frac{\text{distance}}{\text{time}}$$

$$\text{Density} = \frac{\text{mass}}{\text{volume}}$$

HOW TO

(1) Read the question carefully and calculate using your knowledge of measures.

(2) Include units in your answer and give your answer to an appropriate degree of accuracy.

EXAMPLE

A measurement is given as 3.8 cm, correct to the nearest 0.1 cm.

Pritesh writes:

Lower bound = 3.75 cm

Upper bound is a bit less than 3.85 cm, say 3.849 99

Is he correct?

(1) The lower bound is correct; 3.75 is the largest value that is less than or equal to any possible value of the measurement.

(2) The upper bound is incorrect. It should be 3.85.

EXAMPLE

A pile of 16 sheets of card is 7.3 mm thick.

Calculate the thickness of one sheet.

(1) 7.3 mm ÷ 16 = 0.456 25 mm = 0.46 mm (2 sf)

(2) 0.456 25 mm implies that you could measure the thickness of a piece of card to the nearest 0.000 01 mm — unlikely!

EXAMPLE

A model boat travels 3.9 metres in 7.3 seconds. Both measurements are correct to 1 dp. Find the upper and lower bounds of the speed of the boat in metres per second.

(1) 3.9 m is a rounded value in the range 3.85 m to 3.95 m, and 7.3 s is in the range 7.25 s to 7.35 s.

(2) To get the maximum value of the speed, divide the largest possible distance by the shortest possible time:

The upper bound of the speed is
3.95 ÷ 7.25 = 0.544 827 ... metres per second.

To get the minimum value of the speed, divide the smallest possible distance by the greatest possible time.

The lower bound of the speed is
3.85 ÷ 7.35 = 0.523 809 ... metres per second.

$$\text{Speed}_{Max} = \text{Distance}_{Max} \div \text{Time}_{Min}$$

$$\text{Speed}_{Min} = \text{Distance}_{Min} \div \text{Time}_{Max}$$

Exercise 9.3A

1 Work these out, giving your answers to a suitable degree of accuracy.

   a The total weight of two people weighing 68 kg and 73 kg.

   b The area of a rectangle with sides 2.2 m and 3.8 m.

   c The cost of 2.37 m of material at £5.75 per metre.

   d The time taken to travel 1 mile at a speed of 80 mph.

2 A completed jigsaw puzzle measures 24 cm by 21 cm (both to the nearest centimetre). Find the maximum and minimum possible area of the puzzle. Show your working.

3 A bag contains 98 g of sugar; then 34 g of sugar are taken out. Find the upper and lower bounds for the amount of sugar remaining in the bag, if the measurements are all correct to the nearest gram.

4 This trapezium has area 450 cm² to 2 sf. It has parallel sides 18 cm and 22 cm, each to the nearest centimetre.

Calculate the lower bound of the height, h, of the trapezium.

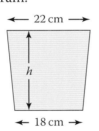

22 cm

h

18 cm

5 Unladen, a lorry weighs 2.3 tonnes, measured to the nearest 100 kg. The lorry is loaded with crates, each weighing 250 kg, correct to the nearest 10 kg. On its journey the lorry crosses a bridge with a maximum safe load of 15 tonnes. What is the maximum number of crates that the driver can safely load onto the lorry?

6 The maximum load a van can carry is 450 kg. The van is used to carry boxes that weigh 30 kg to the nearest 1 kg.
Find the maximum number of boxes that the van can safely carry. Show your working.

7 A crane has a maximum working load of 670 kg to 2 sf. It is used to lift crates that weigh 85 kg, to the nearest 5 kg.
What is the greatest number of crates that the crane can safely lift at one time?

8 A lift can carry a maximum of five people, and the total load must not exceed 440 kg. Five members of a judo team enter the lift. Each person weighs 87 kg to the nearest kilogram. Is it possible that the total weight of the group exceeds the maximum load of the lift? Show your working.

9 The length of Eva's stride is 86 cm, to the nearest centimetre.

   a Write the upper and lower bounds of Eva's stride.

   b The length of a path is 28 m, to the nearest metre.

   Starting at the beginning of the path, Eva takes 32 strides in a straight line along the path. Explain, showing all your working, why Eva may not reach the end of the path.

10 A model car was rolled down a track. The length of the track was measured as 2.55 m, to the nearest centimetre. The time for the journey was 1.7 seconds, measured to the nearest tenth of a second.

Find the upper and lower bound of the speed of the model car in metres per second.

11 A metal rod in the shape of a cuboid is measured to weigh 3925 kg. The dimensions of the rod are measured as 25 cm × 25 cm × 8.0 m.

Marina is trying to decide what metal the rod is made from. The densities of five different possible metals are shown in the table.

Metal	Density kg/m³
Titanium	4500
Tin	7280
Steel	7850
Brass	8525
Copper	8940

Marina claims that the rod must be made of steel.
Do you agree?
Explain your answer fully.

Summary

Checkout
You should now be able to...

	Test it Questions
✓ Use approximate values obtained by rounding to estimate calculations.	**1, 2**
✓ Use an estimate to check an answer obtained using a calculator.	**2**
✓ Use, and convert between, standard units of length, mass, capacity and other measures including compound measures.	**3 – 7**
✓ Solve problems involving compound measure such as speed and density.	**6, 7**
✓ Find upper and lower bounds on the value of a quantity that has been rounded.	**8, 9**
✓ Find upper and lower bounds on expressions that involve quantities that have been rounded.	**10, 11**

Language	Meaning	Example
Approximation	A stated value of a number that is close to but not equal to the true value of a number. Usually obtained by rounding.	$\pi = 3.141592654...$ $\quad = 3.1$ (1 dp) $\quad = 3 \quad$ (1 sf)
Estimate	A simplified calculation based on approximate values or a judgement of a quantity.	$1.9^2 \times \pi \simeq 2^2 \times 3 \simeq 12$ Exact $= 11.34114....$
Length	Length is a measure of linear extent.	A standard ruler is 30 cm long.
Mass	A measure of the amount of matter in an object. Weight measures the force of gravity acting on a mass.	The mass of a bag of sugar is 1 kg.
Volume	A measure of the amount of 3D space occupied by an object.	Volume = 2000 cm³ Capacity = 2 litres
Capacity	A measure of the amount of fluid that a 3D shape will hold.	
Speed	A measure of the distance travelled in a unit of time.	The speed of the car is 65 km/h.
Density	A measure of the amount of mass in a unit of volume.	The density of iron is 7.87 g/cm³.
Accuracy	How close a measured or calculated quantity is to the true value.	$1.239648 = 1.2$ to 1 dp $\quad\quad\quad\quad = 1.24$ to 2 dp
Implied accuracy	The accuracy of a value implied by the number of significant figures or decimal places given.	1.3 has implied accuracy of 1 dp. 1.30 has implied accuracy of 2 dp
Upper bound	The largest value a quantity can take.	$1.45 \leqslant 1.5$ (1dp) < 1.55 ↗ ↗ Lower bound Upper bound
Lower bound	The lowest value a quantity can take.	
Error interval	The range of values between the lower and upper bounds.	

Review

1 Round each of these numbers to

    i 2 decimal places

    ii 2 significant figures.

    a 93021   **b** 27.941

    c 0.00625   **d** 0.895

2 **a** Estimate the answers to these calculations.

    i $3.8 + 27.3 \times 2.1$

    ii $\dfrac{9.81^2 + 112}{\sqrt{3.7}}$

    iii $\dfrac{2.2 + 0.6 \times 7.1}{11.2 - 3.6^2}$

    iv $\sqrt{\dfrac{43 + 67}{7.6 - 3.4}} - 2.9$

    b Find the answers to the calculations in part **a** using a calculator. Give your answers to 2 decimal places.

3 State two metric units for each of these quantities.

    a length   **b** mass

    c volume   **d** capacity.

4 Convert these measurements to litres.

    a 672 cl   **b** 205 ml

    c $3500\,\text{cm}^3$

5 Daryl took 97 minutes to complete his homework. He finished at 21:52, at what time did he start?

6 Scott leaves Kent town at 13:52. He travels at 15 m/s to Clare Valley which is 137.7 km away. When does he arrive in Clare Valley?

7 A block of silver has mass 3.147 kg and density 10.49 g/cm³. Find its volume.

8 For each of these measurements give its

    i upper bound   **ii** lower bound.

    a 12.5 cm (1 dp)

    b 11.5 kg (3 sf)

    c 1.00 m (2 dp)

    d 0.025 km (2 sf)

9 Hugh cuts a piece of wood of length x which is measured as 85.0 cm to the nearest millimetre. Write an inequality to describe the range of values x could take.

10 $X = 10$ (1 sf) $Y = 3.5$ (1 dp)

Find the maximum and minimum values of these expressions.

    a $X - Y$ **b** $\dfrac{Y}{X}$

    c X^2 **d** $\sqrt{X + Y}$

11 The triangle shown below has been measured to the nearest cm.

    a What is the lower bound for the area of the triangle?

    b What is the upper bound for the area of the triangle?

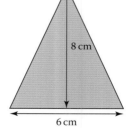

8 cm

6 cm

What next?

Score		
0 – 4		Your knowledge of this topic is still developing.
		To improve look at MyMaths: 1005, 1006, 1043, 1057, 1067, 1121, 1246, 1736, 1737
5 – 9		You are gaining a secure knowledge of this topic.
		To improve look at InvisiPens: 09Sa – f
10 – 11		You have mastered these skills. Well done you are ready to progress!
		To develop your exam technique looks at InvisiPens: 09Aa – d

Assessment 9

1 David makes the following incorrect statements. Rewrite the statements correctly.

   a 3.234 to 3 sf is 3.234   [1]   **b** 29.482 to 2 sf is 29.48   [1]

   c 0.203762 to 1 sf is 0   [1]   **d** 312 to the nearest 10 is 31   [1]

   e 5678 to the nearest 100 is 5600   [1]

   f 256 133 to the nearest 1000 is 256   [1]

2 Ahmed, Bart and Christian make these estimates for the following calculations. For each calculation, decide who makes the best estimate.

		Ahmed	Bart	Christian	
a	5.8×6.5	30	36	40	[2]
b	20.2×5.6	100	110	120	[2]
c	7.9×8.8	56	66	72	[2]
d	$6.6 \div 1.3$	5	6	9	[2]
e	$7.9 \div 0.91$	7	8	9	[2]

3 On average, a person blinks about 6 times per minute. Esther says that a good estimate for the number of times a person blinks in one day is 9000. Use your calculator to explain how she got that estimate.   [2]

4 Estimate the square root of 186. Give your answer to the nearest integer.   [2]

5 **a** Patrick estimates the value of the following calculation to be $1\frac{1}{3}$, find his mistakes and give a sensible estimation of the value.

$$\frac{5.8^2 - 1.5 \times 1.8^2}{4.59 + 9.21} \approx \frac{6^2 - 2 \times 2^2}{5 + 10} = \frac{36 - 16}{15} = 1\frac{1}{3}$$ [3]

   b Seth estimates that $\dfrac{28.9 + \sqrt{0.51}}{(25.2 - 14.7)^2} \approx 0.0775$

      Use your calculator to find the difference between the exact value and Seth's estimate. Say why this difference is large.   [2]

6 At a wedding reception, guests ate from a chocolate fountain. Everyone ate some chocolate and none was left over! The chocolate in the fountain weighed 3.56 kg and each person ate, on average, 88 g of chocolate. Estimate the number of people at the reception.   [2]

7 The diagram shows the plan of a garden in the shape of a rectangle 19.5 m by 10.5 m. Four identical triangular flowerbeds have been cut from the corners, each with shorter edges 5.6 m and 2.8 m. There is a square fishpond, side length 6.6 m, in the centre of the garden. The rest of the garden is laid to lawn.

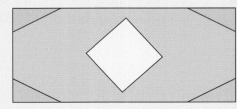

Estimate the percentage of the garden that is taken up by the lawn. Give your answers to the nearest 1%.   [6]

8 Put brackets into each of the expressions to make them correct.

   a $4.6 + 5.3 \times 2.6 = 18.38$ [1]

   b $14.9 - 6.8 \div 2.5 = 12.18$ [1]

   c $3.4 \times 1.6 + 5.9 - 2.8 = 8.54$ [2]

   d $2.6 + 7.56 \div 1.8 - 0.72 = 6.08$ [2]

   e $12.3 - 5.2 \times 1.6 + 3.4 \times 2 = 14.76$ [2]

9

tonne	kilogram	gram	kilometre	metre	centimetre	millimetre	litre	centilitre	millilitre

Estimate the following values using an appropriate unit from the list in each case.

a The capacity of a paddling pool. [1]

b The mass of a full suitcase going on holiday. [1]

c The mass of an eyeliner pencil. [1]

d The distance from your classroom to the headteacher's room. [1]

e The length of a rugby pitch. [1]

f The mass of an aeroplane. [1]

g The amount in a dose of cough mixture. [1]

h The thickness of a DVD case. [1]

i The amount of coffee consumed in a café during a day. [1]

j The distance run in a marathon race. [1]

k The length of a hairbrush. [1]

l The amount of tea in one cup. [1]

10 a Bradley cycles 165 km in $2\frac{1}{2}$ hours. What is his average speed? [2]

b Lewis' racing car travels 9500 metres in 3.8 minutes.
What is its average speed in km/h? [2]

c A train travels for $2\frac{3}{4}$ hours at an average speed of 175 km/h. How far does it go? [2]

d An aeroplane travels for $7\frac{1}{3}$ hours at an average speed of 465 mph.
How many miles does it fly? [2]

e A snail travels 14 m at 47 m/h. How long does his journey take? [2]

11 a Water has a density of 1 g/cm³. A cricket ball has a mass of 158 grams and
volume of 195 cm³. Would it float in water? Explain your answer. [3]

b A gold bracelet has a mass of 44 g and a density of 19.3 g/cm³.
What is its volume? [2]

c A concrete girder measuring 25 m by 15 m by 6 m has a density of 2.4 g/cm³.
What is the mass of the bar in tonnes? [3]

12 Write down the upper and lower limits of the following measurements:

a The shortest distance from the moon to the earth is 221 460 miles to 5 sf. [1]

b Farakh's leg is 90 cm long to the nearest 10 cm. [1]

c A box of chocolates weighs 455 g to the nearest 5 g. [1]

d In July 2013 Mo Farah beat Steve Cram's 28 year old British 1500 m record
with a time of 3 minutes 28.8 s to the nearest 0.1 s. [1]

e The number of tea bags in a box is 240 to the nearest 4 bags. [1]

f Sami is 28 years old to the nearest year. [1]

g A walking stick has a length of 586 mm to the nearest mm. [1]

13 A rectangular tray measures 45 cm by 24 cm by 0.3 cm correct to the nearest cm.
Its mass is 76 g correct to the nearest gram.

a Calculate the biggest and smallest possible values of its volume. [2]

b Calculate the biggest and smallest possible values of its density. [5]

10 Equations and inequalities

Introduction

Maths is not all about whether something is equal to something else. Sometimes it can be useful to know when a quantity needs to be less than or more than a particular value. An example could be a recipe for a soft drink, where the proportion of cane sugar needs to be within certain bounds to conform to standards and to customers' appetites. Some companies are very secretive about the formula for their particular branded soft drink. All you are allowed to know is that the proportions of ingredients lie within a certain range.

Inequalities are statements that allow us to work mathematically with this kind of restriction.

What's the point?

In the food and drink industries it is important to be able to work within restrictions on particular ingredients, which are often imposed by health legislation. In the pharmaceutical industry, it can be critical as too much of one particular chemical can have harmful effects.

Objectives

By the end of this chapter, you will have learned how to …

- Solve linear equations including when the unknown appears on both sides.
- Solve quadratic equations using factorisation, completing the square and the quadratic formula.
- Solve a pair of linear or linear plus quadratic simultaneous equations.
- Use iterative processes to find approximate solutions to equations.
- Solve inequalities and display your solution on a number line or a graph.

Check in

1 Evaluate these, expressing your answers in their simplest form.

    a $\frac{12}{30}$ **b** $\frac{1}{5} + \frac{2}{9}$ **c** $\frac{3}{4} - \frac{1}{4}$ **d** $2\frac{1}{6} + 3\frac{2}{5}$

    e $\frac{5}{6} \times \frac{7}{10}$ **f** $\frac{5}{8} \div \frac{1}{5}$

2 What value must the □ represent in each case?

    a $\square + 3 = 12$ **b** $\square \times 5 = 20$ **c** $\square - 11 = 19$ **d** $\square^2 = 100$

3 Rewrite these expressions, by expanding any brackets and collecting like terms.

    a $5x + 7y - 4x + 9y$ **b** $11x + 9x^2 - 8x$ **c** $7p - 9$

    d $3(2x - 1) + 6(3x - 4)$ **e** $2(4y - 8) - 3(y - 9)$ **f** $x(3x - 2y)$

    g $(x + 9)(x - 7)$ **h** $(2w - 8)(3w - 4)$ **i** $(p - q)^2$

4 Factorise these expressions into double brackets.

    a $x^2 + 5x + 6$ **b** $x^2 - 2x - 24$ **c** $x^2 - 6x + 9$ **d** $x^2 - 100$

Chapter investigation

This L shape is drawn on a 10×10 grid numbered from 1 to 100. It has 5 numbers inside it. We can call it L_{35} because the largest number inside it is 35.

What is the sum total of the numbers inside L_{35}?

Find a connection between n and L_n. Can you write a formula to describe the relationship? What if the L shape is extended in each direction?

1	2	3	4	5	6	7	8	9	10
11	12	13	14	15	16	17	18	19	20
21	22	23	24	25	26	27	28	29	30
31	32	33	34	35	36	37	38	39	40
41	42	43	44	45	46	47	48	49	50
51	52	53	54	55	56	57	58	59	60
61	62	63	64	65	66	67	68	69	70
71	72	73	74	75	76	77	78	79	80
81	82	83	84	85	86	87	88	89	90
91	92	93	94	95	96	97	98	99	100

10.1 Solving linear equations

- An **equation** is a statement with an equals sign.
- To **solve** an equation, do the same **operation** to both sides.

> Undo the operations in reverse order.

EXAMPLE

Solve these equations.

a $\dfrac{3x - 5}{2} = 8$

b $5 - 8x = 45$

a $\dfrac{3x - 5}{2} = 8$ Multiply both sides by 2.

$3x - 5 = 16$ Add 5 to both sides.

$3x = 21$ Divide both sides by 3.

$x = 7$

b $5 - 8x = 45$ Add $8x$ to both sides.

$5 = 45 + 8x$ Subtract 45.

$-40 = 8x$ Divide both sides by 8.

$x = -5$

- When solving an equation with brackets, **expand** the brackets first.

- To solve an equation with unknowns on both sides, start by subtracting the smaller unknown from both sides.

EXAMPLE

Solve $4(x + 2) = 7(x - 2)$

$4(x + 2) = 7(x - 1)$ Expand the brackets.

$4x + 8 = 7x - 7$ Subtract $4x$ from both sides.

$8 = 3x - 7$ Add 7 to both sides.

$15 = 3x$ Divide both sides by 3.

$x = 5$

- To **solve equations** involving fractions, first clear the fractions by multiplying both sides of the equation by the denominator.

For equations with fractions on both sides, multiply both sides of the equation by the product of the denominators.

EXAMPLE

Solve $\dfrac{2x - 1}{5} = \dfrac{3x - 4}{8}$

$\dfrac{2x - 1}{5} = \dfrac{3x - 4}{8}$ Multiply both sides by 8×5.

$8 \times 5 \times \dfrac{2x - 1}{5} = 8 \times 5 \times \dfrac{3x - 4}{8}$ Cancel any common factors.

$8(2x - 1) = 5(3x - 4)$ Expand the brackets.

$16x - 8 = 15x - 20$ Subtract $15x$ from both sides.

$x - 8 = -20$ Add 8 to both sides.

$x = -12$

> This method is called 'cross-multiplying'.

Algebra   Equations and inequalities

Exercise 10.1S

1 Solve these equations.

   a $c + 13 = 21$ **b** $d - 6 = 35$

   c $7f = 63$ **d** $\dfrac{g}{4} = 6$

2 Solve these equations.

   a $3a - 5 = 25$ **b** $2b + 9 = 27$

   c $\dfrac{f}{5} + 3 = 6$ **d** $\dfrac{g}{7} - 8 = 2$

   e $\dfrac{i}{6} + 4 = 9$ **f** $\dfrac{j}{5} - 8 = 4$

   g $\dfrac{2x - 8}{4} = -3$ **h** $\dfrac{3x - 5}{2} = 10$

3 Expand and solve

   a $3(a + 4) = -6$ **b** $2(b - 5) = -12$

   c $-4(6 - c) = -16$ **d** $3(2d - 3) = -21$

   e $4(e + 1) = 10$ **f** $3(f - 2) = -4$

   g $8(2 - g) = 10$ **h** $-4(h - 6) = 26$

4 Solve

   a $5x - 4 = 17$ **b** $3(2x - 4) = 17$

   c $4x - 9 = 27$ **d** $10 - 4x = 11$

   e $3(2 - x) = 19$ **f** $15 = 9 - 5y$

5 Solve

   a $10 + 2x = 7x - 9$

   b $5x - 4 = 12 - 2x$

   c $2 - 3y = 9 - 8y$

   d $5 - y = 10 - 2y$

   e $3w + 9 = -2w - 8$

   f $4(1 - 2x) = 2(3 - x)$

   g $2(3y - 1) = 3(y - 1)$

   h $3z + 2(4z - 2) = 5(3 - z)$

   i $6(p - 1) + 5(4 - p) = 6p$

   j $10q - (2q + 4) = 15$

   k $2(r - 7) - (2r - 3) = 9(r - 2)$

6 Solve these equations.

   a $2x + 4 = x + 3$

   b $10 - 3x = 7x - 10$

   c $8 - 3x = 5 - 2x$

   d $4(7 + 2z) = 15 - 8z$

7 Solve these equations.

   a $3x + 9 = 2x + 15$ **b** $5p - 9 = 3p + 7$

   c $10 - 3t = 8t - 12$ **d** $3 - 7b = 9 - 10b$

8 Solve these equations.

   a $3x + 7x - 3 + 9 = 17$

   b $3(2y - 1) = 4(7y + 6)$

   c $2(3 - 8z) = 4(5 - 6z)$

   d $2a + 3(2a - 7) = 20$

   e $3a - (9 - 7a) = 34$

   f $6 - (3x - 9) = -2(5 - x)$

   g $(x + 3)(x + 4) = (x + 7)(x - 2)$

   h $(y - 7)^2 = (y + 5)^2$

9 Solve these equations by cross-multiplying.

   a $\dfrac{7}{x} = 21$ **b** $15 = \dfrac{5}{x}$

   c $\dfrac{4}{y} = 3$ **d** $\dfrac{7}{p} = 8$

   e $\dfrac{10}{x} = -2$ **f** $11 = \dfrac{5}{y}$

   g $-3 = \dfrac{7}{y}$ **h** $-\dfrac{3}{x} = -9$

10 Solve these equations.

   a $\dfrac{5}{x} + 9 = 10$ **b** $\dfrac{10}{p} + 7 = 8$

   c $\dfrac{x}{4} + 3 = 10$ **d** $-2 = 1 + \dfrac{3}{x}$

11 Solve these equations.

   a $\dfrac{16}{x} + 4 = 2$ **b** $\dfrac{12}{2y} - 3 = 5$

   c $\dfrac{6}{3p} - 1 = 10$ **d** $\dfrac{15}{2x} + 4 = -2$

12 Solve these equations.

   a $\dfrac{x + 1}{3} = \dfrac{x - 1}{4}$ **b** $\dfrac{2y - 1}{3} = \dfrac{y}{2}$

   c $\dfrac{5}{w + 5} = \dfrac{15}{w + 7}$ **d** $\dfrac{3}{x - 1} = \dfrac{9}{2x - 1}$

13 Solve these equations by clearing the fractions first.

   a $\dfrac{x + 7}{3} = \dfrac{2x - 4}{5}$ **b** $\dfrac{5y - 9}{2} = \dfrac{3 - 2y}{6}$

   c $\dfrac{3z + 1}{5} = \dfrac{2z}{3}$ **d** $\dfrac{p}{2} + \dfrac{p}{4} = 7$

   e $\dfrac{4q}{3} - \dfrac{2q}{5} = 9$ **f** $\dfrac{3m}{4} + \dfrac{7m}{6} = \dfrac{1}{3}$

Q 1182, 1319, 1928, 1929   SEARCH

10.1 Solving linear equations

- To solve an equation do the same operation to both sides.
- Simplify the equation by expanding brackets or cross-multiplying to clear fractions.
- To solve an equation with unknowns on both sides, start by subtracting the smaller unknown from both sides.

HOW TO

1. Read the question and form an equation.
2. Solve the equation by simplifying and applying inverse operations.
3. Give your answer in the context of the question.

EXAMPLE

The square and the rectangle have the same area.

x

$x + 5$

$x - 2$

Find the value of x.

1. Area of square = Area of rectangle

$x^2 = (x + 5)(x - 2)$ 2. Remember brackets first (FOIL).

$x^2 = x^2 + 5x - 2x - 10$ Collect like terms.

$x^2 = x^2 + 3x - 10$ Subtract x^2 from each side.

$0 = 3x - 10$ Add 10 to both sides.

$10 = 3x$ Divide both sides by 3.

3. $x = 3\frac{1}{3}$

EXAMPLE

Three numbers are arranged in numerical order.

The mean of these three numbers is equal to the median. Find the three numbers.

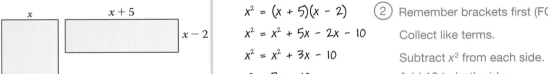

$\dfrac{x + 5}{2}$ $x + 1$ $2(x - 4)$

$\dfrac{\frac{x + 5}{2} + x + 1 + 2(x - 4)}{3} = x + 1$ 1. Mean = median

$\dfrac{x + 5}{2} + x + 1 + 2(x - 4) = 3(x + 1)$ Multiply both sides by 3.

$\dfrac{x + 5}{2} + x + 1 + 2x - 8 = 3x + 3$ 2. Expand the brackets and simplify.

$\dfrac{x + 5}{2} + 3x - 7 = 3x + 3$ The $3x$ terms cancel.

$\dfrac{x + 5}{2} - 7 = 3$

$\dfrac{x + 5}{2} = 10$

$x + 5 = 20$ Multiply both sides by 2.

$x = 15$

3. Substitute $x = 15$ into the expression to find the three numbers.

$\dfrac{x + 5}{2} = \dfrac{15 + 5}{2} = \dfrac{20}{2} = 10$

$x + 1 = 15 + 1 = 16$

$2(x - 4) = 2(15 - 4) = 2 \times 11 = 22$

Algebra Equations and inequalities

Exercise 10.1A

1 The perimeter of this shape is 30 mm. Find the length of each side.

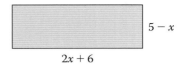

$5 - x$

$2x + 6$

2 The diagram shows an isosceles triangle. The equal angles are 10 less than double the third angle.

Find the size of each angle.

x

3 In each case, use the information to form an equation and solve it.

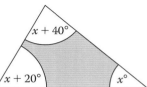

$x + 40°$

$x + 20°$ $x°$

a The angles in a triangle are $x°$, $x + 20°$ and $x + 40°$. Find the angles of the triangle.

b The perimeters of these two shapes are equal. What are the dimensions of each shape?

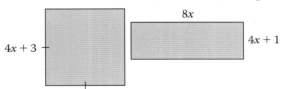

$8x$

$4x + 3$

$4x + 1$

4 Write an equation and solve it to find the starting number in each case.

a I think of a number, add 4, divide it into 12 and get 7.

b I think of a number, take away 3 and divide it into 11. This gives me the same answer as when I take the same number and divide it into 8.

5 Find the length of the rectangle:

$5(7 - 3x)$

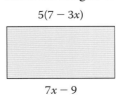

$7x - 9$

6 The means of each set of expressions are equal. Find x and the value of each expression.

Set 1

$2x - 1$	$3x + 2$
$5x + 4$	$7x$
$6x - 4$	$11 - 2x$

Set 2

$3x - 7$	$5x + 8$
$13 - x$	
$2(3x+1)$	$12 + 4x$

7 a In 10 years' time, my age will be double what it was 11 years ago. How old am I?

b Find the length of a square of side $2x + 3$ with an area equal to the area of a rectangle measuring $x + 6$ by $4x - 2$.

8 A number is doubled and divided by 5. Three more than the number is divided by 8. Both calculations give the same answer. What is the number?

9 The sum of one-fifth of Lucy's age and one-seventh of her age is 12. Use algebra to work out Lucy's age.

10 These trapeziums have equal areas. What are the lengths of the parallel sides in each?

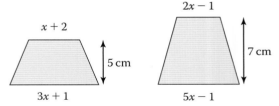

$x + 2$

$2x - 1$

5 cm

7 cm

$3x + 1$

$5x - 1$

11 The mean of these three numbers is 9.

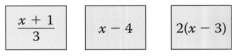

$\dfrac{x + 1}{3}$ $x - 4$ $2(x - 3)$

Find the three numbers.

12 Three numbers are arranged in numerical order.
The mean of these three numbers is equal to the median.

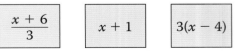

$\dfrac{x + 6}{3}$ $x + 1$ $3(x - 4)$

Find the three numbers.

Q 1182, 1319, 1928, 1929 SEARCH

10.2 Quadratic equations

- **Quadratic** equations contain a squared term as the highest power, for example $2x^2 + 7x - 9 = 0$.

Quadratic equations can have 0, 1 or 2 solutions.

For example	$x^2 = 100$	$x^2 = 0$	$x^2 = -25$
Solution(s)	$x = 10$ or -10	$x = 0$	Impossible
Number of solutions	2	1	0

- Many quadratic equations can be solved by rearranging so that one side equals zero and then **factorising**.

EXAMPLE

Solve $x^2 = 2x + 15$

$x^2 - 2x - 15 = 0$ Make the equation equal to zero.

$(x + 3)(x - 5) = 0$ Find two numbers with sum -2 and product -15.

Either $x + 3 = 0$ or $x - 5 = 0$

$x = -3$ $x = 5$

- You can solve a quadratic equation by **completing the square**.

You could write the answer as $y = 5 \pm \sqrt{5}$.

EXAMPLE

Solve $y^2 - 10y + 20 = 0$ by completing the square.

$y^2 - 10y + 20 = 0$ Complete the square.

$(y - 5)^2 - 25 + 20 = 0$

$(y - 5)^2 - 5 = 0$

$(y - 5)^2 = 5$ Add 5 to both sides.

5 has positive and negative square roots.

$y - 5 = \sqrt{5}$ or $y - 5 = -\sqrt{5}$

Leave your answers in surd form.

$y = 5 + \sqrt{5}$ or $y = 5 - \sqrt{5}$

- You can use the quadratic formula $x = \dfrac{-b \pm \sqrt{b^2 - 4ac}}{2a}$ to solve a quadratic equation $ax^2 + bx + c = 0$.

EXAMPLE

Solve the quadratic equation $3x^2 - 5x = 1$.

Give your answers to 3 sf.

$3x^2 - 5x = 1$

$3x^2 - 5x - 1 = 0$

In the formula:
$a = 3, b = -5$ and $c = -1$

$x = \dfrac{5 \pm \sqrt{(-5)^2 - 4 \times 3 \times (-1)}}{2 \times 3}$

$x = \dfrac{5 \pm \sqrt{25 + 12}}{6}$

$x = \dfrac{-b \pm \sqrt{b^2 - 4ac}}{2a}$

$x = \dfrac{5 + \sqrt{37}}{6} = 1.847127088\ldots = 1.85$ (to 3 sf)

Or $x = \dfrac{5 - \sqrt{37}}{6} = -0.180460421\ldots = -0.180$ (to 3 sf)

Exercise 10.2S

1 First factorise these quadratic equations by using the common factor, then solve them.

a $x^2 - 3x = 0$ **b** $x^2 + 8x = 0$

c $2x^2 - 9x = 0$ **d** $3x^2 - 9x = 0$

e $x^2 = 5x$ **f** $x^2 = 7x$

g $6x - x^2 = 0$ **h** $9y - 3y^2 = 0$

2 Solve these equations by factorising.

a $x^2 + 7x + 12 = 0$ **b** $x^2 + 5x - 14 = 0$

c $x^2 - 4x - 5 = 0$ **d** $x^2 - 5x + 6 = 0$

e $2x^2 + 7x + 3 = 0$ **f** $3x^2 + 7x + 2 = 0$

g $2x^2 + 5x + 2 = 0$ **h** $6y^2 + 7y + 2 = 0$

i $x^2 = 8x - 12$ **j** $2x^2 + 7x = 15$

k $0 = 5x - 6 - x^2$ **l** $x(x + 10) = -21$

3 Solve these quadratic equations by factorising using the difference of two squares.

a $x^2 - 16 = 0$ **b** $x^2 - 64 = 0$

c $y^2 - 25 = 0$ **d** $9x^2 - 4 = 0$

e $4y^2 - 1 = 0$ **f** $x^2 = 169$

g $4x^2 = 25$ **h** $36 = 9y^2$

4 Solve these quadratic equations.

a $3x^2 - x = 0$ **b** $x^2 - 2x - 15 = 0$

c $3x^2 - 11x + 6 = 0$ **d** $9y^2 - 16 = 0$

e $25 = 16x^2$ **f** $x^2 = x$

g $20x^2 = 7x + 3$ **h** $8x = 12 + x^2$

5 Solve these quadratic equations by completing the square.

a $x^2 - 12x + 20 = 0$ **b** $x^2 + 2x - 15 = 0$

c $x^2 - 4x - 5 = 0$ **d** $x^2 + 2x + 1 = 0$

e $x^2 + 2x - 63 = 0$ **f** $x^2 - 14x + 49 = 0$

g $x^2 - 8x = 0$ **h** $y^2 = 1 - 12y$

i $p^2 = 3p + 2$

> Don't forget the two square roots.

6 Solve these quadratic equations using the quadratic formula. Where necessary, give your answers to 3 significant figures.

a $3x^2 + 10x + 6 = 0$

b $5x^2 - 6x + 1 = 0$

c $2x^2 - 7x - 15 = 0$

6 **d** $x^2 + 4x + 1 = 0$

e $2x^2 + 6x - 1 = 0$

f $6y^2 - 11y - 5 = 0$

g $3x^2 + 3 = 10x$

h $6x + 2x^2 - 1 = 0$

i $20 - 7x - 3x^2 = 0$

7 Solve these equations using the quadratic formula. Give your answers to 3 sf.

a $x(x + 4) = 9$

b $2x(x + 1) - x(x + 4) = 11$

c $(3x)^2 = 8x + 3$

d $y + 2 = \dfrac{14}{y}$

e $\dfrac{3}{x + 1} + \dfrac{4}{2x - 1} = 2$

8 Solve these equations by factorising.

a $x^2 + 7x + 10 = 0$ **b** $x^2 + 4x - 12 = 0$

c $x^2 - 49 = 0$ **d** $x^2 - 8x = 0$

e $(x + 2)^2 = 16$ **f** $3y^2 - 7y + 2 = 0$

9 Use the quadratic formula to solve these equations. Give your answers to 3 sf where necessary.

a $x^2 + 8x + 6 = 0$ **b** $7x^2 + 6x + 1 = 0$

c $x^2 - 2x - 1 = 0$ **d** $10y^2 - 2y - 3 = 0$

e $4x^2 + 3x = 2$ **f** $(2x - 3)^2 = 2x$

*10 Solve these equations.

a $5 = x + 6x(x + 1)$

b $(x + 1)^2 = 2x(x - 2) + 10$

c $10x = 1 + \dfrac{3}{x}$

d $\dfrac{2}{x - 2} + \dfrac{4}{x + 1} = 0$

e $x^4 - 13x^2 + 36 = 0$

*11 Solve these, by writing them as quadratic equations.

a $2x^2 = 5x - 1$

b $(y - 5)^2 = 20$

c $p(p + 10) + 21 = 0$

d $10x + 7 = \dfrac{3}{x}$

e $\dfrac{x^2 + 3}{4} + \dfrac{2x - 1}{5} = 1$

Q 1160, 1169, 1181, 1185, 1950   SEARCH

10.2 Quadratic equations

- A quadratic equation has the form $ax^2 + bx + c = 0$

$$x = \frac{-b \pm \sqrt{b^2 - 4ac}}{2a}$$

- You can solve a quadratic equation by factorising, completing the square or using the quadratic formula.

- If the question asks for an answer as a square root, or 'to 3 sf' then it is likely you will need the formula or completing the square. Factorising usually gives integer or fraction answers.

Quadratic equations can have 2, 1 or no solutions.

HOW TO

(1) Use the information in the question to form a quadratic equation.

(2) Rearrange the quadratic so that it equals zero.

(3) Solve the quadratic by factorising, completing the square or using the quadratic formula.

(4) Check that your answers make sense. You may need to reject one solution, depending on the context.

EXAMPLE

Two numbers have a product of 105 and a difference of 8.

If the larger number is x, form and solve an equation to find the two numbers.

(1) The two numbers are x and $x - 8$.
So $x(x - 8) = 105$
(2) $x^2 - 8x - 105 = 0$
(3) $(x + 7)(x - 15) = 0$
Either $x + 7 = 0$ or $x - 15 = 0$
So $x = -7$ and $x - 8 = -15$ or $x = 15$ and $x - 8 = 7$.
(4) Two answers for x lead to two answers for $x - 8$.
The two numbers are -7 and -15 or 7 and 15.

EXAMPLE

If the area of the trapezium is $400 \, cm^2$, show that $x^2 + 10x = 400$ and find the value of x correct to 3 dp.

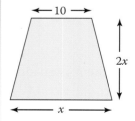

(1) Area of trapezium $= \frac{(a + b)}{2}h$
$A = \frac{1}{2}(x + 10) \times 2x = x(x + 10)$
$400 = x(x + 10)$
$x^2 + 10x = 400$ (as required)
(2) $x^2 + 10x - 400 = 0$
(3) The equation does not factorise, so use
$$x = \frac{-b \pm \sqrt{b^2 - 4ac}}{2a}$$
$$x = \frac{-10 \pm \sqrt{10^2 - 4 \times 1 \times -400}}{2}$$
$= 15.61552...$ or $-25.61552...$
(4) Since x is a length, it must be positive, so $x = 15.616$ to 3 dp.

Exercise 10.2A

1 A rectangle has a length that is 7 cm more than its width, w. The area of the rectangle is 60 cm².

w

    a Write an algebraic expression for the area of the rectangle.

    b Show that $w^2 + 7w - 60 = 0$.

    c Find the dimensions of the rectangle.

2 Prove that $p^2 + 6p + 9$ can never be negative.

3 Use the quadratic formula to solve $2x^2 + 7x + 9 = 0$. What do you notice?

4 Give answers to 2 decimal places where appropriate.

    a Two numbers which differ by 4 have a product of 117. Find the numbers.

    b The length of a rectangle exceeds its width by 4 cm. The area of the rectangle is 357 cm². Find the dimensions of the rectangle.

    c The square of three less than a number is ten. What is the number?

    d Three times the reciprocal of a number is one less than ten times the number. What is the number?

    e The diagonal of a rectangle is 17 mm. The length is 7 mm more than the width. Find the dimensions of the rectangle.

    f If this square and rectangle have equal areas, find the side length of the square.

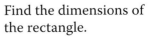

$3x - 2$

$7x - 3$

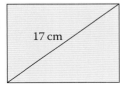

x

5 The perimeter of a rectangle is 46 cm and its diagonal is 17 cm.

Find the dimensions of the rectangle.

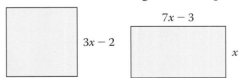

17 cm

6 In each part form an equation and solve it to answer the problem.

    a The diagonal of this rectangle is 13 cm. What is the perimeter of the rectangle?

$2x$

$x - 1$

    b The area of this hexagon is 25 m². Find the perimeter of the hexagon.

$2y + 5$

$3y - 2$

2

7 The mean of these three numbers is 24.

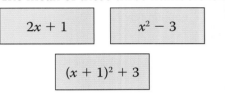

$2x + 1$ $x^2 - 3$

$(x + 1)^2 + 3$

Find the two possible sets of three numbers.

8 Alexa and Briana have tried to solve a quadratic equation by completing the square.

They have each made one mistake.

Identify each mistake and find the correct solutions.

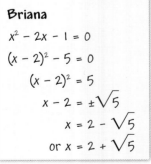

Alexa

$x^2 - 6x + 5 = 0$

$(x - 3)^2 - 4 = 0$

$(x - 3)^2 = 4$

$x - 3 = \pm 4$

$x = -1$

or $x = 7$

Briana

$x^2 - 2x - 1 = 0$

$(x - 2)^2 - 5 = 0$

$(x - 2)^2 = 5$

$x - 2 = \pm \sqrt{5}$

$x = 2 - \sqrt{5}$

or $x = 2 + \sqrt{5}$

*9 By completing the square on $ax^2 + bx + c = 0$, prove the quadratic equation formula

$$x = \frac{-b \pm \sqrt{b^2 - 4ac}}{2a}$$

1160, 1169, 1181, 1185, 1950   SEARCH

10.3 Simultaneous equations

- **Simultaneous** equations are true at the same time. They share a solution.

For example $x + y = 10$ ← x and y add to make 10 and their difference is 4.
$x - y = 4$ ← The shared solution must be $x = 7$ and $y = 3$.

- You can solve simultaneous equations by **eliminating** one of the two **variables**.

- Multiply one or both equations to get equal numbers of one variable.
- Add or subtract the equations to eliminate the variable.

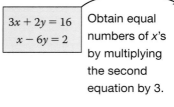

| $3x + 2y = 16$ $x - 6y = 2$ | Obtain equal numbers of x's by multiplying the second equation by 3. | $3x + 2y = 16$ (1) $3x - 18y = 6$ (2) | Subtract (1) − (2) to eliminate x. | $20y = 10$ $y = \frac{1}{2}$ | Substitute for y in (1) to find x. | $3x + 1 = 16$ $x = 5$ |

> Check the solution by substituting in one of the original equations.

EXAMPLE

Solve the simultaneous equations $\quad 2x - 4y = 8$
$\qquad\qquad\qquad\qquad\qquad\qquad\qquad 3x + 3y = -15$

$2x - 4y = 8$ (1)   multiply by 3 $6x - 12y = 24$ (3)

$3x + 3y = -15$ (2)   multiply by 4 $12x + 12y = -60$ (4)

Add (3) + (4): $18x = -36$

$\qquad\qquad\qquad x = -2$

Substituting in (2): $-6 + 3y = -15$ so $3y = -9$

$\qquad\qquad\qquad\qquad\qquad y = -3$

- You can use substitution to solve **simultaneous** equations where one is **linear** and one **quadratic**.

> A linear equation contains no square or higher terms. A quadratic equation contains a square term, but no higher powers.

- Rearrange the linear equation to make one unknown the subject.
- Then substitute this expression into the quadratic equation and solve.

EXAMPLE

Solve the simultaneous equations

$x + y = 7$

$x^2 + y = 13$

$x + y = 7$ (1)

$x^2 + y = 13$ (2)

Equation (1) is linear. Rearrange it to make y the subject: $y = 7 - x$

Substitute in (2): $x^2 + (7 - x) = 13$

$\qquad\qquad\qquad x^2 - x - 6 = 0$

$\qquad\qquad\quad (x + 2)(x - 3) = 0$

Either $x = -2$ and $y = 7 - (-2) = 9$

or $\qquad x = 3$ and $y = 7 - 3 = 4$

Algebra Equations and inequalities

Exercise 10.3S

1 Which pairs of equations have the solution $x = 2$ and $y = 7$?

a
$$x + y = 9$$
$$x - y = -5$$

b
$$2x + y = 11$$
$$3x - y = -1$$

c
$$2x + 2y = 15$$
$$4x - y = 1$$

d
$$x + 2y = 16$$
$$x - 2y = 8$$

2 a Find three pairs of solutions for the equations

    i $2x + y = 11$ **ii** $6x + y = 27$

b Solve these simultaneous equations.

$$2x + y = 11$$
$$6x + y = 27$$

3 Solve these pairs of simultaneous equations by subtracting one equation from the other.

a $3x + y = 15$
    $x + y = 7$

b $6x + 2y = 6$
    $4x + 2y = 2$

c $x + 5y = 19$
    $x + 7y = 27$

d $5x + 2y = 16$
    $x + 2y = 4$

e $m + 3n = 11$
    $m + 2n = 9$

f $4x + 3y = -5$
    $7x + 3y = -11$

4 Solve these pairs of simultaneous equations by adding one equation to the other.

a $3x + 2y = 19$
    $8x - 2y = 58$

b $5x + 2y = 16$
    $3x - 2y = 8$

c $7a - 3b = 24$
    $2a + 3b = 3$

d $2x + 3y = 19$
    $-2x + y = 1$

e $4x - 7y = 15$
    $2x + 7y = 4\frac{1}{2}$

f $6p - 2q = -2$
    $6p + 2q = 26$

5 Solve these pairs of simultaneous equations by either adding or subtracting in order to eliminate one variable.

a $x + y = 3$
    $3x - y = 17$

b $5x - 2y = 4$
    $3x + 2y = 12$

c $5a + b = -7$
    $5a - 2b = -16$

d $3v + w = 14$
    $3v - w = 10$

e $20p - 4q = 32$
    $7p + 4q = 22$

f $3x - 2y = 11$
    $3x + 4y = 23$

6 Solve these simultaneous equations.

a $2x + y = 8$
    $5x + 3y = 12$

b $3x + 2y = 19$
    $4x - y = 29$

c $8a - 3b = 30$
    $3a + b = 7$

d $2v + 3w = 12$
    $5v + 4w = 23$

e $9p + 5q = 15$
    $3p - 2q = -6$

f $3x - 2y = 11$
    $2x - y = 8$

7 a $2x - 3y = 13$
    $3x + 5y = -28$

b $9x + 2y = 0$
    $3x - 5y = 17$

c $5x + 3y = -3$
    $10x - 6y = -10$

d $6x - 7y = 4$
    $3x + 14y = -5.5$

8 Solve these simultaneous equations. Remember to clear the fractions first.

a $\dfrac{x}{3} - \dfrac{y}{4} = \dfrac{3}{2}$
    $2x + y = 14$

b $\dfrac{a}{2} + 3b = 1$
    $5a - 7b = 47$

c $p - \dfrac{2q}{3} = \dfrac{26}{3}$
    $\dfrac{p}{4} + 3q + 1 = 0$

d $\dfrac{5x}{6} + \dfrac{y}{4} = 8$
    $\dfrac{2x}{5} + \dfrac{y}{10} = 4$

9 Solve these simultaneous equations.

a $2x + y = 18$
    $x - 2y = -1$

b $5x + 2y = -30$
    $3x + 4y = -32$

c $14(c + d) = 14$
    $5d - 3c = -11$

d $5x = 7 + 6y$
    $8y = x + 2$

e $24a + 12b + 7 = 0$
    $6a + 12b - 5 = 0$

f $4p - 1\frac{1}{2}q = 5\frac{1}{2}$
    $6p - 2q = 21$

10 Solve these simultaneous equations.

a $x^2 + y = 55$
    $y = 6$

b $x + y^2 = 32$
    $x = 7$

c $x^2 - 3y = 73$
    $y = 9$

d $2y^2 - x = 13$
    $x = 5$

11 Solve these simultaneous equations.

a $y = x^2 - 2x$
    $y = x + 4$

b $y = x^2 - 1$
    $y = 2x - 2$

c $x = 2y^2$
    $x = 9y - 4$

d $x = y^2 - 4$
    $x = 2x - 1$

12 Solve these simultaneous equations.

a $y = x^2 - 3x + 7$
    $y - 5x + 8 = 0$

b $p^2 + 3pq = 10$
    $p = 2q$

Q 1174, 1177, 1236, 1319   SEARCH

10.3 Simultaneous equations

- You can solve simultaneous linear equations by eliminating one of the two variables. Your answer will be one pair of solutions.

- You can use substitution to solve simultaneous equations where one is linear and one quadratic. Your answer will be one or two pairs of solutions.

You can solve **simultaneous** equations graphically.
A solution is at a point of **intersection**.
For example, for the equations $3x - y = 2$ and $2x + y = 8$,
the lines intersect at $(2, 4)$ so the solution is $x = 2$ and $y = 4$.

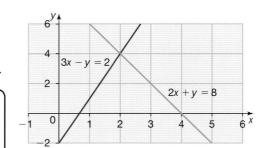

HOW TO

① Use the information in the question to form a pair of simultaneous equations.

② Solve the simultaneous equations using elimination, substitution or by drawing a graph.

③ Give your answers and check that they make sense.

EXAMPLE

The perimeter of this isosceles triangle is 50 cm.

Find the length of its base.

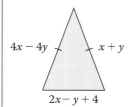

① There are two unknowns, so you need to set up two equations to find them.

Isosceles: $4x - 4y = x + y \rightarrow 3x - 5y = 0$ (1)

Perimeter: $7x - 4y + 4 = 50 \rightarrow 7x - 4y = 46$ (2)

② $7 \times (1)$ $21x - 35y = 0$ (3)

    $3 \times (2)$ $21x - 12y = 138$ (4)

    $(3) - (4)$ $-23y = 138$ Equal terms have the **S**ame **S**igns so **S**ubtract.

                 $y = 6$

Substituting in (1): $3x - 30 = 0$

                   $3x = 30$, so $x = 10$

③ The base is $(2 \times 10) - 6 + 4 = 18$ cm.

EXAMPLE

Solve the equation $5x - x^2 = 2x - 1$ graphically.

① Split the problem into two simultaneous equations.

② Plot the graphs of $y = 5x - x^2$ and $y = 2x - 1$.

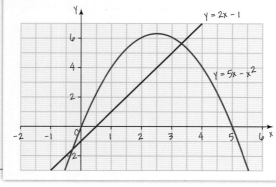

③ The answers are approximate, since you read them off the graph, rather than finding them by a direct algebraic approach.

The solutions are $x \approx -0.3$ and $x \approx 3.3$.

Exercise 10.3A

1 Solve these problems by using simultaneous equations.

3 lemons	4 oranges
4 oranges	5 lemons
£1.27	£1.61

a How much does a lemon cost?

b The perimeter of this triangle is 30 cm.

$2x - 1$ y

x

How long is the base?

2 Write a pair of simultaneous equations to solve each problem.

a Two numbers have a sum of 41 and a difference of 7.
What numbers are they?

b One number is 6 more than another.
Their mean average is 20.
What numbers are they?

c 230 students and 29 staff are going on a school trip.
They travel by large and small coaches.
The large coaches seat 55 and the small coaches seat 39.
There are no spare seats and five coaches make the journey.
How many of each coach are used?

d In an isosceles triangle, the largest angle is 30° more than double the equal angles.
Find the angles in the triangle.

e Two numbers have a difference of 6.
Twice the larger plus the smaller number also equals 6.
What numbers are they?

3 Use simultaneous equations to find the value of each symbol in the puzzle.

☾	✹	☾	✹	92
✹	☾	✹	✹	104

4 Use a graphical method to solve these problems.

a Twice one number plus three times another is 4. Their difference is 2. What are the numbers?

b The sum of the ages of James and Isla is 4. The difference between twice Isla's age and treble James' age is 3. How old are they?

5 Explain why the simultaneous equations $y = 2x - 1$ and $y = 2x + 4$ have no solution.

6 Two lines intersect. One has gradient 4 and y-axis intercept 3. The other has gradient 6 and cuts the y-axis at $(0, 1)$.

Find the point of intersection of the lines.

7 Solve simultaneously $xy^5 = -96$
$$2xy^3 = -48$$

8 Solve simultaneously $2^{p+q} = 32$
$$\frac{3^q}{3^{2p}} = 6561$$

9 **a** Use these graphs to find the approximate solutions of these equations.

  i $x^2 - x - 2 = 2$

  ii $x^2 - x - 2 = -1$

  iii $x^2 - x - 2 = x + 1$

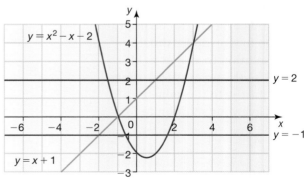

b Where would you find the solutions to $x^2 - x - 2 = 0$? Approximately what are these? Confirm your answer algebraically.

c Which graph would you need to add to the diagram to solve $2 - x = x^2 - x - 2$?

10.4 Approximate solutions

Some equations don't have exact solutions and can't be solved by an algebraic method.

EXAMPLE

Solve the equation $x^3 + x = 145$.
By following this process:

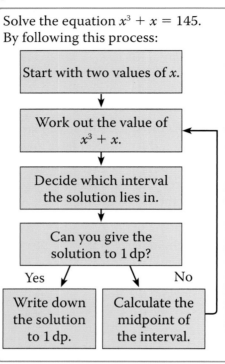

Start with two values of x.

↓

Work out the value of $x^3 + x$.

↓

Decide which interval the solution lies in.

↓

Can you give the solution to 1 dp?

Yes ↓   No ↓

Write down the solution to 1 dp.   Calculate the midpoint of the interval.

Use a table to set out your workings.

x	$x^3 + x$	
5	130	Too small
6	222	Too big
$\dfrac{5 + 6}{2} = 5.5$	171.875	Too big
$\dfrac{5 + 5.5}{2} = 5.25$	149.95...	Too big
$\dfrac{5 + 5.25}{2} = 5.125$	139.73...	Too small
$\dfrac{5.125 + 5.25}{2} = 5.1875$	144.78...	Too small

Start with $x = 5$ and $x = 6$.

The solution is between 5 and 5.5.

The solution is between 5 and 5.25.

The solution is between 5.125 and 5.25.

The solution is between 5.1875 and 5.25, so $x = 5.2$ (1 dp).

You can find approximate solutions using an **iteration** method.

● Iteration involves repeating the same instructions until you find an accurate solution.

EXAMPLE

This process can be used to find an approximate solution to $x^3 + x = 145$.
Use the process four times.
Start with the value $x = 5$.
Give your solution to 5 dp.

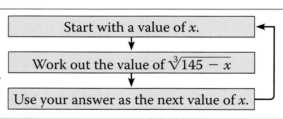

Start with a value of x.

↓

Work out the value of $\sqrt[3]{145 - x}$

↓

Use your answer as the next value of x.

You can also use the recurrence formula $x_{n+1} = \sqrt[3]{145 - x_n}$ to describe the process.

$x_0 = 5$

$x_1 = \sqrt[3]{145 - 5} = 5.192494...$

$x_2 = \sqrt[3]{145 - 5.19249...} = 5.190113...$

$x_3 = \sqrt[3]{145 - 5.19011...} = 5.19014265...$

$x_4 = \sqrt[3]{145 - 5.19014...} = 5.190142292$

So $x = 5.19014$

$5.19014^3 + 5.19014 = 144.9998125$ which is very close to 145.

x_0 means the starting value of x.

x_1 means the result of the first iteration.

Substitute x_1 into the expression to find x_2.

Stop when you find x_4.

Round to 5 dp.

Exercise 10.4S

1 a Show that the equation $x^3 - x^2 = 200$ can be rearranged to $x = \sqrt[3]{200 + x^2}$

b This process can be used to find an approximate solution to $x^3 - x^2 = 200$.

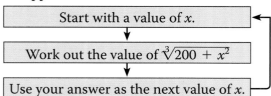

Use the process four times.
Start with $x = 6$.
Give your answer to 3 dp.

c Substitute your answer to part **c** into $x^3 - x^2$.
Comment on the accuracy of your solution to the equation $x^3 - x^2 = 200$.

2 a Show that the equation $p(p + 1) = 100$ can be rearranged to

   i $p = \sqrt{100 - p}$ **ii** $p = 100 - p^2$

b Use this process to find an approximate solution to $p(p + 1) = 100$.

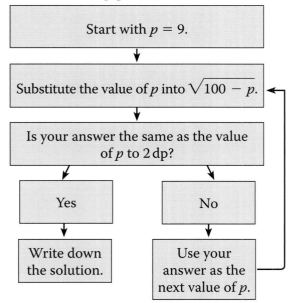

c Try to repeat the process by substituting into $100 - p^2$. What do you notice?
Is your value of p a solution to the equation?

3 John is finding an approximate solution for the equation $x^3 - 5x + 5 = 0$ using this recursive formula.

$$x_{n+1} = \sqrt[3]{5x_n - 5}$$

John starts with $x_1 = -2$ and uses a spreadsheet to generate $x_2, x_3, x_4, \ldots$

x_1	-2
x_2	-2.466212074
x_3	-2.587865565
x_4	-2.617793522
x_5	-2.625052098
x_6	-2.62680652
x_7	-2.627230218
x_8	-2.627332522

a Use John's results to find a solution to 3 decimal places of $x^3 - 5x + 5 = 0$.

b By substituting your answer to part **a** into $x^3 - 5x + 5 = 0$, comment on the accuracy of John's solution to $x^3 - 5x + 5 = 0$.

4 Tracey is trying to find an approximate solution for the equation $x^3 - 4x + 1 = 0$. She has rearranged the equation to form this recursive formula.

$$x_{n+1} = (4x_n - 1)^{\frac{1}{3}}$$

She starts with
$x_1 = -2 \; x_2 = ((4 \times \text{-}2) - 1)^{\frac{1}{3}} = -2.0800838$

a Find x_3, x_4, x_5, x_6 and x_7.

b Comment on your results to part **a**.

c Using the same recursive formula $x_{n+1} = (4x_n - 1)^{\frac{1}{3}}$ but starting with $x_1 = 2$, find another solution of $x^3 - 4x + 1 = 0$.

5 a Show that the equation $x^3 - 6x + 3 = 0$ can be rearranged to form this recursive formula.

$$x_{n+1} = \sqrt[3]{6x_n - 3}$$

b Starting with $x_1 = 2$ and using the recursive formula, find an approximate solution of $x^3 - 6x + 3 = 0$.

10.4 Approximate solutions

- You find solutions to equations by using an iterative method.
- Iteration involves repeating the same instructions until you find an accurate solution.
- A formula of the form $x_{n+1} = f(x_n)$ can often be used to obtain an approximate solution of an equation.

HOW TO

① Obtain an equation which needs to be solved.

② Use an approximate method of solution if the equation is too difficult to solve algebraically.

EXAMPLE

Loans are compared using an interest rate known as the annual percentage rate (APR). For a loan of £L that is repaid in 3 equal annual payments of £R, the APR is $100A$% where

$$L(1 + A)^3 = R(3 + 3A + A^2).$$

A car manufacture offers a loan of £10 000 which has to be repaid in 3 annual payments of £4000.

a Prove that that the APR formula for the loan can be rearrnged to $5A^3 + 13A^2 + 9A - 1 = 0$
Hence, show that the APR is 10% to the nearest percentage.

b Amanda, Brian and Carly each suggest an iterative formala to find a more accurate value for the APR.

Amanda: $A_{n+1} = \sqrt[3]{\dfrac{1 - 13(A_n)^2 - 9A_n}{5}}$

Brian: $A_{n+1} = \sqrt{\dfrac{1 - 5(A_n)^3 - 9A_n}{13}}$

Ceri: $A_{n+1} = \dfrac{1 - 5(A_n)^3 - 13(A_n)^2}{9}$

Subsutite $A_1 = 0.1$ into each formula. Use your answers to decide which of the three formula can be used to find a more accurate value for the APR.

a ① Substitute the values into the APR equation.
$$10\,000(1 + A)^3 = 4000(3 + 3A + A^2)$$
$$5(1 + A)^3 = 2(3 + 3A + A^2) \quad \div 2000$$
Expand the brackets.
Use the fact that $(1 + A)^3 = (1 + A)(1 + A)^2$ to expand the left-hand side.
$$5(1 + A)(A^2 + 2A + 1) = 6 + 6A + 2A^2$$
$$5(A^2 + 2A + 1 + A^3 + 2A^2 + A) = 6 + 6A + 2A^2$$
$$5(A^3 + 3A^2 + 3A + 1) = 6 + 6A + 2A^2$$
$$5A^3 + 15A^2 + 15A + 5 = 6 + 6A + 2A^2$$
$$5A^3 + 13A^2 + 9A - 1 = 0$$

② Show that $100A$ lies between 9.5 and 10.5 by showing that the equation has a solution between 0.095 and 0.105.
$A = 0.095$, LHS $= -0.02$
$A = 0.105$, LHS $= 0.09$
The APR is 10% to the nearest percentage.

b Subsutite $A_1 = 0.1$ into each formula.

Amanda $A_2 = \sqrt[3]{\dfrac{1 - 13(0.1)^2 - 9(0.1)}{5}} = -0.18171...$

Brian $A_2 = \sqrt{\dfrac{1 - 5(0.1)^3 - 9(0.1)}{13}} = 0.08548...$

Ceri $A_2 = \dfrac{1 - 5(0.1)^3 - 13(0.1)^2}{9} = 0.09611...$

The solution is between 0.095 and 0.105.

The formula should give a value between 0.095 and 0.105.

Ceri's formula can be used to find a more accurate value.

Exercise 10.4A

1 A loan of £12 000 has to be repaid in 3 annual payments of £6000.

For a loan of £L that is repaid in 3 equal annual payments of £R, the APR is 100A% where $L(1 + A)^3 = R(3 + 3A + A^2)$.

a Show that the APR formula for the loan can be written as $2A^3 + 5A^2 + 3A - 1 = 0$.

Hence, show that the APR is 23% to the nearest percentage.

b Aria tries to use the iterative formula

$$A_{n+1} = \sqrt{\frac{1 - 2(A_n)^2 - 3A_n}{5}}$$

to find a more accurate value for the APR starting with $A_1 = 0.23$

Will Aria's iterative formula converge on the solution? Explain your answer.

c Use the iterative formula

$$A_{n+1} = \frac{1 - 2(A_n)^3 - 5(A_n)^2}{3}$$

to find the value of the APR to 1 dp.

2 Karlie and Taylor each put £1000 into saving accounts.

After n years, Karlie's simple interest account contains £1000$(1 + 0.04n)$. Taylor's compound interest account contains £1000$(1.03)^n$.

a Show that both accounts have the same amount of money when $1 = 1.03^n - 0.04n$

b Copy and complete this iterative process to find the solution to $1 = 1.03^n - 0.04n$.

Start with two values of n.

↓

Work out the value of $1.03^n - 0.04n$ for each value of n.

↓

c Use your iteration with starting values for $n = 8$ and $n = 40$ to find the numbers of years when both accounts have the same amount of money.

3 **a** Use the 'Babylonian' iterative formula

$$x_{n+1} = \frac{x_n}{2} + \frac{2}{2x_n}$$

to find a fraction approximation to $\sqrt{2}$. Use three iterations starting with the estimate $x_1 = 1$.

b What is the result of squaring your answer to **a**?

Did you know...

Some historians believe that 4000 years ago Babylonian mathematicians used iterative formula to find the square roots of numbers.

4 **a** Use the 'Babylonian' iterative formula

$$x_{n+1} = \frac{x_n}{2} + \frac{3}{2x_n}$$

to find a fraction approximation to $\sqrt{3}$. Use two iterations starting with the estimate $x_1 = 2$.

b What happens if you use the much simpler formula $x_{n+1} = \frac{3}{x_n}$?

5 In 2010, a survey of the birds on an island counted approximately 200 kittiwakes.

A conservationist used the logistic equation
$$P_{n+1} = P_n(1.4 - 0.001P_n)$$

to predict the expected population, P_n, n years later.

a What did the equation predict for the size of the colony of kittiwakes each year from 2011 to 2015?

b Describe in your own words how the size of the colony was expected to change.

c Assuming that the size of the colony will stabilise at a roughly constant value, find this equilibrium size.

22.5

22.5

Q 1956   SEARCH

10.5 Inequalities

- An **inequality** is a mathematical statement using one of these signs:

  $<$ less than $\qquad$ $\leqslant$ less than or equal to

  $>$ greater than $\qquad$ $\geqslant$ greater than or equal to

  > $5 > 3$ and $3 < 5$ are inequalities.

- You can solve an inequality by rearranging and using inverse operations, in a similar way to solving an equation.

- If you multiply or divide an inequality by a negative number you need to reverse the inequality sign to keep it true.

  > $4 < 6$ but $-2 > -3$
  >
  > $5 > 2$ but $-15 < -6$

EXAMPLE

a Find the range of values of x that satisfies both $3x \geqslant 2(x - 1)$ and $12 - 3x > 6$.

Represent the solution set on a number line.

b List the **integer** values of x that satisfy both inequalities.

> Use an 'empty' circle for $<$ and $>$. Use a 'filled' circle for $\leqslant$ and $\geqslant$.

a

$3x \geqslant 2(x - 1)$	$12 - 3x > 6$	Combine the two inequalities.
$3x \geqslant 2x - 2$	$12 > 6 + 3x$	$1\frac{1}{3} > x$ is the same as $x < 1\frac{1}{3}$.
$x \geqslant -2$	$6 > 3x$	So $-2 \leqslant x < 2$.
	$2 > x$	

```
    ●───────────○
 ┼──┼──┼──┼──┼──┼──┼
-3  -2  -1  0  1  2  3
```

b The integer values of x that satisfy both inequalities are $-2, -1, 0$ and 1.

You can represent inequalities as regions on a graph.

- Use a solid line for $\leqslant$ and $\geqslant$.
- Use a dashed line for $<$ and $>$.
- Some questions ask you to shade the region required. Some ask you leave it unshaded.

The unshaded region on this graph represents the inequalities $x > 1$ and $y \geqslant 2$.

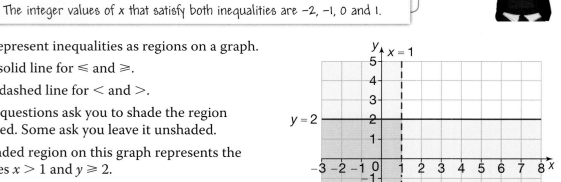

EXAMPLE

Solve the quadratic inequality $x^2 + x - 6 \geqslant 0$

Replace the inequality by an equation and find its roots.

$x^2 + x - 6 = 0 \implies (x + 3)(x - 2) = 0 \implies x = -3$ or $x = 2$

Pick values of x on either side of the roots.

$x = -4$	$(-4)^2 + -4 - 6 = 6 \geqslant 0$	✓
$x = 0$	$0^2 + 0 - 6 = -6 < 0$	✗
$x = 3$	$3^2 + 3 - 6 = 6 \geqslant 0$	✓

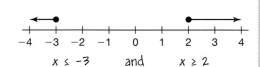

$x \leqslant -3 \qquad$ and $\qquad x \geqslant 2$

You could write the solution using set notation as $\{x \mid x \leqslant -3 \text{ or } x \geqslant 2\}$.

Exercise 10.5S

1 Solve these inequalities and represent the solutions on a number line.

   a $3x \leqslant 21$ **b** $2x - 5 > 17$

   c $\dfrac{p}{2} + 6 \leqslant -2$ **d** $28 < 7x + 49$

   e $5y + 3 \leqslant 2y + 5$ **f** $-3y > 9$

   g $4(x + 2) \leqslant 16$ **h** $-6x < 30$

   i $\dfrac{x}{-5} \geqslant -2$

   j $4p - 3 \leqslant 3(p - 2)$

   k $3(x - 2) < 5(x + 6)$

   l $6x - 4 \geqslant -2x$

2 Represent each set on a number line.

   a $\{x \mid 3x - 5 > 18\}$ **b** $\{p \mid \dfrac{p}{4} + 6 \leqslant -2\}$

   c $\{x \mid 6x + 3 \leqslant 2x - 8\}$

   d $\{y \mid -3y > 12\}$ **e** $\{q \mid \dfrac{q}{-5} \geqslant -2\}$

   f $\{z \mid 4z - 3 \leqslant 3(z - 2)\}$

   g $\{y \mid 3(y - 2) < 8(y + 6)\}$

3 Find the range of values of x that satisfies **both** inequalities.

   a $3x + 6 < 18$ and $-2x < 2$

   b $10 > 5 - x$ and $3(x - 9) < 27$

4 Solve, by treating the two inequalities separately.

   a $y \leqslant 3y + 2 \leqslant 8 + 2y$

   b $z - 8 < 2(z - 3) < z$

   c $4p + 1 < 7p < 5(p + 2)$

> In part **a**, solve $y \leqslant 3y + 2$ and $3y + 2 \leqslant 8 + 2y$.

5 Draw suitable diagrams to show these inequalities. You should leave the required region *unshaded* and label it R.

   a $x \geqslant 4$ **b** $y \leqslant 3$

   c $x > -2$ **d** $y < -3$

   e $-1 \leqslant x \leqslant 6$ **f** $1 < y < 8$

   g $x > 8$ or $x \leqslant -2$

   h $y < 2$ or $y > 9$

   i $2x \geqslant 5$ **j** $8 > y$

   k $-2.5 < x < 7$ and $0 \leqslant y \leqslant 5\frac{1}{4}$

6 For each diagram, write the inequalities shown by the *shaded* region.

   a

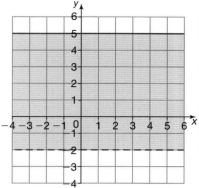

   b

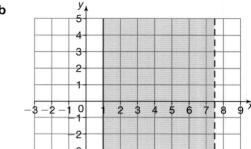

7 List all the points with integer coordinates that satisfy both the inequalities $-2 < x \leqslant 1$ and $2 > y \geqslant 0$.

8 Solve these quadratic inequalities.

   a $x^2 < 64$ **b** $x^2 > 1$

   c $x^2 + 2x > 0$ **d** $x^2 - 6x \leqslant 0$

   e $x^2 + 6x + 8 < 0$ **f** $x^2 + x < 12$

   g $2x^2 - 5x - 3 \leqslant 0$ **h** $3x^2 + 2 \leqslant 0$

9 Find a quadratic inequality that is represented by this solution set.

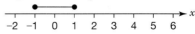

10 George thinks the solution for the quadratic inequality $x^2 \geqslant 49$ is $-7 \leqslant x \geqslant 7$.

   Do you agree with George?

   Explain your reasoning.

10.5 Inequalities

- You can represent inequalities as regions on a graph.
  - Use a solid line for $\leqslant$ and $\geqslant$ to show that points on the line are included.
  - Use a dashed line for $<$ and $>$ to show that points on the line are not included.

HOW TO

1. Read the question carefully, you may need to write inequalities to model the situation.
2. Draw each line on the graph — you could use a table of values to help you. Use a solid line for $\leqslant$ and $\geqslant$ and a dashed line for $<$ and $>$.
3. Choose a point in your solution area and check it obeys all the inequalities.

EXAMPLE

Neha is pricing items to sell in her cafe. She wants the cost of 2 sandwiches to be less than £5.

Neha wants the cost of 3 cupcakes and 2 sandwiches to be less than £9.

x is the price of a sandwich and y is the price of a cupcake.

Show the region on a graph that meets Neha's requirements.

1. 2 sandwiches cost less than £5. $2x < 5$ so $x < \dfrac{5}{2}$

   3 cupcakes and 2 sandwiches cost less than £9. $2x + 3y < 9$

   The sandwiches and cupcakes are not free. $x > 0$ $y > 0$

2. Draw dashed lines at $y = 0$ (x-axis), $x = 0$ (y-axis) and $x = \dfrac{5}{2}$.

   Plot the graph of $3y + 2x = 9$.

   Find the y-values for three values of x and plot the points.

   Draw the line dashed.

   $2x + 3y = 9$ so $y = \dfrac{9 - 2x}{3}$

x	-3	0	3
y	5	3	1

State which region is the required region.

The required region is unshaded.

3. Choose a point in your unshaded solution area and check the coordinates obey all the inequalities.

   $(1, 1)$ $y = 1$ and $1 > 0$

          $x = 1$ and $1 > 0$

          $x = 1$ so $2 \times 1 = 2$ and $2 < 5$

          $x = 1$, $y = 1$ so $(3 \times 1) + (2 \times 1) = 5$ and $5 < 9$

   $(1,1)$ obeys all the required inequalities.

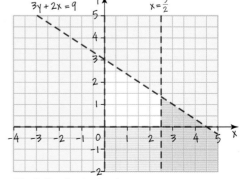

Each line is dashed.

Exercise 10.5A

1 Shade the regions satisfied by these inequalities.

   a $y \leqslant x + 5$ **b** $y > 2x + 1$

   c $y \geqslant 1 - 3x$ **d** $y < \frac{1}{4}x - 2$

   e $2y \geqslant 3x + 4$ **f** $3y < 9x - 7$

   g $x + y < 9$ **h** $2x + 4y > 7$

2 What inequalities are shown by the shaded region in each of these diagrams?

   a

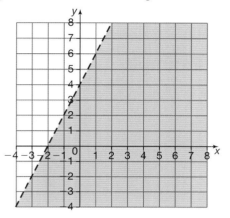

   b

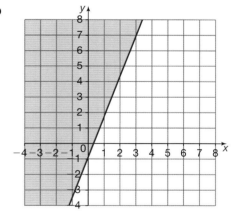

3 Here are two graphs of quadratic equations, with horizontal lines drawn as shown. Write the inequalities represented by the *shaded* regions.

   a

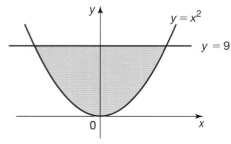

3 **b**

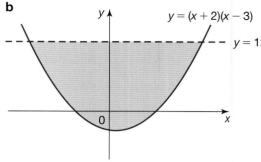

4 On one diagram, show the region satisfied by the three inequalities. List all the integer coordinates that satisfy all three inequalities.

   a $x \geqslant -1$ $y < 2$

         $y \geqslant 3x - 1$

   b $2x + 3y < 10$ $x > 2$

         $2y + 1 > 0$

5 Draw the region satisfied by all three inequalities

   $y \geqslant x^2 - 4, \quad y < 2, \quad y + 1 < x.$

6 300 students are going on a school trip and 16 adults will accompany them. The head teacher needs to hire coaches to take them on the trip. She can hire small coaches that seat 20 people or large coaches that seat 48 people. There must be at least 2 adults on each coach to supervise the students.

   a   If x is the number of small coaches hired and y is the number of large coaches, explain why these inequalities model the situation.
        $x \geqslant 0, y \geqslant 0, 5x + 12y \geqslant 79, x + y \leqslant 8$

   b   Show the region on a graph that satisfies all four inequalities.

7 Draw diagrams and write inequalities that when shaded would give

   a   a shaded, right-angled triangle with right angle at the origin

   b   an unshaded trapezium.

1161, 1162, 1163, 1189   SEARCH

Summary

Checkout

You should now be able to...

✔	Solve linear equations including when the unknown appears on both sides.	**1, 2**
✔	Solve quadratic equations using factorisation, completing the square and the quadratic formula.	**3 – 8**
✔	Solve a pair of linear or linear plus quadratic simultaneous equations.	**7, 8**
✔	Use iterative processes to find approximate solutions to equations.	**9**
✔	Solve inequalities and display your solution on a number line or a graph.	**10, 11**

Language   Meaning   Example

Language	Meaning	Example
Completing the square	Writting a quadratic expression $ax^2 + bx + c$ in the form $p(x + q)^2 + r$.	$2x^2 - 4x - 3 = 2(x - 1)^2 - 5$ $$= 0 \implies x = 1 \pm \sqrt{\frac{5}{2}}$$
Quadratic formula	A formula for the solutions of the quadratic equation $ax^2 + bx + c = 0$. $$x = \frac{-b \pm \sqrt{b^2 - 4ac}}{2a}$$	$2x^2 - 5x + 2 = 0, a = 2, b = -5, c = 2$ $$x = \frac{5 \pm \sqrt{5^2 - 4 \times 2 \times 2}}{2 \times 2} = \frac{5 \pm 3}{4} = 2, \frac{1}{2}$$
Simultaneous equations	Two or more equations that are true at the same time for the same values of the variables.	① $3x - y = 2$ ② $2x + y = 8$ Both true when $x = 2$ and $y = 4$. On a graph the lines intersect at (2, 4).
Elimination	A method of solving simultaneous equations by removing one of the variables.	① + ② $5x = 10 \implies x = 2$
Substitution	Replacing one of the variables in a simultaneous equation with an expression found by rearranging the other equation.	① $\implies y = 3x - 2$ into ② $2x + (3x - 2) = 8 \implies 5x = 10 \implies x = 2$
Inequality	A comparison of two quantities that are not equal.	$5x - 1 < 9$ $5x - 1$ is strictly less than 9.
Iteration	A procedure which is repeated.	$$x_{n+1} = \frac{1}{2}\left(x_n + \frac{2}{x_n}\right)$$ $x_1 = 1 \rightarrow x_2 = 1.5 \rightarrow x_3 = 1.41\dot{6} \rightarrow$
Recursive process	A repeated procedure in which the output of one iteration is used as the input to the next iteration.	$x_4 = 1.4142... \rightarrow \sqrt{2}$

Review

1 Solve these equations.

   a $3a - 7 = 35$ **b** $\dfrac{b}{4} + 3 = 28$

   c $5c + 23 = 8$ **d** $4x + 7 = 3x + 13$

   e $3x - 9 = 8x - 69$

   f $10 - 4x = 24 - 8x$

2 Sam takes x hours to run a marathon and Andy takes 36 minutes longer. The sum of their times is 7 hours.

   a Write an equation in x for the sum of their times.

   b Solve your equation to find Sam and Andy's times. Give your answers in hours and minutes.

3 Solve these quadratic equations by factorising.

   a $x^2 - x - 12 = 0$

   b $x^2 - 12x + 35 = 0$

   c $x^2 + 12x + 36 = 0$

   d $16x^2 - 121 = 0$

   e $6x^2 - 9x = 0$

   f $3x^2 + 7x + 4 = 0$

   g $10x^2 - 19x + 6 = 0$

4 Solve this quadratic equation.

   $2x^2 - 7x = 15$

5 **a** Complete the square for these quadratic expressions.

      i $x^2 - 4x - 1$ **ii** $x^2 + 3x + 2$

   b Solve these quadratic equations by completing the square.

      i $x^2 - 10x + 16 = 0$

      ii $2x^2 + 8x = 24$

6 Use the quadratic formula to solve these equations giving your answers to 3 significant figures.

   a $x^2 - 5x - 7 = 0$ **b** $3x^2 + 9x + 5 = 0$

7 Solve these pairs of simultaneous equations.

   a $3x + 4y = 5$ **b** $5v - 3w = -2$
       $5x - 5y = 20$ $4v - 7w = -8.5$

   c $y - x = 2$ **d** $y = x^2 + 5x$
       $y = x^2$ $y = x + 5$

8 A rectangle has length L and width W. The perimeter of the rectangle is 17 cm and the area is 15 cm².

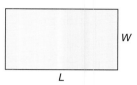

   a Write a pair of simultaneous equations from this information.

   b Solve your simultaneous equations to find the possible lengths W and L.

9 Use an iterative method to find the solutions to these equations correct to 2 decimal places.

   a $x^3 + x = 27$ **b** $\sqrt{x} + 2x = 19$

10 Solve these inequalities and display the results on a number line.

   a $5x - 12 > 13$ **b** $5x + 17 \leqslant 2x + 8$

   c $x^2 + 8x + 7 < 0$ **d** $x^2 + 3x - 18 \geqslant 0$

11 Draw suitable diagrams to show these inequalities. You should leave the required region *unshaded* and label it **R**.

   a $-1 \leqslant x < 3$ **b** $y \leqslant 4$ or $y \geqslant 5$

   c $1 < x < 4, y > 0$ and $x + y \leqslant 4$

What next?

Score		
0 – 4		Your knowledge of this topic is still developing.
		To improve look at MyMaths: 1057, 1160, 1161, 1162, 1163, 1169, 1174, 1177, 1181, 1182, 1185, 1189, 1236, 1319, 1928, 1929
5 – 9		You are gaining a secure knowledge of this topic.
		TTo improve look at InvisiPens: 10Sa – r
10 – 11		You have mastered these skills. Well done you are ready to progress!
		To develop your exam technique looks at InvisiPens: 10Aa – g

Assessment 10

1 **a** Jacques and Gilles went up the hill to fetch a pail of water. The mass of the water was 22 kg more than the pail. In total the mass was 27.5 kg. How heavy was the pail? [2]

b Oliver is three times as old as his brother Albert. Oliver is 6 years older than Albert. How old are Oliver and Albert? [4]

c Alice cut a cake into 3 pieces. Tweedledum's piece was 35 g heavier than Tweedledee's. Tweedledee's piece was 22 g lighter than Alice's. The total mass was 454 g. Calculate the mass of Alice's piece. [4]

d Brendan, Arsene and José go on holiday. Brendan takes x Euros, Arsene takes half as much as José, and José takes 150 Euros more than Brendan. Altogether, they took 2000 Euros spending money. How much money did each person take? [4]

e The ages of Milo and Fizz are in the ratio 3:4. In 6 years time they will be in the ratio 9:10. How old are they now? [4]

f Lewis drove on the M6, averaging 75 mph. He then left the M6 and drove on the M1 averaging 40 mph. He drove a total of 163 miles in 2.5 hours. How far did he drive along the M6? [6]

2 Mario attempted to kayak 30 km on Lake Garda to raise money for charity. Sophia gave him €5 for each *complete* kilometre he covered. Mario gave Sophia €2 for each *complete* kilometre he failed to kayak. The balance was given to charity. Mario gave €115 to the charity.

a Mario kayaked k kilometres. Write down an equation for k and solve it. [4]

b Find the minimum number of kilometres that Mario needed to complete to be sure of making a contribution to charity. [3]

3 **a** A square has side length s cm. Another square has a side 2 cm shorter than the first. The total area of the squares is 200 cm². Find the *exact* side length of the first square. [6]

b Two consecutive prime numbers, which have a difference of 6, have a product of 3127. Find the numbers. [6]

c The formula $S = \dfrac{n(n + 1)}{2}$ represents the sum of the numbers $1 + 2 + 3 + \cdots\cdots + n$. The total of the numbers 1 to n is 5050. What is the largest number in the sequence? [4]

4 Eoin hits a cricket ball straight upwards. The formula $h = 20t - 5t^2$ represents its height above the ground, t seconds after he throws it.

a Find the times when the height of the ball is 15 m above the ground. Say why there are two possible answers. [5]

b Find the time/times when the height of the ball is 20 m above the ground. Say why there is only one answer. [5]

c Form a quadratic equation to find the time when the ball is 25 m above the ground. Give two reasons why there is no solution to this equation. [5]

d Find the time when the ball next hits the ground. Explain your answer. [3]

5 The formula $2d = n(n - 3)$ represents the number of diagonals, d, in a polygon with n sides. A dodecagon has 54 diagonals. How many sides are there in a dodecagon? [5]

6 Greg says that the solutions to the equation $b^2 + 9b + 4 = 0$ are $b = 13$ or $b = 5.04$. Solve the equation by using the quadratic formula. Give your answers to 2 dp. Was Greg correct? [4]

7 **a** The straight line $y = x + 2$ cuts the curve $xy = 24$ in two places.
Find the coordinates of the points of intersection. [6]

   b A circle has the equation $x^2 + y^2 = 25$. Find the coordinates of the points of
intersection of the circle with the straight line $x + y = 1$. [6]

8 **a** On a coach trip, an adult ticket is £a and a child ticket £c. Tickets for 2 adults
and 3 children cost £26.50; tickets for 3 adults and 5 children cost £42.

      How much is   **i** an adult's ticket   [4]   **ii** a child's ticket? [1]

   b A magic goose lays both brown and golden eggs. 5 brown eggs and 2 golden
eggs have a total mass of 200 g, while 8 brown and 3 golden eggs have a total
mass of 310 g. Find the weight of a golden egg. [5]

   c 5 packets of 'Doggibix' and 3 packets of 'Cattibix' cost £5.49
3 packets of 'Doggibix' and 1 packet of 'Cattibix' cost £2.79
Find the cost of one packet of   **i** Doggibix   [4]   **ii** Cattibix? [1]

   d A DVD costs £d and a CD costs £c. 3 DVDs and 2 CDs cost £45.83 in total.
1 DVD and 3 CDs cost £26.92 in total.
What is the cost of
      i a CD   [4]   **ii** a DVD? [1]

   e Farmer Scott can buy 2 sheep and 6 cows, or 10 sheep and 2 cows, for £3 500.
What is the price of   **i** a sheep   [4]   **ii** a cow? [1]

   f Titus Lines is going fishing and needs bait. He can buy 5 maggots and 6 worms
for 38 pence or 6 maggots and 12 worms for 60 pence.
What is the price of
      i a maggot   [4]   **ii** a worm? [1]

9 Shaun represents the inequality $-1 < x \leqslant 3$ on a number line as shown.

   Correct his mistakes. [3]

10 Megan says that the region indicated by the inequalities
$2y - x < 5$, $x + y > 4$ and $x \leqslant 5$ is shown on the diagram.
Correct her diagram and shade the correct region. [4]

11 **a** A lorry averages 22 mpg. It will travel for 345 miles on a full
tank of diesel. The driver has g gallons in his tank. Write
down an appropriate inequality for g and solve it. [2]

   b David and Sophie are getting married. Food has been
ordered for 100 guests. 120 guests have agreed to come but
at least 15% of these will not come. On the day, there is not
enough food for everyone. How many guests attend the
wedding? Find all possible solutions. [4]

   c A busy station has 12 platforms. There are f freight trains and
p passenger trains at platforms. There are 2 more passenger trains than freight trains. There
are at least 3 freight trains at platforms. By writing and solving appropriate inequalities
answer the following questions.

      i Find all possible solutions for f and p. [4]

      ii Are any times when all 12 platforms are being used? [1]

      iii Find the greatest possible number of empty platforms. [1]

Life skills 2: Starting the business

Abigail, Mike, Juliet and Raheem, have completed their business plan and decide they want to locate their restaurant in an area near the railway station in Newton-Maxwell. To choose the ideal location, they need to set boundaries based on proximity to the high street and competitor restaurants. They also start planning other key considerations: designing promotional material, what tables they need to buy and how many, and contacting potential suppliers.

Task 1 – Location

The friends make a list of conditions that the location must meet (see below). Raheem draws a scale map.

a What is the scale of the map?

b Draw an accurate copy of the map and shade the areas that meet all three conditions.

They decide on a location that is on a bearing of 315° from T and 014° from A.

c On your copy of the map, label their desired location with R.

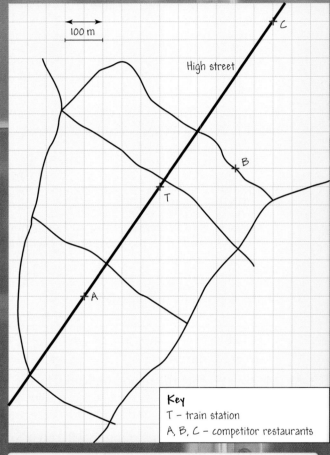

```
Conditions for location

•  No more than 500m from the railway
   station

•  At least 200m from each competitor
   restaurant

•  No more than 150m from the high
   street
```

Task 2 – The restaurant logo

The diagram shows the dimensions used for a logo to appear on the restaurant's business cards.

a Find the area of the logo.

For use on A5 flyers, the logo is enlarged by a scale factor of 2.5

b Find the area of the enlarged logo.

The logo is the only part of the business card and the flyer that uses coloured ink.

c A colour ink cartridge lasts long enough to print 4000 business cards. How many A5 flyers would you expect to print using one colour ink cartridge?

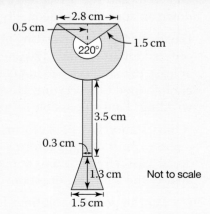

▲ Juliet's logo for use on their promotional material. It consists of a triangle, a circular sector, a rectangle and a trapezium, as shown in the second figure.

Task 3 – Safety regulations

The restaurant must adhere to fire safety regulations.

a x is the number of customers (non-staff) in the restaurant and y is the number of staff in the restaurant. Write down two inequalities in x and y to express the fire safety regulations.

b On graph paper, draw x and y axes from $x = 0$ to 40 and from $y = 0$ to 40. Shade the region where both your inequalities are satisfied.

c Use your inequalities and/or graph to find the maximum number of customers allowed in the restaurant at any time.

FIRE

Fire regulations

* Total number of staff and non-staff must not exceed 32.

* Ratio of non-staff to staff must not exceed 8:1.

Table	Shape of top	Dimensions
Style A	Circular	Diameter = 1.4 m
Style B	Regular octagon	Side length = 58 cm
		Length between opposite sides = 1.4 m

▲ Possible styles of table for restaurant.

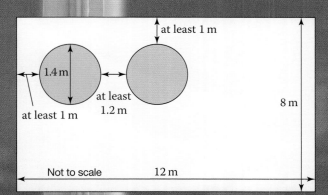

Not to scale

Caller	Total number of calls made	Number of unanswered calls
Abigail	30	10
Mike	22	8
Juliet	40	18
Raheem	18	8

▲ Number of calls made in the first week.

Task 4 – Choosing tables

They narrow down their choice of table based on the maximum number of customers they could have.

a Without doing any calculations, which table top (style A or style B) has the larger area?

b Calculate the area of each table top.

Style A is cheaper, so they decide to buy that one.

c Is this a reasonable decision? Give reasons for your answer.

Task 5 – Number of tables

The dining space is rectangular, and measures 12 m by 8 m. Abigail draws a sketch to indicate the gap required from each wall, and between each table.

a How many tables of Style A can fit in the dining space?

b If the measurements of 12 m and 8 m are accurate to the nearest 10 cm, find upper and lower bounds for the area of the rectangular space.

c Do you still agree with your answer to part **a**? Why?

Task 6 – Calling potential suppliers

Each member of the team makes telephone calls to potential suppliers. They logged how many calls they made in the first week, and how many went unanswered.

a Which person had the highest proportion of unanswered calls?

b Estimate how many unanswered calls you would expect out of the next 50 calls made by the team.

c What assumptions have you made in answering part **b**?

11 Circles and constructions

Introduction

The invention of the wheel was most definitely a landmark event in human technological development, giving people the ability to travel at speed. However it was the use of gears and cogs on a massive scale during the Industrial Revolution that really accelerated advancement, not just in technology but also in social and economic development.

What's the point?

Without mankind's understanding of circles, and how their properties can be exploited in marvellous ways, we would still be living in largely agricultural communities in a pre-industrial state, with no computers, mobile phones, cars, ...

Objectives

By the end of this chapter, you will have learned how to ...

- Find the area and circumference of a circle and composite shapes involving circles.
- Calculate arc lengths, angles and areas of sectors.
- Prove and apply circle theorems.
- Use standard ruler and compass constructions and solve problems involving loci.

Check in

1 Work out the missing angles.

a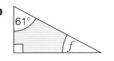

b

c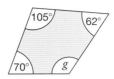

d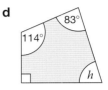

2 Work out the missing angles in these shapes.

a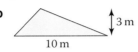

b

c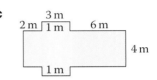

d

3 Find the area of these shapes.

a

2.5 cm

7 cm

b

3 m

10 m

c

3 m

2 m 1 m 6 m

4 m

1 m

Chapter investigation

Sketch a circle, and draw a straight line through it. How many pieces have you divided the circle into?

Draw a second straight line thought the circle. What is the maximum number of pieces you can divide the circle into?

Continue drawing straight lines through the circle (the lines can cut at any angle, and the circle can be as big as you want it to be). Is there a relationship between the number of cuts and the maximum number of pieces? Investigate.

11.1 Circles 1

- The **diameter**, d, is the distance across the circle through the centre.
- The **radius**, r, is the distance from the centre to the edge.
- The **circumference**, C, is the perimeter – the distance around the edge.

The circumference of a circle is in **proportion** to its diameter: $C \approx 3d$

The actual proportion is not a whole number. You use a symbol, π (pi).

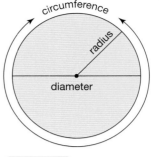

- $C = \pi \times d$ or $C = 2 \times \pi \times r$

$d = 2 \times r$

- Area of a circle $= \pi \times r^2$ or $A = \pi \times r^2$

π is about 3.14

EXAMPLE

Find **i** the circumference **ii** the area of each circle.

a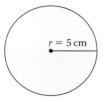

$r = 5\,cm$

b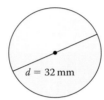

$d = 32\,mm$

> Round your answers to a sensible degree of accuracy. 3 sf is usually good practice.

a i $C = 2 \times \pi \times r = 2 \times \pi \times 5$
$= 31.415...$
$= 31.4\,cm$ (to 3 sf)

ii $A = \pi \times r^2 = \pi \times 5^2$
$= 78.539...$
$= 78.5\,cm^2$ (to 3 sf)

b i $C = \pi \times d = \pi \times 32$
$= 100.530...$
$= 101\,mm$ (to 3 sf)

ii $A = \pi \times r^2 = \pi \times 16^2$
$= 804.247...$
$= 804\,mm^2$ (to 3 sf)

- Area of a semicircle $= \dfrac{1}{2} \times$ area of whole circle.

- Perimeter of a semicircle $= \dfrac{1}{2} \times$ circumference of whole circle + diameter.

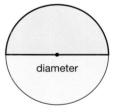

diameter

EXAMPLE

Calculate the perimeter of this semicircle.

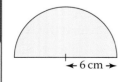

$\leftarrow 6\,cm \rightarrow$

Circumference of whole circle $= 2 \times \pi \times r$
$= 37.699 ...\, cm$

Perimeter of semicircle $= \dfrac{1}{2} \times$ circumference of whole circle + diameter
$= (\dfrac{1}{2} \times 37.699...) + (2 \times 6)$
$= 18.849 ... + 12$
$= 30.8\,cm$ (to 1 dp)

> In terms of π, Perimeter $= 12 + 6\pi$

Exercise 11.1S

1 Calculate the circumferences of these circles. State the units of your answers.

a
diameter = 10 cm

b
diameter = 8 m

c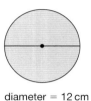
diameter = 12 cm

d
diameter = 20 m

2 Find the circumferences of these circles.

a $r = 12$ mm   **b** $r = 23$ cm

c $d = 105$ mm   **d** $d = 1.2$ cm

e $r = 3.6$ cm   **f** $d = 125$ cm

3 Find the areas of the circles in question **1**.

4 Find the areas of the circles in question **2**.

5 Calculate the diameter of a circle, if its circumference is

a 18.84 cm   **b** 15.7 m

c 28.26 cm   **d** 47.1 m

6 Find the area of each semicircle.

a
5 cm

b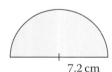
7.2 cm

c
24 mm

d
18 cm

e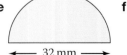
32 mm

f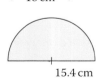
15.4 cm

7 Find the area of each semicircle.

a $r = 12$ cm   **b** $r = 2.3$ cm

c $d = 12.9$ m   **d** $d = 22.3$ cm

e $r = 9.5$ mm   **f** $d = 3.39$ cm

8 Work out the perimeter of each semicircle in question **6**.

9 Work out the perimeter of each semicircle in question **7**.

10 Shamin is cutting out circles for an art project. She has squares of card that are 4.2 cm wide.

4.2 cm

a What is the area of the biggest circle she can cut out?

b What area of card is left?

11 Viaduct arches have straight sides 50 m high. The arch at the top is a semicircle with diameter 8 m.

Find the perimeter of the arch.

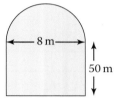
8 m
50 m

Give your answer in terms of π and to 1 dp.

12 A bathroom window is a semicircle with an internal diameter of 80 cm.

Work out the area of the glass in the window.

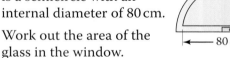
80 cm

*13 Calculate the diameter of a circle, if its area is

a 18.84 cm^2

b 15.7 m^2

c 28.26 cm^2

d 47.1 m^2

e 314 cm^2

14 A round hole has circumference 44 cm. Work out the radius of the hole, to 1 decimal place.

r

15 A circle has area 200 cm^2. Find the circumference of the circle. Give your answer to the nearest cm.

11.1 Circles 1

RECAP

- Use the formula $C = \pi d$ or $C = 2\pi r$ to find the circumference of a circle.
- Use $A = \pi r^2$ to find the area.
- Area of a semicircle $= \dfrac{1}{2} \times$ area of whole circle.
- Perimeter of a semicircle
  $= \dfrac{1}{2} \times$ circumference of whole circle + diameter.

▲ A circle is the shape with the largest area for a fixed perimeter. The cross-section of a tree trunk is a circle. This allows trees to store more nutrients and also minimise damage to the trunk.

HOW TO

① Decide which formulae to use.

② Calculate the perimeters or areas needed.

③ Answer the question, giving units where necessary.

EXAMPLE

A semicircle has perimeter 32 cm. Find the area of the semicircle to the nearest cm².

① Use the formula Perimeter $= \pi r + 2r$ to find the radius.
Then use $A = \dfrac{1}{2}\pi r^2$ to find the area.

$\pi r + 2r = 32$

$\quad\quad r = 32 \div \pi + 2$

$\quad\quad r = 6.223\ldots$

② Calculate the area.

$Area = \dfrac{1}{2}\pi \times 6.223\ldots^2$

$\quad\quad = 60.844\ldots$

③ $\quad\quad = 61\,cm^2$

EXAMPLE

Jake wants to give the front of his dog's kennel two coats of paint.

He has enough paint to cover $\dfrac{1}{2}$ m².

Does Jake need any more paint? Show your working.

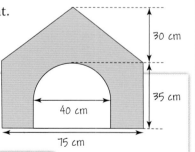

30 cm

35 cm

40 cm

75 cm

① Use the area formulae for a triangle, rectangle and semi-circle. Work in metres.

② Find the area of the shapes on the front, ignoring the hole.

$Area\ of\ triangle = \dfrac{1}{2} \times 0.75 \times 0.3 = 0.1125\,m^2$

$Area\ of\ rectangle = 0.75 \times 0.35 = 0.2625\,m^2$

Area of triangle $= \dfrac{1}{2}bh$

② Find the area of the shapes making the hole.

$Area\ of\ semi\text{-}circle = \dfrac{1}{2} \times \pi r^2 = \dfrac{1}{2} \times \pi \times 0.2^2 = 0.02\pi\,m^2$

$Area\ of\ rectangle = 0.4 \times 0.15 = 0.06\,m^2$ Height $= 0.35 - 0.2$

$Area\ to\ be\ painted = 0.1125 + 0.2625 - 0.02\pi - 0.06 = 0.252168\ldots\,m^2$

$Area\ for\ 2\ coats\ of\ paint = 2 \times 0.252168\ldots = 0.504336\ldots\,m^2$

Or you could work in cm and the convert to m using 10 000 cm² = 1 m².

③ $0.5043\ldots\,m^2$ is more than $\dfrac{1}{2}$ m², so Jake does need more paint.

Geometry   Circles and constructions

Exercise 11.1A

1 A circular helipad (for landing a helicopter) has a radius of 14 m. The cost of building the helipad is £85 per square metre. Is £50 000 enough? Explain your answer.

2 A flowerbed in the park is semicircular. It has a radius of 2 m.

Percy the park keeper wants to plant flowers that each need an area of 0.3 m².

   a How many of these flowers can Percy plant in the flowerbed?

   b What space does he have left?

3 A standard running track is 400 m in length.
Design a running track which includes

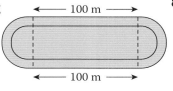

   ● two straight sections of 100 m each

   ● two semicircular sections at opposite ends.

Give your answer as a scale drawing, with dimensions clearly marked.

4 Two circles have the same centre. One has a radius of 3.5 cm and the other has a radius of 2.5 cm. Find the area of the shaded region.

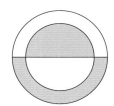

5 The diagram shows the dimensions of a flowerbed in Yusuf's garden.

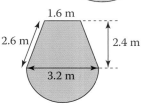

   a Yusuf says that 12 metres of edging will be enough to go around the flowerbed. Is Yusuf correct? Show your working.

   b Yusuf also wants to buy fertiliser to feed the flowers six times in the summer. He uses 35 grams per square metre each time. Is a 2 kilogram bag of fertiliser enough? Explain your answer

6 This shape is formed by 3 semi-circles. Show that

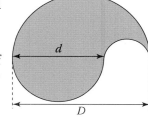

   a the perimeter of the shape $= \pi D$

   *b the area of the shape $= \frac{1}{4}\pi Dd$

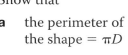

7 Imogen says that a circle with radius r has the same area as a semicircle with radius $2r$. Do you agree with Imogen? Explain your answer.

8 The diagram shows a regular hexagon inscribed in a circle. Write the perimeter of the hexagon as a fraction of the circumference of the circle. Give your answer in terms of π.

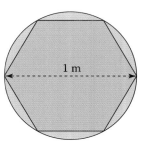

9 A rectangular sheet of metal is 1.2 metres long and 80 centimetres wide.

A badge-making machine cuts circles of metal from this sheet.

   a **i** Find the maximum number of circles of diameter 4 cm that can be cut from the sheet.

      ii Sally says that less than 25% of the metal is wasted. Is Sally correct? Show your working.

   *b The diameter 4 cm is only correct to the nearest millimetre. What difference could this make to the answers to part **a**?

*10 The area of this aircraft window must be less than 100 square inches. Find the maximum value for the height h.

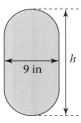

An **arc** is a fraction of the circumference of a circle.

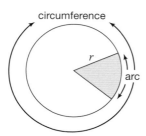

A **sector** is a fraction of a circle, shaped like a slice of pie.

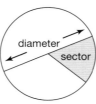

▲ The arc of a rainbow is one of the closest things to a perfect circle that can be found in nature.

To calculate the area and arc length of a sector, you use the angle at the centre of the sector.

> θ is the Greek letter 'theta'.

● Arc length $= \dfrac{\theta}{360} \times$ circumference of whole circle

$= \dfrac{\theta}{360} \times 2\pi r$

● Sector area $= \dfrac{\theta}{360} \times$ area of whole circle

$= \dfrac{\theta}{360} \times \pi r^2$

EXAMPLE

Find the arc length and area of this sector.

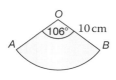

Substitute the values in the formula.

Arc length of sector $= \dfrac{106°}{360°} \times 2 \times \pi \times 10$

$= 18.50049...$ using a calculator

$= 18.5\,\text{cm}$ (to 3 sf)

Area of sector $= \dfrac{106°}{360°} \times \pi \times 10^2$

$= 92.502...$ using a calculator

$= 92.5\,\text{cm}^2$ (to 3 sf)

EXAMPLE

A circle has a radius of 28 cm. A sector of this circle has an area of 200 cm².
Calculate the angle of the sector to the nearest degree.

The area of the sector is proportional to the size of the angle.

$\dfrac{\text{Sector area}}{\text{Area of whole circle}} = \dfrac{\text{angle}}{360}$

Rearrange the formula to make the angle the subject.

Angle $= \dfrac{\text{Sector area}}{\text{Area of whole circle}} \times 360$ Area of whole circle $= \pi \times 28^2 = 2463.0086..$

$= \dfrac{200}{2463.0086} \times 360$

$= 29.232..$

Angle $= 29°$ (nearest degree)

Exercise 11.2S

1 This **quarter-circle** has a radius of 8 cm. Calculate

a the length of arc *AB*

b the area of sector *OAB*.

2 **a** Simplify

    i $\dfrac{150°}{360°}$ **ii** $\dfrac{200°}{360°}$ **iii** $\dfrac{144°}{360°}$

  b Simplify, leaving each in terms of π

    i $\dfrac{60°}{360°} \times 2\pi \times 24$

    ii $\dfrac{120°}{360°} \times 2\pi \times 15$

    iii $\dfrac{80°}{360°} \times \pi \times 6^2$

    iv $\dfrac{135°}{360°} \times \pi \times 12^2$

3 Find the area of each sector. Give your answer to 1 dp.

a

b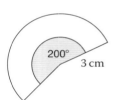

c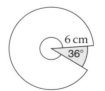

d

e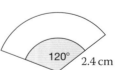

f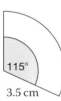

4 Find the arc length of each sector in question **3**.

5 Copy and complete the table for four sectors.

	angle	radius	arc length	area of sector
a	40°	18 mm		
b	135°	1.6 m		
c	252°	3.5 cm		
d	312°	0.46 km		

6 Find the perimeter of each sector.

a

b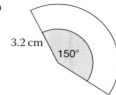

c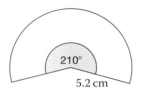

7 The radius of a circle is 30 cm.

An arc of this circle is 75 cm long.

Calculate the angle θ.

8 The area of a sector is 8 m².

The radius of the circle is 2 m.

Calculate the angle of the sector.

9 A square card has sides of length 24 cm.

A quarter-circle of radius 24 cm is cut from the square.

Find the area that is left.

10 Find the area of this incomplete annulus.

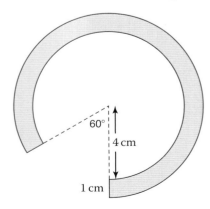

An annulus is the region between two concentric circles, shown here as the purple ring.

11.2 Circles 2

- Arc length $= \dfrac{\theta}{360} \times$ circumference of whole circle $= \dfrac{\theta}{360} \times 2\pi r$
- Sector area $= \dfrac{\theta}{360} \times$ area of whole circle $= \dfrac{\theta}{360} \times \pi r^2$

▲ Circle irrigation is used in dry climates to water crops by rotating a sprinkler around a pivot. This forms crop circles and also conserves water.

HOW TO

(1) Draw a diagram if it helps. Remember that a sector is a fraction of a whole circle.

(2) Decide which area, arc length or perimeter formula to use.

(3) Work out the answer to the question, giving units where necessary.

EXAMPLE

Show that the shaded area is given by $A = \dfrac{1}{8}\pi(R + r)(R - r)$.

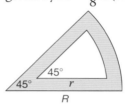

(2) Use area of **sector**, $A = \dfrac{\theta}{360°} \times \pi r^2$

Area of large sector $= \dfrac{45°}{360°} \times \pi \times R^2$ Cancel the fraction

$= \dfrac{1}{8}\pi R^2$

Area of small sector $= \dfrac{1}{8}\pi r^2$ The fraction is the same

Shaded area, $A = \dfrac{1}{8}\pi R^2 - \dfrac{1}{8}\pi r^2$

$A = \dfrac{1}{8}\pi(R^2 - r^2)$ Taking out the common factor

(3) $A = \dfrac{1}{8}\pi(R + r)(R - r)$ Factorising the difference of two squares

EXAMPLE

Three identical pencils are held together with a band.

The radius of each pencil is 4 mm.

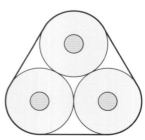

Show that the length of the band is $8(\pi + 3)$ mm.

The band has three straight sections and three arcs.

(1) Joining the pencil centres gives an equilateral triangle. Drawing other radii from the centres to the ends of the arcs gives three congruent rectangles.

(2) The arc length formula needs an angle

$\angle PAU = 360° - 60° - 90° - 90°$

$= 120°$

Arc $PU = \dfrac{120°}{360°} \times 2\pi \times 4$

$= \dfrac{1}{3} \times 8\pi$

Simplify where possible.

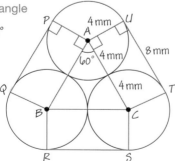

(3) Length of a straight section

$UT = AC = 8$ mm

Total length $= 3 \times \dfrac{1}{3} \times 8\pi + 3 \times 8$

$= 8\pi + 24$

$= 8(\pi + 3)$ mm

Exercise 11.2A

1 Jan wants to lay a patio in a corner of her garden. The patio is shaped like the sector of a circle.

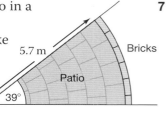

a What is the area of the patio?

b She wants to lay a brick path around the curved edge of the patio. Each brick is 30 cm long.

Calculate the number of bricks she needs.

*2 A circle has diameter 20 cm. The arc length of a sector of this circle is 10 cm. Calculate the area of the sector.

3 A continuous belt is needed to go around four identical oil drums as shown. The diameter of each drum is 24 inches.

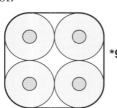

Work out the length of the belt. Give your answer in terms of π.

4 A sector of radius 3 cm is cut from one corner of an equilateral triangle.

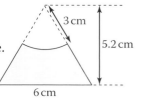

Find the area of the remaining shape.

5 Two quarter-circles are used to draw the leaf on this square tile.

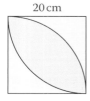

Show that

a the perimeter of the leaf is 20π cm

b the area of the leaf is 200(π − 2) cm².

6 The diagrams show the water in two pipes.

Pipe A   **Pipe B**

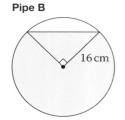

Show that the difference between the area of the water in these pipes is 128(π + 2) cm².

7 **a** Show that the area of this shape is given by
$$A = \frac{1}{12}\pi(R + r)(R - r)$$

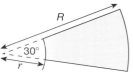

b Find the area when $R = 4\frac{1}{2}$ m and $r = 1\frac{1}{2}$ m. Give your answer in terms of π.

8 Find the *exact* area of this shape.

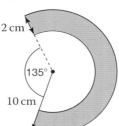

*9 The diagram shows the entrance to a road tunnel.

a Calculate the perimeter.

b The area needs to be at least 24 m². Does the area satisfy this requirement? Explain your answer.

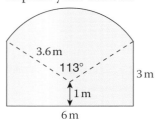

*10 Two sectors are cut from an isosceles triangle as shown.

Find, in terms of π, the area of the remaining shape.

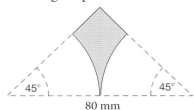

11.3 Circle theorems

Theorem 1

● The angle at the centre of a circle is twice the angle at the circumference from the same arc.

EXAMPLE

Draw the **radius** OC and extend it to D.

In $\triangle AOC$, $AO = OC$ Both **radii**.

$\qquad \angle OAC = \angle OCA = x$ Isosceles triangle.

$\qquad \angle COA = \angle 180° - 2x$

$\qquad \angle AOD = 2x$ Angles on a straight line.

Similarly using $\triangle COB$ you can prove that

$\qquad\qquad \angle DOB = 2y$

Now $\angle ACB = x + y$ and $\angle AOB = 2x + 2y$ as required.

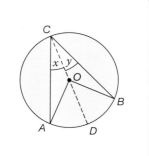

Theorem 2

● Angles from the same arc in the same segment are equal.

EXAMPLE

$\qquad\qquad \angle AOB = 2 \times \angle AXB$ Angle at centre is twice

$\qquad\qquad \angle AOB = 2 \times \angle AYB$ angle at circumference.

so $\angle AXB = \angle AYB$ as required.

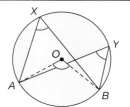

Theorem 3

● The angle in a semicircle is a right angle.

EXAMPLE

Angle at the centre = $180°$

Angle at centre = $2 \times$ angle at circumference

so angle at circumference = $90°$

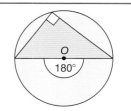

● All four vertices of a **cyclic quadrilateral** lie on the circumference of a circle.

Theorem 4

● The opposite angles of a cyclic quadrilateral add to $180°$.

EXAMPLE

Prove that the opposite angles of a cyclic quadrilateral add up to $180°$.

Draw the two diagonals AC and BD.

$\angle ABD = \angle ACD = x$ Angles on the same arc

$\angle CBD = \angle CAD = y$ are equal.

$\angle ADC = 180° - x - y$ Angles in a triangle.

$\qquad\quad = 180° - (x + y)$

But $x + y = \angle ABC$

Therefore $\angle ADC + \angle ABC = 180°$ as required.

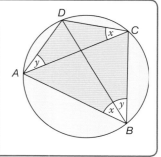

Geometry   Circles and constructions

Exercise 11.3S1

1 Work out the missing angles. Give a reason for each answer.

a

b

c

d

e
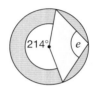

f

g

h

2 Work out the missing angles. Give a reason for each answer.

a

b

c

d

e

f

g

h

3 Work out the missing angles. Give a reason for each answer.

a

b

c

d

e
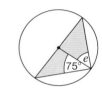

f

g
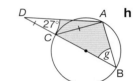

h
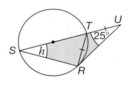

4 Work out the missing angles. Give a reason for each answer.

a

b

c

d

e

f

g

h
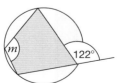

Q 1087, 1142, 1321   SEARCH

11.3 Circle theorems

You need to know some facts about **tangents** to circles.

Theorem 5

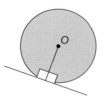

- The angle between a tangent and the **radius** at the point where the tangent touches the circle is a right angle.

Theorem 6

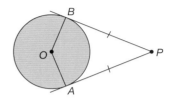

- Two tangents drawn from a point to a circle are equal.

Theorem 7

- The **perpendicular** line from the centre of a circle to a **chord** bisects the chord.

You need to know the alternate segment theorem.

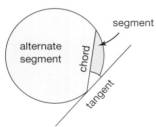

The **chord** divides the circle into two **segments**. The acute angle between the **tangent** and the chord is in the minor segment. The major segment is the **alternate segment**.

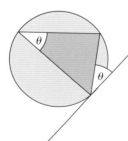

Theorem 8

- The angle formed between a tangent and a chord is equal to the angle from that chord in the alternate segment of the circle.

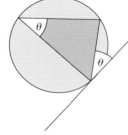

EXAMPLE

Draw a diameter where the tangent touches the circle (T).
Join this diameter (TX) to the other end of the chord (P) to form a triangle PTX.

$\angle XPT = 90°$ Angle in a semicircle.
The angle between tangent and radius is a right-angle, so
$\angle PTX = 90° - \theta$.
$\angle TXP + 90° + (90° - \theta) = 180°$ Angles in a triangle add up to 180°.
so $\angle TXP = 180° - 180° + \theta = \theta$
Angles from the same arc in the same segment are equal, so
the angle in the alternate segment, $\angle PYT = \theta$.

Exercise 11.3S2

1 Work out the missing angles. Give a reason for each answer.

a
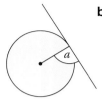

b
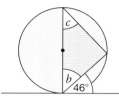

c
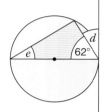

d
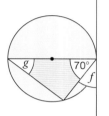

e
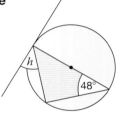

f
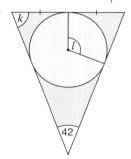

g
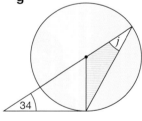

h
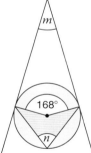

i
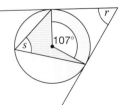

j
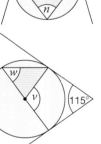

k
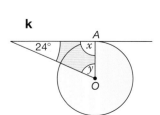

l
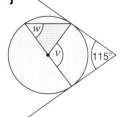

2 Work out the missing angles. Give a reason for each answer.

a
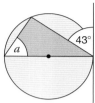

b
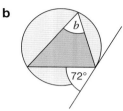

c
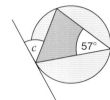

d
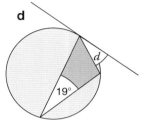

e
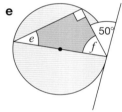

f
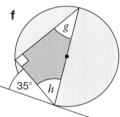

g
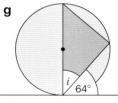

h
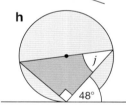

i
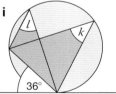

j
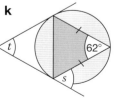

k
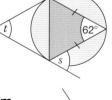

l
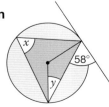

m
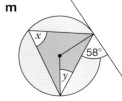

n
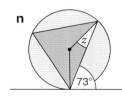

1087, 1142, 1321   SEARCH

11.3 Circle theorems

You must be able to recall and apply all of the theorems from the Skills lessons.

HOW TO

① Draw a diagram (or use one that is given).

② Decide which theorems may apply.

③ Construct a chain of reasoning to find the angles you need or prove the required results.

EXAMPLE

In a **cyclic quadrilateral** *ABCD* prove that an interior angle is equal to the opposite exterior angle, $\angle DAB = \angle BCE$.

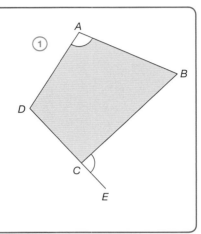

② You know that in a cyclic quadrilateral, opposite angles add up to 180°, so look at the opposite angles of *ABCD*.

③ $\angle DCB = 180° - \angle DAB$ Opposite angles in a cyclic quadrilateral.

$\angle BCE = 180° - \angle DCB$ Angles on a line

or

$\angle DCB = 180° - \angle BCE$

Therefore

$\angle DAB = \angle BCE$ as required.

EXAMPLE

PQR is a triangle drawn inside a circle.
Angle *PQR* = 100°.
Is *PR* a diameter?

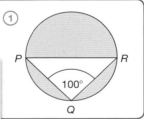

② If *PR* were a diameter, angle *PQR* would be an angle in a semicircle, which is 90°.

③ Angle *PQR* ≠ 90°, so *PR* is not a diameter.

Suppose that *PR* is a diameter, and see if the rest of the information fits. As it does not, your supposition must be wrong.

EXAMPLE

In the diagram *AB* = *ST*.

Prove   **a** $\angle AOB = \angle SOT$.

      b   Angles at the **circumference** from equal **arcs** are equal.

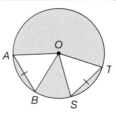

a ② *AO* = *BO* = *SO* = *TO*   Radii
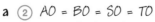
AB = *ST* (given)

③ So △*AOB* and △*SOT* are congruent (SSS)
and $\angle AOB = \angle SOT$

b ② Arcs *AB* and *ST* are equal.
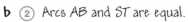
$\angle AOB = 2 \times \angle APB$ Angle at centre is twice
$\angle SOT = 2 \times \angle SQT$ angle at circumference.

③ So $\angle APB = \angle SQT$

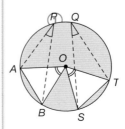

(given) means a fact you have been told.

Exercise 11.3A

1 In the quadrilateral *PQRS* ∠*PQR* = 38° and ∠*RSP* = 138°.

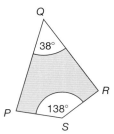

Explain why *PQRS* cannot be a cyclic quadrilateral.

2 *ABCD* is a parallelogram and it is also a cyclic quadrilateral.
Use circle theorems and properties of quadrilaterals to explain which special type of quadrilateral *ABCD* must be.

3 *D*, *E* and *F* are three points on the circumference of a circle of radius 68 mm. *DE* = 120 mm and *EF* = 64 mm.

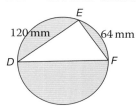

Show that *DF* is a diameter of the circle.

4 Explain why *C* is not the centre of the circle.

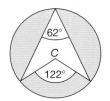

5 *PQRS* is a cyclic quadrilateral.
X lies outside the circle such that *XQR* and *XPS* are straight lines.
XQ and *XP* are equal in length.
Use the 'angles in cyclic quadrilaterals' theorem to show that *PQRS* is an isosceles trapezium.

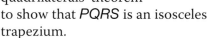

6 *X* and *Y* lie on the circumference of a circle with centre *O*.
P is a point outside the circle.
Angle *PXY* = 62°
Reflex angle *XOY* = 256°

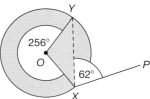

Explain why *PX* is not a tangent to the circle.

7 Prove that two tangents drawn from a point to a circle are equal in length.

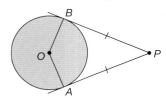

8 *XS* and *XT* are equal chords of a circle with centre *O*.

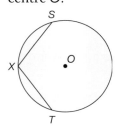

Using congruent triangles, prove that *OX* bisects angle *SXT*.

9 Prove that the perpendicular line from the centre of a circle to a chord bisects the chord.

*10 Two chords *AB* and *AC* in a circle are equal. Prove that the tangent at *A* is parallel to *BC*.

*11 Points *P*, *Q* and *R* lie on a circle.

The tangent at *P* is parallel to *QR* and the tangent at *Q* is parallel to *PR*.

Prove that triangle *PQR* is equilateral.

11.4 Constructions and loci

You can **construct** a unique **triangle** when you know

| two sides and the angle between them (SAS) | or | two angles and a side (ASA) | or | right angle, the hypotenuse and a side (RHS) | or | three sides (SSS). |

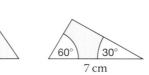

> Any two triangles constructed using any one of these four sets of information will be congruent.

5 cm, 30°, 6 cm

60°, 30°, 7 cm

5 cm, 3 cm

R, 6 cm, 8 cm, P, 10 cm, Q

You will need a ruler and a protractor for SAS, ASA and RHS triangles.

You will need a ruler and compasses for SSS triangles.

- **Bisect** means cut into two equal parts.

> Use the same compass radius throughout the constructions. Start at the red dots.

You can use a straight edge and compasses to construct an angle bisector.

- To bisect angle *ABC*

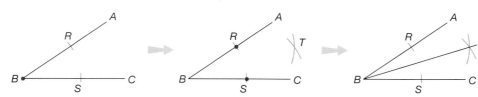

- All points on the angle bisector are **equidistant** from the arms of the angle.

- The **perpendicular** bisector of a line bisects the line at right angles.

> Equidistant means equal distance from.

- To construct the perpendicular bisector of line *AB*

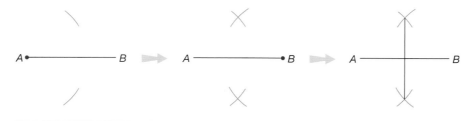

- All points on the perpendicular bisector of *AB* are equidistant from *A* and *B*.

- To construct the perpendicular from a point *X* to a line *YZ*.

Start at the red dots.

Keep the same compass radius throughout the construction.

Exercise 11.4S

1 Trace these lines and the points marked *X*.
For each, use ruler and compasses to construct a perpendicular from the point *X* to the line.
Check your constructions using a protractor.

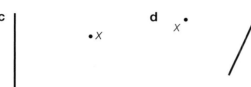

2 Trace these lines and mark the point *X*.
For each, use ruler and compasses to construct the perpendicular from the point *X* on the line. Show all construction lines.
Check your constructions using a protractor.

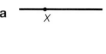

3 Trace and extend these angles.
Construct the angle bisector of each angle, using ruler and compasses.

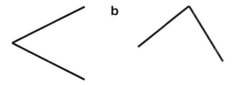

You will need to extend the lines.

4 Use a protractor to draw these angles.

a	70°	**b**	110°	**c**	90°
d	130°	**e**	50°	**f**	85°

Using compasses, construct the angle bisector for each angle.

Use a protractor to check each angle bisector.

5 Use a straight edge and compasses or a protractor to construct these triangles.

a Sides 8 cm, 4 cm, 7 cm   (SSS)
b 3 cm, 30°, 4 cm   (SAS)
c Sides 10 cm, 7.5 cm, 6 cm   (SSS)
d 8 cm, 2 cm, 90°   (RHS)
e Sides 6 cm, 9 cm, 5 cm   (SSS)
f 45°, 4 cm, 45°   (ASA)

It helps to draw a rough sketch first.

6 Construct isosceles triangles with sides

a 7 cm, 7 cm, 5 cm   **b** 5 cm, 5 cm, 7 cm

7 **a** Construct an equilateral triangle with sides 5 cm.

b Construct the angle bisector of each angle of the triangle.

c What do you notice about the three angle bisectors?

8 Follow these steps to construct an angle of 30°.

a Construct an equilateral triangle with sides 4 cm.

b Bisect one of the base angles.

9 Draw lines *AB* for these lengths and construct their perpendicular bisectors.

a 6 cm   **b** 9 cm   **c** 5.6 cm
d 10 cm   **e** 11.2 cm

Check by measuring that each bisector intersects the line *AB* at its midpoint.

10 **a** Construct an equilateral triangle with sides 5 cm.

b Construct the perpendicular bisectors of each side of the triangle.

c Compare your diagram with the one for question 7.

Write down what you notice.

 1089, 1090, 1147   SEARCH

11.4 Constructions and loci

- The locus of a point which is a constant distance from another point is a circle.

- The locus of a point that is **equidistant** from two other fixed points is the **perpendicular bisector** of the line joining the fixed points.

- The locus of a point at a constant distance from a fixed line is a parallel line.

- The locus of a point equidistant from two intersecting lines is the angle bisector of the lines.

HOW TO

① Draw a diagram or copy a given diagram.

② Decide what you need to construct.

③ Use a pencil, pair of compasses and ruler to carry out the construction.

④ Answer any questions asked.

EXAMPLE

P and Q are two points 2.5 cm apart on a line.

P Q

Shade in the region that satisfies all these conditions:

- Right of the perpendicular to the line PQ at point P.

- Closer to P than to Q.

- More than 1 cm from P.

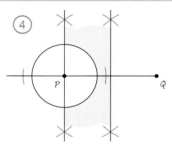

② ③ Construct the perpendicular to the line at point P.

Construct the perpendicular bisector of PQ.
Points to the left are nearer to P than Q.

Draw a circle radius 1 cm, centre P. Points outside are more than 1 cm from P.

EXAMPLE

$ABCD$ is the plan of a garden.

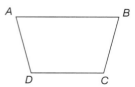

A tree is to be planted in the garden so that it is

- nearer to BC than to BA

- nearer to AD than to AB.

Shade the region where the tree may be planted.

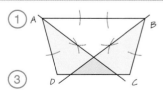

② Construct the angle bisector of ∠ABC. Shade the region between this line and BC.

Construct the angle bisector of ∠BAD. Shade the region between this line and AD.

④ The tree can be planted in the region where the shadings overlap.

Geometry Circles and constructions

Exercise 11.4A

1 a Draw points *A* and *B*, 6 cm apart.

b Shade in the region that satisfies both these conditions:

    i closer to *A* than to *B*

    ii less than 4 cm from *B*.

2 a Draw a rectangle *PQRS* where *PQ* = 5 cm and *QR* = 3 cm.

b Shade the region of the rectangle that is within 4 cm of *P* and within 2.5 cm of *R*.

3 a Construct a right-angled triangle *ABC* where angle *ABC* = 90°, *AB* = 6 cm, *BC* = 4.5 cm.

b Shade the region that satisfies all three of these conditions.

    i closer to *A* than to *B*

    ii less than 5 cm from *A*

    iii less than 5 cm from *C*.

4 a Construct two triangles with sides 8 cm, 5 cm, 7 cm. Label them triangle *A* and triangle *L*.

b On triangle *A* construct the angle bisector of each internal angle of the triangle.

c On triangle *L* construct the perpendicular bisector of each side of the triangle.

d Compare and comment on your answers to **b** and **c**.

5 a Draw a line *AB*, 8 cm long, and construct its perpendicular bisector.

b Construct an angle of 45° where the perpendicular bisector intersects *AB*.

c What other angles have you created in this construction?

6 A lifeboat *L* is 10 km from another lifeboat *K* on a bearing of 045°. They both receive a distress call from a ship. The ship is within 7 km of *K* and within 5 km of *L*.

Draw a scale drawing to show the positions of *K* and *L*. Shade on your diagram the area in which the ship could be.

7 The diagram shows the rectangular garden of a house. There are two trees, *T*, in the garden. A radio mast is to be placed in the garden. It must be more than 5 m from the rear of the house. It must be more than 3 m from any tree. Using a scale of 1 cm : 2 m, draw a scale diagram and shade the possible site for the radio mast.

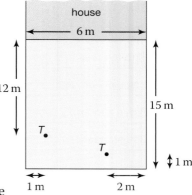

8 Pat wants to tether her goat on this grass using a 5 foot long chain. One end of the chain will be attached to the goat and the other end to a ring that can slide along an 18 foot long rail.

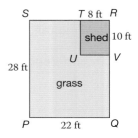

a Calculate the area of the grass that the goat can reach if Pat puts the rail along the sides *TU* and *UV* of the shed.

b Where on the perimeter of the grass should Pat put the rail to allow the goat to reach the greatest area of grass? Explain your answer.

*9 a Jamie says that the construction of an angle bisector works because of congruent triangles.

Is this true? Give your reasons.

b Give geometric reasons why each of the following constructions work.

    i 60° angle.

    ii Perpendicular bisector of a line segment.

Summary

Checkout
You should now be able to...

Test it
Questions

✔ Find the area and circumference of a circle and composite shapes involving circles.	1 – 3
✔ Calculate arc lengths, angles and areas of sectors.	4
✔ Prove and apply circle theorems.	5, 6
✔ Use standard ruler and compass constructions and solve problems involving loci.	7 – 9

Language	Meaning	Example
Circle	A closed curve in a flat surface which is everywhere the same distance from a single fixed point.	
Diameter	A chord that passes through the centre of the circle.	
Radius **Radii (plural)**	A straight line segment drawn from the centre of the circle to the perimeter.	
Circumference	The distance around the edge of a circle.	
Arc	A continuous section of the circumference of a circle.	
Chord	A straight line segment with endpoints on the circumference of a circle.	
Tangent	A straight line which touches the circle at one point only and does not cut the circle.	
Segment	The 2D shape enclosed by an arc and a chord.	
Sector	The shape enclosed by two radii and an arc.	
Bisect	Cut into two parts of the same shape and size.	
Perpendicular bisector	A line which bisects another line at right angles.	
Construct	Draw something accurately using compasses and a ruler.	
Construction lines	Lines drawn during a construction that are not part of the final object.	
Locus **Loci (plural)**	A set of points which satisfy a given set of conditions. The path followed by a moving point.	The set of points less than 1 cm from P is the interior of a circle.

Review

1 For this circle calculate the

    a area

    b circumference.

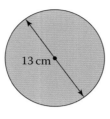

13 cm

2 The area of a circle is 160 cm². What is the diameter of the circle to 1 decimal place?

3 The shape of a swimming pool consists of a rectangle with a semi-circle at one end as shown.

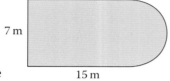

7 m

15 m

    Calculate

    a the area

    b the perimeter of the swimming pool.

4 Calculate

    a the area

    b the arc length

    c the perimeter of the sector.

82°

8 cm

5 Calculate the size of angles a, b, c and d. Give reasons for your answers.

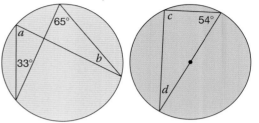

65°
a
33°
b
c
54°
d

6 Calculate the size of angles a and b.

Give reasons for your answers.

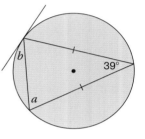

b
39°
a

7 **a** Draw a line exactly 6.5 cm long.

    b Use a ruler and pair of compasses to accurately construct the perpendicular bisector of this line.

8 A dog is attached to a lamp post by a rope which is 2.4 m long and can rotate freely around the post.

Use a scale of 60 : 1 to draw to scale the locus of the furthest points the dog can reach.

9 Use a pencil, ruler and pair of compasses to construct a triangle with sides of lengths 8 cm, 7 cm and 5 cm.

What next?

Score	0 – 4		Your knowledge of this topic is still developing.
			To improve look at MyMaths: 1083, 1087, 1088, 1089, 1090, 1118, 1142, 1147, 1321
	5 – 8		You are gaining a secure knowledge of this topic.
			To improve look at InvisiPens: 11Sa – i
	9		You have mastered these skills. Well done you are ready to progress!
			To develop your exam technique looks at InvisiPens: 11Aa – g

Assessment 11

1 **a** Paul calculates the radii of three different circles from their circumferences.
Match the correct radius to the correct circumference and find the missing radius.
Show your working.

  Circumferences: **i** 61.58 cm **ii** 314.16 cm **iii** 2.01 cm

  Radii: **a** x cm **b** 0.320 cm **c** 50.0 cm [3]

  b He then calculates the radii of three different circles from their areas. Find the missing
radius and match the correct radius to the correct area. Show your working.

  Areas: **i** 176.7 m² **ii** 81.1 m² **iii** 0.407 m²

  Radii: **a** 5.08 m **b** 7.50 m **c** y m [3]

2 A reel of electrical flex contains 50 m of flex. The diameter of the wheel is 25 cm.
An electrician completely unwinds the reel.
Calculate the number of times the reel rotates (ignore the thickness of the hose). [4]

3 **a** A Polo mint is 19 mm in diameter with an 8 mm hole in the middle.
What is the area of the face to the nearest mm²? [3]

  b Find the circumference of the outer and inner holes [3]

4 John is doing a 50 mile cycle ride for charity. The diameter of his bike wheels
is 27.56 in. 1 mile = 63360 in. How many times, to the nearest whole number,
do the wheels rotate on his ride? [6]

5 The London Eye's wheel has a diameter of 120 m. The pods rotate at
26 cm per second. How long, to the nearest minute, does the wheel take
to make a complete revolution? [6]

6 The diagram shows a heart shape.

  There are two semicircles at the top each 6 cm in
diameter.

  The sides are made of two circular arcs. The right hand
arc, YZ has centre X and the left hand arc, XZ has
centre Y. The distance from the middle of the two
semicircles at the top to the point Z is 10.39 cm.
Calculate

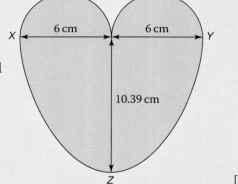

  a the total perimeter of the heart diagram [8]

  b the area of the heart diagram. [10]

7 A sector has a radius equal to its arc length.

  Jacob says that the angle at the centre is 57.3° to 1 dp.

  Show how he worked out the value of this angle. [3]

8 This is a diagram of the centre circle of an ice hockey pitch.
There are players at the centre point O and on the
circumference at points A, B and C. Angle $AOB = 69°$

  Write down the values of the angles p, q and r.
Give reasons for your answers. [6]

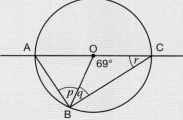

9   The diagram shows a surveyor's trundle wheel.
    The handle is inclined at 34° to the vertical.
    O is the centre of the circle and the diameter through
    O is horizontal.
    Write down the values of the angles *BOC*, *BCO* and *CAB*.
    Give reasons for your answers.   [6]

10 The diagram shows the framework housing a ball on a
    tenpin bowling rink.
    O is the centre of the circle and the diameter through it
    is vertical.
    The line at the bottom of the framework is the horizontal
    runner on which the ball rests. Write down the values of
    the angles *v*, *w* and *x*. Give reasons for your answers.   [6]

11 Thomas says that *x* = *y*.
    Without using the Alternate Segment Theorem prove
    that he is correct.   [8]

    O is the centre of the circle.

    The diagram includes the tangent to the circle at *B*.

    A, *B* and *C* are points on the circumference.

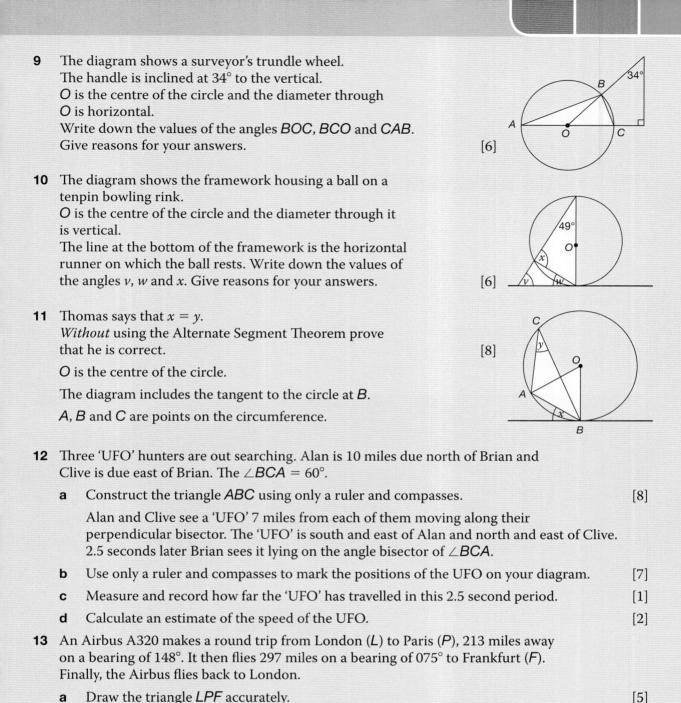

12 Three 'UFO' hunters are out searching. Alan is 10 miles due north of Brian and
    Clive is due east of Brian. The ∠*BCA* = 60°.

    a   Construct the triangle *ABC* using only a ruler and compasses.   [8]

        Alan and Clive see a 'UFO' 7 miles from each of them moving along their
        perpendicular bisector. The 'UFO' is south and east of Alan and north and east of Clive.
        2.5 seconds later Brian sees it lying on the angle bisector of ∠*BCA*.

    b   Use only a ruler and compasses to mark the positions of the UFO on your diagram.   [7]

    c   Measure and record how far the 'UFO' has travelled in this 2.5 second period.   [1]

    d   Calculate an estimate of the speed of the UFO.   [2]

13 An Airbus A320 makes a round trip from London (*L*) to Paris (*P*), 213 miles away
    on a bearing of 148°. It then flies 297 miles on a bearing of 075° to Frankfurt (*F*).
    Finally, the Airbus flies back to London.

    a   Draw the triangle *LPF* accurately.   [5]

    b   Find the distance and bearing of the journey from Frankfurt to London.   [2]

12 Ratio and proportion

Introduction

Colour theory is based on the idea that all possible colours can be created from three 'primary colours'. Video screens use the additive colours red, green and blue. Artists often use red, yellow and blue as the basis for mixing paints. Whilst the printing industry uses the subtractive colours cyan, magenta and yellow. The ratio in which these primary colours are mixed determines the colour of the result.

What's the point?

An understanding of ratio is essential for artists in mixing colours on a palette to achieve a desired result. Furthermore, and understanding of proportion is essential for artists in ensuring that the elements of their artwork are in the 'right proportions'.

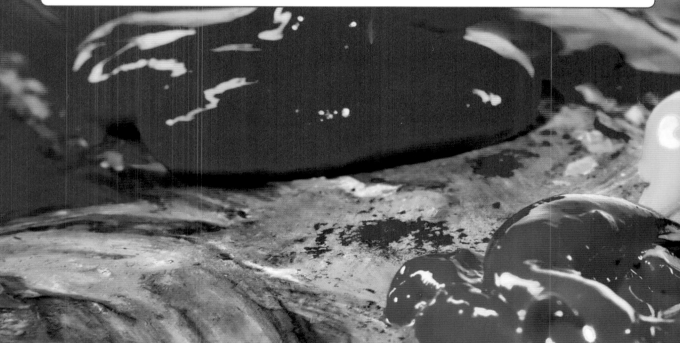

Objectives

By the end of this chapter, you will have learned how to ...

- Express proportions of amounts as fractions or percentages.
- Divide a quantity in a given ratio.
- Use scale factors to convert between lengths on maps and scale diagrams and the distances they represent.
- Calculate percentage increases and decreases using multiplication.
- Find the original value of a quantity that has undergone a percentage increase or decrease.

Check in

1 Calculate each of these expressions.

a $\frac{2}{3}$ of 120   **b** 0.35×60 **c** 28% of 70   **d** $\frac{3}{5}$ of 280

2 Carry out each of these conversions.

a Convert $\frac{3}{5}$ to a percentage.   **b** Convert 0.35 to a fraction.

c Convert 65% to a fraction.   **d** Convert $\frac{7}{20}$ to a decimal.

3 Write these ratios in their simplest form.

a $4:12$ **b** $5:30$ **c** $6:9$ **d** $15:6$

Chapter investigation

The ratio of height to head circumference for the average person is said to be around $3:1$.
Investigate.

12.1 Proportion

- A **proportion** is a part of the whole. It can be written using a fraction, a decimal or a percentage.

EXAMPLE

Find **a** $\dfrac{3}{5}$ of 85 **b** 120% of 45 **c** $\dfrac{7}{8}$ of 86

a $\dfrac{3}{\cancel{5}_1} \times \cancel{85}^{17} = \dfrac{3}{1} \times 17 = 51$ Cancel the 5s before multiplying.

b Work out 20% of 45 and add it to the original amount.

120% of 45 = (100% of 45) + (20% of 45)

20% of 45 = $\dfrac{1}{5} \times 45 = 9$ ⟹ 120% of 45 = 45 + 9 = 54

c Work out $\dfrac{1}{8}$ of 86 and then subtract from the original amount.

$\dfrac{1}{8}$ of 86 = 86 ÷ 8 = 10 + $\dfrac{6}{8}$ = $10\dfrac{3}{4}$ ⟹ $\dfrac{7}{8}$ of 86 = 86 − $10\dfrac{3}{4}$ = 76 − $\dfrac{3}{4}$ = $75\dfrac{1}{4}$

You can express one quantity as a percentage of another in three steps:

1. Write the first quantity as a fraction the second.
2. Convert the fraction to a decimal by division.
3. Convert the decimal to a percentage by multiplying by 100.

Percentage is per-cent which means parts per hundred.

EXAMPLE

What proportion is **a** 4 of 32 **b** 5 of 35 **c** 12 of 4.8?
Write your answers as percentages.

a $\dfrac{4}{32} = \dfrac{1}{8}$

so 4 is $\dfrac{1}{8}$ of 32

$\dfrac{1}{8} = \dfrac{1}{8} \times 100\%$

= $12\dfrac{1}{2}\%$

b $\dfrac{5}{35} = \dfrac{1}{7}$

so 5 is $\dfrac{1}{7}$ of 35

$\dfrac{1}{7} = \dfrac{1}{7} \times 100\%$

= 14.3%

c $\dfrac{12}{100} = \dfrac{3}{25}$

so 12 is $\dfrac{3}{25}$ of 100

$\dfrac{3}{25} = \dfrac{3}{25} \times 100\%$

= 12%

EXAMPLE

Skye took two tests. In German she scored 35 out of 50, and in French she scored 60 out of 80. Write these scores as percentages.

German

35 out of 50 = $\dfrac{35}{50}$

= 35 ÷ 50

= 0.7

= 70%

French

60 out of 80 = $\dfrac{60}{80}$

= 60 ÷ 80

= 0.75

= 75%

- You can compare proportions by converting them to percentages or decimals.

Ratio and proportion   Ratio and proportion

Exercise 12.1S

1 One tenth ($\frac{1}{10}$) of the weight of a soft drink is sugar. Find the amount of sugar in these weights of drink.

   a 750 g   **b** 45 g

   c 1 kg   **d** 1250 g

2 Three fifths ($\frac{3}{5}$) of the volume of a fruit cocktail is orange juice. Find the amount of orange juice in these volumes of fruit cocktail. You should show all of your working.

   a 150 ml   **b** 380 ml

   c 2 litres   **d** 280 cm³

3 Calculate

   a 120% of 50 g   **b** 90% of 40 mm

   c 95% of 400 g   **d** 80% of 39 km

4 Write these percentages as fractions and as decimals.

   a 26%   **b** 71%   **c** 2%   **d** 102%

5 Write these percentages as fractions in their simplest form and decimals.

   a 38%   **b** 46%   **c** 80%   **d** 7%

6 What proportion of

   a 15 is 5

   b 20 is 10

   c 10 cm is 3 cm

   d 12 kg is 3 kg

   e £30 is £12

   f £8 is £16

   g 200 is 25

   h 6.4 is 0.4

   i £1.44 is 12p

   j 1.2 m is 30 cm

7 Write your answers from question **6** as percentages.

8 Write the proportion of each of these shapes that is shaded. Write each of your answers as

   i a fraction in its simplest form

   ii a percentage (to 1 dp as appropriate).

   a   **b**

   c   **d**

9 Work out these proportions, giving your answers as

   i fractions in their lowest terms

   ii percentages.

   a 7 out of every 20

   b 8 parts in a hundred

   c 6 out of 20

   d 75 in every 1000

   e 18 parts out of 80

   f 9 parts in every 60

   g 20 out of every 3

   h 20 out of 8

   i 412 parts in a hundred

   j 6400 in every thousand

10 Use an appropriate method to work out these percentages. Show your method each time.

   a 50% of 270 kg   **b** 27.9% of 115 m

   c 37.5% of £280   **d** 25% of 90 cm³

   e 19% of 2685 g   **f** 27.5% of £60.00

11 Four candidates stood in an election.

   A received 19 000 votes

   B received 16 400 votes

   C received 14 800 votes

   D received 13 200 votes

   Write each of these results as a percentage of the total number of votes.

   Give your answers to 1 d.p.

12.1 Proportion

EXAMPLE

A 1 kg bag of 'Grow Up' fertiliser contains 45 grams of phosphate. A 500 gram packet of 'Top Crop' fertiliser contains 20 grams of phosphate. What is the proportion of phosphate in each fertiliser?

1. Write the whole.

The whole for 'Grow Up' fertiliser is 1 kg = 1000 g.
The whole for 'Top Crop' fertiliser is 500 g.

2. Express the proportions as fractions.

Proportion of phosphate in 'Grow Up'
$$= \frac{45}{1000} = \frac{9}{200}$$
Proportion of phosphate in 'Top Crop'
$$= \frac{20}{500} = \frac{4}{100}$$

3. Convert the fractions to percentages.

Percentage of phosphate in 'Grow Up'
$$= \frac{45}{1000} \times 100 = 4.5\%$$
Percentage of phosphate in 'Top Crop'
$$= \frac{20}{500} \times 100 = 4\%$$

EXAMPLE

Tom has £2800. He gives $\frac{1}{5}$ to his son and $\frac{1}{4}$ to his daughter.
How much does Tom keep himself? You must show all of your working.

1. Write the whole.

The whole is £2800.

2. Multiply by the fraction to find the size of the part.

Tom's son receives
$$\frac{1}{5} \text{ of } £2800 = \frac{2800}{5} = £560$$
Tom's daughter receives
$$\frac{1}{4} \text{ of } £2800 = \frac{2800}{4} = £700$$
Tom keeps £2800 − (£560 + £700) = £1540

Exercise 12.1A

1 A 250 ml glass of fruit drink contains 30 ml of pure orange juice. What proportion of the drink is pure orange juice?
Give your answer as

   a a fraction in its lowest terms

   b a percentage.

2 Samantha wins £4500 in a competition.
She gives $\frac{1}{3}$ to her mother and $\frac{1}{5}$ to her sister.

   a How much does she keep?
Show your working.

   b What proportion of the prize money does she give away? Give your answer as a fraction and as a percentage.

3 A dairy farmer takes 200 cheeses to sell at a market. In the first hour she sells 20% of the cheeses. In the second hour she sells 15% of those that are left.

   a How many cheeses has she sold in total?

   b What percentage of the original number of cheeses did she sell in the first 2 hours?

4 Sunita took three tests. In Maths she scored 48 out of 60, in English she scored 39 out of 50 and in Science she scored 55 out of 70.

   a In which subject did she do

     i the best

     ii the worst?

   b Suggest a way in which you could give Sunita a single overall score for her three tests.

5 20% of the cars in a car park are red.
40 cars are not red.
How many cars are in the car park?

6 Peter is driving 220 miles from Bristol to York.
Peter drives at an average speed of 45 mph for the first hour.
He drives at an average speed of 60 mph for the next two hours.
Peter says that
"After three hours I will have completed three-quarters of the journey."
Do you agree with Peter?
Explain your answer.

7 Jada compares the amount of sugar in three different types of fruit juice.
  ● Tropical 22.4 g of sugar in 180 ml
  ● Pineapple 28.8 g of sugar in 250 ml
  ● Blackcurrant 23.4 g of sugar in 200 ml
Jada says that tropical juice has the highest proportion of sugar. Is she correct? Show your workings.

8 **a** Complete these equivalent fractions.

$$\frac{16}{\boxed{?}} = \frac{18}{45} \qquad \frac{18}{45} = \frac{20}{\boxed{?}}$$

   b A company has 45 employees at the start of 2015.
18 employees are female.
By 2016, two more female employees have joined the company.
The **proportion** of female employees has not changed.
Use your answer to part **a** to find the number of total employees in 2016.

9 Jenni compares the proportion of shaded area of two shapes.

Shape A   **Shape B**

She says that shape A has the largest proportion of shaded area.
Do you agree with Jenni?
Explain your answer.

10 What proportion of the shape is shaded?

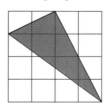

12.2 Ratio

You can compare the size of two quantities using a **ratio**.
2:3 means '2 parts of one quantity, compared with 3 parts of another'.

Check that the measurements have the same units.

- You can simplify a ratio by dividing both parts by the same number. When a ratio cannot be simplified any further it is in its **simplest form**.

EXAMPLE

Henry the snake is only 30 cm long. George the snake is 90 cm long. Express this as a ratio.

Compare the two lengths.

= Henry's length : George's length

= 30 cm : 90 cm

= 30 : 90

Express the ratio 30:90 in its simplest form.

$$÷10 \Big(\begin{array}{c} 30:90 \\ \\ = 3:9 \\ \\ = 1:3 \end{array} \Big) ÷10$$

$$÷3 \qquad\qquad ÷3$$

The ratio 1:3 means that 90 is three times bigger than 30.

You can divide a quantity in a given ratio.

EXAMPLE

Sean and Patrick share £355 in the ratio 3:7. How much money do they each receive?

Find the total number of parts of the whole amount.

Total number of parts = 3 + 7 = 10 parts

Find the value of one part.

Each part = £355 ÷ 10 = £35.50

Sean receives 3 parts = 3 × £35.50 = £106.50

Patrick receives 7 parts = 7 × £35.50 = £248.50

Check your answer by adding up all the parts. They should add up to the amount being shared!

£106.50 + £248.50 = £355

A simplified ratio does not contain any units in the final answer.

- You can express a **ratio** in the form 1 : n using division. This is often called a **scale**.

Maps and plans are drawn to scale. You can solve problems involving scales by multiplying or dividing by the scale.

In this scale diagram, 1 cm on the drawing represents 100 cm on the real house.

Corresponding lengths are multiplied by the same **scale factor**. You can write the scale factor as a ratio.

You can write this scale factor as 1 cm represents 100 cm or 1:100.

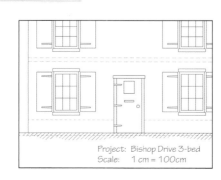

Project: Bishop Drive 3-bed
Scale: 1 cm = 100cm

Exercise 12.2S

1 Write each of these ratios in its simplest form.

a	2:6	**b**	15:5
c	6:18	**d**	4:28
e	5:50	**f**	30:6
g	24:8	**h**	2:30

2 Write the number of blue squares to the number of red squares as a ratio in its simplest form for each of these shapes.

a   **b**

c   **d**

3 Simplify each of these ratios.

a	4:2	**b**	3:9
c	21:14	**d**	18:16
e	45:60	**f**	48:36
g	80:72	**h**	96:120

4 Solve each of these problems.

a Divide £90 in the ratio 3:7.

b Divide 369 kg in the ratio 7:2.

c Divide 103.2 tonnes in the ratio 5:3.

d Divide 35.1 litres in the ratio 5:4.

e Divide £36 in the ratio 1:2:3.

5 Solve each of these problems.
Give your answers to 2 decimal places where appropriate.

a Divide £75 in the ratio 8:7.

b Divide £1000 in the ratio 7:13.

c Divide 364 days in the ratio 5:2.

d Divide 500 g in the ratio 2:5.

e Divide 600 m in the ratio 5:9.

6 Write each of these ratios in its simplest form.

a	40 cm:1 m	**b**	55 mm:8 cm
c	3 km:1200 m	**d**	4 m:240 cm
e	700 mm:42 cm	**f**	12 mins:450 secs

7 Divide the amounts in the ratios indicated.

a	£500	2:5:3
b	360°	4:1:3
c	2 km	9:4:7

> Hint for **c**:
> Convert to metres.
> 1 km = 1000 m

8 Express each of these ratios as a ratio in the form 1:n (a scale).

a	2:6	**b**	3:12
c	10:20	**d**	8:40
e	3:6	**f**	5:15
g	4:20	**h**	12:36
i	30:60	**j**	9:45
k	45:90	**l**	20:120

9 A map has a scale of 1:50000 or 1 cm represents 50000 cm. Calculate in metres the actual distance represented on the map by these measurements.

a	2 cm	**b**	8 cm
c	10 cm	**d**	0.5 cm
e	14.5 cm

10 A map has a scale of 1:5000.

a What is the distance in real life of a measurement of 6.5 cm on the map?

b What is the distance on the map of a measurement of 30 m in real life?

11 A map has a scale of 1:500.

a What is the distance in real life of a measurement of 10 cm on the map?

b What is the distance on the map of a measurement of 20 m in real life?

12 A map has a scale of 1:5000.

a What is the distance in real life of a measurement of 4 cm on the map?

b What is the distance on the map of a measurement of 600 m in real life?

13 A map has a scale of 1:2000.

a What is the distance in real life of a measurement of 5.8 cm on the map?

b What is the distance on the map of a measurement of 3.6 km in real life?

12.2 Ratio

HOW TO

1. Set out the problem to show the ratio or proportion.
2. Multiply or divide to keep the ratios the same.
3. Write the answer.

EXAMPLE

A paint mix uses red and white paint in a ratio of 3:8.
How much red paint will be needed to mix with 2.4 litres of white paint?

1. Set out the problem as equivalent ratios.

   red : white

   ×0.3 (3:8) ×0.3   2.4 ÷ 8 = 0.3

   ?:2.4

2. 3 × 0.3 = 0.9

3. 0.9 ml of red paint is needed.

EXAMPLE

Kara and Mia share money in the ratio 3:7.
Mia receives £168 more than Kara.
How much money did they share?

1. Mia gets 4 more parts than Kara.

   4 parts = £168

2. 1 part = £42

   Total number of parts = 3 + 7 = 10

   10 parts = £420

3. They shared £420.

EXAMPLE

In this scale drawing 1 cm represents 3 m.
Calculate the area of the front.
State your units.

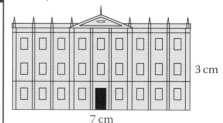

3 cm

7 cm

1. Change 3 m to cm.   3 m = 300 cm
   Write the scale.   ×300

   1cm   300 cm

   ÷300

2. Multiply by the scale.   3 cm × 300 = 900 cm = 9 m
                              7 cm × 300 = 2100 cm = 21 m

3. Calculate the answer.   Area of front = 9 m × 21 m = 189 m²

Exercise 12.2A

1 The ratio of the length of a car to the length of a van is $2:3$.
The car has a length of 240 cm.

   a Express the length of the car as a percentage of the length of the van.

   b Calculate the length of the van.

2 The ratio of the weight of Dave to Morgan is $6:5$. Morgan has a weight of 85 kg.

   a Express the weight of Dave as a percentage of the weight of Morgan.

   b Calculate the weight of Dave.

3 A metal alloy is made from copper and aluminium.
The ratio of the weight of copper to the weight of aluminium is $5:3$.

   a What weight of the metal alloy contains 45 grams of copper?

   b Work out the weight of copper and the weight of aluminium in 184 grams of the metal alloy.

4 Siobhan and Ralph shared £700 in the ratio $2:3$.
Siobhan gave a quarter of her share to Karen.
Ralph gave a fifth of his share to Karen.
What fraction of the £700 did Karen receive?

5 Using this scale drawing of the Eiffel Tower, calculate

   a the height

   b the width of the base.

Scale: 1 cm represents 60 m

6 On the plan of a house, a door measures 3 cm by 8 cm.
If the plan scale is 1 cm represents 25 cm.
Calculate the dimensions of the real door.

7 This is a scale drawing of a volleyball court. The drawing has a scale of $1:200$. Calculate

   a the actual distances marked x, y and z

   b the area of the court.

State the units of your answers.

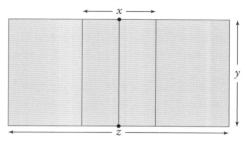

8 The ratio of red counters to blue counters in a bag is $3:7$. What is the probability of choosing a blue counter?

9 Nick and Jess share money in the ratio $2:5$.
Nick gets £120 less than Jess.

   a How much money did they share?

   b How much money did they both receive?

10 Marnie, Hannah and Jessa share money between them.
Hannah gets twice as much as Marnie.
Jessa gets one-third of Hannah's amount.

   a Write the ratio of Marnie: Hannah: Jessa in the form $m:h:j$ where m, h and j are whole numbers and the ratio is in its simplest form.

   b Hannah receives $120. How much more money did Marnie receive than Jessa?

11 The ratio of the internal angle to the external angle in a regular polygon is $3:1$.
How many sides does the polygon have?

12 The angles P and Q in this trapezium are in the ratio $11:25$.
Find the size of angles P and Q.

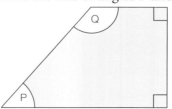

12.3 Percentage change

You often need to calculate a percentage of a quantity.

- A quick method, especially when using a calculator, is to multiply by the appropriate decimal number.

To find 38% of a quantity, multiply by 0.38.

Banks and building societies pay **interest** on money in an account. The interest is always written as a **percentage**.

People can have the interest they earn at the end of each year paid out of their bank account. This is called **simple interest**.

- To calculate simple interest you multiply the interest earned at the end of the year by the number of years.

EXAMPLE

Calculate the simple interest on £3950 for 4 years at an interest rate of 5%.

Calculate the interest for one year.

> Don't forget to estimate.
> 5% of £3950
> ≈ 5% of £4000
> = £200

Interest each year = 5% of £3950

$$= \frac{5}{100} \times 3950$$

$$= 0.05 \times 3950 = £197.50$$

Total amount of simple interest after 4 years = 4 × £197.50
$$= £790$$

- To calculate a **percentage increase**, work out the increase and add it to the original amount.

- To calculate a **percentage decrease**, work out the decrease and subtract it from the amount.

Percentages are used in real life to show how much an amount has increased or decreased.

You can calculate a percentage increase or decrease in a single calculation.

EXAMPLE

a In a sale all prices are reduced by 16%. A pair of trousers normally costs £82. What is the sale price of the pair of trousers?

b Last year, Leanne's Council Tax bill was £968. This year the local council have raised the bill by 16%. How much is Leanne's new bill?

a Sale price = (100 − 16)% of the original price

= 84% of £82

$$= \frac{84}{100} \times 82$$

= 0.84 × 82

= 68.88

= £68.88

−16%   +16%

84%   100%   116%

b New bill = (100 + 16)% of the original bill

= 116% of £968

$$= \frac{116}{100} \times £968$$

= 1.16 × £968

= £1122.88

Exercise 12.3S

1 Write a decimal number equivalent to each percentage.

   a 50%   **b** 60%   **c** 25%

   d 8.5%   **e** 0.15%   **f** 0.01%

2 Calculate these amounts without using a calculator.

   a 10% of £400   **b** 10% of 2600 cm

   c 5% of 64 kg   **d** 25% of 80 m

   e 50% of 380p   **f** 5% of £700

   g 25% of 12 kg   **h** 20% of £31

3 Calculate these percentages, giving your answer to 2 decimal places where appropriate.

   a 45% of 723 kg   **b** 25% of $480

   c 23% of 45 kg   **d** 21% of 28 kg

   e 17.5% of £124   **f** 34% of 230 m

4 Calculate these percentages, giving your answer to two decimal places where appropriate.

   a 7% of £3200   **b** 12% of £3210

   c 27% of €5400   **d** 3.5% of £2200

   e 0.3% of €4450   **f** 3.7% of £12 590

5 Which is larger:
10% of £350 or 15% of £200?

6 Which is larger: 5% of £40 or 8% of £28?

7 Calculate the simple interest paid on £4580

   a at an interest rate of 4% for 3 years

   b at an interest rate of 11% for 5 years

   c at an interest rate of 4.6% for 4 years

   d at an interest rate of 8.5% for 3 years.

8 Calculate the simple interest paid on

   a an amount of £3950 at an interest rate of 10% for 2 years

   b an amount of £6525 at an interest rate of 8.5% for 2 years

   c an amount of £325 at an interest rate of 2.4% for 7 years

   d an amount of £239.70 at an interest rate of 4.25% for 13 years.

9 Write the decimal number you must multiply by to find these percentage increases.

   a 20%   **b** 30%   **c** 45%

10 Write the decimal number you must multiply by to find these percentage decreases.

   a 40%   **b** 60%   **c** 35%

11 Calculate these percentage changes.

   a Increase £450 by 10%

   b Decrease 840 kg by 20%

   c Increase £720 by 5%

   d Decrease 560 km by 30%

   e Increase £560 by 17.5%

   f Decrease 320 m by 20%

12 Calculate these amounts.

   a Increase £250 by 10%

   b Decrease £2830 by 20%

   c Increase £17 200 by 5%

   d Decrease £3600 by 30%

   e Increase £3.60 by 17.5%

   f Decrease £2500 by 20%

13 Calculate each of these using a mental or written method.

   a Increase £350 by 10%

   b Decrease 74 kg by 5%

   c Increase £524 by 5%

   d Decrease 756 km by 35%

   e Increase 960 kg by 17.5%

   f Decrease £288 by 25%

14 Calculate these percentage changes. Give your answers to 2 decimal places as appropriate.

   a Increase £340 by 17%

   b Decrease 905 kg by 42%

   c Increase £1680 by 4.7%

   d Decrease 605 km by 0.9%

   e Increase $2990 by 14.5%

   f Decrease 2210 m by 2.5%

Q 1060, 1073, 1237, 1302, 1934 SEARCH

12.3 Percentage change

RECAP

- You can find a percentage of an amount by multiplying the quantity by an equivalent decimal or fraction.
- You can calculate a percentage increase or decrease by calculating the percentage increase or decrease and then adding it to, or subtracting it from, the original amount.

HOW TO

To write one number as a percentage of another

① Write the first number as a fraction of the other.

② Convert the fraction to a percentage.

EXAMPLE

Find

a 13 as a percentage of 20 **b** 8 as a percentage of 23

a ① $\frac{13}{20} = \frac{13 \times 5}{20 \times 5} = \frac{65}{100} = 65\%$ ② Convert to a fraction with a denominator of 100.

b ① $\frac{8}{23} = (8 \div 23) \times 100\% = 34.8\%$ ② Use a calculator to convert to a percentage.

If a price was reduced from £20 to £13, them the discount would be $(100 - 65)\% = 35\%$.

- In a **reverse percentage** problem, you are given an amount after a percentage change, and you have to find the original amount.

HOW TO

To calculate the original amount after a percentage change

① Write the percentage change as a decimal.

② Find the original amount by dividing by the decimal.

EXAMPLE

Find the original price of a denim jacket reduced by 15% to £32.30.

① $100\% - 15\% = 85\%$

$85\% = 0.85$

original price = £32.30 ÷ 0.85 = £38

② You can test your answer by finding 85% of £38

85% of £38 = 0.85 × £38 = £32.30

EXAMPLE

Following a 5% price increase, a car radio costs £168.
How much did it cost before the increase?

① $100\% + 5\% = 105\%$

$105\% = 1.05$

original price = £168 ÷ 1.05 = £160

② 105% of £160 = 1.05 × £160 = £168

Exercise 12.3A

1 Use a written method to find

  a 12 as a percentage of 20

  b 36 as a percentage of 75

  c 24 as a percentage of 40

2 Use a calculator to find

  a 19 as a percentage of 37

  b 42 as a percentage of 147

  c 8 as a percentage of 209

3 **a** A t-shirt is reduced from £25 to £20. Find the percentage discount.

  b A coat is reduced from £80 to £64. Find the percentage discount.

4 **a** Jenna's savings increase from £400 to £450. Calculate the interest rate.

  b Amanda's antique vase increases in value from £600 to £720. Calculate the percentage increase in value.

5 A car costs £9750 when new. 5 years later it is sold for £4500. What is the average percentage loss each year?

6 Carys is trying to find the original price of an item that is on sale. Its sale price is £56 and it was reduced by 20%.

> 20% of 56 = 11.2
> 56 + 11.2 = 67.2
> The answer is £67.20

Carys' working is shown here.

What is wrong with her working?

7 A book costs £4 after a 20% price reduction. How much did it cost before the reduction? Show your working.

8 These numbers are the results when some amounts were increased by 10%.
For each one, find the original number.

  a 55   **b** 44

  c 88   **d** 121

9 Find the original cost of the following items.

  a A vase that costs £7.20 after a 20% price increase.

  b A table that costs £64 after a 20% decrease in price.

10 **a** Francesca earns £350 per week. She is awarded a pay rise of 3.75%. Frank earns £320 per week. He is awarded a pay rise of 4%. Who gets the bigger pay increase? Show all your working.

  b Bertha's pension was increased by 5.15% to £82.05. What was her pension before this increase?

11 Calculate the original cost of these items, before the percentage changes shown. Show your method. You may use a calculator.

  a A hat that costs £46.50 after a 7% price cut.

  b A skirt that costs £32.80 after a price rise of 6%.

12 A car manufacturer increases the price of a Sunseeker sports car by 6%.
The new price is £8957.
Calculate the price before the increase.

13 During 2005 the population of Camtown increased by 5%. At the end of the year the population was 14 280.

What was the population at the beginning of the year?

*14 To decrease an amount by 8%, multiply it by 0.92.
For a further decrease of 8%, multiply it by 0.92 again, and so on.
Use this idea to calculate

  a the final price of an item with an original price of £380, which is given two successive price cuts of 8%.

  b the final price of an item with an original price of £2400, which is given three successive price cuts of 10%.

Q 1060, 1073, 1237, 1302, 1934   SEARCH

Summary

Checkout

You should now be able to...

Test it

Questions

	Test it Questions
✔ Find fractions and percentages of amounts and express one number as a fraction or percentage of another.	**1, 2, 11**
✔ Divide a quantity in a given ratio and reduce a ratio to its simplest form.	**2 – 6**
✔ Use scale factors, scale diagrams and maps.	**6 – 8**
✔ Solve problems involving percentage change.	**9 – 12**

Language   Meaning   Example

Language	Meaning	Example
Proportion	A proportion is a part of the whole. Two quantities are in proportion if one is always the same multiple of the other.	If there are 6 eggs in a carton Total number of eggs = 6 × number of full cartons
Ratio	A ratio compares the size of one quantity with the size of another.	 Ratio of blue squares to yellow squares = 2:6 = 1:3
Simplify (ratio)	Divide both parts by common factors.	
Scale	The ratio of the length of an object in a scale drawing to the length of the real object.	The Ordnance survey produce maps with scales such as: 1 : 100 000, 1 : 50 000 and 1 : 25 000.
Scale drawing	An accurate drawing of an object to a given scale.	
Percentage	A type of fraction in which the value given is the number of parts in every hundred.	$\frac{33}{100} = 33\%$
Interest	A fee paid to somebody for the use of their money. This is a percentage of the loan amount that must be paid to the lender in addition to the loan itself.	A loan with an interest rate of 4% means that the borrower has to pay back an extra 4% of the amount.
Simple interest	Interest that is calculated on the original amount only and not on any extra interest that has built up.	£100 saved in a bank account at 4% Amount after 3 years = 100 x (1 + 3 x 0.04) = £112
Compound interest	Interest that is calculated on the original amount plus any interest that has built up previously.	£100 saved in a bank account at 4% Amount after 3 years = 100 x (1 + 0.04)^3 = £112.49
Percentage increase/ decrease	An increase/decrease by a percentage of the original amount.	4.20 increased/decreased by 25% = 1.25 × 4.20 = 0.75 × 4.20 = 5.25 = 3.15
Reverse percentage problem	Calculating the original value of a quantity using the value after a percentage change.	Quantity after a 15% decrease = 544 85% of original quantity = 544 100% of original quantity = $\frac{544}{0.85}$ = 640

Review

1 A class has 16 boys and 9 girls.

    a What fraction of the class are boys?

    b What percentage of the class are girls?

    c This class has girls and boys in the same proportion as the overall year group. There are 125 students in the year group, how many of them are girls?

2 Laura spends 8 hours at work each day.

    a What fraction of the day does she spend at work?

    b Write the ratio of the amount of time she spends at work to not at work.

3 Write these ratios in their simplest form.

    a $22:33$ **b** $45:27$ **c** $2\,cm:5\,m$

4 Share

    a £42 in the ratio $5:1$

    b £99 in the ratio $2:4:3$

5 Janet is twice as old as Jake.

    a Write a ratio of Janet's age to Jake's age.

    b What fraction of Janet's age is Jake?

6 The ratio of the mass of milk to flour in a recipe for batter is $2:3$.

    a How much milk and how much flour will be needed to make 1 kg of batter?

    b How much milk will be needed to mix with 450 g of flour?

    c What fraction of the mass of the mixture is milk?

7 This circle is enlarged by scale factor 5.

    a What is the diameter of the enlarged circle?

7 m

7 **b** What is the area of the enlarged circle?

8 A map is drawn using the scale $1:400\,000$.

    a How far in real life is a length of 2 cm on the map?

    b A road is 12.5 km long, how long will it be on the map?

9 **a** Increase 60 by 34%.

    b Decrease 192 by 58%.

10 A bank pays 1.9% per annum interest on a savings account. £6000 is paid in at the beginning of the year and no further deposits or withdrawals are made.

    How much is in the account after 1 year?

11 Mel has a mortgage of £200 000. He pays £1100 a month. This covers only the interest on his mortgage so he still owes the bank £200 000.

    a What is the annual interest rate?

    After 20 years of paying the interest of £1100 each month the mortgage ends and he pays back in £200 000.

    b How much money has he paid in total over the 20 years?

    c What is the total amount paid as a percentage of the original amount borrowed?

12 **a** The price of a loaf of bread rose 8% to £1.34, what was it before the increase?

    b A car is reduced by 17% and now costs £7520, what was the original price of the car?

What next?

Score			
	0 – 5		Your knowledge of this topic is still developing. To improve look at MyMaths: 1015, 1036, 1037, 1038, 1039, 1060, 1073, 1103, 1237, 1302, 1934
	6 – 10		You are gaining a secure knowledge of this topic. To improve look at InvisiPens: 12Sa – j
	11 – 12		You have mastered these skills. Well done you are ready to progress! To develop your exam technique look at InvisiPens: 12Aa – c

Assessment 12

1   There are some Brazil nuts, walnuts and chestnuts in a bowl. 3 out of 5 of the nuts are Brazils, W out of N are walnuts and 3 out of every 20 are chestnuts.

   a   Work out the values of W and N in their simplest form.   [2]

   b   Write all these proportions as fractions.   [3]

   c   Write all these proportions as percentages.   [3]

   d   Which nut represents the smallest proportion of the bowl's contents?   [1]

   e   Which nut represents the biggest proportion of the bowl's contents?   [1]

   f   What percentage of the nuts in the bowl are either Brazil nuts or chestnuts?   [2]

   g   What fraction of the nuts are <u>not</u> walnuts?   [2]

   h   What fraction of the nuts in the bowl are neither chestnuts nor Brazil nuts?   [2]

2   **a**   **i**   Nomsah received $\frac{13}{25}$ of the prize money from a competition.

            Write this proportion as a percentage and a decimal.   [1]

      ii   The total prize was 225 000 rand. How much did Nomsah win?   [2]

   b   Jeremy bought a car for £17 250. A year later its value had decreased by $\frac{6}{25}$.
   How much was the car worth a year later?   [3]

3   A tin contains 63 biscuits, some made with chocolate and the rest 'plain', in the ratio 3:4. Calculate the numbers of chocolate and plain biscuits in the tin.   [3]

4   Debs makes cheese straws, using flour, butter and cheese in the ratio 24:6:9.

   a   Write this ratio in its simplest form.   [1]

   Debs uses 160 g of flour in her mixture.

   b   Work out the number of grams of butter and cheese Debs uses.   [3]

5   A café serves cups of coffee in three sizes: small, medium and large.
   The volumes are in the ratio 2:5:8.

   a   A medium coffee is 375 ml. Find the volumes of the other two sizes.   [3]

   b   The café serves pots tea in three sizes, the ratio of volumes small:medium:large is the same as it is for coffee. The volume of a large pot of tea is 1 litre.
   Find the volumes of the medium and small pots.   [3]

6   **a**   The cast of a school play has 24 boys and 20 girls.
   Write this ratio in its simplest form.   [1]

   b   1 more boy joins the cast but 5 girls leave. Write the new ratio in the form $n:1$.   [3]

7   The ratio of girls to boys in a cycling squad is 4:7. There are 88 people in the squad.

   a   How many of each gender are in the squad?   [4]

   b   The number of girls goes up by 12 and the number of boys goes up by 37.5%.
   Find the number of girls and boys in the squad after the change and show that the ratio of girls to boys stays the same.   [5]

8   Two circles have diameters of 15 m and 20 m. Find the ratio of their

   a   circumferences   [2]   **b**   areas.   [3]

   Write all answers in their simplest form.

9 It takes Dwain 24.5 seconds to run 200 m. At the same pace, how long will he take to run

  a 74 m   [3]   **b** 172 m?   [1]

  Give both answers to the nearest 0.1 s.

10 A road atlas' scale is 1 inch to 5 miles.

  a Gatwick airport to Heathrow is 8.1 inch on map. How far apart are they? [2]

  b Liverpool FC and Newcastle United FC are 128 miles apart as the crow flies. How far apart are they on the map? [2]

  c The straight-line distance from John O' Groats to Lands End on a map is 120.4 in. The actual distance by road is 838 miles. Calculate the difference in mileage between the actual distance and the straight-line distance. [3]

11 On a street map of London the scale is written as 1 : 20 000.

  a How many metres in London is represented by 1 cm on the map? [1]

  b On the map, Buckingham Palace is 5.85 cm from the statue of Eros in Piccadilly Circus. How far apart are they in reality? Give your answer in km. [2]

  c Trafalgar Square is 0.96 km from Big Ben. How far apart are they on the map? [2]

  d A sick child is taken from Paddington Station to Great Ormond Street Children's Hospital. On the map, the distance is represented by 20.75 cm. The ambulance travels at 80 km/h. Show that the ambulance journey takes less than 200 seconds. [4]

12 a Ben says that to increase a value by 29.1% you multiply by 29.1. Gavin says you multiply by 1.291. Who is correct? [1]

  b Explain how to decrease 550 ml by 45.4%. [2]

In the following questions give your answers to 2 dp wherever appropriate.

13 a A station has 45 steps up to the platform bridge. The first section has 19 steps. What percentage of the total number of steps does the first section take? [2]

  b Kate buys material for her wedding dress. The shop assistant cut her material from a 25 m roll, leaving 17.25 m on the roll. What percentage of the roll has Kate bought? [3]

14 a Cowell bought a painting as an investment for £2 500 000. Some years later it was revalued as being worth 8.5% more than originally. What is its new value? [2]

  b In 2000 the average price of a new house was £86 100. In 2010 this average was 164 300. By what percentage had the 2010 value risen above the 2000 value? [3]

  c 'FALSEPRINT', a film processor, sells prints various in sizes. They claim that the area 7″ by 5″ print is more than 45% bigger than their 6″ by 4″ size. Are they correct? [4]

15 A river contains 250 crocodiles. After two years of hunting, the number of crocodiles decreases by 8% in the first year and 5.5% in the second. How many crocodiles, to the nearest crocodile, are there in the river after

  a 1 year   [2]   **b** 2 years? [2]

16 a Dani bought a motorcross bike for £4 776. Dani sold the bike at a loss of 11.3%. What price did Dani receive for the bike? [2]

  b After a pay rise of 3.5% Osborne's monthly salary is £2 566.80. What was his monthly salary before his increase? [2]

  c 'Titus Lines', the angler, buys a new rod for £251.34 which the dealer had sold him making himself a profit of 6%. What price had the dealer originally paid for the rod? [2]

Revision 2

1 Andrea draws a quadrilateral with corners at the coordinates (1,2), (3,4), (4,7) and (3,10). Find the area of this quadrilateral. [8]

2 The diagram shows a company logo.
$BC = CD = 8$ cm,
$AC = 15$ cm and
$AE = 9$ cm.

Calculate the area of the logo. [6]

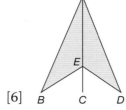

3 Town B is 5 km from Town A on a bearing of 135°. Town C is 12 km from Town A on a bearing of 045°.

  a Sketch the triangle *ABC*. [2]

  b Find angle *BAC*. [1]

  c Calculate the area of the triangle enclosed by the three towns. [1]

4 A field in the shape of a trapezium has two parallel sides, 150 m and 176 m. The perpendicular distance between them is 243 m. Calculate the area of the field in

  a m^2 [2]

  b hectares. (1 hectare $= 10\,000\,m^2$) [1]

5 **a** Triangle *ABC* has been transformed on to triangle *PQR* as shown. Describe this transformation fully. [3]

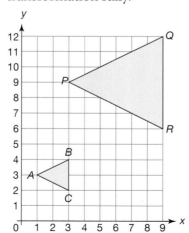

  b Reflect triangle *ABC* in the line $y = -x$ and label the image *A'B'C'*. [2]

6 Triangle *I* has vertices at (2, 2), (6, 2) and (4, 6). Triangle *O* has vertices at (3, −3), (9, −6) and (3, −9). Draw both triangles on a coordinate grid and fully describe the **two** transformations needed to map *I* onto *O*. [6]

7 Pepys's bookshelf can hold 38 books at an average thickness of 2.9 cm.

  a Estimate the length of the bookcase. [1]

  The average weight of a book is 1100 g.

  b Estimate the total mass of the books on Pepys's bookshelf. [1]

  c Use your calculator to find exact answers to parts **a** and **b**. [2]

  d Find your percentage error for the estimates in parts **a** and **b**. [4]

8 A shape is made up of a cylinder, radius 1.5 m with height 2.3 m, and a cuboid measuring 5 m × 3 m × 4 m. All measurements are correct to the nearest cm.

  a Find the upper and lower bounds for the volume. Show your working. [9]

  b Calculate the upper and lower bounds of the mass of the cylinder if the density of the material is 2.8 g/cm³ [5]

9 Rose is planting flowers, Purple Lupins which cost £3 each and White Stocks which cost £4 each. She must spend at least £10 and has room for not more than 6 Lupins and 4 Stocks. She wants at least twice as many Stocks as Lupins.

  a Taking the *y* axis as Lupins and the *x* axis as Stocks, draw 4 straight lines to illustrate the information above. [5]

  b By shading OUT inappropriate areas identify the area which satisfies these conditions. [1]

  c Use your graph to find the minimum number of each flower Rose can buy given that she wants at least one flower of each type. [2]

10 A length of wire is 58 cm long. It can be bent into a rectangle *x* cm long. The area of the rectangle is 100 cm². Find the dimensions of the rectangle. [7]

11 A coffee machine takes only 10p and 50p coins. When emptied, it had 41 coins in it totalling £10.10. How many of each value coin did the machine have? [7]

12 A triangle has angles p, q and r, where $p > 21$, $q > 59$. Find an appropriate inequality for r. [2]

13 A drinks manufacturer makes circular bottle tops. A rectangular metal sheet, of dimensions 35 cm by 30 cm is cut into 42 squares. A circle of the largest possible size is then cut from each square.

   a Find the percentage of metal sheet discarded in this process. [8]

   b The remainder of the sheet is recast. How many bottle tops could be made from it

      i using the arrangement described in part **a**

      ii using any arrangement? [7]

14 A man is trying to move a heavy roller of radius r, using a straight crowbar, BDF. The angle between the crowbar and the ground, ABC, is 60°, AE is the diameter of the circle and O is the centre.

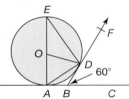

   a Find angles DEA and EDF. Explain your answers. [11]

   b Prove that $AD = r$. [5]

   c Calculate, in terms of r, the length of arc AD and the area of the sector AOD. [5]

15 A house has a rectangular garden, $ABCD$, which is 12 m by 9 m, CD is the side of the house and is 9 m long. A window at the centre of CD has a modem in it which gives a WiFi signal range of 7.5 m.

   a Use a ruler and compasses to construct a diagram of the garden. [7]

   b Shade in the area of the garden within range of the WiFi. [3]

15 c Ida has a deckchair 7.2 m from A and 5.2 m from B. Is she within WiFi range? Explain your answer. [5]

   d Draw and measure a line on your diagram to find the WiFi range needed to make all of the garden accessible. [2]

16 A tennis club has the ratio men : women as 7 : 6. There are 104 people in the club.

   a How many of each sex are there in the club? [3]

The number of men and the number of women each rises by 25%.

   b Does the ratio of men to women change? Explain your answer. [2]

   c What are the new numbers of women and men in the squad? [1]

   d The same number of men leave as women join. The ratio is now 1 : 1. How many people of each gender are in the club now? [1]

17 On a map 1 inch represents 3 miles on the ground.

   a From York to Leeds, as the crow flies, is 22 miles. How far apart are they on the map? [2]

18 a A daily newspaper claims an average daily sale of 1 456 739. What percentage increase would take its daily average sales to 1 500 000? [2]

   b The value of a new car is £18 000. It depreciates by 16% yearly. How much is the car worth at the end of 3 years? [3]

19 Sunetra has 6 cards: 1, 3, 5, 7, 9 and 11. She selects two cards and puts the smaller number on top of the larger to make a fraction.

   a Show there are 15 possible outcomes. [4]

   b Work out the probability her number is

      i $\frac{1}{3}$ [3]   **ii** less than $\frac{1}{4}$ [3]

13 Factors, powers and roots

Introduction

Cryptography is the study of codes, with the aim of communicating in a secure way without messages being deciphered. Modern cryptography is often based on prime numbers. In particular finding very large numbers that can be written as the product of two not-quite-as-large prime factors. Finding what these two prime numbers are is a challenge, even with modern computers; but you need to find them in order to crack the code.

What's the point?

In the modern day, illegal computer-based syndicates use increasingly sophisticated techniques to access sensitive digitally-held information, including bank accounts. Prime number encryption increases the security of stored data, making it harder for the hackers.

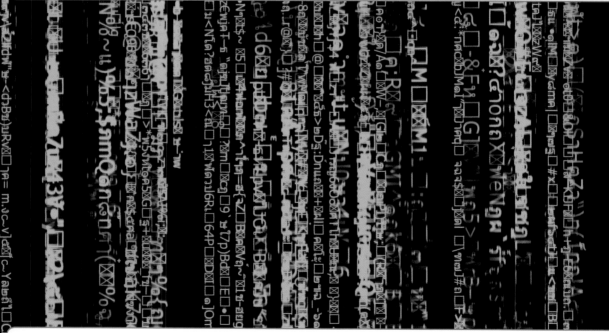

Objectives

By the end of this chapter, you will have learned how to ...

- Know and use the language of prime numbers, factors and multiples.
- Write a number as a product of its prime factors.
- Find the HCF and LCM of a pair of integers.
- Estimate the square or cube root of an integer.
- Find square and cube roots of numbers and apply the laws of indices.
- Simplify expressions involving surds including rationalising fractions.

Check in

1 Write all the factors of each number.

    a 12   **b** 30   **c** 120   **d** 360

2 Write a list of all of the prime numbers up to 100. (There are 25 of them.)

3 Find the results of these calculations, giving the answers in index form.

    a $2^3 \times 2^4$ **b** $3^5 \div 3^2$ **c** $5^2 \times 5^3 \times 5^2$

    d $6^6 \div 6^4$ **e** $7^8 \div (7^2 \times 7^3)$ **f** $(4^6 \times 4^2) \div (4^3 \times 4)$

4 Write the value of each number.

    a 4^0 **b** 6^0 **c** 5^1 **d** 2^{-1}

Chapter investigation

77 is the product of two prime numbers: 7 and 11.

That is, $7 \times 11 = 77$

Is 702 the product of two prime numbers? If not, why not?

What about 703?

13.1 Factors and multiples

You can write a number as a **product** of **factors** in different ways.

This means 1 is not a prime number

- A **prime** number is a number with only two factors – itself and 1.
  $17 = 17 \times 1$

- Any number greater than 1 can be written as a unique product of its prime factors. This is called the **prime factor decomposition** of the number.

A factor tree helps you keep track of all the prime factors, even if you don't find them in ascending order.

EXAMPLE

Find the prime factor decomposition of 990.

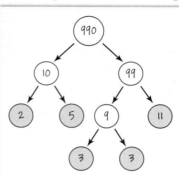

$990 = 2 \times 3^2 \times 5 \times 11$

Write the number as the product of two smaller numbers.

Keep breaking the numbers down until you reach a prime.

- The **highest common factor (HCF)** of two numbers is the largest number that is a factor of both of them.

Factors of 18 = {1, 2, 3, **6**, 9, 18}

Factors of 24 = {1, 2, 3, 4, **6**, 8, 12, 24}

HCF of 18 and 24 = 6

- The **lowest common multiple (LCM)** of two numbers is the smallest number that is a **multiple** of both of them.

Multiples of 18 = 18, 36, 54, **72**, 90, ...

Multiples of 24 = 24, 48, **72**, 96, ...

LCM of 18 and 24 = 72

You can find the HCF and LCM of two numbers by writing their **prime factors** in a Venn diagram.

- The HCF is the product of the numbers in the intersection.
- The LCM is the product of all the numbers in the diagram.

EXAMPLE

Find the HCF and LCM of 60 and 280.

$60 = 2^2 \times 3 \times 5 \qquad 280 = 2^3 \times 5 \times 7$

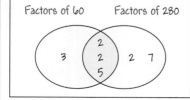

The HCF is the product of the numbers in the intersection.

$HCF = 2^2 \times 5 = 20$

An alternative is to write:

$60 = 2^2 \times 3 \times 5$

$280 = 2^3 \times 5 \times 7$

Identify the highest power of each factor.

$LCM = 2^3 \times 3 \times 5 \times 7 = 840$

The LCM is the product of all the numbers in the diagram.

$LCM = 2^3 \times 3 \times 5 \times 7 = 840$

Exercise 13.1S

1 Copy and complete these calculations to show the different ways that 24 can be written as a product of its factors.

  a $24 = \square \times 2$ **b** $24 = 3 \times \square$

  c $24 = 2 \times 3 \times \square$ **d** $24 = 4 \times \square$

2 Each of these numbers has just two prime factors, which are not repeated. Write each number as the product of its prime factors.

  a 77   **b** 51

  c 65   **d** 91

  e 119   **f** 221

3 Copy and complete the factor tree to find the prime factor decomposition of 18.

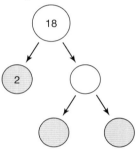

4 Work out the values of each of these expressions.

  a $3^2 \times 5^2$ **b** $2^3 \times 5$

  c $3^2 \times 7$ **d** $2^3 \times 3^2 \times 5$

  e $3^2 \times 7^2$ **f** $2^2 \times 3 \times 5^2$

5 For each number, find its prime factors and write it as the product of powers of its prime factors.

  a 36   **b** 120   **c** 34

  d 25   **e** 48   **f** 90

  g 27   **h** 60   **i** 105

  j 99   **k** 37   **l** 91

6 Write the prime factor decomposition for each of these numbers.

  a 1052   **b** 2560   **c** 630

  d 825   **e** 715   **f** 1001

  g 219   **h** 289   **i** 2840

  j 2695   **k** 1729   **l** 3366

  m 9724   **n** 11830   **o** 2852

7 The diagram below shows how a student found the lowest common multiple (LCM) of 8 and 6.

> Multiples of 8 = 8, 16, ⟨24,⟩ 32, 40, 48 ...
> Multiples of 6 = 6, 12, 18, ⟨24,⟩ 30, 36 ...

  a List the multiples of 12 and 9 in the same way.

  b The lowest common multiple (LCM) of 12 and 9 is the smallest number that is in both lists.
  Write the LCM of 12 and 9.

8 Using the method from question 7, find the LCM of these pairs of numbers.

  a 4 and 5   **b** 12 and 18

  c 5 and 30   **d** 12 and 30

  e 14 and 35   **f** 8 and 20

9 Find the highest common factor (HCF) of each pair of numbers, by drawing a Venn diagram or otherwise.

  a 35 and 20   **b** 48 and 16

  c 21 and 24   **d** 25 and 80

  e 28 and 42   **f** 45 and 60

10 Find the lowest common multiple (LCM) of each pair of numbers, by drawing a Venn diagram or otherwise.

  a 24 and 16   **b** 32 and 100

  c 22 and 33   **d** 104 and 32

  e 56 and 35   **f** 105 and 144

11 Find the LCM and HCF of each pair of numbers.

  a 180 and 420   **b** 77 and 735

  c 240 and 336   **d** 1024 and 18

  e 762 and 826   **f** 1024 and 1296

12 a Charlie says 33 105 is divisible by 15. Is Charlie correct? Explain your answer.

  b Amy says 262 262 is divisible by 1001. Is she correct? Explain your answer.

  c Luisa says that 14 355 is divisible by 45. Is she correct? Explain your answer.

🔍 1032, 1034, 1044   SEARCH

13.1 Factors and multiples

- The **highest common factor** (HCF) of two numbers is the largest number that is a factor of them both.
- The **lowest common multiple** (LCM) of two numbers is the smallest number that they both divide into.
- You can find the HCF and LCM of two numbers by writing their **prime factors** in a Venn diagram.
- The HCF is the product of the numbers in the intersection.
- The LCM is the product of all the numbers in the diagram.

> The prime factor decomposition for any number is always unique.

HOW TO

1. Read the question carefully; decide how to use your knowledge of factors and multiples.

2. Be systematic – use listing, factor trees and/or Venn diagrams to help you to find multiples and factors.

3. Answer the question. Make sure that you explain your answers.

EXAMPLE

Explain why the square of a prime number has exactly three factors.

1. A prime number, p, has two factors, 1 and p (itself).

2. List the factors of the square of a prime number.
   $p^2 = 1 \times p^2$ and $p^2 = p \times p$

3. A prime number squared has exactly three factors because the only factors are: 1, the prime number (p) and the number itself (p^2).

EXAMPLE

The highest common factor of two numbers is 6.
The lowest common multiple is 1080.
Billie says that the two numbers must be 54 and 120.
Show that there is another possibility.

1. Find another pair of numbers with the same HCF and LCM.

2. Write the prime factors of 54 and 120 in a Venn diagram.
   $54 = 2 \times 3^3$ $120 = 2^3 \times 3 \times 5$

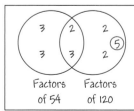

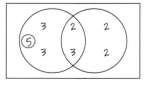

Factors   Factors
of 54   of 120

> Rearrange the factors to find the other pair of numbers.
> The numbers in the intersection must stay the same.

3. Find the other pair of numbers.
   The other possible pair of numbers is: $2 \times 3^3 \times 5 = 270$ and $2^3 \times 3 = 24$

Exercise 13.1A

1 Emily checks her phone every 10 minutes to see if her friend Charlotte is available to chat. Charlotte checks hers every 8 minutes. They both decide to check at 9:00 am. When is the first time after this that they are both checking?

2 Two hands move around a dial. The faster hand moves around in 24 seconds, and the slower hand in 30 seconds. If the two hands start together at the top of the dial, how many seconds does it take before they are next together at the top?

3 A wall measures 234 cm by 432 cm. What is the largest size of square tile that can be used to cover the wall, without needing to cut any of the tiles?

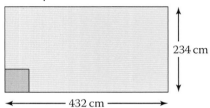

4 The number 18 can be written as $2 \times 3 \times 3$. You can say that 18 has three prime factors.

   a Find three numbers with exactly three prime factors.

   b Find five numbers with exactly four prime factors.

   c Find four numbers between 100 and 300 with exactly five prime factors.

   d Find a two-digit number with exactly six prime factors.

5 A metal cuboid has a volume of 1815 cm³. Each side of the cuboid is a whole number of centimetres, and each edge is longer than 1 cm.

> Volume of a cuboid = length × width × height.

   a Find the prime factor decomposition of 1815.

   b Use your answer to part **a** to find all the possible dimensions of the cuboid.

6 Explain why the cube of a prime number has exactly four factors.

7 **a** Amos says "all odd numbers are prime numbers". Give two examples that show he is wrong.

7 **b** Aya says "all prime numbers are odd numbers". Give an example to show she is wrong.

   c Arik says "take one off any multiple of 6 and you always get a prime number".

      i Give 3 examples where this is true.

      ii Give one example where it is false.

8 The highest common factor of two numbers is 30. The lowest common multiple is 900. Omar says that the two numbers must be 150 and 180. Show that there is another possibility.

9 Two numbers have HCF = 15 and LCM = 90. One of the numbers is 30. What is the other number?

10 Look at this statement.

> The product of any two numbers is equal to the product of their HCF and their LCM.

   a Test the statement. Do you think it is true?

   b Use a Venn diagram to justify your answer to part **a**.

11 Find the HCF and LCM of each set of three numbers.

   a 30, 42, 54

   b 90, 350 and 462

   c 462, 510 and 1105

*12 A perfect number is one where if you add all the factors which are less than the number itself, the total is that number. For example 6 is a perfect number, because its factors are 1, 2, 3, 6 and $1 + 2 + 3 = 6$.

   a Show that 28 is a perfect number.

   b Show that 120 is not a perfect number.

   c Show that 496 is a perfect number.

> **Perfect Numbers**
> No one knows how many perfect numbers there are, but less than 50 have been found so far, even with the help of powerful computers. The fourth one is 8128, and the fifth is 33 550 336. After that they get really, really big!

Q 1032, 1034, 1044   SEARCH

13.2 Powers and roots

- A **square root** is a number that when multiplied by itself gives a result equal to a given number. $\sqrt{25} = 5$

Use a calculator to find a square root using the $\boxed{\sqrt{x}}$ function key.

- A **cube root** is a number that when multiplied by itself and then multiplied by itself again gives a result equal to a given number. $\sqrt[3]{27} = 3$

Use a calculator to find a cube root using the $\boxed{\sqrt[3]{x}}$ function key.

You can use **index** notation to describe **powers** of any number.
$4.6^5 = 4.6 \times 4.6 \times 4.6 \times 4.6 \times 4.6$

Some calculators may have a different key, for example $\boxed{x^y}$ or $\boxed{\wedge}$ or $\boxed{\text{EXP}}$.

EXAMPLE

Use your calculator to find the missing powers.

a $18^{\square} = 5832$ **b** $1.7^{\square} = 8.3521$ **c** $(-1.2)^{\square} = -1.728$

For each question test the different powers.

a $18^2 = 324$

Try again.

$18^3 = 5832$

Correct.

b $1.7^2 = 2.89$

Try again.

$1.7^3 = 4.913$

$1.7^4 = 8.3521$

Correct.

c Square numbers are always positive.

$(-1.2)^3 = -1.728$

Correct.

Powers of the same number can be multiplied and divided.

- To multiply powers of the same base, add the indices. $5^3 \times 5^2 = 5^5$
- To divide powers of the same base, subtract the indices. $5^8 \div 5^2 = 5^6$
- To raise a power, multiply the indices. $(5^2)^4 = 5^8$
- For all values of x except $x = 0$, $x^0 = 1$ $5^0 = 1$

EXAMPLE

Simplify these expressions, giving your answers in index form.

a $7^2 \times 5^3 \times 7^3 \times 5^4$

b $(2^5 \times 3^4) \div (2^3 \times 3^2)$

a $7^2 \times 5^3 \times 7^3 \times 5^4 = 7^{(2 + 3)} \times 5^{(3 + 4)} = 7^5 \times 5^7$

b $(2^5 \times 3^4) \div (2^3 \times 3^2) = 2^{(5 - 3)} \times 3^{(4 - 2)} = 2^2 \times 3^2$

EXAMPLE

Write the value of

a $(16^3 - 81 \times 17)^0$

b $(4.8)^1$

a There is no need to evaluate the expression in the brackets (except to note that it is non-zero).
$(16^3 - 81 \times 17)^0 = 1.$

b $x^1 = x$ for any value of x (including decimal numbers).
$(4.8)^1 = 4.8.$

Exercise 13.2S

1 Use your calculator to work out each of these squares.

   a $(-3.9)^2$ **b** 2.1^2

   c $(-0.7)^2$ **d** 13.25^2

2 Use your calculator to find the positive and negative square roots of these numbers. Give your answers to 2 decimal places as appropriate.

   a 1345   **b** 38.6

   c 7093   **d** 234.652

3 Use your calculator to work out each of these. Give your answers to 2 decimal places as appropriate.

   a $(-5.4)^3$ **b** 9.9^3

   c $(-0.1)^3$ **d** 16.85^3

4 Calculate these using a calculator, giving your answers to 2 dp as appropriate.

   a $\sqrt[3]{12\,167}$ **b** $\sqrt[3]{-216}$

   c $\sqrt[3]{-70}$ **d** $\sqrt[3]{0.015\,625}$

5 Write

   a 81 as a power of 9

   b 125 as a power of 5

   c 128 as a power of 2

   d 100 000 as a power of 10

   e 81 as a power of 3

   f 343 as a power of 7

6 Use your calculator to work out these powers. In each case copy the question and fill in the missing numbers.

   a $24^\square = 576$ **b** $1.5^\square = 3.375$

   c $(-2.25)^\square = 5.0625$ **d** $(-0.5)^\square = -0.125$

   e $5^\square = 3125$ **f** $(-3)^\square = 729$

7 Simplify these expressions, giving your answers in index form.

   a $6^2 \times 6^3$ **b** $4^5 \times 4^4$

   c $2^6 \times 2^7$ **d** $11^5 \times 11^2$

   e $1^{17} \times 1^{13}$ **f** $7^8 \times 7^4$

8 Simplify these expressions, giving your answers in index form where appropriate.

   a $7^8 \div 7^6$ **b** $8^6 \div 8^2$ **c** $3^3 \div 3^2$

   d $9^{11} \div 9^8$ **e** $4^7 \div 4^1$ **f** $2^9 \div 2^9$

9 Simplify these expressions, giving your answers in index form.

   a $8^6 \times 8^2 \div 8^3$ **b** $5^7 \times 5^2 \div 5^4$

   c $2^8 \times 2^3 \div 2^5$ **d** $9^6 \times 9^3 \div 9^7$

   e $8^5 \times 8^5 \div 8^2$ **f** $7^6 \times 7^5 \div 7^4$

   g $4^6 \times 4^8 \div 4^4$ **h** $11^2 \times 11^2 \div 11^3$

10 Simplify these expressions, giving your answers in index form.

   a $3^4 \times 3^2 \div (3^3 \times 3^2)$

   b $(5^6 \div 5^2) \times 5^4 \times 5^2$

   c $(4^5 \div 4^2) \div (4^6 \div 4^5)$

   d $(7^9 \div 7^2) \div (7^2 \times 7^3)$

11 Simplify these expressions, giving your answers in index form.

   a $\dfrac{4^2 \times 4^2}{4^2}$ **b** $\dfrac{6^3 \times 6^4}{6^5}$ **c** $\dfrac{9^8}{9^2 \times 9^4}$

   d $\dfrac{8^6 \div 8^3}{8^2}$ **e** $\dfrac{5^9 \times 5^4}{5^3 \times 5^7}$ **f** $\dfrac{6^3 \times 6^4}{6^5 \div 6^3}$

12 Simplify these expressions as far as possible, giving your answers in index form.

   a $4^2 \times 3^3 \times 4^2$

   b $6^2 \times 5^3 \times 6^2 \times 5^3$

   c $5^4 \times 2^3 \div 5^2$

   d $8^2 \times 5^6 \times 8^3 \div 5^3$

   e $9^3 \times 2^5 \div 2^3 \times 9^2$

13 Simplify these expressions, giving your answers in index form.

   a $\dfrac{5^2 \times 8^5}{8^2}$ **b** $\dfrac{6^5 \times 7^2}{6^3}$

   c $\dfrac{6^4 \times 5^4}{6^2 \times 5^2}$ **d** $\dfrac{7^8 \times 5^6}{5^3 \times 7^2}$

   e $\dfrac{8^7 \times 8^5}{3^2 \times 8^5}$ **f** $4^3 \times \dfrac{4^5 \times 5^9}{4^3 \times 5^7}$

   g $\dfrac{6^9 \times 7^5}{6^7 \times 7^3} \times 6^2$ **h** $4^3 \times \dfrac{7^6 \times 4^5}{4^4 \times 7^3} \times 7^2$

1033, 1053, 1301, 1924   SEARCH

13.2 Powers and roots

- Add indices when multiplying powers of the same number.
- Subtract indices when dividing powers of the same number.
- Multiply indices when finding the power of a power.

> Any number (except 0) to the power of 0 is 1.

HOW TO

① Read the question carefully; decide how to use your knowledge of powers and roots.

② Apply the index laws and look out for chances to use known facts about squares and cubes.

③ Answer the question; make sure that you leave your answer in index form if the question asks for it.

EXAMPLE

Find three whole numbers n, such that $\sqrt{5 + 4n}$ is a whole number.

① If $\sqrt{5 + 4n}$ is a whole number, then $5 + 4n$ must be a square number.

$5 + 4n$ = square number

② Find a square number that is 5 more than a multiple of 4.

Square numbers: 1, 4, 9, 16, 25, 36, 49, 64, 81, 100 ...

Multiples of 4: 4, 8, 12, 16, 20, 24, 28, 32, 36, 40, 44, 48 ...

$5 + 4 \times 1 = 5 + 4 = 9$

$5 + 4 \times 5 = 5 + 20 = 25$

$5 + 4 \times 11 = 5 + 44 = 49$

③ Three possible values for n are 1, 5 and 11.

> This method is more efficient than trying different values of n until you find a correct solution.

EXAMPLE

For the calculation $2^3 \times 5^4$, Jack wrote:

$2 \times 5 = 10$, and $3 + 4 = 7$.

So, the answer is 10^7.

Explain why Jack is incorrect. Find the correct answer.

① The index laws only apply when both numbers have the same base.

② For this calculation you need to evaluate each term and then multiply them together:

③ $2^3 = 8$ and $5^4 = 625$ $\qquad 2^3 \times 5^4 = 8 \times 625 = 5000$

Did you know...

French mathematician and philosopher Rene Descartes was the first person to use index form to write powers of numbers.

EXAMPLE

$2^{12} = 4^a$. Find the value of a.

① Rewrite 2^{12} as a power of 4. This will give you the value of a.

② Use the index law for the power of a power and the fact that $4 = 2^2$.

$2^{12} = (2^2)^6 = 4^6$

③ $a = 6$

Exercise 13.2A

1 Use your calculator to work out these problems.

   a $\sqrt{7} = 2.645751$
Calculate $(2.645751)^2$.
Explain why the answer is not 7.

   b   Two consecutive numbers are multiplied together.
The answer is 3192.
What are the two numbers?

2 The cube of a particular integer lies between 3000 and 4000. Without using a calculator, find the integer. Show your workings.

3 In a school the teachers never reveal their ages but always give clues.
Mr Earle, the History teacher, gave three clues.

My age

- is a multiple of 9
- is a cube number
- is less than 40

   a   What is Mr Earle's age?

   b   Can you work out Mr Earle's age from two clues?
Explain your answers.

4 Some cube numbers are also square numbers.

   a   Show that 64 is one of these.

   b   Show that 3^6 is another.

   c   What is the next number which is a square and a cube?

5 A scale is split into five boxes, each box starting on a multiple of 2.

Copy the diagram, and put these numbers into the correct boxes.

6 Find a pair of prime numbers a, b such that $\sqrt{a^2 - b}$ is a whole number.

7 Find a pair of whole numbers p, q such that $\sqrt{p^3 - 3q}$ is a whole number.

8 For the calculation $3^2 + 3^2 + 3^2$, Kira wrote:
$3^2 + 3^2 + 3^2 = (3 + 3 + 3)^2 = 9^2$
Explain why Kira is incorrect.
Find the correct answer.

9

$6^0 = 1$	$6^1 = 6$
$6^2 = 36$	$6^3 = 216$
$6^4 = 1296$	$6^5 = 7776$
$6^6 = 46656$	

Use the table to

   a   explain why $36 \times 216 = 7776$

   b   work out $46656 \div 36$.

10 Computer memory is measured in bytes. For example one byte is needed to hold one character of text. Humans think in tens, so we often approximate large units of memory to powers of ten. Computers work in powers of two. Four common units are in this table:

Unit	Approximate	Exact
Kilobyte (KB)	A thousand	2^{10}
Megabyte (MB)	A million	2^{20}
Gigabyte (GB)	1000 million	2^{30}
Terabyte (TB)	A million million	2^{40}

   a   Write the numbers in the second column as powers of ten.

   b   Work out the exact values of a KB and a MB.

   c   What is $2^{10} - 10^3$?

   d   What is $2^{20} - 10^6$?

   e   What is the percentage error in the megabyte approximation?

   f   Bill says that a megabyte is a kilobyte squared.
Is he correct? Explain your answer.

11 Find the value of the letter in each equation.

   a $3^a \times 9 = 3^{18}$ **b** $5^b = 25^{b-1}$

   c $4^c = 16^{2c+1}$ **d** $9^{2d-1} = 27^d$

12 Show that $\dfrac{2 \times 4^{x+3}}{2^{2x+4}} = 8$

1033, 1053, 1301, 1924   SEARCH

13.3 Surds

- Some numbers cannot be written as fractions. These **irrational** numbers have no repeating decimal patterns.
- Irrational numbers such as $\sqrt{2}$ are called **surds**.

$\sqrt{2} = 1.414213562...$

EXAMPLE

Say whether each of these numbers is **rational** or irrational.

  a $\dfrac{3}{17}$ **b** $\sqrt{16}$ **c** $\sqrt{17}$

a	$\dfrac{3}{17}$	rational	$\dfrac{3}{17}$ is an exact fraction.
b	$\sqrt{16} = 4$	rational	16 is a square number.
c	$\sqrt{17}$	irrational	17 is not a square number.

- The square root of a number is the product of the square roots of the number's factors.

To simplify a surd, look for square factors of the number under the square root.

For example, $\sqrt{20} = \sqrt{4 \times 5} = \sqrt{4} \times \sqrt{5} = 2\sqrt{5}$.

EXAMPLE

  a   Simplify $\sqrt{12} + \sqrt{48}$.

  b   Write $6\sqrt{2}$ in the form $\sqrt{n}$, where n is an integer.

  a $\sqrt{12} + \sqrt{48} = \sqrt{4} \times \sqrt{3} + \sqrt{16} \times \sqrt{3} = 2\sqrt{3} + 4\sqrt{3} = 6\sqrt{3}$

  b $6\sqrt{2} = \sqrt{6^2 \times 2} = \sqrt{36 \times 2} = \sqrt{72}$

Decimal **approximations** are useful in practical contexts.

Leaving an answer in **surd form** is more accurate than a decimal approximation.

If you have an expression with an irrational number in the **denominator**, you should **rationalise** it.

If you wanted to mark out a square of area $5\,m^2$, you would give the side length as $2.24\,m$, not $\sqrt{5}\,m$.

- To rationalise the denominator in a fraction, multiply numerator and denominator by the denominator.

EXAMPLE

Rewrite each of these expressions without roots in the denominator.

  a $\dfrac{5}{\sqrt{2}}$ **b** $\dfrac{7}{\sqrt{8}}$

  a $\dfrac{5}{\sqrt{2}} = \dfrac{5}{\sqrt{2}} \times \dfrac{\sqrt{2}}{\sqrt{2}}$ **b** $\dfrac{7}{\sqrt{8}} = \dfrac{7}{\sqrt{8}} \times \dfrac{\sqrt{8}}{\sqrt{8}}$

      $= \dfrac{5\sqrt{2}}{2}$ $= \dfrac{7\sqrt{8}}{8}$

Exercise 13.3S

1 Explain whether each of these numbers is rational or irrational.

a $\dfrac{1}{3}$ **b** 2π **c** $\dfrac{3}{19}$

d $\sqrt{3}$ **e** $\sqrt{25}$ **f** $\sqrt[3]{10}$

2 Write each of these expressions as the square root of a single number.

a $\sqrt{2} \times \sqrt{3}$ **b** $\sqrt{5} \times \sqrt{3}$

c $\sqrt{3} \times \sqrt{7} \times \sqrt{11}$

3 Write these expressions in the form $\sqrt{a}\sqrt{b}$, where a and b are prime numbers.

a $\sqrt{14}$ **b** $\sqrt{33}$

c $\sqrt{21}$ **d** $\sqrt{35}$

4 Write these expressions in the form $a\sqrt{b}$, where b is a prime number.

a $\sqrt{20}$ **b** $\sqrt{27}$

c $\sqrt{98}$ **d** $\sqrt{25}$

5 Write these expressions in the form $\sqrt{n}$, where n is an integer.

a $4\sqrt{3}$ **b** $5\sqrt{2}$

c $4\sqrt{5}$ **d** $10\sqrt{2}$

6 Without using a calculator, evaluate these expressions.

a $\sqrt{12} \times \sqrt{3}$ **b** $\sqrt{18} \times \sqrt{8}$

c $\sqrt{33} \times \sqrt{132}$

7 Simplify these expressions.

a $3\sqrt{5} \times \sqrt{20}$ **b** $\sqrt{28} + 5\sqrt{7}$

c $7\sqrt{12} - 2\sqrt{27}$

8 **a** Use a calculator to evaluate your simplified expressions from question 7, giving your answers to 3 decimal places.

     b Use a calculator to evaluate the non-simplified expressions from question 7.

     c Compare your answers to parts **a** and **b**.

9 Simplify these expressions.

a $\sqrt{3} + \sqrt{3}$ **b** $\sqrt{5} + \sqrt{5}$

c $\sqrt{9} + \sqrt{4}$ **d** $\sqrt{7} + \sqrt{7} + \sqrt{7}$

10 Simplify these expressions, giving your answers in surd form where necessary.

a $\sqrt{2} \times \sqrt{2}$ **b** $\sqrt{5} \times \sqrt{5}$

c $\sqrt{3}(\sqrt{3} + 3)$ **d** $\sqrt{4}(\sqrt{3} + 4)$

e $\sqrt{15}(3 + \sqrt{5})$ **f** $\pi(2^3 - \sqrt{20})$

g $(1 + \sqrt{5})(2 + \sqrt{5})$

h $(4 - \sqrt{7})(6 - 2\sqrt{7})$

11 Use a calculator to find an approximate decimal value for each of these expressions. Give your answers to 2 decimal places.

a $4\sqrt{2}$ **b** $\sqrt{5} + 1$

c $2 + \sqrt{5}$ **d** $36\pi - 7$

12 Simplify these expressions, and then use a calculator to find approximate decimal values for each one.

Give your answers correct to 3 significant figures.

a $4(3 + \sqrt{5})$ **b** $\sqrt{5}(5^2 - 5)$

c $\sqrt{7}(2 + \sqrt{7})$ **d** $\pi(2^3 + \sqrt{5})$

13 Simplify these expressions.

a $\sqrt{16} + \sqrt{3}$

b $\sqrt{49} + \sqrt{2} - \sqrt{16}$

c $3\sqrt{7} + \sqrt{49} - \sqrt{7}$

d $17 + \sqrt{17} - \sqrt{9}$

14 Rewrite each of these fractions without roots in the denominator.

a $\dfrac{1}{\sqrt{2}}$ **b** $\dfrac{1}{\sqrt{3}}$

c $\dfrac{1}{\sqrt{7}}$ **d** $\dfrac{1}{\sqrt{6}}$

e $\dfrac{1}{\sqrt{5}}$ **f** $\dfrac{2}{\sqrt{8}}$

g $\dfrac{2}{\sqrt{10}}$ **h** $\dfrac{3}{\sqrt{12}}$

i $\dfrac{5}{\sqrt{30}}$ **j** $\dfrac{8}{\sqrt{40}}$

15 Simplify the following expressions.

a $\dfrac{\sqrt{18} \times \sqrt{27}}{\sqrt{54}}$ **b** $\dfrac{\sqrt{98} \times \sqrt{12}}{\sqrt{48} \times \sqrt{128}}$

c $\dfrac{\sqrt{8}}{\sqrt{10}} \div \dfrac{2\sqrt{45}}{\sqrt{18}}$ **d** $\dfrac{3\sqrt{2}}{10} + \dfrac{\sqrt{8}}{\sqrt{50}}$

13.3 Surds

- Irrational numbers such as $\sqrt{2}$ are called surds.
- For any pair of surds $\sqrt{a} \times \sqrt{b} = \sqrt{a \times b}$
- Surds are written in the simplest form when the smallest possible integer is written inside the square root sign.

You can simplify brackets containing surds.

$$(1 + \sqrt{3})^2 = 1 + 2\sqrt{3} + (\sqrt{3})^2$$
$$= 4 + 2\sqrt{3}$$

Write the rational number first then the surd.

To remove the surd, multiply by the surd with the opposite sign.

$$(3 + \sqrt{5})(3 - \sqrt{5})$$
$$= 9 - 3\sqrt{5} + 3\sqrt{5} - 5$$
$$= 9 - 5$$
$$= 4$$

HOW TO

1. Read the question carefully; decide how to use your knowledge of surds.
2. Use $\sqrt{a} \times \sqrt{b} = \sqrt{a \times b}$ and algebraic techniques to simplify expressions containing surds.
3. Answer the question; make sure that you leave your answer in simplified surd form if the question asks for it.

EXAMPLE

Write $\dfrac{1}{1 + \sqrt{3}}$ without irrational numbers in the denominator.

1. Multiply by an expression that will remove the surd.
2. $(1 + \sqrt{3})(1 - \sqrt{3}) = 1 + \sqrt{3} - \sqrt{3} - 3$ Use FOIL to expand the brackets.
   $$= 1 - 3 = -2$$ The surds cancel.
3. Write the fraction with a rational denominator.

$$\frac{1}{1 + \sqrt{3}} = \frac{1}{1 + \sqrt{3}} \times \frac{1 - \sqrt{3}}{1 - \sqrt{3}}$$

$\dfrac{1 - \sqrt{3}}{1 - \sqrt{3}} = 1$ so the expressions are equivalent (the same).

$$= \frac{1 - \sqrt{3}}{-2} = \frac{-1 + \sqrt{3}}{2}$$

EXAMPLE

The diagram shows two similar rectangles. Find the length of side x.

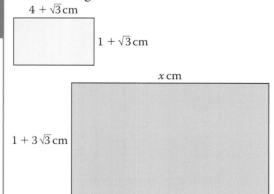

4 + √3 cm

1 + √3 cm

x cm

1 + 3√3 cm

1. The rectangles are similar so $\dfrac{\text{width}}{\text{height}}$ will be the same.

$$\frac{x}{1 + 3\sqrt{3}} = \frac{4 + \sqrt{3}}{1 + \sqrt{3}}$$

2. Solve for x by multiplying both sides by $1 + 3\sqrt{3}$.

$$x = \frac{(4 + \sqrt{3})}{1 + \sqrt{3}} \times (1 + 3\sqrt{3})$$

$$= \frac{4 + 12\sqrt{3} + \sqrt{3} + 9}{1 + \sqrt{3}}$$ $\sqrt{3} \times 3\sqrt{3} = 3 \times 3$

$$= \frac{13 + 13\sqrt{3}}{1 + \sqrt{3}} = \frac{13(1 + \sqrt{3})}{1 + \sqrt{3}}$$ 3. Simplify the fraction.

$$= 13 \text{ cm}$$

Exercise 13.3A

1 You are told that $\sqrt{2} \approx 1.414$, $\sqrt{3} \approx 1.732$ and $\sqrt{5} \approx 2.236$. Use this information to estimate the value of each of these. Show your working, and give your answers to 4 significant figures.

a $\sqrt{2} + \sqrt{3}$ **b** $\sqrt{10}$

c $\sqrt{125}$ **d** $\sqrt{120}$

2 Write down four more surds whose value can be estimated using the information given in question **1**, and estimate the values of your surds.

3 Find the perimeter and area of each shape. Give your answers in simplified surd form.

a

$1 + \sqrt{5}$

$\sqrt{5}$

b

$\sqrt{3}$

$2 - \sqrt{3}$

4 Expand the brackets and simplify.

a $(1 + \sqrt{5})(2 + \sqrt{5})$

b $(\sqrt{3} - 1)(2 + \sqrt{3})$

c $(4 - \sqrt{7})(6 - 2\sqrt{7})$

d $(5 + \sqrt{2})(5 - \sqrt{2})$

e $(6 + 2\sqrt{5})(6 - 2\sqrt{5})$

f $(7 + 5\sqrt{3})(7 - 5\sqrt{3})$

5

$\sqrt{5}$	$2 - \sqrt{5}$	$\sqrt{64}$	$8\sqrt{2}$
$\sqrt{2}$	$\sqrt{20}$	$3 + \sqrt{5}$	$2 + \sqrt{5}$
$3 - \sqrt{5}$	$3\sqrt{2}$	$\sqrt{10}$	$\sqrt{100}$

From this table, find

a two numbers which are rational

b a pair which add up to give 4

c a pair where one number is double the other

d a pair where one is the other squared

e a pair which multiply to give 4

f a pair which multiply to give 48

g the largest value

h the smallest value.

6 Rewrite these fractions without surds in the denominators.

a $\dfrac{1}{1 - 2\sqrt{3}}$ **b** $\dfrac{\sqrt{5}}{1 + \sqrt{5}}$

c $\dfrac{1 - \sqrt{2}}{1 + \sqrt{2}}$ **d** $\dfrac{4 + \sqrt{2}}{3 - 2\sqrt{2}}$

7 Simplify these expressions.

a $\dfrac{6 + \sqrt{27}}{3} - \dfrac{2 + \sqrt{3}}{4}$

b $\dfrac{20 + 3\sqrt{7}}{3} - (5 + \sqrt{28})$

c $\dfrac{6 + \sqrt{8}}{2} + \dfrac{3 + \sqrt{2}}{9}$

d $\dfrac{8 + \sqrt{45}}{3} - \dfrac{4 + \sqrt{20}}{9}$

8 Find the size of the shaded area. Give your answer in the form $p(1 + \sqrt{3})$ where p is an integer.

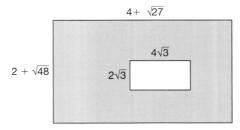

$4 + \sqrt{27}$

$4\sqrt{3}$

$2 + \sqrt{48}$ $2\sqrt{3}$

9 The diagram shows two similar triangles.

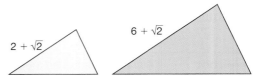

$2 + \sqrt{2}$ $6 + \sqrt{2}$

Find the scale factor of enlargement. Give your answer in simplified surd form.

10 The numbers X, Y and Z are such that $Y^2 = XZ$.

a Find the number Z if $X = \sqrt{3}$ and $Y = 1 - \sqrt{3}$. Give your answer in the form $p + q\sqrt{3}$.

b Explain why it would not be possible to find Y if $X = \sqrt{5}$ and $Z = 1 - \sqrt{5}$.

Summary

Checkout
You should now be able to...

Test it
Questions

	Test it Questions
✓ Know and use the language of prime numbers, factors and multiples.	1
✓ Write a number as a product of its prime factors.	2, 3
✓ Find the HCF and LCM of a pair of integers.	3
✓ Estimate the square or cube root of an integer.	4
✓ Find square and cube roots of numbers and apply the laws of indices.	5, 6
✓ Simplify expressions involving surds including rationalising fractions.	7, 8

Language   Meaning   Example

Language	Meaning	Example
Multiple	The original number multiplied by an integer.	Multiples of 6 include 12 (2 × 6) and 18 (3 × 6).
Factor	A whole number that divides exactly into a given integer.	The factors of 18 are 1, 2, 3, 6, 9 and 18.
Prime	A number that has only two factors, itself and one.	$13 = 1 \times 13$
Prime factor decomposition	Writing a number as a product of its prime factors.	$52 = 2^2 \times 13$
Highest common factor (HCF)	The largest factor that is shared by two or more numbers.	Factors of 30 Factors of 70
Lowest common multiple (LCM)	The smallest multiple that is shared by two or more numbers.	$HCF = 2 \times 5 = 10$ $LCM = 3 \times 2 \times 5 \times 7 = 210$
Square root	A number that when multiplied by itself two times is equal to the given number.	$\sqrt{64} = 8$ (8 × 8 = 64)
Cube root	A number that when multiplied by itself three times is equal to the given number.	$\sqrt[3]{64} = 4$ (4 × 4 × 4 = 64)
Surd	The root of a number which cannot otherwise be written exactly.	$\sqrt{2}$ $\sqrt[3]{5}$
Rationalise	To rewrite a fraction so that it does not contain any surds in the denominator.	$\dfrac{1}{\sqrt{2}} = \dfrac{\sqrt{2}}{2}$

Review

1

1	2	3	15	37	63	101	105

List the numbers in the panel that are

a prime

b factors of 105

c multiples of 21.

2 Write each number as the product of its prime factors. Use index notation where appropriate.

a 105   **b** 37

c 300   **d** 126

3 For each pairs of numbers find the

i lowest common multiple

ii highest common factor.

a 5 and 7   **b** 13 and 39

c 60 and 36   **d** 30 and 108

4 Estimate the value of these roots to 2 decimal places.

a $\sqrt{30}$ **b** $\sqrt[3]{45}$

5 Calculate the value of these expressions.

a $\sqrt[3]{64}$ **b** $\sqrt[3]{125}$

c 4^3 **d** 3^4

6 Simplify these expressions giving your answer in index form.

a $7^2 \times 7^5 \div 7^3$

b $(3^5 \div 3^2)^3$

c $\dfrac{3^{11} \div 3^2}{3^6}$

d $(7^{12} \div 7^3) \times 7^4 \times 7^8$

e $3^4 \times 5^3 \times 3^{-6} \div 5^2 \times 3^2$

7 Simplify these expressions involving surds.

a $\sqrt{108}$ **b** $5\sqrt{3} - \sqrt{27} + \sqrt{8}$

c $2\sqrt{5} \times \sqrt{5}$ **d** $\sqrt{6} \times \sqrt{2}$

e $3\sqrt{8} \div \sqrt{2}$ **f** $5\sqrt{10} \div 10\sqrt{5}$

g $(\sqrt{3} - 2)(\sqrt{3} - 5)$

h $\dfrac{\sqrt{8} + \sqrt{2}}{\sqrt{8} - \sqrt{2}}$

8 Rationalise these fractions.

a $\dfrac{1}{\sqrt{7}}$ **b** $\dfrac{3}{\sqrt{6}}$

c $\dfrac{5}{2\sqrt{5}}$ **d** $\dfrac{4 + \sqrt{6}}{\sqrt{6}}$

e $\dfrac{3\sqrt{2} - 5}{\sqrt{128} - 4\sqrt{2}}$ **f** $\dfrac{4\sqrt{3} + 5\sqrt{2}}{6\sqrt{3} + \sqrt{27}}$

What next?

Score			
	0 – 3		Your knowledge of this topic is still developing. To improve look at MyMaths: 1032, 1033, 1034, 1044, 1053, 1064, 1065, 1924
	4 – 7		You are gaining a secure knowledge of this topic. To improve look at InvisiPens: 13Sa – f
	8		You have mastered these skills. Well done you are ready to progress! To develop your exam technique looks at InvisiPens: 13Aa – h

Assessment 13

1 Isa and Josh both try to write 19 800 as a product of its prime factors.

Isa writes $19\,800 = 2^3 \times 3^2 \times 5^2 \times 11$
Josh writes $19\,800 = 2 \times 3^3 \times 5 \times 11$

Who is correct? Write 19 800 as a product of its prime factors using index notation.
Show your working. [3]

2 Tope says that all square numbers over 1 can be written as the sum of two prime numbers.

  a Show there are two ways of doing this for the number 16. [2]

  b Show there are five ways of doing this for the number 64. [3]

  c Show that there is only one possible way in which any *odd* square number
     can be written as the sum of two prime numbers. [2]

3 Ms Connell took her class to the monkey house at the zoo, but a naughty child
opened the cage door and let the monkeys out with the children. Ms Connell
counted about 70 heads and tails and the ratio of heads to tails was roughly $3:2$.
She also knew that the number of children was a prime number.
How many children were in Ms Connell's class? [4]

4 Arthen thinks that all prime numbers can be found by substituting $p = 0, 1, 2$, etc
into the formula $P = 41 - p + p^2$.

  a Confirm Arthen's assertion for $p = 0, 3$ and 6. [4]

  b Write down one value of p which shows that Arthen is *not* correct.
     Explain your reasoning. [2]

5 **a** Hans says that the highest common factor of 54 and 270 is 27.
     Is he correct? If not, show how to work out the highest common factor
     of these two numbers. [2]

  b Vanessa says that the lowest common multiple of 96 and 270 is 4 320.
     Show that this is true by using prime factors. [3]

6 MegaBurgers sell frozen burgers. The burgers are first packed into boxes of a
uniform size and then the boxes are packed into crates of various sizes.
A large crate contains 48 burgers, a medium crate contains 30 burgers and a
small crate contains 18 burgers.
What is the largest number of burgers that could be in a box? [4]

7 Henry knows an alternative method to find the HCF of two numbers.

    1. Write down the numbers side by side

    2. Cross out the largest and write underneath it the 'difference' between the
       largest and smallest.

    3. Repeat step 2 until the numbers left are the same.

    4. The remaining number is the HCF of the two original numbers.

  a Try Henry's method with these numbers.

    i 30 and 66   [3]   **ii** 252 and 588 [3]

  b Does Henry's method work with more than two numbers? [4]

8 Trains from Birmingham go to Glasgow, Sheffield and Leeds. Trains to Glasgow leave once an hour, to Sheffield every 45 minutes and to Leeds every 25 minutes. At 6 am three trains leave for all these destinations. When is the next time trains to all these destinations will leave simultaneously? [6]

9 **a** Kerry says that to evaluate $3^3 \times 3^2$ you multiply the indices. Saqib says you add the indices. Who is correct? Use the correct rule to evaluate $3^3 \times 3^2$. [2]

b Gino says that to evaluate $14^8 \div 14^2$ you subtract the indices. Jonas says you divide the indices. Who is correct? Use the correct rule to evaluate $14^8 \div 14^2$. [2]

10 Using the equations below, Thomas evaluates the values of p and q as $p = 5$ and $q = 2$. Show how he worked out these values.

a $6^{(p + 5)} = 36^p$ [3] **b** $8^{(2q - 4)} = 4^{(q - 2)}$ [4]

11 A cuboid gift box has length $5 + 2\sqrt{7}$ width $5\sqrt{3}$ and depth $\sqrt{27}$ as shown.

a Find the area of

   i face A [2] **ii** face B [2]

   iii face C. [2]

b Find the total surface area. [2]

c A piece of ribbon will be used to wrap around the gift in both directions.
What is the minimum length of ribbon needed? [2]

d Find the volume of the gift box. [2]

12 Niamh has attempted to simplify the following expressions fully but has made several errors. In each case explain her error and give the correct solution.

a $\sqrt{3} + \sqrt{3} = \sqrt{6}$ [3] **b** $4\sqrt{3} \times 5\sqrt{12} = 20\sqrt{36}$ [4]

c $(\sqrt{3} + 4)^2 = 3 + 16 = 19$ [4] **d** $(\sqrt{6} - 2)^2 = 6 - 4 = 2$ [4]

e $(3 + \sqrt{3})(7 - \sqrt{3}) = 21 - 3 = 18$ [3]

f $(\sqrt{7} + \sqrt{5})(\sqrt{7} - \sqrt{5}) = 2$ [2]

g $(4\sqrt{8} + 2\sqrt{11})(3\sqrt{8} - 5\sqrt{11}) = 12\sqrt{64} - 110$ [4]

13 Rationalise the denominator in each of the following expressions. Leave the fraction in its simplest form.

a $\dfrac{4}{\sqrt{5}}$ [2] **b** $\dfrac{8}{\sqrt{8}}$ [2]

c $\dfrac{1 - \sqrt{2}}{3\sqrt{2}}$ [2] **d** $\dfrac{12 - 3\sqrt{18}}{\sqrt{45}}$ [3]

e $\dfrac{15 - \sqrt{21}}{\sqrt{28} + \sqrt{63}}$ [3] **f** $\dfrac{\sqrt{44}}{\sqrt{704} - \sqrt{176}}$ [3]

14 A right angled triangle has sides as shown. Mariah says that

a $h = 6$ [3]

b The perimeter of the triangle is numerically double its area. [5]

Show that both of Mariah's statements are correct.

14 Graphs 1

Introduction

Kinematics is the topic within maths that deals with motion. By writing equations and drawing graphs that describe the relationship between distance, speed and acceleration, it is possible to calculate things such as the speed of a vehicle at a particular time during its journey.

What's the point?

Having the mathematical tools to describe how objects move means that we can better understand our world, which is in continual motion.

Objectives

By the end of this chapter, you will have learned how to ...

- Find and interpret the gradient and y-intercept of a line and relate these to the equation of the line in the form $y = mx + c$.
- Identify parallel and perpendicular lines using their equations.
- Draw line graphs and quadratic curves.
- Identify roots, intercepts and turning points of quadratic curves using graphical and algebraic methods.
- Use graphs to solve problems involving distance, speed and acceleration.

Check in

1 Evaluate these expressions for **i** $x = 3$ **ii** $x = -2$.

   a x^2 **b** x^3 **c** $2x^2$ **d** $x^3 + x$

   e $3x^2 - x$ **f** $2x^3 + 2x$ **g** $x^3 + 4x^2$ **h** $2x^3 - 4x^2 - x$

2 Plot these coordinate groups on one set of axes. Join each group of coordinates with a straight line.

   a $(3, 1), (3, 2), (3, 3)$ **b** $(1, -2), (2, -2), (3, -2)$

   c $(1, 3), (2, 5), (3, 7)$ **d** $(1, 4), (2, 3), (3, 2)$

3 Find the value of the unknown in each equation.

   a $9 = 3n + 3$ **b** $6 = 1 + 5m$ **c** $10p - 4 = 1$

4 For each line give its **i** gradient   **ii** y-axis intercept   **iii** direction.

   a $y = 3x + 4$ **b** $y = 10 - 4x$ **c** $2y = 8x + 10$

   d $2y - 4x = 15$ **e** $y = 7$ **f** $x = 2y - 4$

Chapter investigation

The points $(2, 4)$ and $(5, 13)$ both lie on the same straight line.

Find the equation of this line.

The point $(3, -2)$ lies on a line that is perpendicular to this line.
Find the equation of the perpendicular line.

14.1 Equation of a straight line

- The equation of a straight line is of the form $y = mx + c$, where m is the **gradient** and c is the **y-intercept**.

The y-intercept is the y-value where the graph cuts the y-axis.

The **gradient** of a line segment is calculated as

$$\frac{\text{Change in the } y\text{-direction}}{\text{Change in the } x\text{-direction}}.$$

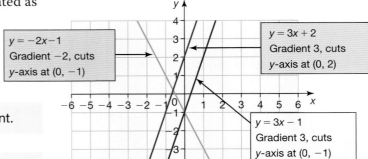

$y = -2x - 1$
Gradient -2, cuts y-axis at $(0, -1)$

$y = 3x + 2$
Gradient 3, cuts y-axis at $(0, 2)$

$y = 3x - 1$
Gradient 3, cuts y-axis at $(0, -1)$

- **Parallel** lines have the same gradient.

- If line A has gradient m, any line perpendicular to line A has gradient $-\dfrac{1}{m}$.

If you know the gradient and a point on the line you can find the equation of the line.

EXAMPLE

Find the equation of the line perpendicular to $y = 2x - 1$ that passes through $(4, 5)$.

$y = 2x - 1$ has gradient 2, so a line perpendicular to it has gradient $-\frac{1}{2}$.

$$y = -\frac{1}{2}x + c$$

At $(4, 5)$: $5 = \left(-\frac{1}{2}\right) \times 4 + c$

$$c = 7$$

The equation is $y = -\frac{1}{2}x + 7$.

If the equation is not in the form $y = ...$, rearrange it first, for example

$$3x + 2y = 12 \implies 2y = -3x + 12 \implies y = -\frac{3}{2}x + 6$$

Now you can see that the gradient is $-\dfrac{3}{2}$ and the intercept is 6.

EXAMPLE

Find the gradient and intercept of the lines

  a $x + y = 4$ **b** $2x - 5y = 10$

a $x + y = 4$
       $y = 4 - x$
       $y = -x + 4$

   gradient $= -1$, intercept $= 4$

b $2x - 5y = 10$
       $2x - 10 = 5y$ Divide by 5.
       $\frac{2}{5}x - 2 = y$

   gradient $= \frac{2}{5}$, intercept $= -2$

Algebra   Graphs 1

Exercise 14.1S

1 For each graph

    i write its gradient and intercept

    ii write the equation of the line.

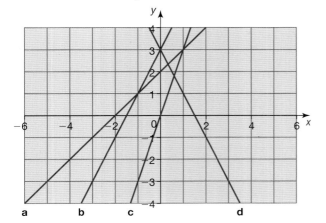

    a   **b**   **c**   **d**

2 Copy and complete the table.

Gradient	Intercept	Equation
3	5	
5	−2	
−2	7	
$\frac{1}{2}$	9	
$-\frac{1}{4}$	−3	
0	4	
1	0	

3 Rearrange these equations in the form $y = mx + c$.

    a $x + y = 5$ **b** $y - x = 3$

    c $x + y = -2$ **d** $x - y = 3$

    e $2x + y = 6$ **f** $5x + y = 9$

    g $3x + y = -2$ **h** $y - 2x = 5$

    i $2y + x = 4$ **j** $2y - x = 8$

    k $2x + 4y = 16$ **l** $6x + 2y = 8$

4 Find the equations of these lines.

    a Gradient 6, passes through $(0, 2)$

    b Gradient −2, passes through $(0, 5)$

    c Gradient −1, passes through $(0, \frac{1}{2})$

    d Gradient −3, passes through $(0, -4)$

5 For each of these lines, give the equation of a line parallel to it.

    a $y = 2x - 1$ **b** $y = -5x + 2$

    c $y = -\frac{1}{4}x + 2$ **d** $y = 7 - 4x$

    e $y = 6 + \frac{3}{4}x$ **f** $2y = 9x - 1$

6 Find the equations of these lines.

    a A line parallel to $y = -4x + 3$ and passing through $(-1, 2)$

    b A line parallel to $2y - 3x = 4$ and passing through $(6, 7)$.

7 Give the equation of a line perpendicular to each of these lines.

    a $y = 2x - 1$ **b** $y = -5x + 2$

    c $y = -\frac{1}{4}x + 2$ **d** $y = 7 - 4x$

    e $y = 6 + \frac{3}{4}x$ **f** $2y = 9x - 1$

8 If these lines are arranged in perpendicular pairs, which is the odd one out?

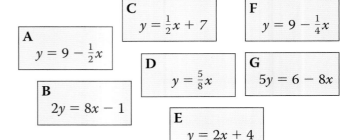

A $y = 9 - \frac{1}{2}x$

B $2y = 8x - 1$

C $y = \frac{1}{2}x + 7$

D $y = \frac{5}{8}x$

E $y = 2x + 4$

F $y = 9 - \frac{1}{4}x$

G $5y = 6 - 8x$

9 Find the equation of a line

    a parallel to $y = 3 - x$ and passing through $(9, 10)$

    b perpendicular to $y = 2x + 4$ and passing through $(3, 7)$

    c perpendicular to $2y = 8 - x$ and crossing the y-axis at the same point

    d perpendicular to $y = -\frac{2}{3}x - 5$ and passing through $(2, 8)$.

10 Find the gradients of the line segments joining these pairs of points.

    a $(4, 9)$ to $(8, 25)$ **b** $(5, 6)$ to $(10, 16)$

    c $(2, 1)$ to $(5, 2)$

11 Find the equations of the line segments joining each pair of points in question **10**.

Q 1153, 1311, 1314, 1957   SEARCH

14.1 Equation of a straight line

RECAP

RECAP

- You can write the equation of any straight line in the form $y = mx + c$, where m is the gradient of the line and c is the y-intercept.
- Parallel lines have the same gradient.
- The gradients of two perpendicular lines multiply together to give -1.

HOW TO

① Find the gradient of the line. Use the graph in the question or draw a sketch.

② Find the y-intercept either using the graph or by substituting a known point into the equation $y = mx + c$.

③ Give your answer in the context of the question.

EXAMPLE

Interpret the line of best fit on this scatter diagram of 18 students' heights and weights.

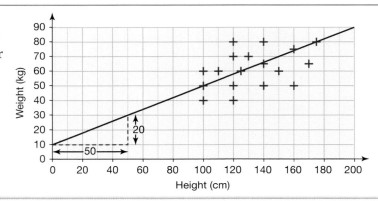

① The gradient, $m = \frac{20}{50} = \frac{2}{5}$

② The y-axis intercept, $c = 10$

The equation of the graph is $y = \frac{2}{5}x + 10$.

③ The gradient tells you that for every 5 cm you grow you gain 2 kg.

The intercept tells you that at 0 cm height, you weigh 10 kg.

It does not always make sense to interpret the intercept on a straight line graph for a big range of values.

EXAMPLE

What is the equation of the perpendicular **bisector** of the line segment passing through $(4, 8)$ and $(6, 16)$?

A perpendicular bisector goes through the midpoint of a line segment and is at right angles to it.

① The midpoint is (mean of x-coordinates, mean of y-coordinates).

The midpoint of $(4, 8)$ and $(6, 16)$ is $\left(\frac{4 + 6}{2}, \frac{8 + 16}{2}\right) = (5, 12)$.

The gradient of this line segment is $\frac{16 - 8}{6 - 4} = 4$.

So, the perpendicular bisector has gradient $-\frac{1}{4}$ and passes through $(5, 12)$.

$$y = -\frac{1}{4}x + c$$

② At $(5, 12)$: $12 = \left(-\frac{1}{4}\right) \times 5 + c$

$$c = 13\frac{1}{4}$$

③ The equation is $y = -\frac{1}{4}x + 13\frac{1}{4}$

Algebra Graphs 1

Exercise 14.1A

1 Here are the equations of several lines.

A $y = 3x - 2$	**B** $y = 4 + 3x$
C $y = x + 3$	**D** $y = 5$
E $2y - 6x = -3$	**F** $y = 3 - x$

a Which three lines are parallel to one another?

b Which two lines cut the y-axis at the same point?

c Which line has a zero gradient?

d Which line passes through (2, 4)?

e Which pair of lines are reflections of one another in the y-axis?

2 For each graph, find its equation in the form $y = mx + c$ and interpret the meaning of m and c, deciding if it is sensible to interpret c.

a
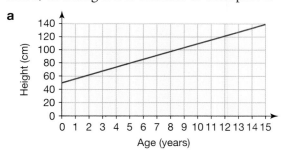

b
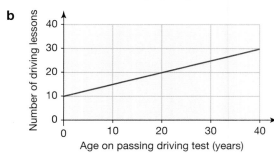

3 Mrs Harman gives her class a geography test. The results (%) for Paper 1 and Paper 2 for 10 students are shown.
Mrs Harman plots a scatter diagram and draws the line of best fit. The points (50, 40) and (60, 50) lie on the line of best fit.

3
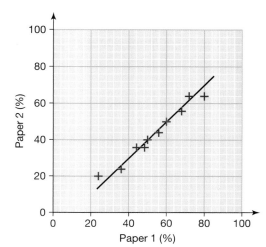

Interpret the line of best fit on the scatter diagram.

4 Find the equation of the perpendicular bisector of the segment joining each pair of points.

a (3, 10) and (7, 12)

b (2, 20) and (5, 18)

c $(-2, 7)$ and $(4, -10)$

5 Write an expression for the gradient of a line perpendicular to the line segment joining $(3t, 9)$ to $(2t, 12)$.

> Use gradients to make your decision.

6 The triangle formed by joining the point (5, 12) to (14, 24) and (2, 40) is right-angled. True or false?

7 Ben solved a pair of simultaneous equations graphically. His solution was $x = 2$ and $y = 3$. Use this diagram to help you find the second simultaneous equation that he solved.

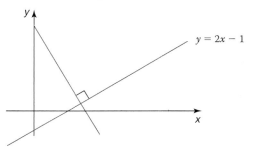

14.2 Linear and quadratic functions

- The graph of a **linear equation** $y = mx + c$ is a straight line with gradient m and y-intercept $(0, c)$.

> If you plot three points you can tell if you have made a mistake.
>
> One point must be wrong.

To plot a graph of a function:

- Draw up a table of values
- Calculate the value of y for each value of x
- Draw a suitable grid
- Plot the (x, y) pairs and join them with a straight line.

A straight line can be **diagonal**, **vertical** or **horizontal**.

The x-coordinate of every point on this vertical line is 2.

The y-coordinate can have any value.

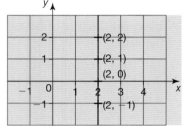

The equation of the line is $x = 2$.

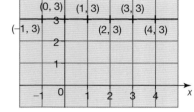

The equation of the line is $y = 3$.

The y-coordinate of every point on this horizontal line is 3.

The x-coordinate can have any value.

- Horizontal lines have equations of the form $y = c$.
- Vertical lines have equations of the form $x = c$.

> c stands for a number.

- A **quadratic function** has an x^2 term. It has a general equation $y = ax^2 + bx + c$ and is a U-shaped curve. Its gradient changes as x changes and its y-intercept is $(0, c)$.

EXAMPLE

Plot the curve $y = x^2 - x$ for $-2 \leqslant x \leqslant 2$ and find the coordinates of its minimum point.

Make a table of values.

x	-2	-1	0	1	2
x^2	4	1	0	1	4
$-x$	2	1	0	-1	-2
$y = x^2 - x$	6	2	0	0	2

Draw the x-axis from -2 to 2 and the y-axis from -1 to 7.

Plot the points $(-2, 6)$, $(-1, 2)$, ..., $(2, 2)$.

Join the points in a smooth curve.

The **minimum** point is $\left(\dfrac{1}{2}, -\dfrac{1}{4} \right)$.

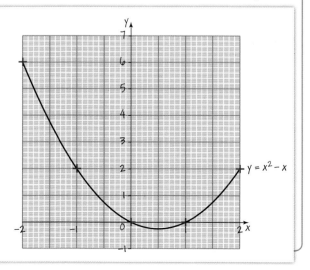

Exercise 14.2S

1 On one set of axes labelled from -6 to $+6$, plot these graphs.

   a $x = 5$ **b** $y = 2$

   c $x = 1.6$ **d** $y = -3$

   e $y = 1$ **f** $x = -1\frac{1}{4}$

2 Draw the graphs of these functions.

   a $y = 3x - 2$ **b** $y = -2x + 4$

   c $y = \frac{1}{2}x + 3$ **d** $y = 5 - x$

3 Draw the graphs of these functions.

   a $x + y = 4$ **b** $2x - y = 3$

   c $2x + y = 4$ **d** $3x + 6y = 9$

4 Draw the graphs of these functions.

   a $y = 5x - 3$ **b** $4y - x = 2$

   c $y = 3 - 4x$ **d** $2x - 3y = 6$

5 **a** The point $(2, 5)$ lies on the graph $y = 2x + 1$. Does $(3, 8)$ lie on this graph? Explain your answer.

   b The point $(2, 5)$ lies on the graph $y = 2x + 1$. Name another point that lies on this line.

6 **a** Plot the graphs $y = 3x - 1$ and $y = 3x + 2$ on the same axes.

   b Explain why there is no point that lies on both of these graphs.

   c Name the equation of another line that would have no points in common with these two.

7 **a** Copy and complete this table to show if each graph will be a straight line or a parabola.

   b Add an equation of your own in each column.

Straight line	Parabola

$y = 3x - 2$

$y = x^2 - 2$

$3x + 2y = 8$

$y = 10 + x^2$ $y = x^2 + 2x + 1$ $y = x$

8 **a** Draw axes labelled from -4 to $+4$ on the x-axis and -5 to $+15$ on the y-axis.

   b Copy and complete this table for $y = x^2 - 2$.

x	-4	-3	-2	-1	0	1	2	3	4
x^2	16							9	
$y = x^2 - 2$	14							7	

   c Plot the points that you have found in part **b** on your axes from part **a**. Join them to form a smooth parabola.

9 For each equation

   i make a table with x-values from -4 to 4 and find the corresponding y-values

   ii draw an x-axis from -4 to 4 and a suitable y-axis

   iii plot the points and join them to form a parabola

   iv write the coordinates of the minimum point of each parabola.

   a $y = x^2 + 3$

   b $y = 2x^2$

   c $y = 3x^2 - 1$

   d $y = x^2 + x$

> In part **b**, square before you multiply by 2.

10 **a** Copy and complete the table of values for the graph $y = x^2 + x + 1$.

x	-3	-2	-1	0	1	2	3
x^2	9						
y	7				3		

   b Plot the points for x and y and join them to form a smooth parabola.

   c What is the approximate minimum value of $x^2 + x + 1$ and for what value of x does it occur?

11 Imagine that these quadratic graphs are drawn.

   A $y = 4 + 3x - x^2$ **B** $y = x^2 + 2x - 4$

   C $y = 2x^2 - 4$ **D** $y = x^2 - 3x - 4$

   Which graphs have the same y-intercept?

🔍 1180, 1312   SEARCH

14.2 Linear and quadratic functions

- Linear functions ($y = mx + c$) are straight lines with gradient m and y-intercept $(0, c)$.
- A quadratic function ($y = ax^2 + bx + c$) is a U-shaped curve. Its gradient changes as x changes and its y-intercept is $(0, c)$.

HOW TO

To use a graph to solve a real-life problem

① Draw up a table of values. Calculate the values of y for at least three x-values. You may decide to write a formula for the situation in the question first.

② Draw a suitable grid, label the axes and plot the graph.

③ Use your graph to give the answer in the context of the question.

EXAMPLE

The parabola
$y = 1.2x - 0.02x^2$
models a javelin throw.

y = the height of a javelin

x = the horizontal distance.

a How far was the javelin thrown?

b Find the maximum height of the javelin.

① Make a table of values.

x	0	10	20	30	40	50	60
y	0	10	16	18	16	10	0

② Plot the points and draw a smooth curve to join the points.

③ The javelin lands when $y = 0$.

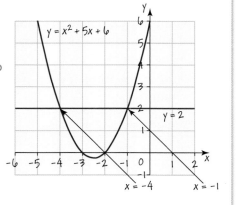

Maximum point

a The javelin was thrown 60 m.

b The maximum height was 18 m.

EXAMPLE

Solve $x^2 + 5x + 6 = 2$

① Make a table of values.

x	-5	-4	-3	-2	-1	0
x²	25	16	9	4	1	0
5x	-25	-20	-15	-10	-5	0
y	6	2	0	0	2	6

② Plot the points and draw a smooth curve to join the points.

Since $y = x^2 + 5x + 6$, then $x^2 + 5x + 6 = 2$ when $y = 2$.
Draw the line $y = 2$ on the graph.

③ The line $y = 2$ crosses the curve $y = x^2 + 5x + 6$ when $x = -1$ and $x = -4$;

$x^2 + 5x + 6 = 2$ when $x = -1$ and $x = -4$.

$y = x^2 + 5x + 6$

$y = 2$

$x = -4$ $x = -1$

Exercise 14.2A

1 Sketch graphs of these functions showing clearly all their main features.

a $y = 10 + 3x - x^2$

b $y = x^2 - 6x + 9$

*c $y = 2x^2 - 3x - 5$

2 A ball is thrown into the air. The formula $y = 20x - 4x^2$ shows its height, y metres, above the ground x seconds after it is thrown.

a Copy and complete the table of values to show the height of the ball during its first five seconds.

Time (x)	0	1	2	3	4	5
20x						
4x²						
Height (y)						

b Use the table to plot a graph to show the ball's height against time.

c Use your graph to find

   i the maximum height reached by the ball and the time at which it reaches this height

   ii two times when the ball is 12 metres above the ground

   iii the interval of time when the ball is above 15 metres.

3 The height above ground of an arrow is modelled by the function $h = 120 + 58d - d^2$ where h = height in centimetres and d = distance travelled in metres.

a Complete the table and plot the graph of the function.

d	0	10	20	30	40	50	60
h							

b Find the maximum height of the arrow.

c State the distance the arrow travelled.

4 Some graphs are drawn on these axes.

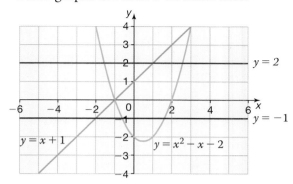

Use the graphs to find the approximate solutions of

a $x^2 - x - 2 = 2$

b $x^2 - x - 2 = -1$

c $x^2 - x - 2 = x + 1$

5 The graph shows $y = x^2 - 2x - 3$.

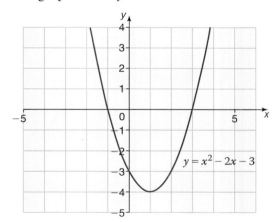

Copy the graph and, by adding lines, use it to find the approximate solutions of

a $x^2 - 2x - 3 = 1$

b $x^2 - 2x - 3 = -3$

c $x^2 - 2x - 3 = -4$

d $x^2 - 2x - 3 = x - 2$

e $x^2 - 2x - 3 = 1 - x$

f $x^2 - 2x - 3 = 0$

*6 Draw appropriate graphs to find the approximate solutions of

a $x^2 - 2 = 5$ **b** $x^3 + x = 2x - 1$

c $2x^2 - x = 0$ **d** $x^3 - x^2 = 2$

   Q 1180, 1312   SEARCH

14.3 Properties of quadratic functions

A parabola has a line of symmetry. The point where the line of symmetry crosses the curve is called a **turning point**.

- The turning points of a quadratic function will either be a **maximum** or a **minimum**.
- An input which gives an output of zero is called a **root** of the function.

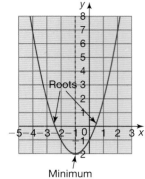

The **x-intercepts** of the graph are roots of the function.

EXAMPLE

14.2

For the function $y = x^2 - 3$, find

a the coordinates of the turning point

b the type of turning point

c the value of the y-intercept

d the roots.

Plot the function $y = x^2 - 3$.

a $(0, -3)$

b Minimum

c -3

d $x = -1.75$ and $x = 1.75$

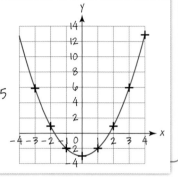

EXAMPLE

18.1

Use algebra to find the roots of these functions.

a $y = x^2 - 2x$ **b** $y = (x + 1)(x - 7)$ **c** $y = x^2 + 8x + 15$ **d** $y = x^2 - 4$

The roots of the function are values that have an output of 0.

a $x^2 - 2x = 0$ **b** $(x + 1)(x - 7) = 0$ **c** $x^2 + 8x + 15 = 0$ **d** $x^2 - 4 = 0$

$x(x - 2) = 0$ $x = -1$ and $x = 7$ $(x + 3)(x + 5) = 0$ $(x + 2)(x - 2) = 0$

$x = 0$ and $x = 2$ $x = -3$ and $x = -5$ $x = -2$ and $x = 2$

Not all quadratic functions have two roots.

EXAMPLE

a Write the function $y = x^2 + 8x + 18$ in the form $y = (x + a)^2 + b$.

b Find the coordinates of the turning point of the function $y = x^2 + 8x + 18$.

c Explain why there are no roots of the function $y = x^2 + 8x + 18$.

a $x^2 + 8x = (x + 4)^2 - 16$

$x^2 + 8x + 18 = (x + 4)^2 - 16 + 18$

$y = (x + 4)^2 + 2$

b The x^2 term of the function is positive. Therefore the turning point is a minimum.

$y = (x + 4)^2 + 2$ has a minimum output when $x = -4$ and $y = 2$.

The turning point of the function is at $(-4, 2)$.

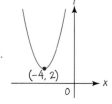

c The quadratic has a minimum at $(-4, 2)$. Therefore the curve does not cross the x-axis and has no x-intercepts. This means that there are no roots of the function $y = x^2 + 8x + 18$.

This information can be used to sketch the graph.

Exercise 14.3S

1 a Sketch the graph of these functions on the same axes and label each line.

$$y = x^2 \qquad y = x^2 - 2 \qquad y = -x^2$$

b Which function has a maximum?

c Which function has two roots?

2 The roots of a quadratic function are $x = -2$ and $x = 8$.

The graph has a turning point at $(3, 5)$.

a Sketch the graph of this function.

b Is the turning point a minimum or maximum?

3 The roots of a quadratic function are $x = -3$ and $x = 1$.

The y-intercept of the function is $y = 6$.

a Sketch the graph of this function.

b Does the function have a maximum or a minimum?

4 A quadratic function has a turning point at $(-3, 0)$ and a y-intercept at $(0, 4)$.

a Is the turning point a maximum or a minimum?

b How many roots does the function have?

5 The y-intercept of a quadratic function is $y = 1$.

The graph has a minimum at the point $(4, -3)$.

a Sketch the graph of this function on squared paper.

b Use your sketch to estimate the roots of the function.

6 Use algebra to find the roots of these functions.

a $y = x^2 + 7x$ **b** $y = x^2 - 8x$

c $y = x^2 + 5x$ **d** $y = x^2 + 2x$

e $y = 2x^2 + 8x$ **f** $y = 2x^2 + 5x$

g $y = 4x^2 + 12x$ **h** $y = 5x^2 + 3x$

7 Use algebra to find the roots of these functions.

a $y = x^2 + 7x + 6$ **b** $y = x^2 - 3x + 2$

c $y = x^2 + 5x + 6$ **d** $y = x^2 + x - 12$

e $y = x^2 - x - 6$ **f** $y = x^2 - 5x + 4$

g $y = x^2 - 2x - 15$ **h** $y = x^2 + 9x + 14$

8 Use algebra to find the roots of these functions.

a $y = x^2 + 4x + 4$ **b** $y = x^2 - 16$

c $y = x^2 - 4$ **d** $y = x^2 + 10x + 25$

9 a Express the function $y = x^2 + 6x + 3$ in the form $y = (x + a)^2 + b$.

b Write the coordinates of the y-intercept.

c Find the coordinates of the turning point.

d Sketch the graph of $y = x^2 + 6x + 3$.

e How many roots does the function have?

10 For each of these functions

i express the function in completed square form

ii find the coordinates of the turning point

iii sketch the graph of the function

iv write down the number of roots of the function.

a $y = x^2 + 2x - 3$ **b** $y = x^2 + 2x + 3$

c $y = x^2 + 8x + 12$ **d** $y = x^2 - 6x + 12$

e $y = x^2 + 5x - 3$ **f** $y = x^2 - 6x + 9$

g $y = -x^2 + 10x - 1$ **h** $y = -x^2 + 3x + 7$

> Parts **g** and **h**, remove a factor of -1 before completing the square. Multiply through again afterwards.

*11 a Write the function $y = 3x^2 + 6x - 2$ in the form $y = p(x + q)^2 + r$.

b Write the coordinates of the y-intercept.

c Find the coordinates of the turning point.

d Sketch the graph of $y = 3x^2 + 6x - 2$.

e How many roots does the function have?

12 Calculate the discriminant of the function $y = x^2 + 4x + 9$. Explain how this shows that the function has no roots.

13 Use the discriminant to show that the function $y = 3x^2 + 2x - 8$ has two roots.

14 A cricket ball is hit into the air. Its height (in metres) above ground after t seconds is given by the formula $h = 1 + 8t - t^2$. What is the greatest height that the ball reaches?

Q 1180, 1185, 1959, 1960   SEARCH

14.3 Properties of quadratic functions

- A parabola has a line of symmetry. The point where the line of symmetry crosses the curve is called a turning point.
- The turning points of a quadratic function will either be a maximum or a minimum.
- An input which gives an output of zero is called a root of the function.

HOW TO

Here is a general strategy for solving problems involving quadratic graphs.

① Identify the quadratic function.

② Sketch a graph.

③ Identify whether the roots or turning point help solve the problem.

④ Complete the necessary algebra and interpret the solution.

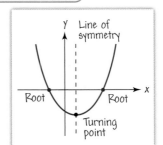

EXAMPLE

A company manufactures and sells packs of batteries. If the selling price of their batteries is too low they make no profit. If the selling price is too high they do not sell enough to make a profit. The company works out that their profit is a function of price:

$$P = -s^2 + 7s - 10 \text{ where } P = \text{profit (p) and } s = \text{selling price (£).}$$

a At what selling price will the company start losing money?

b What selling price will maximise their profit? How much profit will they make in this case?

6.4

① The quadratic function will form a 'negative parabola' since the s^2 term is negative. The graph could be plotted, but it is quicker to factorise and use the roots.

② $P = -s^2 + 7s - 10 \Rightarrow P = -(s^2 - 7s + 10) \Rightarrow P = -(s - 2)(s - 5)$
The roots of the function are $s = 2$ and $s = 5$.

③ Since a parabola is symmetrical the maximum of the quadratic occurs halfway between the roots: $(5 + 2) \div 2 = 3.5$.

④ **a** More than £5.

    b £3.50, $P = -3.5^2 + 7 \times 3.5 - 10 = 2.25$. Their profit is 2.25p.

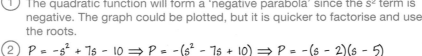

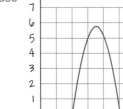

EXAMPLE

A second company also sells batteries. Their model uses the profit function $P = -s^2 + 8s - 6$ where $P = \text{profit (p) and } s = \text{selling price (£).}$

    a What selling price will maximise their profit?

    b How much profit will they make in this case?

    c Comment on the profit for a selling price of £0.

10.2

a $P = -s^2 + 8s - 6 \Rightarrow P = -(s^2 - 8s + 6)$

This does not factorise easily so complete the square instead.

$s^2 - 8s + 6 = (s - 4)^2 - 16 + 6 = (s - 4)^2 - 10 \Rightarrow P = -(s - 4)^2 + 10$

The optimum selling price is £4.

b $s = 4 \Rightarrow P = 10$. Their profit is 10p.

c $P = -0^2 + 8 \times 0 - 6 = -6$. The company loses 6p per battery.
The model does not make sense at this point.

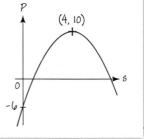

Algebra Graphs 1

Exercise 14.3A

1 A company establishes a profit function of $P = 6s - s^2 - 5$ where $P = $ profit (p) and $s = $ selling price (£)

 a Sketch the graph of the profit function.

 b At what prices would the profit be zero?

 c What selling price will maximise their profit? How much profit will they make in this case?

2 A company uses a profit function of $P = -2s^2 + 900s - 100\,000$ where $P = $ profit (£) and $s = $ selling price (£)

 a What selling price will maximise their profit? How much profit will they make in this case?

 b At what prices would the profit be zero?

3 The height above ground of a javelin is modelled by the function $h = 100 + 48t - t^2$ where $h = $ height in centimetres and $\quad t = $ length of throw in metres.

 a Sketch the graph of the function.

 b Find the maximum height of the javelin.

 c State the length of the throw.

4 The jet of water in a fountain is modelled by the function $y = -\frac{1}{10}x\,(x - 50)$ where $x = $ distance from source (cm) and $y = $ height (cm)

 a At what distance from the source does the jet enter the water again?

 b What is the maximum height reached by the jet?

5 Shot balls used to be made by dropping molten lead from the top of a tower into a pool of water. After lead is dropped, its height (in metres) above ground is given by the formula $h = 44.1 - 4.9t^2$, where t is time in seconds.

 a Sketch the graph of the function

 b What height is the lead dropped from?

 c After how many seconds does the lead land in the pool of water?

6 A quadratic function has roots at $x = -2$ and $x = 8$ and a turning point at $(3, 5)$.

 a Sketch the graph of this function.

 b Is the turning point a minimum or maximum?

7 A quadratic function has roots $x = -3$ and $x = 1$ and intercept $y = 6$.

 a Sketch the graph of this function.

 b Does the function have a maximum or a minimum?

8 A quadratic function has a turning point at $(-3, 0)$ and a y-intercept at $(0, 4)$

 a Is the turning point a maximum or a minimum?

 b How many roots does the function have?

9 A quadratic function has a minimum at $(4, -3)$ and a y-intercept at $y = 1$.

 a Sketch the graph of this function on squared paper.

 b Use your sketch to estimate the roots of the function.

*10 A parabolic trough solar collector captures energy from sunlight. Every cross-section focuses the sun's rays onto a single point.

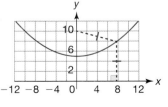

A parabolic trough is designed using the function $y = \frac{1}{20}x^2 + 5$. The single point is $(0, 10)$. A feature of this point is that every position on the parabola is equidistant from it and the x-axis.

 a Verify that the point $(10, 10)$ is on the curve.

 b Find another point with integer coordinates that lies on the curve.

 c Use graphing software to verify your solution.

 d Investigate this feature of the parabola.

14.4 Kinematic graphs

A distance–time graph shows information about a journey.

The gradient of a straight line in a distance–time graph is the speed of the object.

Velocity–time graphs also give information about a journey.

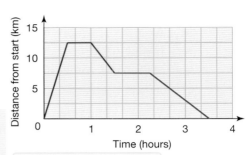

- The gradient of a straight line in a velocity–time graph is the **acceleration** of the object.
- The area under a line in a velocity–time graph is the distance travelled by the object.

Velocity is speed in a certain direction.

EXAMPLE

Debbie is running.
At the end of her training session her GPS app shows this graph.

a What is Debbie's speed during the first 45 minutes?

b What is Debbie's average speed for the whole journey?

22.1

To get answers in km/h use 45 mins = 0.75 hrs.

a speed = $\dfrac{\text{distance}}{\text{time}}$ = $\dfrac{5}{0.75}$ = 6.666... km/h

= 6.7 km/h (1 dp)

b For average speed use total distance and total time.

average speed = $\dfrac{\text{total distance}}{\text{total time}}$ = $\dfrac{12.5 \text{ km}}{1.75 \text{ hours}}$ = 7.142857... km/h

= 7.1 km/h (1 dp)

Remember the gradient is 'change in y' over 'change in x'.

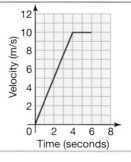

EXAMPLE

The velocity–time graph describes the journey of a radio-controlled car.

a Describe the car's journey in words.

b What is the acceleration of the car during the first four seconds?

c What is the distance travelled over the first six seconds?

a The car starts at rest and reaches a speed of 10 m/s after 4 seconds.
Initially the car accelerates steadily.
Then, between 4 and 6 seconds, the car travels at a constant speed.

The graph is a *straight* line.
The line is *horizontal*.

b Calculate the gradient.

Acceleration = $\dfrac{\text{change in } y}{\text{change in } x}$ = $\dfrac{10 \text{ m/s}}{4 \text{ seconds}}$ = 2.5 m/s²

c Calculate the area below the line.

From $t = 0$ to $t = 4$ Area = $\frac{1}{2}$ × 4 × 10 = 20

From $t = 4$ to $t = 6$ Area = 2 × 10 = 20

The distance travelled is 40 metres.

Or as a trapezium

Area = $\frac{1}{2}$ (2 + 6) × 10

= 40 m

Algebra   Graphs 1

Exercise 14.4S

1 This distance–time graph shows Lisa's coach journey.

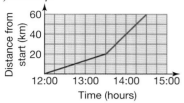

a What is the speed between 12:00 and 13:30?

b What is the speed between 13:30 and 14:30?

2 Tamera is riding a bike. Information about her journey is shown in this graph.

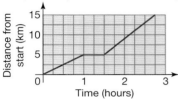

Tamera's journey starts at 8 am.

a What is her average speed for the whole ride?

b What is her speed between 10:00 and 10:30?

3 Mark sets off from home in his car at 2 pm. He stops to get petrol and then continues on his journey.

Mark returns home later in the afternoon.

The distance–time graph shows more information about this journey.

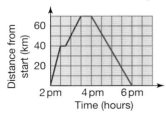

a Between which times was Mark travelling fastest? Explain how you know.

b What was Mark's overall average speed on the outward journey?

c What was Mark's average speed on the homeward journey?

d The gradient of the line after 4 pm is negative. What does this tell you about the velocity?

4 The velocity–time graph shows information about a runner in a race.

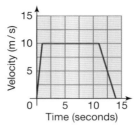

a What is the speed at 12 seconds?

b At what times is the speed 6 m/s?

c What is the acceleration during the first second?

d Find the overall distance travelled.

5 The diagram shows a velocity–time graph.

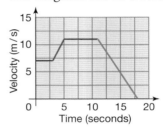

a What is the acceleration between 3 and 5 seconds?

b What is the distance travelled between 3 and 5 seconds?

c What is the overall distance travelled?

*6 Describe the journey shown by this velocity–time graph.

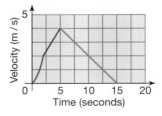

7 A train leaves a station at 11 am and travels at a constant speed of 65 km/h.

A second train leaves the same station 30 minutes later. It travels at 150 km/h.

a Construct a distance–time graph to show this.

b At what time are the two trains the same distance from the starting station?

Q 1322, 1323   SEARCH

14.4 Kinematic graphs

- In a distance–time graph the gradient of a straight line is the speed of the object.
- Velocity is speed in a certain direction.
- In a velocity–time graph
  - the gradient of a straight line is the acceleration of the object
  - the area under a line is the distance travelled by the object.

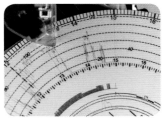

▲ The tachograph on a lorry plots a distance–time graph and a speed–time graph every 24 hours.

HOW TO

To sketch a distance–time or acceleration–time graph given a velocity–time graph

1. Split the total time into distinct sections.
2. Consider how the speed varies in each section and the effect that this would have on distance or acceleration.
3. Sketch each section of the distance–time graph in turn.

EXAMPLE

For this velocity–time graph

a   sketch the distance–time graph for the same data

b   construct a graph to show how acceleration varies over time.

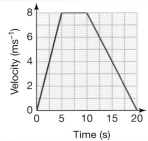

1. Consider 0–5 s, 5–10 s and 10–20 s separately.

2. Distance = area under graph
   Acceleration = gradient of graph

<u>0 to 5</u>   Speed: steady increases $0 \rightarrow 8$ m/s.

    Total distance $= \dfrac{1}{2} \times 5 \times 8 = 20$ m.

3. Rate of covering distance increases as speed increases.

    Acceleration $= \dfrac{8}{5} = 1.6$ m/s^2.

<u>5 to 10</u>   Speed: constant at 8 m/s.

    Total distance $= 5 \times 8 = 40$ m.

    Distance increases at a steady rate: straight line, positive gradient.

    Acceleration $= 0$ m/s^2.

<u>10 to 20</u>   Speed: steady decreases, $8 \rightarrow 0$ m/s.

    Total distance $= \dfrac{1}{2} \times 10 \times 8 = 40$ m.

3. Rate of covering distance decreases as speed decreases.

    Acceleration is $\dfrac{-8}{10} = -0.8$ m/s^2

    (Deceleration is 0.8 m/s^2).

a

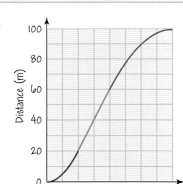

b

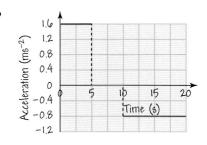

Exercise 14.4A

1 For each of the velocity–time graphs

i sketch a distance–time graph

ii construct an acceleration–time graph.

a

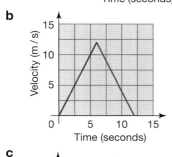

b

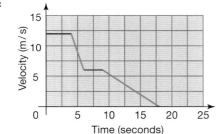

c

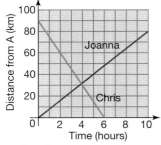

2 Two cyclists start a training ride at 7:30 am.

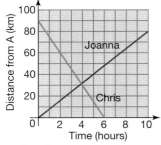

a What happens at 11:20 am?

b What is the average speed of each cyclist?

c What is the average velocity of each cyclist?

3 This velocity–time graph gives information about two cars.

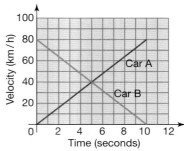

a What is the distance travelled by each car?

b Sketch, on the same axes, a distance–time graph for each car.

c Plot, on the same axes, an acceleration–time graph for each car. Clearly state the units.

4 Explain why each of these graphs is impossible.

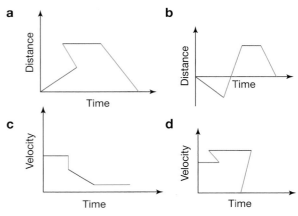

5 Becky and Sara run a 400 m race. This graph shows the race.

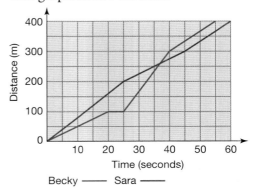

Becky ——— Sara ———

Describe the race.
Ensure that you state who won.

Summary

Checkout
You should now be able to...

Test it
Questions

✔ Find and interpret the gradient and y-intercept of a line and relate these to the equation of the line in the form $y = mx + c$.	1 – 3
✔ Identify parallel and perpendicular lines using their equations.	4
✔ Draw line graphs and quadratic curves.	5, 6
✔ Identify roots, intercepts and turning points of quadratic curves using graphical and algebraic methods.	7, 8
✔ Use graphs to solve problems involving distance, speed and acceleration.	9

Language · Meaning · Example

Language	Meaning	Example
Gradient	A measure of the slope of a line on a graph found by dividing the change in y by the change in x.	y-intercept, $c = 3$ Gradient, $m = -\frac{1}{2}$ $y = -\frac{1}{2}x + 3$
y-intercept	A point at which a graph crosses the y axis.	
$y = mx + c$	The standard form for a straight lime graph. m = gradient, c = y-intercept	
Quadratic function	A function of the form $f(x) = ax^2 + bx + c$ where a, b and c are numbers and $a \neq 0$.	$f(x) = 6x^2 - 7x - 3$ $a = 6, b = -7, c = -3$
Parabola	The shape of a quadratic curve.	Maximum $(-2,3)$ Root Minimum $(2,0)$ $y = 3 - (x + 2)^2$ $\qquad$ $y = 4(x - 2)^2$
Turning point	A point on a curve where the gradient changes from increasing to decreasing, ∩ – a **maximum** decreasing to increasing, ∪ – a **minimum**.	
Root	An x-intercept of a graph. If $y = f(x)$, a root is a solution to the equation $f(x) = 0$, that is, an input to the function $f(x)$ that gives output 0.	
Kinematics	A branch of mathematics relating to the motion of objects.	Acceleration $= \dfrac{\text{change in velocity}}{\text{change in time}} = \dfrac{10}{5} = 2\,\text{m/s}^2$
Speed	The gradient of a distance–time graph.	
Acceleration	The gradient of a speed–time graph.	

Review

1 **a** What is the gradient of this line?

   b What is the equation of this line?

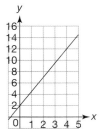

2 What is the equation of a line that has gradient -3 and passes through $(0, 5)$?

3 State the gradients and y-intercepts of the lines with these equations.
   a $y = 7 - 2x$ **b** $y = x + 9$
   c $x + y = 3$ **d** $2x + 3y - 5 = 0$

4 Given a line with the equation $y = 5x + 2$, write down the equation of the line that is
   a parallel and passes through $(2, 16)$
   b perpendicular and passes through $(10, -1)$.

5 Draw the graphs of these functions.
   a $y = 5x + 2$ **b** $y = 14 - 3x$

6 Draw the graphs of these functions.
   a $y = 4x^2 - 16$ **b** $y = x^2 + 10x + 25$
   c $y = 2x^2 - 5x - 3$
   d $y = -x^2 + x + 6$

7 Here is a graph of the function $y = f(x)$. Write down

   a the y-intercept

   b the x-intercepts

   c the roots

   d the x-coordinate of the turning point

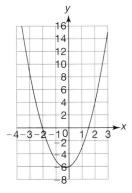

8 **a** Complete the square for each of these functions.
      i $f(x) = x^2 + 2x - 5$
      ii $f(x) = 2x^2 - 12x + 5$
      iii $f(x) = 5x - x^2$

   b Work out the coordinates of the turning point for each of the functions and state if it is a maximum or a minimum point.

9 The graph shows the distance covered by a sprinter.

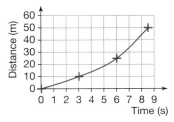

   a What is the total distance run?

   b What was the speed of the sprinter in the first 3 seconds?

   c What is the average speed, in km/h, of the sprinter?

   d What is happening to the speed of the sprinter during the run?

   At the end of the run the sprinter takes 5 seconds to slow down to a complete stop.

   e What is the sprinter's deceleration, assuming it is constant?

What next?

Score			
	0 – 4		Your knowledge of this topic is still developing. To improve look at MyMaths: 1153, 1169, 1180, 1185, 1311, 1312, 1314, 1322, 1323, 1396
	5 – 8		You are gaining a secure knowledge of this topic. To improve look at InvisiPens: 14Sa – k
	9		You have mastered these skills. Well done you are ready to progress! To develop your exam technique looks at InvisiPens: 14Aa – f

Assessment 14

1 **a** State the gradient of each line (A–D) on the grid. [4]

b Match each equation to one of the lines A, B, C and D drawn on the graph. [4]

i $5x + 3y = 20$ **ii** $2y - 3x = 9$

iii $2x - y = 0$ **iv** $2x + 3y = 0$

c State the y-intercept for each equation. [4]

2 Work out the gradient and y-intercept for each of the following straight lines.

a $y = 2x + 7$ [2]   **b** $y = 9 + 4x$ [2]

c $y = 6x - 11$ [2]   **d** $y = 12 - 4x$ [2]

e $4x - 7y = 14$ [3]   **f** $15x + 14y = 35$ [3]

3 Given the gradient and a point on the line, find the equation of each line in the form $y = mx + c$.

a Gradient $= -5$, point $(0, -61)$ [2]   **b** Gradient $= 3$, point $(1, 2\frac{1}{2})$ [2]

c Gradient $= \frac{1}{3}$, point $(3, -1)$ [2]   **d** Gradient $= \frac{1}{4}$, point $(0, -\frac{3}{4})$ [2]

4 Calculate the gradients of the straight lines which pass through each pairs of points.

a $(1, 5)$ and $(5, 9)$ **b** $(6, 7)$ and $(8, -9)$

c $(-3, 5)$ and $(-4, 6)$ **d** $(-11, 0)$ and $(5, -8)$ [8]

5 Find the equation of each line in the form $y = mx + c$.

a Line parallel to $y - 7x - 9 = 0$ that intercepts the y-axis at $(0, 5)$.   [3]

b Line parallel to $3x + y - 77 = 0$ that intercepts the y-axis at $(0, -7.25)$.   [3]

c Line parallel to $2x + 6y = 15$ that intercepts the y-axis at $(0, 2\frac{2}{3})$ [3]

6 Here are the equations of ten lines.

A: $y = 4x - 8$ B: $x + 1 = 0$ C: $4x = 3y$ D: $4y + x - 15 = 0$ E: $y = x$

F: $3x + y + 8 = 0$ G: $4 - x = 0$ H: $2y - 8x + 1 = 0$ I: $y - 7 = 0$ J: $x - \frac{1}{4}y + 3 = 0$

a Find the three lines that are parallel to one another.   [3]

b Find the two pairs of lines that are perpendicular to one another.   [3]

c Which line has a zero gradient?   [1]

d Which two lines have an infinite gradient?   [2]

e Which two lines pass through the point $(1, 3\frac{1}{2})$?   [2]

f Which two lines pass through the origin?   [1]

7 **a** Find the mid-point and gradient of the line segment joining $(-1, 5)$ and $(7, 3)$.   [4]

b Find the equation of the perpendicular bisector of this line segment.   [4]

8 A town planner investigated the correlation between the population density and distance from the urban centre in a large city. These are the results.

Distance (km)	2	4	6	8	10	12	14	16	18	20	22	24	26
Population density (people per square km)	95	93	90	33	78	66	92	59	48	40	33	27	27

  a The line of best fit passes through (6, 90) and (20, 41). Find the equation of the line of best fit. [4]

  b Complete this sentence.
As the distance from the urban centre increases by 1 km, the population density _____ by _____ people per square km. [2]

  c Is the reading at 8 km an outlier? Explain your answer. [2]

9 **a** Draw $y = x^2 - x - 6$ and $y = x$ on the same grid for values of x from -4 to 3. [5]

  b Estimate the coordinates of the points where the graphs intersect. [1]

  c Use your graph to estimate the solutions of the equation $x^2 - x - 6 = 0$ [2]

  d By factorising, solve the equation $x^2 - x - 6 = 0$.
Compare your answer with your estimate in part **d**. [3]

10 A metal spring stretches when a mass of m grams hung on the end.

The distance stretched, d cm, is given by the formula $d = 8 + \dfrac{m}{15}$.

  a Complete the table of values for m and d.
Write your answers to 1 dp where appropriate. [3]

m	10	20	30	40	50	60	70	80	90	100
d										

  b Plot the graph of d against m for values of m from 0 to 100. [3]

  c Use your graph to
    i find the stretch, d, when a mass of 75 g is hung on the spring [1]
    ii find the mass hung on the spring when $d = 9$ cm. [1]

  d What is the length of the spring when no weights are hung on it? [1]

11 The average safe stopping distance for cars d metres, is given by the equation

$d = \dfrac{3v^2}{125} + \dfrac{v}{2.7}$, where v is the speed of the car in km/h.

  a Complete the table of values for d and v. Write your distances to the nearest metre. [4]

v	0	10	20	30	40	50	60	70	80	90	100
d											

  b Draw the graph of $d = \dfrac{3v^2}{125} + \dfrac{v}{2.7}$. [4]

  c Use your graph to find the safe stopping distance when a car travels at
    i 25 km/h   **ii** 42 km/h   **iii** 77 km/h. [3]

15 Working in 3D

Introduction

In the Cave of Crystal Giants in Mexico, vast crystals of selenite grow to lengths of up to 11 metres, weighing up to 55 tons – the largest crystals to have yet been discovered. The cave was only discovered because there was a mine nearby. It is likely that there are even more awesome geological structures lying somewhere undiscovered beneath our feet!

What's the point?

We live in a three-dimensional world. 3D geometry allows us to describe the wonderful things that we can see in the natural world, as well as helping us to devise increasingly sophisticated structures in the manmade world.

Objectives

By the end of this chapter, you will have learned how to ...

- Draw and interpret plans and elevations of 3D shapes.
- Calculate the volume of cuboids and right prisms.
- Calculate the surface area and volume of spheres, pyramids, cones and composite shapes.
- Know and apply the relationship between lengths, areas and volumes of similar shapes.

Check in

1 **a** Draw a net of this cube.

   b Hence calculate its surface area.

   c Calculate its volume.

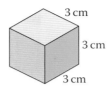
3 cm
3 cm
3 cm

2 For each circle work out its

   i area

   ii circumference.

a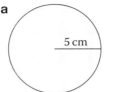
5 cm

b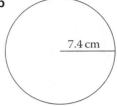
7.4 cm

Chapter investigation

A manufacturer wants to create a drinks container with a capacity of 1 litre. In order to minimise the cost, they also want the container to have the least surface area.

Advise the manufacturer on suitable designs for the container.

15.1 3D shapes

A **face** is a flat surface of a solid.
An **edge** is the line where two faces meet.
A **vertex** is a point at which three or more edges meet.

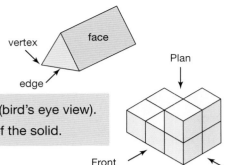

- A **plan** of a solid is the view from directly overhead (bird's eye view).
- An **elevation** is the view from the front or the side of the solid.

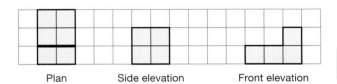

Plan   Side elevation   Front elevation

Notice the extra bold line in the plan, when the level of the cubes alters.

- A **net** is a 2D shape that can be folded to form a 3D shape.

EXAMPLE

The nets of two solids are shown. Name the solid that can be made from each net.

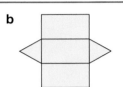

a cuboid

b triangular prism

- The **surface area** of a 3D shape is the total area of its **faces**.

EXAMPLE

Find the surface area of this triangular **prism**.

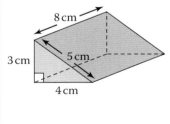

Triangle: Area = $\frac{1}{2}$ × 4 × 3 = 6 cm²

Triangle: Area = $\frac{1}{2}$ × 4 × 3 = 6 cm²

Side: Area = 3 × 8 = 24 cm²

Bottom: Area = 4 × 8 = 32 cm²

Sloping side: Area = 5 × 8 = 40 cm²

Surface area
= 6 + 6 + 24 + 32 + 40 = 108 cm²

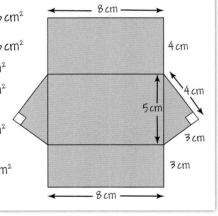

- Volume of a prism = area of cross-section × length

EXAMPLE

Calculate the volume of this triangular prism. State the units of your answer.

Area of triangle = $\frac{1}{2}$ × 4 × 2 = 4 m²

Volume of prism = area of triangle × 8

= 4 × 8 = 32 m³

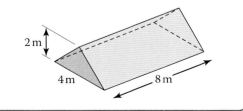

Geometry   Working in 3D

Exercise 15.1S

1 On square grid paper, draw the plan (P), the front elevation (F) and the side elevation (S) for each solid.

a

b

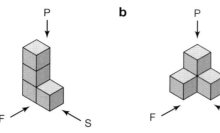

2 The plan, front elevation and side elevation are given for these solids made from cubes. Draw a 3D sketch of each solid and state the number of cubes needed to make it.

a

b

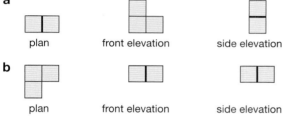

plan   front elevation   side elevation

3 For these solids, draw

i the plan

ii the front elevation from the direction marked with an arrow.

a   **b**

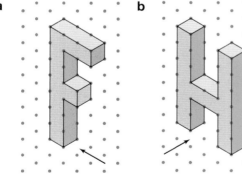

c   **d**

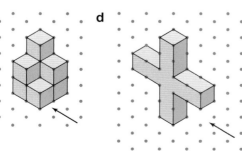

4 a On square grid paper, draw the net for each cuboid.

i   **ii**

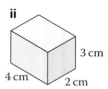

2 cm
3 cm   5 cm
3 cm
4 cm   2 cm

b Find the surface area of each cuboid.

c Find the volume of each cuboid.

5 Work out the surface areas of these prisms.

a

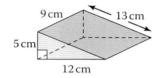

9 cm   13 cm
5 cm
12 cm

b

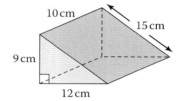

10 cm   15 cm
9 cm
12 cm

6 a Draw the net of this square-based pyramid.

b Use the net to find the surface area.

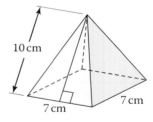

10 cm
7 cm   7 cm

7 a Find the volume of this shape.

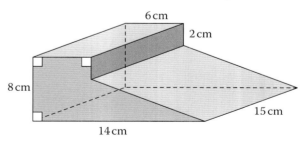

6 cm   2 cm
8 cm
14 cm   15 cm

b Draw the net of the shape.

c Use the net to find the surface area.

Q 1078, 1098, 1106   SEARCH

15.1 3D shapes

- A 3D shape is a solid shape.
- A net is a 2D shape that you can fold to make a 3D shape.
- A plan of a solid is the view from directly overhead.
- An elevation is the view from the front or the side of the solid.
- The area of the net is called the surface area.
- Volume of a prism = area of cross-section × length

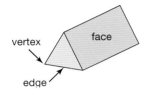

vertex
face
edge

HOW TO

(1) Use your knowledge of 3D solids to draw a net, the plan and elevations or a sketch of the solid.

(2) Work out the answer, remember to include units and give an explanation if the question asks for it.

Each face is a cross-section.

EXAMPLE

A cube has surface area $150\,cm^2$.
Find the **volume** of the cube.

(1) A cube has six faces.
Each face has the same area.

Area of one face is $150 \div 6 = 25\,cm^2$

Each face is a square, so length of each side is $\sqrt{25} = 5\,cm$

(2) Volume of cube = area of cross-section × length
$= 25 \times 5$
$= 125\,cm^3$

EXAMPLE

A 3D solid has this net.

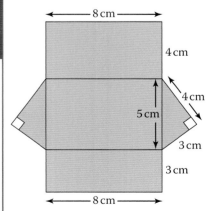

8 cm
4 cm
4 cm
5 cm
3 cm
3 cm
8 cm

Find the volume of the solid.

(1) Folding up the net gives a triangular prism.

The cross-section is a triangle.

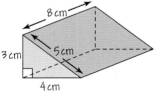

8 cm
3 cm
5 cm
4 cm

Area of triangle $= \frac{1}{2} \times$ base $\times$ height

$= \frac{1}{2} \times 3 \times 4$

$= 6\,cm^2$

(2) Volume = area of triangle × length

Volume of solid $= 6 \times 8$

$= 48\,cm^3$

Exercise 15.1A

1 A solid is made from eight isosceles triangles and one octagon. For this solid, write

    a the mathematical name

    b the number of faces

    c the number of edges

    d the number of vertices.

2 Sketch **six** different nets of a cube.

3 Find the volumes of the cubes with these surface areas.

    a Surface area $54\,cm^2$

    b Surface area $294\,cm^2$

    c Surface area $96\,cm^2$

    d Surface area $1.5\,m^2$

4 Find the surface areas of the cubes with these volumes.

    a Volume $512\,cm^3$

    b Volume $1000\,cm^3$

    c Volume $216\,cm^3$

    d Volume $1\,m^3$

5 The diagrams show the plan and the front elevation of different solids.

    i Draw a sketch of each solid.

    ii Find the volume of each solid.

    iii Find the surface area of each solid.

    a

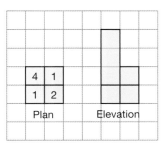

    b

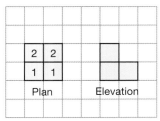

The numbers on the plan tell you the number of cubes in each column.

6 A cuboid has volume $200\,cm^3$. Write down three different possible dimensions of the cuboid.

7 Draw and label a sketch of a triangular prism with volume $120\,cm^3$.

8 A 3D solid has this net.

Find the volume of the solid.

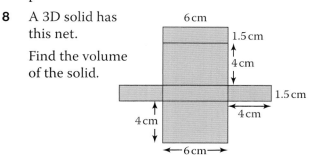

9 The diagrams show the plan and the front elevation of different solids.

Find the volume of each solid.

    a

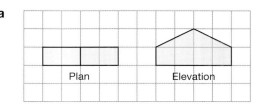

    b

*10 Ben wants to use this card to make an open-topped box in the shape of a cube.

He wants the box to be as big as possible.

    a Draw a sketch of a net he could use.

    b How long will the sides of the box be?

11 Harriet says that you can use this formula to find the surface area of a prism.
Surface area $= 2a + pl$,
where a = area of cross-section,
p = perimeter of cross-section and
l = length.
Explain why Harriet's formula works.

Q 1078, 1098, 1106　SEARCH

15.2 Volume of a prism

● Volume of a prism = area of cross-section × length

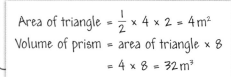

cross section

length

A

h

A

h

A

h

2πr

h

EXAMPLE

Calculate the volume of this triangular prism. State the units of your answer.

2 m

4 m

8 m

Area of triangle = $\frac{1}{2}$ × 4 × 2 = 4 m²

Volume of prism = area of triangle × 8

= 4 × 8 = 32 m³

A cylinder is a prism with circular cross-section.

● Volume of a **cylinder** = area of circle × height

3D solids such as cylinders and spheres have curved surfaces.

The curved face of a **cylinder** opens out to make a rectangle.

● Surface area of a cylinder = 2 × area of circle + area of curved face
= 2πr² + 2πrh.

EXAMPLE

Find the volume and surface area of this cylinder. Give your answers to 3 significant figures.

3 cm

7 cm

Area of circle = π × 3² = 28.274 ... cm²

Volume = 28.274 ... × 7 = 198 cm³ (3 sf)

Surface area = 2 × π × 3² + 2 × π × 3 × 7

= 188.495... cm² = 188 cm² (3 sf)

Do not round intermediate steps of the calculation.

A **sphere** is like a ball. It has one curved surface, no edges and no vertices.

For a sphere with radius = r cm

● **Volume of a sphere** $V = \frac{4}{3}\pi r^3$ cm³

● **Surface area of a sphere** $SA = 4\pi r^2$ cm²

EXAMPLE

A tennis ball has radius 4.2 cm.

Find, giving your answers to 1 dp

a   its volume

b   its surface area.

a   Volume of tennis ball = $\frac{4}{3}\pi r^3$

= $\frac{4}{3}$ π × 4.2³ = 310.3 cm³ (1 dp)

b   Surface area = 4πr²

= 4π × 4.2² = 221.7 cm² (1 dp)

Exercise 15.2S

1 Calculate the area of cross-section and the volume for each prism.

a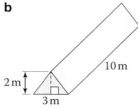

6 cm
5 cm
8 cm

b

10 m
2 m
3 m

c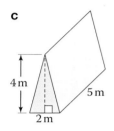

4 m
5 m
2 m

d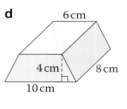

6 cm
4 cm
8 cm
10 cm

2 Find the volume of this shape.

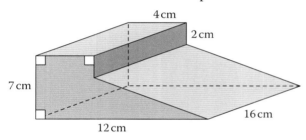

4 cm
2 cm
7 cm
16 cm
12 cm

3 Use the formula:

> curved surface area of a cylinder = $2\pi rh$

to find the curved surface area of cylinders with radius (r) and height (h).
Give your answers to 1 decimal place and include the correct units.

a radius = 4 cm, height = 9 cm

b radius = 5 cm, height = 7.5 cm

c $r = 16$ cm, $h = 8$ cm

d $r = 11.3$ cm, $h = 26$ cm

e $r = 13.9$ cm, $h = 2.9$ cm

f $r = 13$ mm, $h = 2.36$ mm

4 Use the formula:

> total surface area of a cylinder = $2\pi r^2 + 2\pi rh$

to find the total surface area of each cylinder in question **3**.

Give your answers to 1 decimal place and include the correct units.

5 Find the volume and total surface area of each cylinder.

> Be careful with units in part **d**.

a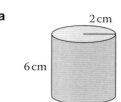

2 cm
6 cm

b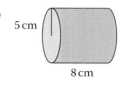

5 cm
8 cm

c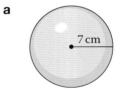

4 cm
4 cm

d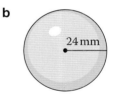

32 mm
5 cm

6 Use the formula:

> volume of a sphere = $\dfrac{4}{3}\pi r^3$

to calculate the volume of each sphere.

a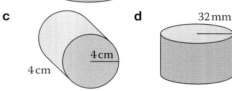

7 cm

b

24 mm

c radius = 13 cm **d** radius = 11 cm

e radius = 25 cm **f** radius = 31 cm

g radius = 17.3 cm **h** radius = 8.1 cm

i radius = 13.3 mm **j** radius = 1.26 m

7 Use the formula:

> surface area of a sphere = $4\pi r^2$

to find the surface areas of the spheres in question **6**.

15.2 Volume of a prism

RECAP

- Volume of a prism = cross-sectional area × height
- Volume of a sphere = $\frac{4}{3}\pi r^3$
- Total surface area of a cylinder = $2\pi r^2 + 2\pi rh$
- Surface area of a sphere = $4\pi r^2$

The volume of liquids are usually measured in litres and millilitres.

$1\,m^3 = 1000$ litres

1 litre $= 1000\,cm^3$

$1\,ml = 1\,cm^3$

HOW TO

① Decide which face is the cross-section of the prism.

② Use the formula for the volume of a prism.

③ Answer the question, remember to include units and give an explanation if needed.

EXAMPLE

20 kg of molten metal is made into prisms. The density of the metal is 8 g/cm³. How many prisms are made?

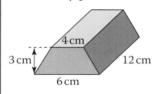

4 cm
3 cm
12 cm
6 cm

② Volume = mass ÷ density

Volume of metal $= 20\,000 \div 8 = 2500\,cm^3$

② Area of trapezium $A = \frac{1}{2}(a + b)h$

Area of trapezium $= \frac{1}{2}(6 + 4) \times 3 = 15\,cm^2$

② Volume of prism $V = Al$

Volume of prism $= 15 \times 12 = 180\,cm^3$

③ Number of prisms $= 2500 \div 180 = 13.88...$

④ Number of prisms $= 13$ Round down.

EXAMPLE

A drinks manufacturer wants a can to hold 330 ml of cola. The diameter of the can must be 7 cm.

a How tall does the can need to be?

The cans are to be packed in boxes of 12.

b Give the dimensions of a box that could be used.

c What advice can you give about the use of your answers in practice?

a ①

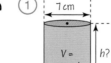

7 cm

$V = 330\,ml$

h?

② For a cylinder $V = \pi r^2 h$ $1\,ml = 1\,cm^3$

③ Substitute the values.

$330 = \pi \times 3.5^2 \times h$ Radius $= 7 \div 2$

$330 = 12.25\pi \times h$ Rearrange to find h.

$h = \frac{330}{12.25\pi} = 8.57\,cm$ (3 sf)

b ① A possible box.

② Length of box $= 3 \times 7\,cm = 21\,cm$

Width of box $= 2 \times 7\,cm = 14$ cm

Height of box $= 2h = 17.1\,cm$ (3 sf)

c ④ h and the box dimensions will need to be a little larger. This is because the answers have been rounded downwards and the thickness of the can and box have been neglected.

Exercise 15.2A

1 A block has a length and width of 22.50 mm, and a height of 3.15 mm. It has a mass of 9.50 g.

   a Find the density of the metal from which the block is made, giving your answer in g/cm³.

$$\text{Density} = \frac{\text{mass}}{\text{volume}}$$

   b How many blocks can be made from 1 kg of the material?

2 A steel cable weighs 2450 kg.

   The cable has a uniform circular cross-section of radius 0.85 cm.

   The steel from which the cable is made has a density of 7950 kg/m³.

   Find the length of the cable.

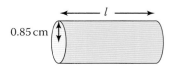

3 This metal cuboid is melted and made into triangular prisms.

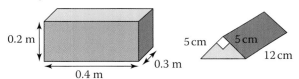

   How many prisms are made? State any assumptions you make.

4 Marcus is planning to go camping, but requires a tent with a capacity greater than 5 m³. He can borrow the tent shown. Does it fulfil Marcus' needs?

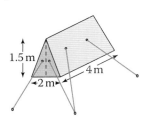

5 A trough with this cross-section must hold 50 litres of water. Find the shortest possible length of the trough.

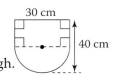

6 The surface area of a sphere is 616 cm². Work out the radius of the sphere.

7 Rachel bought a cylindrical tube containing three power balls.

Each ball is a sphere of radius 5 cm.

The balls touch the sides of the tube.

The balls touch the top and bottom of the tube.

Work out the volume of empty space in the tube.

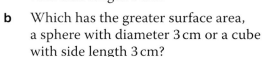

8 **a** Which has the greater volume, a sphere with diameter 3 cm or a cube with side length 3 cm?

   b Which has the greater surface area, a sphere with diameter 3 cm or a cube with side length 3 cm?

9 A sphere with radius 6 cm fits exactly inside a cylinder.

   a Write **i** the radius
             ii the height of the cylinder.

   b Work out the surface area of the sphere.

   c Work out the curved surface area of the cylinder.

   d For a sphere radius r that fits exactly inside a cylinder, find an expression for:

      i the surface area of the sphere

      ii the curved surface area of the cylinder.

   e Explain why your answers to **b** and **c** must be the same. You may want to use your answer to part **d** to help you.

*10 Unsharpened pencils have a cylinder of graphite surrounded by wood. The diameter of the graphite is 2 mm. The density of the graphite is 640 kg/m³. The density of the wood is 420 kg/m³.

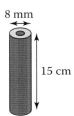

   a Find the mass of 500 pencils.

   b **i** Suggest dimensions for a box to pack the 500 pencils in.

      ii Find the percentage of the box that is filled by the pencils.

Q 1137, 1138, 1139, 1246   SEARCH

15.3 Volume and surface area

Regular tetrahedron

Square-based pyramid

A pyramid is a 3D shape with sides that taper to a point.

● Volume of a pyramid = $\frac{1}{3}$ of base area × vertical height

EXAMPLE

Find the volume of this solid.

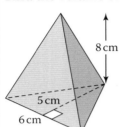

8 cm

5 cm

6 cm

Volume of pyramid = $\frac{1}{3}$ base area × height

$= \frac{1}{3} \times (\frac{1}{2} \times 6 \times 5) \times 8$

$= 40 \text{ cm}^3$

Area of triangle

$= \frac{1}{2} \times$ base × height

$= \frac{1}{2} bh$

Surface area is the total area of all the surfaces of a 3D solid.

● For 3D solids with flat surfaces, you work out the area of each surface and add them together.

EXAMPLE

19.1

Find the surface area of this pyramid.

Give your answer to 1 dp.

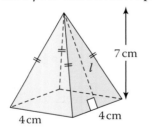

7 cm

4 cm   4 cm

Each face is an isosceles triangle.

The sloping side of this triangle is l, the height of the triangular face of the pyramid.

Base area 4 × 4 = 16 cm²

4 cm

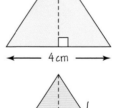
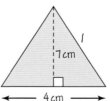

l

7cm

4 cm

Using Pythagoras: $7^2 + 2^2 = l^2$

$l = \sqrt{7^2 + 2^2} = \sqrt{53} = 7.28 \text{ cm}$ (3 sf)

Area of each triangular face

$= \frac{1}{2} \times 4 \times 7.28 = 14.56 \text{ cm}^2$

Surface area

= base area + area of 4 triangular faces

$= 16 + (4 \times 14.56) = 74.2 \text{ cm}^2$ (1 dp)

You are only given the base measurement.

To find the height of each side triangle, imagine a vertical slice through the pyramid:

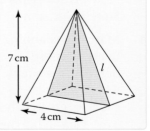

7 cm

l

4 cm

Exercise 15.3S1

1 Use the formula:

> volume of a pyramid
> $= \dfrac{1}{3}$ of base area × vertical height

to calculate the volumes of pyramids with these base areas and vertical heights.

Include the correct units in your answer.

a base area = 18 cm², vertical height = 8 cm

b base area = 25 m², vertical height = 9 m

c base area = 14 cm², vertical height = 12 cm

d base area = 18 mm², vertical height = 8 mm

2 Find the volume of each solid.

a
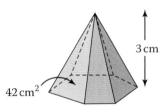
3 cm
42 cm²

b
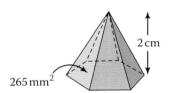
2 cm
265 mm²

c
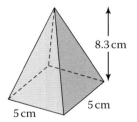
8.3 cm
5 cm
5 cm

d
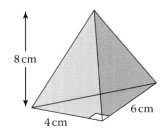
8 cm
4 cm
6 cm

3 Find the surface area of each 3D solid.

a
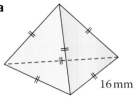
16 mm

b
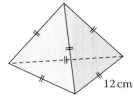
12 cm

c
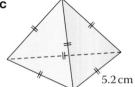
5.2 cm

d
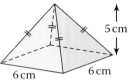
5 cm
6 cm
6 cm

e
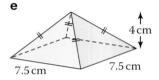
4 cm
7.5 cm
7.5 cm

f
6 cm
85 mm
85 mm

g
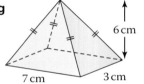
6 cm
7 cm
3 cm

h
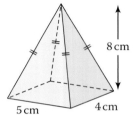
8 cm
5 cm
4 cm

i
72 mm
65 mm
35 mm

> Take care with parts **g–i**.
> The triangular faces are not all the same.

4 Calculate the volume and surface area of the solid.
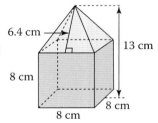
6.4 cm
13 cm
8 cm
8 cm
8 cm

5 Find the height of the solid.
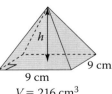
h
9 cm
9 cm
$V = 216$ cm³

6 Two square-based pyramids are joined. The total volume is 2700 mm³. Find x.
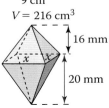
16 mm
x
20 mm

15.3 Volume and surface area

● Volume of a pyramid $= \frac{1}{3}$ of base area $\times$ vertical height

EXAMPLE

Find the volume of the cone.

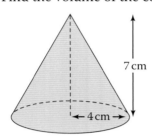

7 cm

4 cm

Volume of cone

$= \frac{1}{3}$ base area $\times$ height

$= \frac{1}{3} \times \pi \times 4^2 \times 7$

$= 117.3 \text{ cm}^3$ (1 dp)

Area of base $= \pi r^2$

Volume of cone

$= \frac{1}{3} \pi r^2 h$

Round your answer to a sensible degree of accuracy.

A **cone** is a solid with a circular base.
You form the curved surface by folding a sector of a circle.
The **radius** of the sector, l, is the sloping side of the cone.

Arc length of the sector = **circumference** of the base of the cone so $\frac{\theta}{360} \times 2\pi l = 2\pi r$

Radius of cone $r = \frac{\theta}{360} \times l$

Area of sector = **curved surface area** of cone

$= \frac{\theta}{360} \times \pi l^2$

$= \frac{\theta}{360} \times l \times \pi \times l$

$= r \times \pi \times l$

Remember this formula.

$r = \dfrac{\theta}{360} \times l$

● Curved surface area of cone $= \pi r l$

● Surface area of cone = curved surface area + area of base

$= \pi r l + \pi r^2$

Base radius, vertical height and slant side make a right-angled triangle, so use Pythagoras.

EXAMPLE

This sector is folded to form a cone.
Find, giving your answers to 1 dp:

a the radius of the cone

b the curved surface area of the cone

c the total surface area of the cone

d the vertical height of the cone.

300°
6 cm

a $\theta = 300°, l = 6 \text{ cm}$

$r = \dfrac{\theta}{360} \times l = \dfrac{300}{360} \times 6 = 5$ so $r = 5 \text{ cm}$

b Curved surface area $= \pi r l = \pi \times 5 \times 6 = 94.2 \text{ cm}^2$ (1 dp)

c Base area $= \pi r^2 = \pi \times 5^2 = 78.5 \text{ cm}^2$ (1 dp)

Total surface area $= 94.24 + 78.53 = 172.8 \text{ cm}^2$ (1 dp)

d Vertical height

$h^2 = 6^2 - 5^2$

$h = \sqrt{6^2 - 5^2} = 3.3 \text{ cm}$ (1 dp)

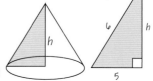

Exercise 15.3S2

1 Use the formula:

> volume of a cone
> $= \dfrac{1}{3}$ of base area $\times$ vertical height

to calculate the volumes of cones with these base areas and vertical heights.

Include the correct units in your answer.

a base area $= 15\,\text{cm}^2$,
vertical height $= 10\,\text{cm}$

b base area $= 22\,\text{m}^2$,
vertical height $= 12\,\text{m}$

c

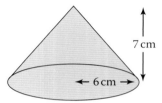

7 cm

← 6 cm →

d

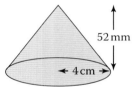

52 mm

← 4 cm →

2 Use the formula:

> curved surface area of cone $= \pi r l$

to find the curved surface area of cones with radius (r) and slant height (l).

Give your answers to 1 decimal place and include the correct units.

a base radius 5 cm
slant height 8.2 cm

b base radius 3 cm
slant height 6 cm

c base radius 45 mm
slant height 30 mm

d base radius 2.5 cm
slant height 30 mm

e base radius 6.7 cm
slant height 10.5 cm

f base radius 135 mm
slant height 18.5 cm

3 Use the formula:

> total surface area of cone $= \pi r l + \pi r^2$

to find the total surface area of each cone in question **2**.

Give your answers to 1 decimal place and include the correct units.

4 Find the total surface area of each cone.

a

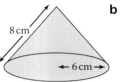

8 cm

← 6 cm →

b

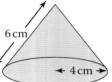

6 cm

← 4 cm →

5 Find the curved surface area of each cone.

a

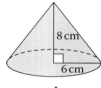

8 cm

6 cm

> Remember to find the slant height first.

b

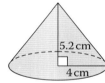

5.2 cm

4 cm

c

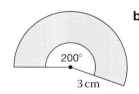

12 cm

7.2 cm

6 These sectors are folded to form cones. Find the curved surface area of each cone.

a

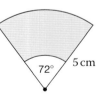

200°

3 cm

b

72°

5 cm

c

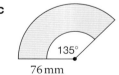

135°

76 mm

d

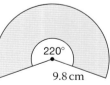

220°

9.8 cm

e

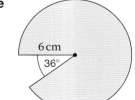

6 cm

36°

f

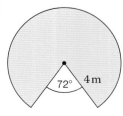

72°

4 m

Q 1107, 1122, 1136   SEARCH

15.3 Volume and surface area

- Volume is the amount of space that a 3D solid occupies.
- Volume of a pyramid = $\frac{1}{3}$ of base area $\times$ vertical height
- Surface area is the total area of all of the surfaces of a 3D solid.
- Total surface area of a cone = $\pi r l + \pi r^2$

HOW TO

1. Decide which shapes make up the 3D solid in the question.
2. Use one of the formulae for surface area or volume.
3. Give your answer and include units.

Cone

Frustum

When you cut the top off a **cone** with a cut parallel to the base, the part that is left is called the **frustum**.

The cone removed is similar to the whole cone.

- Volume of a frustum =
  volume of the whole cone − volume of the smaller cone

EXAMPLE 19.1

Find **a** the volume
b the surface area of this frustum.

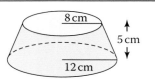
8 cm
5 cm
12 cm

Use similar triangles to find the height of the whole cone.

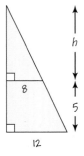
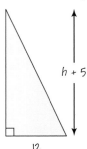

$\dfrac{h}{8} = \dfrac{h + 5}{12}$

$12h = 8(h + 5)$

$4h = 40$

$h = 10$

The triangles are similar, so the ratio of the sides is constant.

You can leave π in your working to avoid rounding. That is more accurate.

a Volume of whole cone: $\frac{1}{3}\pi \times 12^2 \times 15 = 720\pi$

Volume of small cone: $\frac{1}{3}\pi \times 8^2 \times 10 = 213\frac{1}{3}\pi$

Volume of frustum: $720\pi - 213\frac{1}{3}\pi = 1592\,\text{cm}^3$ (4 sf)

b Curved surface area of whole cone

$\pi \times 12 \times \sqrt{369} = 12\pi\sqrt{369}$

Curved surface area of small cone

$\pi \times 8 \times \sqrt{164} = 8\pi\sqrt{164}$

Surface area of frustum

= section of curved surface + area top circle + area bottom circle

$= (12\pi\sqrt{369} - 8\pi\sqrt{164}) + \pi \times 8^2 + \pi \times 12^2 = 1056\,\text{cm}^2$ (4 sf)

Use Pythagoras to find the sloping length l for the whole cone:

$l^2 = 12^2 + (h + 5)^2$

$l^2 = 12^2 + 15^2$

$l = \sqrt{369}$

For the small cone:

$l^2 = 8^2 + h^2$

$l^2 = 8^2 + 10^2$

$l = \sqrt{164}$

Exercise 15.3A

1 James is going to fill a paper cone with sweets.

He can choose between cone X and cone Y.
Cone X has top radius 4 cm and height 8 cm.
Cone Y has top radius 8 cm and height 4 cm.

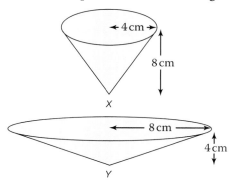

a Which cone has the greater volume?

b What is the difference in volume between cone X and cone Y?

2 A square-based pyramid has base side length x cm and vertical height 2x cm.
The volume of the pyramid is 18 cm³.
Work out the value of x.

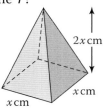

3 A pencil is in the shape of a regular hexagonal prism.
Each side of the hexagon is 6 mm.

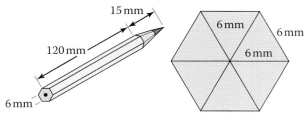

a Find the area of the base of the prism.

The pencil is sharpened at one end to form a pyramid.

The pyramid has height 15 mm and sides 6 mm.

The prism that makes the remainder of the pencil is 120 mm long.

b Find the total surface area of the pencil.

4 Laura made a model rocket from a cylinder with height 6 cm and a cone with vertical height 2.4 cm.
The radius of the cylinder and the cone is 1.8 cm.

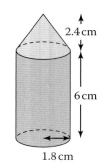

a Find the volume of the rocket.

b Find the surface area of the rocket.

5 Work out

i the volume

ii the surface area of each of these frustums.

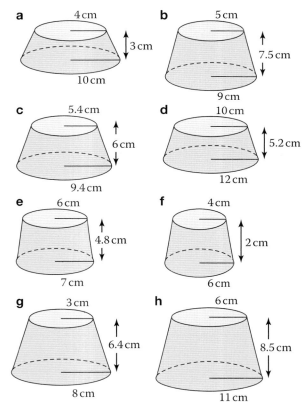

6 The diagram shows a frustum.
The base radius, 2r cm, is twice the radius of the top of the frustum, r cm.
The height of the frustum is h cm.
Write down an expression for

a the volume

b the surface area of this frustum.

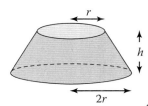

 Q 1107, 1122, 1136 SEARCH

Summary

Checkout
You should now be able to...

Test it
Questions

✔ Draw and interpret plans and elevations of 3D shapes.	1, 2
✔ Calculate the volume of cuboids and right prisms.	3, 4
✔ Calculate the surface area and volume of spheres, pyramids, cones and composite shapes.	5, 6
✔ Know and apply the relationship between lengths, areas and volumes of similar shapes.	7

Language	Meaning	Example
Face	A flat surface of a solid enclosed by edges.	Vertex, Edge, Face
Edge (solid)	A line along which two faces meet.	
Vertex **Vertices (plural)**	A point at which two or more edges meet.	
Plan	A drawing of a 3D object looking straight down at the object from directly overhead.	Plan, Front, Side
Elevation	A 2D drawing of a 3D object looking straight at the object from the front or side.	
Net	A 2D shape that can be folded to make a 3D solid.	4 cm, 4 cm, 4 cm
Surface area	The total area of all the faces of a 3D solid.	Surface area of cube = 6 × 4 × 4 = 96 cm²
Volume	The amount of space occupied by, or inside, a 3D shape.	Volume of cube = 4 × 4 × 4 = 64 cm³
Cross-section	The 2D shape formed when a solid shape is cut through in a specified direction, usually parallel to one of its faces.	Cross section, Cylinder, Triangular prism
Prism	A 3D solid with a constant cross-section.	
Pyramid	A 3D solid with a polygon as its base. All the other faces are triangular in shape and meet at a single vertex.	Regular tetrahedron, Square-based pyramid, Irregular pyramid
Cylinder	A prism with a circular cross-section.	Cylinder, Cone, Sphere
Cone	A solid with a circular base and one vertex.	
Sphere	A 3D shape with every point on its surface the same distance from the centre.	
Frustum	The part of a cone which remains when the top part is cut off with a cut parallel to the base	Frustum

Review

1 Draw

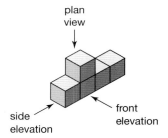

plan view

side elevation

front elevation

   a   the front elevation

   b   the plan view of the solid.

2 Here are three elevations of a 3D solid made from cubes.

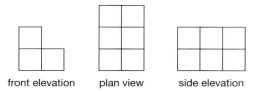

front elevation   plan view   side elevation

Draw the 3D solid these views come from.

3 A cuboid has dimensions 14 m, 3.5 m and 5 m.

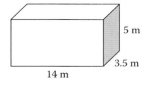

5 m

3.5 m

14 m

Calculate

   a   the volume

   b   the surface area of the cuboid.

The cuboid is made from a material with a density of $1.5 \, \text{kg/m}^3$.

   c   What is the mass of the cuboid?

4 A cylinder has a radius of 13 cm and a length of 18 cm.

Calculate

   a   the volume

   b   the surface area assuming the cylinder is closed at both ends.

5 Calculate the volume of these 3D shapes.

   a

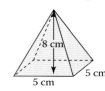

15 cm

   b

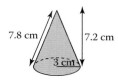

8 cm

5 cm

5 cm

6 Calculate the volume and surface area of this shape

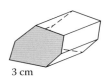

7.8 cm   7.2 cm

3 cm

7 A prism of volume $20 \, \text{cm}^3$ is enlarged to a solid of volume $160 \, \text{cm}^3$.

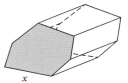

3 cm x

   a   What will be the length x?

The area of the cross section of the larger solid is $40 \, \text{cm}^2$.

   b   What is the area of the cross section of the smaller solid?

What next?

Score			
	0 – 3		Your knowledge of this topic is still developing. To improve look at MyMaths: 1078, 1098, 1106, 1107, 1122, 1136, 1137, 1138, 1139, 1246
	4 – 6		You are gaining a secure knowledge of this topic. To improve look at InvisiPens: 15Sa – j
	7		You have mastered these skills. Well done you are ready to progress! To develop your exam technique looks at InvisiPens: 15Aa – g

Assessment 15

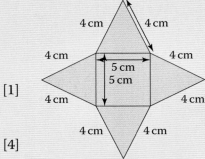

1 a Claire has a shape with 4 faces that are all triangles.
She says that her shape is a prism.
Write the correct name of the shape. [1]

b Graham draws the net of a square based right
pyramid with base length 4 cm and longest edge 5 cm.
Draw the correct net for this shape. [4]

2 David draws two isometric (3D) drawings for the shapes below.
Which drawing matches which shape? Give reasons for your answer. [2]

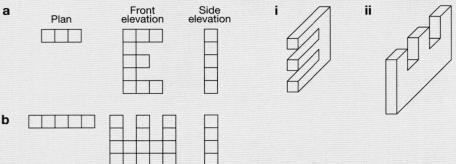

3 Bob has exactly enough 1 cm cubes to make a solid cube of side length 8 cm.

a How many 1 cm cubes has Bob got in total? [2]

b How many cubes with side length 2 cm can Bob make? [3]

c How many cubes with side length 4 cm can he make? [3]

d Bob tries to make cubes with side length 3 cm.

   i How many can he make? [2]

   ii How many 1 cm cubes are left over? [3]

4 Liz and Sue run a jewellery shop. Sue has a cuboid box, 24 cm × 30 cm × 36 cm.

a Liz packs earrings in small boxes 6 cm × 5 cm × 2 cm. How many of Liz's boxes
will fit into Sue's? [6]

b Sue says that the ratio of her surface area to Liz's surface area is the same as
the ratio of the volumes. Is Sue correct? Show units in your explanation. [7]

5 'Our Plaice' fish and chip shop sells chips in two sizes, 'Slim Jims' which are 13.5 cm long
with a square cross section of side 3 mm, or 'Chunky Dunkies' which are 6 cm long with
a square cross section of side 4.5 mm.

a Does each chip have the same volume? Show your workings. [4]

b Healthier chips have a smaller surface area so they absorb fat.
Which of 'Our Plaice's' chips are healthier to eat? [3]

6 The chocolate bar in the diagram consists of a slab of milk
chocolate with a cylinder of cream filling.
The whole bar is then covered with chopped nuts.

a Calculate the volume of

   i cream filling   [2]   **ii** chocolate.   [5]

b Calculate the surface area of the chocolate bar. [8]

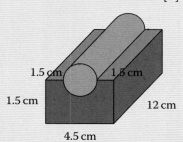

7 a A slab of cheese is in the shape of a prism with a right angled triangular cross section. The triangle has sides of 1.5, 3.6 and 3.9 inches. The length of the prism is 5 inches.
Calculate **i** its total surface area [6] **ii** its volume. [2]

b A child's ice cream cornet is in the shape of a cone of height 12 cm with a circular top of outer radius 4 cm. The inner radius of the top is 3.9 cm. Work out the volume of the biscuit that makes up the cone. [6]

c A golden right pyramid has a rectangular base with dimensions 55 cm × 40 cm. The perpendicular height of the pyramid is 65 cm.

   i Calculate its volume. [4]

   ii The gold is melted down and recast as a number of spheres of diameter 10 cm. How many complete spheres can be made? [4]

d A glass tube is 8 cm long, with an outer radius of 0.5 cm and a central hole of diameter 0.5 cm. Calculate

   i the volume of the glass [6]

   ii the mass of the tube if the density of the glass is 2.6 g/cm^3. [2]

8 A child's toy consists of a hemisphere and a cone with the same radius, as shown.

Calculate the **a** volume [6] and **b** surface area of the toy. [6]

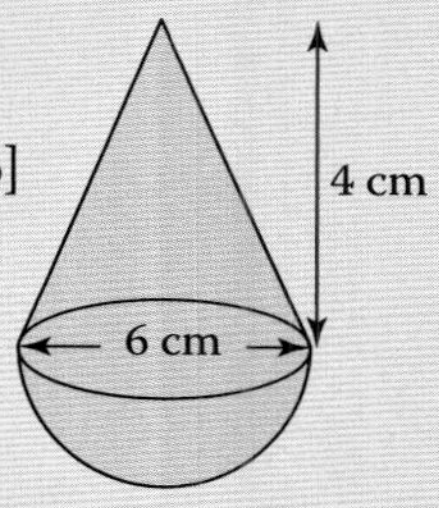

9 An unsharpened wooden pencil is in the shape of a hexagonal prism. Each side of the hexagon is 4 mm and the pencil is 6.5 cm long. The graphite core is a cylinder of radius 0.6 mm. Calculate the

a cross sectional area [2] **b** total volume [2]

c volume of wood in the pencil. [4]

10 A café sells ice cream and has a plastic three sectional advertising stand outside as shown in the diagram. The plastic has thickness 1.25 inches.

a Calculate the volume of the complete model. [8]

b Calculate the volume of the plastic in the hemisphere. [2]

c Calculate the total surface area. [11]

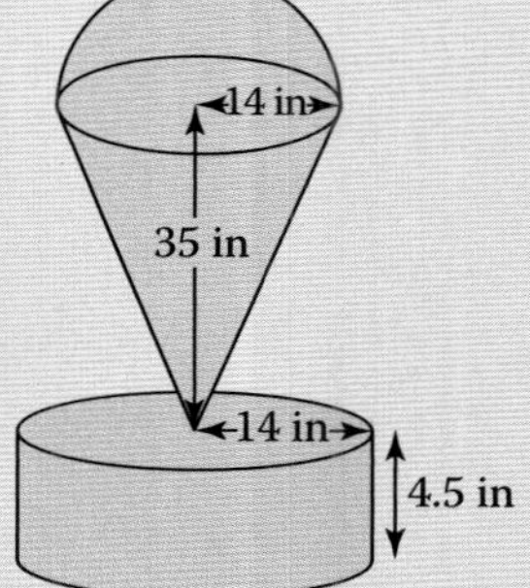

11 The clepsydra, or water clock, was invented in the 14th century BC. Water slowly drips out of the hole at the bottom and time is measured by the level of water remaining.

The clepsydra is filled at midnight and water drips out a a rate of 3 ml/min. Find the following.

a The volume, to the nearest ml, of the full cone. [2]

b The time taken, to the nearest minute, to empty the clepsydra. [2]

c The radius of the water's surface when its height is 15 cm. [2]

d The volume of water when the height is 15 cm. [2]

e The volume of water lost before its height drops to 15 cm. [2]

f The actual time when the height of water is 15 cm. [2]

The 'hour' marks on the side of the clepsydra were evenly spaced.

g Explain why this was not sensible and suggest a better shape. [2]

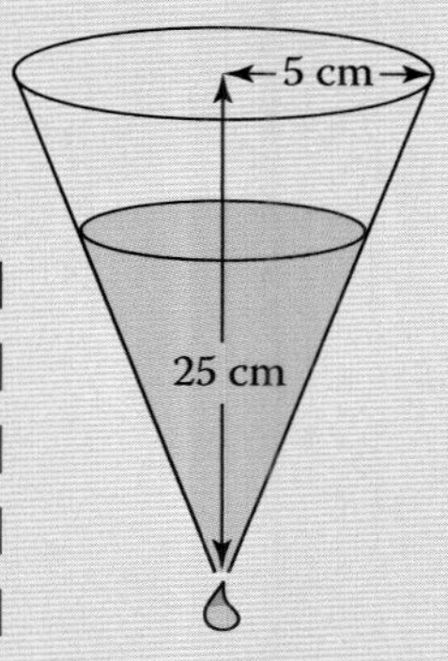

Abigail, Mike, Juliet and Raheem are now ready to start making the building suitable for their restaurant. They have also chosen some suppliers to work with, and are considering what stocks of non-perishable food and drink they will need. They analyse the data that they have gathered from their market research to help set their menu and price list.

Task 1 – The ingredients

One of the starters on the menu is an old family recipe that Mike borrowed from his grandmother: *Artichokes with anchovies*.

The friends are keen to avoid waste, and want to use ingredients efficiently. Each tin of anchovy fillets contains 24 anchovy fillets.

Each jar of artichoke hearts contains 20 artichoke hearts.

How many portions of *Artichokes with anchovies* will they have to make in order to use up complete tins of anchovy fillets and complete jars of artichoke hearts with no leftovers?

Artichokes with anchovies – serves 1

2 artichoke hearts
3 anchovy fillets
1 tablespoon of olive oil
8 tablespoons of butter, softened
2 garlic cloves
1 pinch each, salt and pepper
Juice of 1 lemon

Task 2 – Price list

They decide to look again at the data they collected from their small pilot survey (*Life skills 1: The business plan*).

p.104

a   On graph paper, plot separate male and female scatter graphs of amount prepared to spend against age.

b   For each of these graphs, describe the correlation and explain what it suggests in each case.

c   Should they use these graphs alone to decide a specific age group to target for their restaurant? Why?

Task 3 – Supplier meeting

Raheem drives from the restaurant to the vegetable suppliers, 9 miles away. Each friend keeps a log of all business-related travel and time spent in meetings so that they can claim back costs on expenses and keep a record of how much time they are investing.

a   Draw a distance-time graph showing Raheem's journey.

b   At what time does he arrive back at the restaurant?

c   How long was he away from the restaurant?

Veggie-r-us supplier meeting 3, journey log

Total distance to supplier, 9 miles.

1. Departed restaurant at 1 pm.
2. 6-minute drive down high street (2 miles).
3. 7 miles at 60mph (miles per hour) on motorway.
4. Reach supplier, stay there for 10 minutes.
5. 3 miles on motorway at 60 mph.
6. Stop at motorway services (5 minutes).
7. 4 miles on motorway at 40mph (bad traffic).
8. 2 miles through town (4 minutes).

Task 4 – Maintenance

They need to hire a plumber to fix some pipes in the restaurant.

There are two local plumbers who have been recommended by friends.

a On the same axes, draw graphs showing the cost (excluding parts) of hiring each plumber against the time taken.

b For what range of times is Bill cheaper than Alan?

Both quote that the job will take about 6.5 hours.

c Who should they get to do the job? Give your reasons.

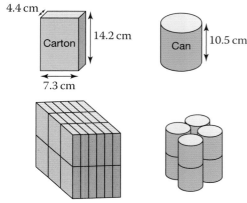

▲ Top: dimensions of one carton and one can.
Bottom: stacked arrangement of cartons and an example of how cans are able to stack.

Annual amount spent eating out (£)	Number of people
0–400	36
Over 400–600	21
Over 600–800	38
Over 800–1000	35
Over 1000–1200	26
Over 1200–1400	18
Over 1400–1700	17
Over 1700–2000	9
Total	200

▲ Annual spend on eating out in Newton-Maxwell.

Task 5 – Stock keeping

The friends need to decide between buying cartons or cans of tomatoes. The cartons are stacked in 2 layers in a cardboard box. Each layer is 6 cartons deep and 3 cartons across (see diagram).

a Find the volume of a cardboard box. Add an extra 5% for the cardboard.

The boxes are stored in a cuboid cupboard. Its dimensions are:

70 cm deep, 115 cm wide and 180 cm high.

b Calculate the volume of the cupboard in cm³ and m³. Explain how these figures are related.

c What percentage of the space in the cupboard do 30 boxes take up?

d Find the surface area of one carton.

The cans are cylindrical and have the same volume as a carton.

e What is the radius of the can?

f The cans can be stacked as shown to the left. Work out how many cans could fit in the same space taken up by one cardboard box.

g The cans and the cartons cost the same amount to buy. Which is better value for money? Why?

Task 6 – Repeat business

The friends decide to investigate the results of a larger survey into peoples' spending on eating out.

a Calculate an estimate of the mean annual amount spent eating out.

b Draw a histogram showing the data.

c Draw a cumulative frequency graph showing the data.

d Find the median and interquartile range of the annual amount spent eating out.

e Draw a box plot showing the data.

f What conclusions can you make from the different representations of the data? Do they support the mean?

Juliet assumes that a given person will visit all four local restaurants equally often, and spend the same amount at each one.

g Based on Juliet's assumption, find the mean amount the friends should expect one customer to spend at their restaurant in a year.

16 Handling data 2

Introduction

In the UK, literacy is almost taken for granted. By the time people reach adulthood, they can generally read and write. In many other countries, this is not the case. Statistics show that there is a correlation, or link, between the wealth per person of a country (as measured in 'GDP per capita') and the adult literacy rate. In 2013, the UK GDP and literacy rate $36 000 and 99% respectively; by comparison, in Sierra Leone they were $2000 and 35%.

The branch of statistics that deals with the relationship between variables is called correlation.

What's the point?

Understanding the relationship between quantities helps us to make informed decisions on a global scale. Literacy problems will not be resolved effectively unless poverty is also tackled.

Objectives

By the end of this chapter, you will have learned how to …

- Use frequency tables to represent grouped data.
- Construct histograms with equal or unequal class widths.
- Calculate summary statistics from a grouped frequency table.
- Plot scatter graphs and recognise correlation.
- Draw lines of best fit and use them to make predictions.

Check in

1 Work out:
  a $\frac{1}{2}$ of 124   **b** 50% of 120
  c $\frac{1}{2}$ of $(36 + 1)$ **d** 25% of 60
  e $\frac{1}{4}$ of $(27 + 1)$ **f** 25% of 120
  g $\frac{3}{4}$ of 144   **h** 75% of 200

2 This graph shows the cost per day to hire a power tool.

  a How much does it cost to hire the power tool for
    i 3 days   **ii** 5 days?

  b Mike has £40.
    What is the maximum number of days he can hire the power tool?

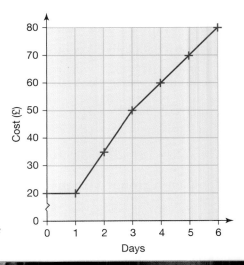

Chapter investigation

Imogen and Toby are trying to work out if there is a correlation between the size of an animal and how long it lives.

Toby says, 'Larger animals live longer than smaller ones. This must be true because elephants live much longer than mice.'

Imogen replies, 'Well, I read that the oldest animal ever found was a small clam that lived for over 400 years. Imagine living 400 years as a clam!'

Is there any correlation between animal size and longevity.

16.1 Frequency diagrams

Some surveys produce data with many different values.

You can **group** data into **class intervals** to avoid too many categories.

EXAMPLE

The exam marks of class 10A are shown:

35	47	63	25	31	8	19	55	47	14
24	36	56	61	15	43	22	50	66	10
36	45	18	20	53	31	40	60	44	47

Complete a **grouped frequency table**.

Mark	Tally	Frequency
1–20	ЖІІ	7
21–40	ЖІІІІ	9
41–60	ЖЖІ	11
61–80	ІІІ	3

Ж = 5

Check that the frequencies add to 30.

● **Discrete** data can only take exact values (usually collected by counting).

● **Continuous** data can take any value (collected by measuring). Continuous data cannot be measured exactly.

The number of students in a class is discrete data. The heights of the students is continuous data.

You must be careful if the data is **grouped**.

You can use a **bar chart** to display grouped discrete data.

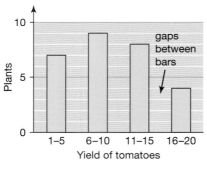

You can use **a histogram** to display grouped continuous data.

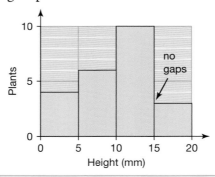

EXAMPLE

The times taken, in seconds, to run 100 m are shown in the table.

Time (seconds)	Number of people
$0 < t \le 10$	0
$10 < t \le 20$	4
$20 < t \le 30$	6
$30 < t \le 40$	3

Draw a frequency diagram to illustrate this information.

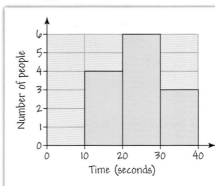

Draw and label the axes.

Draw a bar to show the frequency of each category.

Statistics   Grouped and bivariate data

Exercise 16.1S

1 **a** Copy and complete the frequency table using these test marks.

8 14 21   4 15 22 25 24 15 11

10 17 24 20 13 16 12   9   3 14

20 10 16 15   7 23 23 14 15 16

8   2   9 19 12 10 10 20 13 13

15 17 11 14 19 20 23 23 24   5

Test mark	Tally	Frequency
1–5		
6–10		
11–15		
16–20		
21–25		

b Calculate the number of people who took the test.

2 **a** Copy and complete the frequency table using these weights of people, given to the nearest kilogram.

48 63 73 55 59 61 70 63 58 67

46 45 57 58 63 71 60 47 49 51

53 61 68 65 70 60 52 59 50 49

48 47 63 61 58 71 53 51 60 70

Weight (kg)	Tally	Number of people
$45 \leqslant w < 50$		
$50 \leqslant w < 55$		
$55 \leqslant w < 60$		
$60 \leqslant w < 65$		
$65 \leqslant w < 70$		
$70 \leqslant w < 75$		

b Calculate the number of people in the sample.

3 **a** Copy and complete the frequency table using these heights of people.

153 134 155 142 140 163 150 135

170 156 171 161 141 153 144 163

140 160 172 157 136 160 134 154

176 154 173 179 160 152 170 148

151 165 138 143 147 144 156 139

3

Height (cm)	Tally	Number of people
$130 < h \leqslant 140$		
$140 < h \leqslant 150$		
$150 < h \leqslant 160$		
$160 < h \leqslant 170$		
$170 < h \leqslant 180$		

b Copy and complete the histogram on graph paper.

4 The depth, in centimetres, of a reservoir is measured daily throughout April.

Draw a histogram to show the depths.

Depth (cm)	Number of days
$0 < d \leqslant 5$	1
$5 < d \leqslant 10$	5
$10 < d \leqslant 15$	14
$15 < d \leqslant 20$	8
$20 < d \leqslant 25$	2

5 The exam marks of 36 students are shown in the frequency table.

Draw a bar chart to show the exam marks.

Exam mark (%)	Number of students
1 to 20	4
21 to 40	8
41 to 60	13
61 to 80	6
81 to 100	5

6 The histogram shows the best distances, in metres, that athletes threw a javelin in a competition.

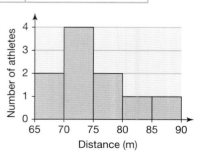

a State the number of athletes who threw the javelin

  i between 65 and 70 metres

  ii between 80 and 85 metres.

b In which class interval was the winner?

c What is the modal class interval?

d Calculate the total number of athletes who threw a javelin.

Q 1196, 1197 SEARCH

16.1 Frequency diagrams

- You can group data into class intervals when there are many values.
- Discrete data can only take exact values (usually collected by counting), for example the number of students in a class.
- Continuous data can take any value (collected by measuring), for example the heights of the students in a class.
- You can use a bar chart to display grouped discrete data.
- You can use a histogram to display continuous data.

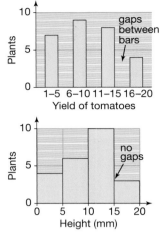

The bars in a histogram can have different widths.

The area of each bar represents the frequency.

The height of the bar represents the **frequency density**.

- Frequency density $= \dfrac{\text{frequency}}{\text{class width}}$

① Find the class width of each interval.

② Calculate the frequency density.

③ Draw a histogram. There are no gaps between the bars.

Ursula collected data on the time taken to complete a simple jigsaw.

Draw a histogram to represent the data.

Time, t seconds	$40 \leqslant t < 60$	$60 \leqslant t < 70$	$70 \leqslant t < 80$	$80 \leqslant t < 90$	$90 \leqslant t < 120$
Frequency	6	6	10	7	6

Time, t seconds	$40 \leqslant t < 60$	$60 \leqslant t < 70$	$70 \leqslant t < 80$	$80 \leqslant t < 90$	$90 \leqslant t < 120$
Class width	20	10	10	10	30
Frequency	6	6	10	7	6
Frequency density	0.3	0.6	1	0.7	0.2

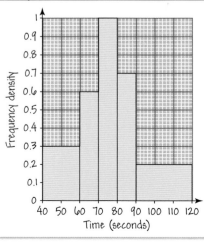

Add rows to the table to calculate class width and frequency density.

Exercise 16.1A

For each set of data in questions **1**–**3**

a Copy and complete the table.   **b** Draw a histogram to represent the data.

1 Reaction times of a sample of students.

Time, *t* seconds	$1 \leqslant t < 3$	$3 \leqslant t < 4$	$4 \leqslant t < 5$	$5 \leqslant t < 6$	$6 \leqslant t < 9$
Class width					
Frequency	12	17	19	11	18
Frequency density					

2 Amounts spent by the first 100 customers in a shop one Saturday.

Amount spent, £*a*	$0 \leqslant a < 5$	$5 \leqslant a < 10$	$10 \leqslant a < 20$	$20 \leqslant a < 40$	$40 \leqslant a < 60$	$60 \leqslant a < 100$
Class width						
Frequency	6	10	23	29	24	8
Frequency density						

3 Times dog owners spend on daily walks.

Time, *t* minutes	$10 \leqslant t < 20$	$20 \leqslant t < 40$	$40 \leqslant t < 60$	$60 \leqslant t < 90$	$90 \leqslant t < ?$
Class width					30
Frequency	8				9
Frequency density		0.8	1.4	1.3	

4 The histogram shows the times a sample of students spent watching TV one evening.

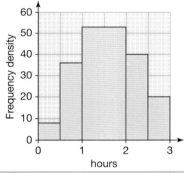

a How many students spent longer than $2\frac{1}{2}$ hours watching TV?

b Copy and complete the frequency table for the data.

Time, *t* hours	$0 \leqslant t < 0.5$	$0.5 \leqslant t < 1$	$1 \leqslant t < 2$	$2 \leqslant t < 2.5$	$2.5 \leqslant t < 3$
Frequency					

c How many students were in the sample?

5 The incomplete table and histogram give some information about the weights, in grams, of a sample of apples.

a Use the information in the histogram to work out the missing frequencies in the table.

b Copy and complete the histogram.

Weight, *g* grams	Frequency
$120 \leqslant g < 140$	8
$140 \leqslant g < 150$	6
$150 \leqslant g < 155$	
$155 \leqslant g < 160$	
$160 \leqslant g < 165$	
$165 \leqslant g < 175$	16
$175 \leqslant g < 185$	12
$185 \leqslant g < 200$	6

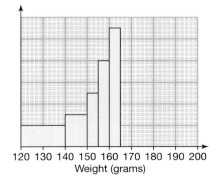

  Q 1196, 1197   SEARCH

16.2 Averages and spread 2

- You can put large amounts of continuous data into a **grouped frequency table**.

A grouped frequency table does not tell you the actual data values so you can only find estimates of the averages.

- You use **estimates** of averages to summarise the data.

- For grouped data in a frequency table, you can calculate
  - the **estimated mean**
  - the **modal class**
  - the **class interval** in which the median lies.

> The modal class is the class with the greatest frequency.

EXAMPLE

The table shows the time taken, to the nearest minute, by a group of students to solve a crossword puzzle.

Time, t, minutes	Frequency
$5 < t \leqslant 10$	2
$10 < t \leqslant 15$	14
$15 < t \leqslant 20$	13
$20 < t \leqslant 25$	6
$25 < t \leqslant 30$	1

For these data, work out

a the **modal class**

b the class containing the **median**

c an estimate for the **mean**

a Modal class is $10 < t \leqslant 15$

b Class containing median is $15 < t \leqslant 20$

c

Time t minutes	Frequency	Midpoint	Midpoint × frequency
$5 < t \leqslant 10$	2	7.5	$7.5 \times 2 = 15$
$10 < t \leqslant 15$	14	12.5	$12.5 \times 14 = 175$
$15 < t \leqslant 20$	13	17.5	$17.5 \times 13 = 227.5$
$20 < t \leqslant 25$	6	22.5	$22.5 \times 6 = 135$
$25 < t \leqslant 30$	1	27.5	$27.5 \times 1 = 27.5$
Total	**36**		**580**

Total number of students

Total time

Estimated mean = $\dfrac{\text{Estimated total time}}{\text{Total number of students}}$

Estimated mean = $\dfrac{580}{36} = 16.1$ (1dp)

> Put two extra columns in the table.

> Find the midpoint of each interval.

> Find the totals of the Frequency and Midpoint × frequency columns.

Exercise 16.2S

1 The weights, to the nearest kilogram, of 25 men are shown.

69	82	75	66	72
73	79	70	74	68
84	63	69	88	81
73	86	71	74	67
80	86	68	71	75

Note:
You can use a scientific calculator to work out the mean of grouped data. You should find out how to do this on your calculator.

a Copy and complete the frequency table.

Weight (kg)	Tally	Number of men
60 to 64		
65 to 69		
70 to 74		
75 to 79		
80 to 84		
85 to 89		

b State the modal class.

c Find the class interval in which the median lies.

2 The speeds of 10 cars in a 20 mph zone are shown in the frequency table.

Speed (mph)	Mid-value	Number of cars	Mid-value × Number of cars
11 to 15		1	
16 to 20		6	
21 to 25		2	
26 to 30		1	

a Calculate the number of cars that are breaking the speed limit.

b Copy the frequency table and calculate the mid-values for each class interval.

20

c Complete the last column of your table and find an estimate of the mean speed.

3 The grouped frequency tables give information about the time taken to solve four different crosswords.

For each table, copy the table, add extra working columns and find

i the modal class

ii the class containing the median

3 **iii** an estimate of the mean.

a

Time, t, minutes	Frequency
$5 < t \leqslant 10$	2
$10 < t \leqslant 15$	14
$15 < t \leqslant 20$	13
$20 < t \leqslant 25$	6
$25 < t \leqslant 30$	1

b

Time, t, minutes	Frequency
$0 < t \leqslant 10$	3
$10 < t \leqslant 20$	6
$20 < t \leqslant 30$	4
$30 < t \leqslant 40$	5
$40 < t \leqslant 50$	2

c

Time, t, minutes	Frequency
$5 < t \leqslant 10$	8
$10 < t \leqslant 15$	5
$15 < t \leqslant 20$	7
$20 < t \leqslant 25$	4
$25 < t \leqslant 30$	0
$30 < t \leqslant 35$	1

d

Time, t, minutes	Frequency
$5 < t \leqslant 15$	3
$15 < t \leqslant 25$	9
$25 < t \leqslant 35$	7
$35 < t \leqslant 45$	8
$45 < t \leqslant 55$	2
$55 < t \leqslant 65$	1

4 The heights of 50 Year 10 students were measured. The results are shown in the table.

Height, h, cm	Number of students
$150 \leqslant h < 155$	3
$155 \leqslant h < 160$	5
$160 \leqslant h < 165$	15
$165 \leqslant h < 170$	25
$170 \leqslant h < 175$	2

a What is the modal group?

b Estimate the mean height.

c Which class interval contains the median?

Q 1201, 1202, 1254, 1255 SEARCH

16.2 Averages and spread 2

● For large amounts of data presented as grouped data in a frequency table, you cannot calculate the exact mean, mode or median. Instead, you can calculate:

 ● the estimated mean

 ● the modal class

 ● the class interval in which the median lies.

> The mean, mode and median are measures of average. The range is a measure of spread.

HOW TO

Compare grouped data

① Compare a measure of average such as the modal class, median class or estimated mean.

② Compare the ranges of the data sets.

EXAMPLE

The tables show the ages of people attending concerts to see the bands Badness and Cloudplay.

Badness	
Age, a, years	Frequency
$20 \leqslant a < 30$	1600
$30 \leqslant a < 40$	4300
$40 \leqslant a < 50$	2100
$50 \leqslant a < 60$	1000

Cloudplay	
Age, a, years	Frequency
$10 \leqslant a < 20$	2800
$20 \leqslant a < 30$	4600
$30 \leqslant a < 40$	3300
$40 \leqslant a < 50$	700
$50 \leqslant a < 60$	500

Compare the ages of the people attending the two concerts.

> You could also compare the median class or the estimated mean.

① Compare a measure of average.

The modal age is greater at Badness concerts then Cloudplay concerts.

The highest frequency for Badness is in the class interval 30–40 years old, whereas the highest frequency for Cloudplay is in the interval 20–30 years old.

② Compare the range in ages.

There is more variation in the ages of the people at Cloudplay concerts than Badness converts.

The largest possible range for Cloudplay is 60 – 10 = 50 years, whereas for Badness it is 60 – 20 = 40 years.

Exercise 16.2A

1 Jayne kept a daily record of the number of miles she travelled in her car during two months.

	December	January
Miles travelled, m	**Frequency**	**Frequency**
$0 < m \leqslant 20$	3	0
$20 < m \leqslant 40$	8	5
$40 < m \leqslant 60$	10	12
$60 < m \leqslant 80$	6	8
$80 < m \leqslant 100$	4	6

  a Estimate the mean number of miles for each month.

  b Find the modal class and median class for each month.

  c Compare the number of miles Jayne travelled in December and January.

2 David carried out a survey to find the time taken by 120 teachers and 120 office workers to travel home from work.

Teachers

Time taken, t, minutes	Frequency
$0 < t \leqslant 10$	12
$10 < t \leqslant 20$	33
$20 < t \leqslant 30$	48
$30 < t \leqslant 40$	20
$40 < t \leqslant 50$	7

Office workers

Time taken, t, minutes	Frequency
$10 < t \leqslant 20$	2
$20 < t \leqslant 30$	21
$30 < t \leqslant 40$	51
$40 < t \leqslant 50$	28
$50 < t \leqslant 60$	18

  a Estimate the mean number of minutes for the teachers and office workers.

  b Find the modal class and median class for the teachers and office workers.

  c Make comparisons between the time taken by the teachers and office workers to travel home from work.

3 The weights of some apples are shown in the table.

Weight, w (g)	Frequency
$30 \leqslant w < 40$	6
$40 \leqslant w < 50$	28
$50 \leqslant w < 60$	21
$60 \leqslant w < 70$	25

Granny Smith apples have a mean weight of 45 g and a range of 39 g.

Is the data in the table about Granny Smith apples?

4 A company produces three million packets of crisps each day. It states on each packet that the bag contains 25 g of crisps. To test this, a sample of 1000 bags are weighed.

The table shows the results.

Weight, w (g)	Frequency
$23.5 \leqslant w < 24.5$	20
$24.5 \leqslant w < 25.5$	733
$25.5 \leqslant w < 26.5$	194
$26.5 \leqslant w < 27.5$	53

Is the company justified in stating that each bag contains 25 g of crisps?

Show your workings and justify your answer.

5 Two machines are each designed to produce paper 0.3 mm thick.

The tables show the actual output of a sample from each machine.

	Machine A	Machine B
Thickness, t (mm)	**Frequency**	**Frequency**
$0.27 \leqslant t < 0.28$	2	1
$0.28 \leqslant t < 0.29$	7	50
$0.29 \leqslant t < 0.30$	32	42
$0.30 \leqslant t < 0.31$	50	5
$0.31 \leqslant t < 0.32$	9	2

Compare the output of the two machines using suitable calculations.

Which machine is producing paper closer to the required thickness?

1201, 1202, 1254, 1255 SEARCH

16.3 Box plots and cumulative frequency graphs

The heights of 120 boys are given in the table.

Height, h, cm	$145 \leqslant h < 150$	$150 \leqslant h < 155$	$155 \leqslant h < 160$	$160 \leqslant h < 165$	$165 \leqslant h < 170$
Frequency	8	27	48	31	6

a Use the table to draw a cumulative frequency diagram.

b Draw a box plot for the data.

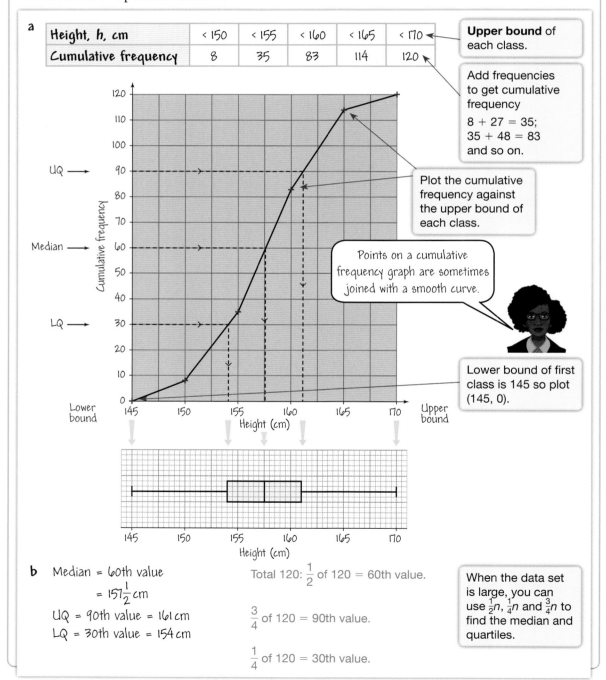

a

Height, h, cm	< 150	< 155	< 160	< 165	< 170
Cumulative frequency	8	35	83	114	120

Upper bound of each class.

Add frequencies to get cumulative frequency

$8 + 27 = 35$;
$35 + 48 = 83$
and so on.

Plot the cumulative frequency against the upper bound of each class.

Points on a cumulative frequency graph are sometimes joined with a smooth curve.

Lower bound of first class is 145 so plot (145, 0).

b Median = 60th value
$= 157\frac{1}{2}$ cm
UQ = 90th value = 161 cm
LQ = 30th value = 154 cm

Total 120: $\frac{1}{2}$ of 120 = 60th value.

$\frac{3}{4}$ of 120 = 90th value.

$\frac{1}{4}$ of 120 = 30th value.

When the data set is large, you can use $\frac{1}{2}n$, $\frac{1}{4}n$ and $\frac{3}{4}n$ to find the median and quartiles.

Exercise 16.3S

1 Mr Cheong summarised the test results for a group of students.

Lowest mark 41% Lower quartile 54%
Median 60% Highest mark 84%
Upper quartile 70%

Draw a box plot for these results.

2 In a science experiment, Davinder recorded reaction times to the nearest tenth of a second. He summarised the data.

Quickest time 3.2 Lower quartile 3.7
Median 4.4 Slowest time 9.6
Upper quartile 7.5

Draw a box plot to represent these results.

For each of the data sets in questions **3–8**

a draw a cumulative frequency diagram

b draw a box plot

c use the cumulative frequency diagram to make each estimate.

3 The heights of 100 girls.

Height, h, cm	Frequency
$145 \leqslant h < 150$	7
$150 \leqslant h < 155$	25
$155 \leqslant h < 160$	46
$160 \leqslant h < 165$	17
$165 \leqslant h < 170$	5

c Estimate the number of girls with height

i < 152 cm **ii** > 163 cm.

4 The ages of teachers in a school.

Age, A, years	Frequency
$20 \leqslant A < 30$	18
$30 \leqslant A < 40$	37
$40 \leqslant A < 50$	51
$50 \leqslant A < 60$	28
$60 \leqslant A < 70$	16

c Estimate the number of teachers who are

i < 35 **ii** > 55.

5 The times taken to complete a crossword puzzle.

Time, t, minutes	Frequency
$0 \leqslant t < 10$	4
$10 \leqslant t < 20$	11
$20 \leqslant t < 30$	29
$30 \leqslant t < 40$	37
$40 \leqslant t < 50$	27
$50 \leqslant t < 60$	12

c Estimate the number of people who took

i < 25 min **ii** > 45 min

to complete the puzzle.

6 The weights of a sample of cats and kittens.

Weight, w, grams	Frequency
$1500 \leqslant w < 2000$	9
$2000 \leqslant w < 2500$	22
$2500 \leqslant w < 3000$	37
$3000 \leqslant w < 3500$	20
$3500 \leqslant w < 4000$	12

c Estimate the number of cats and kittens that weighed

i < 2200 g **ii** > 3600 g.

7 The heights of sunflowers growing in one field.

Height, h, cm	Frequency
$40 \leqslant h < 60$	2
$60 \leqslant h < 80$	17
$80 \leqslant h < 100$	28
$100 \leqslant h < 120$	39
$120 \leqslant h < 140$	24
$140 \leqslant h < 160$	10

c Estimate the number of sunflowers that were

i < 130 cm **ii** > 90 cm.

8 The total spent by 100 shoppers at Tesbury's superstore.

Amount, p, £	Frequency
$0 \leqslant p < 10$	16
$10 \leqslant p < 20$	14
$20 \leqslant p < 30$	23
$30 \leqslant p < 50$	17
$50 \leqslant p < 70$	15
$70 \leqslant p < 100$	15

c Estimate the number of shoppers who spent

i $< £20$ **ii** $> £80$.

Q 1194, 1195, 1333 SEARCH

16.3 Box plots and cumulative frequency graphs

- You can represent grouped data on a cumulative frequency diagram.
- You use a box plot to show the median and IQR of a set of data.

HOW TO

1. Compare a measure of average, such as the median.
2. Compare a measure of spread, such as the IQR.

EXAMPLE

These cumulative frequency graphs summarise the weights of a sample of 100 men and 100 women.

Make two comparisons between the weights of the men and women.

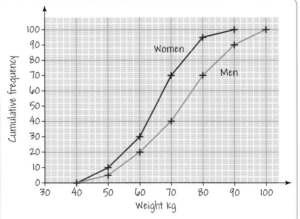

1. Median weight of women = 65 kg
   Median weight of men = 73 kg
   On average, the women are lighter than the men.

2. IQR of women's weights = 73 − 58 = 15 kg
   Range of men's weights = 84 − 62 = 22 kg
   The men's weights vary more than the women's weights.

EXAMPLE

These box plots summarise the heights of samples of 13- and 14-year-old boys and girls.

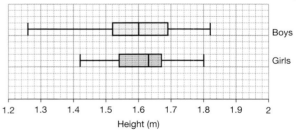

Compare like measures such as the medians, IQR, etc.

Write two comparisons between the heights of the boys and the girls.

1. **Median** height of girls = 1.63 m   Median height of boys = 1.60 m
   On average, the girls are taller than the boys.

2. **IQR** for girls = 1.67 − 1.54 = 0.13 m   IQR for boys = 1.69 − 1.52 = 0.17 m

   The IQR for the boys is greater than for the girls so the middle half of the heights is more varied for the boys.

Exercise 16.3A

1 Write three comparisons between the test results of a group of girls and a group of boys.

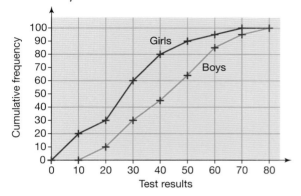

2 Write three comparisons between the heights of samples of sunflowers grown by two farmers.

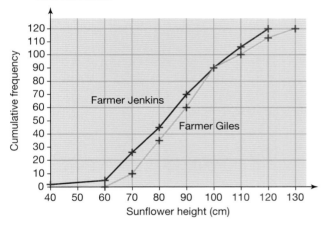

3 Write three comparisons between the mobile phone bills paid by samples of boys and girls.

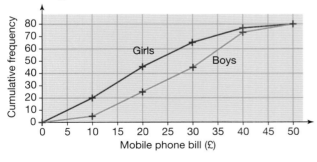

Write four comparisons between each pair of box plots.

4 The box plots summarise the waiting times, to the nearest minute, of a group of patients at the doctor and the dentist.

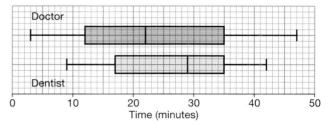

5 The box plots summarise the reaction times, of a group of boys and girls.

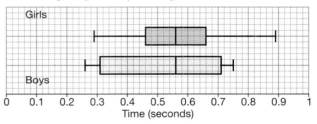

6 The box plots summarise the French and English test results of a group of students.

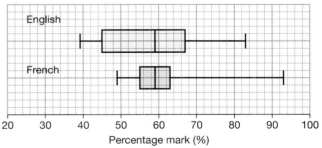

7 The box plots summarise the average length of a phone call, to the nearest minute, made by two groups of girls aged 13 and 17.

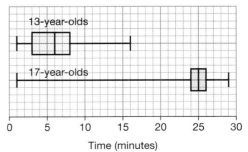

16.4 Scatter graphs and correlation

You can use a **scatter graph** to compare two sets of data, for example, height and weight.

● The data is collected in pairs and plotted as coordinates.

● If the points lie roughly in a straight line, there is a **linear relationship** or **correlation** between the two **variables**.

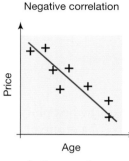

As height increases, weight also increases.

As the age of a car increases, the price decreases.

There is no linear relationship between height and exam mark.

The red straight line is the **line of best fit**.

You cannot draw a line of best fit for no correlation.

If the points lie close to the line of best fit, the correlation is strong.

EXAMPLE

The exam results (%) for Paper 1 and Paper 2 for 10 students are shown.

Paper 1	56	72	50	24	44	80	68	48	60	36
Paper 2	44	64	40	20	36	64	56	36	50	24

a Draw a scatter graph and line of best fit.

b Describe the relationship between the Paper 1 results and Paper 2 results.

Points that are an exception to the general pattern of the data are called **outliers**.

a Plot the exam marks as coordinates. The line of best fit should be close to all the points, with approximately the same number of crosses on either side of the line.

b There is a positive correlation. Students who did well on Paper 1 did well on Paper 2. Students who did not do well on Paper 1 did not do well on Paper 2 either.

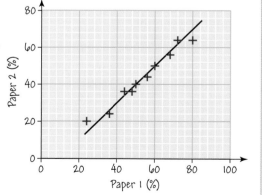

Performing well on Paper 1 does not cause you to perform well on Paper 2. The tests could be on completely different subjects.

● If two variables are correlated then it does not always mean that one **causes** the other.

The events could both be a result of a common cause, or there could be no connection at all between the events.

Exercise 16.4S

1 Match each scatter graph to the correct type of correlation.

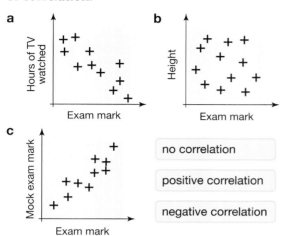

no correlation

positive correlation

negative correlation

2 Describe the points A, B, C, D and E on each scatter graph.

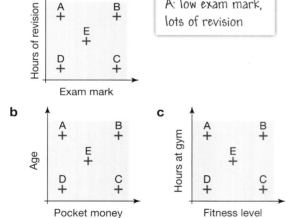

A: low exam mark, lots of revision

3 Match each scatter graph to the correct type of correlation.

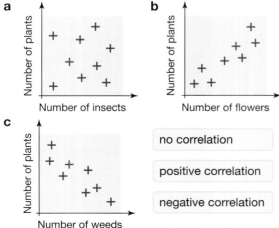

no correlation

positive correlation

negative correlation

4 The table shows the amount of water used to water plants and the daily maximum temperature.

Water (litres)	25	26	31	24	45	40	5	13	18	28
Maximum temperature (°C)	24	21	25	19	30	28	15	18	20	27

a Copy and complete the scatter graph for this information.

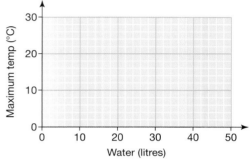

b State the type of correlation shown in the scatter graph.

c Copy and complete these sentences:

   i As the temperature increases, the amount of water used _____.

   ii As the temperature decreases, the amount of water used _____.

5 The times taken, in minutes, to run a mile and the shoe sizes of ten athletes are shown in the table.

Shoe size	10	$7\frac{1}{2}$	5	9	6	$8\frac{1}{2}$	$7\frac{1}{2}$	$6\frac{1}{2}$	8	7
Time (mins)	9	8	8	7	5	13	15	12	5	6

a Draw a scatter graph to show this information.

   Use 2 cm to represent 1 shoe size on the horizontal axis.

   Use 2 cm to represent 5 minutes on the vertical axis.

b State the type of correlation shown in the scatter graph.

c Describe, in words, any relationship that the graph shows.

Q 1213, 1250   SEARCH

16.4 Scatter graphs and correlation

- You can compare two sets of data on a scatter graph.
- The data is collected in pairs and plotted as coordinates.
- If the points lie roughly in a straight line, there is a correlation between the two variables.

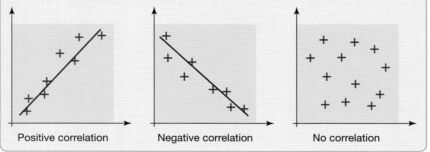

Positive correlation   Negative correlation   No correlation

You can use the line of best fit to make predictions for a value that falls in the range of the data – this is called **interpolation**. Interpolation is reliable, especially if the correlation is strong.

You can use the line of best fit to make predictions for a value that falls outside the range of the data – this is called **extrapolation**. Extrapolation is not always reliable as you cannot be sure that pattern holds for values outside of the data collected.

HOW TO

1. Describe and interpret the relationship shown by the diagram.

2. Make predictions based on the correlation shown.

EXAMPLE

The scatter graph shows the number of goals scored by 21 football teams in a season plotted against the number of points gained.

a   Describe the relationship between the goals scored and the number of points.

b   Describe the goals and points for team A.

c   If a team scored 45 goals, how many points would you expect it to have?

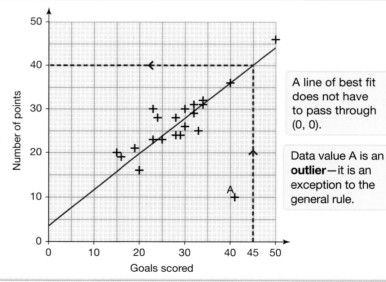

A line of best fit does not have to pass through (0, 0).

Data value A is an **outlier**—it is an exception to the general rule.

1. Describe and interpret the relationship shown by the diagram.

2. Make predictions based on the correlation shown.

a   The graph shows a positive correlation: the more goals scored, the more points gained.

b   Team A has scored lots of goals but has gained very few points.

c   Reading from the graph, and using the line of best fit as a guide, you could expect a team that scored 45 goals to gain 40 points.

Exercise 16.4A

1 The scatter graph shows the exam results (%) for Paper 1 and Paper 2 for 11 students.

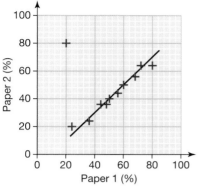

a Describe the relationship between Paper 1 and Paper 2 results.

b Identify an outlier and describe how this student performed in the papers.

c If a student scored 40 in Paper 1, what score would you expect them to gain in Paper 2?

d Jenna extends the line of best fit on the graph and tries to predict the score that a student who scored 100 on Paper 2 will score on Paper 1.
Will her estimate make sense?

2 The graph shows the marks in two papers achieved by ten students.

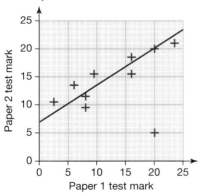

Use the line of best fit to estimate

a the Paper 2 mark for a student who scored 13 in Paper 1

b the Paper 1 mark for a student who scored 23 in Paper 2.

c Describe the outlier in the scatter graph. How well did this student perform on the two papers?

3 The table shows the age and diameter, in centimetres, of trees in a forest.

Age (years)	10	27	6	22	15	25	11	16	21	19
Diameter (cm)	20	78	9	65	38	74	25	44	59	50

a Draw a scatter graph for the data.

b State the type of correlation between the age and diameter of the trees.

c Draw a line of best fit.

d If the diameter of a tree is 55 cm, estimate the age of the tree.

e Use your graph to estimate the diameter of a tree that is one year old.

f Explain why the estimate in part **d** is more reliable than the estimate in part **e**.

4 The table shows the number of hot water bottles sold per month in a chemist's shop and the average temperature for each month.

Month	Average monthly temperature °C	Sales of hot water bottles
Jan	2	32
Feb	4	28
Mar	7	10
April	10	4
May	14	6
June	19	0
July	21	2
Aug	20	3
Sept	18	7
Oct	15	15
Nov	11	22
Dec	5	29

a Draw a scatter graph for the data.

b Describe the correlation shown and the relationship between the results.

c Draw in a line of best fit.

d Predict the average temperature in a month 20 hot water bottles were sold.

e The weather forecast predicts an average January temperature of −5°C. Could you use your graph to find how many hot water bottles would be sold?

Q 1213, 1250   SEARCH

16.5 Time series

You can use a **line graph** to show how data changes as time passes.

The data can be discrete or continuous.

The temperature of a liquid is measured every minute.

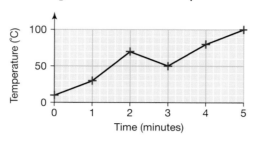

Time could be seconds, minutes, hours, days, weeks, months or years.

Time is always the **horizontal** axis.

This is an example of a **time series graph**.

- A time series graph shows
  - how the data changes over time, or the **trend**
  - each individual value of the data.

The table shows the average monthly rainfall, in centimetres, in Sheffield over the last 30 years.

Month	Jan	Feb	Mar	Apr	May	Jun	Jul	Aug	Sep	Oct	Nov	Dec
Rainfall (cm)	8.7	6.3	6.8	6.3	5.6	6.7	5.1	6.4	6.4	7.4	7.8	9.2

Draw a time series graph to show this information.

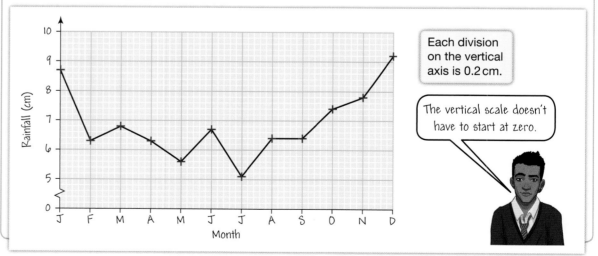

Each division on the vertical axis is 0.2 cm.

The vertical scale doesn't have to start at zero.

Exercise 16.5S

1 The number of photographs taken each day during a 7-day holiday is given.

Sunday	Monday	Tuesday	Wednesday	Thursday	Friday	Saturday
8	12	11	16	19	2	13

Copy and complete the line graph to show this information.

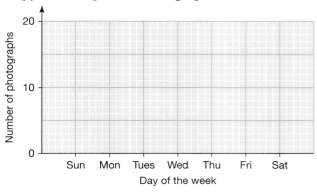

2 The numbers of DVDs rented from a shop during a week are shown.

Sunday	Monday	Tuesday	Wednesday	Thursday	Friday	Saturday
18	9	7	11	15	35	36

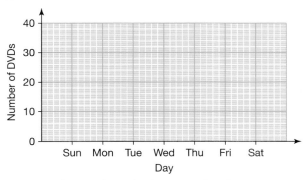

Copy and complete the line graph, choosing a suitable vertical scale.

3 Every year on his birthday, Peter's weight in kilograms is measured.

Age	2	3	4	5	6	7	8	9	10	11	12	13
Weight (kg)	14	16	18	20	22	25	28	31	34	38.5	40	45

Draw a line graph to show the weights.

4 The hours of sunshine each month are shown in the table.

Jan	Feb	Mar	Apr	May	Jun	Jul	Aug	Sep	Oct	Nov	Dec
43	57	105	131	185	176	194	183	131	87	53	35

Draw a line graph to show the hours of sunshine.

5 The daily viewing figures, in millions, for a reality TV show are shown.

Day	Sat	Sun	Mon	Tue	Wed	Thu	Fri
Viewers (in millions)	3.2	3.8	4.3	4.5	3.1	5.2	7.1

Draw a line graph to show the viewing figures.

 1198, 1939   SEARCH

16.5 Time series

- You can use a line graph to show how data changes over time.
- This is sometimes called a time series graph. It shows:
  - how the data changes over time, or the trend
  - each individual value of the data.
- This graph shows the how the temperature changes over time.

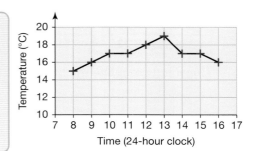

HOW TO

① Draw axes on graph paper and label with suitable scales.

② Plot the data from the table as coordinate pairs. Join the plotted points with straight lines.

③ Discuss any short term trends, seasonal variation and any longer term trends.

EXAMPLE

Jenny's quarterly gas bills over a period of two years are shown in the table.

	Jan–March	April–June	July–Sept	Oct–Dec
2003	£65	£38	£24	£60
2004	£68	£42	£30	£68

Plot the data on a graph and comment on any pattern in the data.

① Draw axes on graph paper with time on the horizontal axis. Plot time on the horizontal axis, J–M means Jan–March.

② Plot the coordinates as crosses on the grid. Join them up with straight lines.

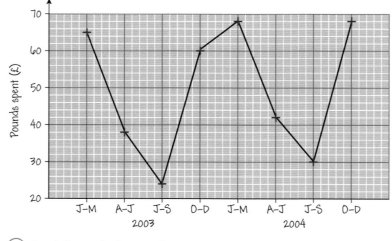

③ Gas bills are highest in the winter months and lowest in the summer months. This annual pattern appears to repeat itself.

There is a slight trend for the bills to rise from year to year.

The graph shows seasonal variation.

Exercise 16.5A

For each of questions **1–4**

- **a** Plot the data on a graph
- **b** Comment on any patterns in the data.

1 The table shows Ken's monthly mobile phone bills.

Jan	Feb	Mar	April	May	June	July	Aug	Sept	Oct	Nov	Dec
£16	£12	£15	£18	£16	£18	£12	£10	£12	£15	£16	£20

2 The table shows Mary's quarterly electricity bills over a two-year period.

	Jan–March	April–June	July–Sept	Oct–Dec
2004	£45	£20	£15	£48
2005	£54	£24	£18	£50

3 The table shows monthly ice-cream sales at Angelo's shop during one year.

Jan	Feb	Mar	April	May	June	July	Aug	Sept	Oct	Nov	Dec
£16	£12	£15	£18	£38	£48	£52	£58	£18	£15	£16	£40

4 A town council carried out a survey over a number of years to find the percentage of local teenagers who used the town's library. The table shows the results.

year	1998	1999	2000	2001	2002	2003	2004	2005
%	14	18	24	28	25	20	18	22

5 This news report was written about sales representatives of a small firm.

> On average, sales representatives at the firm travel 77 km per day. The range of distances travelled is 116 km.

These two graphs were drawn to summarise the distances travelled daily by representatives at two different firms.

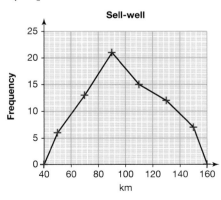

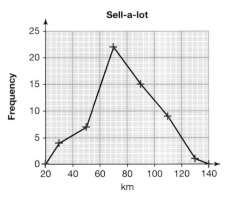

Use the graphs to identify which firm was being reported on. Give a reason for your choice.

Q 1198, 1939   SEARCH

Summary

Checkout
You should now be able to...

✓ Use frequency tables to represent discrete or continuous grouped data.	1 – 4
✓ Construct bar charts and construct histograms with equal or unequal class widths.	1
✓ Calculate summary statistics from a grouped frequency table.	2
✓ Construct and interpret cumulative frequency curves and box plots.	3
✓ Plot scatter graphs and recognise correlation.	4
✓ Use tables and line graphs to represent time series data.	5

Language | Meaning | Example

Language	Meaning	Example
Discrete data	Data that can only take values from a set whose values can be listed.	Shoe sizes 4, $4\frac{1}{2}$, 5, $5\frac{1}{2}$, 6, ... Discrete data is usually counted
Continuous data	Data that can take any value within an interval.	Heights 1.63 m, 1.42 m, 1.565 m ..... Continuous data is usually measured.
Histogram	A way of presenting grouped continuous data. The areas of the bars are proportional to the frequencies of each class.	See lesson 16.1
Frequency density	The height of the bar in a histogram represents the frequency density.	Frequency density = frequency ÷ class width
Box plot	A drawing which displays the median, quartiles and greatest and least value of a set of data.	See lesson 16.3
Cumulative frequency	The total of all the frequencies of a set of data up to a particular value of data.	See lesson 16.3
Scatter graph	A graph that displays bivariate data – data points which involve two variables.	See lesson 16.4
Line of best fit	A straight line through the points on the scatter graph which shows the trend of the data.	
Correlation	A measure of how strongly two variables appear to be related.	
Time series graph	A scatter graph with time as one variable.	
Trend (time series)	The long term behaviour of the data, 'averaging out' short term fluctuations.	

Room temperature

Review

1 The table shows the lengths of the cars in a car park. Represent this data in a histogram.

Length, L (m)	Frequency
$3.4 < L \leqslant 3.8$	20
$3.8 < L \leqslant 4.2$	30
$4.2 < L \leqslant 4.4$	18
$4.4 < L \leqslant 4.6$	20
$4.6 < L \leqslant 5$	10

2 The table gives data about the annual salaries in a company.

Salary, s (£1000s)	Frequency
$10 < s \leqslant 20$	5
$20 < s \leqslant 30$	25
$30 < s \leqslant 40$	26
$40 < s \leqslant 50$	17
$50 < s \leqslant 60$	5
$60 < s \leqslant 70$	2

  a What is the modal class?

  b In which class does the median lie?

  c Estimate the mean salary.

3 Use the data given in question **2** to complete the following questions.

  a Draw a cumulative frequency graph for this data.

  b Use your graph to estimate the

    i lower quartile **ii** upper quartile

    iii median **iv** IQR

    v percentage of people who earn less than £45,000 a year.

  c Draw a box plot for this data.

4 The table shows the engine size and average CO_2 emissions of a number of cars.

Engine Capacity (cc)	CO_2 g/km
2200	175
1950	160
1900	150
1400	125
1500	145
1350	130
1200	125
900	100

  a Draw a scatter diagram to display this data.

  b Describe the correlation between engine capacity and CO_2 emissions.

  c Draw a line of best fit on your diagram.

  d Estimate the emissions for a car with an engine size of 2000 cc.

  e Can you use this graph to estimate the engine size of a car which emits 200 g/km?

5 a Draw a line graph for this time series data.

Time	Temperature (°C)
06:00	15
09:00	19
12:00	20
15:00	24
18:00	25
21:00	20

  b Comment on any pattern(s) present in the data.

What next?

Score			
	0 – 2		Your knowledge of this topic is still developing. To improve look at MyMaths: 1194, 1195, 1196, 1197, 1198, 1201, 1202, 1213, 1250, 1254, 1255, 1333
	3 – 4		You are gaining a secure knowledge of this topic. To improve look at InvisiPens: 16Sa – k
	5		You have mastered these skills. Well done you are ready to progress!. To develop your exam technique look at InvisiPens: 16Aa – i

Assessment 16

1 The table shows the age distribution of the members of an athletics club.

Age in years (Y)	Frequency
10 ≤ Y < 15	14
15 ≤ Y < 20	18
20 ≤ Y < 25	26
25 ≤ Y < 35	33
35 ≤ Y < 45	28
45 ≤ Y < 60	24
60 ≤ Y < 80	7

   a Draw a histogram to illustrate this data. [10]

   b How many members are younger than 20 years old? [1]

   c How many members are at least 45? [1]

   d The club's rules only allow athletes between 20 and not more than 35 years old to represent them in competitions. How many athletes do they have to choose from? [1]

2 As part of a statistics project, the ages of all the people in a hotel were recorded. There were 50 guests present who were 10 years old or younger.

The histogram shows these results. The classes are 0 < Age ≤ 10, ...

   a How many people aged 10 to 20 were in the hotel? [2]

   b How many people aged 50 to 80 were in the hotel? [1]

   c How many people were in the hotel altogether? [2]

3 A police speed checkpoint is situated beside the M6 and the speeds of 250 vehicles noted.

Speed (mph)	No. of Vehicles	Frequency Density
0 < V ≤ 30	6	
30 < V ≤ 60	36	
60 < V ≤ 70	94	
70 < V ≤ 90	105	
90 < V ≤ 115	6	
115 < V ≤ 130	3	

   a The national speed limit is 70 mph. How many motorists were breaking the speed limit? [1]

   b Draw a histogram showing this data. [11]

4 Using all the sets of data in questions **1–3**, for each set:

   a Write down the modal class. [3]

   b Write down the class interval which contains the median. [3]

   c Calculate an estimate for the mean. [15]

5 A group of key stage 2 children were given tests in Science and in English. The results are shown in the table.

Science	22	25	31	33	37	45	48	54	56	61	68	70	75	80	86	89	90	93	96	98
English	44	93	47	58	12	48	76	55	82	60	75	38	59	52	71	66	80	15	72	68

   a Plot a scatter diagram showing the above data. [6]

   b Draw the line of best fit. [2]

5 **c** Comment on the type of correlation. [1]

   d Comment on the two outliers. [1]

   e Use your line to estimate

     i a Science Mark corresponding to an English mark of 52. [1]

     ii an English Mark corresponding to a Science mark of 52. [1]

6 The number of patients in the waiting room of a health centre was recorded at hourly intervals and the results recorded on this time graph.

   a How many patients were in the waiting room at 10:00? [1]

   b How many patients were in the waiting room at 12:00? [1]

   c How many patients were in the waiting room at 14:30? [1]

   d Why is your answer to **c** only an approximation? [1]

   e What is the maximum number of patients recorded in the waiting room? [1]

   f Give a possible reason for the three peaks in the time series. [1]

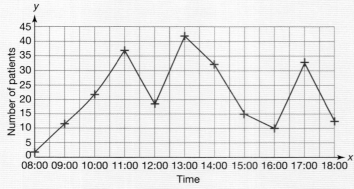

7 'Lickalot' ice cream manufacturers have issued their sales figures for the past 2 years. (Figures are to the nearest £1000)

	Jan	Feb	Mar	Apr	May	Jun	Jul	Aug	Sep	Oct	Nov	Dec
Year 1	100	95	125	150	176	203	251	266	204	131	101	88
Year 2	70	70	105	135	165	176	215	189	217	145	133	110

   a Draw a time series diagram of these figures with Year 2 superimposed on top of Year 1. [6]

   b Compare the summer and winter sales for these periods. [2]

8 The following table gives the waiting time for a fairground ride at 'Thorpe Towers'.

Waiting time (mins)	$0 < t \leqslant 5$	$5 < t \leqslant 10$	$10 < t \leqslant 15$	$15 < t \leqslant 20$	$20 < t \leqslant 25$	$25 < t \leqslant 30$
Frequency	3	5	12	16	7	2

   a Draw the cumulative frequency diagram for this information. [8]

   b Use your diagram to answer the following.

     i How many people waited 10 minutes or less? [1]

     ii How many people waited more than 15 minutes? [1]

   c Use your cumulative frequency diagram to estimate the values of

     i the median [2]

   and **ii** the interquartile range. [4]

   d Use your cumulative frequency curve to draw a box plot for the data. [7]

17 Calculations 2

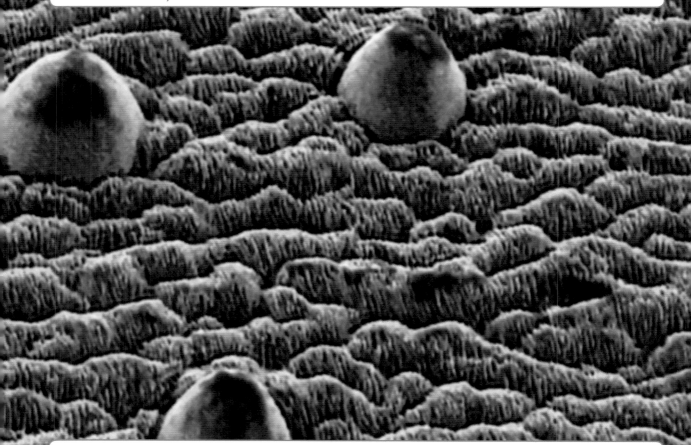

Introduction

An electron microscope is much more powerful than a normal microscope, and is used to look at very small objects. Biologists use electron microscopes to look at micro-organisms such as viruses. Materials scientists might use one to analyse crystalline structures. Nowadays there are electron microscopes that can detect objects that are smaller than a nanometre (0.000000001 m), so they can 'see' molecules and even individual atoms.

What's the point?

Mathematics needs to be able to describe very small quantities, such as lengths measured in nanometres. Without this ability, researchers could not analyse microscopic organisms like viruses.

Objectives

By the end of this chapter, you will have learned how to …

● Perform calculations involving roots and indices, including negative and fractional indices.

● Perform exact calculations involving fractions, surds and π.

● Work with numbers in standard form.

Check in

1 Simplify each expression.

    a $4\pi + 3\pi$ **b** $\pi(3^2 + 4)$ **c** $4\pi(5^2 - 4^2)$

    d $\sqrt{(5^2 - 3^2)}$ **e** $\sqrt{2} + 2\sqrt{2}$ **f** $5\sqrt{3} - \sqrt{3}(4^2 - 3 \times 4)$

2 **a** List all of the factors of these numbers.

    i 6   **ii** 12   **iii** 28   **iv** 36

    b List all of the prime numbers between 1 and 50.

3 Write these multiplications in power form. For example, $4 \times 4 \times 4 = 4^3$.

    a 3×3 **b** $4 \times 4 \times 4 \times 4 \times 4$ **c** $6 \times 6 \times 6$ **d** $5 \times 5 \times 5 \times 5$

4 Evaluate these expressions.

    a 2^4 **b** 3^3 **c** 4^2 **d** 5^3 **e** 2^7 **f** 7^1

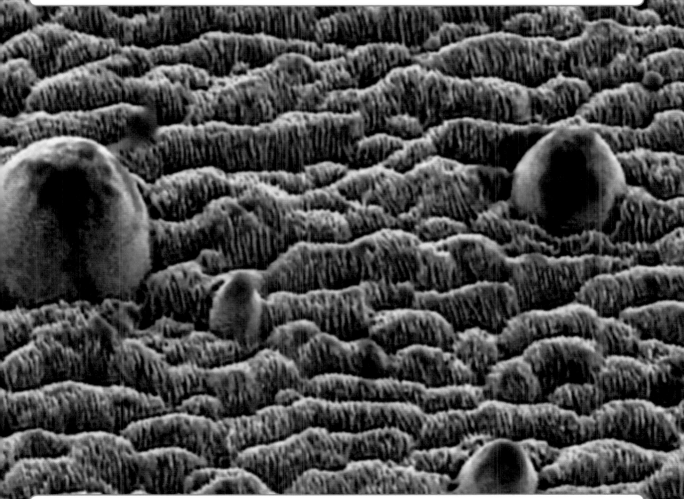

Chapter investigation

A picometre is a unit of length that is used to describe atomic distances.
It is 0.000000000001 m in length.

See if you can find out about the smallest metric unit of length that has a name, and if it has any possible real-world uses. What about the largest unit of length?

17.1 Calculating with roots and indices

Indices can be fractions as well as whole numbers.

2.2

They obey the same index laws as positive integer powers.

Index laws
$x^a \times x^b = x^{a+b}$
$x^a \div x^b = x^{a-b}$
$(x^a)^b = x^{a \times b}$

- $x^{\frac{1}{2}} = \sqrt{x}$ for any value of x.
  In general, $x^{\frac{1}{n}} = \sqrt[n]{x}$ (the nth root of x).

13.2

- $x^{-1} = \frac{1}{x}$, for any value of $x \cdot 0$.
  This is the **reciprocal** of x.

- In general, $x^{-n} = \frac{1}{x^n}$.
  This is the **reciprocal** of x^n.

$5^{\frac{1}{2}} \times 5^{\frac{1}{2}} = 5^{(\frac{1}{2}+\frac{1}{2})} = 5^1 = 5$
but $\sqrt{5} \times \sqrt{5} = 5$,
so $5^{\frac{1}{2}}$ means $\sqrt{5}$

$\frac{1}{5} = 1 \div 5 = 5^0 \div 5^1 = 5^{0-1} = 5^{-1}$,
so 5^{-1} means $\frac{1}{5}$

$\frac{1}{5^2} = 1 \div 5^2 = 5^0 \div 5^2 = 5^{0-2} = 5^{-2}$,
so 5^{-2} means $\frac{1}{5^2}$

Zero has no reciprocal as 0^{-1} is not defined.

EXAMPLE

Evaluate

a $16^{\frac{1}{2}}$ **b** 8^{-1} **c** 10^{-2}

A **fractional** index means a root.

A **negative** index means a reciprocal.

a $16^{\frac{1}{2}} = \sqrt{16} = 4$ **b** $8^{-1} = \frac{1}{8}$ **c** $10^{-2} = \frac{1}{10^2} = \frac{1}{100} = 0.01$

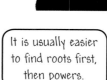

You can give meaning to more complex fractional powers using the index laws.

- $x^{\frac{m}{n}} = (x^{\frac{1}{n}})^m = (\sqrt[n]{x})^m$ or $x^{\frac{m}{n}} = (x^m)^{\frac{1}{n}} = \sqrt[n]{x^m}$

EXAMPLE

Find the value of

a $4^{-\frac{1}{2}}$ **b** $9^{\frac{3}{2}}$ **c** $8^{\frac{2}{3}}$ **d** $4^{-\frac{5}{2}}$

It is usually easier to find roots first, then powers.

a $4^{-\frac{1}{2}} = \frac{1}{4^{\frac{1}{2}}} = \frac{1}{\sqrt{4}} = \frac{1}{2}$

b $9^{\frac{3}{2}} = (9^{\frac{1}{2}})^3 = (\sqrt{9})^3 = (3)^3 = 27$

c $8^{\frac{2}{3}} = (8^{\frac{1}{3}})^2 = (\sqrt[3]{8})^2 = 2^2 = 4$

d $4^{-\frac{5}{2}} = \frac{1}{4^{\frac{5}{2}}} = \frac{1}{(4^{\frac{1}{2}})^5} = \frac{1}{(\sqrt{4})^5} = \frac{1}{(2)^5} = \frac{1}{32}$

You can use the index laws with fractional and negative indices.

EXAMPLE

Simplify these expressions.

a $2^{\frac{3}{4}} \times 2^{\frac{1}{2}}$ **b** $2^{\frac{1}{2}} \div 2^{\frac{1}{6}}$ **c** $(2^{-\frac{1}{2}})^4$ **d** $(2^{-2} \times 2^{\frac{3}{2}})^2$

a $2^{\frac{3}{4}} \times 2^{\frac{1}{2}} = 2^{\frac{3}{4}+\frac{1}{2}}$
$= 2^{\frac{5}{4}}$

b $2^{\frac{1}{2}} \div 2^{\frac{1}{6}} = 2^{\frac{1}{2}-\frac{1}{6}}$
$= 2^{\frac{1}{3}}$

c $(2^{-\frac{1}{2}})^4 = 2^{-\frac{1}{2} \times 2}$
$= 2^{-2}$

d $(2^{-2} \times 2^{\frac{3}{2}})^2 = (2^{-\frac{1}{2}})^2$
$= 2^{-\frac{1}{2} \times 2} = 2^{-1}$

Exercise 17.1S

1 Evaluate

   a $100^{\frac{1}{2}}$ **b** $16^{0.5}$ **c** $49^{\frac{1}{2}}$

   d $4^{0.5}$ **e** $121^{\frac{1}{2}}$ **f** $144^{0.5}$

   g $8^{\frac{1}{3}}$ **h** $27^{\frac{1}{3}}$ **i** $100^{0.5}$

   j $81^{\frac{1}{2}}$ **k** $9^{\frac{1}{2}}$ **l** $125^{\frac{1}{3}}$

   m $0^{\frac{1}{2}}$ **n** $1000^{\frac{1}{3}}$ **o** $64^{\frac{1}{3}}$

2 Write these numbers in index form.

   a $\dfrac{1}{2}$ **b** $\dfrac{1}{5}$ **c** $-\dfrac{1}{7}$

   d 0.5 **e** 0.1 **f** $0.\dot{3}$

3 Write these expressions in index form.

   a $\dfrac{1}{7^2}$ **b** $\dfrac{1}{9^2}$ **c** $\dfrac{1}{2^2}$

   d $\dfrac{1}{2^5}$ **e** $\dfrac{1}{3^4}$ **f** $\dfrac{1}{6^4}$

4 Write these expressions as fractions.

   a 8^{-2} **b** 7^{-3} **c** 5^{-2}

   d 9^{-4} **e** 3^{-2} **f** 9^{-3}

5 Write these expressions

   i   in fraction form

   ii   as decimals.

   a 3^{-2} **b** 2^{-3} **c** 10^{-5}

   d 1^{-7} **e** 8^{-1} **f** 2^{-4}

6 Evaluate these expressions.

   a 4^{-2} **b** 4^{-1} **c** 4^{0}

   d $4^{0.5}$ **e** 4^{1} **f** 4^{2}

   g 4^{3} **h** $4^{-0.5}$ **i** 4^{-3}

7 Write these expressions in index form.

   a $\dfrac{1}{\sqrt{3}}$ **b** $\dfrac{1}{\sqrt{5}}$ **c** $\dfrac{1}{\sqrt{11}}$

8 Evaluate these expressions.

   a $25^{\frac{1}{2}}$ **b** $25^{-\frac{1}{2}}$ **c** 25^{0}

   d $25^{\frac{3}{2}}$ **e** 25^{-1} **f** $25^{-\frac{3}{2}}$

   g $16^{\frac{3}{2}}$ **h** $27^{\frac{2}{3}}$ **i** $4^{-\frac{3}{2}}$

   j $81^{-0.25}$ **k** $125^{-\frac{2}{3}}$ **l** $100^{-\frac{7}{2}}$

   m $4^{\frac{3}{2}}$ **n** $8^{\frac{2}{3}}$ **o** $9^{\frac{5}{2}}$

   p $100^{-\frac{1}{2}}$ **q** $16^{-\frac{3}{2}}$ **r** $1000^{\frac{2}{3}}$

   s $400^{-\frac{1}{2}}$ **t** $169^{\frac{3}{2}}$ **u** $4^{-\frac{3}{2}}$

9 Simplify these expressions.

   a $2^{\frac{1}{3}} \times 2^{\frac{1}{6}}$ **b** $2^{\frac{1}{3}} \div 2^{\frac{1}{6}}$

   c $(2^{\frac{3}{2}})^3$ **d** $2^{\frac{2}{5}} \times 2^{\frac{1}{5}} \div 2$

   e $(2^{\frac{3}{4}} \times 2^{\frac{1}{2}})^2$ **f** $(2^3 \times 2^{\frac{5}{2}}) \div (2^{\frac{7}{2}} \times 2^2)$

10 Simplify these expressions.

   a $3^{-2} \times 3^3$ **b** $3^{-4} \div 3^2$

   c $(3^{-1})^2$ **d** $3^{-2} \times 3^{-1} \div 3^3$

   e $(3^{-1} \div 3^{-2})^{-1}$ **f** $(3^{-2} \div 3^{-1})^2 \div 3^{-6}$

11 Simplify these expressions.

   a $(5^{\frac{1}{2}})^2$ **b** $(5^{\frac{1}{3}})^2$ **c** $(5^2)^{\frac{1}{3}}$

   d $(5^{\frac{1}{3}})^{\frac{1}{2}}$ **e** $(5^{-2})^2$ **f** $(5^3)^{-2}$

   g $(5^{-1})^{-3}$ **h** $(5^{-\frac{5}{4}})^{-2}$ **i** $(5^{-2})^{\frac{3}{4}}$

   j $(5^{-\frac{2}{5}})^{\frac{3}{2}}$ **k** $(5^{\frac{6}{5}})^{-\frac{1}{3}}$ **l** $(5^{-\frac{2}{3}})^{\frac{5}{4}}$

12 Simplify these expressions.

   a $2^{-\frac{3}{2}} \times 2^{\frac{1}{4}}$ **b** $2^{-\frac{1}{6}} \div 2^{-\frac{1}{3}}$

   c $(2^{\frac{1}{3}} \div 2^{-\frac{1}{2}})^3$ **d** $2^{\frac{3}{2}} \times 2^{-2} \div 2^{\frac{1}{4}}$

   e $(2^{\frac{5}{4}} \times 2^{-1})^{-\frac{3}{2}}$ **f** $(2^3 \div 2^{-\frac{1}{2}})^{-2} \times (2^{\frac{3}{2}} \div 2^{-1})$

13 Write these expressions as

   i   powers of 4   **ii**   powers of 16

   a $\dfrac{1}{4}$ **b** 16 **c** $\dfrac{1}{16}$ **d** $\dfrac{1}{2}$

14 Write these expressions as powers of 10.

   a $\dfrac{1}{10}$ **b** $\sqrt{10}$ **c** $(\sqrt{10})^3$ **d** $\dfrac{1}{(\sqrt{10})^5}$

15 Write these expressions as powers of 5.

   a 0.04 **b** 0.008 **c** $\dfrac{1}{\sqrt[3]{5}}$

   d $\sqrt[3]{25}$ **e** $\dfrac{1}{\sqrt{125}}$ **f** $\dfrac{1}{\sqrt[3]{625}}$

16 a Fill in the missing numbers.

      i $125 = 25^{\square}$

      ii $25 = 125^{\square}$

   b   **i** $32 = 4^{\square}$

      ii $4 = 32^{\square}$

   c   **i** $81 = 27^{\square}$

      ii $27 = 81^{\square}$

   d   **i** $0.125 = 16^{\square}$

      ii $16 = 0.125^{\square}$

17.1 Calculating with roots and indices

- Fractional indices represent **roots**.
  - $x^{\frac{1}{2}} = \sqrt{x}$ for all values of x
  - $x^{\frac{1}{n}} = \sqrt[n]{x}$ the nth root of x
  - $x^{\frac{m}{n}} = (x^{\frac{1}{n}})^m = (\sqrt[n]{x})^m$, or $x^{\frac{m}{n}} = (x^m)^{\frac{1}{n}} = \sqrt[n]{x^m}$
- Negative indices represent **reciprocals**.
  - $x^{-n} = \dfrac{1}{x^n}$

(1) Read the question carefully; decide how to use connections between powers and roots.

(2) Apply the index laws and look out for chances to use known facts about squares and cubes.

(3) Answer the question; make sure that you leave your answer in index form if the question asks for it.

Rather than relying on learning the rules for fractional and negative indices 'by heart', make sure you understand the underlying relationships:

- $x^{\frac{1}{2}} \times x^{\frac{1}{2}} = x \Rightarrow x^{\frac{1}{2}} = \sqrt{x}$
- $x^{-n} = x^0 \div x^n = 1 \div x^n = \dfrac{1}{x^n}$

Write 27 as a power of 9.

(1) 27 and 9 are both powers of 3. $27 = 3^3$ and $9 = 3^2$

(2) Rewrite 3 as a power of 9. $3 = \sqrt{9} = 9^{\frac{1}{2}}$

    Substitute $3 = 9^{\frac{1}{2}}$ into $27 = 3^3$. $27 = (9^{\frac{1}{2}})^3$

(3) Simplify the expression. $27 = 9^{\frac{3}{2}}$ Multiply the powers.

$a^{\frac{3}{2}} = 1000$. Find the value of a.

(1) Use inverse operations to make a the subject.

(2) $a^{\frac{3}{2}}$ means 'cube' and 'square root'. The inverse is 'cube root' and 'square'.

    $a = 1000^{\frac{2}{3}} = (\sqrt[3]{1000})^2 = 10^2 = 100$

(3) $a = 100$

Find the missing powers in these equations.

a $(2^{-2x} \div 2^{-5})^2 = 2^{14}$ b $2^{\frac{1}{3}} \times 2^{\frac{1}{2}-x} = \dfrac{1}{2^{\frac{1}{6}}}$ c $\dfrac{2^{\frac{1}{2}} \times 4^x \div 2^{-\frac{1}{4}}}{8^{2x}} = 2$

(1) Simplify the expressions using the index laws.

a $(2^{-2x} \div 2^{-5})^2 = 2^{14}$ (2)

   $(2^{-2x--5})^2 = 2^{14}$

   $2^{2(5-2x)} = 2^{14}$

   Compare indices.

   $2(5-2x) = 14$

   $5 - 2x = 7$

     $x = -1$ (3)

b $2^{\frac{1}{3}} \times 2^{\frac{1}{2}-x} = \dfrac{1}{2^{\frac{1}{6}}}$ (2)

   $2^{\frac{1}{3}+\frac{1}{2}-x} = 2^{-\frac{1}{6}}$

   $2^{\frac{5}{6}-x} = 2^{-\frac{1}{6}}$

   $\dfrac{5}{6} - x = -\dfrac{1}{6}$

     $x = 1$ (3)

c   First write 4 and 8 as powers of 2.

   $\dfrac{2^{\frac{1}{2}} \times (2^2)^x \div 2^{-\frac{1}{4}}}{(2^3)^{2x}} = 2$

   $2^{\frac{1}{2}+2x--\frac{1}{4}-6x} = 2$ (2)

   $2^{\frac{3}{4}-4x} = 2^1$

   $\dfrac{3}{4} - 4x = 1 \Rightarrow 4x = -\dfrac{1}{4}$

     $x = -\dfrac{1}{16}$

Exercise 17.1A

1 Write these expressions as powers of 4.

 a $\dfrac{1}{4}$ **b** 16

 c $\dfrac{1}{16}$ **d** $\dfrac{1}{2}$

2 Write each of the numbers given in question **1** as a power of 16.

3 Write these expressions as powers of 10.

 a $\dfrac{1}{10}$ **b** $\sqrt{10}$

 c $(\sqrt{10})^3$ **d** $\dfrac{1}{(\sqrt{10})^5}$

4 Write these expressions as powers of 5.

 a 0.04 **b** 0.008

 c $\dfrac{1}{\sqrt[3]{5}}$ **d** $\sqrt[3]{25}$

 e $\dfrac{1}{\sqrt{125}}$ **f** $\dfrac{1}{\sqrt[3]{625}}$

5 **a** The volume of a cube is $x\,\text{cm}^3$.
 Find an expression for the surface area.
 Give your answer in the form ax^b, where a and b are rational numbers.

 b The surface area of a cube is $54\,\text{cm}^2$.
 Find the volume of the cube.

 c A cube with volume $216\,\text{cm}^3$ has surface area $216\,\text{cm}^2$.
 Are there any other cubes that have the same numerical value for the volume and surface area? Explain your answer.

6 **a** Copy and complete the table to show the value of 9^x for various values of x.

x	-1	$-\dfrac{1}{2}$	0	$\dfrac{1}{2}$	1
9^x					

 b Use your table to plot a graph to show the value of 9^x, for values of x in the range -1 to $+1$.

 c On the same set of axes, plot the graph of 4^x for the same values of x. Comment on any similarities and differences between the two graphs.

7 **a** **i** Write 125 as a power of 25.

 ii Write 25 as a power of 125.

 b **i** Write 32 as a power of 4.

 ii Write 4 as a power of 32.

 c **i** Write 81 as a power of 27.

 ii Write 27 as a power of 81.

 d **i** Write 0.125 as a power of 16.

 ii Write 16 as a power of 0.125.

8 Find the value of the letter in each of these equations.

 a $a^{\frac{2}{3}} = 81$ **b** $b^{\frac{4}{5}} = 16$

 c $c^{-\frac{1}{3}} = 2$ **d** $d^{-\frac{3}{2}} = 0.008$

9 Evaluate $\left(2\dfrac{1}{4}\right)^{-\frac{1}{2}}$

10 Solve these equations.

 a $(2^{-3x} \div 2^{-2})^2 = 2^{10}$

 b $\dfrac{2^{3x} \times 2^4}{2^3 \times 2^{4x}} = \dfrac{1}{2^2}$

 c $(2^{4+x} \div 2^{2+2x})^2 = (2^{3x} \times 2^{x-2})^{-1}$

 d $2^{\frac{3}{2}} \times 2^x \div 2^{\frac{2}{3}} = 2$

 e $(2^{\frac{1}{2}} \times 2^x)^{\frac{1}{2}} = \dfrac{1}{(2^{\frac{1}{3}})^3}$

11 Copy and complete these number pyramids. The number in each cell is the product of the numbers in the two cells immediately below it.

 a
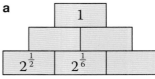

 b
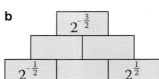

 c Create your own number pyramid containing powers of 2. Challenge a partner to find three missing entries.

17.2 Exact calculations

- A fraction will have a **terminating** decimal equivalent if the only prime factors of the denominator are 2 or 5. $\frac{3}{20} = \frac{3}{(2 \times 2 \times 5)}$

- All other fractions give **recurring** decimals. $\frac{1}{18} = \frac{1}{(2 \times 3 \times 3)} = 0.055\,55\ldots = 0.05\dot{5}$

When you calculate with fractions, you should work with the numbers in fraction form as far as possible.

EXAMPLE

Calculate

a $\frac{1}{2} + \frac{2}{3}$

b $\frac{3}{4} \times \frac{2}{9}$

c $\frac{4}{5} \div \frac{3}{10}$

a $\frac{1}{2} + \frac{2}{3} = \frac{3}{6} + \frac{4}{6} = \frac{7}{6} = 1\frac{1}{6}$

b $\frac{\cancel{3}^{1}}{\cancel{4}_{2}} \times \frac{\cancel{2}^{1}}{\cancel{9}_{3}} = \frac{1}{2} \times \frac{1}{3} = \frac{1}{6}$

c $\frac{4}{5} \div \frac{3}{10} = \frac{4}{\cancel{5}_{1}} \times \frac{\cancel{10}^{2}}{3} = \frac{4}{1} \times \frac{2}{3} = \frac{8}{3} = 2\frac{2}{3}$

- Some numbers cannot be written as fractions. These **irrational** numbers have no repeating decimal patterns.

- Irrational numbers such as $\sqrt{2}$ are called **surds**.

$\sqrt{2} = 1.414\,213\,562\ldots$

You can work with combinations of rational numbers, surds and other irrational numbers such as π.

$\dfrac{1 + \sqrt{5}}{2}$ or $\dfrac{1}{2\pi}\sqrt{\dfrac{7}{9.8}}$ Work with the numbers in surd form as far as possible.

EXAMPLE

Calculate
$\dfrac{7 + \sqrt{20}}{4} - \dfrac{3 + \sqrt{5}}{5}$

$\dfrac{7 + \sqrt{20}}{4} - \dfrac{3 + \sqrt{5}}{5} = \dfrac{35 + 5\sqrt{20}}{20} - \dfrac{12 - 4\sqrt{5}}{20}$

$= \dfrac{35 + 5\sqrt{20} - 12 - 4\sqrt{5}}{20}$

$= \dfrac{23 + 10\sqrt{5} - 4\sqrt{5}}{20}$

$= \dfrac{23 + 6\sqrt{5}}{20}$

Decimal **approximations** are useful in practical contexts.

EXAMPLE

Evaluate
$2\pi(4^2 + 4 \times 6)$

First simplify, leaving π in the expression.

$2\pi(4^2 + 4 \times 6) = 2\pi(16 + 24) = 2\pi \times 40 = 80\pi$

You can now find a decimal approximation:

$\pi = 3.14$ to 2 dp so $80\pi \approx 251$ to 3 sf.

Using π you can give exact answers to problems involving areas of circles.

Leaving an answer in **surd form** is more accurate than a decimal approximation.

Exercise 17.2S

1 Evaluate exactly

a $\dfrac{1}{3} + \dfrac{1}{5}$ **b** $\dfrac{2}{5} + \dfrac{3}{7}$

c $\dfrac{3}{8} - \dfrac{2}{7}$ **d** $\dfrac{8}{15} + \dfrac{4}{9}$

e $6\dfrac{1}{2} - 1\dfrac{5}{8}$ **f** $\dfrac{2}{5} + \dfrac{1}{3} + \dfrac{1}{4}$

g $\dfrac{3}{8} + \dfrac{1}{2} - \dfrac{2}{5}$ **h** $7\dfrac{2}{9} + 2\dfrac{1}{4}$

2 Evaluate exactly

a $\dfrac{2}{3} \times \dfrac{3}{4}$ **b** $\dfrac{5}{9} \div \dfrac{1}{3}$

c $2\dfrac{1}{2} \times \dfrac{5}{8}$ **d** $\dfrac{8}{9} \div 1\dfrac{2}{3}$

e $3\dfrac{1}{2} \div 2\dfrac{1}{4}$ **f** $5\dfrac{1}{5} \times 2\dfrac{3}{4}$

g $8\dfrac{2}{5} \div 3\dfrac{1}{7}$ **h** $3\dfrac{1}{2} \times 7\dfrac{5}{9}$

3 For each answer from questions **1** and **2**, either give an exact decimal equivalent of the answer, or explain why it is not possible to do so.

4 Evaluate exactly

a $\dfrac{2}{5} \times \left(\dfrac{3}{4} + \dfrac{2}{3} \right)$

b $\left(\dfrac{2}{3} + \dfrac{4}{5} \right) \div \left(\dfrac{2}{5} + \dfrac{3}{7} \right)$

c $\dfrac{5}{6} \div \dfrac{3^2 + 4^2}{10}$

d $\left(\dfrac{3}{7} \right)^2 \times \left(\dfrac{4}{5} - \dfrac{1}{7} \right)$

e $\left(\dfrac{1}{2} + \dfrac{5}{9} \right)^2 + \left(\dfrac{2}{3} \right)^3$

f $\left(\dfrac{5}{6} + \dfrac{1}{2} \right) - \left(\dfrac{4}{7} \times 3 \right)$

5 Julia says that all these six calculations give the same answer. Is she correct?
(Show your working.)

a $\dfrac{3}{4} + \dfrac{1}{2}$ **b** $\dfrac{1}{2} \times 2\dfrac{1}{2}$

c $2\dfrac{3}{4} - 1\dfrac{1}{2}$ **d** $3 \div 2\dfrac{2}{5}$

e $\sqrt{50} \div \sqrt{32}$ **f** $\dfrac{1}{4}(3 + \sqrt{2})(3 - \sqrt{2})$

6 Simplify these expressions.

a $\pi + \pi$ **b** $2\pi + \pi$

c $4 + \pi - 2 + \pi$ **d** $\pi(4^2 - 6)$

e $2\pi(4^2 - 7)$ **f** $4\pi(2 + \sqrt{4})$

7 Simplify these expressions.

a $\pi(6^2 - 4)$ **b** $\sqrt{49} + \pi(7 - 5)$

c $4(7 + \sqrt{2})$ **d** $\pi(8^2 - \sqrt{4})$

8 Use a calculator to find an approximate decimal value for each of these expressions. Give your answers to 2 decimal places.

a $4\sqrt{2}$ **b** $\sqrt{5} + 1$

c $2 + \sqrt{5}$ **d** $36\pi - 7$

9 Simplify these expressions.

a $\sqrt{20} + 3\sqrt{5}$

b $(5 + 3\sqrt{3}) - (2 + 4\sqrt{3})$

c $\dfrac{6 + \sqrt{27}}{3} - \dfrac{2 + \sqrt{3}}{4}$

d $\dfrac{20 + 3\sqrt{7}}{3} - (5 + \sqrt{28})$

e $\dfrac{6 + \sqrt{8}}{2} + \dfrac{3 + \sqrt{2}}{9}$

f $\dfrac{8 + \sqrt{45}}{3} - \dfrac{4 + \sqrt{20}}{9}$

10 Evaluate, giving your answers in surd form.

a $\sqrt{5}(1 + \sqrt{5})$ **b** $\sqrt{3}(5 - \sqrt{3})$

c $(2 + \sqrt{3})(3 + \sqrt{3})$ **d** $(3 - \sqrt{2})^2$

e $(5 - \sqrt{7})(5 + \sqrt{7})$

f $(4 + 3\sqrt{5})(6 - \sqrt{5})$

11 Simplify these expressions, writing your solutions without irrational numbers in the denominators.

a $(1 + \sqrt{5}) \div (1 + \sqrt{3})$

b $(2 + \sqrt{7}) \div (2 + \sqrt{3})$

c $(5 - \sqrt{11}) \div (2 + \sqrt{11})$

d $(1 + 2\sqrt{3}) \div (2 - \sqrt{3})$

e $(5 + 2\sqrt{3}) \div (3 - 2\sqrt{3})$

f $(7 - 2\sqrt{5}) \div (8 - 2\sqrt{5})$

17.2 Exact calculations

● When you calculate with fractions, you should write the numbers as exact fractions.

● Irrational numbers such as $\sqrt{2}$ are called surds. For any pair of surds $\sqrt{a} \times \sqrt{b} = \sqrt{a \times b}$.

● Surds are written in the simplest form when the smallest possible integer is written inside the square root sign.

● The area of circles and volume of cylinders, cones and spheres can be written in terms of π.

HOW TO

① Read the question carefully; decide how to use your knowledge of surds. You may need to use your knowledge of geometric formulae.

② Simplify the expressions by calculating with fractions and collecting like terms containing π.

Use $\sqrt{a \times b}$ to simplify expressions containing surds.

EXAMPLE

10.3

Solve these simultaneous equations.

$2p + q = 6\sqrt{3}$ (A)

$2p - 2q = 3\pi$ (B)

① Subtract (B) from (A).

$2p + q - (2p - 2q) = 6\sqrt{3} - 3\pi$

② $3q = 6\sqrt{3} - 3\pi$ Keep your calculations exact.

$q = 2\sqrt{3} - \pi$

Substitute q into (A).

$2p + 2\sqrt{3} - \pi = 6\sqrt{3}$

$2p = 4\sqrt{3} + \pi$

$p = 2\sqrt{3} + \dfrac{\pi}{2}$

EXAMPLE

The blue cylinder has diameter 4 cm and height 3 cm. The red cylinder has diameter 3 cm and height 4 cm. The cuboid has a square base with side 3 cm, and is 5 cm high.

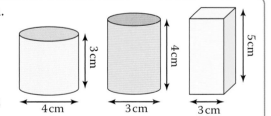

a What is the exact total volume of the shapes?

b The blue cylinder is enlarged by scale factor 1.5 what is the new total of the volumes?

a Blue volume $= \pi \times \left(\dfrac{4}{2}\right)^2 \times 3 = 12\pi$

Red volume $= \pi \times \left(\dfrac{3}{2}\right)^2 \times 4 = \pi \times \dfrac{9}{4} \times 4 = 9\pi$

Yellow volume $= 3 \times 3 \times 5 = 45$

Total volume $= 45 + 21\pi$ cm²

b Volume scale factor $= \left(\dfrac{3}{2}\right)^3 = \dfrac{27}{8}$.

New blue volume $= \dfrac{27}{8} \times 12\pi = \dfrac{81}{2}\pi = 40\frac{1}{2}\pi$

New total volume $= 40\frac{1}{2}\pi + 9\pi + 45$

$= 49\frac{1}{2}\pi + 45$ cm³

① Volume of cylinder $= \pi r^2 h$.

① Volume of a cuboid $= l \times w \times h$

② Terms involving π are like terms.

③ Cancel common factors

Leaving π in your answers actually makes working easier!

Exercise 17.2A

Do not use a calculator for this exercise.

1 A bathroom window is a semicircle with an internal diameter of 80 cm. Work out the exact area and perimeter of the glass in the window.

2 Shamin is cutting out circles for an art project. She has squares of card that are 4 cm wide.
If she cuts out the largest possible circle, what is the exact area of the card remaining?

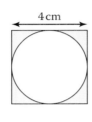
4 cm

3 Find the exact solutions to these simultaneous equations.

   a $x + y = 3\sqrt{5}$
      $2x - y = 3\pi$

   b $3x + y = 9\sqrt{5} + \pi$
      $x + y = 5\sqrt{5} - \pi$

   c $3x + 2y = 2(\pi - \sqrt{5})$
      $2x + 3y = 3\sqrt{5} - 2\pi$

4 A circle of radius 10 cm has the same area as a square with sides x cm.
Find the exact value of x.

5 A cube with side length 5 cm has the same volume as a cone with vertical height 3 cm.
Show that the exact radius of the cone is $\dfrac{5\sqrt{5}}{\sqrt{\pi}}$ cm.

6 The diagram shows a cuboid with volume 25 cm³.
What is the exact surface area of the cuboid?

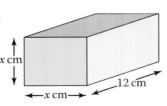
x cm
x cm
12 cm

7 An orange has diameter 12 cm. Once peeled, its volume is $\dfrac{500\pi}{3}$ cm³. How thick was the skin?

8 A hemisphere has surface area 60π mm². Find its volume.

9 The areas of two squares are in the ratio $1:3$.
The side of the larger square is 12 cm.
What is the side of the smaller square?
Give your answer in the form $a\sqrt{b}$ where a and b are integers.

10 An equilateral triangle has sides 2 cm.

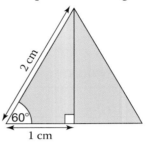
2 cm
60°
1 cm

   a Prove that the height of the triangle is $\sqrt{3}$ cm.

The triangle is enlarged and the area of the image is $3\sqrt{3}$ cm².

   b Find the height of the enlarged triangle.

   c Find the side length of the enlarged triangle.

11 Rachel bought a cylindrical tube containing three exercise balls. Each ball is a sphere of radius 5 cm.
The balls touch the sides of the tube.
The balls touch the top and bottom of the tube.

Work out the exact volume of empty space in the tube.

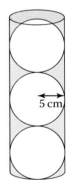
5 cm

15.3

> You saw this question in a previous exercise. Here you are calculating the exact volume.

*12 Lindsey and Laura are arguing about the square root of two. Lindsey says, 'you can write it as a fraction, $\sqrt{2} = \dfrac{m}{n}$'. Laura says, 'that means both m and n must be even so your fraction can never be in its lowest terms'.

Who is right? Give your reasons.

17.3 Standard form

You can use **standard form** to represent large and small numbers.

- In standard form, a number is written as $A \times 10^n$, where $1 \leqslant A < 10$ and n is an integer.

EXAMPLE

Write these numbers in standard form.

a   235   b 0.23×10^6

c   0.45   d $0.000\,000\,416$

a $235 = 2.35 \times 10^2$

b $0.23 \times 10^6 = 2.3 \times 10^5$

c 4.5×10^{-1}

d 4.16×10^{-7}

First order the powers of 10: $10^{-1} < 10^2 < 10^3 < 10^4$.

Then compare numbers with the same powers of 10: $5.44 < 6.35$.

EXAMPLE

Write these numbers in order, starting with the smallest.
6.35×10^4, 5.44×10^4, 6.95×10^3, 7.075×10^2, 9.9×10^{-1}

The correct order is:

9.9×10^{-1}, 7.075×10^2, 6.95×10^3, 5.44×10^4, 6.35×10^4

- You can multiply and divide in standard form. Multiply or divide the numbers and then multiply or divide the powers.

EXAMPLE

Calculate

a $(3.6 \times 10^5) \div (1.2 \times 10^3)$

b $(5.4 \times 10^4) \times (2 \times 10^{-3})$

a $(3.6 \times 10^5) \div (1.2 \times 10^3) = (3.6 \div 1.2) \times (10^5 \div 10^3)$
$= 3 \times 10^{(5-3)} = 3 \times 10^2$

b $(5.4 \times 10^4) \times (2 \times 10^{-3}) = (5.4 \times 2) \times (10^4 \times 10^{-3})$
$= 10.8 \times 10^{4-3} = 1.08 \times 10^2$

- You can add and subtract in standard form. First rewrite the numbers with the same power of 10, and then add or subtract the numbers.

EXAMPLE

Calculate $3.2 \times 10^5 + 7.1 \times 10^4$.
Give your answer in standard form.

$3.2 \times 10^5 + 7.1 \times 10^4 = 3.2 \times 10^5 + 0.71 \times 10^5$
$= (3.2 + 0.71) \times 10^5 = 3.91 \times 10^5$

- You need to know how to enter standard form calculations into your calculator and how to interpret the display.

EXAMPLE

Use your calculator to work out $(6.43 \times 10^6) \div (4.21 \times 10^{-2})$.

$\boxed{1.5273\,159\,14\,^{08}}$

$(6.43 \times 10^6) \div (4.21 \times 10^{-2}) = 1.53 \times 10^8$ (to 3 sf)

Exercise 17.3S

1 Convert to ordinary numbers.

   a 10^3 **b** 10^6 **c** 10^5

2 Convert to ordinary numbers.

   a 10^0 **b** 10^{-2} **c** 10^{-5}

3 Write these numbers in standard form.

   a 9000   **b** 650   **c** 6500

   d 952   **e** 23.58   **f** 255.85

4 Write these numbers in standard form.

   a 0.00034   **b** 0.1067

   c 0.0000091   **d** 0.315

   e 0.0000505   **f** 0.0182

   g 0.00845   **h** 0.000000000306

5 Write these numbers in order of size, smallest first.

   4.05×10^4 4.55×10^4 9×10^3

   3.898×10^4 1.08×10^4 5×10^4

6 Write these numbers as ordinary numbers.

   a 6.35×10^4 **b** 9.1×10^{17}

   c 1.11×10^2 **d** 2.998×10^8

7 Convert to ordinary numbers.

   a 4.5×10^{-3} **b** 3.17×10^{-5}

   c 1.09×10^{-6} **d** 9.79×10^{-7}

8 Although they are written as multiples of powers of 10, these numbers are not in standard form. Rewrite each of them correctly in standard form.

   a 60×10^1 **b** 45×10^3

   c 0.65×10^1 **d** 0.05×10^8

   e 21.5×10^3 **f** 0.7×10^{14}

   g 122.516×10^{18} **h** 0.015×10^9

   i 28×10^{-2} **j** 0.4×10^{-1}

   k 13.5×10^{-4} **l** 12×10^{-8}

9 Evaluate these calculations, giving your answers in standard form.
Do not use a calculator.

   a $(5 \times 10^3) \times (5 \times 10^4)$

   b $(8 \times 10^7) \times (3 \times 10^5)$

   c $(2.5 \times 10^{-3}) \times (2 \times 10^2)$

   d $(4.6 \times 10^{-6}) \times (2 \times 10^{-2})$

10 Evaluate these calculations, showing your working. Do not use a calculator; give your answers in standard form.

   a $(2 \times 10^6) \div (4 \times 10^4)$

   b $(3 \times 10^5) \div (4 \times 10^2)$

   c $(4 \times 10^4) \div (2 \times 10^6)$

   d $(8.4 \times 10^{-2}) \div (2 \times 10^6)$

11 Work out these calculations, giving your answers in standard form.

   a $2 \times 10^3 \times 3 \times 10^4$

   b $(8 \times 10^{15}) \div (2 \times 10^3)$

   c $7.5 \times 10^3 \times 2 \times 10^5$

   d $(3.5 \times 10^3) \div (5 \times 10^2)$

   e $5 \times 10^5 \times 3 \times 10^4$

   f $4 \times 10^3 \times 5 \times 10^2 \times 6 \times 10^4$

12 Work out these calculations without using a calculator.

   a $(4 \times 10^6) \div (2 \times 10^8)$

   b $5 \times 10^4 \times 2 \times 10^{-6}$

   c $(3 \times 10^{-3}) \div (2 \times 10^5)$

   d $(4 \times 10^5) \div (2 \times 10^2)$

   e $5 \times 10^{-3} \times 5 \times 10^{-4}$

   f $(9.3 \times 10^{-2}) \div (3 \times 10^{-6})$

13 Work out these calculations without using a calculator, giving your answers in standard form.

   a $(5 \times 10^{-1}) + (2 \times 10^{-2})$

   b $(4 \times 10^{-2}) + (6 \times 10^{-3})$

   c $(2 \times 10^{-2}) + (9 \times 10^{-4})$

   d $(1.5 \times 10^{-2}) - (2 \times 10^{-3})$

14 Use a calculator to find the value of these in standard form.

   a $(6.4 \times 10^{-4}) + (7.1 \times 10^{-3})$

   b $(9.9 \times 10^5) - (2.7 \times 10^4)$

   c $(4.8 \times 10^{-6}) + (3.9 \times 10^{-5})$

   d $(3.3 \times 10^2) - (7.5 \times 10^1)$

   e $(9.8 \times 10^5) - (6.4 \times 10^5)$

   f $(3.5 \times 10^{-2}) + (9.7 \times 10^{-3})$

15 Use your calculator to check your answers to questions **11–13**.

Q 1049, 1050, 1051   SEARCH

17.3 Standard form

RECAP

- You can write a number in standard form as $A \times 10^n$, where n is a positive or negative integer and $1 \leqslant A < 10$.

HOW TO

1. Read the question carefully and decide which calculation you need to carry out.
2. Calculate using your knowledge of standard form.
3. Give your answer in standard form and check that your answer is a sensible order of magnitude.

EXAMPLE

Rosie wrote

$(7.2 \times 10^{-4}) \div (3.6 \times 10^{-8}) = 2 \times 10^{-12}$

Is she correct?

1. No. She has multiplied 10^{-4} and 10^{-8}, instead of dividing.
2. The correct answer is:

   $(7.2 \div 3.6) \times (10^{-4} \div 10^{-8}) = 2 \times 10^{-4 - -8} = 2 \times 10^4$

   A quick check shows Rosie's answer is incorrect.
   Since you are dividing one number by another smaller number, the answer must be **greater** than 1.

EXAMPLE

The mass of a carbon atom is 2×10^{-23} g.

How many atoms are there in one gram of carbon?

1. Divide 1 g by the mass of one carbon atom.
2. Write both numbers in standard form.

   $1 \div (2 \times 10^{-23}) = 1 \times 10^0 \div (2 \times 10^{-23})$

   Divide the numbers and then divide the powers of 10.

   $= (1 \div 2) \times (10^0 \div 10^{-23})$
   $= 0.5 \times 10^{0 - -23}$
   $= 0.5 \times 10^{23}$

3. Give your answer in standard form.

   There are 5×10^{22} atoms in one gram of carbon.
   You would expect there to be a very large number of atoms, so the answer is sensible.

EXAMPLE

Mercury is 5.79×10^7 km from the Sun.

Venus is 1.08×10^8 km from the Sun.

Find the minimum distance between Mercury and Venus.

1. Draw a sketch of the orbits. The minimum distance is the difference.

   Minimum distance $= (1.08 \times 10^8) - (5.79 \times 10^7)$

2. Write both numbers with the same power of 10.

   $= (10.8 \times 10^7) - (5.79 \times 10^7)$
   $= (10.8 - 5.79) \times 10^7$
   $= 5.01 \times 10^7$ km   Include the units.

3. You would expect the answer to be of magnitude 10^7.

Exercise 17.3A

1 The mass of the Sun is 2×10^{30} kg.
The mass of the Earth is 6×10^{24} kg.
How many times heavier than the Earth is the Sun?

2 The mass of one atom of the element mercury is 3.3×10^{-22} g.
The mass of the planet Mercury is 3.3×10^{23} kg.
How many mercury atoms are there in 3.3×10^{23} kg?

3 The width of a plant cell is 60 micrometres.
A micrometre is 1×10^{-6} m (one millionth of a metre).
The diagram of a plant cell in a science textbook has width 3 cm.
How many times bigger is the diagram than the real plant cell?

4 A bumblebee weighs 5.2×10^{-5} kg.
An adult man weighs 70 kg.
A bumblebee can carry 75% of its weight.
How many bumblebees would it take to lift a man?
Give your answer in standard form to 3 sf.

5 The diameter of the Earth is 1.3×10^4 km.
The diameter of the Moon is 3.5×10^3 km.
The diameter of the Sun is 400 times greater than the diameter of the Moon.
How many times smaller is the diameter of the Earth than the diameter of the Sun?

6 The Earth is 1.496×10^8 km from the Sun.
Mars is 2.279×10^8 km from the Sun.
Find the minimum and maximum distances between Earth and Mars.

7 Light travels about 3×10^8 metres per second.

   a Find the time it takes for light to travel 1 metre.

   b Find the distance light travels in 1 year.
Give your answers in standard form.

8 The mass of a proton is approximately 1800 times greater than the mass of an electron.
The mass of an electron is 9.11×10^{-31} kg.
A hydrogen atom contains one proton and one electron.
Find the mass of a hydrogen atom.
Give your answer in standard form.

9 The masses of the eight planets in our solar system are listed in the table.
The masses are given in kg.

Mercury	3.30×10^{23}
Venus	4.87×10^{24}
Earth	5.97×10^{24}
Mars	6.42×10^{23}
Jupiter	1.90×10^{27}
Saturn	5.68×10^{26}
Uranus	8.68×10^{25}
Neptune	1.02×10^{26}

Carrie calculated the total mass of the planets. Her answer is $7.686\,12 \times 10^{26}$ kg.

   a Which planet did Carrie forget to include in her total?
Calculate the correct total mass.

   b What percentage of the total mass does Earth account for?

Give your answer to two significant figures.

10 A gas giant is a large planet that is made up of gas and liquid.
The largest gas giant planets in our solar system are Jupiter and Saturn.
Jupiter has a radius of 6.99×10^4 km and a mass of 1.90×10^{27} kg.
Saturn has a radius of 5.82×10^4 km and a mass of 5.68×10^{26} kg.
Lamar claims that Jupiter is twice as dense as Saturn.
Do you agree? Explain your answer.

11 You are given these facts about the planet Venus.

- Venus orbits the Sun at a speed of 1.26×10^5 km per hour.
- Venus is 1.08×10^8 km from the Sun.
- Venus' orbit is approximately circular.
- A planet's year is the time it takes for the planet to orbit the Sun.
- Venus' day is 243 Earth days.

   a Prove that Venus' day is longer than its year.

   b How does the assumption that Venus' orbit is circular affect your answer to part **a**?

Summary

Checkout

You should now be able to...

	Test it Questions
✔ Perform calculations involving roots and indices, including negative and fractional indices.	**1 – 4**
✔ Perform exact calculations involving fractions, surds and π.	**5 – 8**
✔ Work with numbers in standard form.	**9 – 11**

Language	Meaning	Example
Index **Base** **Power**	In index notation, the index or power shows how many times the base has to be multiplied by itself. The plural of index is indices.	Power or index $5^3 = 5 \times 5 \times 5$ Base
Fractional index	A fractional index indicates that a root needs to be found.	$64^{\frac{1}{3}} = \sqrt[3]{64} = 4$
Negative index	The reciprocal of the base to the positive power of the index.	$2^{-3} = \frac{1}{2^3} = \frac{1}{8}$ $\quad 64^{-\frac{1}{3}} = \frac{1}{\sqrt[3]{64}} = \frac{1}{4}$
Reciprocal	The reciprocal of a number is one divided by that number. A number multiplied by its reciprocal equals 1.	The reciprocal of 5 is $\frac{1}{5} = 1 \div 5 = 0.2$ $5 \times \frac{1}{5} = 1$
Root	The nth root of a number x is the number that when multiplied by itself n times gives the number x: $(\sqrt[n]{x})^n = x$	$\sqrt[5]{243} = 3$ $(\sqrt[5]{243})^5 = 3^5 = 3 \times 3 \times 3 \times 3 \times 3 = 243$
Surd	Surds are roots that can only be written accurately using fractional indices or a root symbol.	$\sqrt{2}$ and $\sqrt{3}$ are surds. $\sqrt{4} = 2$ so $\sqrt{4}$ is not a surd.
Approximation	A stated value of a number that is close to but not equal to the true value of a number.	$\pi = 3.14$ (3 sf)
Surd form	Writing an exact answer using surds instead of finding a decimal approximation.	$\sqrt{3} + \sqrt{3} = 2\sqrt{3}$
Exact calculation	A calculation that does not involve truncated decimals or other approximations. Exact answers are given in terms of integers, fractions, surds and π.	$\frac{4}{3}\pi \times \left(1\frac{1}{2}\right)^3 = \frac{4}{3} \times \pi \times \frac{3^3}{2^3} = \frac{9}{2}\pi$ $\sqrt{7^2 - 3^2} = \sqrt{40} = 2\sqrt{10}$
Standard form	Writing a number in the form of a number greater than or equal to 1 and less than 10 multiplied by an integer power of 10.	$498000 = 4.98 \times 10^5$ $0.0056 = 5.6 \times 10^{-3}$

Review

1 Write these numbers in index form

   a $0.\dot{1}$ **b** $\dfrac{1}{5^7}$ **c** $\sqrt[3]{12}$ **d** $\dfrac{1}{\sqrt[3]{8}}$

2 Write down the value of these expressions.

   a $10^5 \times 10^2$ **b** $5^6 \div 5^3$ **c** 7^0

   d $(3^2)^3$ **e** $\sqrt{100}$ **f** $64^{\frac{1}{2}}$

   g $125^{-\frac{1}{3}}$ **h** 6^{-2} **i** $4^{\frac{3}{2}}$

   j $(13^{\frac{1}{2}})^2$ **k** $27^{\frac{2}{3}}$ **l** $121^{-0.5}$

3 Simplify these expressions.

   a $2^{\frac{3}{7}} \times 2^{\frac{2}{7}}$ **b** $2^{\frac{3}{7}} \div 2^{\frac{2}{7}}$

   c $(2^{\frac{3}{7}})^2$ **d** $2^{-3} \times 2^{-4}$

   e $2^{-3} \div 2^{-4}$ **f** $(2^{-3})^{-4}$

   g $2^{-\frac{5}{4}} \times 2^{\frac{1}{2}} \div 2^{-\frac{3}{4}}$ **h** $(2^{-2} \times 2^{\frac{5}{6}})^2 \div 2^{-\frac{2}{3}}$

4 Find the value of the letter in each of these equations.

   a $27 = 3^a$ **b** $\sqrt{6} = 6^b$

   c $\sqrt[3]{4} = 2^c$ **d** $\dfrac{1}{1000} = 10^d$

5 Evaluate these expressions exactly.

   a $\dfrac{5}{14} \times 8$ **b** $\dfrac{5}{9} \times \dfrac{3}{20}$ **c** $\dfrac{7}{15} + \dfrac{2}{3}$

   d $\dfrac{6}{7} - \dfrac{3}{8}$ **e** $5\frac{1}{5} - 1\frac{1}{3}$ **f** $8\frac{1}{3} \times \dfrac{3}{5}$

   g $\dfrac{6}{11} \div 9$ **h** $\dfrac{3}{14} \div \dfrac{9}{7}$ **i** $2\frac{1}{4} \div 3\frac{1}{3}$

6 Simplify these expressions.

   a $7 + 2\pi - 3 - 5\pi$

   b $1 - \sqrt{12} + \sqrt{3}$

   c $11\pi(\sqrt{64} - 5)$

   d $\dfrac{6 + \sqrt{27}}{2} + \dfrac{8 + \sqrt{75}}{3}$

   e $(9 + 2\sqrt{7})(1 - \sqrt{7})$

   f $\dfrac{5 - \sqrt{5}}{3\sqrt{5}}$

7 Find the missing sides in these shapes.

   a   **b**

Perimeter of rectangle $= \sqrt{3}(4 + \sqrt{2})$

19.1

8 **a** Calculate the exact diameter of the circle and the semi-circle.

   i   **ii**

   Area $= 2\pi$ Area $= 18\pi$

   b Calculate the exact circumference of the circle and the perimeter of the semi-circle.

9 Write these approximations in standard form.

   a The UK population in 2014: 63 million.

   b The average distance from Earth to the sun: 149 600 000 km.

   c The width of a transistor: 0.000 022 mm.

   d 30 g in kg.

10 Write these as ordinary numbers.

   a 2.18×10^9 **b** 3.1×10^5

   c 5×10^{-7} **d** 9.92×10^{-5}

11 Calculate these and leave your answer in standard form.

   a $(4 \times 10^3) \times (2.1 \times 10^8)$

   b $(2.4 \times 10^8) \div (5 \times 10^{-3})$

   c $(4.5 \times 10^7) + (3 \times 10^5)$

   d $(8.4 \times 10^{-6}) - (2 \times 10^{-4})$

What next?

<table>
<tr><td rowspan="3">Score</td><td>0 – 4</td><td></td><td>Your knowledge of this topic is still developing.
To improve look at MyMaths: 1033, 1045, 1049, 1050, 1051, 1065, 1074, 1301, 1924</td></tr>
<tr><td>5 – 9</td><td></td><td>You are gaining a secure knowledge of this topic.
To improve look at InvisiPens: 17Sa – f</td></tr>
<tr><td>10 – 11</td><td></td><td>You have mastered these skills. Well done you are ready to progress!
To develop your exam technique looks at InvisiPens: 17Aa– g</td></tr>
</table>

Assessment 17

1 Peter says that

    a $0^2 = 1$ **b** $49^{\frac{1}{2}} = 7$ **c** $(-3)^2 = -9$

  Soraya says that

    a $0^2 = 0$ [1]   **b** $49^{\frac{1}{2}} = \frac{1}{49}$ [1]   **c** $(-3)^2 = 9$ [1]

  Without using your calculator, for each value who is correct?
Show your working.

2 Some students have attempted indices calculations. In each case explain whether the student has got it right, if it is wrong give the correct answer as a single power.

  a Paige says that to work out $15^7 \times 15^5$ you multiply the indices. [1]

  b Hayley says that to work out $(3^4)^5$ you multiply the indices. [1]

  c Matt says that $(3^4)^0 = 3^4$. Liz says that $(3^4)^0 = 1$. Lois says that $(3^4)^0 = 3$. [2]

  d Mike says that $7^7 \times 7^2 \div 7^6 = 7^8$. [1]

3 Joe tries to work out the values of a set of expressions, his solutions are shown below. In each case explain whether Joe got the answer right. Correct the answer if he is wrong.

  a $25^{\frac{3}{2}} = 125$ [2]   **b** $216^{\frac{2}{3}} = 144$ [2]   **c** $49^{-\frac{1}{2}} = -7$ [2]

  d $196^{\frac{3}{2}} = 941\,192$ [2]   **e** $9^{-\frac{5}{2}} = 243$ [2]   **f** $156^0 = 156$ [2]

4 Find the values of a, b and c that make these expressions correct.

  a $\sqrt{7} = 7^a$ [2]   **b** $\sqrt[5]{40} = 40^b$ [2]

  c $\sqrt{81} = 9^c$ [2]   **d** $5 = 625^d$ [2]

5 Write the following expressions as powers of   **i** 3   **ii** 9.

  a 3 [2]   **b** 9 [2]   **c** 81 [2]

  d $\frac{1}{3}$ [2]   **e** $\frac{1}{9}$ [2]   **f** $\frac{1}{81}$ [2]

  g 27 [2]   **h** $\frac{1}{27}$ [2]   **i** $\sqrt{3}$ [2]

6 **a** The volume of a cone, whose radius is the same as its vertical height, x, is given by the formula $V = kx^3$. Show that $k = \dfrac{\pi}{3}$. [2]

  b The curved surface area of this cone, A, is given by the formula $A = px^q$ where q can be written as a fraction or integer and p cannot. Find the values of p and q. [4]

  c The curved surface area of a similar cone is $16\pi\sqrt{2}\,\text{m}^2$.
Find the volume of this cone. [4]

7 An unsharpened wooden pencil is in the shape of a hexagonal prism.
The side of the hexagon is $\sqrt{3}$ inches. The pencil is 2π inches long.

  The graphite core is a cylinder with radius $0.1\sqrt{2}$ inches. Calculate the following exact values.

  a The area of the hexagonal cross section. [3]

  b The total volume of pencil and core. [2]

  c The volume of the core. [2]

  d The volume of wood in the pencil. [2]

8 **a** The areas of six oceans and seas are given below in miles² (mi²).
Convert them to ordinary numbers and write them in increasing order of size.

Malay Sea:	3.14×10^6 mi²	Bering Sea:	8.76×10^5 mi²
Indian Ocean:	2.84×10^7 mi²	Caribbean Sea:	1.06×10^6 mi²
English Channel:	2.9×10^4 mi²	Baltic Sea:	1.46×10^5 mi²

[4]

b Which is bigger: 1×10^9 or 999 999 999 and by how much? [2]

c Find the value of n in each of the following equations.

i $4.7 \times 10^n = 47\,000$ [2]   **ii** $6.81 \times 10^n = 681$ [1]

iii $3.467 \times 10^n = 3\,467\,000$ [1]   **iv** $27.5 \times 10^n = 0.0275$ [2]

d Rewrite each of these sentences using ordinary numbers.

i The distance from Mexico City to Moscow is 1.0763×10^4 km. [1]

ii The energy released by the wingbeat of a honey bee is 8×10^{-4} Joules/second. [1]

e **i** Rewrite the following sentence using standard form.

The average length of a bedbug is $\frac{4}{1\,000\,000}$ of a kilometre thick. [1]

ii Write the length given in part **i** in millimetres. [1]

9 **a** The Wright Brothers 'Flyer 1', the world's first aircraft, had a mass of 3.4×10^2 kg.
The Saturn V Rocket had a mass of 2.96×10^6 kg.
How many Flyer's have the same mass as a Saturn? [2]

b 'Flyer 1' attained a speed of 3.04×10^0 ms⁻¹ on its first flight and the supersonic airliner, Concorde, attained a speed of 2.179×10^3 kmh⁻¹.
How much faster was the Concorde than 'Flyer 1'? [2]

c There are approximately 4.336×10^9 stars in our galaxy and about 5.776×10^3 stars visible to the naked eye. What fraction of the galaxy can we see?
Write your fraction in the form $\frac{1}{x}$. [3]

d A triathlon has 3 stages, the largest triathlon has a 3.8×10^0 swim, 1.8×10^2 km cycle ride and 0.42195×10^2 km run.
How far is the race in full?
Give your answer as an ordinary number to 3sf and in standard form. [5]

10 The highest point on the earth's surface, Mount Everest, has a recorded height of 8.848×10^3 m.
The deepest point in the ocean, the Challenger Deep, has a depth of 1.1×10^4 m.
How far is it from the bottom of Challenger to the top of Everest?
Give your answer in standard form and as an ordinary number. [4]

11 **a** The "**Americas**" consists of Canada, area 3.852×10^6 mi²; the USA, area 3.676×10^6 mi²; Central America, area 2.022×10^5 mi² and South America, area 6.879×10^6 mi².

i Calculate the total area of the Americas. Give your answer in standard form and as an ordinary number. [3]

ii Calculate the difference in area between the largest and smallest areas in the list. [2]

b The total surface area of the earth is 1.969×10^8 mi². The surface area of the world's land masses is 5.73×10^7 mi².

i Find the ratio (area of the Americas) : (total area of the land mass of the earth).

Give your answer in the form $1:m$ [2]

ii What area of the earth's surface is covered by water? [2]

iii What percentage of the world's surface is covered by water? [4]

18 Graphs 2

Introduction

When you look at a ball in flight, you already know certain things about its path, or trajectory. It will travel in a smooth curve. It will reach a maximum point. Its downward path will tend to be a mirror image of the upwards path. In football a goalkeeper knows this as well, so a striker might apply spin to the ball to make its flight less predictable.

The path of the ball can be modelled mathematically by a type of equation called a quadratic equation, which is described in this chapter.

What's the point?

An appreciation of quadratic equations and their graphs enables us to understand how an object moves under gravity, and tells us where it's likely to land!

Objectives

By the end of this chapter, you will have learned how to ...

- Recognise and draw graphs of cubic and reciprocal functions.
- Recognise and draw the graphs of exponential functions.
- Recognise and sketch the graphs of trigonometric functions.
- Recognise and sketch translations and reflections of graphs.
- Draw and interpret graphs of non-standard functions and use them in real-life problems.
- Approximate the gradient of a curve at a given point and the area under a graph. Interpret these values in real-life problems including kinematice graphs.
- Recognise and use simple equations of circles and find the tangent to a circle at a point.

Check in

1 Sketch a graph to show each situation.

    a Your height as you age from a baby to a twenty year old (x-axis time, y-axis height).

    b The temperature of a cup of tea as it is left to stand for half an hour (x-axis time, y-axis temperature).

2 Give the gradient of each line segment.

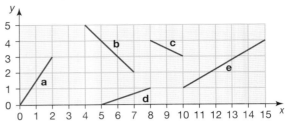

3 Solve these using factorisation or the quadratic formula.

    a $x^2 - 7x + 12 = 0$ **b** $y^2 - 8y = 0$ **c** $x^2 - 3x - 2 = 0$

    d $2x^2 + 7x + 3 = 0$ **e** $y^2 = 11y - 24$ **f** $3x^2 - 2x - 1 = 0$

4 Evaluate each expression for the given value of x.

    a $3x^2 - 2x$ $x = 4$ **b** $x^2 - x$ $x = -2$ **c** $x^2(2x + 3)$ $x = -4$

Chapter investigation

It takes 20 people 18 days to build an extension to a sports centre.

How long would it take one person? State any assumptions that you have made.

Find out how long the job would take for different numbers of people.
Draw a graph to show this information. Try to find an equation for your graph.

18.1 Cubic and reciprocal functions

- A **cubic** equation contains a term in x^3. It has a distinctive **S-shaped** graph.

EXAMPLE

Draw the graph of $y = x^3 - 1$ and use it to estimate the value of y when $x = 2.5$.

First make a table of x and y values.

x	-3	-2	-1	0	1	2	3
x^3	-27	-8	-1	0	1	8	27
$y = x^3 - 1$	-28	-9	-2	-1	0	7	26

$(-3)^3 = -27$

Draw the x-axis from -3 to 3 and the y-axis from -30 to 30.
Plot the points $(-3, -28)$, $(-2, -9)$ and so on, and join them in a smooth curve.

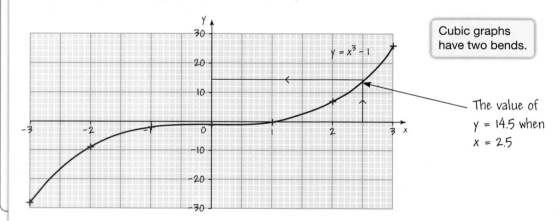

$y = x^3 - 1$

Cubic graphs have two bends.

The value of $y = 14.5$ when $x = 2.5$

A **reciprocal** equation contains a term in $\frac{1}{x}$.
It has a different-shaped curve.

EXAMPLE

Draw the graph of $y = \frac{1}{x}$.

Use the x-axis from -3 to $+3$.

x	-3	-2	-1	0	1	2	3
y	$-\frac{1}{3}$	$-\frac{1}{2}$	-1		1	$\frac{1}{2}$	$\frac{1}{3}$

Draw the x-axis from -3 to 3 and
the y-axis from -1 to 1.

Plot the points.
Note that you cannot plot $x = 0$ as
$\frac{1}{0}$ is not a defined value.

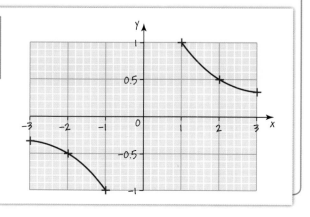

Algebra Graphs 2

Exercise 18.1S

1 **a** Draw an x-axis from -3 to 3 and a y-axis from -30 to $+30$.

b Copy and complete the table for $y = x^3 + 1$.

x	−3	−2	−1	0	1	2	3
y	−26						28

when $x = -3$, $y = (-3)^3 + 1 = -27 + 1$

c Plot the coordinates that you have found in part **b** on your axes from part **a**. Join them to form a smooth, S-shaped curve.

> When using your calculator, remember brackets.

d Use your graph to estimate the value of y when

  i $x = 1.5$ **ii** $x = 0.5^3 + 1$

2 For each equation, copy and complete the table of values. Plot these points on suitable axes and join them to form a smooth curve.

a $y = x^3 - 4$

x	−2	−1	0	1	2	3
x³ − 4					4	

b $y = \dfrac{2}{x}$

x	−2	−1	1	2	3
y				1	

c $y = x^3 + x + 1$

x	−2	−1	0	1	2	3
y					11	

d $y = \dfrac{3}{x} + 1$

x	−2	−1	1	2	3
y	−$\frac{1}{2}$				

3 Draw graphs of these functions for the range of x-values given.

a $y = x^3 + 3x$, for $-3 \leqslant x \leqslant 3$

b $y = x^3 + x - 2$, for $-3 \leqslant x \leqslant 3$

c $y = x^3 + x - 4$, for $-2 \leqslant x \leqslant 3$

d $f(x) = x^3 - x^2 + 3x$, for $-3 \leqslant x \leqslant 3$

4 **a** Plot the graph $f(x) = x^3 - 2x^2 + x + 4$ for $-3 \leqslant x \leqslant 3$.

b Use your graph to find

  i the value of x when $y = -20$

  ii the value of y when $x = 1.7$.

5 Plot these functions for the range of x-values given.

a $y = \dfrac{12}{x - 2}$, for $-2 \leqslant x \leqslant 6$

b $y = \dfrac{x}{x + 4}$, for $-4 \leqslant x \leqslant 4$

c $f(x) = \dfrac{6}{x + x} - 2$, for $-3 \leqslant x \leqslant 3$

6 The graph here has had the axes removed.

The four functions in the sketch are

a $y = 3x + 12$ **b** $y = x^3 - 2x$

c $y = \dfrac{1}{x}$ **d** $y = x^2 + 6x + 7$

Work out which colour graph matches each function.

Did you know...

People have been interested in solving equations for a long time. The Babylonians new how to solve quadratic equations as far back as 2000 BC. However it was not until 1545 AD that a method for solving cubic equations was published by Gerolamo Cardano.

This proved quite controversial at the time as mathematicians challenged one another to equation solving competitions and so jealously guarded their methods.

Q 1071, 1172, 1958   SEARCH

18.1 Cubic and reciprocal functions

RECAP

- A cubic function will always include a power of 3 and no higher power.
- In general, a cubic function will be of the form $y = ax^3 + bx^2 + cx + d$. a, b, c and d are numbers, and some of them may be zero.
- A reciprocal function will always involve dividing by the input.
- In general, a reciprocal function will be of the form $y = \frac{a}{x}$.
- The graph of a reciprocal function includes asymptotes and is always in two sections.

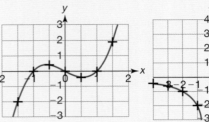 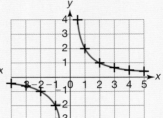

HOW TO

Here is a general strategy for plotting more complex cubic functions.

① Create a table of values that allows you to substitute into each term in turn.
② 'Collect' the answers together to find the y-values.
③ Use enough points to enable a smooth cubic curve to be drawn.
④ Create axes on 2 mm graph paper and plot points carefully.

EXAMPLE

a Plot the graph of $y = 2x^3 - 3x^2 - 4x + 1$. Use values of x from $x = -1.5$ to $x = 2$.

b Use your graph to show that there are three solutions to the equation $2x^3 - 3x^2 - 4x + 1 = x - 1$.

a ①/③ This table of values increases in steps of 0.5 to give eight points in total

x	-1.5	-1	-0.5	0	0.5	1	1.5	2
$2x^3$	-6.75	-2	-0.25	0	0.25	2	6.75	16
$-3x^2$	-6.75	-3	-0.75	0	-0.75	-3	-6.75	-12
$-4x$	+6	+4	+2	0	-2	-4	-6	-8
$+1$	+1	+1	+1	+1	+1	+1	+1	+1
Y	-6.5	0	2	1	-1.5	-4	-5	-3

② The last row in the table gives the corresponding y-values for each value of x.

④ Notice that $(-0.5, 2)$ and $(1.5, -5)$ are not quite turning points. The curve needs to be smooth and 'flow' through the points plotted.

b The line $y = x - 1$ intersects $y = 2x^3 - 3x^2 - 4x + 1$ in three places. Therefore there are three solutions to the equation $2x^3 - 3x^2 - 4x + 1 = x - 1$.

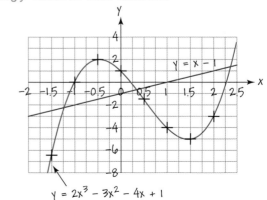

Exercise 18.1A

1 **a** Copy and complete this table of values for the function $y = x^3 - 2x^2 + 3$.

x	−1.5	−1	−0.5	0	0.5	1	1.5	2
x^3								
$-2x^2$								
$+3$								
y								

b Plot the graph of the function $y = x^3 - 2x^2 + 3$.

c Use your graph to show that there is one solution to the equation $x^3 - 2x^2 + 3 = 2 - x$.

d Estimate the solution to the equation in **c**. Give your solution to 2 decimal places.

2 **a** Copy and complete this table of values for the function $y = 3x^3 - 4x + 3$.

x	−1.5	−1	−0.5	0	0.5	1	1.5
$3x^3$							
$-4x$							
$+3$							
y							

b Plot the graph of the function $y = 3x^3 - 4x + 3$.

c On the same axes, plot the graph of the function $y = -\dfrac{2}{x}$.

d Explain how you know that there are two solutions to the equation $3x^4 - 4x^2 + 3x = -2$.

e Estimate the solutions to the equation in **d**. Give your solutions to 2 decimal places.

3 **a** Create, and complete, a table of values for the function $y = 2x^3 - 4x^2 + 3x - 1$. Use values of x from -0.5 to 1.5, increasing in steps of 0.25.

b Plot the graph of $y = 2x^3 - 4x^2 + 3x - 1$.

c Solve the equation $2x^3 - 4x^2 + 3x - 1 = 2 - 2x$. Explain how you know you have found all the solutions.

4 **a** Plot the graph of the function $y = -2x^3 + x^2 + 5x - 1$. Use values of x from -1.5 to 2, increasing in steps of 0.5.

b Plot the graph of $y = \dfrac{1}{x}$ on the same axes.

c Write the equation that can be solved using these graphs.

d Find all the solutions to your equation in **c**.

5 The graph has had the axes removed.

The equations of the three functions are

$$y = x^3 + 7x^2 + 14x + 11$$
$$y = -x^3 + 6x^2 - 8x + 1$$
$$y = x^3$$

Match each function to the colour of its graph. Explain your reasoning.

6 A cubic function has roots at $x = -2$, $x = 1$ and $x = 2$. The y-intercept is 4. Sketch the graph of this function.

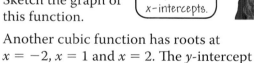

The roots of a function show where the graph has x-intercepts.

7 Another cubic function has roots at $x = -2$, $x = 1$ and $x = 2$. The y-intercept is -4. Sketch the graph of this function.

*8 A cubic function has roots at $x = 0$ and $x = 3$. The x^3 term is positive. Sketch the two possible graphs of this function.

> Use graphing software for question **9**.

9 **a** Plot the graphs of $y = x^2 + x - 9$ and $y = \dfrac{9}{x}$ on the same axes.

b Write the equation that can be solved using these graphs.

c Use your graph to find the solutions to the equation in **b**.

d Plot the graph of the function $y = x^3 + x^2 - 9x - 9$. Write the roots of the function.

e Use algebra to explain why the answers to **c** and **d** are the same.

Q 1071, 1172, 1958   SEARCH

18.2 Exponential and trigonometric functions

- **Exponential** functions include a term with a variable **index**, for example $f(x) = 2^x$. Graphs of exponential functions have a characteristic shape.

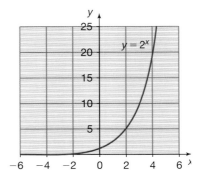

As x gets larger, y gets larger and vice versa.
Since $2^x (= y)$ can never be zero for any value of x, the x-axis is an **asymptote**, the curve gets near but never actually touches it.

- The trigonometric functions are $y = \sin x$,
  $y = \cos x$ and $y = \tan x$.

You can find the value of sin and cos for any value of x.
The graphs of sin and cos are **periodic**, they repeat every $360°$.

Sine graph

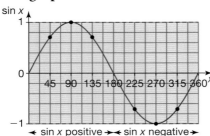

Cosine graph

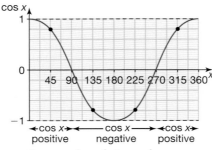

The graph of $y = \cos x$ is the same as the graph of $y = \sin x$ translated $90°$ along the x-axis.

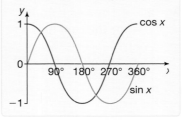

The graph of $y = \tan x$ is different from the graphs of sin and cos.
The graph is periodic, it repeats every $180°$.
The graph has asymptotes at $90°$, $270°$, $450°$, ...

Tangent graph

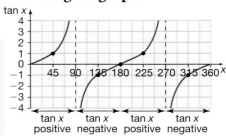

- Functions and their graphs can be translated or reflected.
- $y = f(x) + a$ is a translation of $\begin{pmatrix} 0 \\ a \end{pmatrix}$ on $y = f(x)$
- $y = f(x + a)$ is a translation of $\begin{pmatrix} -a \\ 0 \end{pmatrix}$ on $y = f(x)$
- $y = -f(x)$ is a reflection of $y = f(x)$ in the x-axis
- $y = f(-x)$ is a reflection of $y = f(x)$ in the y-axis

EXAMPLE

The graph shows the function $y = g(x)$.
What will be the position of the point $(2, 5)$ on the transformed graph $y = g(x + 3)$?
Sketch the transformed graph.

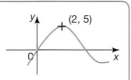

The y-coordinate remains unchanged as the function is only moved left.

The transformation $g(x + 3)$ translated the graph of $g(x) - 3$ units parallel to the x-axis, $\begin{pmatrix} -3 \\ 0 \end{pmatrix}$.

The point $(2, 5)$ will be translated to $(-1, 5)$.

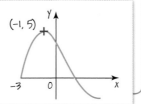

Algebra Graphs 2

Exercise 18.2S

1 Match each graph with its equation.

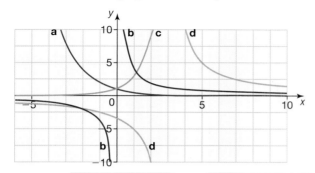

i $y = \dfrac{4}{x}$ **ii** $y = 3^x$

iii $y = \dfrac{10}{x-3}$ **iv** $y = 2^{-x}$

2 a Without using a calculator, copy and complete this table of values for the function $g(x) = 2^{x+1}$.

x	−4	−3	−2	−1	0	1	2	3	4
g(x)		$\dfrac{1}{4}$					8		

b Draw suitable axes and plot the graph of $g(x) = 2^{x+1}$.

c Approximate the value of x for which $g(x) = 25$.

3 Plot these functions for the range of x-values given.

a $f(x) = 4^{x-2}$, for $-2 \leqslant x \leqslant 6$

b $y = 3^{-x} - 1$, for $-4 \leqslant x \leqslant 4$

4 Use the graph of $y = \sin x$ to solve these equations for angles between $0°$ and $360°$.

Use a calculator to check the accuracy of your answers.

a $\sin x = 0.7$ **b** $\sin x = 0.8$

c $\sin x = -0.8$ **d** $\sin x = -0.5$

5 Use the graph of $y = \sin x$ to find the sine of these angles. Use a calculator to check the accuracy of your answers.

a $\sin 30°$ **b** $\sin 150°$

c $\sin 45°$ **d** $\sin 225°$

6 Use the graph of $y = \cos x$ to solve these equations for angles between $0°$ and $360°$.

6 Use a calculator to check the accuracy of your answers.

a $\cos x = 0.7$ **b** $\cos x = 0.8$

c $\cos x = -0.8$ **d** $\cos x = -0.5$

7 Use the graph of $y = \cos x$ to work out the cosine of each angle. Use a calculator to check the accuracy of your answers.

a $\cos 60°$ **b** $\cos 140°$

c $\cos 45°$ **d** $\cos 225°$

8 Use the graph of $y = \tan x$ to solve these equations for angles between $0°$ and $360°$.

Use a calculator to check the accuracy of your answers.

a $\tan x = 0.5$ **b** $\tan x = -4.2$

c $\tan x = 29$ **d** $\tan x = -0.1$

9 Sketch the graphs of these functions on the same axes. Label the y-intercept in each case.

$y = 3^x$ $y = 3^x + 2$ *f(x) is a way of writing 'a function of x'*

$y = 3^x - 5$ $y = 3^x - 1$

Copy and complete this statement.

$f(x) + a$ is a vertical ____ of $f(x)$ by ___ units.

10 Plot the graph of these functions on the same set of axes. Label the minimum point in each case.

$y = x^2$ $y = (x + 1)^2$

$y = (x + 4)^2$ $y = (x - 3)^2$

Copy and complete this statement.

$f(x + a)$ is a _____ _____ of $f(x)$ by ___ units.

11 Plot the graphs of

a $y = 7^x$ and $y = -7^x$

b $y = \sin x$ and $y = -\sin x$

c $y = x^2 + 2$ and $y = -(x^2 + 2)$

Copy and complete this statement.

$-f(x)$ is a _____ of $f(x)$ in the _____.

12 Plot the graphs of

a $y = 8^x$ and $y = 8^{-x}$

b $y = \tan x$ and $y = \tan(-x)$

c $y = x^2 + 1$ and $y = (-x)^2 + 1$

Copy and complete this statement.

$f(-x)$ is a _____ of $f(x)$ in the _____.

18.2 Exponential and trigonometric functions

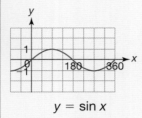

$y = \sin x$

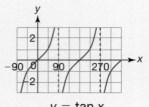

$y = \cos x$

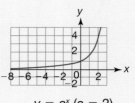

$y = \tan x$

$y = a^x \ (a = 2)$

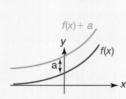

$f(x) + a$ is a vertical translation of $f(x)$ by a units.

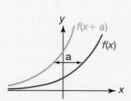

$f(x + a)$ is a horizontal translation of $f(x)$ by $-a$ units.

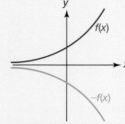

$-f(x)$ is a reflection of $f(x)$ in the x-axis.

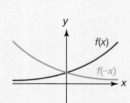

$f(-x)$ is a reflection of $f(x)$ in the y-axis.

HOW TO

To transform a graph

① Identify which transformation is needed.

② Sketch the graph in its new position.

③ Label any maximums, minimums or points of intersection if possible.

EXAMPLE

The graph shows the function $y = f(x)$.

a Sketch the graph of $y = f(x) + 2$.

b Label any intercepts with the axes if possible.

c The point P on the graph $y = f(x)$ is at $(-1, 4)$.
What point does P move to on the curve $y = f(x) + 2$?

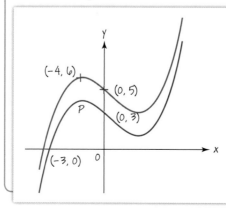

① $y = f(x) + 2$ is a vertical translation of 2 units up.

a ② The graph keeps its shape but moves up by 2 units.

b ③ The y-intercept, $(0, 3)$, moves to $(0, 3 + 2)$, that is, $(0, 5)$.

c $(-1, 6)$ $6 = 4 + 2$

Exercise 18.2A

1 Use the graph of $y = f(x)$ in the example.

   a Sketch the graph of $y = f(x + 3)$.

   b Label any intercepts with the axes if possible.

   c What is the position of point P when $y = f(x + 3)$?

2 The graph shows a function $y = f(x)$.
The point A is at $(3, 2)$.

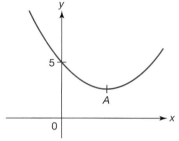

For each of the following functions, find the new position of point A.

   a $y = f(x + 1)$ **b** $y = f(x) - 2$

   c $-f(x)$ **d** $f(-x)$

3 The graph shows an exponential function $y = f(x)$. There is an asymptote at $y = 2$.

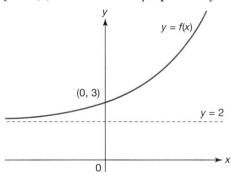

For each of the following functions, sketch the graph including its main features.

   a $y = f(x) + 5$ **b** $y = f(x - 2)$

   c $-f(x)$ **d** $f(-x)$

4 Are these statements true or false.
In each case give your reasons.

   a $\cos(x) = \cos(-x)$

   b $\sin(x) = \sin(x + 180°)$

   c $\tan(x) = \tan(x + 180°)$

   d $\sin(x + 90°) = \cos(x)$

   e $\cos(90° - x) = \sin(x)$

   f $\tan(x) = -\tan(-x)$

5 The graph shows the function $y = \sin(x - p)°$.

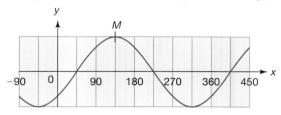

   a What are the coordinates of the point M?

   b Work out the value of p. | There are lots of possible answers to part **b**, keep it simple.

   c Josh says,
'The graph could be of another function $y = \cos(x - q)°$ instead.'

   Josh is correct.
Work out a possible value of q.

6 The graph shows $y = \tan x$.

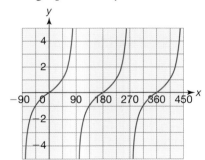

   a Write the value of $\tan 45°$

   b Use the graph to work out two more angles such that $\tan x = 1$.

   c Explain how you could continue finding angles with a tangent of 1.

   d Show how you could find different angles with a sine of $\frac{1}{2}$.

   e Repeat **d** for $y = \cos x$.

Use graphing software for question **7**.

7 **a** Investigate what happens with $y = a \sin x$, and $y = \sin(ax)$, when a is varied.

   b Research to find out where these kinds of waves are used in the real world.

Q 1070, 1955   SEARCH

18.3 Real-life graphs

● The shape of a graph shows the **trend**.

The graph shows the numbers of video recorders sold over a 10-year period.

The trend is that the number of video recorders sold is **decreasing**.

You can read information from a graph, but read the axis labels and scale carefully.

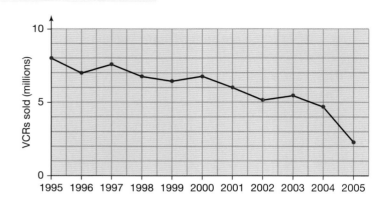

EXAMPLE

The graph shows the amount of rainwater in a barrel over a few days.

a On day 1 it rained heavily. What happened to the amount of water in the barrel?

b On which day was 25 litres poured out of the barrel?

c What happened to the amount of water on day 2? Suggest a reason for this.

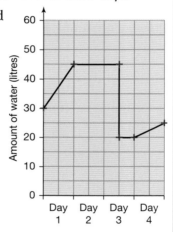

a The amount of water increased.

b Day 3, as the amount suddenly reduced by 25 litres.

c Amount of water stayed the same. It probably did not rain on day 2 and no water was poured out.

Think what could affect the amount of water.

● A straight line shows that a quantity is changing at a steady rate.

The steeper the slope, the faster the change.

quantity increasing   no change   quantity decreasing

The graphs show the water level as two tanks fill with water. Water is poured into both tanks at a steady rate.

The first tank fills more slowly, as it is wider. The second tank fills more quickly, as it is narrower.

The steeper the slope, the faster the change in water level.

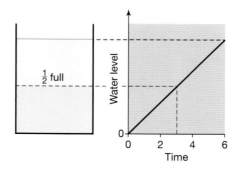

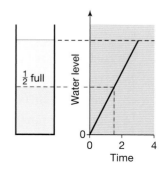

Exercise 18.3S

1 The graph shows sales of 'Time 2 Chat' mobile phones.

  a How many phones were sold in March?

  b How many phones were sold in June?

  c How many more phones were sold in June than in January?

  d Here are the sales figures for the next three months.

Month	October	November	December
Number of phones sold	400	325	400

  Copy the graph and complete it for this information.

  e What happened to sales in November and December?

  Suggest a reason for this.

  f What overall trend does the graph show?

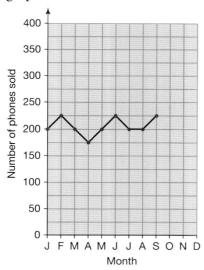

2 The graph shows how a bean plant grew from a seed over several weeks.

  a How tall was the plant after 6 weeks?

  b How much did the plant grow between weeks 8 and 10?

  c How tall did the plant grow in total?

  d How much did the plant grow between 15 and 20 weeks.

  e Is the plant likely to reach a height of 3 metres? Explain your answer.

2

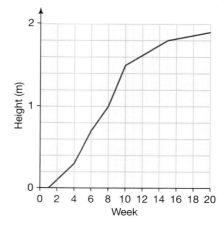

3 Match the four sketch graphs with the containers.

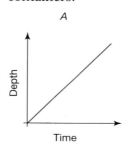

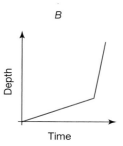

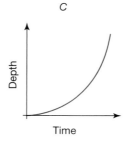

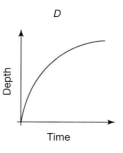

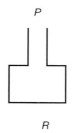

1184, 1322 SEARCH

18.3 Real-life graphs

RECAP

- There are many occasions when the situation can be modelled by a graph.
- Axes should be clearly labelled so that the reader can interpret the data.
- Scales should be chosen carefully so that the picture of the data is clear.
- The gradient of a line shows how fast a quantity is changing.
  - a straight line implies a constant rate of change.
  - a horizontal line means there is no change.

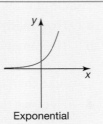

Exponential

EXAMPLE

Water is poured into this container at a constant rate.
Sketch the depth of water versus time.

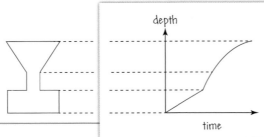

Container widens-rate of filling slows.

Container narrow-fills quickly.

Container wide-falls slowly.

HOW TO

Here is a general strategy for identifying the type of function that can be used to model some data.

① Plot the points on a graph.

② Compare the shape of the data with the common types of function. It may be that just a section of the curve fits.

③ Use your graph to answer the question.

EXAMPLE

Beth is investigating the features of an engine. The table shows the power generated by different amounts of torque.

Torque (N/m)	1	2	3	4	5	6	7
Power (W)	22	42	53	63	62	55	40

> Torque is a turning force. You can still complete the question without knowing that detail though.

a Plot the data on a graph.

b What type of function could be used to model the data?

c Estimate the power when a torque of 5.5 N/m is applied.

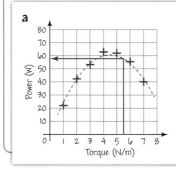

a

① Choose axes that allow you to plot all the data and give you a roughly square graph.

b ② $y = -x^2$ has the right type of shape- ⌒.
A quadratic function

c ③ Read up from 5.5 N/m and then across.
58 W.

Exercise 18.3A

1 Georgina is playing a computer game.

Describe the type of function that created the path of the bird.

2 The diagrams show four different types of beaker.

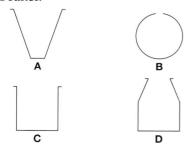

Liquid is poured into a beaker at a constant rate. A graph is plotted to show 'time' on the horizontal axis and 'depth of liquid' on the vertical axis.

a Match each beaker to the type of function that best describes the graph.

Cubic	Quadratic
No standard function	Linear

b Sketch the graph for each beaker.

3 A lottery offers a jackpot of £1000. Entrants choose any three numbers from a set of seven numbers. It is likely that the jackpot will have to be shared.

a Copy and complete the table to show the prizes available in different cases.

Winners	1	2	3	4	5	6
Prize (£)						

b Plot a graph to show this information.

c What type of function describes the relationship?

4 Steve is a scientist developing energy-efficient LCD displays. Power is required to update his display. When this happens, the 'pulse size' (in volts) is connected to the temperature. Steve has this set of data.

°C	−10	−5	0	10	23	40
Volts	20.5	17.7	16	13.6	12.5	12.1

a Plot the data on a graph.

b What type of function best describes the data?

c Steve has worked out a function that connects temperature and pulse size. He uses his formula to work out that if temperature = −20°C, then pulse size = 30 V. He also knows that the pulse size cannot be less than 11.8 V. Do these facts confirm your reasoning in **b**? Explain why.

> Use graph plotting software for question **5**.

*5 During the 1960s British mathematicians developed radar that could track the position of artillery fire. From this it could be worked out where the fire had been launched.

Artillery fire follows a parabolic path. The diagram shows a simplified example.

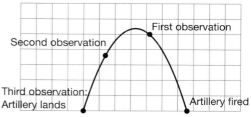

a Plot the points (6, 12), (2, 8) and (0, 0). Use these as three observations. The artillery lands at (0, 0).

b Plot the graph of $y = -a(x + b)^2 + c$.

c Vary the values of a, b and c. Find a quadratic curve that passes through all three points.

d Write the coordinates of the point where the artillery is fired from.

e Work with a partner. Choose three points, including (0, 0). Challenge them to find a quadratic curve that passes through those points.

Q 1184, 1322 SEARCH

18.4 Gradients and areas under graphs

The gradient of a straight line, $\dfrac{\text{change in } y}{\text{change in } x}$, is the **rate of change** of y with respect to x.

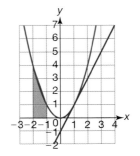

14.1

● If two points on a straight line are (x_1, y_1) and (x_2, y_2) then
$$\text{gradient} = \frac{y_2 - y_1}{x_2 - x_1}.$$

It is also useful to be able to find the area under a graph between two points.

EXAMPLE

Find the gradient of the line that passes through the points

a $(-2, 3)$ and $(7, 6)$ **b** $(3, 5)$ and $(6, -7)$

a $\text{Gradient} = \dfrac{y_2 - y_1}{x_2 - x_1} = \dfrac{6 - 3}{7 - -2} = \dfrac{3}{9} = \dfrac{1}{3}$ **b** $\text{Gradient} = \dfrac{y_2 - y_1}{x_2 - x_1} = \dfrac{-7 - 5}{6 - 3} = \dfrac{-12}{3} = -4$

This rate of change varies for a curve, such as $y = x^2$. Therefore the gradient varies too.

You can find the gradient at a given point on a curve.

● The gradient at a given point on a curve is the gradient of the **tangent** at that point.

EXAMPLE

Find the gradient of the curve $y = x^2$ when $x = 2$.

Draw an accurate graph.
Use your ruler to draw a line that just touches the curve at $(2, 4)$.
Find two points on the tangent line.

$(1, 0)$ and $(2, 4)$ are points on the tangent.

$\text{Gradient} = \dfrac{4 - 0}{2 - 1} = \dfrac{4}{1} = 4$

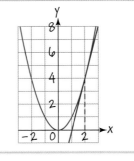

It is useful to be able to find the area under a curve between two points.
The area under a speed-time graph gives the distance travelled.

EXAMPLE

Estimate the area under the curve $y = x^2$ between the points when $x = -3$ and $x = 0$.

You can estimate this area by splitting it into two trapezia and a triangle.

$\text{Area of large trapezium} = \dfrac{9 + 4}{2} \times 1 = 6.5$

$\text{Area of small trapezium} = \dfrac{4 + 1}{2} \times 1 = 2.5$

$\text{Area of triangle} = \dfrac{1 \times 1}{2} = 0.5$

$\text{Estimate for whole area} = 6.5 + 2.5 + 0.5 = 9.5$

This is an over-estimate of the actual area as each of the three shapes is slightly too large.

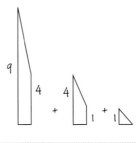

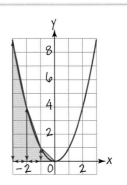

Algebra   Graphs 2

Exercise 18.4S

1 Find the gradient of the straight line through these points.

  a (1, 1) and (5, 9)

  b (2, 6) and (8, 2)

  c (−2, −1) and (1, 5)

  d (−2, 1) and (1, −2)

  e (1, −2) and (2, 4)

  f (0, −3) and (−2, 5)

  g (−4, 3) and (2, 1)

  h (3, −1) and (7, 1)

  i (−5, −2) and (3, 6)

  j (−1, 5) and (7, 2)

> You may use graph plotting software to help with questions **2** to **5**.

2 **a** Plot the graph of $y = x^2$ for values of x from −4 to 4. If using pencil and paper, aim to use a single side of A4 graph paper.

  b Find the gradient of the tangent to the curve $y = x^2$ at these points.

    i $x = 1$ **ii** $x = 3$ **iii** $x = -1$

    iv $x = -2$ **v** $x = 0$ **vi** $x = -3$

3 Describe your results from question **2** and the second example. What do you notice about the value of x and the gradient at that point?

4 **a** Plot the graph of the function $y = \frac{1}{2}x^2 + 3$ for $-4 \leqslant x \leqslant 4$.

  b Find the gradient of the tangent to the curve $y = \frac{1}{2}x^2 + 3$ at these points.

    i (2, 5)   **ii** (−3, 7.5)   **iii** $x = -2$

    iv $x = 0$ **v** $x = 1$ **vi** $x = 3$

  Use your results to predict the gradient at the point $x = 5$. Give your reasons.

5 The graph shows the function $y = x^3 + 2$.

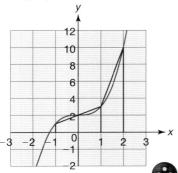

5 **a** Use the trapezia to estimate the area under the curve between the points where $x = -1$ and $x = 2$.

  b Is your answer an over-estimate or an under-estimate? Give your reasons.

*6 The graph shows the function $y = x^2 - x + 2$.

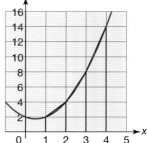

  a Use the trapezia to estimate the area under the curve between $x = 1$ and $x = 4$.

  b A more accurate estimate would be found if each trapezium had a 'width' of $\frac{1}{2}$. Work out the value of this estimate.

7 Estimate the distance travelled by an object in this velocity – time graph.

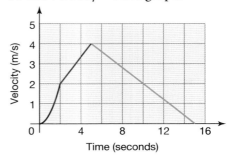

8 **a** Find the area under the line $y = 2x$ for values of x between

    i 0 and 1   **ii** 0 and 2

    iii 0 and 3   **iv** 0 and 4.

  b Plot the areas you found in part **a** as a function of the upper bound of the x range and join them by a smooth curve. Comment on what you notice.

9 Draw the graph of $y = 3x^2$ for $0 \leqslant x \leqslant 4$. Repeat question **8** finding the areas under the curve $y = 3x^2$.

Q 1128, 1312, 1944, 1953   SEARCH

18.4 Gradients and areas under graphs

- The gradient of a straight line is the **rate of change** of y with respect to x.
- If two points on a straight line are (x_1, y_1) and (x_2, y_2) then the gradient $= \dfrac{y_2 - y_1}{x_2 - x_1}$
- The gradient at a point on a curve is the gradient of the **tangent** at that point.
- The area under a curve can be estimated by splitting it into simple shapes.

HOW TO

To solve problems involving gradients and areas of curves

① Use a clear diagram that has large enough scales to show values accurately.

② Draw a tangent at the relevant point or split the area into simple shapes.

③ Choose suitable points to substitute into the formula or calculate the total area of the simple shapes.

④ Interpret the solution in the context of the question.

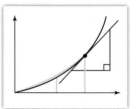

EXAMPLE

The speed–time graph shows information about a runner during the first 2 seconds of a race.

What was the runner's **acceleration** at 1.25 seconds?

Acceleration is the rate of change of speed with respect to time.

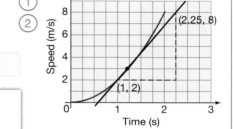

① ②

③ Use the points (2.25, 8) and (1, 2).

$$\text{Gradient of tangent} = \frac{8 - 2}{2.25 - 1} = \frac{6}{1.25} = 4.8$$

④ Acceleration = 4.8 m/s².

- In a velocity-time graph the area under the curve gives the distance travelled.

EXAMPLE

Estimate the distance travelled by the runner in the first two seconds.

①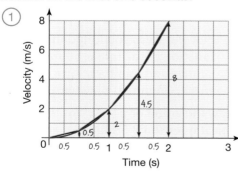

② Split the area at every 0.5 seconds.

③ Read the required lenghts off the graph.

$$\text{Area of large trapezium} = \frac{8 + 4.5}{2} \times 0.5 = 3.125$$

$$\text{Area of medium trapezium} = \frac{4.5 + 2}{2} \times 0.5 = 1.625$$

$$\text{Area of small trapezium} = \frac{2 + 0.5}{2} \times 0.5 = 0.625$$

$$\text{Area of triangle} = \frac{1}{2} \times 0.5 \times 0.5 = 0.125$$

$$\text{Estimate for whole area} = 3.125 + 1.625 + 0.625 + 0.125$$
$$= 5.5$$

④ The estimated distance travelled is 5.5 metres.

Exercise 18.4A

1 Here is a speed–time graph.

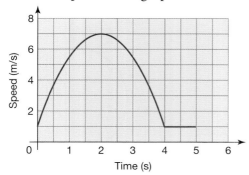

a Work out the acceleration at 1 second.

b Find the acceleration at 4.5 seconds.

c Work out the deceleration at 3.5 seconds.

d Estimate the distance travelled over the 5-second period.

2 John's grandmother put £1000 into a savings account when he was born. She adds money every year so that the total increases by 10%. The (red) graph shows this information.

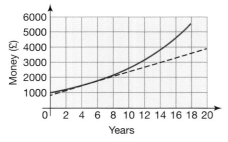

a How much money will be in the account when John is 18?

b The tangent at 5 years is shown. The tangent goes through the points (14, 3000) and (−3, 400). Find the gradient of the tangent.

c What is the rate of change of money in the account at 5 years? State the units of your answer.

d Work out the rate of change of money in the account at

   i 10 years   **ii** 15 years.

3 Rory borrows £150 from a short-term loan company. The money should be paid back in 18 days with interest of £28. This is equivalent to an interest rate of 365% per annum.

If Rory does not pay the money back, the interest grows very quickly as shown in the graph.

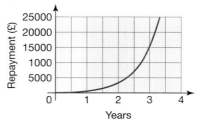

The function for the graph is
$y = 150 \times 4.65^x$ where $x = $ years and
$y = $ repayment.

Work out

a The repayment after 3 years.

b The rate of change of the repayment at 3 years.

4 Georgina is conducting an experiment. She measures the temperature of a substance once a minute for 10 minutes. The results are shown in the table.

Time	0	1	2	3	4	5	6	7	8	9	10
°C	100	50	33	26	20	17	14	13	11	10	9

Estimate the rate of change in the temperature at 3 minutes.

5 A radio-controlled model car travels a distance of 14 metres, and its speed-time graph looks like this.

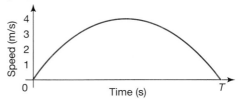

Estimate the time T that the car takes to come to rest again.

*6 Estimate the area between these curves.

a $x \geqslant 0$, $y = 8 - x^2$ and $y = 2x$

b $y = -x^2$, $y = x^3$ and $x = 2$

c $x^2 + y^2 = 8$ and $y \geqslant x^2 - 2$

   Q 1128, 1312, 1944, 1953   SEARCH

18.5 Equation of a circle

A circle is the infinite set of points that are a fixed distance from a fixed point. That distance is the radius.

11.4 A circle can be thought of as a locus.

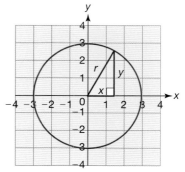

▲ By Pythagoras $x^2 + y^2 = r^2$. The radius is the hypotenuse of a right-angled triangle.

- The equation $\mathbf{x^2 + y^2 = r^2}$ describes a circle with radius r and centre at the origin (0, 0).

EXAMPLE

A circle has equation $x^2 + y^2 = 9$. Write

a the coordinates of the centre of the circle

b the radius of the circle.

    a Centre = (0, 0)   **b** Radius = $\sqrt{9}$ = 3

EXAMPLE

What is the equation of this circle?

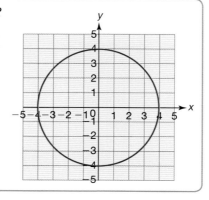

$x^2 + y^2 = r^2$ describes a circle with radius r and centre (0, 0).

The centre is (0, 0).

The radius is 4.

$$x^2 + y^2 = 4^2$$
$$\text{or } x^2 + y^2 = 16.$$

Understanding the equation of a circle enables you to solve more complex algebraic problems.

EXAMPLE

Solve the simultaneous equations $x^2 + y^2 = 25$ and $y = 2x - 2$

The graph shows that there are two points of intersection and therefore two solutions to find.

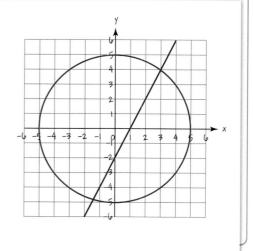

$$y = 2x - 2$$
$$x^2 + y^2 = 25$$
$$x^2 + (2x - 2)^2 = 25 \quad \text{Eliminate } y.$$
$$x^2 + 4x^2 - 8x + 4 = 25$$
$$5x^2 - 8x - 21 = 0$$
$$(5x + 7)(x - 3) = 0 \quad \text{Try to factorise if possible.}$$
$$x = 3 \text{ or } x = -\frac{7}{5}$$

Next find the corresponding values for y.

$$x = 3 \implies y = 2 \times 3 - 2 = 4$$
$$x = -\frac{7}{5} \implies y = 2 \times -\frac{7}{5} - 2 = -\frac{24}{5}$$

The solutions are $x = 3$, $y = 4$ and $x = -\frac{7}{5}$, $y = -\frac{24}{5}$

Exercise 18.5S

1 Write the equation of a circle with centre at the origin when the radius is

   a 5   **b** 6

   c 11   **d** 14

   e 2.5   **f** 4.5

2 For each of these circles write

   i the centre

   ii the radius.

   a $x^2 + y^2 = 1$ **b** $x^2 + y^2 = 81$

   c $x^2 + y^2 = 100$ **d** $x^2 + y^2 = 60$

3 Sketch the graph of each of these equations. Label any intersections with the axes.

   a $x^2 + y^2 = 49$ **b** $x^2 + y^2 = 64$

   c $x^2 + y^2 = 2$ **d** $x^2 + y^2 = 20$

   e $y^2 = 4 - x^2$ **f** $y^2 = 16 - x^2$

4 Work out the equation of each of these circles.

   a 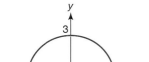   **b**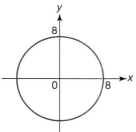

   c 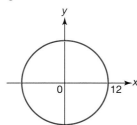   **d**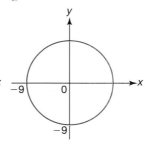

5 The equation of a circle is $x^2 + y^2 = 25$. Find the coordinates of the points where

   a $x = 0$ **b** $y = 4$

   c $y = -3$ **d** $x = 5$

   e $y = 2$ **f** $x = -1$

   g $x = \dfrac{1}{2}$ **h** $y = -\dfrac{2}{3}$

6 The diagram shows the circle $x^2 + y^2 = 25$. The radius of the circle at the point $(3, 4)$ has been drawn. The dotted line is the tangent to the circle at the point $(3, 4)$.

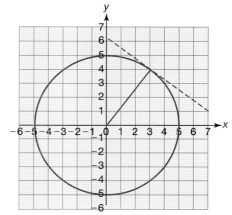

   a Find the gradient of the radius shown.

   b Write the equation of the line through $(0, 0)$ and $(3, 4)$.

   c Find the gradient of the tangent at the point $(3, 4)$.

   d Work out the y-intercept of the tangent.

   e Write the equation of the tangent to the circle at the point $(3, 4)$.

7 A circle has equation $x^2 + y^2 = 100$. Find the equation of the tangent to the circle at the point

   a $(6, 8)$ **b** $(8, 6)$

8 A circle has equation $x^2 + y^2 = 17$. Find the equation of the tangent to the circle at the point

   a $(-1, 4)$ **b** $(4, 1)$ **c** $(-1, -4)$

9 Find the exact solutions to these simultaneous equations.

   a $x^2 + y^2 = 25$ and $y = x + 1$

   b $x^2 + y^2 = 25$ and $y = 2x - 5$

   c $x^2 + y^2 = 100$ and $y = -\dfrac{3}{4}x$

   d $x^2 + y^2 = 169$ and $y = 3x - 3$

  *e $x^2 + y^2 = 36$ and $y = x - 2$

  *f $x^2 + y^2 = 4$ and $y = 2x + 1$

Q 1152   SEARCH

18.5 Equation of a circle

- The equation of a circle with centre (0, 0) and radius r is $x^2 + y^2 = r^2$.
- You may also have to use results from other topics.
  - A tangent to a circle is perpendicular to the radius at that point.
  - Two tangents from an external point to a circle are of equal length.
  - A radius that is perpendicular to a chord also bisects that chord.
  - The gradient of the line joining (x_1, y_1) and (x_2, y_2) is $\dfrac{y_2 - y_1}{x_2 - x_1}$.
  - Two perpendicular lines have gradients with a product of -1.

11.3

14.1

HOW TO

To solve problems involving circles
1. Draw a diagram and add known points.
2. Consider whether any circle theorems or Pythagoras' theorem are relevant.
3. Fully answer the question.

EXAMPLE

The point (5, 6) lies on a circle with centre (0, 0). Find the equation of the circle.

2. Use Pythagoras' theorem to find the radius.
   This is the same as substituting $x = 5$ and $y = 6$ into $x^2 + y^2 = r^2$ to find r^2.

   $5^2 + 6^2 = r^2$
   $25 + 36 = r^2$
   $r^2 = 61 \Rightarrow r = \sqrt{61}$

   You don't really need to find r since the equation of the circle is $x^2 + y^2 = r^2$

3. $x^2 + y^2 = 61$

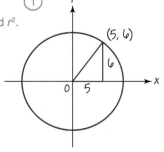

EXAMPLE

A circle has equation $x^2 + y^2 = 20$.
Find the equation of the tangents to the circle at the points where $x = 4$.

$x = 4 \Rightarrow 16 + y^2 = 20 \Rightarrow y = 2$ or $y = -2$.

2. A tangent is perpendicular to the radius at that point.

At (4, 2)
gradient of radius = $\dfrac{2}{4} = \dfrac{1}{2}$

$\Rightarrow$ gradient of tangent = $\dfrac{-1}{\frac{1}{2}} = -2$

Equation of tangent, $y = -2x + c$

Use the point on the tangent line to find the constant c.

$2 = -2 \times 4 + c \Rightarrow c = 10$

$y = -2x + 10$

At (4, −2)
gradient of radius = $-\dfrac{2}{4} = -\dfrac{1}{2}$

$\Rightarrow$ gradient of tangent = $\dfrac{-1}{\frac{1}{2}} = 2$

Equation of tangent, $y = 2x + c$

$-2 = 2 \times 4 + c \Rightarrow c = -10$

$y = 2x - 10$

3. There are two tangents.

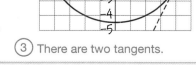

Algebra Graphs 2

Exercise 18.5A

1 Find the equation of a circle with centre $(0, 0)$ if the point $(2, 7)$ lies on the circle.

2 A circle has equation $x^2 + y^2 = 20$. Find the equation of the tangents to the circle at $x = -4$.

3 Each of these points lie on a different circle with centre $(0, 0)$. Find the equation of each circle.

   a $(1, 6)$ **b** $(4, 4)$

   c $(12, 23)$ **d** $(-3, 7)$

   e $(6, -2)$ **f** $(-4, 10)$

   g $(-5, -12)$ **h** $(-1, -3)$

4 Find the equations of the tangents to the given circle at the points with the given value of x or y.

   a $x^2 + y^2 = 40$ $x = 6$

   b $x^2 + y^2 = 34$ $y = 5$

   c $x^2 + y^2 = 13$ $y = -2$

   *d $x^2 + y^2 = 10$ $x = -2$

   *e $x^2 + y^2 = 12$ $y = 2$

5 Two tangents to a circle meet at the point $(13, 0)$. The angle between the two tangents is $46°$. Find the equation of the circle.

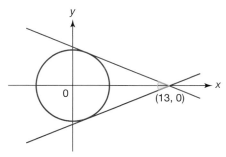

6 Two tangents to a circle meet at the point $(0, -12)$. The angle between the two tangents is $68°$. Find the equation of the circle if its centre is $(0, 0)$.

> Use graph plotting software for questions **7** to **9**.

7 **a** Plot the circle with equation $x^2 + y^2 = 29$.

   b Find the equation of a tangent that is perpendicular to the tangent at $(-2, 5)$.

8 The tangent to a circle with centre $(0, 0)$ passes through the points $(11, 2)$ and $(-1, 8)$. Find the equation of the circle.

*9 Plot the graph of this equation
$$(x - 2)^2 + (y - 4)^2 = 9$$
Write down a description of your graph in words.
Experiment with graphs of similar equations. Describe what do you notice?

10 Find the gradient of the green chord. Give your reasons.

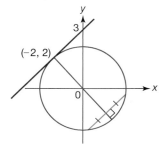

11 Explain why the simultaneous equations $x^2 + y^2 = 4$ and $y = \frac{1}{2}x + 5$ have no solution.

12 The simultaneous equations $x^2 + y^2 = 4$ and $y = ax + b$ have one solution. Suggest possible values for a and b.

13 A and B are two points on the x-axis, each a distance r from the origin. P is an arbitary point (x, y).

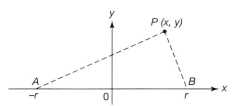

   a Write down expressions for the gradients of these lines.

      i AP **ii** BP

   b If the point P is chosen so that angle APB is a right angle, write down an equation that must be satisfied by x and y.

   c Simplify the equation and comment on your result.

Q 1152   SEARCH

Summary

Checkout

You should now be able to...

✓ Recognise and draw graphs of cubic and reciprocal functions.	1, 2
✓ Recognise and draw the graphs of exponential functions.	2, 7
✓ Recognise and sketch the graphs of trigonometric functions.	3–5
✓ Recognise and sketch translations and reflections of graphs.	5, 6
✓ Draw and interpret graphs of non-standard functions and use them in real-life problems.	7, 8
✓ Approximate the gradient of a curve at a given point and the area under a graph. Interpret these values in real-life problems including kinematic graphs.	7, 8
✓ Recognise and use simple equations of circles and find the tangent to a circle at a point.	9, 10

Language | Meaning | Example

Language	Meaning	Example
Quadratic function	A function of the form $ax^2 + bx + c$. They have a characteristic ∪- or ∩-shape.	
Cubic function	A function of the form $ax^3 + bx^2 + cx + d$. They have a characteristic S-shape.	
Reciprocal function	A function of the form $\frac{c}{x}$. They have two parts for negative and positive x.	
Exponential function	A function of the form a^x.	
Asymptote	A straight line which a curve gets as close as you like to but never touches.	$x = 0$ is a vertical asymptote to $y = \frac{2}{x}$. $y = 0$ is a horizontal asymptote to both $y = \frac{2}{x}$ and $y = 2^{-x}$.
Trigonometric function	The functions $y = \sin x$, $y = \cos x$ and $y = \tan x$.	
Periodic	A function is periodic if it can be represented by a graph which repeats itself identically at regular intervals.	$\sin x$ and $\cos x$ have period 360°, $\tan x$ has period 180°.
Tangent to a curve	The tangent to a curve at a point P on the curve is a straight line that touches the curve at P without crossing it at P.	

Review

1   Which of the graphs has equation

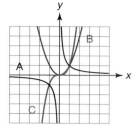

   a $y = x^3$

   b $y = \dfrac{1}{x}$?

2   Draw the graphs of these functions for the range of x-values given

   a $y = x^3 + 2x^2 - x - 2$ for $-3 \leqslant x \leqslant 3$

   b $y = 3^x$ for $-2 \leqslant x \leqslant 2$

3   Sketch the graphs of

   a $y = \cos x$ b $y = \tan x$

   for values of x between 0 and 360°.

4   Use the trigonometric graphs from 18.2 Skills to

   a   solve these equations for angles between 0° and 360°.

      i $\sin x = 0.5$ ii $\cos x = -0.5$

   b   evaluate

      i $\tan 135°$ ii $\cos 300°$.

5   Sketch graphs of these functions.

   a $y = \cos(x - 30)$ b $y = -\sin x$

   for values of x between $-180°$ and $180°$.

6   These graphs are all transformations of the graph with equation $y = x^2$.

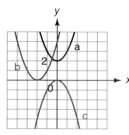

   i   Describe the transformation of the graph of $y = x^2$ in each case.

   ii   Write the equation of each curve.

7   The table shows the number of bacteria present at the end of each hour of an experiment.

End of hour	0	1	2	3	4
Bacteria Present	5	10	20	40	80

   a   Predict how many bacteria will be present after 6 hours?

   b   Plot a graph to show the values from the table.

   c   Describe the type of growth shown by the bacteria.

   d   Write down the equation of the curve.

   e   Find the gradient of the curve at the end of the second hour.

8   Use the speed–time graph to find

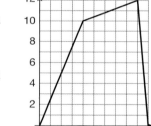

   a   the acceleration during the first 3 seconds

   b   the acceleration during the final second

   c   the total distance traveled.

9   What is the equation of a circle with centre (0, 0) and radius 9?

10   A circle has equation $x^2 + y^2 = 8$.

   a   Write down

      i   the coordinates of the centre

      ii   the radius of the circle.

   b   Work out the equation of the tangent to the circle at the points where $x = 2$.

What next?

Score			
	0 – 4		Your knowledge of this topic is still developing. To improve look at MyMaths: 1070, 1071, 1126, 1128, 1152, 1172, 1184, 1188, 1312, 1322
	5 – 8		You are gaining a secure knowledge of this topic. To improve look at InvisiPens: 18Sa – f
	9 – 10		You have mastered these skills. Well done you are ready to progress! To develop your exam technique looks at InvisiPens: 18Aa – c

Assessment 18

1 Match the graphs **i–iv** to their equations:

A $y = \dfrac{1}{x}$

B $y = 3^x$

C $y = \dfrac{3}{x}$

D $y = x(x - 2)(x + 3)$ [4]

2 Match the graphs **A–E** to one of the five statements **i–v**. [5]

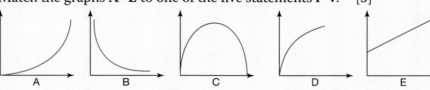

i The cost of gas is a set amount added to a price per unit.

ii The motion of a cricket ball after being thrown.

iii The time taken to drive a set distance compared to the average speed.

iv The height of water in a conical flask being filled at a steady rate.

v The height of water in a hemispherical flask being filled at a steady rate.

3 The product of x and y is 36.

    a Write down a formula for y in the form $y = \ldots..$ [1]

    b Draw a suitable graph of the equation in part *a* for x values between -12 and 12. [5]

    c Use your graph to find a pair of numbers that multiply to give 36 and add
       together to give 13. [2]

4   **a** Taking values of x from -3 to 3, draw the graph of $y = x^3 + x^2 - 12x$. [6]

    b On the same grid, draw the graph of $y = 4 - 8x$ [2]

    c Use your graph to solve the equation $4 - 8x = 0$ [1]

    d Use your graph to solve the equation $x^3 + x^2 - 12x = 0$ [3]

    e Find and record the coordinates of the points where the two graphs intersect. [3]

    f Complete the sentence the solutions to $x^3 + x^2 -$ _____ $x -$ _____ $= 0$ are the points
       of intersection of the graphs of $y = x^3 + x^2 - 12x$ and $y = 4 - 8x$ [2]

5   **a** Draw the graph of $y = x^2 - \dfrac{12}{x}$ for values of x from -5 to 5. [6]

    b Write down the values of y when   **i** $x = -2.5$ **ii** $x = 2.5$ **iii** $x = 4$ [3]

    c Use your graph to solve, to 1 dp, the following equations

       i $x^2 - \dfrac{12}{x} = 12$ **ii** $x^2 - \dfrac{12}{x} - 20 = 0$ **iii** $x^2 - \dfrac{12}{x} = -10$

       iv $x^2 - \dfrac{12}{x} + 15 = 0$ **v** $x^2 - \dfrac{12}{x} = 0$ [5]

    d Draw the graph of $y = 3 - 4x$ on the same axes. [4]

    e Find the values of x at the points of intersection of $y = 3 - 4x$ with $y = x^2 - \dfrac{12}{x}$. [3]

    f Explain why the points of intersection of $y = 3 - 4x$ and $y = x^2 - \dfrac{12}{x}$ are the solutions
       of the equation $x^3 + 4x^2 - 3x - 12 = 0$. [4]

6 The height of a point on the tooth of a cogwheel above its centre can be modelled by the equation $h = 3 \sin x$, where x is the angle turned through from the horizontal.

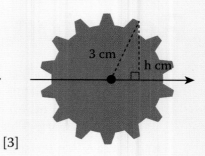

3 cm

h cm

a Complete the table to show how h varies as the angle changes. Give your values of h to 1 dp.

x	0	30	45	60	90	135	180	270	360
h									

[3]

b Draw a graph of h for values of x between 0 and 360. [4]

c State the range of values that h can take: ___ $\leq h \leq$ ___. [2]

d The cog is 2 cm above the centre. Estimate the possible values of x. [4]

7 A circle has equation $x^2 + y^2 = 169$.

a Find the area enclosed by the circle. Give your answer in terms of π. [3]

b Show that the point (5, 12) lies on the circle. [2]

c Find the gradient of the line that passes through the centre of the circle and the point (5, 12). [3]

d Show that the equation of the tangent to the circle at the point (5, 12) has equation $y = -\frac{5}{12}x + 14\frac{1}{12}$. [5]

8 a Sketch the graph of $y = \cos(x)$ for $0 \leq x \leq 360$. [3]

b How many solution does the equation $\cos(x) = \tan(x)$ have in the range $0 \leq x \leq 360$? Explain your answer fully. [2]

9 A new lake on a nature reserve is initially stocked with 250 fish. The owners predicted that the population, p, would rise by 12% per year.

a Calculate, using the form $250 \times a^n$, where a is a constant, and n is the number of years, the population of the lake after 1, 2, 3 and 4 years. [4]

b Explain why the population after n years is 250×1.12^n. [2]

c Draw the graph of $p = 250 \times 1.12^n$ for the first 10 years. [6]

d Find the number of years before the average population rise

    i doubles   **ii** triples. [2]

e Estimate the **rate** of change in the population after 5 years. [4]

f If the population increased at a steady rate after 5 years, what would the population be after 10 years? [3]

10 The curve $y = f(x)$ passes through the points A(−2, −4), B(0, 1), C(2, 0) and D(3, 1).

What coordinates do the points A, B, C and D move to after the following transformations?

a $y = f(x) + 4$ [4]   **b** $y = f(x - 1)$ [4]

c $y = f(-x)$ [4]   **d** $y = -f(x)$ [4]

11 The velocity of a car at the start of a journey can be modelled using the equation $v = 1.9^t - 1$, where v is the velocity in m/s and t is the time in seconds.

a Draw a velocity time graph for the first 3 seconds of the journey. [4]

b Show that the car travels approximately 6 m in the first 3 seconds of the journey. [4]

Revision 3

1 a Mo says that the highest common factor of the numbers 4410 and 3300 is 2310, and that their lowest common multiple is 485100. Correct his mistakes. [7]

b Mo says that the smallest positive integers that 3300 and 4410 must be multiplied by to make them square numbers are 33 and 10. Explain why he is correct. [3]

2 a A rectangle with sides $(4 + \sqrt{3})$ and $(p - \sqrt{3})$ has an area q, where p and q are integers. Find p and q. [6]

b The sides of an equilateral triangle are $(\sqrt{3} - a)$, $(6 + a\sqrt{3})$ and $(b\sqrt{3} - 4\frac{1}{2})$. Find the values of a and b. [10]

3 a Arrange these numbers from smallest to largest. [8]

　i 4^2 　　　　**ii** $\sqrt{2.25}$

　iii 1^7 　　　　**iv** $(-3)^3$

　v 3.2^3 　　　**vi** $\sqrt[3]{46.656}$

　vii $\sqrt[3]{(-592.704)}$ 　**viii** 5^0

b Which two adjacent numbers have the biggest difference? [1]

c Which two adjacent numbers have the smallest difference? [1]

4 Helen has two consecutive even numbers that when squared and added together equal 100. What are the numbers? [2]

5 David has the following equations. He says that $a = 3$, $c = 1$ and $d = 2$. Is David correct? For the values that are not correct, give all of the possible correct values.

a $(5^a)^2 = 625$ [2]

b $6^{b-1} \times 36 = 1296$ [2]

c $c^2 = 2^c$ [2]

6 Hari draws a chord joining the points (2, 1) and (6, 5) on a circle.

a Calculate the equation of the perpendicular bisector of this chord. [8]

b He then draws a chord joining the points (2, 9) and (6, 5) on the same circle. Repeat part **a** for this chord. [6]

c Solve your equations simultaneously to find the centre of the circle. [3]

d Sketch the circle. [3]

7 The graphs $y = (\frac{1}{2}x - 1)^2$ and $y = \frac{1}{2}x + 3$ intersect at points A and B.

a Use a graphical method to estimate the x-coordinates of A and B. [12]

b Use an algebraic method to find the exact values of the x-coordinates of A and B. [6]

8 Take the Earth as a sphere of radius 6370 km. It has a mean density of $5.5\,\text{g/cm}^3$.

a i Calculate the volume of the Earth. [2]

　ii Calculate Earth's mass. [5]

b During a storm water falls at a rate of 1460 litres per second. Assume the earth is spherical and none of the water is absorbed. How long would it take for the entire earth to be covered in water to a depth of 10 cm? [9]

9 The table shows the number of words per sentence in 50 sentences in a book.

Words per sentence	Number of sentences
$1 < W \le 5$	3
$5 < W \le 10$	15
$10 < W \le 15$	21
$15 < W \le 20$	9
$20 < W \le 25$	2

a Pablo calculates the mean number of words per sentence to be 11.73. Show how he worked out this value. [5]

b Draw a cumulative frequency diagram. Estimate the median and interquartile range. [18]

10 10 students preparing for a music exam recorded how many hours per day, on average, they practised. The data was compared with the mark they scored in their Music exam.

Student	A	B	C	D	E	F	G	H	I	J
Hours of practice	4	7	9	7	8	6	10	10	9	5
Test mark	92	65	76	45	74	40	95	30	45	30

a Draw a scatter graph and draw the line of best fit. [5]

b Which two students do not fit the trend? Give possible reasons why. [3]

c Another student missed the exam but had practised, on average, for 5.5 hours daily. What mark should he have achieved? [1]

11 UK unemployment rates for the period October 2012 to September 2014 are shown.

Month	Oct 2012	Nov 2012	Dec 2012	Jan 2013	Feb 2013	Mar 2013
Rate (%)	7.8	7.7	7.8	7.8	7.9	7.8
Month	Apr 2013	May 2013	Jun 2013	Jul 2013	Aug 2013	Sep 2013
Rate (%)	7.8	7.8	7.8	7.7	7.7	7.6

Month	Oct 2013	Nov 2013	Dec 2013	Jan 2014	Feb 2014	Mar 2014
Rate (%)	7.6	7.4	7.1	7.2	7.2	6.9
Month	Apr 2014	May 2014	Jun 2014	Jul 2014	Aug 2014	Sep 2014
Rate (%)	6.8	6.6	6.5	6.4	6.2	6.0

a Draw a time series diagram of these figures with Oct 13 – Sep 14 superimposed on top of Year Oct 12 – Sep 13 . [3]

b Compare unemployment rates in these periods. [1]

12 a Evaluate the following without using a calculator

i 6^3 [1]

ii 10^5 [1]

iii $\sqrt[3]{125}$ [1]

iv $\sqrt[3]{512}$ [1]

v $\sqrt[4]{0}$ [1]

vi $\sqrt[3]{-343}$ [1]

12 b Kia has attempted to write each of the following as a single power. In each case explain whether Kia is right and give the correct answer if she is wrong.

i $12^4 \times 12^5 = 144^{20}$ [1]

ii $(6^2)^3 = 6^5$ [1]

iii $210^3 \div 210^2 = 210^{1.5}$ [1]

iv $14^{12} \div 14^{13} = \frac{1}{14}$ [1]

v $4 \div 4^5 = 4^{-5}$ [1]

vi $\left(\dfrac{9^8 \times 9^8}{9^5 \times 9^7}\right)^2 = 9^{20}$ [3]

c Find the value of x when:

i $5^2 \times 125 = 5^x$ [2]

ii $16^5 = 4^x$ [2]

iii $4^8 = 16^x$ [2]

iv $(3^2)^3 = 9^x$ [2]

13 a A gym has 1.2×10^2 members, each of whom use, on average, 4.7×10^3 units of electricity per year. How many units does the gym use in a year? [3]

b The mass of a nitrogen atom is 2.326×10^{-23} g. One litre of air contains 4.3602×10^{-1} g of nitrogen. How many nitrogen atoms are there in one litre of air? [2]

14 The fuel consumption in a car engine is modelled by the function $C = \dfrac{240}{v} + \dfrac{v}{8} + 10$, where C is the consumption in litres per hour and v is the speed in mph.

a Taking values of v, in tens, from 10 to 100, draw up a table, calculate the values of C and draw the graph. [7]

b Find the consumption when

i $v = 17.5$ [1]

ii $v = 67.5$ mph. [1]

c Find the speed when

i $C = 30$ l/h [1]

ii $C = 23$ l/h. [2]

d At what speed is the car running most efficiently? Explain your answer. [2]

19 Pythagoras and trigonometry

Introduction

The highest mountain in the world is Mount Everest, located in the Himalayas. Its peak is now measured to be 8848 metres above sea level.

The mountain was first climbed in 1953, by Edmund Hilary and Sherpa Tenzing, almost 100 years after its height was first measured as part of the Great Trigonometrical Survey of India in 1856. The original surveyors, who included George Everest, obtained a height of 8840 metres by measuring the distance and angle of elevation between Mount Everest and a fixed location.

What's the point?

Once a right-angled triangle is seen in a particular problem then a mathematician only needs two pieces of information to be able to calculate all the other lengths and angles.

Objectives

By the end of this chapter, you will have learned how to ...

- Use Pythagoras' theorem to find a missing side in a right-angled triangle.
- Use trigonometric ratios to find missing lengths and angles in triangles.
- Find the exact values of $\sin\theta$, $\cos\theta$ and $\tan\theta$ for key angles.
- Use the sine and cosine rules to find missing lengths and angles.
- Express vectors in terms of simple base vectors.

Check in

1 Work out each of these.

    a 7^2 **b** $4^2 + 6^2$ **c** $3^2 + 5^2$ **d** $8^2 - 4^2$ **e** $7^2 - 2^2$ **f** $\sqrt{17^2 - 15^2}$

2 Rearrange these equations to make x the subject.

    a $y = \dfrac{x}{6}$ **b** $y = \dfrac{x}{5}$ **c** $y = \dfrac{x}{10}$ **d** $y = \dfrac{2}{x}$ **e** $y = \dfrac{5}{x}$ **f** $y = \dfrac{8}{x}$

Chapter investigation

An engineering company is building a ski lift.
The height of the lift is exactly 45 m.
An engineer suggests using a cable of length 200 m.

The maximum angle of incline is 12°.
Does the engineer's lift meet the criteria?
Design a lift that meets the criteria using the smallest possible length of cable.

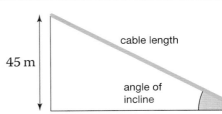

19.1 Pythagoras' theorem

In a right-angled triangle the **hypotenuse** is the longest side. It is opposite the right angle.

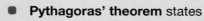

hypotenuse

- **Pythagoras' theorem** states

  For any right-angled triangle, $c^2 = a^2 + b^2$
  where c is the hypotenuse.

You can use Pythagoras' theorem to find any side given two other sides.

- To find a shorter side use $a^2 = c^2 - b^2$ or $b^2 = c^2 - a^2$

EXAMPLE

Work out the missing sides in these triangles.

a

c
7.6 cm
4.2 cm

b
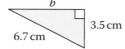
b
3.5 cm
6.7 cm

c
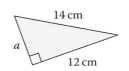
14 cm
a
12 cm

To find the hypotenuse: 'square, add and square root'.

a $\quad c^2 = a^2 + b^2$
$\quad\quad = 4.2^2 + 7.6^2$
$\quad\quad = 17.64 + 57.76$
$\quad\quad = 75.4$
$\quad c = \sqrt{75.4}$
$\quad\quad = 8.7\,\text{cm} \;(1\,\text{dp})$

b $\quad c^2 = a^2 + b^2$
$\quad\quad b^2 = c^2 - a^2$
$\quad\quad\quad = 6.7^2 - 3.5^2$
$\quad\quad\quad = 44.89 - 12.25$
$\quad\quad\quad = 32.64$
$\quad\quad b = \sqrt{32.64}$
$\quad\quad\quad = 5.7\,\text{cm} \;(1\,\text{dp})$

c $\quad c^2 = a^2 + b^2$
$\quad\quad a^2 = c^2 - b^2$
$\quad\quad\quad = 14^2 - 12^2$
$\quad\quad\quad = 196 - 144$
$\quad\quad\quad = 52$
$\quad\quad a = \sqrt{52}$
$\quad\quad\quad = 7.2\,\text{cm} \;(1\,\text{dp})$

To find a shorter side: 'square, subtract and square root'.

You can use **Pythagoras' theorem** to find the length of a line joining two points on a grid.

EXAMPLE

Find the midpoint and length of the line joining A and B.

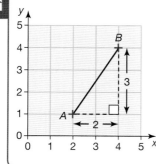

Midpoint

$$\left(\frac{2+4}{2}, \frac{1+4}{2}\right) = (3, 2.5)$$

Length

Draw a right-angled triangle and label the lengths of the two shorter sides.

$AB^2 = 2^2 + 3^2$

$AB^2 = 13$

$AB = 3.61 \;(2\,\text{dp})$

PITAGORA

Exercise 19.1S

1 Calculate the unknown area for these right-angled triangles.

a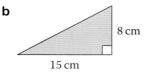

Area = ?
Area = 12 cm²
Area = 8 cm²

b

Area = ?
Area = 4 cm²
Area = 4 cm²

c

Area = 10 cm²
Area = ?
Area = 6 cm²

d

Area = 20 cm²
Area = 5 cm²
Area = ?

2 Find the length of the hypotenuse in each of these right-angled triangles.

> Give answers in this exercise to 1 dp where appropriate.

a
3 cm
4 cm

b
8 cm
15 cm

c
5 cm
12 cm

d
5 cm
9 cm

e
4 cm
10 cm

f
7 cm
7 cm

g
6.2 cm
10.9 cm

h
3.5 cm
6.4 cm

3 In some of the triangles in question **2**, all three sides have integer (whole number) values. Such sets of three numbers are called Pythagorean triples. List the Pythagorean triples from question **2**.

4 Find the length of the missing side in each of these right-angled triangles.

a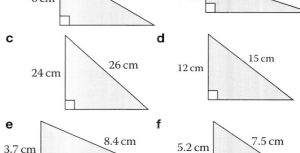
6 cm
10 cm

b
7 cm
21 cm

c
24 cm
26 cm

d
12 cm
15 cm

e
3.7 cm
8.4 cm

f
5.2 cm
7.5 cm

g
4.8 cm
7.3 cm

h
10 cm

5 Some of the triangles in question **4** are Pythagorean triples.

a Write down the Pythagorean triples in question **4**.

b Compare these with your answers to question **3**.

c Comment on anything you notice.

6 Use Pythagoras' theorem to find the length of the line segment joining each pair of points.

a $(2, 5)$ and $(6, 8)$

b $(7, 1)$ and $(2, 8)$

c $(0, 7)$ and $(1, -3)$

d $(-4, 5)$ and $(4, -2)$

7 Find the length of the diagonal in the rectangle.

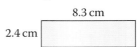
8.3 cm
2.4 cm

8 A rectangle has one side 4 cm and diagonal 10.4 cm. Find the length of the other side.

4 cm
10.4 cm

9 Find the length of the diagonal of a square with side length 8 cm.

19.1 Pythagoras' theorem

- **Pythagoras' theorem** states

  For any right-angled triangle, $c^2 = a^2 + b^2$
  where c is the hypotenuse (longest side).

- To find the shorter side use $a^2 = c^2 - b^2$ or $b^2 = c^2 - a^2$

HOW TO

To solve a problem using Pythagoras' theorem
1. Sketch a diagram. Label the right angle and the sides a, b and c.
2. Substitute the values into the formula.
3. Round your answer to a suitable degree of accuracy and include any units.

EXAMPLE

Find the length AD.

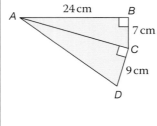

First find AC $AC^2 = 7^2 + 24^2$
$AC^2 = 625$
$AC = 25\,cm$

In triangle ACD $AD^2 = 9^2 + 25^2$
$AD^2 = 706$
$AD = 26.57\,cm$ (2 dp)

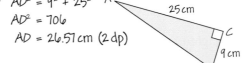

AC is the hypotenuse of the right-angled triangle ABC.

AD is the hypotenuse of the right-angled triangle ACD.

In 3D, a point has x-, y- and z-coordinates.

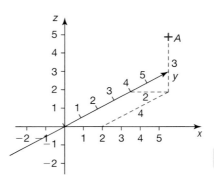

- You can use **Pythagoras' theorem** to find the distance d between two points in 3D.
  $d^2 = x^2 + y^2 + z^2$

A is the point $(2, 4, 3)$

EXAMPLE

Find the distance from $A(3, 4, 1)$ to $B(1, 3, 5)$.

$AB^2 = (3-1)^2 + (4-3)^2 + (1-5)^2$
$= 2^2 + 1^2 + (-4)^2 = 21$
$AB = \sqrt{21} = 4.6$ (1 dp)

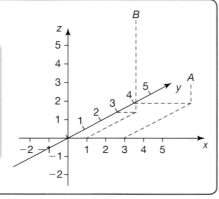

Exercise 19.1A

1 Find the length of the side of a square with diagonal length 8 cm.

2 An isosceles triangle has base 10 cm.

The other two sides of the triangle are each 13 cm.

   a Find the height of the triangle.

   b Find the area of the triangle.

3 The top of a 4-metre ladder leans against the top of a wall. The wall is 3.8 metres high.

How far from the wall is the bottom of the ladder?

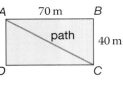

4 The diagram shows a path across a rectangular field.

How much further is it from A to C along the sides of the field than along the path?

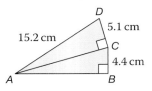

5 Find the distances between these points.

   a $(2, 4, 5)$ and $(3, 7, 10)$

   b $(1, 5, 3)$ and $(6, 2, 6)$

   c $(-1, 9, 2)$ and $(-3, 0, 5)$

   d $(4, -2, 3)$ and $(2, -4, 7)$

6 Work out the length of the diagonal d in this cuboid.

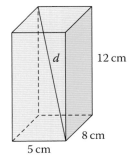

7 PQR and PRS are right-angled triangles. Find the length PS.

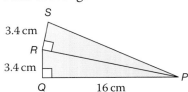

8 ABC and ACD are right-angled triangles. Find the length AB.

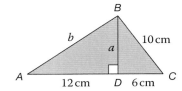

9 Find the missing lengths.

   a

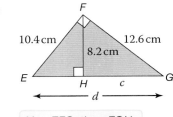

> Use BCD, then ABD.

   b

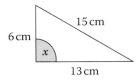

> Use EFG, then FGH.

10 Prove that a triangle with sides 20 cm, 21 cm and 29 cm is right-angled.

11 Jeremy draws this triangle.

   a Explain why angle x cannot be $90°$.

   b Is angle x acute or obtuse? Use a sketch to explain your answer.

12 A Pythagorean triple is three whole numbers (a, b, c) that satisfy $a^2 + b^2 = c^2$.

$(3, 4, 5)$ and $(6, 8, 10)$ are Pythagorean triples.

$(3, 4, 5)$ is a primitive Pythagorean triple.

$(6, 8, 10)$ is not a primitive Pythagorean triple, as it is a multiple of $(3, 4, 5)$.

There are seven primitive Pythagorean triples with $c < 50$. Can you find them all?

19.2 Trigonometry 1

In a right-angled triangle the longest side is the **hypotenuse**.

The side next to the labelled angle is called the **adjacent**.

The side opposite the labelled angle is called the **opposite**.

For a right-angled triangle with angle θ, the ratio of each pair of sides is constant.

> You can find tan, sin and cos values of any angle using your calculator.

- $\sin\theta = \dfrac{\text{opposite}}{\text{hypotenuse}}$ $\cos\theta = \dfrac{\text{adjacent}}{\text{hypotenuse}}$ $\tan\theta = \dfrac{\text{opposite}}{\text{adjacent}}$

EXAMPLE

Find the missing sides.

a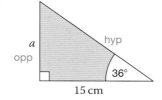

b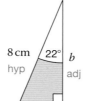

c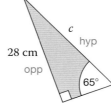

a $\tan 36° = \dfrac{a}{15}$

$15 \times \tan 36° = a$

$a = 10.9\,\text{cm}\ (1\,dp)$

b $\cos 22° = \dfrac{b}{8}$

$8 \times \cos 22° = b$

$b = 7.42\,\text{cm}\ (3\,sf)$

c $\sin 65° = \dfrac{28}{c}$

$c = \dfrac{28}{\sin 65°}$

$c = 30.9\,\text{cm}\ (3\,sf)$

> To find the hypotenuse, you will always need to divide by either sin or cos.

- You can use the inverse functions $\sin^{-1}$, $\cos^{-1}$ and $\tan^{-1}$ to find the angle if you know two sides.

EXAMPLE

Find the missing angles. Give your answers to the nearest degree.

a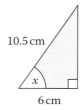

b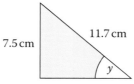

c

a You have adjacent and hypotenuse, so use cosine.

$\cos x = \dfrac{6}{10.5}$

$x = \cos^{-1}\left(\dfrac{6}{10.5}\right)$

$x = 55°$ (to the nearest degree)

b You have opposite and hypotenuse, so use sine.

$\sin y = \dfrac{7.5}{11.7}$

$y = \sin^{-1}\left(\dfrac{7.5}{11.7}\right)$

$y = 40°$ (to the nearest degree)

c You have opposite and adjacent, so use tan.

$\tan z = \dfrac{5.2}{11.1}$

$z = \tan^{-1}\left(\dfrac{5.2}{11.1}\right)$

$z = 25°$ (to the nearest degree)

Exercise 19.2S

1 Find the missing side in each triangle.
Give your answers to 3 significant figures.

a
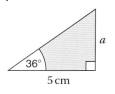

b
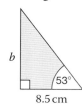

c
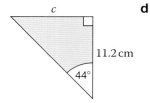

d
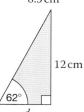

e
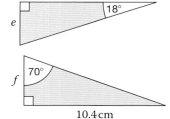

f

2 Find the missing side in each of these right-angled triangles.
Give your answer to 3 significant figures.

a
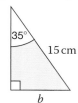

b

c
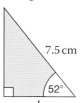

d
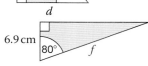

e
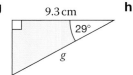

f
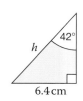

g

h

3 Find the missing angle in each triangle.
Give your answers to 3 significant figures.

a
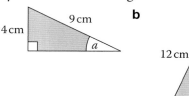

b

c

d
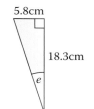

e

f
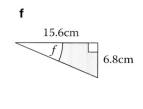

g
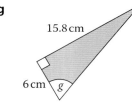

h
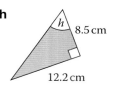

i
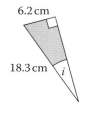

j
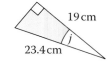

k
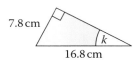

l
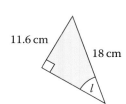

m

n
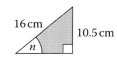

1131, 1133, 1943 SEARCH

Trigonometry 1

RECAP

- You can use **sine**, **cosine** and **tangent** ratios in a right-angled triangle:

$$\sin\theta = \frac{opp}{hyp} \qquad \cos\theta = \frac{adj}{hyp} \qquad \tan\theta = \frac{opp}{adj}$$

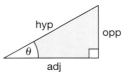

Angle	sin	cos	tan
30°	$\dfrac{1}{2}$	$\dfrac{\sqrt{3}}{2}$	$\dfrac{1}{\sqrt{3}}$
45°	$\dfrac{1}{\sqrt{2}}$	$\dfrac{1}{\sqrt{2}}$	1
60°	$\dfrac{\sqrt{3}}{2}$	$\dfrac{1}{2}$	$\sqrt{3}$

HOW TO

① Draw a copy of the triangle. Label the sides adjacent, hypotenuse and opposite.

② Decide which ratio to use.

③ Give your answer; include units or an explanation if the question asks for it.

EXAMPLE

ABCD is a parallelogram.

$AB = 8.2\,cm$, $BC = 6.6\,cm$ and angle $ABC = 53°$.
Work out the area of the parallelogram.

① Sketch a diagram.

Area of a parallelogram = $b \times x$

You need to find the perpendicular height, x.

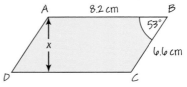

② $\sin 53° = \dfrac{x}{6.6}$

$x = 6.6 \times \sin 53°$

$x = 5.27...$

The height x must be at right angles to the base *DC*.

Finding x is an intermediate step. Do not round any values until the end of the calculation.

③ Area = $b \times x$ = $8.2 \times 5.27...$ = $43.2\,cm^2$ (3 sf)

EXAMPLE

Draw an equilateral triangle with sides 2 cm.
Use the triangle to find the exact values of cos 60° and sin 60°.
Show that $(\cos 60°)^2 + (\sin 60°)^2 = 1$

① The angles in an equilateral triangle are all 60°. Draw a vertical line to divide the triangle into two right-angled triangles.

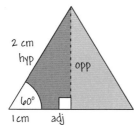

Label the hypotenuse and the opposite and adjacent sides.

② $\cos 60° = \dfrac{adj}{hyp} = \dfrac{1}{2}$

Use Pythagoras' theorem to find the length of the opposite side.

$(opp)^2 = 2^2 - 1^2 = 3 \qquad opp = \sqrt{3}$

$\sin 60° = \dfrac{opp}{hyp} = \dfrac{\sqrt{3}}{2}$

③ $(\cos 60°)^2 + (\sin 60°)^2 = \left(\dfrac{1}{2}\right)^2 + \left(\dfrac{\sqrt{3}}{2}\right)^2$

$= \dfrac{1}{4} + \dfrac{3}{4} = 1$

Exercise 19.2A

1 Find the missing angles.

a

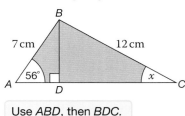

Use *ABD*, then *BDC*.

b

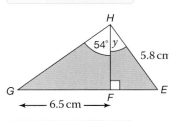

Use *GFH*, then *FEH*.

2 *JKL* and *JLM* are right-angled triangles.

a Find *JL*.

b Find angle *JML*.

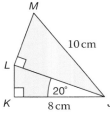

3 *PQR* and *PRS* are right-angled triangles.

a Find *PR*.

b Find *RQ*.

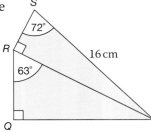

4 Find the area of a parallelogram with side lengths 5 cm and 11 cm and smaller angle 64°.

5 Find the area of a rhombus with side lengths 9 cm and smaller angle 52°.

6 Jenny walks 4 km on a bearing 052°. She changes direction and walks a further 5 km to finish due east of her starting point.

Find how far Jenny is from her starting point.

7 Liz leaves home and cycles 16 km on a bearing 215° to a lake. She changes direction and cycles 12 km to a wood which is due south of her home.

a On what bearing does she cycle from the lake to the wood?

b How far does she have to cycle home?

8 Edina and Patsy are estimating the height of the same tree. Edina stands 20 m from the tree and measures the angle of elevation of the treetop as 52°. Patsy stands 28 m from the tree and measures the angle of elevation of the treetop as 44°.

Can they both be correct? Explain your reasoning.

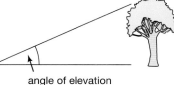

angle of elevation

9 a Find *PQ*.

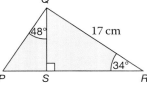

b Find *AB*.

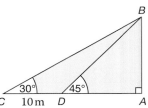

10 Draw an equilateral triangle with sides 2 cm. Use the triangle to find the exact values of cos 30° and sin 30°.

Show that $(\cos 60°)^2 + (\sin 60°)^2 = 1$

11 Draw a right-angled isosceles triangle with equal sides 1 cm.

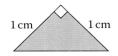

Use the triangle to find the exact values of cos 45° and sin 45°.

What can you say about $(\cos 45°)^2 + (\sin 45°)^2$?

12 a Use your calculator to work out cos 0° and sin 0°.

b Use the diagram to investigate what happens to lengths of the sides of the triangle as the angle θ gets close to 0°.

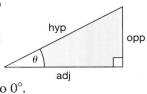

c Use your calculator to work out cos 90° and sin 90°.

d Use the triangle to investigate what happens to lengths of the sides of the triangle as the angle θ gets close to 90°.

Q 1131, 1133, 1943   SEARCH

19.3 Trigonometry 2

The **sine** and **cosine** rules let you find angles and sides in triangles that are not right-angled.

Label the angles as A, B and C.

Label the sides opposite each angle as a, b, and c.

Angle A is directly opposite side a.

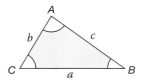

> To use the sine rule you need:
> - an angle and the side opposite to it
> - one other angle or side

● This is the **sine rule**: $\dfrac{a}{\sin A} = \dfrac{b}{\sin B} = \dfrac{c}{\sin C}$

EXAMPLE

a Find the length PQ.

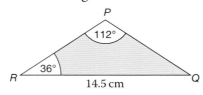

b Find the angle X.

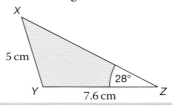

To find a side, use $\dfrac{a}{\sin A} = \dfrac{c}{\sin C}$

a $\dfrac{PQ}{\sin 36°} = \dfrac{14.5}{\sin 112°}$

$PQ = \sin 36° \times \dfrac{14.5}{\sin 112°}$

$PQ = 9.2\,cm$ (2 sf)

To find an angle, use $\dfrac{\sin A}{a} = \dfrac{\sin C}{c}$

b $\dfrac{\sin X}{7.6} = \dfrac{\sin 28°}{5}$

$\sin X = 7.6 \times \dfrac{\sin 28°}{5}$

$X = \sin^{-1}\left(7.6 \times \dfrac{\sin 28°}{5}\right) = 46°$ (nearest degree)

● This is the cosine rule: $a^2 = b^2 + c^2 - 2bc \cos A$

● It can be rearranged to give: $\cos A = \dfrac{b^2 + c^2 - a^2}{2bc}$

> You use the cosine rule when a problem involves:
> - all three sides and one angle
> - two sides and the angle between them

EXAMPLE

a Work out the length EF.

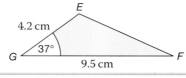

b Work out the angle P.

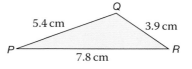

a Use the cosine rule:

$a^2 = b^2 + c^2 - 2bc \cos A$

$EF^2 = 9.5^2 + 4.2^2$
$\qquad - 2 \times 9.5 \times 4.2 \times \cos 37°$

$\qquad = 90.25 + 17.64 - 63.73...$

$EF = 6.6\,cm$ (1 dp)

b Use the rearranged cosine rule:

$\cos A = \dfrac{b^2 + c^2 - a^2}{2bc}$

$\cos P = \dfrac{7.8^2 + 5.4^2 - 3.9^2}{2 \times 7.8 \times 5.4}$

$P = \cos^{-1}\left(\dfrac{7.8^2 + 5.4^2 - 3.9^2}{2 \times 7.8 \times 5.4}\right)$

$\qquad = 27°$ (nearest degree)

Geometry Pythagoras and trigonometry

Exercise 19.3S

1 Find the sides marked *x*.

Use $\dfrac{a}{\sin A} = \dfrac{b}{\sin B} = \dfrac{c}{\sin C}$

a
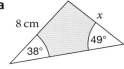

b
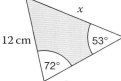

c
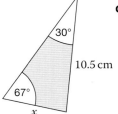

d
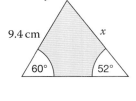

e
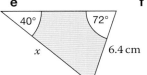

f
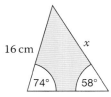

2 Find the angles marked θ

Use $\dfrac{\sin A}{a} = \dfrac{\sin B}{b} = \dfrac{\sin C}{c}$

a
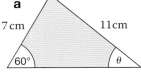

b

c
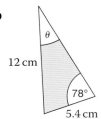

d

e
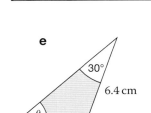

f
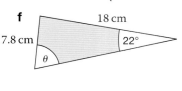

3 Find the sides marked *x*.

Use $a^2 = b^2 + c^2 - 2bc \cos A$
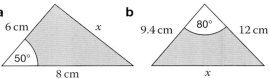

a
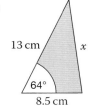

b
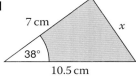

c
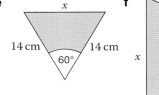

d
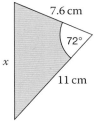

e
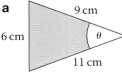

f
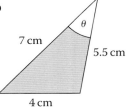

4 Find the angles marked θ.

Use $\cos A = \dfrac{b^2 + c^2 - a^2}{2bc}$

a
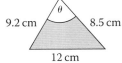

b

c

d
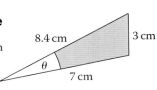

e

Q 1094, 1095, 1120, 1943 SEARCH

19.3 Trigonometry 2

- You use the sine rule and the cosine rule to solve problems in triangles that are not right-angled.
- You may need to use a combination of rules to solve a problem.

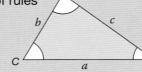

$$\frac{a}{\sin A} = \frac{b}{\sin B} = \frac{c}{\sin C}$$

$$a^2 = b^2 + c^2 - 2bc \cos A$$

▲ The ancient Egyptians used an early form of trigonometry for building pyramids.

① Draw a sketch of the problem. Label all the known sides and angles.

② Use the sine rule and/or the cosine rule to find missing angles and sides.

③ Give your answer; include units or an explanation if the question asks for it.

Peter walks 5 km from **S**, on a **bearing** of 063°.
At **C** he changes direction and walks a further 3.2 km on a bearing of 138°, to **F**.
Find the distance, **SF**, from where he began.

① Make a sketch.

$x = 180° - 63° = 117°$ Angles in parallel lines.

$\angle SCF = 360° - 117° - 138° = 105°$ Angles at a point.

You know two sides and the included angle.

② Use the cosine rule.

$SF^2 = 5^2 + 3.2^2 - 2 \times 5 \times 3.2 \cos 105°$

③ $SF = 6.6$ km (1 dp)

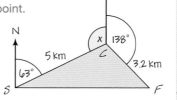

The diagram shows a bicycle frame.
PQ is parallel to **SR**.
Work out the length **QR**.

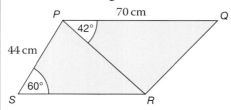

① Draw a sketch.

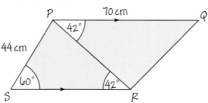

② PQ is parallel to RS so $\angle SRP = \angle QPR = 42°$.

Use the sine rule to find the length PR.

In triangle PRS you know an angle, the side opposite it, one other angle and one other side.

$$\frac{PR}{\sin 60°} = \frac{44}{\sin 42°} \qquad PR = \frac{44 \times \sin 60°}{\sin 42°} \qquad PR = 56.947...$$

Use the cosine rule to find QR.

In triangle PQR you know two sides and the included angle.

$QR^2 = 56.947...^2 + 70^2 - 2 \times 56.947... \times 70 \cos 42°$

③ $QR = 47$ cm (2 sf)

Exercise 19.3A

1 Use the sine rule to find angles *B* and *C*. Then use the cosine rule to find the side *x*.

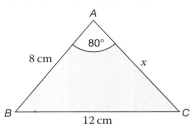

2 Debbie runs 8 km on a bearing of 310°.

She stops, changes direction and continues running for 10 km on a bearing of 055°.

Find the distance and bearing on which Debbie should run to return to her starting point.

3 Clare cycles 12 km on a bearing of 050°.

She stops, changes direction and continues cycling 10 km on a bearing of 120°.

Find the distance and bearing on which Clare should cycle to return to her starting point.

4 *AB* is parallel to *DC*. Work out the length *BC*.

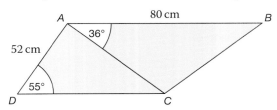

5 *JK* is parallel to *ML*. Work out the length *JM*.

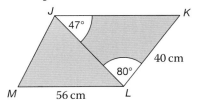

6 *PQ* is parallel to *SR*. Work out the length *PS*.

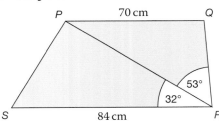

7 a Find *x*.

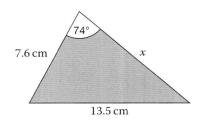

b Find *y*.

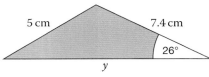

8 Two sides of a triangle are 15.4 cm and 12 cm. The angle between them is 72°. Work out the perimeter of the triangle.

9 Adjacent sides of a parallelogram are 6 cm and 8.3 cm. The shorter diagonal is 7 cm. Work out the length of the other diagonal.

10 Adjacent sides of a parallelogram are 8 cm and 11 cm. The longer diagonal is 15.2 cm. Work out the length of the other diagonal.

11 a Use triangles *ADC* and *BDC* to write two different expressions for *h*.

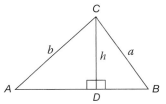

b Use your answer to part **a** to show that
$$\frac{a}{\sin A} = \frac{b}{\sin B}$$

c Show that the area of triangle *ABC* can be found from $\frac{1}{2} bc \sin A$ or $\frac{1}{2} ca \sin B$

*12 a Show that $a^2 = b^2 + c^2 - 2bc \cos A$

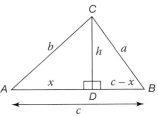

b Rearrange this result to show that
$$\cos A = \frac{b^2 + c^2 - a^2}{2bc}$$

c Show that these results also apply when angle *A* is obtuse.

Q 1094, 1095, 1120, 1943 SEARCH

19.4 Pythagoras and trigonometry problems

The same theorems and rules that you use to solve problems in two dimensions can be used to solve problems in three dimensions.

You can use these rules for right-angled triangles only.

- **Pythagoras' Theorem** $O^2 + A^2 = H^2$
- **Trigonometry** $\sin \theta = \dfrac{O}{H}$ $\cos \theta = \dfrac{A}{H}$ $\tan \theta = \dfrac{O}{A}$

You can use these rules for any triangle.

- **Sine rule** $\dfrac{a}{\sin A} = \dfrac{b}{\sin B} = \dfrac{c}{\sin C}$
- **Cosine rule** $a^2 = b^2 + c^2 - 2bc \cos A$ or $\cos A = \dfrac{b^2 + c^2 - a^2}{2bc}$
- **Area of triangle** $= \dfrac{1}{2} ab \sin C$

EXAMPLE

The base *BCDE* of this pyramid is a square with sides of length 8 cm.
The length of each slant edge is 10 cm.
F is the centre of the base and *M* is the mid-point of *CD*.
Find   **a**   the height of the pyramid
   b   ∠*ABF*, the angle between a slant edge and the base
   c   ∠*AMF*, the angle between a triangular face and the base.

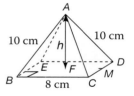

a   *h* is one of the short sides of the right-angle triangle *ABF* and *BF*
extended is the hypotenuse of right-angle triangle *BCD*.
In triangle *BCD*

By Pythagoras $BD^2 = 8^2 + 8^2 = 128$
$BD = \sqrt{128} = 8\sqrt{2}$ or $11.3137... \text{ cm}$
$BF = BD \div 2 = 4\sqrt{2}$ or $5.656... \text{ cm}$

In triangle *ABF*

By Pythagoras $h^2 = 10^2 - BF^2 = 100 - 32 = 68$
$h = \sqrt{68} = 2\sqrt{17}$ or $8.2462... \text{ cm}$
The height of the pyramid $= 2\sqrt{17}$ or 8.2 cm (2 sf)

b   In triangle *ABF*

$\cos \angle ABF = \dfrac{BF}{AB} = \dfrac{4\sqrt{2}}{10} = 0.5656...$
$\angle ABF = \cos^{-1} 0.5656... = 55.550...°$
The angle between a slant edge and the base is 56° (1°).

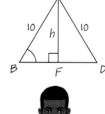

You could use the cosine rule in triangle *ABD* to find angle *ABF* but it is easier to use right-angled triangles.

c   In triangle *AMF*

$\tan \angle AMF = \dfrac{2\sqrt{17}}{4} = 2.0615...$
$\angle AMF = \tan^{-1} 2.0615... = 64.123...°$
The angle between a triangular face and the base is 64° (1°).

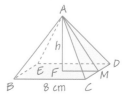

Exercise 19.4S

1 Calculate

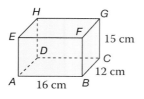

  a *AC*

  b *AG*

  c ∠*GAC*, the angle between *AG* and the base

  d ∠*FAB*, the angle between plane *AFGD* and the base.

2 **a** Find the height of this cone.

  b **i** Use triangle *SPR* to find angle *SPR*.

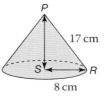

    ii Use the cosine rule in triangle *PQR* to check your answer to part **i**.

3 For this square-based pyramid, find

  a the height

  b ∠*PSU*, the angle between a slant edge and the base

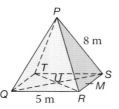

  c ∠*PMU*, the angle between a triangular face and the base.

4 This cube has sides of length 20 cm.

  a Find the following lengths, giving each answer as a surd.

    i *QV*   **ii** *PV*

  b Calculate

    i ∠*VQR* (between *VQ* and the base)

    ii ∠*VPR* (between *VP* and the base).

5 For this triangular prism, work out

  a the height, *h*

  b *AC*

  c *EC*

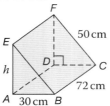

  d ∠*ECA* (between *EC* and the base)

  e ∠*EBA* (between *BCFE* and the base).

6 In this prism, two faces are isosceles trapezia and the other faces are rectangles.

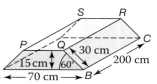

  a Use triangle *AQB* to calculate

    i *AQ*   **ii** ∠*QAB*

    iii the area of *AQB*.

  b Use different methods to check your answers.

  c Calculate   **i** *AR*   **ii** ∠*RAC*.

7 This solid is made from two square-based pyramids. Calculate

  a the lengths of the sloping edges

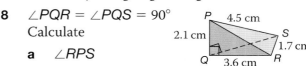

  b the angle between a small triangular face and an adjoining large triangular face.

8 ∠*PQR* = ∠*PQS* = 90°

  Calculate

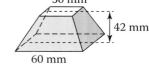

  a ∠*RPS*

  b the area of triangle *RPS*

  c ∠*RQS*

  d the area of triangle *RQS*.

*9 The top and bottom of this frustum are squares. The other faces are congruent isosceles trapezia.

  Find

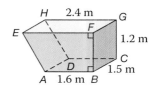

  a the length of a sloping edge

  b the angle between a sloping edge and the base

  c the angle between a trapezium and the base.

10 The diagram shows the dimensions of a waste skip.

  Calculate

  a *AE*

  b the angle the face *ADHE* makes with the horizontal.

Q 1095, 1112, 1120, 1144 SEARCH

Pythagoras and trigonometry problems

RECAP

For *right-angled* triangles only
- **Pythagoras' Theorem** $O^2 + A^2 = H^2$
- **Trigonometry** $\quad \sin\theta = \dfrac{O}{H} \quad \cos\theta = \dfrac{A}{H} \quad \tan\theta = \dfrac{O}{A}$

For *any* triangle
- **Sine rule** $\qquad \dfrac{a}{\sin A} = \dfrac{b}{\sin B} = \dfrac{c}{\sin C}$
- **Cosine rule** $\qquad a^2 = b^2 + c^2 - 2bc\cos A \quad$ or $\quad \cos A = \dfrac{b^2 + c^2 - a^2}{2bc}$
- **Area of triangle** $\quad = \dfrac{1}{2}ab\sin C$

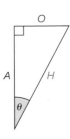

HOW TO

To solve 2D problems involving triangles

Bearings are measured clockwise from North

① Draw a sketch and include all the known values.

② Decide which trigonometric rule(s) to use and substitute values.

③ Work out the value(s) required. Check whether they look reasonable.

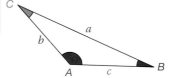

EXAMPLE

Two observers, Kelly and Liam, are on a straight coastline with Kelly 16 km due north of Liam. Kelly observes a ship on bearing 224°. Liam observes the same ship on bearing 302°. Kelly says the ship is more than 10 km from the coastline.
Is Kelly correct? Show how you decide and state any assumptions made.

① $\angle SKL = 224° - 180° = 44°, \quad \angle SLK = 360° - 302° = 58° \quad \angle KSL = 180° - 44° - 58° = 78°$

Assume Kelly and Liam are at sea level.

② Using the sine rule in triangle KSL.

$\dfrac{SL}{\sin 44°} = \dfrac{16}{\sin 78°} \qquad$ A side and opposite $\angle$ are known.

③ $SL = \dfrac{16 \times \sin 44°}{\sin 78°} = 11.362...\text{ km}$

② In right-angled triangle STL

$\sin 58° = \dfrac{d}{11.362...}$

③ $d = 11.362... \times \sin 58° = 9.636...\text{ km} \qquad < 10\text{ km}$

This is less than 10 km, so Kelly is not correct.

You can also use triangle KST to find d, or to check your answer.

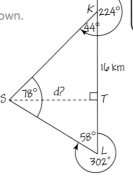

EXAMPLE

Given that θ is acute and $\cos\theta = \dfrac{7}{25}$, find as fractions $\quad$ **a** $\quad \sin\theta \qquad$ **b** $\quad \tan\theta$

① Sketch a right-angled triangle in which $\cos\theta = \dfrac{7}{25}$.

② Use Pythagoras to find the unknown side x, then sin and tan can be found.

③ $x^2 = 25^2 - 7^2 = 576$

$x = \sqrt{576} = 24$

a $\quad \sin\theta = \dfrac{O}{H} = \dfrac{24}{25} \qquad$ **b** $\quad \tan\theta = \dfrac{O}{A} = \dfrac{24}{7}$

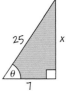

Geometry Pythagoras and trigonometry

Exercise 19.4A

State any assumptions you make.
Check that each answer seems reasonable.

1 A guidebook says this tower is over 50 m tall.
Is it correct?

(diagram showing angles 27° and 46°, base 50 m, height h?)

2 The diagram shows 2 sections *AB* and *BC* of a cable car route.
Find angle *ABC*

(diagram: 75 m, 58 m, 64 m, 42 m, points A, B, C)

3 From a port *P*, a ship sails 46 km on a bearing of 104° followed by 32 km on a bearing of 310°.

  a Find the distance and bearing of the ship from *P* after this journey.

  b The ship travels west until it is due north of *P*. The captain says they are now less than 10 km from *P*.
Is he correct?

4 In each part θ is an acute angle.

  a Given that $\tan \theta = \frac{3}{4}$, find as fractions

    i $\sin \theta$ **ii** $\cos \theta$.

  b Given that $\sin \theta = \frac{12}{13}$, find as fractions

    i $\cos \theta$ **ii** $\tan \theta$.

  c Given that $\cos \theta = 0.8$, find as decimals

    i $\sin \theta$ **ii** $\tan \theta$.

5 Prove that, for any acute angle θ

  $(\sin \theta)^2 + (\cos \theta)^2 = 1$

6 A regular *n*-sided polygon has its vertices on a circle, radius *r*.
Show that the area of the polygon is given by

$$A = \frac{1}{2}\,nr^2 \sin\left(\frac{360°}{n}\right)$$

7 For this building, calculate

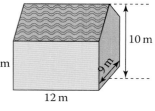

  a the angle that the roof makes with the horizontal

  b the cost of tiling the roof at £49 per m² plus 20% VAT.

8 The diagram shows a rectangular flowerbed. The radius of the large circle is 50 cm. The radius of each small circle is 18 cm. Find the area of the flowerbed in square metres.

9 A rugby cross-bar, *AB*, is 5.6 m wide and 3 m above the ground.

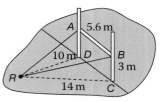

The rugby ball, *R*, is placed on the ground, 14 m from the bottom of one post, *C*, and 10 m from the bottom of the other post, *D*. Calculate angle *ARB*.

10 A hat is shaped as a cone with radius 8 cm and height 15 cm. Find, in terms of π, the area of card needed to make the hat.

*11 Find, in terms of π, the mass of this pedestal given that the density is 500 kg/m³.

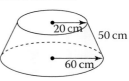

*12 The diagram shows the cross-section of a pipe with diameter 30 mm. The speed of the water in the the pipe is 1.5 metres per second. How long will it take to fill a 40 litre tank?

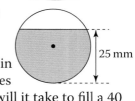

*13 The top of a table is made in the shape of a regular octagon by cutting congruent isosceles triangles from the corners of a 1 m square piece of wood. Show that the perimeter of the octagon is $8(\sqrt{2} - 1)$ metres.

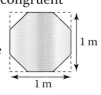

Q 1095, 1112, 1120, 1144   SEARCH

19.5 Vectors

You can find the length of the vector $\begin{pmatrix} 4 \\ 3 \end{pmatrix}$ using Pythagoras' theorem.

length $= \sqrt{3^2 + 4^2} = 5$

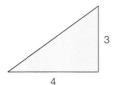

- A vector has a length and a direction.

You can draw a vector as an arrowed line. Its orientation gives the direction of movement; its length gives the distance.

These lines are all parallel and the same length.

They all represent the same vector **a**.

The arrow shows the direction is from A to B.

You can tie a vector to a starting point.

The line PQ represents the vector $\overrightarrow{PQ} = \mathbf{p}$.

Note that $\overrightarrow{QP} = -\mathbf{p}$

The vector $-\mathbf{a}$ is parallel, the same length and in the opposite direction to **a**.

- You can multiply a vector by a number.

The vector $2\mathbf{p}$ is parallel to the vector **p** and twice the length.

The vector $3\mathbf{p}$ is parallel to the vector **p** and three times the length.

The vector $-\mathbf{p}$ is parallel to the vector **p** and the same length, but in the opposite direction.

The pairs of letters in the sequence fit together.

- Vectors represented by parallel lines are multiples of each other.

- Vectors can be described using the notation $\overrightarrow{AB}$ or bold type, **a**.
- In handwriting, vectors can be shown with an underline, <u>a</u>.

You can add and subtract vectors by putting them 'nose to tail'.

The result of the addition or subtraction is called the **resultant** vector.

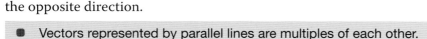

$$\overrightarrow{AB} + \overrightarrow{BC} + \overrightarrow{CD} + \overrightarrow{DE} + \overrightarrow{EF} = \overrightarrow{AF}$$

$\mathbf{m} = \begin{pmatrix} 3 \\ 2 \end{pmatrix}$ $\mathbf{n} = \begin{pmatrix} 2 \\ -1 \end{pmatrix}$

Calculate $\mathbf{m} + \mathbf{n}$ and $\mathbf{m} - \mathbf{n}$.

a $\begin{pmatrix} 3 \\ 2 \end{pmatrix} + \begin{pmatrix} 2 \\ -1 \end{pmatrix} = \begin{pmatrix} 3 + 2 \\ 2 + -1 \end{pmatrix} = \begin{pmatrix} 5 \\ 1 \end{pmatrix}$ 5 right, 1 up

$\begin{pmatrix} 3 \\ 2 \end{pmatrix} - \begin{pmatrix} 2 \\ -1 \end{pmatrix} = \begin{pmatrix} 3 - 2 \\ 2 - -1 \end{pmatrix} = \begin{pmatrix} 1 \\ 3 \end{pmatrix}$ 1 right, 3 up

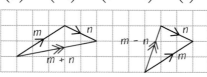

Geometry   Pythagoras and trigonometry

Exercise 19.5S

1 a On squared paper draw the vectors

$\begin{pmatrix} 1 \\ -2 \end{pmatrix}$ and $\begin{pmatrix} -1 \\ 2 \end{pmatrix}$.

b Write what you notice about these two vectors.

2 *ABCDEF* is a regular hexagon.
X is the centre of the hexagon.
$\overrightarrow{XA} = \mathbf{a}$ and $\overrightarrow{AB} = \mathbf{b}$.

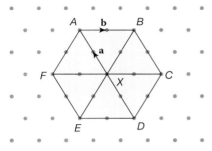

Write all the vectors that are equal to

a **a**   **b** **b**   **c** −**a**   **d** −**b**

3 *JKLMNOPQ* is a regular octagon.

$\overrightarrow{OJ} = \mathbf{j}$ $\overrightarrow{OM} = \mathbf{m}$ $\overrightarrow{OP} = \mathbf{p}$

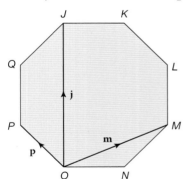

Write all the vectors that are equal to

a **j**   **b** **m**

c **p**   **d** −**j**

e −**m**   **f** −**p**

4 The diagram shows vectors
s and **t**. On square grid
paper draw the vectors that
represent

a **s** + **s**   **b** **s** + **t**

c **t** + **s**   **d** **s** − **t**

e **t** − **s**   **f** **t** + **t** − **s**

5

OPQR is a rectangle.
$\overrightarrow{OP} = \mathbf{p}$ and $\overrightarrow{OR} = \mathbf{r}$.

Work out the vector, in terms of **p** and **r**, that represents

a $\overrightarrow{PQ}$ **b** $\overrightarrow{OQ}$ **c** $\overrightarrow{QO}$ **d** $\overrightarrow{RP}$

6 $\overrightarrow{OP} = \mathbf{p}$ $\overrightarrow{OQ} = \mathbf{q}$

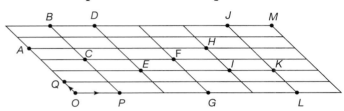

Write, and simplify, the vectors

a $\overrightarrow{OG}$ **b** $\overrightarrow{OL}$ **c** $\overrightarrow{OK}$

d $\overrightarrow{OJ}$ **e** $\overrightarrow{OA}$ **f** $\overrightarrow{OC}$

g $\overrightarrow{OB}$ **h** $\overrightarrow{OF}$ **i** $\overrightarrow{PE}$

j $\overrightarrow{PD}$ **k** $\overrightarrow{EF}$ **l** $\overrightarrow{CA}$

m $\overrightarrow{JK}$ **n** $\overrightarrow{JI}$ **o** $\overrightarrow{JE}$

p $\overrightarrow{FD}$ **q** $\overrightarrow{DF}$ **r** $\overrightarrow{DC}$

s $\overrightarrow{ME}$ **t** $\overrightarrow{HG}$ **u** $\overrightarrow{KD}$

7 The diagram shows vectors **x** and **y**.
On square grid paper draw the vectors

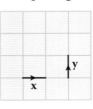

a 2**x**   **b** 3**y**

c 2**x** + 3**y**   **d** 3**x** − **y**

e **y** − 2**x**   **f** $1\frac{1}{2}\mathbf{x} + 1\frac{1}{2}\mathbf{y}$

g 2(**x** + **y**)   **h** $\frac{1}{2}(2\mathbf{x} - 3\mathbf{y})$

i 3**x** + 4**y**

Q 1134, 1135   SEARCH

19.5 Vectors

- A vector has a fixed length and direction.
- You describe a translation with a vector.
- You can add and subtract vectors by putting them 'nose to tail'.
- Vectors represented by parallel lines are multiples of each other.
- An equal vector in the opposite direction is negative.
- Vectors can be described using the notation $\overrightarrow{AB}$ or bold type, **a**.
- In handwriting, vectors can be shown with an underline, <u>a</u>.

The arrow shows the direction is from A to B.

HOW TO

① Draw a diagram and mark the known vectors. Use the diagram to find the vectors you need.
② Use your knowledge of vector addition and subtraction. Look out for parallel vectors.
③ Make sure you use the correct notation for vectors. Always end any proofs with a clear statement.

EXAMPLE

OAB is a triangle.

M is the midpoint of OA.

N is the midpoint of OB.

$\overrightarrow{OA} = \mathbf{a}$ and $\overrightarrow{OB} = \mathbf{b}$.

Show that MN is parallel to AB.

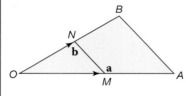

① find the vectors $\overrightarrow{AB}$ and $\overrightarrow{MN}$

$$\overrightarrow{AB} = \overrightarrow{AO} + \overrightarrow{OB}$$
$$= -\mathbf{a} + \mathbf{b}$$
$$\overrightarrow{MN} = \overrightarrow{MO} + \overrightarrow{ON}$$
$$= -\tfrac{1}{2}\mathbf{a} + \tfrac{1}{2}\mathbf{b}$$
$$= \tfrac{1}{2}(-\mathbf{a} + \mathbf{b})$$

To prove that lines are **parallel**, you show that the vectors they represent are multiples of each other.

② $\overrightarrow{AB}$ is a multiple of $\overrightarrow{MN}$

$$(-\mathbf{a} + \mathbf{b}) = 2 \times \tfrac{1}{2}(-\mathbf{a} + \mathbf{b})$$

③ So MN is parallel to AB.

EXAMPLE

$PQRS$ is an isosceles trapezium.

PQ and SR are parallel sides with $PQ = 2 \times SR$.

$\overrightarrow{PQ} = 2\mathbf{p}$ $\overrightarrow{QR} = \mathbf{q}$

X lies on QR such that $QX:XR = 2:1$.

Y lies on PQ extended such that $PQ:QY = 1:1$.

Prove that S, X and Y are collinear.

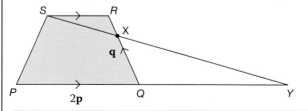

① $PQ = 2SR$, so $\overrightarrow{SR} = \mathbf{p}$

$$\overrightarrow{RX} = \tfrac{1}{3}\overrightarrow{RQ} = -\tfrac{1}{3}\mathbf{q}$$

② First show that SX and XY are parallel.

$$\overrightarrow{SX} = \overrightarrow{SR} + \overrightarrow{RX} = \mathbf{p} - \tfrac{1}{3}\mathbf{q}$$
$$\overrightarrow{XY} = \overrightarrow{XQ} + \overrightarrow{QY}$$
$$= -\tfrac{2}{3}\overrightarrow{RQ} + \overrightarrow{QY} = -\tfrac{2}{3}\mathbf{q} + 2\mathbf{p}$$
$$= 2(-\tfrac{1}{3}\mathbf{q} + \mathbf{p}) = 2(\mathbf{p} - \tfrac{1}{3}\mathbf{q})$$
$$= 2\,\overrightarrow{SX}$$

③ So $\overrightarrow{SX}$ and $\overrightarrow{XY}$ are parallel.

SX and XY have the point X in common.

So S, X and Y are collinear.

To prove that points are collinear, you show that the vectors joining pairs of the points are parallel and that the vectors share a common point.

Exercise 19.5A

1 *WXYZ* is a square.
The diagonals *WY* and
XZ intersect at *M*.
$\overrightarrow{WX} = \mathbf{w}$ and $\overrightarrow{WZ} = \mathbf{z}$.
Write these vectors in
terms of **w** and **z**.

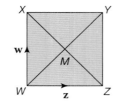

   a $\overrightarrow{XY}$ **b** $\overrightarrow{WY}$ **c** $\overrightarrow{WM}$

   d $\overrightarrow{MY}$ **e** $\overrightarrow{MX}$ **f** $\overrightarrow{XM}$

2 *ORST* is a rhombus.
$\overrightarrow{OR} = \mathbf{r}$ and $\overrightarrow{OT} = \mathbf{t}$.
P lies on *OS* such that
OP:*PS* = 3:1.
Work out the vector,
in terms of **r** and **t**,
that represents

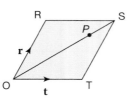

   a $\overrightarrow{RS}$ **b** $\overrightarrow{OS}$ **c** $\overrightarrow{OP}$

   d $\overrightarrow{SP}$ **e** $\overrightarrow{TP}$

3 *OAB* is a triangle.
X is the point on
AB for which
AX:*XB* = 1:4.
$\overrightarrow{OA} = \mathbf{a}$ and $\overrightarrow{OB} = \mathbf{b}$.

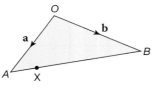

   a Write, in terms of **a** and **b**, an
      expression for $\overrightarrow{AB}$.

   b Express $\overrightarrow{OX}$ in terms of **a** and **b**.
      Give your answer in its simplest form.

4 In the diagram, $\overrightarrow{OA} = 2\mathbf{a}$,
$\overrightarrow{OB} = 2\mathbf{b}$, $\overrightarrow{OC} = 3\mathbf{a}$ and
$\overrightarrow{BD} = \mathbf{b}$. Prove that AB
is parallel to *CD*.

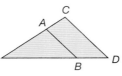

5 *RSTU* is a rectangle.
M is the midpoint of
the side *RS*.
N is the midpoint of
the side *ST*.
$\overrightarrow{RS} = 2\mathbf{r}$ $\overrightarrow{ST} = 2\mathbf{s}$

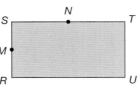

   a Work out the vector, in terms of **r** and **s**,
      that represents

      i $\overrightarrow{RT}$ **ii** $\overrightarrow{RM}$ **iii** $\overrightarrow{SN}$ **iv** $\overrightarrow{MN}$.

   b Show that *MN* is parallel to *RT*.

6 *OPQR* is a trapezium.
OP is parallel to *RQ* and
$OP = \frac{1}{3}RQ$.
$\overrightarrow{OP} = \mathbf{p}$ and
$\overrightarrow{OR} = \mathbf{r}$.

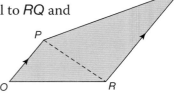

   a Work out the vector, in terms of **p** and
      r, that represents

      i $\overrightarrow{RQ}$ **ii** $\overrightarrow{OQ}$ **iii** $\overrightarrow{RP}$ **iv** $\overrightarrow{PQ}$

   b The point *X* lies on *PR* such that
      PX:*XR* = 1:3.
      Show that *O*, *X* and *Q* lie on the same
      straight line.

7 *OJKL* is a parallelogram.
$\overrightarrow{OJ} = \mathbf{j}$ $\overrightarrow{JK} = \mathbf{k}$

   a Express in terms of **j** and **k** these
      vectors.

      i $\overrightarrow{OK}$ **ii** $\overrightarrow{OL}$ **iii** $\overrightarrow{JL}$

   b *M* is the point on *KL* extended such that
      KL:*LM* = 1:2.
      X is a point on *OL* such that $OX = \frac{1}{3}OL$.

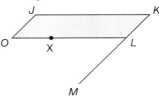

      Show that *J*, *X* and *M* lie on the same
      straight line.

*8 *ABCD* is a quadrilateral.
The mid-points of its
sides are *P*, *Q*, *R* and *S*.
Use vectors to prove that
PQRS is a parallelogram.

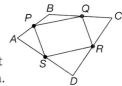

*9 Use vector methods to prove that the
diagonals of a parallelogram bisect
each other.

Summary

Checkout

You should now be able to...

	Test it Questions
✔ Use Pythagoras' theorem to find a missing side in a right-angled triangle or the length of a line segment on a coordinate grid.	**1, 3, 4**
✔ Use trigonometric ratios to find missing lengths and angles in triangles.	**2 – 5**
✔ Find the exact values of $\sin\theta$ and $\cos\theta$ for key angles.	**6**
✔ Use the sine and cosine rules to find missing lengths and angles.	**7**
✔ Use the sine formula for the area of a triangle.	**8**
✔ Calculate with vectors and use them in geometric proofs.	**9, 10**

Language	Meaning	Example
Hypotenuse	The side opposite the right angle in a right-angled triangle.	hypotenuse
Pythagoras' theorem	For a right-angled triangle, the area of the square drawn along the hypotenuse is equal to the sum of the areas of the squares drawn along the other two sides.	$c^2 = a^2 + b^2$
Adjacent	A side next to the labelled angle in a right-angled triangle.	
Opposite	The side opposite the labelled angle in a right-angled triangle.	
Sine ratio	The ratio of the length of the opposite side to the hypotenuse in a right-angled triangle.	$\sin\theta = \frac{3}{5} = 0.6$
Cosine ratio	The ratio of the length of the adjacent side to the hypotenuse in a right-angled triangle.	$\cos\theta = \frac{4}{5} = 0.8$
Tangent ratio	The ratio of the length of the opposite side to the adjacent side in a right-angled triangle.	$\tan\theta = \frac{3}{4} = 0.75$
Scalar	A quantity with just size.	Mass, temperature, speed
Vector	A quantity with both size and direction.	
Resultant	The vector that is equivalent to adding or subtracting two or more vectors.	
Multiple	The original vector multiplied by a scalar.	$3\mathbf{a} = \mathbf{a} + \mathbf{a} + \mathbf{a}$
Collinear	Points are collinear if they lie on the same straight line.	

Review

1 Calculate the lengths a and b in these right—angled triangles. Give your answers to 1 decimal place.

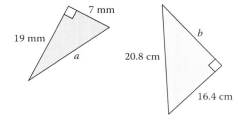

2 Calculate the lengths a, b and c to 3 s.f.

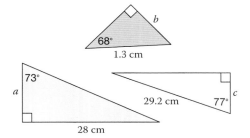

3 Calculate the size of angles A and B.

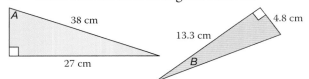

4 A cuboid is shown below.

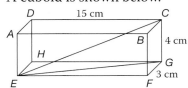

   a Calculate these lengths.
     i EG **ii** EC
   b Calculate these angles.
     i FEG **ii** HDF

5 A ship starts at a port then sails 120 km due south then 75 km due west.

   a How far is the ship from port?

   b Give the bearing of the ship from the port.

6 Without using a calculator, write down the exact value of these expressions.

   a $\sin 90°$ **b** $\cos 30°$

7 Calculate the size of angle θ and length x.

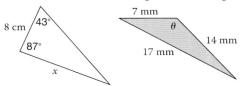

8 Calculate the area of the triangle ABC where $AB = 8$ cm, $BC = 11$ cm and angle $ABC = 35°$.

9 Work out these vector sums and draw the resultant vectors.

$$\mathbf{u} = 5\begin{pmatrix} 2 \\ -1 \end{pmatrix} + 4\begin{pmatrix} -1 \\ 0 \end{pmatrix} \quad \mathbf{v} = 3\begin{pmatrix} 3 \\ 2 \end{pmatrix} - 4\begin{pmatrix} 3 \\ -1 \end{pmatrix}$$

10 AB is parallel to DC and the ratio of the lengths DC to AB is $3:2$.

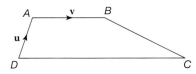

   a Write these vectors in terms of $\mathbf{u}$ and $\mathbf{v}$.
     i $\overrightarrow{DB}$ **ii** $\overrightarrow{DC}$ **iii** $\overrightarrow{BC}$

   M is the midpoint of AB and N is the midpoint of DC.

   b Write $\overrightarrow{MN}$ in terms of $\mathbf{u}$ and $\mathbf{v}$.

What next?

Assessment 19

1 a Hannah says incorrectly that the hypotenuse of this triangle is 181 m.
Describe her error and work out the correct hypotenuse. [3]

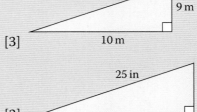

b Pawel says incorrectly that the missing side in this triangle is 34.66 in to 2 dp.
Describe his error and work out the correct value of the missing side. [3]

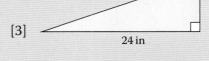

2 a Angelina says that 2, 3, 13 is a Pythagorean triple.
Say why she is wrong. [2]

b Darshna says that 20, 99 and 100 is a Pythagorean triple. Say why she is wrong and find a Pythagorean triple with highest value 100. [3]

3 Laura plots a pairs of points, $(-3, -5)$ and $(6, -1)$.
Find the distance between the points. [3]

4 Douglas draws six triangles with the following sides. Which of these are right angled triangles? Give reasons for your answers.

a 5, 12, 13 km [2]   **b** 9, 12, 15 cm [2]

c 9, 14, 17 m [2]   **d** 1.6, 3.0, 3.4 in [2]

e 11, 19, 22 yds [2]   **f** 3.6, 7.7, 8.5 ft [2]

5 A tangent, PQ, is drawn to a circle of radius 5.875 cm.
Q is 11.6 m from the centre of the circle.

a What sort of triangle is OPQ? [1]

Calculate

b the distance PQ [3]

c angle θ. [3]

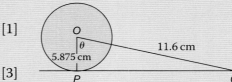

6 Ainslie is yachting. He sails from the harbour H for 2.5 km on a bearing of 062° and then goes round a buoy B. He then sails for 3.6 km on a bearing of 152° until he reaches lighthouse L and then back to the Harbour, as shown.

a Prove that the angle x is a right angle. [4]

b Calculate the distance the yacht sails in the final leg from L back to H. [3]

c Hence calculate the total distance travelled. [1]

d Find angle y to the nearest degree. [3]

e Find the bearing of the Lighthouse at L from the start line at H. [2]

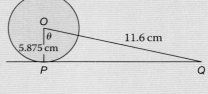

7 A vertical TV mast is 300 m high. The mast is secured by ropes leading from the top of the mast to a peg the ground. The angle between each rope and the ground is 69.7°.

    a How long is each rope? [3]

    b How far is horizontal distance from the peg to the base of the mast? [3]

8 **a** A plane is approaching an airport at an altitude of 1000 m. It is 4 km short of the airport horizontally. At what angle to the horizontal must it constantly descend at in order to land safely at the airport? [3]

    b A bank of seats at a stadium has a length of 40 m from back to front. Its vertical height is 25 m. At what angle to the horizontal does the bank of seats slope? [3]

9 **a** The base of a ladder is on horizontal ground and is leaning against a vertical wall. The base is 1.75 m from the wall and the ladder makes an angle of 12.5° with the wall. How long is the ladder? [3]

    b How far up the wall does the ladder reach? [2]

10 Tweedledum and Tweedledee are playing conkers. Tweedledum's conker, C, is tied to the end of a string 59.5 cm long. He pulls it back from the vertical until it is 28 cm horizontally from its original position. Calculate

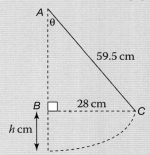

    a the vertical distance, h, that the conker has risen [6]

    b angle θ. [3]

11 **a** Draw these vectors on squared paper.

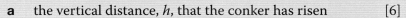

    i $\begin{pmatrix} 3 \\ 4 \end{pmatrix}$ [2] **ii** $\begin{pmatrix} -5 \\ 12 \end{pmatrix}$ [2] **iii** $\begin{pmatrix} 3 \\ -1 \end{pmatrix}$ [2] **iv** $\begin{pmatrix} -2 \\ -3 \end{pmatrix}$ [2]

    b $x = \begin{pmatrix} 4 \\ 6 \end{pmatrix}$ and $y = \begin{pmatrix} -2 \\ 7 \end{pmatrix}$. Draw these vectors on squared paper.

    i $x + y$ [2] **ii** $y - x$ [2] **iii** $x - y$ [2]

    iv $2x$ [2] **v** $-2y$ [2] **vi** $2x - 3y$ [2]

    c Find the length of each vector in part **b**. Leave your answers in surd form. [12]

    d Compare your drawings to parts **b ii** and **iii**. Write down two things you notice. [2]

12 Greg has a square grid. In his grid a vector one square to the right is given by x and one square vertically upwards by y. He writes down the vectors shown in terms of the vectors x and y.

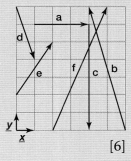

    a $3x$ **b** $2x + 7y$ **c** $6y$

    d $x - 3y$ **e** $3x + 2y$ **f** $3x + 7y$

Which vectors has he written down correctly?
For the vectors he has written incorrectly, write down the correct vector in terms of x and y. [6]

13 $x = \begin{pmatrix} 4 \\ 1 \end{pmatrix}$, $y = \begin{pmatrix} 7 \\ -2 \end{pmatrix}$ and $z = \begin{pmatrix} -8 \\ 5 \end{pmatrix}$. Write down these vectors.

    a $3x$ [1] **b** $5y$ [1] **c** $-2z$ [1]

    d $y + z$ [2] **e** $y - x$ [2] **f** $x + y + z$ [2]

    g $4z - 2x$ [3] **h** $4y + 3x - 2z$ [3] **i** $3z - (x + y)$ [3]

20 Combined events

Introduction

There is an old British myth, dating back hundreds of years, that says if it rains on St Swithin's Day (15th July) it will rain for 40 days afterwards. Surprisingly the myth retains its popularity despite statistics showing that it is untrue. Long-scale weather forecasting is fairly unreliable, particularly in the UK where the summer weather is largely determined by the 'jet stream'. However people will always look for tell-tale signs to help their predictions. In probability language, if a particular event occurs (rain on St Swithin's Day), does this increase the probability of another event occurring (a wet summer)?

What's the point?

Quite often, the occurrence of one event will significantly increase the probability of another event occurring. For example, lung cancer appears unpredictably, but its occurrence is greatly increased if a particular person is a smoker. Understanding the probabilities attached to linked events helps us to evaluate everyday risks.

Objectives

By the end of this chapter, you will have learned how to …

● Use Venn diagrams to represent sets.

● Use a possibility space to represent the outcomes of two experiments and to calculate probabilities.

● Use a tree diagram to show the outcomes of two experiments.

● Calculate conditional probabilities.

Check in

1 Work out each of these expressions.

a $1 - 0.45$ **b** $1 - 0.96$ **c** $1 - 0.28$

d $1 - 0.375$ **e** $0.2 + 0.4$ **f** $0.3 + 0.04$

g $0.65 + 0.25$ **h** 0.5×0.36 **i** 0.25×0.68

j 0.64×0.3 **k** $1 - 0.125 - 0.64$ **l** $1 - 0.125 \times 0.64$

2 Work out each of these expressions.

a $1 - \frac{5}{6}$ **b** $1 - \frac{1}{5}$ **c** $1 - \frac{7}{9}$ **d** $\frac{1}{5} + \frac{2}{3}$

e $\frac{3}{4} + \frac{1}{6}$ **f** $\frac{2}{3} \times \frac{5}{6}$ **g** $\frac{2}{9} \times \frac{4}{5}$ **h** $1 - \frac{3}{4} \times \frac{4}{5}$

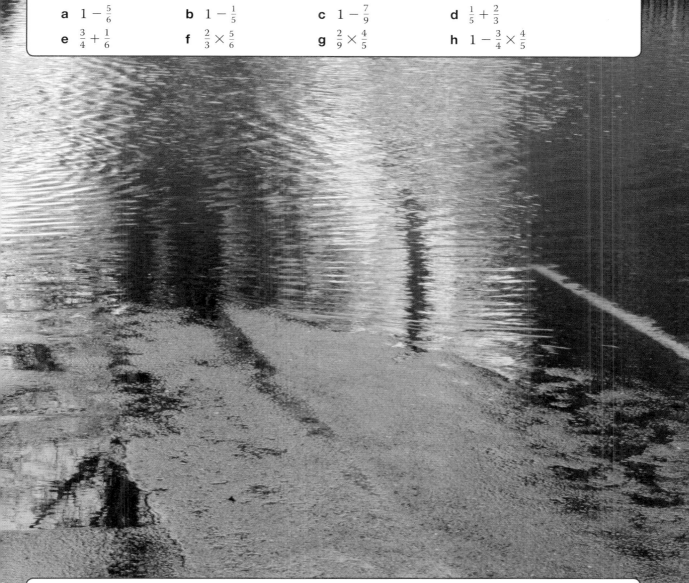

Chapter investigation

Two six-sided dice, labelled 1 to 6, are thrown. The score is given by adding the two numbers that appear uppermost.

Which result is most likely to occur or are all possible results equally likely?

What if, instead of adding the two numbers, you multiply them together?

20.1 Sets

● A **set** is a collection of numbers or objects.

● The objects in the set are called the **members** or **elements** of the set.

If the set X is 'the factors of 6', then you can write $X = \{1, 2, 3 \text{ and } 6\}$.

$3 \in X$ means that 3 is an element (member) of the set X.

If the set Y is 'the even numbers', then you can write $Y = \{2, 4, 6, 8, ...\}$.

● The universal set, which has the symbol ξ, is the set containing all the elements.

● The empty set, $\varnothing$, is the set with no elements.

You can use a Venn diagram to show the relationship between sets.

● The **intersection** of two sets, $A \cap B$, consists of the elements common to both sets.

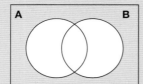

● The **union** of two sets, $A \cup B$, consists of the elements which appear in at least one of the sets.

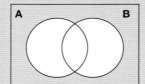

● The **complement** of a set, A', consists of the elements which are not in A.

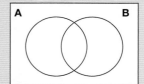

EXAMPLE

$\xi = \{1, 2, ... 11, 12\}$
$A = \{\text{factors of 12}\}$
$B = \{2, 3, 5, 6, 11\}$.

Find   **a** $A \cap B$

b $A \cup B$

c $(A \cup B)'$

a $A = \{1, 2, 3, 4, 6, 12\}$
$A \cap B = \{2, 3, 6\}$

b $A \cup B = \{1, 2, 3, 4, 5, 6, 11, 12\}$

c $(A \cup B)' = \{7, 8, 9, 10\}$

● You list the elements of each set   or   show the number of elements in each region

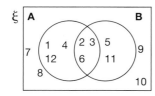

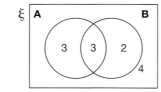

In the example,

$P(A) = \dfrac{6}{12} = \dfrac{1}{2}$

● You can use Venn diagrams to work out probabilities.

● $P(A) = \dfrac{\text{number of elements in set A}}{\text{total number of elements in } \xi}$

Probability   Combined events

Exercise 20.1S

1 List the elements of these sets.

   a P – the first ten square numbers

   b R – countries in North America

   c S – the first ten prime numbers

   d T – factors of 36

2 Using the sets in question **1**, give the sets

   a P ∩ T   **b** S ∩ T

   c P ∩ S   **d** P ∪ S

3 Give a precise description of each set.

   a {1, 2, 5, 10}

   b {2, 4, 6, 8, 10, 12, ...}

   c {a, e, i, o, u}

   d {HH, HT, TH, TT}

   e {1p, 2p, 5p, 10p, 20p, 50p, £1, £2}

   f {3, 6, 9, 12, 15, 18, 21, 24, 27, 30}

4 List the elements of these sets.

   a A – the first ten positive integers

   b B – single digit odd numbers

   c C – single digit prime numbers

   d D – single digit square numbers

5 Using the sets in question **4**, give the sets

   a B ∩ C   **b** B ∩ D

   c B ∪ D   **d** C ∪ D

6 Say why B, C and D must be subsets of A for the sets in question **4**.

7 A = {even numbers}

   B = {odd numbers}

   C = {multiples of 5}

   D = {prime numbers}

   E = {multiples of 3}

   F = {square numbers}

   G = {factors of 36}

   For these pairs of sets, state if they have any elements in common, if they do, then list them.

7 **a** A, C   **b** A, D   **c** F, G

   d E, F   **e** A, B   **f** C, G

   g D, F   **h** D, E   **i** D, E and F

8 The Venn diagram shows information about the sport that 25 students are playing in PE this term.

The results are shown on the Venn diagram.

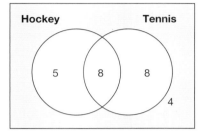

   a How many students are

     i in the intersection of hockey and tennis

     ii playing hockey

     iii not playing tennis

     iv in the union of hockey and tennis.

   b Describe the shaded region in words.

   c What fraction of students are playing either hockey or tennis, but not both?

9 An insurance company surveys 50 customers. The customers are sorted into

   P = {pet insurance}

   H = {home insurance}

   The results are shown on the Venn diagram.

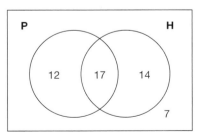

   a A customer is chosen at random. Find

     i P(P)   **ii** P(H')

     iii P(P ∩ H)   **iv** P(P ∪ H)

   b How many customers had home insurance but not pet insurance?

Q 1262, 1921, 1922   SEARCH

20.1 Sets

RECAP

- The **intersection** of two sets, A ∩ B, consists of the elements common to both sets.
- The **union** of two sets, A ∪ B, consists of the elements which appear in at least one of the sets.
- The **universal set**, which has the symbol ξ, is the set containing all the elements.
- The **empty set**, ∅, is the set with no elements.
- **Venn diagrams** can be used to represent the relationships between sets.

HOW TO

① Draw a Venn diagram to show the information.
② Fill in all the regions of the Venn diagram.
③ Use the Venn diagram to calculate probabilities.

> Don't forget to fill in the region outside the circles.

EXAMPLE

Parts made on a production line can have three kinds of defects, A, B or C.
1000 parts were inspected and the following results were obtained.

- 31 had type A defect
- 37 had type B defects
- 42 had type C defect
- 11 had both type A and type B defects
- 13 had both type B and type C defects
- 10 had both type A and type C defects
- 920 had no defects

What is the probability that a part chosen at random has all three types of defect?

① Draw a Venn diagram to show the information.
   Let x = parts with all three defects

11 had both type A and type B defects.
$11 = (11 - x) + x$
So $11 - x$ had only type A and type B.

② Find the number in the other regions in terms of x.

31 had type A defect.
$31 - (11 - x + x + 10 - x)$
$= 31 - (21 - x)$
$= 10 + x$
had only type A defects.

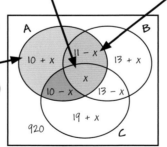

> Try and work out the other parts of the Venn diagram for yourself.

All of the regions add up to 1000.
$1000 = 920 + (10 + x) + (13 + x) + (19 + x) + (11 - x) + (10 - x) + (13 - x) + x$
$1000 = 920 + 76 + x$
$1000 = 996 + x$
$x = 4$

③ $P(A \cap B \cap C) = \dfrac{4}{1000} = 0.004$

Exercise 20.1A

1 The Venn Diagram shows the **numbers** of pupils in Year 11 in a school.

G = {Pupils who are taking biology}

H = {pupils who walk to school}

L = {pupils who are an only child}

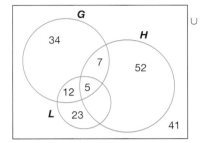

a How many pupils are taking biology?

b How many pupils do not walk to school?

c How many pupils are there in G ∩ H ∩ L?

d What do you know about a pupil in G ∩ H ∩ L?

2 U = {all triangles}
E = {equilateral triangles}
I = {isosceles triangles}
R = {right-angled triangles}

a Sketch a member of I ∩ R.

b Explain why E ∩ R = ∅.

3 Elsie sorts a group of objects into the sets A and B. She draws a Venn diagram to show her results.

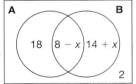

a If A and B are mutually exclusive, find the value of x.

b If $x = 8$, what can you say about the sets A and B?

4 At Newtown School there are 27 students in class 11B.
17 students play tennis, 11 play basketball and 2 play neither game.
How many students play both tennis and basketball?

5 A group of people took a literacy and a numeracy test. 85% of applicants passed the literacy test and 75% passed the numeracy test. What are the maximum and minimum proportions of people who did not pass either test?

6 David is organising a family reunion.
They can take part in two activities, archery or paintball.
David shows the results on a Venn diagram.

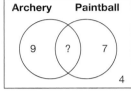

Archery costs £22 and paintball costs £18. David collects £524 to pay for the activities. How many sign up for both activities?

*7 U = {all quadrilaterals}

X = {rhombuses}

Y = {rectangles}

a Sketch a member of X ∩ Y.

b What sort of quadrilaterals are in the set X ∩ Y?

c If X had been the set containing kites, how would your answers to parts **b** and **c** differ?

*8 At Highfield School there are 100 students in year 10.
They all study one or more of these subjects.

- 36 study French
- 42 study Spanish
- 47 study German
- 10 study both French and Spanish
- 12 study both Spanish and German
- 9 study both French and German

What is the probability that a student chosen at random studies all three subjects?

20.2 Possibility spaces

- The list or table of all of the possible outcomes of a trial is called a **possibility space** or **sample space**.

When two ordinary dice are thrown, there are 36 possible outcomes.

If the dice are fair then the 36 outcomes are equally likely.

- You can use a sample space to calculate probabilities.

	1	2	3	4	5	6
1	1,1	1,2	1,3	1,4	1,5	1,6
2	2,1	2,2	2,3	2,4	2,5	2,6
3	3,1	3,2	3,2	3,4	3,5	3,6
4	4,1	4,2	4,3	4,4	4,5	4,6
5	5,1	5,2	5,3	5,4	5,5	5,6
6	6,1	6,2	6,3	6,4	6,5	6,6

EXAMPLE

Jasper throws two dice and adds the results.

Lena throws two dice and multiplies the results.

- **a** Draw a possibility space for Jasper's and Lena's experiments.
- **b** Find the probability that Jasper scores 8.
- **c** Find the probability that Lena scores 6.
- **d** Find the probability that Lena scores 5 or less.

a Jasper

+	1	2	3	4	5	6
1	2	3	4	5	6	7
2	3	4	5	6	7	8
3	4	5	6	7	8	9
4	5	6	7	8	9	10
5	6	7	8	9	10	11
6	7	8	9	10	11	12

Lena

×	1	2	3	4	5	6
1	1	2	3	4	5	6
2	2	4	6	8	10	12
3	3	6	9	12	15	18
4	4	8	12	16	20	24
5	5	10	15	20	25	30
6	6	12	18	24	30	36

b The same sum appears on each diagonal.

$$P(8) = \frac{5}{36}$$

c $P(6) = \dfrac{4}{36} = \dfrac{1}{9}$

d $P(5 \text{ or less}) = \dfrac{10}{36} = \dfrac{5}{18}$

- You can use the possibility space to write the set of all possible outcomes.

EXAMPLE

A fair coin is tossed and a fair die is thrown.

- If a head is seen then the score on the dice is doubled.
- If a tail is seen then the score is just the number on the dice.
- **a** Show the possibility space in a grid.
- **b** Write the set of all possible outcomes.
- **c** Find P(6).

a

	1	2	3	4	5	6
T	1	2	3	4	5	6
H	2	4	6	8	10	12

b The set of all outcomes is {1, 2, 3, 4, 5, 6, 8, 10, 12}

c $P(6) = \dfrac{2}{12} = \dfrac{1}{6}$

Exercise 20.2S

1 a A shop sells five different sandwiches and four different drinks. How many different sandwich and drink combinations are there?

b Lucy spins an eight-sided spinner and throws a dice. How many outcomes are there?

2 Using Jasper's table shown opposite find the probability that the **sum of the scores** seen on two fair dice is

a exactly 10 **b** at least 10

c a square number **d** less than 5.

e Write the set of all possible outcomes.

3 Using Lena's table shown opposite find the probability that the **product of the scores** seen on two fair dice is

a exactly 10 **b** at least 10

c a square number **d** less than 5.

e Write the set of all possible outcomes.

4 Two fair dice are thrown and the difference between the scores showing on the two dice is recorded.

a Make a table to show the possibility space.

b Write the set of all possible outcomes.

c Find the probability that the difference is

i 0 **ii** 3

iii 6 **iv** a prime number.

5 A fair coin is tossed three times and the outcome recorded (for example HHT).

a Write the set of the 8 possible outcomes.

b In how many of these are exactly two heads seen?

c In how many do you see three of the same?

6 Two fair spinners are used. On one the possible scores are 1, 2 and 4, on the other the scores are 1, 3 and 5. The sum of the scores on the two spinners is recorded.

a Make a table to show the possibility space.

b Write the set of all possible outcomes.

c Find the probability that the score is

i 2 **ii** 3 **iii** even

7 The two spinners in question **6** are used again but the score recorded is the product of the two scores.

a Make a table to show the possibility space.

b Write the set of all possible outcomes.

c Find the probability that the score is

i 2 **ii** 3 **iii** even.

8 For the spinners used in questions **6** and **7**, if you wanted to get an even number, would you be better to use the sum or the product of the scores on the two spinners?

9 A pair of unbiased dice are thrown and the sum and product of the scores are recorded in two lists. The dice are thrown 100 times.

a Estimate the number of times a sum of exactly 10 will be seen.

b Estimate the number of times a product of exactly 10 will be seen.

c Would you expect to see 6 in the list of sums more often, less often or about the same number of times as in the list of products?

d Would you expect to see 3 in the list of sums more often, less often or about the same number of times as in the list of products?

e Marin says that 7 is certain to appear more often in the list of sums because you can't score 7 using the product. Is he correct? Give your reasons.

10 Two fair dice are thrown together. One is an ordinary dice with the numbers 1 to 6, and the other has faces labelled 1, 2, 2, 3, 3, 3.

a Make a table to show the possibility space.

b Find the probability that the score is

i 6 **ii** 7 **iii** 9 **iv** 3.

c What other scores are as likely to happen as 6?

d Why are some scores less likely to occur than 6?

Q 1199, 1263   SEARCH

20.2 Possibility spaces

- When dealing with equally likely outcomes for a single or a combined experiment, a list or table showing the outcomes is helpful.

> Once you've drawn a table, the individual cells give you the outcomes.

HOW TO

1. If you are constructing a list, work systematically so you can generate them all in sequence.
   For tossing three coins, THH, HTH, TTT, HHT, HHH, THT are 6 of the 8 possibilities – but what are the other 2?
2. When combining two simple experiments, a table allows you to enter the 'score' in the cell while the row and header column still tells you what each outcome was.

EXAMPLE

Tara has 6 cards numbered from 1 to 6.
She takes two cards without replacement.
A = {product is even} B = {sum is even}
Show that P(A) = 2P(B).

> Could you work out what the probabilities of even sum or product is without finding all the sums and products?

1. Draw a possibility space.

You cannot take the same card twice so there are 30 outcomes.

2. Use the sample space to calculate probabilities.

$P(A) = \dfrac{24}{30} = \dfrac{4}{5}$ There are 24 cells with an even product.

	1	2	3	4	5	6
1		2	3	4	5	6
2	2		6	8	10	12
3	3	6		12	15	18
4	6	8	12		20	24
5	5	10	15	20		30
6	6	12	18	24	30	

$P(B) = \dfrac{12}{30} = \dfrac{2}{5}$ There are 12 cells with an even sum.

$P(A) = 2P(B)$

	1	2	3	4	5	6
1		3	4	5	6	7
2	3		5	6	7	8
3	4	5		7	8	9
4	5	6	7		9	10
5	6	7	8	9		11
6	7	8	9	10	11	

EXAMPLE

A fair coin is tossed and a fair die is thrown. If a head is seen then the score on the dice is squared. If a tail is seen then the score is just the number on the dice.

Show the possibility space in a grid. Explain why 1 and 4 are the most likely scores to be recorded.

1. Draw a possibility space.

	1	2	3	4	5	6
T	1	2	3	4	5	6
H	1	4	9	16	25	36

2. 1 and 4 are the only square numbers that fall within the range 1–6.

Exercise 20.2A

1 Two cards are taken from a set of cards numbered 1 to 6.
Find the probability that the **difference** in the value of the two cards is

a 3

b a factor of 6

c at least 1.

2 A fair coin is tossed until a head is seen. The number of times it has been tossed is recorded as the score on that trial.

What is the probability that the score recorded is at least 3?

> Hint: consider the possible outcomes if a fair coin is tossed twice.

3 A fair dice is thrown and the spinner shown is spun. If the spinner lands on yellow you miss a turn (move 0 squares) otherwise you move the number of squares given by the product of the scores on the spinner and the dice.

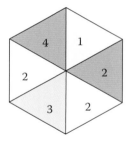

a Using labels R1, Y2 etc for the outcomes on the spinner construct a table to show the possible scores.

b What is the probability that the score recorded is

i 4   **ii** more than 6?

4 Cards numbered 1 to 100 are put in a box and Alessandra is asked to pick one at random. What is the probability that she chooses

a a single digit number

b a two digit number

c a number containing at least one 3?

> Don't write out the whole list, but imagine how many cards satisfy each condition.

5 Two fair spinners are used – one has sections showing the numbers 1, 1, 2, 3, 5 and the other has sections showing 3, 4, 5, 7, 8, 9.

a What is the probability that the total score on the two spinners is

i 6   **ii** even?

b What is the probability that the score on one spinner is at least twice the score on the other spinner?

*6 Jack and Jill each roll a fair dice. Whoever gets the larger score wins the game.

a If a draw is allowed, what is the probability that Jill wins the game?

b A draw is not allowed and if the two dice show the same they roll again until one wins.
What is the probability it will be Jill?

*7 A blue and a red dice are both fair and are thrown together. The following events are defined on the scores seen

A – the dice show the same score

B – the total score is at least 10

C – the total score is odd

D – the high score is a 4

E – the score on one dice is a proper factor of the score on the other (a proper factor is a factor which is not 1 or the number itself)

a Find these probabilities

i $P(A \cap B)$ **ii** $P(E)$ **iii** $P(C \cap D)$.

b Find two pairs of mutually exclusive events.

c How many pairs of mutually exclusive events are there?

20.3 Tree diagrams

● You can use a **frequency tree** to show the outcomes of two events.

EXAMPLE

A factory employs 300 workers. 100 workers are skilled and the rest are unskilled. 25 skilled workers work part time. 120 unskilled workers work full time.

a Draw a frequency tree to show this information.

b How many part-time workers are there altogether?

c If a worker is chosen at random what is the probability they are full-time?

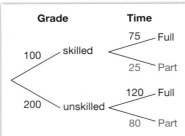

The frequency tree shows the number of workers along each branch.

The probability of each outcome is the relative frequency.

b 25 skilled and 80 unskilled workers are part time. $25 + 80 = 105$

c 75 skilled and 120 unskilled workers are full time. $\dfrac{75 + 120}{300} = \dfrac{13}{20}$

You can use a tree **diagram** to show the probabilities of two events.
● Write the outcomes at the end of each branch.
● Write the probability on each branch.
● The probabilities on each set of branches should add to 1.

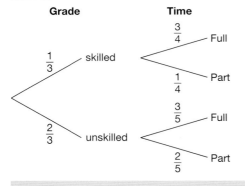

80 out of 200 unskilled workers work part time.
$\dfrac{80}{200} = \dfrac{2}{5}$

When you give a probability as a fraction, try to reduce it to its simplest form.

To find probabilities when an event can happen in different ways
● Multiply the probabilities along the branches.
● Add the probabilities for the different ways of getting the chosen event.

$P(\text{full time}) = P(\text{skilled and full time}) + P(\text{unskilled and full time})$

$$= \left(\frac{1}{3} \times \frac{3}{4}\right) + \left(\frac{2}{3} \times \frac{3}{5}\right)$$

$$= \frac{3}{12} + \frac{6}{15} = \frac{13}{20}$$

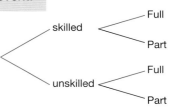

Probability Combined events

Fluency

Exercise 20.3S

1 A university surveys 3250 students.
Approximately 2000 students live at home.
A quarter of the students living at home are worried about debt.
Three in five of students living away from home are worried about debt.

   a Draw a frequency tree showing the number of students on each branch.

   b A student is chosen at random. What is the probability that they are concerned about debt?

2 The MOT test examines whether cars are roadworthy.
10% of cars fail the MOT because there is something wrong with their brakes.
40% of the cars with faulty brakes also have faulty lights.
20% of cars whose brakes are satisfactory have faulty lights.

An MOT centre tested 3000 cars in March.

   a Draw a frequency tree showing the numbers of cars failing with brakes and lights in March.

   b How many cars fail on at least one of brakes and lights?

   c Tommy says that this means the rest of the 3000 cars passed the MOT test. Why is Tommy wrong?

3 A company has 360 employees.
120 employees are male.
80% of the male employees and 168 of the female employees are in the pension scheme.

   a Draw a frequency tree to show this information.

   b Find the probability that an employee chosen at random is in the pension scheme.

   c Find the probability that the employee is male.

   d A female employee is chosen at random. Find the probability that she is in the pension scheme.

3 **e** Add the probabilities of each outcome onto each branch of your frequency diagram.

   f Find the probability that an employee is in the pension scheme. Does your answer agree with your answer in part **a**?

4 Jessica travels to work by car two days a week and by train on the other three.
She is late for work 10% of the time when she travels by car, and late 20% of the time when she travels by train.

   a Draw a probability tree diagram showing the probability of each outcome.

   b Jessica works 150 days during the first 8 months of a year. Draw a frequency tree to show the numbers of days she travels to work by car and train, and on which she is late and on time.

   c Estimate how many days Jessica is late for work during this period.

   d What is the probability Jessica is late for work on a day chosen at random during this period?

5 Athletes are regularly tested for performance enhancing drugs.
If an athlete is taking a drug, the test will give a positive result 19 times out of 20.
However in athletes who are not taking the drug one in fifty tests is also positive.

It is thought that around 20% of athletes in a particular event are taking the drug.
The authorities tests on 500 athletes.

   a Draw a probability tree showing the numbers of athletes expected to be taking or not taking the drug, and the results of the test.

   b Calculate the total number of positive tests expected to be seen.

   c How many of these came from athletes not taking the drug?

   d Find the probability that the test gives an accurate result.

20.3 Tree diagrams

- You can use tree **diagrams** to show the probabilities of two events.
  - Write the outcomes at the end of each branch
  - Write the probability on each branch
  - The probabilities on each set of branches should add to 1
- To find probabilities when an event can happen in different ways
  - Multiply the probabilities along the branches.
  - Add the probabilities for the different ways of getting the chosen event

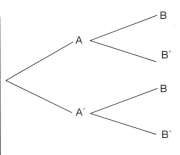

- If the outcome of one event does not affect what happens in another then the events are **independent**.
  If A and B are **independent** events then
  P (A and B) = P(B) × P(A)

> If you can shows that
> P(A and B) = P(B) × P(A)
> then A and B are independent.

HOW TO

To prove that two events are independent

① Draw a tree diagram showing the possible outcomes and probabilities of each outcome

② Use the tree diagram to find the probability of P(A and B), P(A) and P(B).

③ Test to see if P (A and B) = P(B) × P(A) is satisfied.

EXAMPLE

A bag has 5 red and 3 blue balls in it.
A ball is taken at random and not replaced, and then a second ball is taken out.
Show that choosing a red ball on the second attempt is dependent on whether or not the first ball was red.

① Draw a tree diagram.

First ... Second

$\frac{5}{8}$ Red
$\frac{4}{7}$ Red
$\frac{3}{7}$ Blue

$\frac{3}{8}$ Blue
$\frac{5}{7}$ Red
$\frac{2}{7}$ Blue

② Find the probabilities of each event.

$P(\text{first ball red}) = \frac{5}{8}$

$P(\text{second ball red}) = P(R, R) + P(B, R)$

$= \frac{5}{8} \times \frac{4}{7} + \frac{3}{8} \times \frac{5}{7}$

$= \frac{20}{56} + \frac{15}{56} = \frac{35}{56}$

$= \frac{5}{8}$

$P(\text{both balls are red}) = \frac{5}{8} \times \frac{4}{7}$

$= \frac{20}{56} = \frac{5}{14}$

> Don't cancel fractions to lowest form when you multiply the probabilities along the path – you are likely to have to add them!

③ If the events are independent then

$P(\text{first ball red}) \times P(\text{second ball red}) = P(\text{both balls are red})$

$P(\text{first ball red}) \times P(\text{second ball red}) = \frac{5}{8} \times \frac{5}{8} = \frac{25}{64} \neq \frac{5}{14}$

So the events are dependent.

Probability Combined events

Exercise 20.3A

1 In a league, teams are awarded 3 points for a win, one for a draw and none for a loss. Amelie thinks that

- her team has a probability of 0.6 of winning any match.
- her team has a probability of 0.3 for a draw
- the result of any game is independent of other results.

a Find the probability that her team has at least three points after two games.

b How have you used Amelie's assumption that the results of the game are independent in your answer to part **a**?

c Do you think that Amelie's assumption that the results of the game are independent is reasonable? Give a reason for your answer.

2 A Year 13 pupil is taking their driving test. Records show that people taking the test at that age have a 70% chance of passing on the first attempt and 80% on any further attempts needed.

a Show this information on a tree diagram showing up to three attempts.

b Find the probability that

   i the pupil passes at the second attempt

   ii the pupil has still not passed after three attempts.

3 Denzel is going to the airport to catch a flight. He needs to travel on a bus and then catch a train to the airport.
He catches a bus which has a probability of 0.8 of making a connection with a train which always gets to the airport on time. The next train has a probability of 0.7 of getting him to the airport on time.

a Show this information on a tree diagram.

b Find the probability that Denzel gets to the airport in time for his flight.

4 A bag has 4 red and 4 blue balls in it. A ball is taken from it at random and not replaced and then a second ball is taken out.

X = the second ball is blue

Y = the two balls are the same colour

Z = both balls are blue

a Ellie says that $Z = (X \cap Y)$. Is she correct?

b Show that X and Y are independent events.

5 A bag has 10 white and 5 black balls in it. A ball is taken from it at random, the colour noted and the ball is replaced and then a second ball is taken at random.

A = the two balls are different colours

B = at least one ball is black.

Construct a tree diagram and decide if A and B are independent events.

*6 A spinner has a probability of $\frac{1}{6}$ of landing on blue.
Green is three times as likely to occur as blue.
Red, black and yellow are equally likely to occur.

a Calculate the probability that the spinner lands on red.

Sara is playing a game where she spins the spinner and if it lands on red, she takes double the score seen when she throws a fair dice.
Sara needs a 4 to finish.

b Draw a probability tree showing the outcomes of the spinner and the dice.

c What is the probability Sara finishes on her go?

d Is the use of the spinner a help or a hindrance to Sara getting a 4 to finish, or does it not matter? Give a reason.

e How would Sara's experiment change if she needed

   i a 5 to finish?

   ii an 8 to finish?

 Q 1208, 1334, 1935, 1966   SEARCH

20.4 Conditional probability

In life, some events are dependent on other events occurring beforehand.

● In **conditional** probability, the probability of subsequent events depends on previous events occurring.

You can calculate probabilities based on the conditions you are given.

● For two events A and B. The probability of B happening, **given** that A has already happened, is written as P(B|A).
To calculate conditional probabilities, use the formulae

$$P(A \cap B) = P(B|A) \times P(A) \text{ and } P(B|A) = \frac{P(A \cap B)}{P(A)}$$

Your chances of exam success are conditional on whether you revise or not.

If A and B are independent, then P(B|A) = P(B).

EXAMPLE

An insurance company records from 200 accident claims.
The table shows the speed of the car and weather conditions during the accident.

	Wet	Dry	Total
Speeding	11	21	32
Not speeding	77	91	168
Total	88	112	200

A is the event 'it is wet' and
B is the event 'a car is speeding'.
Are A and B independent?
Give a reason why.

$P(B) = \frac{32}{200} = 0.16$

$P(B|A) = \frac{11}{88} = 0.125$ 11 out of the 88 cars were speeding when it was wet

$P(B|A) \neq P(B)$, so A and B are not independent.

You can use Venn diagrams to calculate conditional probabilities.
This Venn diagram shows the probabilities attached to two events A and B.

EXAMPLE

Find P(A|B) and P(B|A) for the probabilities in this Venn diagram.

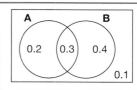

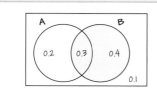

$P(B) = 0.7, P(A \cap B) = 0.3$

So $P(A|B) = \dfrac{P(A \cap B)}{P(B)} = \dfrac{0.3}{0.7}$

$= \dfrac{3}{7}$

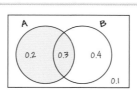

$P(A) = 0.5, P(A \cap B) = 0.3$

So $P(B|A) = \dfrac{P(A \cap B)}{P(A)} = \dfrac{0.3}{0.5}$

$= \dfrac{3}{5}$

You can also use tree diagrams to calculate conditional probabilities.

Probability Combined events

Exercise 20.4S

1 For the sets shown in the Venn diagram, find

   a P(S ∪ R)

   b P(S|R)

   c P(R|S)

   d P(R ∩ S)

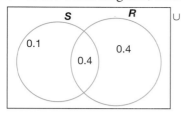

   Are S and R independent?

2 Three sets are defined

   U = {single digit integers}

   P = {prime numbers}

   F = {factors of 6}

   a Draw a Venn diagram showing the probability of a single digit number chosen at random lying in each of the regions created by the sets P and F.

   b Find **i** P(P)   **ii** P(P ∩ F)   **iii** P(P|F)

   c Explain why P and F are not independent.

3

	Walk	Other
Year 7	43	136
Year 11	98	83

   The table shows the way year 7 and year 11 pupils normally get to a particular school.

   a What is the probability a Year 7 pupil chosen at random walks to school?

   b What is the probability a Year 11 pupil chosen at random walks to school?

   c Is the way pupils travel to school independent of the year group they are in?

4 95% of drivers wear seat belts.
44% of car drivers involved in accidents die if they are not wearing a seat belt.
92% of those that do wear a seat belt survive.

   a Draw a tree diagram to show this information.

   b What is the probability that a driver in an accident did not wear a seat belt and survived?

5

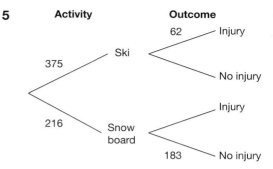

   a Copy and complete the tree diagram showing the number of people in a ski resort who completed a survey about their holiday.

   b If a person is chosen at random from the people **who were snowboarding**, what is the probability they suffered an injury?

   c If a person is chosen at random from those **who were injured**, what is the probability that they were snowboarding?

6 There are eight hundred thousand adults in a large city. A rare illness affects about one in 500 adults.
A new screening test gives a positive result 98% of the time when a person has the illness.
It also gives a positive result on 1% of people who do not have the illness.

   a Draw a tree diagram to show the numbers of adults in the city expected to have and not have the illness, and what the results of the screening test would be.

   b Find the probability that a person with a negative test result has the disease.

   c Find the probability that a person with a positive test result has the disease.

20.4 Conditional probability

- The probability of B happening, **given** that A has already happened, is written as P(B|A).
- To calculate conditional probabilities, use the formulae

P(A ∩ B) = P(B|A) × P(A)

and $P(B|A) = \dfrac{P(A \cap B)}{P(A)}$

If B happening does not depend on A happening then
P(B|A) = P(B).

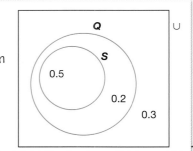

HOW TO

① Decide which events depend on the outcome of previous events – if two events are independent then P(B|A) = P(B).
② Use the information from a two-way table, Venn diagram or tree diagram to calculate the probability of the events.

EXAMPLE

In a series of long jump competitions, athletes have to jump at least 6.5 metres in the first three rounds of a competition to be eligible for three more jumps.
In the series, 70% of the athletes qualify for the extra jumps, and 50% record a best jump of at least 7 metres.

Draw a Venn diagram to represent this information. Calculate the probability that an athlete makes a jump over 7 metres given that she qualified for extra jumps.

Q is the set of athletes qualifying for extra jumps.   P(Q) = 0.7
S is the set of athletes jumping over 7 metres.   P(S) = 0.5
S must be a **subset** of Q, because if you've jumped over 7 metres, then you've definitely jumped over 6.5 metres – so you've qualified.

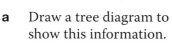

EXAMPLE

95% of drivers wear seat belts. 60% of car drivers involved in serious accidents die if they are not wearing a seat belt, whereas 80% of those that do wear a seat belt survive.

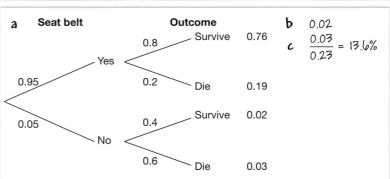

a Draw a tree diagram to show this information.

b What is the probability that a driver in a serious accident did not wear a seat belt and survived?

c What is the probability that a driver who died in a serious accident was not wearing a seat belt? Give your answer as a percentage to 1 decimal place.

Exercise 20.4A

1 Of the employees in a large factory

- $\frac{1}{3}$ travel to work by bus
- $\frac{1}{4}$ by train
- the rest by car.

Those travelling by bus have a probability of $\frac{1}{4}$ of being late.

Those by train will be late with probability $\frac{1}{5}$

Those by car will be late with probability $\frac{1}{10}$.

a Draw and complete a tree diagram, and calculate the probability that an employee chosen at random will be late.

b Calculate the probability that an employee who is late travels by bus.

2 An insurance company classifies drivers in three categories.

P is 'low risk', and they represent 25% of drivers who are insured.

Q is 'moderate risk' and they represent 60% of the drivers.

R is 'high risk'.

The probability that a category P driver has one or more accidents in a twelve month period is 2%.

The corresponding probabilities for Q and R are 6% and 10%.

a Find the probability that a motorist, chosen at random, is assessed as a category Q risk and has one or more accidents in the year.

b Find the probability that a motorist, chosen at random, has one or more accidents in the year.

c If a customer has an accident in a twelve month period, what is the probability that they were a category Q driver?

3 A team plays 38 matches over a season in different conditions with these results.

	Wet	Dry
Win	6	17
Draw	2	5
Lose	4	4

Find the probability that the team

a won a match chosen at random

b won a match that was played in wet conditions

c played the match in wet conditions given that they won the match.

4 If a team wins a match the probability they win their next match is 0.6 and the probability of a draw is 0.2. If the match was a draw the probability of winning the next match is 0.5 and drawing is 0.1. If they lost the match the probability of winning the next match is 0.3 and drawing is 0.4.

a Draw a tree diagram for two matches after the team has drawn a match.

b A win is worth 3 points, a draw 1 point and a loss 0 points.
What is the probability the team scores at least 3 points in the two matches?

c Given that the team scores at least three points what is the probability they scored exactly 4 points?

5 A GP practice encourages elderly people to have a flu vaccination in the autumn.
They claim it reduces the likelihood of having flu over the winter from 50% to 15%.
The practice gives flu vaccination to 60% of the elderly people one year.
What is the probability that an elderly person chosen at random from the practice

a gets flu that winter

b has been vaccinated given that they get flu

c has been vaccinated given that they did not get flu?

Summary

Checkout
You should now be able to...

Test it
Questions

✔ Use tables and Venn diagrams to represent sets.	1
✔ Use a possibility space to represent the outcomes of two experiments and to calculate probabilities.	2, 3
✔ Use a tree diagram to show the outcomes of one or more experiments and to calculate probabilities.	4 – 6
✔ Calculate conditional probabilities.	1, 4, 5

Language | Meaning | Example

Language	Meaning	Example
Set	A set is a collection of objects.	E = {Even numbers ⩽ of 12}
Element **Member**	An object in a set.	= {2, 4, 6, 8, 10, 12}
Universal set, ξ	The set containing all the elements.	ξ = {Positive numbers ⩽ 12}
Empty set, ∅	The empty set has no members.	{all odd numbers divisible by 2} = {} = ∅
Venn diagram	A way of showing sets and their elements.	T = {Multiples of three, less than 12} ξ = {3, 6, 9, 12} ξ: E [2 4] [6] [3] T [8 10] [12] [9] 1 5 7 11
Intersection	For two or more sets, a new set containing the elements found in *all* of the sets.	Intersection, E ∩ T = {6, 12}
Union	For two or more sets, a new set containing the elements found in *any* of the sets.	Union, E ∪ T = {2, 3, 4, 6, 8, 9, 10, 12}
Complement	The complement of a set is all members which are not in that set but are in the universal set. The complement of A is A'.	O = {odd numbers ⩽ 12} = {1, 3, 5, 7, 9, 11} O' = {2, 4, 6, 8, 10}
Disjoint	Two sets are disjoint if their intersection is empty. Mutually exclusive events have disjoint sets.	{Even numbers} ∩ {Odd numbers} = ∅
Tree diagram	A diagram that shows all the outcomes of one or more consecutive events as successive branches. Probabilities are given on the branches.	100 < 60 H < 35 HH / 25 HT ; 40 T < 24 TH / 16 TT
Frequency tree	A tree diagram in which each branch is labelled using the number of times that combination of outcomes occurred.	
Conditional probability	The probability of one event given the probability that another event has occurred. P(A given B) = P(A | B) = P(A ∩ B) / P(B)	For a regular dice, P(2 given rolled an even number) = $\frac{1}{3}$
Independent	Two events are independent if the result of one event does *not* affect the result of the other event.	If A and B are independent then P(A | B) = P(A) P(A ∩ B) = P(A) × P(B)

Probability Combined events

438

Review

1 Draw a Venn diagram to show the multiples of 3, even numbers and the factors of 20 less than 20.
Simon selects an even number from the Venn diagram. What is the probability that it is a factor of 20?

2 Jakub has packs of sweets from 3 different brands. The relative frequency of each colour for each brand is shown in the table.

Colour	Brand 1	Brand 2	Brand 3
Red	0.25	0.2	0.15
Blue	0.25	0.3	0.35
Orange	0.25	0.4	0.45
Yellow	0.25	0.1	0.05

Jakub selects one sweet from a pack of each brand.

Estimate the probability he selects

a 3 sweets of the same colour

b at least 2 red sweets.

3 Jayden can choose fillings of egg, cheese or tuna for his sandwich, and can use white, brown or granary bread.

a Draw a table to show all the possible types of sandwiches.

Jayden is equally likely to pick each filling and equally likely to pick each bread type.

b What is the probability that David chooses

i granary bread

ii egg in brown bread

iii not tuna and not white bread?

4 Madison must pass through two sets of traffic lights on her way to work. The probability the first set is red when she

4 approaches is 0.15 and the probability the second set is red is 0.35.

a Draw a tree diagram to show all the possible outcomes.

b Calculate the probability that

i she doesn't have to stop at all

ii she has to stop at least once.

c On Monday the first light is green when Madison approaches it. What is the probability the second set is also green?

5 A bag contains 10 red and 6 black counters. A counter is removed at random and *not* put back. A second counter is then removed and not replaced.

a Calculate the probability that

i both counters are red

ii both counters are black

iii the counters are different colours.

A third counter is now chosen.

b Calculate the probability that all the counters are the same colour.

c Two black counters are chosen without replacement. A third counter is then chosen. What is the probability that the third counter is also black?

6 The probability that Mia is late to work is $\frac{1}{20}$. If Mia is late to work then the probability that Max is late to work is $\frac{1}{5}$, otherwise it is $\frac{1}{10}$.

a Draw a tree diagram to show all the possible outcomes.

b Calculate the probability that

i neither of them are late to work

ii Max is late to work.

What next?

Assessment 20

1 The table shows the groupings of 80 people in a local Gymnastics club.

	Children under 12		Teenagers 13 to 19		Adults 20 to 30	
	Male	**Female**	**Male**	**Female**	**Male**	**Female**
Number of people	4	9	17	24	15	11

a Use the information in the table to complete
the Venn diagram. [4]

b Calculate these probabilities

   i   P(T)   **ii**   P(M')

   iii   P(T ∪ M)   **iv**   P(T ∪ M) [8]

c Are the events T and M independent? [3]

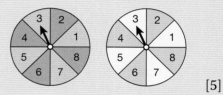

Teenagers (T)   Male (M)

2 ξ = {Whole numbers from 1 to 20}, T = {factors of 18}, F = {factors of n}
P(F) = 0.2   P(T ∪ F) = 0.1 and P(T ∪ F) = 0.45

a Draw a Venn diagram to show the number of elements in each region. [6]

b Find the value of n. [4]

3 Victoria has two unbiased spinners.
Each spinner is divided into eight equal sectors
containing the numbers 1 to 8.
The spinner are spun and their scores are added together.

a Draw a sample space diagram for these spinners. [5]

b Use your diagram to find

   i   P(3)   [2]   **ii**   P(8)   [2]   **iii**   P(12)   [2]

c What is the most likely score? [1]

d How has the assumption that the spinners are unbiased affected your
answers to parts **b** and **c**? [1]

4 32 editors in a publishing company went for a
break in the staff room. The staff room has tea
and coffee and a box of chocolate and plain
biscuits.
17 editors had a cup of tea.
12 of the editors who had tea chose ate a
chocolate biscuit.
22 editors ate chocolate biscuits altogether.

Complete the frequency tree. [4]

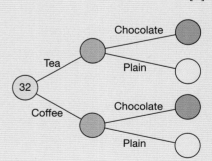

5 The probability that my postman delivers my mail between 10 and 10:30 in the morning is 0.15. The probability it comes before 10 am is 0.1.

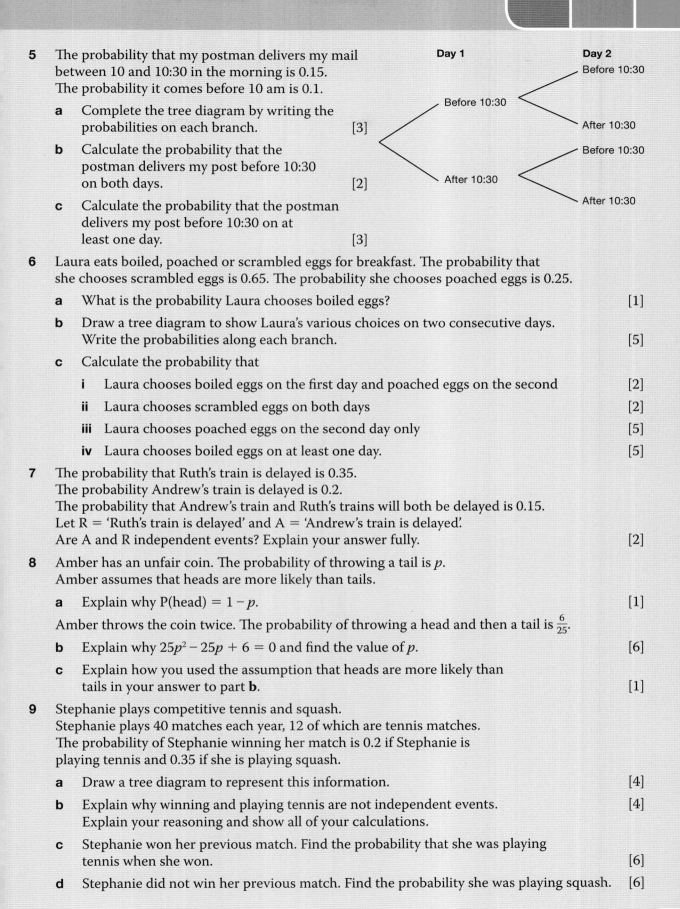

Day 1   Day 2

Before 10:30 — Before 10:30 / After 10:30

After 10:30 — Before 10:30 / After 10:30

  a Complete the tree diagram by writing the probabilities on each branch. [3]

  b Calculate the probability that the postman delivers my post before 10:30 on both days. [2]

  c Calculate the probability that the postman delivers my post before 10:30 on at least one day. [3]

6 Laura eats boiled, poached or scrambled eggs for breakfast. The probability that she chooses scrambled eggs is 0.65. The probability she chooses poached eggs is 0.25.

  a What is the probability Laura chooses boiled eggs? [1]

  b Draw a tree diagram to show Laura's various choices on two consecutive days. Write the probabilities along each branch. [5]

  c Calculate the probability that

    i Laura chooses boiled eggs on the first day and poached eggs on the second [2]

    ii Laura chooses scrambled eggs on both days [2]

    iii Laura chooses poached eggs on the second day only [5]

    iv Laura chooses boiled eggs on at least one day. [5]

7 The probability that Ruth's train is delayed is 0.35.
The probability Andrew's train is delayed is 0.2.
The probability that Andrew's train and Ruth's trains will both be delayed is 0.15.
Let R = 'Ruth's train is delayed' and A = 'Andrew's train is delayed'.
Are A and R independent events? Explain your answer fully. [2]

8 Amber has an unfair coin. The probability of throwing a tail is p.
Amber assumes that heads are more likely than tails.

  a Explain why P(head) = $1 - p$. [1]

Amber throws the coin twice. The probability of throwing a head and then a tail is $\frac{6}{25}$.

  b Explain why $25p^2 - 25p + 6 = 0$ and find the value of p. [6]

  c Explain how you used the assumption that heads are more likely than tails in your answer to part **b**. [1]

9 Stephanie plays competitive tennis and squash.
Stephanie plays 40 matches each year, 12 of which are tennis matches.
The probability of Stephanie winning her match is 0.2 if Stephanie is playing tennis and 0.35 if she is playing squash.

  a Draw a tree diagram to represent this information. [4]

  b Explain why winning and playing tennis are not independent events. Explain your reasoning and show all of your calculations. [4]

  c Stephanie won her previous match. Find the probability that she was playing tennis when she won. [6]

  d Stephanie did not win her previous match. Find the probability she was playing squash. [6]

Life skills 4: The launch party

Now that the business is set up, the restaurant is ready to start receiving customers. Abigail, Raheem, Mike and Juliet plan a grand opening. They expect more people to come than they can fit in the restaurant, so they plan to hire a marquee for the car park. They also continue to plan the future growth of their business.

Task 1 – Number of guests

The friends send emails about the opening night to 128 people.

They ask each person contacted to forward the email to five other people in exchange for entry into a prize draw for a free meal for two.

Based on the assumptions in the box on the right, how many people who received an email would you expect to attend the opening night?

Assumptions

- Half of the 128 people forward the email to 5 others.
- $\frac{1}{4}$ of these forward the email to 5 others.
- $\frac{1}{8}$ of these forward the email to 5 others.
- $\frac{1}{16}$ of these forward the email to 5 others.
- No one receives the email more than once.
- 10% of the people who receive the email go to the opening night.

Task 2 – Marquees

They consider various marquees to hire. One option is shown together with a scale plan. It has two vertical poles which come up through the roof.

Ropes *AP*, *BQ*, *CQ* and *DP* join the points *A*, *B*, *C* and *D*, which are at ground level, to the points *P* and *Q*, which are 8.6 m above the ground near the tops of the poles.

Assuming that the ropes *AP*, *BQ*, *CQ* and *DP* are straight lines, calculate the following.

a The length of the rope *QC*.

b The angle that the rope *QC* makes with the horizontal.

Another marquee is in the shape of a cylinder with a cone on top of it.

c Calculate the volume of the marquee.

d Calculate the surface area of the roof and sides of the marquee.

The marquee company charges a flat fee of £500 for hire of all marquees, plus £1 per cubic metre for marquees larger than the standard size of 8 m × 6 m × 3 m.

e Find which marquee is cheaper to hire using the volume of each marquee.

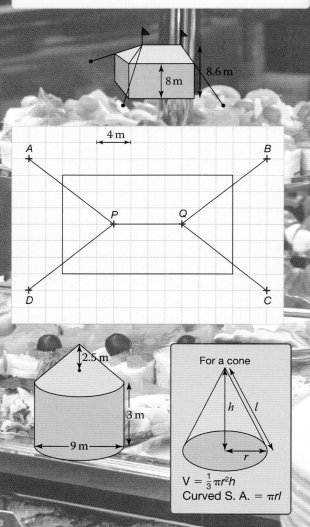

For a cone

$V = \frac{1}{3}\pi r^2 h$

Curved S. A. = $\pi r l$

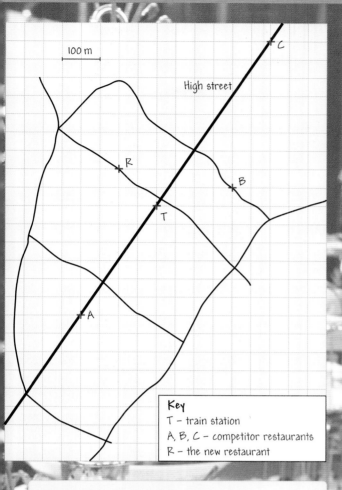

Key

T – train station

A, B, C – competitor restaurants

R – the new restaurant

Results of market research

- P(customer has a starter) = 0.7
- P(customer who has had a starter has an expensive main course) = 0.2
- P(customer who has had a starter has a medium main course) = 0.5
- P(customer who has had a starter has a cheap main course) = 0.3
- P(customer who has not had a starter has an expensive main course) = 0.4
- P(customer who has not had a starter has a medium main course) = 0.4
- P(customer who has not had a starter has a cheap main course) = 0.2
- P(Any customer has a dessert) = 0.4

Projections

Number of customers in first month = 700
Growth of 5% per month

Task 3 – Marketing slogan

The friends make the following claim in their marketing material for the opening night:

'Closest restaurant to the train station!'

Determine if their claim is correct by

a Finding the following as column vectors (using metres as units) $\vec{TB}$ $\vec{TA}$ $\vec{TR}$

b and so finding the distances TB, TA and TR.

c Use the cosine rule and your answers to part **b** to find the angle ARB.

Task 4 – Forecasting

The friends agree to price meals as follows.

All starters £4.

Main courses	£15 (expensive)
	£13 (medium)
	£11 (cheap)

All desserts £3

Assuming all customers have a main course, but not all have a starter or dessert, use the market research and the prices to estimate the probability that a customer spends more than £17.50 on food.

Task 5 – Future growth

The friends have used the outcome of the opening night, and some additional market research, to make some projections about the growth of the business. They create a table as part of a report to the bank who gave them the business loan.

Based on their projections, copy and complete the following table for their report.

Number of customers in one years' time (start of 13th month)	
Total number of customers in the first year	
Month in which 2000 customers first reached	

21 Sequences

Introduction

Musical scales are typically written using eight notes: the C Major scale uses C D E F G A B C. The interval between the first and last C is called an octave.

The pitch of a musical note, measured in Hertz (Hz), corresponds to the number of vibrations per seconds.

The frequencies of the corresponding notes in each octave follow a geometric sequence. If the C in one octave is 130.8 Hz then the C in the next octave is $2 \times 130.8 = 261.6$ Hz, the next C is $2 \times 261.6 = 523.2$ Hz, etc.

What's the point?

Understanding the relationship between terms in a sequence lets you find any term in the sequence and begin to understand its properties.

Objectives

By the end of this chapter you will have learned how to …

- Find terms of a linear sequence using a term-to-term or position-to-term rule.
- Recognise special types of sequence and find terms using either a term-to-term or position-to-term rule.
- Find terms of a quadratic sequence using a term-to-term or position-to-term rule.

Check in

1 Complete the next two values in each pattern.

   a   2, 4, 6, 8, ...   b   100, 94, 88, 82, 76, ...   c   1, 2, 4, 7, 11, ...

   d   10, 7, 4, 1, ...   e   3, 6, 12, 24, ...   f $1, \dfrac{1}{2}, \dfrac{1}{3}, \dfrac{1}{4}, \dfrac{1}{5}, ...$

2 Given that $n = 3$, put these expressions in ascending order.

   $2n + 7$ $4(n - 1)$ $2n^2$ $\dfrac{9}{n} + 15$ $15 - n$

3 This pattern has been shown in three different ways.
   It has a special name.
   What is it called?
   Describe how it got this name.

1	4	9	16
1×1	2×2	3×3	4×4

Chapter investigation

Abi, Bo and Cara are making patterns with numbers.

Abi makes a sequence by adding a fixed number onto a starting number.

1, 4, 7, 10, 13, 16, 19, 22, 25, 28, ...   Start with 1, add 3 each time

Bo makes a second sequence by adding Abi's sequence onto another starting number.

1, 2, 6, 13, 23, 36, 52, 71, 93, 118, ...   Start with 1, then add 1, then add 4, then add 7 then add 10, etc.

Cara takes the first two numbers in Bo's sequence and makes a third term by adding these together, then a fourth term by adding the second and third terms, etc.

1, 2, 3, 5, 8, 13, 21, 34, 55, 89, 144, ... $1 + 2 + 3, 2 + 3 = 5, 3 + 5 = 8$, etc.

How do sequences like these behave?

21.1 Linear sequences

- In a **linear** sequence the terms go up or down by the same amount.

> 12, 15, 18, 21, 24, ...
> goes up in 3s, so it is linear.
>
> 100, 95, 90, 85, 80, ...
> goes down in 5s, so it is linear.

You can find the **nth term** of a linear sequence by comparing the sequence to the times table to which it is related. For example the sequence 10, 17, 24, 31, 38, ... is connected to the 7× table, hence:

Position	1	2	3	4	5
Multiples of 7	7	14	21	28	35
Sequence term	10	17	24	31	38

By comparing the sequence to the 7× table, you can see that you have to add 3 to the multiples of 7 to get the terms of the sequence. Hence nth term T_n = position × 7 + 3.

$$T_n = 7n + 3$$

You can use the nth term **formula** to find any term of the sequence quickly, for example, the 100th term of the above sequence is
$T_{100} = 7 \times 100 + 3$ so $T_{100} = 703$

> The terms go up in 7s.

- To find the general term of a linear sequence
  - work out the common difference
  - write the common difference as the coefficient of n
  - compare the terms in the sequence to the multiples of n.

Find the nth term and, hence, the 100th term of

a   2, 8, 14, 20, 26, ...   **b**   25, 20, 15, 10, 5, ...

a   Compare the sequence to the multiples of 6.

Position	1	2	3	4	5
Multiples of 6	6	12	18	24	30
Term	2	8	14	20	26

> This sequence goes up in 6s, so it must be connected to the 6 times table.

First term = 2 = 6 − 4   Second term = 8 = 12 − 4

Subtract 4 from the multiple of 6 to get each term.

$T_n = 6n - 4$
$T_{100} = 6 \times 100 - 4 = 596$

b   Compare this sequence to the multiples of − 5.

Position	1	2	3	4	5
Multiples of 5	− 5	− 10	− 15	− 20	− 25
Term	25	20	15	10	5

> If a sequence goes down use negative multiples.

First term = 25 = − 5 + 30   Second term = 20 = − 10 + 30

Add 30 to the multiple of − 5 to get each term.

$T_n = -5n + 30$ or $T_n = 30 - 5n$
$T_{100} = 30 - 5 \times 100 = -470$

> On a number line, to get from −5 to 25, you need to add 30.

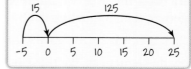

Exercise 21.1S

1 Write the first five terms in each of these sequences.

   a Even numbers

   b Odd numbers larger than 16

   c Multiples of 4

   d Multiples of 6 greater than 20

   e Two more than the 5 times table

   f Square numbers

   g One more than square numbers

   h Powers of 2

2 Write the first five terms of these sequences.

   a 2nd term 7, increases by 3 each time.

   b 2nd term 19, decreases by 6 each time.

   c 3rd term 12, increases by 4 each time.

   d 3rd term 14, decreases by 8 each time.

   e 5th term is 8, increases by 6 each time.

3 Copy each sequence and add the next two terms.

   a 4, 9, 14, 19, 24, ___, ___

   b 100, 93, 86, 79, 72, ___, ___

   c 1, 2, 4, 7, 11, ___, ___

   d 9, 99, 999, 9999, 99 999, ___, ___

   e 1, 1, 2, 3, 5, 8, ___, ___

   f 54, 27, 13.5, 6.75, ___, ___

4 Copy and fill in the missing numbers in each sequence.

   a 4, ___, 10, ___, 16, ...

   b 4, ___, ___, 32, 64, ...

   c 95, ___, ___, ___, 87, ...

   d 1, ___, 27, 64, ___, ...

5 Find the first three terms and the 10th term of these sequences.

   a $5n + 1$ **b** $3n + 8$

   c $8n - 4$ **d** $6n - 8$

   e $24 - 2n$ **f** $15 - 5n$

   g $7n - 20$ **h** $4n - 6$

6 Find the nth term for these sequences.

   a 11, 17, 23, 29, 35, ...

   b 1, 10, 19, 28, 37, ...

   c 15, 22, 29, 36, 43, ...

   d $-10, -6, -2, 2, 6, ...$

   e 20, 17, 14, 11, 8, ...

   f $15, 11, 7, 3, -1, ...$

   g $16, 8, 0, -8, -16, ...$

   h $31, 23, 15, 7, -1, ...$

7 Find the nth term for each of these arithmetic sequences.

   a 7, 11, 15, 19, 23, ...

   b $-6, -2, 2, 6, 10, ...$

   c $32, 23, 14, 5, -4, ...$

   d $15, 9, 3, -3, 9, ...$

8 Find a formula for the nth term, T_n, of each sequence.

   a 4, 9, 14, 19, 24, ...

   b 1, 3, 5, 7, 9, ...

   c 10, 12, 14, 16, 18, ...

   d 1, 1.5, 2, 2.5, 3, ...

   e $-4, -2, 0, 2, 4, ...$

   f 1, 2, 3, 4, 5, ...

   g The multiples of 13

   h Counting up in 10s, starting from 4

   i 10, 8, 6, 4, 2, ...

   j 100, 95, 90, 85, 80, ...

   k $50, 49\frac{3}{4}, 49\frac{1}{2}, 49\frac{1}{4}, ...$

   l Counting down in multiples of 4 from 75

9 Find the nth term of these sequences.

   a $1, -4, -9, -14, -19, ...$

   b 2, 3.5, 5, 6.5, 8, ...

   c 8, 6.5, 5, 3.5, 2, ...

   d 1.4, 2, 2.6, 3.2, 3.8, ...

   e $2, 2\frac{1}{2}, 3, 3\frac{1}{2}, 4, ...$

   f $3, 2\frac{1}{2}, 2, 1\frac{1}{2}, 1, ...$

Q 1165, 1173   SEARCH

21.1 Linear sequences

- To find the general term of a linear sequence
  - work out the common difference
  - write the common difference as the coefficient of n
  - compare the terms in the sequence to the multiples of n.

2, 6, 10, 14, 18, ...

differences all = +4

nth term rule = $4n \pm \square$

$$\begin{array}{ccccc} & 2, & 6, & 10, & 14, & 18 \\ \text{Compare} & 4, & 8, & 12, & 16, & 20 \\ & -2, & -2, & -2, & -2, & -2 \end{array}$$

nth term rule = $4n - 2$

To find the nth term of a pattern

① use the sequence of numbers or diagrams to complete a table for the sequence

② find the common difference and work out the nth term

③ use the formula for the nth term to answer the question.

Here is a sequence of patterns of dots.

a Write the number of dots in the first five patterns.

b Find, in terms of n, an expression for the number of dots in the nth pattern.

c Is there a pattern with 200 dots?

a 5, 7, 9, 11, 13

b Common difference = 2

② Each term is 3 more than a multiple of 2.

Number of dots in nth pattern is $2n + 3$.

③ Look for what changes (the $2n$ term) and what stays the same (the $+3$ term).

c ③ Compare the nth term with the number of dots.

$$2n + 3 = 200$$
$$2n = 197$$
$$n = 98.5 \qquad n \text{ must be a whole number.}$$

There isn't a pattern with 200 dots.

①

n	1	2	3	4	5
$2n$	2	4	6	8	10
Sequence	5	7	9	11	13

Caitlyn tries to find a term that is common to the sequences with nth terms $10n - 4$ and $24 - 3n$

Caitlyn's answer

$10n - 4 = 24 - 3n$
$13n - 4 = 24$
$13n = 27$
$n = 2.07969...$
n is not a whole number.
There isn't a term in both sequences.

Explain why Caitlyn is wrong.

Caitlyn's answer shows that there is not a common term that also has the **same position** in both sequences.

Generate the first few terms of each sequence.

$10n - 4 \quad 6, 16, 26, ...$

All positive numbers that end in 6.

$24 - 3n \quad 21, 18, 15, ...$

Multiples of 3 counting down from 21.

6 ends in 6 and is a multiple of 3, so 6 is in both sequences.

Exercise 21.1A

1 Derive your own formulae for these patterns and explain why they work.

a Connect the number of coloured tiles (B) and the number of white tiles (W).

b **i** Connect the number of white circles (W) with the number of coloured circles (B).

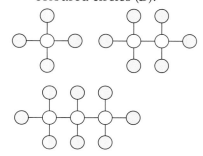

ii Use the same pattern to connect the number of white circles (W) with the number of lines (L).

2 By considering the sequence of patterns, write a formula to connect the quantities given.

a

Relate the number of edges E to the number of hexagons H.

b

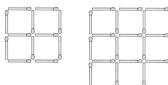

Relate the number of matches M with the length of the square L.

3 Are these statements true or false?

a The 50th term of 2, 5, 8, 11, 14, ... is more than 150.

b The 50th term of 5, 9, 13, 17, 21, ... is even.

c The 100th term of 1000, 990, 980, 970, 960, ... is negative.

4 Here are some terms of a sequence. In each case find the formula for the nth term, T_n.

a The 5th term is 20, the 6th term is 28 and the 7th term is 36.

b The 100th term is 302, 101st term is 305 and the 102nd term is 308.

c The 153rd term is 260, the 154th term is 262 and the 155th term is 264.

5 How could you find the nth term formula for these fractional sequences? See if you can work out what it would be in each case.

a $\dfrac{3}{7}, \dfrac{5}{10}, \dfrac{7}{13}, \dfrac{9}{16}, \dfrac{11}{19}, \cdots$

b $\dfrac{10}{30}, \dfrac{12}{27}, \dfrac{14}{24}, \dfrac{16}{21}, \dfrac{18}{18}, \cdots$

c $\dfrac{7}{1}, \dfrac{8}{4}, \dfrac{9}{9}, \dfrac{10}{16}, \dfrac{11}{25}, \cdots$

d $\dfrac{1}{11}, \dfrac{8}{9}, \dfrac{27}{7}, \dfrac{64}{5}, \dfrac{125}{3}, \cdots$

> Find separate formulae for the numerator and denominator.

6 Is 75 a term in the sequence described by the nth term $5n - 3$?

Explain your reasoning.

7 Find the term that is common to the sequences with nth terms $60 - 12n$ and $7n + 1$.

*8 A sequence has first three terms 140, 133, 126 ...

A second sequence has first three terms $-59, -54, -49$ ...

a Find the nth term of each sequence and prove that the 17th term is the same in both sequences.

b There are 6 terms that are common to both sequences.

Can you find them all without writing out both sequences?

> **Hint:** Think of the numbers in the second sequence as 'one more than a multiple of 5'.

*9 Reya says

> '*The two sequences* $T(n) = 7 + 5n$ *and* $S(n) = 54 - 6n$ *don't have any terms in common because when I solve* $T(n) = S(n)$ *I get* $n = 4\frac{3}{11}$ *which isn't an integer*'.

Give reasons why Reya is wrong.

21.2 Quadratic sequences

Quadratic sequences can be generated and described using a

● 'term-to-term' rule

● 'position-to-term' rule.

> For the sequence
>
> $\quad$ 2, 5, 10, 17, 26...
>
> 'add consecutive odd numbers starting with 3'.
>
> nth term = $n^2 + 1$

> If you plot T(n) against n for a quadratic sequence, then you will get points on a quadratic curve.

EXAMPLE

Find the first five terms of the sequence using this term-to-term rule.

First term $\quad$ 5 $\qquad$ Rule $\quad$ Add consecutive odd numbers

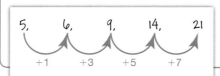

5, $\quad$ 6, $\quad$ 9, $\quad$ 14, $\quad$ 21

$\quad$ +1 $\quad\quad$ +3 $\quad\quad$ +5 $\quad\quad$ +7

● Quadratic sequences are sequences in which the second differences between terms are constant.

EXAMPLE

The nth term of a sequence is $n^2 + n + 1$.

a $\quad$ Find the first five terms of the sequence. $\qquad$ **b** $\quad$ Check the sequence is a quadratic sequence.

a $\quad$ 3, 7, 13, 21, 31, ...

$T(1) = 1^2 + 1 + 1 = 3$
$T(2) = 2^2 + 2 + 1 = 7$
$T(3) = 3^2 + 3 + 1 = 13$
$T(4) = 4^2 + 4 + 1 = 21$
$T(5) = 5^2 + 5 + 1 = 31$

> T(1) is the first term of the sequence.
>
> T(2) is the second term of the sequence.

b $\quad$ 3, $\quad$ 7, $\quad$ 13, $\quad$ 21, $\quad$ 31, ...

$\qquad$ 4 $\quad$ 6 $\quad$ 8 $\quad$ 10 $\qquad$ First difference

$\qquad\quad$ 2 $\quad$ 2 $\quad$ 2 $\qquad$ Second difference

As the second difference is constant, the sequence is quadratic.

EXAMPLE

Find the nth term for these quadratic sequences.

a $\quad$ 3, 8, 15, 24, 35, ... $\qquad\qquad$ **b** $\quad$ 5, 11, 21, 35, 53, ...

a $\quad$ 3, $\quad$ 8, $\quad$ 15, $\quad$ 24, $\quad$ 35, ...

$\quad$ +5 $\quad$ +7 $\quad$ +9 $\quad$ +11 $\quad$ First difference

$\quad$ +2 $\quad$ +2 $\quad$ +2 $\qquad$ Second difference

The coefficient of n^2 is half the value of the second difference.

nth term = $n^2 \pm \square$

Compare the sequence with the first term of the sequence 'n^2'.

$\qquad\quad$ 3, $\quad$ 8, $\quad$ 15, $\quad$ 24, $\quad$ 35
$-n^2 \quad$ 1, $\quad$ 4, $\quad$ 9, $\quad$ 16, $\quad$ 25
$\qquad\quad$ 2, $\quad$ 4, $\quad$ 6, $\quad$ 8, $\quad$ 10

By inspection

Linear sequence, nth term = $2n$

Quadratic sequence, nth term = $n^2 + 2n$

b $\quad$ 5, $\quad$ 11, $\quad$ 21, $\quad$ 35, $\quad$ 53, ...

$\quad$ +6 $\quad$ +10 $\quad$ +14 $\quad$ +18 $\quad$ First difference

$\quad$ +4 $\quad$ +4 $\quad$ +4 $\qquad$ Second difference

nth term = $2n^2 \pm \square$

Compare the sequence with the first terms of the sequence '$2n^2$'.

$\qquad\qquad$ 5, $\quad$ 11, $\quad$ 21, $\quad$ 35, $\quad$ 53
$-2n^2 \quad$ 2, $\quad$ 8, $\quad$ 18, $\quad$ 32, $\quad$ 50
$\qquad\qquad$ 3, $\quad$ 3, $\quad$ 3, $\quad$ 3, $\quad$ 3

Constant difference +3

Quadratic sequence, nth term = $2n^2 + 3$

Exercise 21.2S

1 Find the first four terms of these sequences using the term-to-term rules.

   a First term   2

      Rule Add consecutive odd numbers

   b First term   5

      Rule Add consecutive even numbers

   c First term   10

      Rule Add consecutive whole numbers

   d First term   5

      Rule Add consecutive multiples of 3

   e First term   10

      Rule Subtract consecutive odd numbers

   f First term   8

      Rule Subtract consecutive even numbers

2 Find the next two terms of these quadratic sequences.

   a 3, 7, 13, 21, □, □

   b 4, 6, 10, 16, □, □

   c 1, 5, 10, 16, □, □

   d 1, 4, 9, 16, □, □

   e 10, 7, 6, 7, □, □

   f 8, 18, 24, 26, □, □

   g 10, 7, −1, −14, □, □

3 Find the missing terms in these sequences.

   a 6, 12, 22, □, 54

   b □, 8, 18, 32, 50

   c 2, □, 10, 17, 26

   d 1, 5, □, 19, □

   e −4, 10, 28, □, □

   f 6, 12, 15, □, 12

   g 5, 8, 6, □, −13

4 a Find the first three terms for the sequences described by these rules.

      i $T(n) = n^2 + 3$

      ii $T(n) = n^2 + 2n$

      iii $T(n) = n^2 + n - 2$

      iv $T(n) = 2n^2$

      v $T(n) = 2n^2 + 4$

   b Calculate the 10th term of each sequence.

5 a Using the nth term for each sequence, calculate the first four terms.

      i $n^2 + 1$ **ii** $n^2 - 3$

      iii $n^2 + 2n + 1$ **iv** $n^2 + n - 4$

      v $2n^2 + n$ **vi** $0.5n^2 + n + 1$

      vii $2n^2 + 2n + 2$ **viii** $n(n - 2)$

   b Calculate the second difference in each case to check the sequences are quadratic.

6 Find the nth term of these quadratic sequences.

   a 4, 7, 12, 19, ...   **b** −3, 0, 5, 12, ...

   c 2, 8, 18, 32, ...   **d** 2, 6, 12, 20, ...

   e 5, 12, 21, 32, ...   **f** 3, 8, 15, 24, ...

   g 6, 14, 26, 42, ...   ***h** 3, 4, 3, 0, ...

7 Find the nth term of these sequences.

   a 9, 6, 1, −6, ...   **b** 18, 12, 2, −12, ...

   c 0, −2, −6, −12, ... **d** 10, 8, 4, −2, ...

8 a For the sequence 2, 6, 12, 20, 30, ... predict the position of the first term that will have a value greater than 1000.

   b Find the nth term for the sequence 2, 6, 12, 20, 30, ...

   c Use your answer to part **b** to evaluate the accuracy of your prediction.

9 Is 150 a term in the quadratic sequence described by the nth term $n^2 + 3$?

Explain your reasoning.

10 Kirsty is generating a sequence using the rule $T(n) = 2n^2 - n$. She thinks that every term will be an even number. Do you agree with Kirsty? Explain your reasoning.

11 a Write the first four terms of the sequences with nth term

      i $2n + 1$ **ii** $5n - 3$

   b Use your answers from part **a** to find the nth term of the quadratic sequences

      i 9, 25, 49, 81   **ii** 10, 26, 50, 82

      iii 4, 49, 144, 289   **iv** 9, 54, 149, 294

   c Find the nth term of the sequence 49, 100, 169, 256

21.2 Quadratic sequences

- Know how to generate and describe quadratic sequences using
  - the 'term-to-term' rule and
  - the 'position-to-term' rule using the nth term.

3, 10, 21, 36, 55, ...

second differences all = +4

nth term rule = $2n^2 \pm \square$

		3,	10,	21,	36,	55
Compare:		2,	8,	18,	32,	50
		1,	2,	3,	4,	5

nth term rule = $2n^2 + n$

HOW TO

To describe a quadratic sequence using the nth term

(1) Find the first difference between each term and the next.

(2) Find the difference between the first differences: the sequence is quadratic if the second difference is constant.

(3) The coefficient of n^2 is half the value of the second difference.

(4) Add a linear sequence to adjust the expression for the nth term.

EXAMPLE

Who is right, Ed or Asha? Explain your reasoning.

Ed says

'Since $12 = 3 \times 4$ and there are 13 squares in the 4th pattern then in the 12th pattern there will be $3 \times 13 = 39$ squares.'

Asha says

'There is a pattern in this sequence with 420 squares.'

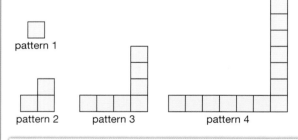

pattern 1 pattern 2 pattern 3 pattern 4

pattern 1 pattern 2 pattern 3 pattern 4

1,	3,	7,	13, ...
	2	4	6
		2	2

(1) First difference

(2) Second difference

(3) $1 = 2 \div 2$

The coefficient of n^2 is 1.

nth term = $n^2 + \square$

1,	3,	7,	13, ...
1	4	9	16
0	-1	-2	-3

(4) By inspection

Linear sequence, nth term = $1 - n$

Quadratic sequence, nth term = $n^2 - n + 1$

The 12th pattern has $12^2 - 12 + 1 = 133$ squares.

2	6	12	20, ...
	4	6	8
		2	2

The coefficient of n^2 is 1.

nth term = $n^2 + \square$

2	6	12	20, ...
1	4	9	16
1	2	3	4

Linear sequence, nth term = n

Quadratic sequence, nth term = $n^2 + n = n(n + 1)$

$n^2 + n = 420$ Is there a positive integer solution?

$n^2 + n - 420 = (n + 21)(n - 20) = 0$

$n = -21$ or 20

The 20th pattern has 420 squares.

Ed is wrong and Asha is correct.

Exercise 21.2A

1 Georgina thinks that the 10th term of the sequence 4, 7, 12, 19, 28, ... will be 56. Do you agree with Georgina? Explain your reasoning.

2 Bob thinks that the nth term of the sequence 5, 7, 11, 17, 25, ... will start with '$2n^2$'.

    a Do you agree with Bob? Explain your reasoning.

    b Find the full expression for the nth term of the sequence.

3   **a** Match each sequence with the correct 'position-to-term' rule.

    b Find the missing cells.

2, 6, 12, 20, ...	$T(n) = n^2 - n$
3, 0, −5, −12, ...	
0, 3, 8, 15, ...	$T(n) = n^2 + 2$
	$T(n) = n(n + 1)$
3, 6, 11, 18, ...	$T(n) = 4 - n^2$

    c Create your own matching puzzle.

4 The first terms of a quadratic sequence are 3 and 7. Find five different quadratic sequences with this property.

5 Sam is making patterns using matches.

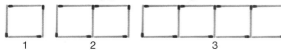

    1   2   3

    a Draw the next pattern.

    b Copy and complete the table.

Pattern (n)	1	2	3	4
Number of matches (m)	4			

    c Find a formula for the number of matchsticks, m, in the nth pattern.

    d Find the pattern that contains 136 matches.

    e How many matches will Sam need for the 50th pattern?

6 Draw a set of repeating patterns to represent the sequence described by the nth term $n^2 + 2$.

7 How many squares are there in the 15th pattern?

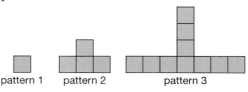

pattern 1   pattern 2   pattern 3

8 Four boys and four girls sit in a row as shown.

B	B	B	B		G	G	G	G

The boys want to swap places with the girls.

G	G	G	G		B	B	B	B

However, they are only allowed to move by either

    – sliding into an empty chair or

    – jumping over one person into an empty chair.

    a How many moves will it take?

    b How many moves will it take if there are 3 boys and 3 girls, 2 boys and 2 girls, ...?

    c Copy and complete the table.

Number of pairs	1	2	3	4
Number of moves				

    d Find a rule to predict the number of moves for 50 pairs of boys and girls.

9 Justify that the nth term of the triangular number sequence 1, 3, 6, 10, ... is $\frac{1}{2}n(n + 1)$.

    a Use an algebraic approach such as analysing the first and second differences to find the nth term.

    b Use a geometric approach by drawing each term of the sequence.

*10 By looking at successive differences, or otherwise, find expressions for the nth term of these *cubic* sequences.

    a 1, 8, 27, 64, 125, 216, ...

    b 2, 16, 54, 128, 250, 432, ...

    c 0, 6, 24, 60, 120, 210, ...

    **d −2, 2, 14, 40, 86, 158, ...

21.3 Special sequences

● Square, cube and triangular numbers are associated with geometric patterns.

Square numbers
1 4 9 16 25 …

Cube numbers
1 8 27 64 125 …

Triangular numbers
1 3 6 10 15 …

EXAMPLE

a Find a square number that is also **i** a triangular number **ii** a cube number.
b Show that 33, 34 and 35 can each be written as the sum of no more than three triangular numbers.

> List the numbers Square numbers   1,   4,   9,   16,   25,   (36,) 49, (64,) …
>
> Triangular numbers 1,   3,   6,   10,   15,   21,   28, (36,) …
>
> Cube numbers   1,   8,   27, (64,) 125, 216, …
>
> **a i** 36   **ii** 64   Look for numbers appearing in both lists.
>
> **b** 33 = 21 + 6 + 6   34 = 21 + 10 + 3   35 = 28 + 6 + 1
>
>   or 15 + 15 + 3   or 28 + 6   or 15 + 10 + 10

● **Arithmetic** (linear) sequences have a constant difference between terms. $T(n + 1) - T(n) = d$
● **Geometric** sequences have a constant ratio between terms. $T(n + 1) \div T(n) = r$

EXAMPLE

Classify these sequences as arithmetic or geometric.
a 5, 8, 11, 14, … **b** 1, −2, 4, −8, … **c** 3, 3√3, 9, 9√3, … **d** 7, 3, −1, −5, …

> **a** $8 - 5 = 11 - 8 = 14 - 11 = 3$ **b** $-2 \div 1 = 4 \div -2 = -8 \div 4 = -2$ $\boxed{\sqrt{3} \times \sqrt{3} = 3}$
>
>   Arithmetic   Constant difference, +3   Geometric   Constant ratio, -2
>
> **c** $3\sqrt{3} \div 3 = 9 \div 3\sqrt{3} = 9\sqrt{3} \div 9 = \sqrt{3}$ **d** $3 - 7 = -1 - 3 = -5 - -1 = -4$
>
>   Geometric   Constant ratio, $\sqrt{3}$ Arithmetic   Constant difference, -4

● In a **Fibonacci**-type sequence each term is the sum of the two previous terms: $T(n + 2) = T(n + 1) + T(n)$.

1, 1, 2, 3, 5, 8, 13, 21, … $2 = 1 + 1, 3 = 2 + 1, 5 = 3 + 2$, etc.

● In a **quadratic** sequence the differences between terms form an arithmetic sequence; the second differences are constant.

EXAMPLE

Find the missing terms in this quadratic sequence.

2, 6, 12, 20, ☐, 42, ☐

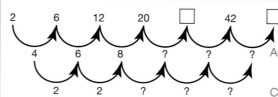

> 2   6   12   20   ☐   42   ☐
>
>   4   6   8   ?   ?   ?   Arithmetic sequence
>
>   2   2   ?   ?   ?   Constant difference +2
>
> Arithmetic sequence 4, 6, 8, 10, 12, 14
> Missing terms 20 + 10 = 30 and 42 + 14 = 56   Check 30 + 12 = 42

Exercise 21.3S

1 Write down the first ten terms of these number sequences.

   a triangular   **b** square   **c** cube

2 Express these numbers as the sum of not more than three triangular numbers.

   a 30   **b** 31   **c** 32

3 Describe these sequences using one of these words.

Arithmetic	Geometric
Quadratic	Fibonacci-type

   a $2, 5, 8, 11, \ldots$ **b** $7, 11, 15, 19, \ldots$

   c $2, 3, 5, 8, 13, \ldots$ **d** $2, 5, 10, 17, \ldots$

   e $2, 6, 18, 54, \ldots$ **f** $18, 15, 12, 9, \ldots$

   g $1, 4, 5, 9, \ldots$ **h** $1, 2, 4, 8, \ldots$

   i $4, 7, 13, 22, \ldots$ **j** $0.5, 2, 3.5, 5, \ldots$

   k $\frac{1}{4}, \frac{1}{2}, 1, 2, \ldots$ **l** $3, 1, \frac{1}{3}, \frac{1}{9}, \ldots$

   m $-2, 4, -8, 16, \ldots$ **n** $2, -2, 2, -2, \ldots$

   o $2, 2 + \sqrt{5}, 2 + 2\sqrt{5}, 2 + 3\sqrt{5}, \ldots$

   p $\sqrt{5}, 5, 5\sqrt{5}, 25, \ldots$

4 Do the triangular numbers form a quadratic sequence? Give your reason.

5 Find the next three terms of the following sequences using the properties of the sequence.

   a Arithmetic $2, 4, \square, \square, \square$

   b Geometric $2, 4, \square, \square, \square$

   c Fibonacci $2, 4, \square, \square, \square$

   d Quadratic $2, 4, \square, \square, \square$

6 Generate the first four terms of a geometric sequence using the following facts.

	First term	Multiplier
a	3	2
b	10	5
c	3	0.5
d	2	-3
e	$\frac{1}{2}$	$\frac{1}{2}$
f	-3	-2
g	4	$\sqrt{3}$
h	$\sqrt{3}$	$\sqrt{3}$
i	$2\sqrt{5}$	$\sqrt{5}$

7 Find the missing term in each sequence, giving your reason in each case.

   a $5, 10, \square, 20, 25, \ldots$

   b $5, 10, \square, 40, 80, \ldots$

   c $5, 10, \square, 25, 40, \ldots$

   d $5, 10, \square, 26, 37, \ldots$

8 The geometric mean of two numbers, x and y, is $\sqrt{xy}$. If a, b and c form a geometric sequence, show that b is the geometric mean of a and c.

9 By considering a number's prime decomposition, or otherwise, find three square numbers that are also cube numbers.

10 Generate the next five terms of these Fibonacci-type sequences starting from the first two terms given.

   a $T(n + 2) = T(n + 1) + T(n)$ $2, 1$

   b $T(n + 2) = T(n + 1) - T(n)$ $1, 1$

   c $T(n + 2) = T(n + 1) + 2T(n)$ $1, 1$

   d $T(n + 2) = 2T(n + 1) - T(n)$ $1, 1$

   e $T(n + 2) = 2T(n + 1) - T(n)$ $0, 1$

11 Iain thinks that the triangular number sequence can be generated using this rule.

$$T(n) = \frac{1}{2}n(n + 1)$$

Do you agree with Iain? Give your reason.

12 Alice thinks that the sequence $1, 4, 9, 16, 25, \ldots$ is a quadratic sequence.

Bob thinks that the sequence $1, 4, 9, 16, 25, \ldots$ is a square sequence.

Who is correct? Give your reason.

13 a Find the first five terms of these sequences.

   i $\dfrac{n}{n + 1}$ **ii** $\dfrac{n}{n^2 + 1}$

   iii $\dfrac{1}{n^2}$ **iv** $\dfrac{n^2}{n + 1}$

   v $(0.9)^n$ **vi** $(1.1)^n$

   b As n increases, comment on the behaviour of $T(n)$.

Q 1920, 1946, 1947   SEARCH

21.3 Special sequences

RECAP

- Square numbers 1, 4, 9, 16, ... n^2 ...
- Cube numbers 1, 8, 27, 64, ... n^3 ...
- Triangular numbers 1, 3, 6, 10, 15, ... – differences 2, 3, 4, 5, ...
- Geometric sequences – constant ratio
- Arithmetic sequence – constant differences
- Quadratic sequence – differences form an arithmetic sequence
- Fibonacci-type sequence – each term is the sum of the previous two terms

HOW TO

① Memorise the square, triangular and cube number sequences.

② Identify arithmetic and quadratic sequences by looking at the difference between terms.

③ Identify geometric sequences by looking at the ratio between terms.

④ Generate Fibonacci-type sequences by adding the previous terms to create the next term.

EXAMPLE

Create two sequences with the following properties.

a Arithmetic sequence with starting term 5   **b** Geometric sequence with starting term 3

c Fibonacci-type sequence with starting term 4   **d** Quadratic sequence with starting term 4

a	5, 8, 11, 14, 17, ...	② the rule is 'add 3'	
	5, 2, −1, −4, −7, ...	the rule is 'subtract 3'	
b	3, 6, 12, 24, 48, ...	③ the rule is '× 2'	
	$3\sqrt{5}$, 15, $15\sqrt{5}$, 75, $75\sqrt{5}$, ...	the rule is '× $\sqrt{5}$'	
c	4, 5, 9, 14, 23, ...	④ 9 = 5 + 4, 14 = 9 + 5, 23 = 14 + 9	
	4, 8, 12, 20, 32, ...	12 = 8 + 4, 20 = 12 + 8, 32 = 20 + 12	
d	4, 7, 12, 19, 28, ...	② first differences 3, 5, 7, 9	second difference is '+2'
	4, 2, −2, −8, −16, ...	first differences −2, −4, −6, −8	second difference is '−2'

EXAMPLE

Karl loves performing his amazing magic trick.

1 Pick any whole number.   **2** Write 4 times the number.

3 Add these two numbers together.   **4** Add the second and third numbers together.

5 Add the third and fourth numbers together.   **6** Keep repeating until you get the 8th number.

7 The 8th number is always 60 times the first number.

a Try the trick out. Does it always work?

b Does it work for negative integers, fractions and decimals?

c Explain how the trick works.

a	Start with 3	3, 4 × 3 = 12, 3 + 12 = 15, 12 + 15 = 27, 15 + 27 = 42, 27 + 42 = 69,	
		42 + 69 = 111, 69 + 111 = 180	Yes 180 = 60 × 3
b	Start with −2	−2, −8, −10, −18, −28, −46, −74, −120	Yes −120 = 60 × −2
	Start with $\frac{1}{2}$	$\frac{1}{2}$, 2, $2\frac{1}{2}$, $4\frac{1}{2}$, 7, $11\frac{1}{2}$, $18\frac{1}{2}$, 30	Yes 30 = 60 × $\frac{1}{2}$
	Start with 1.25	1.25, 5, 6.25, 11.25, 17.5, 28.75, 46.25, 75	Yes 75 = 60 × 1.25
c	Let the first number be N.		
	N, 4N, 5N, 9N, 14N, 23N, 37N, 60N	The 8th term is 60 × N.	

Exercise 21.3A

1 The first term of a sequence is 4. Create five more terms of

  a an arithmetic sequence

  b a geometric sequence

  c a Fibonacci-type sequence

  d a quadratic sequence.

2 Complete the missing cells

	Type	1st term	5th term	Rule
a	Arithmetic	12		$+5$
b	Arithmetic		23	$+4$
c	Arithmetic	0.5		-5
d	Arithmetic			$2n - 3$
e	Geometric	3		$\times 4$
f	Geometric		256	$\times 4$
g	Geometric	$\sqrt{3}$		$\times \sqrt{3}$
h	Geometric		$\dfrac{1}{16}$	$\times \dfrac{1}{4}$

3 Jenny thinks that the triangular number sequence can be created by starting with the number 1 and then adding on 2, adding on 3 and so on.

  a Do you agree with Jenny? Give your reason.

  b What are the pentagonal, hexagonal and tetrahedral numbers? How are they created?

4 a Research and produce a presentation about how Leonardo Fibonacci, also known as Leonardo of Pisa, discovered the Fibonacci sequence 1, 1, 2, 3, 5, …

  b How is the Fibonacci sequence linked to
  i Pascal's Triangle **ii** the Golden Ratio?

5 Hannah and Sam are given three options for a gift to celebrate their birthdays.

  Option 1 £500

  Option 2 £100 for the first month, £200 the next month, £300 the next month until the end of the year

  Option 3 During the month of their birthday 1p on Day 1, 2p on Day 2, 4p on Day 3, 8p on Day 4, …

5 Hannah's birthday is 20th February. Sam's birthday is 5th September.

  a Which option should Hannah choose? Give your reason.

  b Which option should Sam choose? Give your reason.

6 a Describe this geometric sequence.
  $$\frac{1}{2}, \frac{1}{4}, \frac{1}{8}, \frac{1}{16}, \ldots$$

  b A new sequence $S(n)$ is created by adding together the first n terms of the original sequence
  $S(1) = \frac{1}{2}$, $S(2) = \frac{1}{2} + \frac{1}{4} = \frac{3}{4}$ etc.
  Write down the first five terms of the sequence $S(n)$.

  c By considering this diagram, or otherwise, explain what $S(n)$ tends to as n gets large.

7 In the second example, how would you change Karl's magic trick so that

  a the 8th term is 99 times the starting number

  b the 12th term is 500 times the starting number?

  c Invent other magic tricks using sequences.

8 Rearrange each set of terms to make a geometric sequence.

  a $ab^7 \quad ab^3 \quad ab \quad ab^9 \quad ab^5$

  b $c^6d^3 \quad c^2d^7 \quad c^8d \quad c^4d^5 \quad c^{10}d^{-1}$

  c $24x^4 - 48x^3 \quad 48x^5 - 96x^4$
  $3x - 6 \quad 12x^3 - 24x^2 \quad 6x^2 - 12x$

9 The terms of a geometric sequence are given by $T(n) = ar^{n-1}$ where a and r are constants.

  Describe how the terms $T(n)$ behave for different values of r. Use words like diverge, converge, constant and oscillate.

1920, 1946, 1947 SEARCH

Summary

Checkout
You should now be able to...

✓ Generate a sequence using a term-to-term or position-to-term rule.	1, 2, 4
✓ Recognise a linear sequence and find a formula for its nth term.	3
✓ Recognise a quadratic sequence and find a formula for its nth term.	4, 5
✓ Recognise and use special sequences.	6 – 9

Language — Meaning — Example

Language	Meaning	Example
Sequence	An ordered set of numbers or other objects.	Square numbers 1 4 9 16 … $T(1) = 1$ First term, position 1 $T(2) = 4$ Second term, position 2
Term	One of the separate items in a sequence.	
Position	A number that counts where a term appears in a sequence.	
Term-to-term rule	A rule that links a term in a sequence with the previous term.	Sequence: 3, 5, 7, 9, 11, 13 …. Term-to-term rule: 'add 2' $T(n + 1) = T(n) + 2$ Position-to-term rule: $T(n) = 2n + 1$
Position-to-term rule / General term / nth term	A rule that links a term in a sequence with its position in the sequence.	
Linear / Arithmetic sequence	A sequence with a constant difference between terms. A graph of $T(n)$ against n gives points on a straight line.	4, 9, 14, 19, 24, … Constant difference = 5 $T(n + 1) = T(n) + 5$, $T(n) = 5n - 1$
Triangular numbers	The sequence formed by summing the integers: $1, 1 + 2, 1 + 2 + 3, …, \frac{1}{2}n(n + 1), …$	1, 3, 6, 10, 15, …
Geometric sequence	A sequence with a constant ratio between terms.	1, 3, 9, 27, 81, … Constant ratio = 3 $T(n + 1) = 3 \times T(n)$, $T(n) = \frac{1}{3} \times 3^n$
Fibonacci-type sequence	Each term is a sum of previous terms.	1, 1, 2, 3, 5, 8, 13, … $T(n + 2) = T(n + 1) + T(n)$
Quadratic sequence	A sequence in which the differences between terms form an arithmetic sequence; the second differences are constant. A graph of $T(n)$ against n gives points on a quadratic curve.	4, 9, 16, 25, … +5, +7, +9, … First difference 2, +2, … Second difference $T(n + 1) = T(n) + 2n + 3$, $T(n) = n^2 + 2n + 1$

Review

1 a What are the next three terms of these sequences?

    i   8, 17, 26, 35, …

    ii   71, 58, 45, 32, …

    iii 2.8, 4.4, 6, 7.6, …

  b Write the term-to-term rule for each of the sequences in part **a**.

2 Calculate the 13th term for the sequences with these position-to-term rules.

  a $T(n) = 5n + 8$

  b $T(n) = 12n - 15$

3 Write a formula for the nth term of these sequences.

  a 1, 7, 13, 19, …

  b 15, 22, 29, 36, …

  c 51, 39, 27, 15, …

  d $-6.5, -8, -9.5, -11, …$

4 The nth term of a sequence is given by $3n^2 + 4n$.

Calculate

  a the 7th term

  b the 10th term.

5 Work out the rule for the nth term of these sequences.

  a 4, 7, 12, 19, …

  b 2, 10, 24, 44, …

  c 4, 13, 26, 43, …

6 Classify each sequence using one of these words.

Linear	Quadratic
Geometric	Fibonacci-type

  a 2, −6, 18, −54, 162, …

  b 1, 3, 6, 10, 15, …

  c 1, 3, 4, 7, 11, …

  d 0.6, 0.45, 0.3, 0.15, 0, …

  e 4, 6, 10, 16, 24, …

  f 0.1, 0.01, 0.001, 0.0001, 0.00001, …

7 a Write down the next two terms of these sequences.

    i   1, 3, 6, 10, …   **ii** $1, \dfrac{1}{4}, \dfrac{1}{9}, \dfrac{1}{16}$

  b Write down the nth term of the sequences in part **a**.

8 This sequence is formed by doubling the current term to get the next term.

2, 4, 8, 16, …

  a Write down the next three terms of the sequence.

  b Write down the rule for the nth term of the sequence.

9 Write a rule for the nth term of these sequences and use it to work out the 10th term of each sequence.

  a $\sqrt{2}, 3\sqrt{2}, 5\sqrt{2} …$

  b $\dfrac{1}{3}, \dfrac{2}{5}, \dfrac{3}{7}, \dfrac{4}{9}, …$

What next?

<table>
<tr><td rowspan="3">Score</td><td>0 – 4</td><td></td><td>Your knowledge of this topic is still developing.
To improve look at MyMaths: 1054, 1165, 1166, 1173, 1920</td></tr>
<tr><td>5 – 8</td><td></td><td>You are gaining a secure knowledge of this topic.
To improve look at InvisiPens: 21Sa – i</td></tr>
<tr><td>9</td><td></td><td>You have mastered these skills. Well done you are ready to progress!
To develop your exam technique looks at InvisiPens: 21Aa – e</td></tr>
</table>

Assessment 21

1   Chris is given the terms of some sequences and the descriptions of the sequences.
He can't remember which description matches which group of terms.
Can you match each group of terms to the correct description?   [9]

a	The first five even numbers less than 5.	**i**	$-3, -1, 1, 3, 5$
b	The first five multiples of 7.	**ii**	$5, 10, 15, 20, 25, 30$
c	The first five factors of 210.	**iii**	$11, 13, 17, 19, 23, 29$
d	The first five triangular numbers.	**iv**	$7, 14, 21, 28, 35$
e	All the factors of 24.	**v**	$4, 2, 0, -2, -4$
f	Odd numbers from -3 to 5.	**vi**	$1, 3, 6, 10, 15$
g	Prime numbers between 10 and 30.	**vii**	$1, 2, 3, 4, 6, 8, 12, 24$
h	$15 <$ square numbers ≤ 81.	**viii**	$1, 2, 3, 5, 7$
i	Multiples of 5 between 3 and 33.	**ix**	$16, 25, 36, 49, 64, 81$

2   **a**   Nathan is given this sequence.

$$1, \quad 11, \quad 21, \quad 31, \quad 41$$

He says, 'the common difference of this sequence is $+11$.'
Is he correct? If not, work out the common difference.   [1]

   b   He also says, 'the nth term of this sequence is...'
Can you complete his sentence? Show your working.   [4]

3   **a**   For this sequence of patterns find the nth term for the

   i   perimeter   [2]   **ii**   area.   [2]

   b   A pattern in the sequence has an area of $48\,cm^2$. What is its perimeter?   [1]

   c   A pattern in the sequence has an perimeter of $48\,cm$. What is its area?   [1]

4   A very famous sequence starts with these numbers.

$$1, \quad 1, \quad 2, \quad 3, \quad 5, \quad 8, \quad 13, \quad 21, \quad 34, \quad \square, \quad \square, \quad \square$$

   a   Work out the term to term rule that shows how the series works.   [1]

   b   Using your rule work out the next two terms.   [1]

   c   If you are right, the next term after the ones you have just found is a square number.
What is it?   [1]

   d   What is the name of this type of sequence?   [1]

5   Kerry is given some nth terms, $T(n)$, for some sequences. She makes the following
statements. Which of her statements are correct and which are not correct?

For Kerry's statements that are not correct, rewrite the statement correctly.

a	$T(n) = 2n + 7$	The 10th term is 27.	[2]
b	$T(n) = 6n - 5$	The first 3 terms are $-5, 1, 7$.	[2]
c	$T(n) = 13 - 3n$	The 100th term is 287.	[2]
d	$T(n) = n^2 - 10$	The 10th term is 100.	[2]
e	$T(n) = 15 - 3n^2$	The 100th term is $-29\,985$.	[2]

6 Here is a sequence of patterns.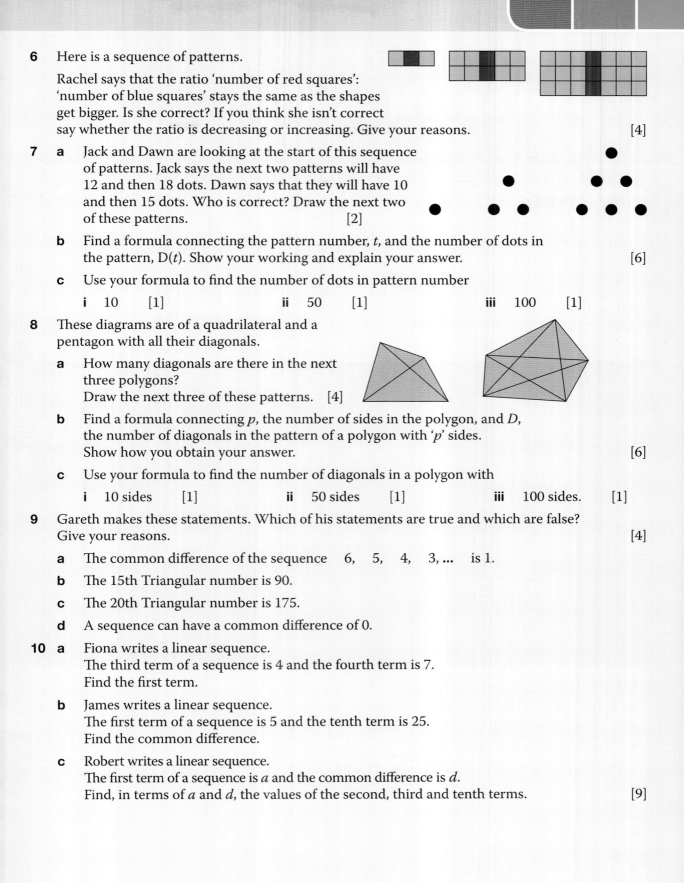

Rachel says that the ratio 'number of red squares':
'number of blue squares' stays the same as the shapes
get bigger. Is she correct? If you think she isn't correct
say whether the ratio is decreasing or increasing. Give your reasons. [4]

7 **a** Jack and Dawn are looking at the start of this sequence
of patterns. Jack says the next two patterns will have
12 and then 18 dots. Dawn says that they will have 10
and then 15 dots. Who is correct? Draw the next two
of these patterns. [2]

b Find a formula connecting the pattern number, t, and the number of dots in
the pattern, $D(t)$. Show your working and explain your answer. [6]

c Use your formula to find the number of dots in pattern number

i 10 [1]   **ii** 50 [1]   **iii** 100 [1]

8 These diagrams are of a quadrilateral and a
pentagon with all their diagonals.

a How many diagonals are there in the next
three polygons?
Draw the next three of these patterns. [4]

b Find a formula connecting p, the number of sides in the polygon, and D,
the number of diagonals in the pattern of a polygon with 'p' sides.
Show how you obtain your answer. [6]

c Use your formula to find the number of diagonals in a polygon with

i 10 sides [1]   **ii** 50 sides [1]   **iii** 100 sides. [1]

9 Gareth makes these statements. Which of his statements are true and which are false?
Give your reasons. [4]

a The common difference of the sequence   6,   5,   4,   3, ...   is 1.

b The 15th Triangular number is 90.

c The 20th Triangular number is 175.

d A sequence can have a common difference of 0.

10 **a** Fiona writes a linear sequence.
The third term of a sequence is 4 and the fourth term is 7.
Find the first term.

b James writes a linear sequence.
The first term of a sequence is 5 and the tenth term is 25.
Find the common difference.

c Robert writes a linear sequence.
The first term of a sequence is a and the common difference is d.
Find, in terms of a and d, the values of the second, third and tenth terms. [9]

22 Units and proportionality

Introduction

The half-life of a radioactive isotope is the time taken for half its radioactive atoms to decay. The number of radioactive isotopes remaining after each half-life forms a geometric sequence. Comparing the proportion of remaining radioactive isotopes to the geometric sequence, allows you to estimate the age of an artefact – even something that is millions of years old.

Scientists can estimate the age of a dinosaur fossil by analysing the proportion of radioactive uranium atoms in the surrounding layers of volcanic rock. The oldest dinosaur fossils are thought to be more than 240 million years old.

What's the point?

Understanding proportion and modelling growth and decay can help you understand the past and make predictions about the future.

Objectives

By the end of this chapter, you will have learned how to …

- Convert between standard units of measure and compound units.
- Use compound measures.
- Compare lengths, areas and volumes of similar shapes.
- Solve direct and inverse proportion problems.
- Describe direct and inverse proportion relationships using an equation.
- Recognise graphs showing direct and inverse proportion and interpret the gradient of a straight line graph.
- Find the instantaneous and average rate of change from a graph.
- Solve repeated proportional change problems.

Check in

1 **a** Increase these amounts by 10%.

    i £45   **ii** 60 mm   **iii** 48 km   **iv** 4 hours

  b Decrease these amounts by a quarter.

    i 6 miles   **ii** 58 minutes   **iii** 14 kg   **iv** £62

2 Calculate the average speed of cars that made these journeys.

  a 480 miles in 12 hours   **b** 85 miles in two hours

  c 27 miles in 30 minutes   **d** 48 miles in 90 minutes

3 Find the volume of these prisms.
  State the units of your answers.

  a   **b** 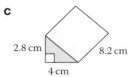   **c**

4 Work out the surface area of the prisms in question **3**.

Chapter investigation

In 1798 Thomas Malthus wrote that human populations typically grow by a fixed percentage each year but that agricultural yield grows by a fixed amount each year. Why does this lead to a 'Malthusian catastrophe'?

	Value in 1950	Growth per year
World population	2.5 billion	1.8%
Number of people who could be fed	4.0 billion	200 million

Using these numbers when would the catastrophe occur?

Could catastrophe be avoided if population growth was held fixed at its rate for the year 2000?

22.1 Compound units

Compound measures describe one quantity in relation to another.

These are examples of compound measures.

- **Speed** = $\dfrac{\text{total distance travelled}}{\text{total time taken}}$ Units such as m/s, km/h

- **Density** = $\dfrac{\text{mass}}{\text{volume}}$ Units such as g/cm³

- **Pressure** = $\dfrac{\text{force}}{\text{area}}$ Units such as N/m²

Use the triangle to work out which calculation to use.

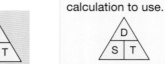

Cover D (for distance)
You multiply
S (speed) × T (time)

EXAMPLE

Kerry jogs at an average speed of 5 km/h for $1\frac{1}{2}$ hours.

What distance does she jog?

Distance = $5 \times 1\frac{1}{2}$

= 7.5 km

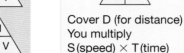

Density = $\dfrac{\text{mass}}{\text{volume}}$

EXAMPLE

Find the density of a piece of wood with cross-section area 42 cm², length 12 cm and mass 693 g.

Volume = 42 × 12 = 504 cm³

Density = 693 ÷ 504 = 1.375 g/cm³

Mass in grams. Volume in cm³.
So density is in g/cm³.

Force is measured in newtons (N).

EXAMPLE

A force of 18 N acts over an area of 5 m².

What is the pressure?

Pressure = $\dfrac{18}{5}$ = 3.6 N/m²

A rate is also a compound unit. It tells you how many units of one quantity there are compared with one unit of another quantity.

- **Rate of pay** = $\dfrac{\text{amount of money}}{\text{time}}$ Units such as £/h

Rate of flow is a compound measure. It is the volume of liquid that passes through a container in a unit of time.

- **Rate of flow** = $\dfrac{\text{volume}}{\text{time}}$ Units such as litres/s

EXAMPLE

Water empties from a tank at a rate of 1.5 litres per second.

It takes 10 minutes to empty the tank.

How much water was in the tank?

Use the triangle to work out which formula to use.

Volume = rate × time

Convert the time to seconds.

10 minutes = 10 × 60 s = 600 s

Volume in tank = 1.5 × 600 = 900 litres

Exercise 22.1S

1 The winners' times in some of the races at a sports day are

a 100 metres in 13 seconds

b 200 metres in 28 seconds

c 400 metres in 58.4 seconds

d 1500 metres in 4 minutes 52 seconds

Calculate the speed of each winner in m/s, correct to 1 dp.

2 Work out the distance travelled in

a 2 hours at 80 km/h

b 7 hours at 23 mph

c 6 seconds at 9 m/s

d 1 day at 12 mph.

3 Work out the time it takes to travel

a 180 kilometres at 60 km/h

b 280 miles at 70 mph

c 8 kilometres at 24 km/h

d 15 miles at 60 mph.

4 A cube with volume 640 cm³ has a mass of 912 g. Find the density of the cube in g/cm³.

5 An emulsion paint has a density of 1.95 kg/litre. Find

a the mass of 4.85 litres of the paint

b the number of litres of the paint that would have a mass of 12 kg.

6 The table shows the densities of different metals.

Metal	Density
Zinc	7130 kg/m³
Cast iron	6800 kg/m³
Gold	19320 kg/m³
Tin	7280 kg/m³
Nickel	8900 kg/m³
Brass	8500 kg/m³

Use the information in the table to find

a the mass of 0.8 m³ of zinc

b the mass of 0.5 m³ of cast iron

c the mass of 3.2 m³ of gold

d the volume of 910 g of tin

e the volume of 220 g of nickel

f the volume of a brass statue that has mass 17 kg.

7 The table shows the pressure, force and area of different materials.
Complete the table. Include the correct units in your answers.

	Pressure	Force	Area
a		12.9 N	10 m²
b		482.5 N	25 cm²
c	2560 N/m²	1200 N	
d	512 N/mm²		14.5 mm²
e	17.8 N/cm²	225 N	
f	24.6 N/m²		2.8 m²

8 If 4 metres of fabric costs £8.40, find the price of the fabric in pounds per metre.

9 Jane is paid £478 a week. Each week she works 40 hours. What is her hourly rate of pay?

10 Find the rate of flow for pipes A and B in litres/s.

a Pipe A: 20 litres of water in 8 seconds.

b Pipe B: 48 litres of water in 30 seconds.

11 Water empties from a tank at a rate of 2 litres/s. It takes 10 minutes to empty the tank.
How much water was in the tank?

12 An electric fire uses 18 units of electricity over a period of 7.5 hours.

a What is the hourly rate of consumption of electricity in units per hour?

b How many units of electricity are used in 24 hours?

13 A car has fuel efficiency of 8 litres per 100 km.

a How far can the car travel on 40 litres of petrol?

b How many litres of petrol would be needed for a journey of 250 km?

c What is the car's rate of fuel consumption in km per litre?

14 Rose received 75 US dollars in exchange for £50.

a Calculate the rate of exchange in US dollars per £.

b How many US dollars would she get for £125?

c If Rose received $120 US dollars, how many pounds did she exchange?

1121, 1246, 1970, 1971 SEARCH

22.1 Compound units

RECAP

Compound units describe one quantity in relation to another.
- The density of a material is its mass divided by its volume.
- Speed is the distance travelled divided by the time taken.
- Pressure is the force divided by the area.
- A formula triangle is a useful way to remember the relationships between the different parts.
- A rate is also a compound unit.

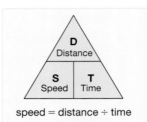

speed = distance ÷ time
distance = speed × time
time = distance ÷ speed

HOW TO

① Draw a formula triangle and write the correct formula.

② Convert units or work out quantities to apply the formula.

③ Work out the answer, making sure the units are correct.

EXAMPLE

A train leaves Norwich at 13:40 and arrives in Cambridge at 15:00.
If the distance is 90 km find the average speed of the train.

① Draw a formula triangle and write the formula for speed.

② Work out the time in hours.

③ Work out the answer using the correct units.

$Speed = \dfrac{distance}{time}$

$Time = 1 \text{ hour } 20 \text{ min} = 1\frac{1}{3}\text{ h} = 1.333... \text{ h}$

$Speed = \dfrac{90}{1.333} = 67.5 \text{ to 3 sf}$

The average speed of the train is 67.5 km/h.

EXAMPLE

Sand was falling from the back of a lorry at a rate of 0.4 kg/s.
It took 20 minutes for all the sand to fall from the lorry.

How much sand was the lorry carrying?

① Draw a formula triangle and write the formula for mass.

② Convert the time to seconds.

③ Work out the answer using the correct units.

The rate of flow is in kg/s, which is mass divided by time.
Mass = rate × time

20 minutes = 20 × 60 s = 1200 s

Mass = 0.4 × 1200 = 480

The lorry was carrying 480 kg of sand.

EXAMPLE

A metal cuboid has a length of 7 cm, a width of 5 cm and a height of 4 cm.
It has a mass of 1.470 kg. Find its density in g/cm³.

① Draw a formula triangle and write the formula for density.

② Work out the volume of the cuboid and convert mass to grams.

③ Work out the answer using the correct units.

$Density = \dfrac{mass}{volume}$

Volume of cuboid = length × width × height
$= 7 \times 5 \times 4 \text{ cm}^3 = 140 \text{ cm}^3$

Mass = 1.470 kg = 1470 g

$Density = \dfrac{1470}{140} = 10.5 \text{ g/cm}^3 \text{ to 3 sf}$

Ratio and proportion   Units and proportionality

Exercise 22.1A

1 A train leaves Euston at 8:57 am and arrives at Preston at 11:37 am.
If the distance is 238 miles find the average speed of the train.

2 Sand falls from the back of a lorry at a rate of 0.2 kg/s.
It took 25 minutes for all the sand to fall from the lorry.
How much sand was the lorry carrying?

3 A metal cuboid has a length of 9 cm, a width of 5 cm and a height of 4 cm.
It has a mass of 1.53 kg.
Find its density in g/cm³.

4 A car travels 24 miles in 45 minutes.
Find the average speed of the car in miles per hour (mph).

5 Copy and complete the table to show speeds, distances and times for five different journeys.

Speed (kmph)	Distance (km)	Time
105		5 hours
48	106	
	84	2 hours 15 minutes
86		2 hours 30 minutes
	65	1 hours 45 minutes

6 A cube of side 2 cm weighs 40 grams.

   a Find the density of the material from which the cube is made. Give your answer in g/cm³.

> Volume of cube = length³.

   b A cube of side length 2.6 cm is made from the same material.
Find the mass of this cube, in grams.

7 A solid block has a length and width of 22.50 mm, and a height of 3.15 mm. It has a mass of 9.50 g.

   a Find the density of the metal from which the block is made.
Give your answer in g/cm³.

   b How many blocks can be made from 1 kg of the material?

8 In this question, give your answers in kg/m³.

   a The volume of 31.5 g of silver is 3 cm³.
Work out the density of silver.

   b The volume of 18 g of titanium is 4 cm³.
Work out the density of titanium.

   c A sheet of aluminium foil has volume 0.4 cm³ and mass 1.08 g. Work out the density of aluminium foil.

9 Grace earns £340 per week for 40 hours work.
If Grace works overtime, she is paid 1.5 times her standard hourly rate.

   a How much is Grace paid for 7 hours of overtime work?

   b Grace earned £531.25 last week. How many hours of overtime did she work?

10 The toll charged for a car travelling on a motorway was £33.60 for a journey of 420 km.
Cars with trailers are charged double.
How much would it cost for a car with a trailer to travel 264 km?

11 A yacht race has three legs of 8 km, 6 km and 10 km.
The average speed for the winning yacht was 6.2 km/h.
The second yacht finished 8 minutes after the winner.
How long did it take the second place yacht to finish the race?

12 a Julia is wearing high-heeled shoes.
Each heel has an area of 1 cm². Julia weighs 55 kg.
How much pressure does Julia's heel exert when she has one heel on the ground?

   b An elephant's foot is 45 cm across is approximately circular.
An elephant walks with two feet on the ground at a time.
An elephant weighs 5500 kg.
How much pressure does an elephant's foot exert when the elephant has two feet on the ground?

22.2 Converting between units

The relationships between the metric units of area are

- **1 cm² = 10 × 10 mm²**
  = **100 mm²**

- **1 m² = 100 × 100 cm²**
  = **10 000 cm²**

$1 m^2 = 100 cm \times 100 cm$

EXAMPLE

Change $5\,m^2$ to cm^2.

$5\,m^2 = 5 \times 10\,000\,cm^2$
$= 50\,000\,cm^2$

> You expect a larger number, so multiply.

The relationships between the metric units of **volume** are

- **1 cm³ = 10 × 10 × 10 mm³**
  = **1000 mm³**

- **1 m³ = 100 × 100 × 100 cm³**
  = **1 000 000 cm³**

$1 m^3 = 100 cm \times 100 cm \times 100 cm$

In **similar** shapes, corresponding sides are in the same **ratio**.

You can use the ratio to work out **areas** and **volumes** in similar shapes and solids.

- For similar shapes with length ratio $1:n$
  - the ratio of the areas is $1:n^2$
  - the ratio of the volumes is $1:n^3$

Area 6

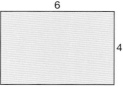

Area 24

Length ratio is $1:2$
Area ratio is $6:24 = 1:4 = 1:2^2$

Volume 12

Volume 96

Length ratio is $1:2$
Volume ratio is $12:96 = 1:8 = 1:2^3$

EXAMPLE

Two similar Russian dolls are on display.

a The surface area of the smaller doll is $7.2\,cm^2$. Work out the surface area of the larger doll.

b The volume of the larger doll is $145.8\,cm^3$. Work out the volume of the smaller doll.

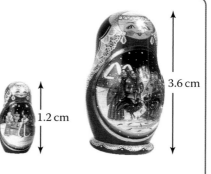

3.6 cm

1.2 cm

Ratio of lengths, smaller : larger = $1.2 : 3.6 = 1 : 3$

a Area ratio is $1 : 3^2 = 1 : 9$

Surface area of larger doll = $9 \times 7.2 = 64.8\,cm^2$

b Volume ratio is $1 : 3^3 = 1 : 27$

Volume of smaller doll = $145.8 \div 27 = 5.4\,cm^3$

Ratio and proportion Units and proportionality

Exercise 22.2S

1 Convert these measurements to millimetres.

  a 5 cm   **b** 8 cm

  c 15 cm   **d** 6 cm 7 mm

  e 19 cm 3 mm   **f** 4.5 cm

  g 4.3 cm   **h** 10.6 cm

  i 80 cm   **j** 1 m

2 Convert these measurements to centimetres.

  a 60 mm   **b** 85 mm

  c 240 mm   **d** 63 mm

  e 4 mm   **f** 4 m

  g 10 m   **h** 3.5 m

  i 1.6 m   **j** 1.63 m

3 Convert these measurements to metres.

  a 400 cm   **b** 450 cm

  c 475 cm   **d** 470 cm

  e 50 cm   **f** 1 km

  g 4 km   **h** 0.5 km

  i 3.5 km   **j** 18 km

4 Here are two identical rectangles, A and B.

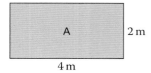

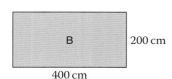

  a Calculate the area of rectangle A in m^2.

  b Calculate the area of rectangle B in cm^2.

5 **a** Calculate the area of this rectangle in m^2.

  b Convert your answer to cm^2.

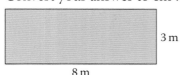

6 Convert these areas to mm^2.

  a $4 cm^2$

  b $7.3 cm^2$

  c $10.9 cm^2$

  d $2.5 cm^2$

  e $400 cm^2$

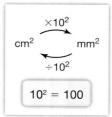

$\times 10^2$

$cm^2 \quad mm^2$

$\div 10^2$

$10^2 = 100$

7 Convert these areas to cm^2.

  a $600 mm^2$ **b** $1200 mm^2$

  c $850 mm^2$ **d** $6500 mm^2$

  e $10000 mm^2$

8 Convert these areas to m^2.

  a $40000 cm^2$

  b $85000 cm^2$

  c $1000000 cm^2$

  d $125000 cm^2$

  e $5000 cm^2$

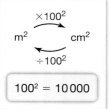

$\times 100^2$

$m^2 \quad cm^2$

$\div 100^2$

$100^2 = 10000$

9 Convert these areas to cm^2.

  a $5 m^2$ **b** $10 m^2$ **c** $6.5 m^2$

  d $7.75 m^2$ **e** $0.6 m^2$

10 P and Q are two similar cuboids.

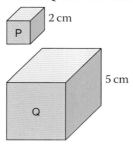

  a The surface area of cuboid P is $37.2 cm^2$. Work out the surface area of cuboid Q.

  b The volume of cuboid P is $12.4 cm^3$. Work out the volume of cuboid Q.

22.2 Converting between units

RECAP

- $1\,cm = 10\,mm$
- $1\,cm^2 = 10\,mm \times 10\,mm = 100\,mm^2$
- $1\,cm^3 = 10 \times 10 \times 10\,mm^3 = 1000\,mm^3$
- $1\,m = 100\,cm$
- $1\,m^2 = 100\,cm \times 100\,cm = 10\,000\,cm^2$
- $1\,m^3 = 100 \times 100 \times 100\,cm^3 = 1\,000\,000\,cm^3$
- If the scale factor for length is n then the scale factor for area is $n \times n = n^2$, and the scale factor for volume is $n \times n \times n = n^3$.

> In similar shapes, the corresponding sides are in the same ratio.

HOW TO

1. Compare corresponding lengths, areas or volumes.
2. Find the ratio between lengths, areas or volumes.
3. Use the ratio to answer the question.
   Make sure that you include units in your answer.

EXAMPLE

P and Q are similar shapes.

Calculate the surface area of shape Q.

 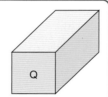

Volume 260 cm³   Volume 877.5 cm³
Surface area 244 cm²   Surface area = ?

1. Compare the volumes of the shapes.

    Volume ratio

    $$\frac{V_Q}{V_P} = \frac{877.5}{260} = \frac{27}{8}$$

2. Find the length ratio first.

    Length ratio

    $$\frac{L_Q}{L_P} = \frac{\sqrt[3]{27}}{\sqrt[3]{8}} = \frac{3}{2}$$

    Then work out the area ratio.

    Area ratio

    $$\frac{A_Q}{A_P} = \frac{3^2}{2^2} = \frac{9}{4}$$

3. Multiply the ratio by the surface area of shape P.

    Surface area of

    $$Q = 244 \times \frac{9}{4} = 549\,cm^2$$

> Always sense check your answer – are you expected a larger or a smaller value?

Exercise 22.2A

1 A soft drink comes in a bottle.
The bottle is 20 cm tall and contains 330 ml of juice.
As an advertising stunt, a similar bottle is made the size of a man.
The bottle is 1.8 m tall.

   a The label has an area of 28 cm². What is the area of the label on the larger bottle?

   b What is the capacity of the larger bottle?

2 J and K are two similar boxes.

The volume of J is 702 cm³.

The volume of K is 208 cm³.

The surface area of J is 549 cm².

Calculate the surface area of K.

3 X and Y are two similar solids.

The total surface area of X is 150 cm².

The total surface area of Y is 216 cm².

The volume of Y is 216 cm³.

Calculate the volume of X.

4 C and D are two cubes.

   a Explain why any two cubes must be similar.

   b Explain why any two cuboids are not necessarily similar.

The ratio of the side lengths of cube C and cube D is 1:7.

   c Write down the ratio of

     i their surface areas

     ii their volumes.

5 Two model cars made to different scales are mathematically similar.

The overall widths of the cars are 3.2 cm and 4.8 cm respectively.

   a What is the ratio of the radii of the cars' wheels?

The cars are packed in mathematically similar boxes so that they just fit inside the box.

   b The surface area of the larger box is 76.5 cm².

     Work out the surface area of the smaller box.

   c The volume of the smaller box is 24 cm³.

     Work out the volume of the larger box.

6 At the local pizzeria Gavin is offered two deals for £9.99.

> Deal A: One large round pizza with radius 18 cm
>
> Deal B: Two smaller round pizzas each with radius 9 cm

Which deal gives the most pizza?

7 A frustum is created by removing a cone of height 10 cm from a cone of height 15 cm.

Show that the volume of the frustum is in the ratio 19:8 to the volume of the cone removed.

8 Cone A has surface area 260.3 cm² and volume 188.5 cm³.
Cone B has surface area 1017.9 cm² and volume 1352.2 cm³.
Are the cones similar?
Explain your answer.

22.3 Direct and inverse proportion

● Numbers or quantities are in **direct proportion** when the ratio of each pair of corresponding values is the same.

One litre of a shade of purple paint is made by mixing 200 ml of red paint and 400 ml of blue paint. The ratio of red paint to blue paint is 1:2.

● When you multiply (or divide) one of the variables by a certain number, you have to multiply or divide the other variable by the same number.

250 ml of red paint has to be mixed with $2 \times 250 = 500$ ml of blue paint.

You can always find the amount of white paint by multiplying the amount of red paint by a fixed number, the constant of proportionality, in this case 4.

● You write 'y is proportional to x' as $y \propto x$. This can also be written as $y = kx$, where k is the **constant** of proportionality.

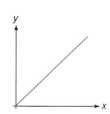

The graph of $y = kx$ is a straight line that passes through the origin.

EXAMPLE

If 35 metres of steel cable weigh 87.5 kilograms, how much do 25 metres of the same cable weigh?

If x = length of the cable (metres) and w = mass (kilograms), then $w = kx$

The constant of proportionality k represents the mass of one metre of cable.

$87.5 = k \times 35 \Rightarrow k = 87.5 \div 35 = 2.5$ so $w = 2.5x$

Substitute $x = 25$ into the formula: $w = 2.5 \times 25 = 62.5$ kg

The graph of $y = \frac{k}{x}$ is a reciprocal graph.

● When variables are in **inverse proportion**, one of the variables increases as the other one decreases, and vice-versa.

● 'y is inversely proportional to x' can be written as $y \propto \frac{1}{x}$, or $y = \frac{k}{x}$, where k is the **constant** of proportionality.

● If two variables are in inverse proportion, the **product** of their values will stay the same.

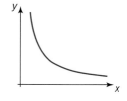

EXAMPLE

A pot of paint, which will cover an area of $12\,\text{m}^2$, is used to paint a garage floor. Find the width of the floor when the length is

   a   1 m   **b**   3 m   **c**   6.5 m

Use l for the length of the floor and w for the width

$lw = 12 \Rightarrow w = \frac{12}{l}$

Substitute the given values into the formula.

   a $w = \frac{12}{1} = 12$ m   **b** $w = \frac{12}{3} = 4$ m   **c** $w = \frac{12}{6.5} = 1.85$ m

Exercise 22.3S

1 The table shows corresponding values of the variables w, x, y and z.

w	3	6	9	15
x	8	14	20	32
y	5	10	15	25
z	4	7	10	16

Which of these statements could be true?

a $w \propto x$ **b** $z \propto x$

c $z \propto w$ **d** $y \propto w$

e $w = ky$, where k is a constant

2 Using the values from the table in question **1**, plot graphs of

a w against x **b** w against y

c x against y **d** x against z.

Use your results to describe the key features of a graph showing the relationship between two variables that are in direct proportion.

3 The weight, w, of a piece of wooden shelving is directly proportional to its length, l.

a Write a formula that connects w and l, including a constant of proportionality, k.

b Given that 2.5 m of the shelving weighs 6.2 kg, find the value of the constant k.

c Calculate the weight of a 2.9 m length of the shelving.

4 You are told that the variable y is directly proportional to the variable x.
Explain what will happen to the value of y when the value of x is

a doubled   **b** halved

c multiplied by 6   **d** divided by 10

e multiplied by a factor of 0.7.

5 You are told that the variable w is inversely proportional to the variable z.
Explain what will happen to the value of w when the value of z is

a doubled   **b** halved

c multiplied by 6   **d** divided by 10

e multiplied by a factor of 0.7.

6 It takes 20 hours for 5 people to decorate 100 cakes.

a How many people are needed to decorate 100 cakes in 10 hours?

b How many hours will it take 4 people to decorate 100 cakes?

c How long will it take 5 people to decorate 200 cakes?

> Take care! Hours and people are inversely proportional.
> People and cakes are directly proportional.

7 You are told that y is inversely proportional to x, and that when $x = 4$, $y = 4$. Find the value of y when x is equal to

a 8   **b** 2   **c** 40

d 1   **e** 100.

8 Given that $y \propto \frac{1}{w}$, and that $y = 10$ when $w = 50$, write an equation connecting y and w.

9 You are told that $y = \frac{k}{x}$, and that $y = 20$ when $x = 40$. Find the value of the constant k.

10 The variables u and v are in inverse proportion to one another. When $u = 6$, $v = 8$. Find the value of u when $v = 12$.

11 The distance, d, that Freya drives travelled while driving at a constant speed is directly proportional to the time spent driving, t.

a Write the statement 'd is directly proportional to t' as a formula including a constant of proportionality, k.

b Given that Freya travels 72 miles in $1\frac{1}{2}$ hours, find the value of the constant k.

c Use your previous answers to calculate the distance travelled in $2\frac{1}{2}$ hours.

d How long would it take Freya to drive 160 miles? Give your answer in hours and minutes.

22.3 Direct and inverse proportion

- Numbers or quantities are in **direct proportion** when the ratio of each pair of corresponding values is the same.
- You write 'y is proportional to x' as $y \propto x$. This can also be written as $y = kx$, where k is the **constant** of proportionality.
- 'y is inversely proportional to x' can be written as 'y is proportional to $\frac{1}{x}$' or $y = \frac{k}{x}$, where k is the constant of proportionality.
- When two quantities are in inverse proportion, then their graph is a reciprocal curve.

- One quantity can be directly proportional to the square, cube, square root, cube root ... of another quantity.
- One quantity can be inversely proportional to the square, cube, square root, cube root ... of another quantity.

1. Write the proportional relationship as an equation.
2. Substitute values into the equation to find the constant, k.
3. Use the formula to answer the question.

A wedding cake is going to be made from two round tiers. The baking time, T minutes, is directly proportional to the square of the individual cake's radius, R mm.
When R = 150, T = 50.
Find T when R = 180.

1. T is directly proportional to R^2.

   $T \propto R^2$ so $T = kR^2$

2. Substitute values into the equation to find k.

   When $R = 150$, $T = 50$, so $50 = k \times 150^2$

   $k = 0.002222\ 2...$

3. Use the formula to find T when R = 180.

   $T = 0.00222... \times 180^2$ $T = 72$

You could leave your value for k as an exact fraction instead of a decimal.

The force of attraction, F, between two magnets is inversely proportional to the square of the distance, d, between them. Two magnets are 0.5 cm apart and the force of attraction between them is 50 newtons.

When the magnets are 3 cm apart what will be the force of attraction between them?

1. F is inversely proportional to d^2.

   $F \propto \dfrac{1}{d^2}$, so $F = \dfrac{k}{d^2}$

2. Substitute values into the formula to find k.

   When F is 50, d is 0.5 $50 = \dfrac{k}{0.5^2}$

   $k = 50 \times 0.5^2 = 12.5$

3. Use the formula to find d.

   So $F = \dfrac{12.5}{d^2}$

   When $d = 3$ $F = \dfrac{12.5}{3^2} = 1.4$ newtons

Ratio and proportion   Units and proportionality

Exercise 22.3A

1 A shop sells drawing pins in two different packs.

> Pack A contains 120 drawing pins and costs £1.45.
>
> Pack B contains 200 of the same drawing pins, and costs £2.30.

Calculate the cost of one drawing pin from each pack, and explain which pack is better value.

2 A store sells packs of paper in two sizes.

Regular	Super
150 sheets	500 sheets
Cost £1.05	Cost £3.85

Which of these two packs gives better value for money? You must show all of your working.

3 R varies with the square of *s*. If *R* is 144 when *s* is 1.2, find

a a formula for *R* in terms of *s*

b the value of *R* when *s* is 0.8

c the value of *s* when *R* is 200.

4 For a circle, explain how you know that the area is directly proportional to the square of its radius.

State the value of the constant of proportionality in this case.

What other formulae do you know that show direct proportion?

5 The surface area of a sphere varies with the square of its radius. If the surface area of a sphere with radius 3 cm is 113 cm², find the surface area of a sphere with radius 7 cm.

6 Imogen carries out a scientific experiment. She sets two magnets at varying distances apart and notices that the force between them decreases as the square of the distance increases. She performs the experiment three times. These are her results.

Distance *d* (cm)	1	2	3	4	5
*d*²	1	4	9	16	25
Force *F* (newtons)	100	25	?	?	4

6 **a** Study Imogen's results. Write a formula for *F* in terms of *d* and use it to complete the rest of the table.

b Plot a graph of *F* against *d* and join the points to form a smooth curve.

7 Match the four sketch graphs with the four proportion statements.

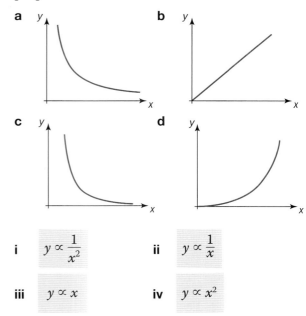

i $y \propto \dfrac{1}{x^2}$ **ii** $y \propto \dfrac{1}{x}$

iii $y \propto x$ **iv** $y \propto x^2$

8 A carpet layer is laying square floor tiles in two identical apartments. The time *t* (mins) that it takes to lay the tiles is inversely proportional to the square of the length of each tile. Given that tiles with a side length of 20 cm take 4 hours and 10 minutes to lay, how long will the second apartment take using tiles with side length 40 cm?

9 Given that *y* is inversely proportional to the square root of *x* and that *y* is 5 when *x* is 4, find

a a formula for *y* in terms of *x*

b the value of

   i *y* when *x* is 100   **ii** *x* when *y* is 12.

10 Given that *P* varies inversely with the cube of *t* and that *P* is 4 when *t* is 2, find

a a formula for *P* in terms of *t*

b without a calculator

   i *P* when *t* is 3   **ii** *t* when *P* is $\frac{1}{2}$.

 Q 1048, 1948, 1949   SEARCH

22.4 Rates of change

- The gradient of a line segment is calculated as $\dfrac{\text{change in the } y\text{-direction}}{\text{change in the } x\text{-direction}}$.

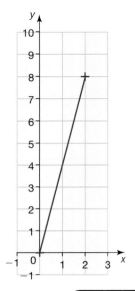

14.1

The chord from $(0, 0)$ to $(2, 8)$ has gradient 4.
The graph of $y = x^3$ also passes through the points $(0, 0)$ and $(2, 8)$.
However, the gradient of the curve changes from point to point.

- To estimate the gradient of a curve at a point P, find the gradient of a chord between points on either side of P.

EXAMPLE

Estimate the gradient of the graph of $y = x^3$ at $(1, 1)$ by finding the gradient of the chord joining $(0.999, 0.999^3)$ to $(1.001, 1.001^3)$.

① Use the formula gradient $= \dfrac{\text{change in the } y\text{-direction}}{\text{change in the } x\text{-direction}}$

$$\frac{1.001^3 - 0.999^3}{1.001 - 0.999} = 3.000001$$

As you zoom in on the graph of $y = x^3$ at $(1, 1)$ and you will see that the curve starts to look more and more like a straight line.

Advanced mathematics can be used to prove that the gradient of $y = x^3$ at $(1, 1)$ is precisely 3.

The straight line through $(1, 1)$ with gradient 3 is called a tangent to the curve.

- The average gradient of a curve **between two points** is the gradient of the chord joining the two points.
- The gradient of a curve **at a point** is the gradient of the tangent at that point.

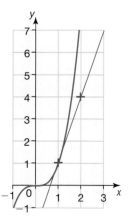

14.1

EXAMPLE

Find the equation of the tangent to the curve $y = x^3$ at the point $(1, 1)$.

① The tangent is a straight line through $(1, 1)$, with gradient 3.

$y = mx + c$ $1 = 3 + c$ so $c = -2$

The equation is $y = 3x - 2$.

Exercise 22.4S

1 a Estimate the gradient of the curve $y = x^3$ at the point (2, 8) by finding the gradient of the chord joining (1, 1) to (3, 27).

b Improve the estimate of part **a** by using a chord closer to (2, 8).

c Use your best estimate to find the equation of the tangent at (2, 8).

2 a Estimate the gradient of the curve $y = x^3 - 9x$ at the point (2, −10) by finding the gradient of the chord joining (1, −8) to (3, 0).

b Improve the estimate of part **a** by using a chord closer to (2, −10).

c Use your best estimate to find the equation of the tangent at (2, −10).

3 a Plot the graph of $y = x^2$ for $-3 \leqslant x \leqslant 3$.

b Draw, as accurately as possible, tangents at the points (±1, 1) and (±2, 4).

c Hence complete the table of gradients.

x	−2	−1	0	1	2
Gradient					

d Suggest a possible formula for the gradient at any point (p, p^2) on the graph of $y = x^2$.

4 A graph consists of two straight line segments; from O to A and from A to B as shown.

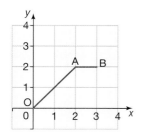

a Find the gradients of

   i OA   **ii** AB.

b What is the average gradient of the graph from O to B?

c Explain why the average gradient cannot be found by simply taking the average of your answers to **ai** and **aii**.

5 a Sketch the graph of $y = x^2 + x - 2$.

b Find the gradients of the chords through

   i (0, −2) and (6, 40)

   ii (1, 0) and (5, 28)

   iii (2, 4) and (4, 18)

c What is the gradient of the curve at (3, 10)?

d Use your sketch of the curve to illustrate the answers to **b** and **c**.

*6 The sketch shows the graph of $y = x^2$ with the tangent drawn at point $P(p, p^2)$.

The tangent crosses the y-axis at point Q.

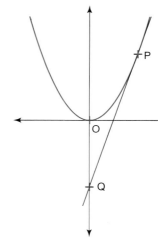

Use your answer to **3d** to find the equation of the tangent at P and show that point Q has coordinates $(0, -p^2)$.

*7 a For the graph of $y = x^2 - 10$, find the gradient of the chord joining the point where $x = 4 - h$ to the point where $x = 4 + h$.

b Use your answer to **a** to explain why the gradient of $y = x^2 - 10$ at the point (4, 6) is precisely 8.

Q 1312, 1953   SEARCH

22.4   Rates of change

- The gradient of a graph at a point can be estimated geometrically by drawing a tangent at that point.
- The gradient of a graph at a point can be estimated numerically by using a chord near that point.

HOW TO

(1) Use the fact that the gradient at a point can represent instantaneous rates of change, especially speed.

(2) Interpret rates of change in different contexts.

EXAMPLE

A ball is thrown vertically upwards with an initial speed of $20\,\text{ms}^{-1}$. Its height – time graph is as shown. The graph has equation $h = 20t - 5t^2$.

  a   How is the initial speed represented on this graph?

  b   Use a chord to estimate the vertical speed of the ball 1 second after being thrown.

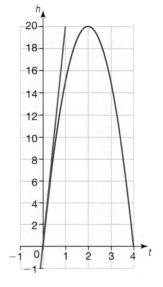

(1)(2)   Each gradient at a point represents vertical speed.

a   The gradient of the tangent at the origin.

     The gradient of the tangent is 20, so initial speed = $20\,\text{ms}^{-1}$.

b   The gradient of the chord from $t = 0.9$ to $t = 1.1$ is

     given by $\dfrac{15.95 - 13.95}{1.1 - 0.9} = 10\,\text{ms}^{-1}$.

(2)   Interpret this in the context of throwing the ball.

     After 1 second, the ball's vertical speed is $10\,\text{ms}^{-1}$.

EXAMPLE

Water is poured at a constant rate into each of these containers.

The containers are initially empty.

Draw graphs of depth of water against time for each container.

  a   **b**   **c**

(2)   The depth of water in the cylinder will increase at a constant rate.

(1)   Therefore the graph of depth against time will have a constant gradient.

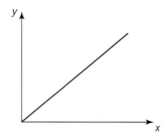 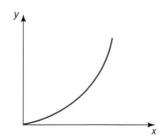 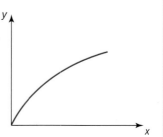

Ratio and proportion   Units and proportionality

Exercise 22.4A

1 A ball is thrown vertically upwards with an initial velocity of $30 \, ms^{-1}$. Its height t seconds later is given by $h = 30t - 5t^2$.

    a Use a chord to estimate its velocity 4 seconds after being thrown.

    b Interpret your answer.

2

The graph shows how the expected weekly profit from the manufacture of a *Choco* bar depends upon its selling price.

    a Explain the meaning of the two points where the profit is zero.

    b What is the importance of the point where the gradient of the graph is zero?

    c What selling price would you recommend and why?

3

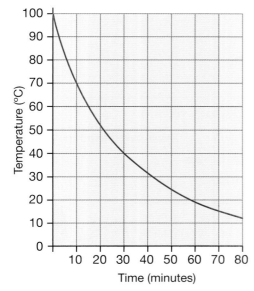

The graph shows a cooling curve for a container of water.

3 **a** Use tangent lines to obtain rough estimates of the gradients of the curve at times 0, 30 and 60 minutes.

    b What are the units of these gradients?

    c Use your answers to **a** and **b** to write a brief description of the cooling of the water.

4 The speed, $v \, ms^{-1}$, of a projectile at time t seconds is given by

$$v = 20 - 10t.$$

    a Sketch a graph of v against t.

    b What is measured by the gradient of this graph?

    c Find the gradient of the graph and state its units.

5 The space shuttle used two solid fuel rockets which burned for just two minutes and were then parachuted back to Earth from a height of 45 kilometres.

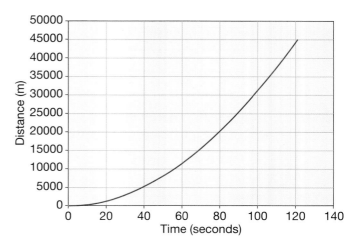

    a Calculate the average speed of the space shuttle during the first two minutes of flight.

    b Use a tangent line to estimate the speed of the space shuttle when the solid fuel rockets detached.

22.5 **Growth and decay**

12.3

- To increase something by r%, multiply it by $\dfrac{100 + r}{100} = 1 + \dfrac{r}{100}$

- To decrease something by r%, multiply it by $\dfrac{100 - r}{100} = 1 - \dfrac{r}{100}$

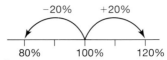

To increase something by 20%, multiply by 1.2
To reduce something by 20%, multiply by 0.8

1.2 and 0.8 are called **multipliers**.

Sometimes percentage increases or decreases are repeated.

If interest is calculated on the interest it is **compound interest** otherwise it is **simple interest**.

EXAMPLE

The number of grey squirrels in a forest is 540. The population increases by 10% each year.

a Draw a graph to illustrate the growth over the next 6 years.

b After how long will the population double?

This is called **exponential growth**.

18.3

a 100% + 10% = 110%, multiplier = 1.1

Find values to plot on a graph.

Years	Population
0	540
1	540 × 1.1 = 594
2	594 × 1.1 = 653
3	653 × 1.1 = 719
4	719 × 1.1 = 791
5	791 × 1.1 = 870
6	870 × 1.1 = 957

Try using the 'Ans' key on your calculator to repeat calculations.

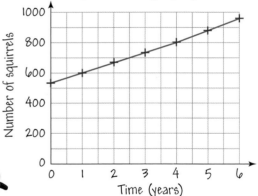

b After 7 years, population = 957 × 1.1 = 1053   Not quite double, 540 × 2 = 1080.

After 8 years, population = 1053 × 1.1 = 1158

It will take just over 7 years for the population to double.

EXAMPLE

The value of a car is £25 000. The value decreases by 22% each year.

a Write a formula for the value of the car after n years.   This is called **exponential decay**.

b Use your formula to find the value of the car after 8 years.

c After how long will the car be worth less than £1000?

a 100% − 22% = 78%, multiplier = 0.78

Value after n years = £25 000 × 0.78^n

You could check this by multiplying by 0.78 eight times.

b Value after 8 years = £25 000 × 0.78^8 = £3425   (nearest £)

c Continuing to multiply £3425 by 0.78 gives

2672, 2084, 1625, 1268, 989

After 13 years the car will be worth less than £1000.

Ratio and proportion   Units and proportionality

Exercise 22.5S

1 Find the decimal multiplier for each percentage change.

    a Increase of 25%   **b** Decrease of 25%

    c Increase of 2.5%   **d** Decrease of 2.5%

2 The number of trout in a lake is 800. The number decreases by 15% each year.

    a Draw a graph to illustrate the fall in the population over the next 8 years.

    b **i** Find a formula for the number of trout after n years.

        ii Use your formula to check three values on your graph.

3 A bacteria population doubles every 20 minutes.

    a Starting with 1 bacterium, draw a graph of the population growth over 3 hours.

    b Estimate the number of bacteria after

        i 150 minutes   **ii** 170 minutes

    c What has happened to the population between 150 minutes and 170 minutes?

4 The table gives information about the population growth of two bacteria colonies.

Colony	Population now	Increase per hour
A	200	50%
B	400	35%

    a Show that the population of Colony A after n hours is 200×1.5^n

    b Find an expression for the population of Colony B after n hours.

    c When does the population of Colony A become bigger than that of Colony B?

5 The value of a new car is £16 000. The car loses 15% of its value at the start of each year.

    a Find a formula for the value of the car after n years.

    b Find the value of the car after 4 years.

    c After how many complete years will the car's value drop below £4000?

6 The population of a town is 52 000. The population increases by 1.5% each year.

    a Find the population after 6 years.

    b When will the population reach 60 000?

7 Sadie invests £2000 in a savings account. The bank adds 4% compound interest at the end of each year. Sadie does not add or take any money from the account for 10 years.

    a Copy and extend this table to show how Sadie's investment grows.

End of year	Amount in the account (£)
1	2000 × 1.04 = 2080
2	

    b Work out the percentage interest that Sadie's investment earns in 10 years.

    c £P is invested with compound interest r% added at the end of each time period,

        i Show that the total amount at the end of n time periods is

$$A = P\left(1 + \frac{r}{100}\right)^n$$

> A time period is usually a year or a number of months.

        ii Use this formula to check the last amount in your table in part **a**.

8 The half-life of a radioactive substance is the time it takes for the amount to go down to half of the original amount.

After how many half-lives will there be less than 1% of the radioactive substance left?

*9 The table shows how the population of the world has grown since 1900.

    a Draw a graph of this data.

    b A growth function $P = 1.65 \times 1.0125^{(y - 1900)}$ where P is the population in billions in year y has been suggested as a model.

        i Show this function on your graph.

        ii What annual percentage increase in world population does the model assume?

Year	Population (billions)
1900	1.65
1910	1.75
1920	1.86
1930	2.07
1940	2.30
1950	2.56
1960	3.04
1970	3.71
1980	4.45
1990	5.29
2000	6.09
2010	6.87

Q 1070, 1238   SEARCH

22.5 Growth and decay

12.3

RECAP

- To increase/decrease something by $r\%$, multiply by $1 \pm \dfrac{r}{100}$
- When a principal amount £P is invested with compound interest $r\%$ added at the end of each time period, the total accrued at the end of n time periods is

$$A = P\left(1 + \frac{r}{100}\right)^n$$

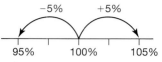

−5%   +5%

95%   100%   105%

▲ To increase something by 5%, multiply by 1.05
To decrease something by 5%, multiply by 0.95

HOW TO

To solve a repeated percentage change problem
1. Find the multiplier.
2. Decide whether to use a single calculation or a step-by-step approach.
3. Calculate the value required and ATQ.

Repeating steps is using an iterative process

EXAMPLE

Harry invests £4000 in an account. Interest of 1.5% is added at the end of every 6 months.
a   How much compound interest is earned in 4 years?
b   How much longer will it take for the amount in the account to reach £5000?
c   What assumptions have you made in answering these questions?

1. The multiplier $= 1 + \dfrac{1.5}{100} = 1.015$

a   Amount after 4 years $= 4000 \times 1.015^8 = £4505.97$ 2. 4 years $= 8 \times 6$ months
Round to the nearest pence.

Interest $= £4505.97 - £4000 = £505.97$ 3. Take away the original amount.

b   4573.56, 4642.16, 4711.80, 4782.47, 4854.21, 4927.02, 5000.93   2. Multiplying by 1.015.
This takes 7×6 months $= 3\frac{1}{2}$ years

c   The interest rate does not change and Harry does not add or take out any money.   3.

EXAMPLE

10.4

18.2

An angling club says that there are 1500 trout in a lake and each year the trout population falls by 20%. The club decides to add 500 trout to the lake at the end of each year.
a   Show that $T_{n+1} = 0.8T_n + 500$ where T_n is the number of trout in the lake after n years.
b   Estimate the number of fish in the lake after 4 years.
c   Sketch a graph to show how the trout population varies in this time. State any assumptions.

a $100\% - 20\% = 80\%$ 1.
Multiplying T_n by 0.8, then adding 500 gives $T_{n+1} = 0.8T_n + 500$

b $T_1 = 0.8 \times 1500 + 500 = 1700$ 2.
$T_2 = 0.8 \times 1700 + 500 = 1860$
$T_3 = 0.8 \times 1860 + 500 = 1988$
$T_4 = 0.8 \times 1988 + 500 = 2090$ 3.
Round to the nearest whole number.

c   Assumes the same conditions each year.   3.

Number of trout

2000

1500

0   1   2   3   4

Time (years)

Decay curves fall more quickly at first.

Exercise 22.5A

1 For each account in the table below, find the compound interest earned.

Acc	Original amount	Compound interest rate	Number of years
a	£250	4% per year	6
b	£840	2.5% per 6 months	5
c	£4500	1.25% per 3 months	3

2 A building society offers two accounts: Karen says that they would give the same interest on an investment. Is Karen correct? Explain your answer.

> **Easy Saver**
> 4% interest added at the end of each year
>
> **Half-yearly saver**
> 2% interest added at the end of every 6 months

3 A road planner uses the formula 2400×1.08^n to estimate the number of vehicles per day that will travel on a new road n months after it opens.

  a Describe two assumptions the planner has made.

  b Sketch a graph to show what the planner expects to happen.

  c Give reasons why the planner's assumptions may not be appropriate.

4 There are 250 rare trees in a forest, but each year the number of trees falls by 30%. A woodland trust aims to plant 60 more trees in the forest at the end of each year.

  a Show that $T_{n+1} = 0.7T_n + 60$ where T_n denotes the number of trees in the forest after n years.

  b Work out the number of trees after 5 years.

  c Sketch a graph to show how the number of trees varies in this time.

5 Ben takes out a loan for £500. Interest of 2% is added to the amount owing at the end of each month, then Ben pays off £90 or all the amount owing when it is less than £90.

  a How long will it take Ben to pay off the loan? Show your working.

5 b Work out the percentage interest that Ben will pay on the loan of £500.

6 Sally invests £8000 in an account that pays 3.5% interest at the end of each year. Sally has to pay 20% tax on this interest. Calculate how much Sally will have in her account at the end of 4 years.

7 Liam finds a formula for the compound interest earned by £P invested for 6 years at a rate of 4.5%. Here is Liam's method.

> Interest in 1 year = $0.045 \times £P$
> Interest for 6 years = $6 \times 0.045 \times £P = £0.27P$

  a Why is Liam's method incorrect?

  b Find a correct formula.

  c After 6 years the interest earned is £1934.46. Find, to the nearest one pound, the original amount £P.

8 Find the minimum rate of interest for an investment of £500 to grow to £600 in 6 years.

9 Tanya measures the temperature of a cup of coffee as it cools.

Time t (min)	0	10	20	30	40	50	60
Temperature T (°C)	85	68	55	45	39	34	31

  a i Use Tanya's data to draw a graph.

  ii Find the rate at which the coffee is cooling after half an hour.

  b Tanya says $T = 20 + 65 \times 0.97^t$ is a good model of the data.

  i Is Tanya correct? Show how you decide.

  ***ii** Explain each term in Tanya's model.

***10** The half-life of caesium-137 is 30 years.

  a Show that when 1 kilogram of caesium-137 decays, the amount left after t years is
  $$f(t) = 2^{-\frac{t}{30}} \text{ kg}$$

  b Sketch a graph of amount against time.

  c Describe how the function and graph would change if $f(t)$ was given in terms of grams instead of kilograms.

p.384

Summary

Checkout
You should now be able to...

	Test it Questions
✔ Use compound measures.	1 – 3
✔ Convert between standard units of measure and compound units.	4
✔ Compare lengths, areas and volumes of similar shapes.	5
✔ Solve direct and inverse proportion problems.	6 – 8
✔ Describe direct and inverse proportion relationships using an equation.	6 – 8
✔ Recognise graphs showing direct and inverse proportion and interpret the gradient of a straight line graph.	9
✔ Find the instantaneous and average rate of change from a graph.	10
✔ Solve repeated proportional change problems.	11

Language	Meaning	Example
Speed	A measure of the distance travelled by an object in a certain time.	The speed limit on the motorway is 70mph..
Density	A measure of the amount of matter in a certain volume.	<table><tr><th>Metal</th><th>Mass of 1 cm³</th><th>Density</th></tr><tr><td>Gold</td><td>19.3 g</td><td>19.3 g/cm³</td></tr><tr><td>Iron</td><td>7.87 g</td><td>7.87 g/cm³</td></tr></table>
Pressure	Pressure is a measure of the amount of force per unit area.	A woman standing on a wooden floor exerts 650 N force over a total footprint of 600 cm². The pressure she exerts is $\frac{650}{600} = 1.08$ N/cm² (2 dp)
Rate	One quantity measured per unit of time or per unit of another quantity.	Total pay for 3 hours work = £16.50 Rate of pay $= \frac{16.50}{3}$ $= £5.50$ per hour
Similar	The same shape but different in size.	(triangles: 4 cm, 5 cm; 8 cm, 10 cm)
Proportional	Two variables are proportional if one is always the same multiple of the other.	$y = kx$ where k is a constant.
Inversely proportional	Two variables are inversely proportional if one is proportional to the reciprocal of the other.	$y = \frac{k}{x}$ where k is a constant.

Review

1   A force of 10 N acts over an area of 8 m². What is the pressure?

2   A box of cereal costs £1.55 for 750 g. What is the cost per 100 g?

3   The nutritional information from a pack of chips is shown in the box.

A suggested serving is 135 g.

per 100 g when baked	
Energy	270 kcal
Protein	4.2 g
Carbohydrate	39 g
Fat	10.5 g
Salt	0.9 g

   a   For a serving, calculate the amount of
      i energy   ii fat   iii salt.
   b   What size serving contains 513 kcal?

4   Convert
   a   6.7 m to mm
   b   2651 s to minutes and seconds
   c   450 cm³ to litres   d   7 m/s to km/h.
   e   67 253 mm² into m².

5   A 3D shape is enlarged by scale factor 3.
   a   The volume of the smaller shape is 19.5 cm³, what is the volume of the larger shape?
   b   The surface area of the larger shape is 360 cm², what is the surface area of the smaller shape?

6   y is proportional to x. When $x = 3.5$, $y = 4.2$
   a   Write a formula linking x and y.
   b   What is the value of
      i y when $x = 7.9$ ii x when $y = 11.6$?

7   y is proportional to the square of x. When $x = 8$, $y = 128$.
   a   Write a formula to link x and y.
   b   What is the value of
      i y when $x = 11$ ii x when $y = 50$?

8   X is inversely proportional to Y. When $X = 12$, $Y = 5.5$
   a   Write a formula to link X and Y.
   b   What is the value of
      i Y when $X = 22$ ii X when $Y = 132$?

9   Sketch graphs to show the relationship
   a   between x and y in question **6**
   b   between X and Y in question **7**.

10   The graph shows the change in the depth of liquid in a container over time.

   Use the graph to estimate
   a   the rate of change of the depth at 3 hours
   b   the average rate of change over the whole 5 hours.

11   Isabella invests £1500 in a bank account that pays 2.5% interest per year.
   a   How much is in the account after
      i   one year   ii   five years?
   b   Write a formula to find the value of the account, V, after t years.

What next?

Score	0 – 4		Your knowledge of this topic is still developing. To improve look at MyMaths: 1036, 1048, 1059, 1061, 1070, 1121, 1238, 1246, 1312, 1329
	5 – 9		You are gaining a secure knowledge of this topic. To improve look at InvisiPens: 22Sa – i
	10 – 11		You have mastered these skills. Well done you are ready to progress! To develop your exam technique look at InvisiPens: 22Aa – j

Assessment 22

1   Light travels at 186 000 miles per second. The Sun is 93 000 000 miles from Earth.
Calculate how long it takes for light from the Sun to reach the Earth.
Give your answer in minutes and seconds. [4]

2   **a**   1 inch is equivalent to 2.54 cm. How many inches are there in 1m? [2]

b   A pack of 8 batteries cost £6.50. How much do 12 batteries cost? [2]

c   Nigel can cook 8 omelettes in 15 minutes. How long does it take to cook 28 omelettes? [2]

3   The Olympic sprinter Usain Bolt can run 100 m in 9.8 seconds.
A wombat can maintain a speed of 40 km/h for 150 m.
Who would win a 100 m race, Usain Bolt or a wombat? [4]

4   **a**   A statue has a mass of 3850 grams and volume of 529 cm³. What is its density? [2]

b   A silver bar has a mass of 250 g and a density of 10.5 g/cm³. What is its volume? [2]

c   A litre of milk has a density of 1.03 g/cm³. What is the mass of the milk? [2]

5   John's dog Muttley eats three tins of dog food in two days.

a   How many tins does Muttley eat in 30 days? Give your answer to the nearest tin. [3]

b   John says that 100 tins will feed Muttley for 67 days. Is John correct? [3]

c   John buys 30 tins of dog food to feed Muttley and Muttley's friend Fido.
Fido eats twice as much dog food as Muttley each day.
Does John have enough food to feed Muttley and Fido for two weeks? [4]

6   Adam Titchswamp has a cylindrical water container in his garden. The container was
full of rainwater, but Adam left the tap running and the container emptied.
It took $17\frac{1}{2}$ minutes for the container to empty.
The rainwater flowed out at an average rate of 150 ml/second.

a   How many litres of rainwater does Adam's container hold when full? [4]

b   The base of the cylinder is 23 cm. How high is it? [4]

7   A cuboid has volume 9600 cm³. The density of the cuboid is
7.2 g/cm³.

a   Find the mass of the cuboid in kg. [3]

The pressure that the base of the cuboid exerts on the floor is
0.7 N/cm².
The force of the cuboid on the floor can be calculated using the
formula: Force = 9.8m, where m is the mass in kg.
The cuboid has depth 40 cm, length x cm and height y cm.

b   Find the values of x and y to the nearest cm. [5]

8   The scale on a map is 1 : 7500. A town has an area of 37.5 cm² on the map.
What is the real area of the town in km²? [4]

9 a The diagram shows two similar cylinders.
   The smaller cylinder has radius r cm.
   The larger cylinder has radius $2r$ cm.
   The cost of making the cylinders depends on the surface area.
   The cost of making the small cylinder is £2.50.
   How much does it cost to make the larger cylinder? [4]

   b The larger cylinder has volume $1.2\,m^3$.
   Find the volume of the smaller cylinder. Give your answer in cm^3. [3]

10 The length of the extension, e cm, when a spring is stretched is directly proportional to T, the tension in Newtons. $T = 10$, when $e = 2.5$.

   a Write a formula connecting e and T. [3]

   b Find the value of e when $T = 17$ N. [2]

   c Find the value of T when $e = 4.5$ cm. [2]

11 The area, A, of a shape is directly proportional to x^2. When $x = 4$, $A = 50.272$.

   a Write a formula connecting A and x. [2]

   b Elsa says that the shape is a circle. Is Elsa correct? [2]

   c Find A when $x = 10$. [1]

   d Find x when $A = 254.502$ [2]

12 Nina believes that monthly rent for a one bedroom flat is inversely proportional to the distance in km to the city centre. Nina collects this data from eight available flats.

	Flat 1	Flat 2	Flat 3	Flat 4	Flat 5	Flat 6	Flat 7	Flat 8
Distance to city centre (km)	2.1	2.8	3.7	4.1	4.7	5.6	6.2	6.5
Monthly rent (£)	900	625	475	450	425	325	300	275

   a Draw a graph to show Nina's data. [5]

   b Which flat does not support Nina's conclusion that the rent and distance are inversely proportional? [1]

   c Nina has a maximum budget of £525 per month to spend on rent.
   How close to the city centre can Nina afford to live? [2]

13 P is inversely proportional to the square root of t. When P is 5, t is 4.

   a Find an expression for P in terms of t. [3]

   b Calculate P when $t = 100$. [2]

14 The height, h metres, of a ball when it is thrown is given by $h = 15t - 5t^2$, where t is the time in seconds.

   a Plot a graph of the height by taking readings at 0.5 second intervals. [6]

   b Estimate the speed of the ball 1 second after being thrown. [3]

15 The value, V, of Jenna's car can be calculated using the formula $V = 27\,000 \times 0.95^t$, where t is the number of years since Jenna first bought the car.

   a Complete this sentence: Jenna originally bought the car for £_____
   and the value of the car decreases by ___% each year. [2]

   b After 10 years the value of the car decreases by 10% each year.
   How much will Jenna's car be worth after 15 years? [5]

Revision 4

1 Three children are standing at points A, B and C as shown and throwing a ball to each other around the triangle. B has the ball and throws it to C. How far does she throw it? [3]

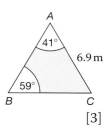

2 Pumba walks 7.6 km through the jungle on a bearing of 200°. How far south is he from his starting point? [3]

3 The sun casts a shadow on Dan 230 cm long. The shadow makes an angle with the ground of 35°. How tall is Dan, to the nearest cm? [2]

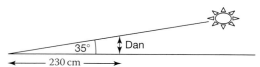

4 A plane coming in to land has 10 miles to go. It is currently 1.5 miles above the ground. Calculate its angle of descent, θ. [2]

5 D is the midpoint of AB in the triangle ABC.

  i Calculate the length, x. [7]

  ii Use the cosine rule in triangle ACD to find the value of θ. [5]

6 A flagpole OP stands at one corner of a rectangular parade ground $ABCO$. The dimensions of the parade ground are 150 m by 100 m where OC and AB are the shorter sides and the distance of P from the corner C is 125 m.

  a Find the height of the flagpole. [4]

  b Find angle PCO. [2]

  c Find lengths BO, and PB. [6]

  d Find angle PBO. [2]

7 Helen says that the vectors $\overrightarrow{RQ}$ and $\overrightarrow{NM}$ are related by $\overrightarrow{RQ} = -\overrightarrow{NM}$. Give the correct relationship, and express the vectors in terms of **a** and **b**. [6]

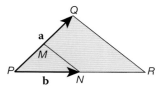

8 Ben writes a sequence of numbers that begins 2 6 12 20 ....

  a He says that the nth term of this sequence is $n^2 + n$. Is he correct? Show your working. [2]

  b Work out the 50th and 100th terms. [2]

  A second sequence is formed of the differences between each term and the next in the first sequence.

  c Work out the nth term of this new sequence. Hence find the 50th and 100th terms. [6]

9 The diagram shows a pattern of black and white triangles.

  a Copy and complete the table. [9]

Rows (r)	1	2	3	4	5
White triangles	1	3			
Black triangles	0	1			
Total number of triangles	1	4			

  b The formula for the number of white triangles is $n = \dfrac{r(r + 1)}{2}$.

  Find the number of white triangles in a

  i 50 row [2]

  ii 100 row pattern. [1]

9 c Find the formula for the total number of triangles in row r. [1]

d Use this formula and the one given in part **c** to derive a formula for the number of black squares in any row. Give your answer in its simplest form. [3]

e Use your formula to find the number of black triangles in the 60th pattern. [2]

f Test your formula by working out the number of black triangles in row 2. [1]

10 a Ellie says that there are 13 ml in 1.3 l. Correct her answer. [1]

b Lewis says that there are 7.6 kg in 76 g. Correct his answer. [1]

11 a James is cooking pancakes for 9 people. A recipe for 6 people uses 4 eggs. How many eggs does he need? [2]

b The supermarkets 'Liddi' and 'Addle' both sell teacakes. 'Liddi' sells a pack of 5 for £1.35 and 'Addle' sells a pack of 4 for £1.10. Which supermarket gives better value for money? [4]

12 Norwich to Harwich is 72 miles.

a Richard drives from Norwich to Harwich in 108 minutes. What is his average speed? Give your answer in mph. [3]

b Jeremy drives from Harwich to Norwich at an average speed of 50 mph. How long, to the nearest minute, did his journey take? [3]

c Jeremy drove from Harwich to Norwich at an average speed 4 mph faster than his return journey. How many minutes did he save on the outward journey? [4]

d Richard and Jeremy both leave on their outward journeys at 10:00. They meet in a village café for coffee. How far from Harwich is the village and at what time do they meet? [6]

13 a An object has a mass of 1.26 kg and volume of 180 cm³. What is its density? [3]

b A cylindrical metal rod with radius 3.5 cm and length 12.5 cm has a density of 11.4 g/cm³. What is the mass of the rod? [4]

c A silver bar has a mass of 30 g and a density of 10.5 g/cm³. What is its volume? [2]

14 a If $a = kb^2$, where k is a constant, we say that a is proportional to b^2 and write $a \propto b^2$. Write the following formulae in the same way.

i $m = \dfrac{45}{d^3}$ [1]

ii $T = \dfrac{2\pi\sqrt{l}}{10}$ [1]

iii $S = 4(r^2 - 3)5$ [1]

b $d \propto \sqrt{h}$. When $d = 80$, $h = 64$. Calculate

i d when $h = 56.25$ [6]

ii h when $d = 22$. [3]

15 Boyle's law states that, under certain conditions, the pressure P exerted by a particular mass of gas is inversely proportional to the volume V it occupies. A volume of 150 cm³ exerts a pressure of 6×10^4 Nm⁻². The volume is reduced to 80 cm³. What is the new pressure? [4]

16 Ally plays tennis, netball or squash every morning. The probability of choosing each sport is:

Tennis 0.28   Netball 0.35   Squash P

a Show that $P = 0.37$ [2]

b Draw a tree diagram showing her possible choices over two consecutive days. [3]

c Find the probability Ally

i plays tennis on day 1 and squash on day 2 [2]

ii plays netball at least once [4]

iii doesn't play tennis on either day. [4]

Formulae

Make sure that you know these formulae.

Quadratic formula

The solutions of $ax^2 + bx + c = 0$ $a \cdot 0$ are $x = \dfrac{-b \pm \sqrt{b^2 - 4ac}}{2a}$

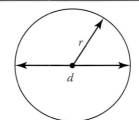

Circles

Circumference of a circle $= 2\pi r = \pi d$

Area of a circle $= \pi r^2$

Pythagoras' theorem

In any right-angled triangle

$\qquad a^2 + b^2 = c^2$

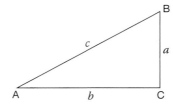

Trigonometry formulae

In any right-angled triangle $\sin A = \dfrac{a}{c}$ $\cos A = \dfrac{b}{c}$ $\tan A = \dfrac{a}{b}$

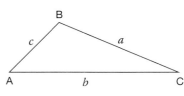

In *any* triangle

sine rule $\dfrac{a}{\sin A} = \dfrac{b}{\sin B} = \dfrac{c}{\sin C}$ cosine rule $a^2 = b^2 + c^2 - 2bc \cos A$ Area $= \frac{1}{2}ab \sin C$

Make sure that you know and can derive these formulae.

Perimeter, area, surface area and volume formulae

Area of a trapezium $= \frac{1}{2}(a + b)h$

Volume of a prism = area of cross section $\times$ length

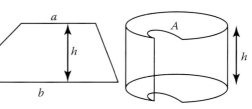

Compound interest

If P is the principal amount, r is the interest rate over a given period and n is number of times that the interest is compounded then

Total accrued $= P\left(1 + \dfrac{r}{100}\right)^n$

Probability

If P(A) is the probability of outcome A and P(B) the probability of outcome B then

P(A or B) = P(A) + P(B) − P(A and B)   P(A and B) = P(A given B)P(B)

These formulae will be provided at the front of each examination paper.

Perimeter, area, surface area and volume formulae

Curved surface area of a cone $= \pi r l$

Surface area of a sphere $= 4\pi r^2$

Volume of a sphere $= \frac{4}{3}\pi r^3$

Volume of a cone $= \frac{1}{3}\pi r^2 h$

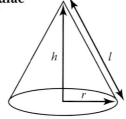

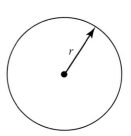

Kinematics formulae

If a is constant acceleration, u is initial velocity, v is final velocity, s is displacement from the position when $t = 0$ and t is time taken then

$v = u + at$ $\qquad\qquad$ $s = ut + \frac{1}{2}at^2$ $\qquad\qquad$ $v^2 = u^2 + 2as$

Key phrases and terms

Use our handy grid to understand what you need to do with a maths question.

Circle	Draw a circle around the correct answer from a list.
Comment on	Give a judgement on a result. This could highlight any assumptions or limitations or say whether a result is sensible.
Complete	Fill in any missing information in a table or diagram.
Construct	Draw a shape accurately.
Construct, using ruler and compasses	Compasses must be used to create angles. Do *not* erase construction lines.
Criticise	Make negative judgement. This could highlight any assumptions or limitations or say why a result is not sensible.
Deduce	Use the information supplied or the preceding result(s) to answer the question.
Describe	Usually describing a graph, one mark per descriptive sentence. When describing transformations give all the relevant information.
Diagram not accurately drawn	Calculate any angles or sides, do *not* measure them on the diagram.
Draw	Accurately plot a straight line, bar chart or transformations.
Draw and label	Draw and mark on values.
Estimate	Simplify a calculation, by rounding, to find an approximate value.
Evaluate	Calculate the value of a numerical or algebraic expression.
Expand	Multiply out brackets.
Factorise fully	Put in brackets with the highest common factor outside the brackets.
Give a reason	Write an explanation of your argument, including the information you use and your reasons or rules used.
Give the exact value	Do not use rounding or approximations in your calculations. You should give your answers as fractions, surds and multiples of π.
Give your answer in terms of π.	Do not use a numerical approximation to π, instead treat it as an algebraic variable (letter).
Give your answer in its simplest form/ simplify	Collect all like terms, any fractions or ratios should be in lowest terms.
Give your answer to an appropriate degree of accuracy	For example, if the numbers in the question are given to 2 decimal places, give your answer to 2 decimal places.
Measure	Use a ruler or protractor to accurately measure lengths or angles.
Prove that	Formally obtain a required result using a sequence of logical steps. Each line of workings should follow from the previous line and any general results used should be quoted.
Rearrange	Change the subject of a formula.
Shade	Use hatching to indicate an area on a diagram.
Show	Present information by drawing the required diagram.
Show that	Obtain a required result showing each stage of your workings.

Show working	Usually asked for to support a decision.
Sketch	Represent using a diagram or graph. This should show the general shape and any important features, such as the position of intercepts, using labels as necessary. It does not require an accurate drawing.
Solve	Find an answer using algebra or arithmetic. This often means find the value of x in an equation.
State	Write one sentence answering the question.
Tick a box	Choose the correct answer from a list.
Work out	Work out the answer and show your working.
Write down	Write down your answer; written working out is not usually required.
Use an approximation	Estimate.
Use a line of best fit	Draw a line of best fit and use it.
Use	Use supplied information or the preceding result(s) to answer the question.
You must show your working.	Marks will be lost for not writing down how you found the answer to the question.

Chapter 1

Check in 1

1 a Four thousand **b** Four hundred
c Four tenths **d** Four thousandths
2 a

b $-3, -2.4, -1.8, 0, +1.5, +5$

1.1S

1 a One thousand three hundred and seven
b Twenty-nine thousand and six
c Three hundred thousand
d Six hundred and five thousand and thirty
2 a 8043 **b** 70 000 000 **c** 200 051 **d** 2010
3 a 0.1, 0.3, 1, 1.3, 2, 3.1
b 6.07.7.06, 27.6, 77.2, 607
4 a 8000.6, 6008, 6000.8, 862.6, 682.8
b 97.4, 94.7, 79.4, 74.7, 49.7, 47.9
5 a 167 **b** 248 **c** 7.16 **d** 10.95
e 2430 **f** 2813
6 a 21.4 **b** 6.73 **c** 410.6 **d** 20.07
e 0.6025 **f** 8.6
7 a 4.52; five tenths is greater than five hundredths
b 5.5; five tenths is greater than five hundredths
c 16.8; eight tenths is eighty hundredths which is greater than seventy-five hundredths
d 16.8; eight tenths is eighty hundredths which is greater than fifteen hundredths
8 a False **b** True **c** True **d** True
e False **f** True
9 a 7.03, 7.08, 7.3, 7.38, 7.8, 7.83
b 2.18, 2.4, 4.18, 4.2, 8.24, 8.4
10 a 18.7, 18.16, 17.6, 17.16, 16.7, 16.18
b 13.2, 13.145, 2.5, 2.38, 1.1, 1.06
11 a i 3050 **ii** 3000 **b i** 1760 **ii** 1800
c i 290 **ii** 300 **d i** 50 **ii** 100
e i 40 **ii** 0 **f i** 740 **ii** 700
12 a 3000 **b** 1000 **c** 0 **d** 25 000
e 16 000 **f** 168 000
13 a i 39.1 **ii** 39.11 **b i** 7.1 **ii** 7.07
c i 5.9 **ii** 5.92 **d i** 512.7 **ii** 512.72
e i 4.3 **ii** 4.26 **f i** 12.0 **ii** 12.01
g i 0.8 **ii** 0.83 **h i** 26.9 **ii** 26.88
14 a i 0.1 **ii** 0.07 **iii** 0.070
b i 15.9 **ii** 15.92 **iii** 15.918
c i 128.0 **ii** 128.00 **iii** 127.998
d i 887.2 **ii** 887.17 **iii** 887.172
e i 55.1 **ii** 55.14 **iii** 55.145
f i 0.0 **ii** 0.01 **iii** 0.007
15 a 480 **b** 1200 **c** 490 **d** 14 000
e 530 **f** 15 000
16 a 0.36 **b** 0.42 **c** 0.057 **d** 0.0047
e 1.4 **f** 0.000 004 2
17 a 200 **b** 2000 **c** 5 **d** 10
e 0.000 5 **f** 100 000
18 a 9.73 **b** 0.36 **c** 147.5 **d** 29
e 0.53 **f** 4.20 **g** 1245.400 **h** 0.004
i 270 **j** 460.0
19 a 1306 **b** 2.085 **c** 1085 **d** 2.487
e 0.000 8 **f** 6.19 **g** 0.045 13 **h** 0.004 5

1.1A

1 a Reece (gold), Sololova (silver), and DeLoach (bronze)
b Yes, she would have won the bronze medal, as DeLoach won bronze with a jump of 6.89 m which is lower than 6.92 m.
2 Irie and Thoman had the closest times, as the difference between their times is 0.05 which is less than all the other differences.
3 a £28 000 **b** 48
4 a 4 **b** 35
5 44.48 and 44.53
6 Jessica is correct because ⩾ and > can be placed between both the first pair and the third pair (giving four points), and ⩽, =, and ⩾ can be placed between the second pair (which gives her the other three points to make seven).
7 a 1111 **b** 26
c $1011 + 1111 = 11\,010$ in binary since $15 + 11 = 26$ in our ordinary number system (see the previous parts of the question).
Yes. Binary numbers can be added and subtracted using the ordinary methods of addition and subtraction, taking special care when carrying or borrowing, which will happen more often than in our ordinary number system.

1.2S

1 a 22 **b** -12 **c** -2 **d** -14
e 12 **f** 12 **g** 19 **h** -15
i 11 **j** -61 **k** 344 **l** 49
2 a -10.8 **b** 0.7 **c** 13.5 **d** -0.7
e -38.2 **f** 112.7
3 a 10 **b** 10.1 **c** 10.3 **d** 11.1
e 16.4 **f** 7.5
4 a 7.77 **b** 5.25 **c** 3.6 **d** 3.9
e 2.13 **f** 13.04
5 a 2.5 **b** 2.1 **c** 2.2 **d** 3.9
e 1.4 **f** 0.7
6 a 0.74 **b** 1.05 **c** 1.51 **d** 7.14
e 1.08 **f** 0.64
7 a 35.96 **b** 34.61 **c** 96.35 **d** 104.42
e 14.931 **f** 10.401 **g** 26.211 **h** 63.836
8 a 10.68 **b** 0.43 **c** 4.097 **d** 0.76
e 2.12 **f** 4.19 **g** 2.04 **h** 5.7
9 a 2.63 **b** 0.45 **c** 24.32 **d** 14.72
10 a 3.97 **b** 0.095 **c** 12.44 **d** 58.34
11 a 2.81 **b** 0.757 **c** 88.01 **d** 1126.72
12 Check appropriate written methods have been used where applicable
a 2.6 **b** 7.86 **c** 5.9 **d** 92.54
e 0.97 **f** 24.27
13 Check appropriate written methods have been used and estimates have been given (estimates in brackets below)
a 13.896 (12) **b** 19.45 (20) **c** 359.79 (360)
d 7.683 (10) **e** 0.326 (0) **f** 11.42 (10)
14 See answers to question 13
15 Check for correct working
a 4.1 **b** 40.2 **c** 11.288 **d** 5.968
e 0.892 **f** 0.469
16 Check for correct working
a 99.88 **b** 28.986 **c** 14.54 **d** 8.6757

1.2A

1 a 248 km **b** 97 **c** 1.102 kg
2 a From top-left to bottom-right, the missing numbers are 12.7, 5.8, 5.4, 3.7, 4.8, and 0.6.
b From top to bottom, the missing numbers are -5, -3, and 5.
c From top-left to bottom-right, the missing numbers are -24, -8, 4, -7, and -1.

3 a There nine possible pairs are (34, 87), (37, 48), (37, 84), (38, 47), (43, 78), (48, 73), (73, 84), (74, 83), and (78, 43).
b 157 (73 + 84 or 74 + 83)

4 a £5.49   **b** £172.38

5 a The correct answer is 100.97; there should have been a carried 1 from the third addition (6 + 4).
b The correct answer is 245.1; the carried 1 should have been added to 4 + 9 in the third addition.
c The correct answer is 161.28; 4 should have been added to (rather than subtracted from) 5 in the third column.
d The correct answer is 87.82; the original 1 in the third column should have been crossed out and replaced by a 2.

6 Check the subtraction is correct and contains each digit once (for example, 612 − 534 = 78).

7 a 1026 and 342
b Yes: 1448 and 724.
c Yes: 1026 + 522 = 2 × (432 + 342).

8 a There are six: 246, 264, 426, 462, 624, and 642.
b The six pairs are (246, 753), (264, 735), (426, 573), (462, 537), (624, 375), and (642, 357).
c (462, 537)

1.3S

1 a −25   **b** −32   **c** −72   **d** −20
e 30   **f** 49   **g** 16   **h** −20
i −18   **j** 26   **k** −42   **l** −48

2 a −2   **b** −5   **c** 5   **d** 4
e −22   **f** −1   **g** 40   **h** 4
i 5   **j** −17   **k** −3   **l** 27

3 a 98   **b** 152   **c** 273   **d** 323
e 308   **f** 609

4 Check appropriate written methods are used and that answers match question **3**.

5 a Given in question   **b** $6 \times 0.2 = 6 \div 5$
c $12 \times 0.2 = 12 \div 5$ **d** $2 \times 0.001 = 2 \div 1000$

6 a Given in question   **b** $6 \div 0.25 = 6 \times 4$
c $16 \div 0.01 = 16 \times 100$ **d** $15 \div 0.05 = 15 \times 20$

7 Check methods have been shown and estimates used (see brackets for examples of estimates).
a 9.3 (9)   **b** 700 (500)   **c** 12.2 (16)   **d** 10.6 (8)
e 463.3̇ (420)   **f** 130 (120)   **g** 570 (510)   **h** 0.65 (0.5)

8 Check written methods have been shown.
a 24.91   **b** 4.284   **c** 105.84   **d** 130.8985
e 42.9442   **f** 369.5328

9 Check written methods have been shown.
a 3.87   **b** 0.775   **c** 0.916   **d** 7.53
e 18.13   **f** 3.45   **g** 4.15   **h** 7.74
i 4.08   **j** 2.35

10 Check written methods have been shown.
a 5.26   **b** 28.88   **c** 1384.29   **d** 175.56
e 28.65   **f** 111.51

11 Check answers match questions **8–10**.

12 a 28.81   **b** 28.81   **c** 4.3   **d** 0.067
e 10

13 a 196   **b** 4   **c** 18   **d** 289

14 a 121   **b** 39   **c** 18   **d** $2\sqrt{7}$ (or $\sqrt{28}$)

15 Check appropriate methods have been chosen.
a 170   **b** 0.58 (2dp)   **c** 1.78 (2dp)

16 a 10   **b** 18

1.3A

1 a £75.82
b She works for 7.5 hours on Saturday and 5.5 hours on Sunday.
c 23 hours

2 £1.37

3 Nancy is wrong since she has tried to use a multiplication law for division (her method gives 30 rather than the correct answer of 7.2).

4 a $5 \times (2 + 1) = 15$ **b** $5 \times (3 − 1) \times 4 = 40$
c No brackets required   **d** $2 + 3^2 \times (4 + 3) = 65$
e No brackets required   **f** $(4 \times 5 + 5) \times 6 = 150$

5 a From top-left to bottom-right, the missing digits are 1, 1, 1, 0, 6, 2, 5, 2.
b From top-left to bottom-right, the missing digits are 1, 9, 2, 7, 7, 5, 9, 1, 9, 7, 5, 9, and 9.

6 a The maximum score is 72 points and the minimum score is −54 points.
b The three different ways to score 12 points are: 3 correct with the rest unanswered; 6 correct and 4 incorrect with the rest unanswered; and 9 correct and 8 incorrect with one unanswered.

7 £32.14 (2dp)

8 a False. Halving negative numbers makes them bigger (half of −2 is −1 which is greater than −2).
b True by definition. $a \div \frac{1}{2} = 2a \div 1 = 2a$.
c False. Multiplying a negative number by 5 makes it smaller ($−1 \times 5 = −5$ which is less than −1).
d False. Multiplying zero by 10 gives zero.

9 a Repeatedly multiplying a number (other than zero) by + 0.9 results in answers tending towards zero without changing from positive to negative (or vice versa).
b Repeatedly multiplying a number (other than zero) by − 0.9 results in answers tending towards zero alternating between positive and negative with each multiplication.

10 Check students' working. Yes this will work for every number between 100 and 999, since $7 \times 11 \times 13 = 1001$ and multiplying any three-digit number by 1001 results in a six-digit number which repeats the original three-digit number.

Review 1

1 a 24.3 < 24.5   **b** −0.5 > −0.9
c 0.5 > 0.06   **d** 1.456 < 1.46

2 a False, $0.85 \div 10 = 85 \div 1000$.   **b** True
c True   **d** False, $3^2 > 3 \times 2$.   **e** True

3 a 45.9   **b** 0.08   **c** 0.085   **d** 2000
e 78 000   **f** 3.5

4 a 8300   **b** 25.9   **c** 310   **d** 5
e 76.4   **f** 5.49   **g** 0.085   **h** 0.008

5 a 3842   **b** 192   **c** 309.32   **d** 53.053
e 7726   **f** 34.6   **g** 917.4   **h** 9.91

6 a 540   **b** 63   **c** 0.56
d 500   **e** 51.522   **f** 2.65335
g 1.9   **h** 290

7 a −11   **b** −241   **c** 12.6   **d** −0.43
e −42   **f** 5.5   **g** −17   **h** 5

8 a 1   **b** 40   **c** 112   **d** 112
e 10.2   **f** 30

9 a 56.8112   **b** −1

Assessment 1

1 a 57.6 < 303; 0.8, 1.9, 3.3, 44, 57.6, 303
b −2.19 < −0.07; −2.19, −0.07, 30, 43.56, 188.0, 194.7

2 Yes, 0.42, 3, 4.236, 51.6, 4200, 216 000.

3 a No, LB = 271.75 cm < 271 cm.
b No, Yao Defen UB = 233.345 cm > 233.341 cm.

4 a Yes   **b** No, Dave 40, Jane 50 (1 sf).
c Dave 45, Jane 53

5 a Abena 13 000, Edward 8100 (2 sf)
b 15 395.8 km
c The estimate would be smaller, the approximation for Abena's x value would decrease more than Edward's value.

Answers

6 2.125 km

7 a 25 sweets **b** 25 000 (nearest 1000)

8 a 30, −20 **b** 7

c Students' answers, for example, 2 correct, 8 wrong; 1 correct, 4 wrong, 5 unanswered; 10 unanswered.

9 Sue 19.9, Clive 27.8, Ben 61.7, Henry 1.83

10 Yes

11 a −2.2, −4.5, −7.8 **b** 7.49, 3.67, −3.82

12

−2	3	−8	5
−7	4	−1	2
7	−6	1	−4
0	−3	6	−5

13 Students' answer, for example, Amanda (more accurate, not enough money), Gardener (less accurate, enough money).

14 a 1 **b** 0, 1 ÷ 0 is undefined.

c Students' answers < 1, for example, $\frac{1}{2} < 2$.

15 a $(3 + 4) \times 5 = 35$ **b** $(4 ÷ 3) + 5 = 6\frac{1}{3}$

c $5(2^3 + 0.4) ÷ (4 − 3 \times −1) = 6$

Chapter 2

Check in 2

1 a 45 **b** 52 **c** −26 **d** 196

e 7 **f** 13 **g** −50 **h** 30

2 a 9 + 6 = 15 **b** 8 + 7 = 15

c 5 × 3 = 15 **d** 27 − 12 = 15

3 a 3 **b** 4 **c** 10 **d** 6

e 4 **f** 25 **g** 33 **h** 7

2.1S

1 a $5w$ **b** $\frac{6}{k}$ **c** y^2 **d** $6ab$

e $8k^3$ **f** $12k^3$

2 a 20 **b** 4 **c** 22 **d** 108

3 a 9 **b** 18 **c** 25 **d** 50

e 16 **f** 32 **g** 36 **h** 72

i 360

4 a −1 **b** −1 **c** 11 **d** 8

e 0 **f** 12 **g** 6 **h** −9

5 a 2 **b** 4 **c** −1 **d** 2

e −4 **f** $\frac{8}{3}$ **g** −2 **h** −6

i −2

6 a $15n$ **b** $12m$ **c** $17p$ **d** $18q$

e x **f** $4w$ **g** $7a$ **h** $5b$

i j **j** $2k$

7 a Given **b** $\frac{x}{3}$ **c** $\frac{y}{7}$ **d** $\frac{t}{9}$

e $\frac{2a}{3}$ **f** $\frac{3n}{4}$ **g** $\frac{5p}{7}$ **h** $\frac{v}{2}$

8 a $5p + 6q$ **b** $9x + 7y$

c $2m + 8n$ **d** $x + 5y$

e $8r − 6s$ **f** $2g − 4f$

g $2a + 6b + 5c$ **h** $5u − 2v + 3w$

i $3x − 4y + 5z$ **j** $6r + 5s + 2t$

9 a $11a + 6b$ **b** $2t + 26$ **c** $x − 12y$ **d** $14p + p^2$

e $20xy$ **f** $7ab$

10 a $28mn$ **b** $12m^2$ **c** $10p$ **d** 2

e $24abc$ **f** $6k^3$ **g** $4b$ **h** $9c$

11 a $9r$ **b** $4m^2$ **c** $4x$ **d** $8tv$

e $10mn$ **f** $6xy^2$ **g** $2x^2 + x$ **h** $9w − 8$

i $z^3 + 3z + 1$

12 a $k^2 + k$ **b** $m^2 + 10m + 2$ **c** $10t + 5$

13 From top-left to bottom-right, the answers are: $9b − 2a$, $12a^2$, $4b$, $2ab$, $10p + 7p^2 + 5p^3$, $13abc$, $5m − 4$, $60m^3$, and $\frac{2}{a^2}$.

2.1A

1 Abdul

2 Paul was right, because indices come before multiplication in BIDMAS.

3 a No, only Cerys is correct. Audrey is incorrect because the equation $b = 3a$ actually means that Billie has three times as much money as Audrey. Cerys is correct because substituting $3a$ for b in the second equation gives the equation $c = 4a$, which means that Cerys does indeed have four times as much money as Audrey.

b £40

4 Top row (left to right): $4p + 7q$, $4p + 7q$, and $4p + 2q$ (the odd one out); Second row (left to right): $6mn$, $10mn$ (the odd one out), and $6mn$; Third row (left to right): $2d$, 2 (the odd one out), and $2d$; Fourth row (left to right): $2n − 8$, $2n$ (the odd one out), and $2n − 8$.

5 a The possible answers are the 16 rearrangements of the following equation, such that there are three terms on one side and one term on the other: $(2x + 3y) + (2x − 4y) − (x − 3y) = (3x + 2y)$.

b Any valid rearrangement of the following equation: $(3a \times 6) + 2a + 10b ÷ 2 − b = (20a + 4b)$

6 a i $16 + 8p$ **ii** $32p$

b i $64p$

ii The two rectangles are joined along one of the sides of length $4p$.

7 $3x$ and $2y$

8 $17xy$ mm^2

9 a $2x + 6y$

b Rectangle. The rectangle has a perimeter of $10y + 8x$ and the trapezium has a perimeter of $10y + 7x$, the difference is x, $x > 0$ as $3x$ is a side length.

10 $4p$

2.2S

1 a y^4 **b** m^6 **c** $6v^2w^3$ **d** $2r^4s$

e $6m^2n$ **f** $8y^3z^2$

2 a $6m^2$ **b** $12p^3$ **c** $6xy^2$ **d** $10r^2s^2$

3 No, because the laws of indices state that the powers should be added in this instance; $a^5 \times a^2 = a^7$.

Alternatively, trying it out with $a = 2$ shows that Kyle is wrong.

4 a n^5 **b** s^7 **c** p^4 **d** t^4

5 a x^7 **b** x^8 **c** x^9 **d** x^7

6 a r^2 **b** r **c** r^5 **d** r^3

7 a m^4 **b** x **c** t^2 **d** y^3

8 a x **b** m^2 **c** s^3 **d** v^3

e q^3 **f** t^4 **g** p **h** y^2

9 a x^{15} **b** x^{12} **c** $x^{12}y^2$ **d** $x^{15}y^5$

e x^8y^3 **f** x^2y^2z

10 No. Tracey has correctly added the powers according to the index rules, but she has then mistakenly added the coefficients 4 and 2 rather than multiplying them.

11 a $3x^7$ **b** $5y^7$ **c** $12b^8$ **d** $10p^{11}$

e $30h^{11}$ **f** $12s^3t^4$

12 No. Andy has correctly subtracted the powers according to the index rules, but he has then mistakenly subtracted the coefficients 3 from 12 rather than dividing 12 by 3.

13 a $2y^4$ **b** $2a^6$ **c** $5k^4$ **d** $3p^5$

e $5x^6$ **f** $\frac{1}{2}x^8y^{-4}$

14 a a^6 **b** y^{12} **c** k^{15} **d** p^{56}

e a^{21} **f** a^{21}

15 a $4a^6$ **b** $729y^{12}$ **c** $25k^6$ **d** $216p^{21}$

e $128a^{21}$ **f** $256a^{16}$

16 a y^2 **b** x^{-2} **c** a^{-6} **d** h^{-6}

e p^4 **f** p^{-1}

17 a g^3 **b** h^{-6} **c** b^{-12} **d** j^{-6}

e t^{10} **f** n^{-2}

18 a $4p^{16}$ **b** $60r^{-1}$ **c** $27h^{-9}$ **d** $3b^8$

e $2m^{-9}$ **f** 2

1 a $32x^2y^3$ b $19a^2b^4$
2 $24p^2q$
3 $4ab^3$
4 a $p^2q \times (pq)^3 = p^5q^4$ b $xy + (xy + xy)^2 = 4x^3y^3$
5 $6x^7y$ is wrong, because the coefficient of the previous term should be multiplied by two to give $8x^7y$.
6 x^{40}
7 The statement is false, because $x^2y^6 \neq x^3y^6$
8 $x^{-1} = x^{0-1} = x^0 \div x^1 = 1 \div x = \frac{1}{x}$
9 Since $(\sqrt{x^2}) = x$, $\sqrt{x} = \sqrt{x^1} = \sqrt{x^{(\frac{1}{2}+\frac{1}{2})}} = \sqrt{x^{\frac{1}{2}} \times x^{\frac{1}{2}}} = \sqrt{(x^{\frac{1}{2}})^2} = x^{\frac{1}{2}}$.
10 $30w^{-9}$
11 $(4p^4)^3 \div (8p^7)^2 = (4^3p^{4\times3}) \div (8^2p^{7\times2}) = 64p^{12} \div 64p^{14} = p^{-2} = \frac{1}{p^2}$
12 False. The index rules say that the correct answer is 3^{x+y}
13 The first expression is the odd one out, as it is equal to $2t^{-8}$, whereas the other two equal $2t^{-8}$.
14 $x = 1$
15 a If $u = 3^x$, then $9^x + 3^{x+1} = (3^2)^x + 3^x \times 3^1 = 3^{2x} + u \times 3$
$= (3^x)^2 + 3u = u^2 + 3u$.
 b $3^{3x} + 3^{2+x}$ c $3^{2x} - \frac{1}{3^x}$ d $u^4 - \frac{u^2}{9}$

1 a $4n + 20$ b $6b - 42$ c $a^2 + 3a$
 d $ab - ac$ e $8x + 12y - 16z$ f $2h^2 + 18h$
2 a $-3k - 27$ b $-2h + 10$ c $-w + 4$
 d $-t + p$ e $-k^2 - 7k$ f $9k - 18m - 36$
 g $-x^2 + x + 8$ h $-2x^2 - 6$ i $3x - 3$
3 a 1 and 2 b 1, 2, and 4 c 1, 2, 5, and 10
 d 1, 2, 3, and 6 e 1 and 3 f 1 and 2
 g 1 and 2 h 1, 2, 4, and 8
4 a 3 b 2 c 2 d 4
5 a y b s c m d $2y$
6 a $2(x + 5)$ b $3(y + 5)$ c $4(2p - 1)$ d $3(m + 2)$
 e $5(n + 1)$ f $6(2 - t)$ g $2(2k + 7)$ h $3(3z - 1)$
7 a $2(x + 2)$ b $3(y - 2)$ c $12(p + 3q)$ d $5(5w - 1)$
 e $x(6y + w)$ f $b(a - 2c)$ g $q(pr + rt - sw)$
 h $x(5y - 1)$ i $2x(y + 3)$ j $2a(3a + 2b)$
 k $5p(5p - 2)$ l $7x(2y + 1)$ m $2a(2a + c - 4)$
 n $5m(2m^2 + 3n - 1)$ o $6p(p^3 - 2)$
8 a $5p + 9$ b $7m + 8$ c $2x + 4$ d $5k + 10$
 e $9t + 10$ f $4r + 7$
9 a $10c + 62$ b $23x + 67$ c $2x^2 + 10x$ d $17t^2 + 32t$
 e $7x - 45$ f $2x - 11$ g $2m - 26$ h $33 - 11g$
 i $p + 14$ j $2q - 7$
10 a $23(x + y)$ b $(a - b)(a - b + 5)$
 c $(q + r)(6 - (q + r))^2$ d $7(pt - w)$
11 $-5x^2 + 44x$
12 a $15x + 27$ b $8p^2 - 16p$ c $15m - 6m^2$
 d $21y + 17$ e $10x^2 + 10xy - 45x$ f $57 - 2t$
 g $4h + 16$ h $x^4 + 3x^3$

1 Clare has confused addition with multiplication; she should have replaced the 0 with a 1.
Ben is acting as though the second term is 3 rather than $3p$.
Vicky has also confused addition with multiplication in her attempted factorization.
2 a $(a + b)(x + y)$ b $(b + c)(d + m)$
 c $(a + 2)(a + b)$ d $(c - m)(d + e)$
3 a 4 b 3 c 4 and 4 d 4 and 5
 e 6 and 4 f 3 and 1
4 a $4(x - 1)$ b $20(b + 2)$
5 a $3(2x - 1)$ b $6x - 3$
 c Using part b we know that $6x - 3 = 15$, so subtracting 15 from both sides of the equation gives $6x - 18 = 0$ as required.

6 a e.g. $8(3x + 2)$
 b e.g. $(3x + 10) + (21x + 6)$
7 The expression is most simply factorised as $2y(y + 1)$; expanding gives $2y^2 + 2y$.
8 a 6 b 16.5 c 58.6 d 33.2
9 $4(3x + 2) - 2(2x - 1) = 12x + 8 - 4x + 2 = 8x + 10$
$= 2(4x + 5)$ as required.
10 Factorise the volume to give any three of the following combinations of dimensions: $10x$, y, and $5x + 2$; $10y$, x, and $5x + 2$; $2x$, $5y$, and $5x + 2$; $5x$, $2y$, and $5x + 2$; 1, $10x$, and $y(5x + 2)$; 1, $10y$, and $x(5x + 2)$; and 1, $10xy$, and $5x + 2$ (other combinations are possible with non-integer coefficients).
11 a Factorise the volume to give any three of the following combinations of dimensions: x, $2x$, and $8y - 27$; 1, $2x^2$, and $8y - 27$; 1, 2, and $x^2(8y - 27)$; 1, $2x$, and $x(8y - 27)$; and 2, x^2, and $8y - 27$ (other combinations are possible with non-integer coefficients).
 b Otherwise the cuboid would have zero (or negative) volume. $2x^2(8y - 27) > 0$ so $8y - 27 > 0$ giving $y > 3.375$
12 $2p + 5$
13 The missing terms are $3x - 6$ and $24x^4 - 48x^3$.

1 a $\frac{1}{3}$ b $\frac{3}{4}$ c $\frac{3}{5}$ d $\frac{4}{9}$
 e $\frac{5}{8}$ f $\frac{1}{3}$
2 a 1 b $\frac{3}{4}$ c $\frac{5}{6}$
3 a $\frac{1}{3}$ b $\frac{1}{3}$ c $\frac{1}{12}$
4 a $\frac{3}{2}$ b $\frac{4}{5}$ c $\frac{7}{16}$ d $\frac{5}{48}$
 e $\frac{9}{2}$ f $\frac{4}{33}$
5 a $\frac{2}{9}$ b $2\frac{1}{2}$ c $\frac{1}{9}$ d $2\frac{1}{4}$
 e $\frac{1}{6}$ f $\frac{2}{3}$
6 a $3w$ b $\frac{b}{3}$ c $2c$ d $\frac{4b}{d}$
 e $4bd^2$ f $\frac{3xy^3}{2}$ g $x + 3$ h $x + 1$
 i $\frac{y - 2}{3}$
7 No. David should perhaps have tried to factorise the numerator and denominator by x to check whether he could cancel the x terms. One way to see that it's an incorrect simplification is to substitute 1 for x, which gives 8, not 10, as a simplified answer.
8 a x b p c $\frac{1}{y}$ d $\frac{6}{y - 1}$
 e 3 f Does not simplify
9 a $y + 2$ b $x - 4$ c $\frac{1}{x + 3}$ d $(p - 1)^2$
 e $y + 4$ f $\frac{1}{(b - 2)^3}$
10 a $\frac{4p}{5}$ b $\frac{4y}{7}$ c $\frac{3}{p}$ d $\frac{11y}{8}$
 e $\frac{p}{15}$ f $\frac{6y - 7x}{xy}$
11 $A = E = \frac{x}{6}$; $C = G = \frac{x}{3}$; $D = F = \frac{x}{12}$; and $B = \frac{5x}{12}$, so B is the odd one out. Check its newly-created pair is equal to $\frac{5x}{12}$.
12 a 3 b $\frac{12a}{7}$ c $\frac{n}{5m}$ d $\frac{15}{g^2}$
 e $\frac{8}{3}$ f $\frac{f}{2p^2}$
13 a $\frac{10}{x^2 + 2x}$ b $\frac{3}{x - 1}$ c $\frac{9}{2x + c}$ d 1
 e $\frac{2}{x + 2}$ f y
14 a $\frac{14x + 3}{20}$ b $\frac{12x + 13}{77}$ c $\frac{y - 1}{15}$ d $\frac{41p - 112}{35}$
 e $\frac{5x - 13}{(x - 7)(x + 4)}$ f $\frac{8x + 9}{(x - 2)(x + 3)}$ g $\frac{11 - y}{(y - 2)(y + 1)}$ h $\frac{7p + 13}{(p + 3)(p - 1)}$
 i $\frac{12(w - 2)}{w(w - 8)}$ j $\frac{4x - 3}{(x - 2)^2}$

1 $\frac{4}{x}$
2 Top-middle = $10x$; bottom-left = $6x$; and bottom-right = $\frac{4x}{x + 1}$.
3 From top to bottom, the answers are $\frac{4}{x + 5}$ and $\frac{x + 5}{4}$.
4 $\frac{x + 7}{6}$
5 Constant difference = $\frac{1}{(x + 1)(x + 2)}$

6 In the first subtraction, Amelia has made her mistake on the second line, by miscalculating -3×-3, which is equal to $+9$, not -9. The correct answer is $\frac{19 - x}{(x - 3)(x + 5)}$.

In the second subtraction, Amelia has forgotten to multiply -2 by 6 to get $6x - 12$ at the start of the Numerator. The correct answer is $\frac{11x + 8}{(x + 4)(x - 2)}$.

7 a $\frac{2(8p - 7)}{(p + 1)(p - 2)}$

b p must be greater than -1 because substituting $p = -1$ into the expression for the perimeter makes the denominator 0, and any value of p less than -1 would give a negative perimeter.

c p cannot equal 2 because that would also make the denominator 0.

8 $\frac{1}{3}\left(\frac{12}{2x + 1} + \frac{2}{7 - x} + \frac{4}{7 - x}\right) = \frac{1}{3}\left(\frac{12}{2x + 1} + \frac{6}{7 - x}\right)$
$= \frac{1}{3}\left(\frac{84 - 12x + 12x + 6}{(2x + 1)(7 - x)}\right) = \frac{30}{(2x + 1)(7 - x)}.$

9 $\frac{75}{x(x + 4)}$

10 From the diagram, the calculation is:
$4\left(\frac{4}{x + 3}\right) + 8\left(5 + \frac{2}{x + 3}\right) - 2\left(\frac{4}{x + 3}\right)$
$= 2\left(\frac{4}{x + 3}\right) + 8\left(5 + \frac{2}{x + 3}\right) = \frac{8}{x + 3} + 40 + \frac{16}{x + 3} = \frac{24 + 40x + 120}{x + 3}$
$= \frac{8(5x + 18)}{x + 3}.$

11 $\frac{2x + 1}{12}$

12 a $\frac{3 \times 2 \times 5(x + 2)(x - 1)(x - 3)(x + 4)}{6 \times 5(x + 2)(x - 1)(x - 3)(x + 4)}$

b From left to right the answers are $4(x + 4)$ and $2 - x$.

Review 2

1 a 3 **b** 7 **c** 1
2 a $13y$ **b** $7xy$ **c** $3x$ **d** y^3
e $\frac{x}{2}$ **f** $5xy$
3 a $6a - 3b$ **b** $2 - 5a + 7b$ **c** $4a + 4a^2$
d $5a^2b - 5ab^2$ **e** $6ab + 3b - 9$ **f** $2a - a^2$
4 a 10 **b** -21 **c** -4 **d** 28
e 12 **f** -14 **g** $1\frac{1}{3}$ **h** 3
5 a 10 **b** $3\frac{1}{2}$
6 a a^7 **b** b^5 **c** c^6 **d** $20d^9$
e $4e^6$ **f** 6
7 a x^5 **b** y^2 **c** $8z^6$ **d** 1
e $28u^4$ **f** $5p^{-2}$ **g** $27r^{-6}$ **h** $\frac{1}{4}s^{-4}t^6$
8 a $10a + 15$ **b** $18b - 9c$ **c** $-32d^2 + 8cd$
d $y - y^2$
9 a $5x(x + 2)$ **b** $7a(3b^2 - 2)$ **c** $15p(2p + q^2 - 3q^3)$
10 a $\frac{x + 2}{4}$ **b** $\frac{x}{x + 1}$
11 a $\frac{3}{a}$ **b** $\frac{b}{40}$ **c** $\frac{3}{2c}$ **d** $\frac{3d - 2}{4d^2}$
e $\frac{3f + 5d}{df}$ **f** $\frac{5}{a}$ **g** $\frac{ab}{3}$ **h** $\frac{a}{(a + 1)(a + 2)}$

Assessment 2

1 a $157\,\text{cm}^2$ **b** $3.58\,\text{m}$ **c** $3.27\,\text{in}$
2 $10z^3$
3 a The sum of the areas of the rectangles should equal the area of the square.
$4y + 28 + 7y + y^2 = y^2 + 11y + 28 > y^2$ as $y > 0$.
This sum is greater than the area of the square so George must be incorrect.
b SADC $= 4y - 28$, ATED $= 28$, DEUB $= 7y - 28$,
CDBV $= y^2 - 11y + 28$
4 a i 80 m **ii** 245 m **b** 8 seconds.
5 a Correct.
b Not complete. $abc^2(ac - b^3 + a^4b^2c^2)$
c Incorrect. $6p^3q^2r^4(2r^5 - 3q^3r - 5p^2q^5)$
d Correct.
e Incorrect. $g(g^2 + g - 1)$
f Incorrect, this cannot be factorised.
g Not complete. $2(p - q)[2 - 3(p - q)]$
h Incorrect. $3(y + 2z)^2[1 + 3(y + 2z)]$
i Not complete. $-2(x - 2)(x - 1)$

6 a $2y^2 + 80y$
b No. The volume is $y \times y \times 20 = 20y^2$ and $20y^2 = 200\,\text{cm}^3$.
Thus $y^2 = 10$ so $y = \sqrt{10}$.
7 a i $5W - 10$ **ii** $5(W - 2)$ **b i** $8W + 4$ **ii** $4(2W + 1)$
8 $w = 5, x = 8\frac{1}{3}$
9 a $32p^2 - 22p$ **b** $2p(16p - 11)$
10 a $\frac{5}{9}$ **b** $\frac{11}{20}$ **c** $\frac{13}{18}$ **d** $\frac{2q}{5}$
e $5x$ **f** $\frac{p}{2q}$ **g** $\frac{27a^2}{2}$ **h** $\frac{z - 4}{6}$
11 a $\frac{10z - 9}{24}$ **b** $\frac{59 - 44y}{35}$ **c** $\frac{7x + 11}{(x + 1)(x + 2)}$
d $\frac{13(p - 2)}{(2p + 3)(3p - 2)}$ **e** $\frac{a(7a + 12)}{(a - 2)(2a + 9)}$ **f** $\frac{6x - 1}{(2x + 1)^3}$
12 a $\frac{4(w + 3)}{5}$ **b** $w = 12$ **c** 4 and 3

Chapter 3

Check in 3

1

2 $a = 56°, b = 112°, c = 235°, d = 160°$
3 $e = 72°, f = 63°, g = 118°, h = 75°$

3.1S

1 a Acute **b** Obtuse **c** Reflex **d** Right angle
2 a Right angle **b** Acute **c** Obtuse **d** Reflex
e Reflex **f** Acute **g** Obtuse **h** Reflex
i Reflex **j** Acute **k** Obtuse **l** Reflex
3 a, b Students' diagrams
4 a 180° **b** 360°
5 a 40°, angles around a point sum to 360°.
b 240°, angles around a point sum to 360°.
c 130°, angles on a line sum to 180°.
d 50°, angles on a line sum to 180°.
6 a 110°, corresponding angles are equal.
b 47°, alternate angles are equal.
c 115°, alternate angles are equal.
d 63°, corresponding angles are equal.
7 a 110°, angles around a point sum to 360°.
b 135°, angles around a point sum to 360°.
c 45°, angles on a straight line sum to 180°.
d 45°, angles on a straight line sum to 180°.
8 $i = 14°$ (angles on a straight line sum to 180°),
$h = 14°$ (vertically opposite angles are equal),
$g = 166°$ (vertically opposite angles are equal),
$j = 37°$ (vertically opposite angles are equal),
$l = 143°$ (angles on a straight line sum to 180°),
$k = 143°$ (vertically opposite angles are equal)
9 a $h = 50°$ (angles on a straight line sum to 180°),
$i = 50°$ (corresponding angles are equal)
b $j = 63°$ (alternate angles are equal),
$k = 63°$ (vertically opposite angles are equal)

3.1A

1 Angles on a straight line add up to 180°, these angles add up to 170°.
2 a $a = 17°$ (angles on a straight line sum to 180°),
$b = 17°$ (alternate angles are equal),
$c = 163°$ (angles on a straight line sum to 180°)
b $d = 125°$ (alternate angles are equal),
$e = 105°$ (vertically opposite angles are equal)

c $f = 134°$ (vertically opposite angles are equal),
$g = 134°$ (corresponding angles are equal),
$h = 24°$ (angles on a straight line sum to 180°, corresponding angles are equal, angles in a triangle sum to 180°)

d $i = 140°$ (vertically opposite angles are equal),
$j = 40°$ (interior angles sum to 180°),
$k = 140°$ (interior angles sum to 180°),
$l = 40°$ (interior angles sum to 180°)

3 a 275°   **b** 68°

4 a 284°   **b** 263°   **c** 117°

5 158° to the right

6 a The pink angle is divided as shown in the first diagram into a (alternate angles are equal) and b (corresponding angles are equal).
b $c = 180° − (a + b)$ (angles on a straight line sum to 180°)
so $c + a + b = 180°$.

7 Starting from the top left vertex moving clockwise let the internal angles be A, B, C, D. $A + B = 180°$ and $B + C = 180°$ (interior angles sum to 180°) so $A = C = 180° − B$.

8 $a = 80°, b = 110°, c = 70°, d = 30°, e = 150°$

3.2S

1 a Right angled triangle   **b** Equilateral triangle
   c Kite   **d** Rhombus

2 a 50°   **b** 70°   **c** 80°   **d** 60°
   e 70°   **f** 125°

3 a a   **b** None

4 a Equilateral **b** Scalene   **c** Isosceles **d** Isosceles

5 a 90°, right-angled   **b** 70°, isosceles
   c 60°, equilateral   **d** 80°, scalene
   e 90°, right-angled   **f** 130°, scalene

6 a $a = 80°, b = 100°$ **b** $c = 55°, d = 125°$
   c $e = 105°$ **d** $f = 110°, g = 70°, h = 110°$

7 a 90°, rectangle   **b** 115°, kite
   c 106°, parallelogram   **d** 108°, rhombus
   e 67°, isosceles trapezium   **f** 130°, kite

3.2A

1 50°, 60°, 70°

2 a $m = 52°, n = 38°$ **b** $o = 59°, p = 61°$
   c $q = 115°, r = 35°$ **d** $s = 69°, t = 39°$

3 a e.g. (1, 1)   **b** e.g. (−1, 3)
   c (−1, 2) or (3, 2)   **d** e.g. (2, 1)
   e (1, 1.5)   **f** e.g. (3, 0) or (−1, 0)

4 a 70°   **b** 50°   **c** 100°

5 a $a = 36°$ (alternate angles are equal),
$b = 63°$ (alternate angles are equal),
$c = 81°$ (angles on a straight line sum to 180°)
b $a = 61°$ (corresponding angles are equal),
$b = 49°$ (corresponding angles are equal),
$c = 70°$ (angles in a triangle sum to 180°)
c $a = 113°$ (alternate angles are equal),
$b = 67°$ (angles on a straight line sum to 180°),
$c = 113°$ (corresponding angles are equal),
$d = 67°$ (interior angles sum to 180°), $e = 113°$ (interior angles sum to 180°)

6 a 75°   **b** 34°

7 a True, two pairs of parallel lines, four right angles.
b False, kites may include two different side lengths.
c False, a rhombus may have a non-90° angle.

8 From the top left vertex moving clockwise let the vertices be A, B, C, D. $\angle BAC = \angle ACD = 30°$ (angles in a triangle sum to 180° and alternate angles are equal). $\angle CAD = 30°$ ($= \angle ACD$) (angles in a triangle sum to 180°).

3.3S

1 a Yes, SSS   **b** No, similar   **c** Yes, SAS

2 B

3 a $a = 40°, b = 50°$ **b** $c = 20°, d = 40°$

4 a (scale factor = 2),   **b** (scale factor = 5),   **e** (scale factor = 4)

5 a scale factor = 2, length = 8 cm
   b scale factor = 3, length = 12 cm

6 $a = 6.4$ cm, $b = 4.5$ cm

7 $c = 9$ cm, $d = 1.9$ cm

3.3A

1 Both triangles have angles 90°, 38° and 52°. Triangles satisfy ASA.

2 Satisfies SSS.

3 a Does not satisfy SSS.   **b** Satisfies SSS.

4 a 6 cm   **b** 12 cm

5 RT = 5 cm, QR = 4 cm, QS = 12 cm

6 a XY = 6.6 cm, VY = 6 cm   **b** 15.5 cm

7 KN = 6 cm, JN = 2.8 cm

8 a EF = HG (opposite sides in a parallelogram are equal),
$\angle EFM = \angle MHG$ (alternate angles are equal),
$\angle FEM = \angle MGH$ (alternate angles are equal). ASA.
b By congruence $EM = GM$.

9 a $RS = RU = UT = TS$, $\angle RSO = \angle RUO$ (isosceles triangle),
$\angle RUO = \angle OST$ (alternate angles are equal), $\angle OST = \angle OUT$ (isosceles triangle) so RUO is congruent to OUT and RSO is congruent to TSO by SAS. $\angle SRT = \angle RTU$ (alternate angles are equal) so OUT is congruent to OSR by ASA.
b $ROS = SOT = TOU = UOR = 360 ÷ 4 = 90°$

3.4S

1 a 3   **b** 4   **c** 5   **d** 6
   e 7   **f** 8

2 a 3   **b** 4   **c** 5   **d** 6
   e 7   **f** 8

3 a 60°   **b** 120°   **c** 360°

4 Exterior angle: 120°, 90°, 72°, 60°, 51.4°, 45°, 40°, 36°
Interior angle: 60°, 90°, 108°, 120°, 128.6°, 135°, 140°, 144°

5 a 156°   **b** $180 − 360 ÷ 15$

6 a 160° e.g. 180 (18 − 2) ÷ 18 or 180 − 360 ÷ 18

7 $180° (8 − 2) = 1080°$.

8 $x = 141°$

9 144°

10 106°

11 a $x = 40°, y = 70°$ **b** $2y = 140°, 180 − 2y = 40°$

3.4A

1 a 18°   **b** 360°   **c** 20   **d** 20

2 a 24°   **b** 156°

3 a 146°   **b** 115°

4 a 45°   **b** 135°   **c** octagon

5 a $x = 125°, y = 50°$ **b** $x = 110°, y = 65°$

6 $x = 135°$, smallest angle = 85°

7 a 12   **b** 8

8 Interior angle = 120°, which is a factor of 360°.

9 No, interior angle = 108° which is not a factor of 360°.

10 a 150°
   b 12-sided regular polygon, or any shapes interior angles totalling 150°.

11 $x = 120°$

12 $\angle ABC = 120°$ (interior angle of a regular hexagon),
$\angle BCA = (180 − 120) ÷ 2 = 30°$ (angles in a triangle sum to 180°, isosceles triangle has equal base angles),
$\angle ACD = 120 − 30 = 90°$.

Review 3

1 $a = 105°, b = 60°, c = 120°$

2 $a = 105°$ (CA), $b = 75°$ (ASL), $c = 65°$ (AA)

3 a 243°   **b** 321°   **c** 025°

4 $40°$

5 $a = 70°$

6 $A(1, 1), B(3, 3), C(5, 1)$ should be plotted and labeled. The triangle is isosceles and right-angled.

7 a Trapezium **b** Rhombus

8 No, in triangle ABC the hypotenuse is $13\,cm$ but in triangle EFD it is one of the shorter sides which is $13\,cm$. Since the triangle is right-angled these sides cannot both be $13\,cm$ so they are not congruent.

9 $x = 2.25\,cm$

10 a $900°$

  b Geometrical proof that the interior angles sum to $540°$
  So each interior angles is $540 \div 5 = 108°$.
  Hence each exterior angle is $180 - 108 = 72°$.
  Hence sum of exterior angles is $72° \times 5 = 360°$

Assessment 3

1 a $182.5°$ or $177.5°$

  b At 1:05 the hour hand is on a bearing of $032.5°$ and the minute hand is on a bearing of $030°$.
  At 1:06 the hour hand is on a bearing of $033°$ and the minute hand is on a bearing of $036°$.
  The minute hand is on a lesser bearing than the hour hand at 1:05 but a greater bearing than the hour hand at 1:06 so the two must be on the same bearing at some point between the two times.

2 $103°$ anti-clockwise.

3 a $130°$ **b** $255°$ **c** $215°$ **d** $310°$

4 Rafa was south of Sunita.

5 a Incorrect. $a = 62°$ (IA).

  b Correct (CA).

  c Correct (CA, ASL).

  d Incorrect. $d = 87°$ (ASL, CA).

  e Incorrect. $e = 88°$ (exterior angles of a polygon).

  f Incorrect. $f = 92°$ (ASL).

  g Correct (isosceles triangle has base angles $52°$, ASL).

  h Correct (ASQ).

6 a False. Two obtuse angles add to more than $180°$ which is the sum of the three angles in a triangle.

  b True.

  c False. One right angle $= 90°$ and one obtuse angle is more than $90°$. This comes to more than $180°$ before the acute angle has been added.

  d True.

7 a $93°$ **b** $170°$ **c** $73°$ **d** $174°$

8 a $108°$ **b** 25 slabs. **c** $72°, 54°, 54°$

9 $120\,cm$

10 a $\angle PTS = \angle PQR$ (CA). $\angle PST = \angle PRQ$ (CA). $\angle SPT = \angle RPQ$ (common angle).

  b The ratio of shorter to longer parallel sides is $\frac{AB}{10}$ in trapezium ABFE and $\frac{AB}{15}$ in trapezium ABFE. $\frac{AB}{10} > \frac{AB}{15}$. The ratios between sides are different so the shapes are not similar.

  c $15\,cm$

11 a $2\,cm$ **b** $18\,cm^2$ **c** $2\,cm^2$

  d $FD = AB = 6\,cm$ and FD is parallel to AB so $ABDF$ is a parallelogram and DB is parallel to FA.
  $\angle DBC$ and $\angle FAB$ are corresponding angles so are equal.

Chapter 4

Check in 4

1 a $22\frac{1}{2}$ **b** 21 **c** 6 **d** 21

2 a A survey of an entire population.

  b In a census, everyone in the population is included. In a sample, only a selection of people from the population are included.

3 It suggests the correct answer is 'yes'.

4.1S

1 Difficult to make sure that the whole population is included – census, disadvantage.
Less data to analyse – sample, advantage.
Unbiased – census, advantage.
Not everyone is represented – sample, disadvantage.

2 a Question gives too many responses.

  b Vague question.

  c Question doesn't specify the time period – weekly/monthly etc

  d Leading question.

  e Leading question.

  f Question doesn't specify the time period – weekly/monthly etc

  g Question gives too many responses.

  h Vague question.

  i Question gives too many responses.

  j Responses mean different things to different people.

3 a Question doesn't ask how much they spend. Assumes you visit cinema. Answers may differ at different times of year so could say on average. No answer choices given

  b On average how many times do you go to the cinema in one month? Once or less often, 2 or 3 times a month, 4 times or more often.

  c On average how much do you spend when you go to the cinema? Less than £5, £5 to £10, more than £10.

4 a i Leading.

  ii What is your favourite sport? Tennis, swimming, football, rugby, cricket, other.

  b On average, how many times a week do you play sport? Never, 1 or 2 times, 3 or 4 times, more than 4 times.

5 Obviously visit cinema so not typical of population.

6 People in an athletics club will probably play sport more often than those who aren't.

7 a It is not representative of the whole school.

  b Pick names out of a hat, or take every 20th person on a list of all the people in the school.

4.1A

1 People listening to MP3 players may be more interested in music than is typical. Does not cover all possible answers

2 May not like either. Not enough choices. Sample size is very small. Year 11 may not be representative of the whole school.

3 a Not true. The raw data may differ as different members of the population may be chosen for the sample.

  b Not true. A sample can be any size.

4 A sample. Taking a census of all the batteries would use up all the stock. A sample is faster and less expensive to carry out.

5 Reception = 19, Year 1 = 21, Year 2 = 20

6 Randomly select 550 women and 250 men.

7 For example, for a sample size of 100, randomly select 14 males 18–30, 14 males 31–50, 12 males over 50, 20 females 18–30, 19 females 31–50 and 21 females over 50.

8 a 10 boys and 8 girls **b** 85

9 The proportion of boys in the sample is much higher than the proportion of boys at the gym club.

10 It would be more natural to choose a sample of size 61 with 9 adults, 26 boys and 26 girls.

4.2S

1 Options for how often respondents go to the cinema must start at 0 and include an open-ended highest option, categories must not overlap, a time frame must be specified. Options for how much spent must start at 0 and have an open-ended highest option, categories should not overlap.

2 Two way table giving gender options and frequency options for playing sport. Frequency options must start at 0 and have an open-ended highest option, categories should not overlap.

3 Two way table including all year options (not grouped) for columns and possible crisp flavours including 'other' for rows (or columns and rows reversed).

4 **a** Two-way table with colour (black/silver) and contract (pay as you go/contract).
   b Check that totals add to 100.

5 Two-way table with saloon/hatchback and part exchange/cash.

6 **a** 78   **b** 20   **c** 43   **d** 55
   e 98

7 **a** 28   **b** 67

8

43.0	0 2 3 8 8 9
44.0	0 1 3 4 5 7
45.0	0 0 1 2 6
46.0	0 1 3 5 5 9 9

Key: 44.0 | 7 means 44.7 seconds

9

40	0 1 2 4 5 6 6 7 7 8 8 9
50	0 3 3 4 5 5 6 6 7
60	0 1 1 3 4 5 5
70	0 0

Key: 50 | 3 means 53

10

200	4 6 7 8 9 9
210	0 1 4 6 6 7
220	5
230	0 0 0 2 3 6 7 8
240	0 1 2 4

Key: 200 | 4 means 204

4.2A

1 8 girls study Spanish.

2 Emily is wrong. There are 17 girls and 19 boys taking part.

3 **a**

Men		Women
	14	8 9
9 8	15	1 3 4 5 5 6 7 8 8
9 8 7 5 0	16	1 2 5 6 7 9
8 8 6 5 4 2 1	17	2 4 8
8 4 3 3 2 0	18	

Key: 2|16|3 means 162 cm for men, 163 cm for women
   b Men are taller than women and heights are evenly spread in both groups

4 **a**

Boys		Girls
8 8 8 6 2 2	3	
7 5 4 2	4	0 3 4 6 8 8
9 9 8 7 6 3 2	5	2 2 5 6 9
8 6 4 2 1	6	2 3 3 5 6 7
2 1	7	2 3 4 6 7
	8	0 2

Key: 5|4|6 means 4.5 minutes for boys, 4.6 seconds for girls

   b Boys have a faster reaction time than girls, variation in times is similar.

5 Agree – out of the 60 red cars, 55 were speeding.

6 **a** Stem-and-leaf diagram displays all the raw data. The table will have too many categories.
   b Advantage – grouping the data makes it easier to spot trends. Disadvantage – the raw data is not recorded.
   c Disadvantage – the data set is large so it is difficult to spot trends. Advantage – all the data is recorded.

1

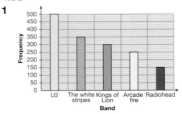

2

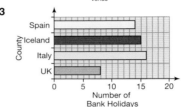

3

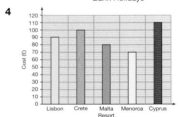

4

5 **a** 30°   **b** Boys 210°, Girls 150°
   c Students pie charts, with the angles from part **b**.

6 **a** 6°
   b Sunny 90°, Cloudy 108°, Rainy 84°, Snowy 18°, Windy 60°
   c Pie chart with angles given in part **b**

7 Pie chart with the following angles – Bat the Rat: 30°, Hook a Duck: 25°, Smash a Plate: 35°, Roll a Coin: 80°, Tombola: 70°, Break 1: 60°, Break 2: 60°

8 **a** 20°   **b i** 7   **ii** 10   **iii** 1

4.3A

1 The vertical scale does not start at 0.

2 Pie chart shows proportion not number of jars sxold. No information about what sauces are in the 'other' category. Angles for BBQ, Chili and Pesto are similar so it is difficult to compare the number of jars sold.

3 **a** Too many categories.
   b Easy to compare the frequency on a bar chart.

4 **a** No, the angle for German is the same as the angle for Spanish.
   b No, Lydia doesn't know how many students in the schools are represented in the pie charts.

5 **a** Monday 60°, Tuesday 100°, Wednesday 60°, Thursday 40°, Friday 80°, Saturday 20°.
   b Monday 80°, Tuesday 40°, Wednesday 0°, Thursday 40°, Friday 80°, Saturday 120°.
   c Pie chart – the angles represent the proportion.
   d Bar chart – the heights of the bars show the frequencies.

4.4S

1 **a** 8   **b** 9   **c** 2   **d** 6
   e 2   **f** 24   **g** 18   **h** 104
   i 15   **j** 5

2 **a** **i** 5 **ii** 10 **iii** 35 **iv** 98
v 2
b Set **ii**, the outlier has not affected the median.
3 **a** Mode = 1, Range = 10 **b** Mode = 8, Range = 3
c Mode = 11, Range = 4 **d** Mode = 25, Range = 15
e Mode = 8, Range = 3 **f** Mode = 5, Range = 3
Parts **a** and **d** contain outliers which increase the range but do not affect the mode.
4 **a, b** 3, 3, 3, 3, 4, 4, 5, 5
Range = 2, Mode = 3, Median = 3.5, Mean = 3.75
5 **a, b** 1, 1, 1, 1, 2, 2, 2, 2, 2, 3, 3, 4, 4, 4, 4, 5, 5, 5, 5, 5, 5, 5
Range = 4, Mode = 5, Median = 4, Mean = 3.28
6 **a** **i** 7 **ii** 6 **iii** 5.82 **iv** 6 **v** 2
b **i** 75 **ii** 63 **iii** 60.1 **iv** 63 **v** 27
c **i** 8 **ii** 96 **iii** 95.6 **iv** 96 **v** 2
d **i** 71 **ii** 22, 37 **iii** 40.4 **iv** 37 **v** 38
e **i** 26 **ii** 88, 89 **iii** 84.2 **iv** 87 **v** 7
f **i** 72 **ii** 27 **iii** 46.9 **iv** 34 **v** 37
g **i** 8 **ii** 105 **iii** 105.2 **iv** 105 **v** 3
7 **a** 1, 6, 8, 2, 8, 5, 6, 9, 3, 5, 7, 4, 4, 5, 5
b **i** 8 **ii** 5 **iii** 5.2 **iv** 5 **v** 3
c Range and IQR stay the same.

4.4A

1 **a** 2, 4
b Number 47's data is more spread out than Number 45's.
2 **a** 1, 3, 4, 1, 1 **b** 1.8, 2, 2
c On average, houses in Ullswater Drive have more cars than those in Ambleside Close.
3 23 or 42
4 −3.1, −2.6, 3.5, 4.1, 4.1
5 e.g. 1, 5, 6, 10, 15 or 4, 6, 6, 11, 18 or 3, 4, 6, 9, 17.
6 Set with the smallest mean is 6, 6, 8, 8, 16 (mean 8.8).
Set with the largest mean is 6, 6, 8, 16, 16 (mean 10.4).
7 **a** Mean = 6, Mode = 4, Median = 6, Range = 7,
Interquartile range = 4
b Mean = 6.2, Mode = 4, Median = 6, Range = 8,
Interquartile range = 4
8 41 years old
9 14 raisins
10 78.6%

Review 4

1 Sample is biased as there probably won't be children in the pub. Not everyone in the pub may be watching the football.
2 11
3 **a** Pie chart with following labeled sectors:
Children's Services (80°), Corporate/Finance (40°),
Adult Social Care (140°), Transport (30°), Resident's Services (70°).
b £648 000
c Heights of bars should be (£1000s):
144, 72, 252, 54, 126
4 **a** **i** 36.1 s **ii** 55 s
b **i** 37 s **ii** 40 s **iii** 29 s
c 11 s
5 **a** **i** 47 **ii** 22
b **i** 37 **ii** 25
c The women are younger in general and their ages vary more.

Assessment 4

1 **a** **i** Yes **ii** Look at the cars in the car park.
iii Take measurements in the middle of day when all teachers are present.
b **i** Yes **ii** Measure the students in class.
iii Have measurements of each student taken multiple times by separate people.

c **i** No
ii Gather data from the Met Office, academic institutions etc.
iii Use data collected from the same locations using the same method each year.
d **i** No
ii Offer subjects a variety of lengths to estimate.
iii Sample subjects randomly from the available population.
2 The villages are different sizes so the smaller ones are overrepresented.
3 **a**

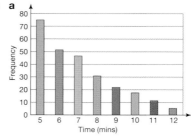

b

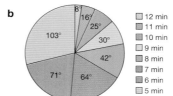

c **i** 5 min **ii** 7 min
d 7 min
4

	Glasses	No glasses
Left-handed	1	4
Right-handed	14	16

5 **a**

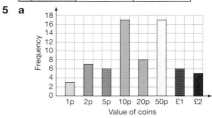

b 50p **c** 20p **d** 47p
6 **a** Walk **b** Car **c** 20% **d** 120
e 300
7 **a** **i** 103 miles **ii** 99.5 miles **iii** 88 miles
b You cannot tell because you do not know how many journeys he makes in one day.
c **i** 107 miles **ii** 52.25 miles
d Mean increases, because the total has increased but not the number of journeys. Median increases, because 98 replaces 90 as one of the two middle numbers. Mode, no mode as each number occurs once. Range, stays the same because the highest and lowest numbers are unaffected.
8 **a** 9 **b** 2 **c** 8 **d** 0
e 1.2

Chapter 5

Check in 5

1 **a** 6 squares shaded **b** 8 squares shaded
c 9 squares shaded **d** 10 squares shaded
e 7 squares shaded
2 **a** $\frac{1}{2}$ **b** $\frac{3}{4}$ **c** $\frac{4}{5}$ **d** $\frac{19}{20}$
e $\frac{3}{4}$
3 **a** 20 × 30 = 600 **b** 360 ÷ 60 = 6
c 1200 − 800 = 400 **d** 7000 + 6000 = 13 000

5.1S

1 **a** $3\frac{1}{2}$ **b** $1\frac{3}{5}$ **c** $3\frac{1}{3}$ **d** $\frac{2}{3}$

   e $2\frac{2}{5}$

2 **a** 10 **b** 21 **c** 12 **d** 13

   e 22

3 **a** 24 m **b** 36 m

4 **a** 15 litres **b** 180 cl **c** 48 cl **d** 250 ml

5 **a** 30 m **b** 8 km **c** 16 mm **d** 90 m

6 **a** $1\frac{2}{3}$ **b** $5\frac{1}{2}$ **c** $4\frac{1}{2}$ **d** $\frac{2}{3}$

   e $9\frac{1}{3}$

7 **a** $2\frac{2}{3}$ miles **b** $33\frac{1}{3}$ miles **c** $12\frac{1}{2}$ miles **d** $3\frac{3}{4}$ miles

8 **a** 6.7 m **b** 5.3 mm **c** 8.7 cm **d** 14.9 km

9 **a** 10.5 **b** 126 **c** 240 **d** 300

   e 132 **f** 99

10 **a** £225 **b** 990 m **c** 2520 kg **d** €144

11 **a** 135 **b** 91 **c** 744 **d** 18

   e 387 **f** 192

12 **a** 1104 mm **b** 1952 kg **c** €1443 **d** £493

13 **a** 0.5 **b** 0.6 **c** 0.25 **d** 0.51

   e 0.64 **f** 0.22 **g** 0.15 **h** 0.7

   i 0.07 **j** 0.085 **k** 0.0015 **l** 0.0001

14 **a** 5.7 **b** 200 **c** 15.93 **d** 99.84

   e 16.81 **f** 20

15 **a** £3.84 **b** £53.55 **c** £13.95 **d** £170.10

   e £28.16 **f** £15.90

5.1A

1 25% of £520 = £130, 30% of 450 = £135, 20% of £640 = £128

2 **a** £104 **b** 35 weeks

3 40.83 m^3

4 £425

5 £73.70

6 649 boys

7 £49090.60

8 £199.17

9 120

10 Merton (£71.25 < £72.25)

11 **a** 48 **b** 40 **c** 2

12 $\frac{2}{5}$

13 59%

5.2S

1 **a** $\frac{3}{5}$ **b** 1 **c** $\frac{2}{3}$ **d** $\frac{3}{5}$

2 **a** 20 **b** 18 **c** 35 **d** 25

3 **a** $\frac{3}{10}$ **b** $\frac{5}{6}$ **c** $\frac{11}{20}$ **d** $\frac{3}{8}$

4 **a** $1\frac{1}{4}$ **b** $1\frac{4}{5}$ **c** $1\frac{5}{8}$ **d** $1\frac{1}{4}$

5 **a** $\frac{7}{4}$ **b** $\frac{23}{16}$ **c** $\frac{14}{9}$ **d** $\frac{18}{7}$

6 **a** $1\frac{3}{4}$ **b** $1\frac{2}{3}$ **c** $3\frac{3}{10}$ **d** $3\frac{1}{8}$

   e $3\frac{14}{15}$ **f** $4\frac{7}{12}$ **g** $7\frac{13}{14}$ **h** $9\frac{19}{53}$

   i $1\frac{7}{20}$ **j** $\frac{3}{4}$ **k** $1\frac{19}{20}$ **l** $4\frac{13}{14}$

7 **a** $\frac{3}{20}$ **b** $\frac{4}{27}$ **c** $\frac{1}{14}$ **d** $\frac{1}{4}$

   e $\frac{20}{63}$ **f** $\frac{1}{12}$ **g** $\frac{12}{65}$ **h** $\frac{1}{15}$

8 **a** $\frac{3}{4}$ **b** $\frac{3}{4}$ **c** $\frac{4}{5}$ **d** $\frac{4}{5}$

   e $\frac{3}{7}$ **f** $\frac{9}{5}$ **g** $\frac{24}{7}$ **h** $\frac{11}{4}$

9 **a** $\frac{5}{32}$ **b** $\frac{1}{8}$ **c** $\frac{2}{21}$ **d** $\frac{1}{48}$

   e 20 **f** 9 **g** $\frac{25}{2}$ **h** $\frac{77}{3}$

10 **a** $\frac{1}{3}$ **b** $\frac{1}{4}$ **c** $\frac{1}{6}$ **d** $\frac{1}{4}$

   e $\frac{8}{9}$ **f** $\frac{45}{56}$ **g** $\frac{2}{9}$ **h** $\frac{8}{21}$

11 **a** $1\frac{1}{8}$ **b** $1\frac{1}{10}$ **c** 3 **d** 3

   e $4\frac{19}{24}$ **f** $1\frac{3}{14}$ **g** $1\frac{21}{22}$ **h** $4\frac{4}{27}$

12 **a** $6\frac{1}{8}$ **b** $\frac{5}{8}$ **c** $10\frac{5}{16}$ **d** $\frac{2}{3}$

13 **a** $3\frac{1}{3}$ **b** $\frac{35}{64}$ **c** $\frac{15}{38}$ **d** $4\frac{57}{160}$

5.2A

1 **a** $\frac{5}{14}, \frac{3}{8}, \frac{3}{7}$ **b** $\frac{2}{7}, \frac{2}{3}, \frac{5}{6}$

2 **a** $\frac{3}{4}, \frac{17}{40}, \frac{2}{5}, \frac{3}{8}$ **b** $\frac{5}{6}, \frac{5}{8}, \frac{7}{12}, \frac{11}{24}$

3 $\frac{7}{24}$

4 **a** 6 inches **b** $\frac{25}{8}$ inches

5 **a** $\frac{3}{8}$ and $\frac{1}{10}$ **b** $\frac{1}{10}, \frac{2}{5}, \frac{1}{2}$,

6 $\frac{2}{3}, \frac{3}{10}, \frac{1}{2}, \frac{3}{20}, \frac{2}{3}, \times$

7 $\frac{4}{7}$

8 $\frac{6}{11}, \frac{6}{11} - \frac{1}{2} = \frac{1}{22} < \frac{1}{2} - \frac{4}{9} = \frac{1}{18}$

9 Many answers possible, e.g. $\frac{1}{3}, \frac{1}{10}, \frac{1}{15}$

10 **a** e.g. $a = 2, b = 3$ **b** Any pair where $a > 0, b < 0$

11 $\frac{3 \times 5 \times 2}{6} = xy$

12 **a** $\frac{7}{4}$ **b** 3.06 – accurate to 1 decimal place

5.3S

1 **a** 0.5 **b** 0.75 **c** 0.4 **d** 0.1

   e 0.2 **f** 0.25

2 **a** 50% **b** 75% **c** 40% **d** 10%

   e 20% **f** 25%

3 **a** 0.43 **b** 0.86 **c** 0.94 **d** 0.455

   e 0.0375 **f** 1.05

4 **a** $\frac{1}{2}$ **b** $\frac{1}{4}$ **c** $\frac{1}{5}$ **d** $\frac{1}{8}$

   e $\frac{3}{4}$ **f** $\frac{9}{10}$

5, 6 **a** 62.5% **b** 80% **c** 87.5% **d** 60%

   e 37.5% **f** 12.5%

7 **a** 55.6% **b** 34.4% **c** 75.8% **d** 51.3%

   e 43.7% **f** 83.9% **g** 83.9% **h** 105.6%

8 **a** 0.22 **b** 0.185 **c** $0.\dot{5}$ **d** $0.3\dot{5}$

   e $0.06\dot{5}$ **f** $0.61\dot{4}9$ **g** $0.544\dot{6}$ **h** $1.5\dot{2}$

9 **a** 80% **b** 15% **c** 87.5% **d** 75%

   e 28% **f** 18.75% **g** 45% **h** 14%

10 **a** $\frac{51}{100}$ **b** $\frac{43}{100}$ **c** $\frac{413}{1000}$ **d** $\frac{719}{1000}$

   e $\frac{91}{100}$ **f** $\frac{871}{1000}$

11 **a** $\frac{49}{100}$ **b** $\frac{53}{100}$ **c** $\frac{73}{100}$ **d** $\frac{81}{100}$

   e $\frac{37}{100}$ **f** $\frac{19}{100}$

12 **a** 0.0625 **b** 0.28 **c** 0.056 **d** 0.075

   e 0.4375 **f** 0.03125

13 **a** 6.25% **b** 28% **c** 5.6% **d** 7.5%

   e 43.75% **f** 3.125%

14 **a** $\frac{8}{25}$ **b** $\frac{11}{20}$ **c** $\frac{11}{25}$ **d** $\frac{31}{200}$

   e $\frac{16}{25}$ **f** $\frac{53}{200}$

15 **a** $\frac{11}{20}$ **b** $\frac{31}{50}$ **c** $\frac{21}{25}$ **d** $\frac{13}{20}$

   e $\frac{18}{25}$ **f** $\frac{37}{200}$

16 $\frac{8}{11}$, this has a prime factor in the denominator other than 2 or 5.

17 **a** $0.\dot{1}$ **b** $0.\dot{5}$ **c** $0.7\dot{5}$ **d** $0.3\dot{4}\dot{6}$

   e $0.7\dot{5}6$ **f** $0.01\dot{2}\dot{6}$

18 **a** $0.\dot{3}$ **b** $0.1\dot{6}$ **c** $0.\dot{6}$ **d** $0.\dot{1}4285\dot{7}$

   e $0.\dot{1}$ **f** $0.8\dot{3}$

19 **a** 33.3% **b** 16.6% **c** 66.6% **d** $14.\dot{2}8571\dot{4}\%$

   e 11.1% **f** 83.3%

21 **a** $0.\dot{4}2857\dot{1}$ **b** 0.1875 **c** 0.2125 **d** $0.\dot{5}$

   e 0.16 **f** $0.\dot{7}1428\dot{5}$

1 a grade 6   **b** grade 7   **c** grade 5   **d** grade 7

2 a Yes   **b** H W Smith

3 a $0.\dot{3}$, 33.3%, $33\frac{1}{3}$%, 33

  b $0.\dot{4}$, 44.5%, 0.45, 0.454

  c 22.3%, 0.232, 23.22%, 0.233, $0.2\dot{3}$

  d $0.6\dot{5}$, 0.66, 66.6%, 0.6666, $\frac{2}{3}$

  e 14%, $14.\dot{1}$%, 0.142, $\frac{1}{7}$, $\frac{51}{350}$

  f $\frac{5}{6}$, $\frac{6}{7}$, 86%, 0.866, $0.8\dot{6}$

4 a 0.1428571429…, 0.2857142857…, 0.4285714286…, 0.5714285714…, 0.7142857143…, 0.8571428571…

  b numbers repeat after every 7th number

5 Shula's calculator has rounded the repeating pattern to 8 decimal places.

6 Abby, $\frac{1}{9} = 0.\dot{1}$ so $1 = \frac{9}{9} = 0.\dot{9}$

7 a $\frac{1}{9}$ **b** $\frac{2}{9}$ **c** $\frac{5}{33}$ **d** $\frac{125}{999}$

  e $\frac{8}{37}$ **f** $\frac{19}{90}$ **g** $\frac{13}{18}$ **h** $\frac{91}{110}$

  i $\frac{421}{666}$ **j** $\frac{1349}{1650}$

8 a $\frac{8}{11}$ **b** $0.\dot{7}$ **c** $0.0\dot{7}\dot{4}$

9 a $100(0.\dot{5}\dot{7}) - (0.\dot{5}\dot{7}) = 57$ so $0.\dot{5}\dot{7} = \frac{57}{99} = \frac{19}{33}$

  b $\frac{59}{165}$

10 They are equally close. $0.5 - 0.\dot{3}\dot{6} = 0.\dot{6}\dot{3} - 0.5 = \frac{3}{22}$

11 π

12 $\frac{123456789}{999999999}$

Review 5

1 a 4   **b** 15   **c** 12   **d** 24

2 a 21   **b** 10.2   **c** 1   **d** 5

3 a $\frac{20}{7}$ **b** $\frac{16}{11}$

4 a $2\frac{2}{9}$ **b** $7\frac{1}{7}$

5 a $\frac{1}{5}$ **b** 3   **c** $\frac{7}{2}$ **d** $\frac{5}{7}$

6 a $\frac{1}{10}$ **b** $\frac{27}{8}$ or $3\frac{3}{8}$ **c** $\frac{2}{3}$ **d** 18

  e $\frac{1}{2}$ **f** $\frac{1}{2}$ **g** $\frac{17}{33}$ **h** $\frac{41}{28}$ or $1\frac{13}{28}$

7 a 0.8   **b** 0.85   **c** 0.03   **d** 0.375

  e 0.22   **f** 0.004   **g** 2   **h** 0.055

8 a $\frac{7}{10}$ **b** $\frac{107}{500}$ **c** $\frac{9}{25}$ **d** $\frac{1}{100}$

  e $\frac{3}{25}$ **f** $\frac{3}{2}$ **g** $\frac{111}{250}$ **h** $\frac{1}{500}$

9 a 60%   **b** 65%   **c** 0.7%   **d** 45%

  e 35%   **f** 80%   **g** 9%   **h** 180%

10 a $0.\dot{2}$ **b** $0.857\dot{1}4\dot{2}$

11 a $\frac{8}{9}$ **b** $\frac{7}{30}$

12 a $\frac{9}{16}$, $\frac{5}{8}$, $\frac{2}{3}$ **b** $\frac{1}{5}$, 22.2%, $0.\dot{2}$

Assessment 5

1 a $\frac{11}{50}$

  b Labour £5.60, materials £2, advertising £8, profit £4.40.

  c £20.09

2 a Blue sky   **b** Space   **c** Chat-chat

  d Ben was incorrect, 8 is 32% of 25 but 40% of 20.

3 a 36.4% (3 sf)   **b** 14%

4 a 0.12   **b** 4th   **c** 3rd

5 $1\frac{1}{2}$ hectares

6 8.5 yards

7 a 540 m².   **b** 252 m²

8 $\frac{10}{33}$

9 Yes, % increase = 60%.

10 $\frac{1}{7}$ is not exactly 14%. Doubling 14% also doubles the error. $\frac{1}{7} = 14.2857…$ % so $\frac{2}{7} = 28.5714……$% = 29% (nearest %).

11 If 102 people represent *exactly* 34% of the total, total = 102 ÷ 0.34 = 300, and 66% of 300 = 198 not 200.

12 $\frac{3}{8}$, 33.3%, $33\frac{1}{3}$%, 0.334, 0.34, $\frac{5}{14}$

13 $\frac{1}{3} = 0.\dot{3} = 0.3333…$
$1 = \frac{1}{3} + \frac{1}{3} + \frac{1}{3}$
$0.333333…+$
$0.333333…+$
$0.333333…$
$\overline{0.999999… = 0.\dot{9}}$

14 a **i** $0.\dot{2}$ **ii** 0.52   **iii** $0.91\dot{6}$ **iv** $0.3\dot{5}$
   v $0.6\dot{3}$ **vi** 0.6875

  b If the denominator of a fraction has a prime factor other than 1, 2 and 5 the decimal expansion will be recurring.

  c $\frac{14}{30}$, $\frac{19}{99}$, $\frac{425}{612}$

15 a $\frac{2}{11}$ **b** $\frac{4}{15}$ **c** $\frac{34}{333}$ **d** $\frac{4165}{9999}$

  e $\frac{5}{22}$ **f** $\frac{3479}{9900}$

Lifeskills 1

1 a

Women		Men
9 8	1	
8 3 2 2 0	2	0 1 3 4 6
8 5 2	3	1 7 9
7 4	4	0 2 7
8 2	5	1 5
1	6	2 6

Key   2|4 means 24 years old.

Overall, the men interviewed were older than the women.

  b Yes, but the difference is small. Women's mean = £28.60, men's mean = £30.

2 a £222 222   **b** $P = R - G - S - C$
  c 52 222   **d** $S = R - G - C - P$
                   $S = £97 778$

3 (16–24) 38, (25–49) 93, (50–64) 36, (65+) 33.

4 a $\frac{7}{40}$

  b Pie chart angles: A 144°, R 90°, J 63°, M 63°.

  c Abigail £20 000, Raheem £12 500, Juliet £8750, Mike £8750.

5 a **i** £21 061.82 **ii** £19 963.55

  b **i** £7121.89 **ii** £7513.69

  c 10.4%

  d $C = \frac{A}{1+i}\left(\frac{1-\left(\frac{1}{1+i}\right)^5}{1-\left(\frac{1}{1+i}\right)}\right) = \frac{A}{1+i}\left(\frac{1-\left(\frac{1}{1+i}\right)^5}{\frac{1+i-1}{1+i}}\right) = \frac{A}{1+i} \times \frac{1-\left(\frac{1}{1+i}\right)^5}{\frac{i}{1+i}}$

      $= \frac{A}{i}\left(1-\left(\frac{1}{1+i}\right)^5\right)$

  e $A = \dfrac{Ci}{1 - \left(\frac{1}{1+i}\right)^5}$

Chapter 6

Check in 6

1 a $4n + 7$ **b** $3(n - 6)$ **c** $10 - n^2$ **d** $\frac{3n - 6}{2}$
  e n^5

2 a 196   **b** 1   **c** -32
  d $\frac{8}{27}$ **e** 0.000008   **f** 10 000 000 000
  g -5 **h** $\frac{2}{5}$

6.1S

1 a 12   **b** 11   **c** -24 **d** 12

2 a 12   **b** 2.75   **c** 128   **d** 144

3 a $C = 20t + 35$ **b** £95

4 a $C = 1.6m + 2$ **b i** £10   **ii** £26

5 a $C = 35t + 75$ **b** €320   **c** 12 days

6 a $m = \frac{y - c}{x}$ **b** $m = wt + k$
  c $m = \sqrt{p - kt}$ **d** $m = (k + l)^3$

7 a $x = 2(y - kw)$ **b** $x = \frac{4m + t^2}{a}$ **c** $x = 4y^2$

d $x = \frac{k - y^2}{l}$ **e** $x = \left(\frac{t - kh}{a}\right)^2$ **f** $x = \frac{p}{m + t}$

g $x = \frac{p}{w - c}$ **h** $x = \frac{y}{a(b + j)}$

8 a $86\,°F$ **b** $C = \frac{5(F + 40)}{9} - 40$ **c** $-35.6\,°C$

9 a $g = L\left(\frac{2\pi}{T}\right)^2$ **b** $9.8\,\text{ms}^{-2}$

10 a $\frac{p(c - qt)}{wr} = m - x^2$ **b** $\frac{1}{b} = \frac{1}{c} - \frac{1}{a}$

$x^2 = \frac{mwr}{wr} - \frac{p(c - qt)}{wr}$ $\frac{1}{b} = \frac{a - c}{ac}$

$x = \sqrt{\frac{mwr + pqt - pc}{wr}}$ $b = \frac{ac}{a - c}$

c $t - qx = 16p^4$

$t - 16p^4 = qx$

$x = \frac{t - 16p^4}{q}$

11 a $f = \frac{uv}{u + v}$ **b** $v = \frac{uf}{u - f}$

12 a $w = \frac{t + r}{q - p}$ **b** $w = \frac{a - k}{c - l}$ **c** $w = \frac{py + qt}{p + q}$

d $w = \frac{t + r}{t + 1}$ **e** $w = \frac{c - gr}{r - 1}$ **f** $w = \frac{5r + t}{r + x}$

g $w = \frac{t + 5k}{k - 1}$ **h** $w = \frac{3q - 4p}{7}$ **i** $w = -\frac{t + 25q}{24}$

6.1A

1 a i $P = 4s$ **ii** $A = s^2$

b i $P = 2y + 26$ **ii** $A = 8y + 5x$

2 a $C = 0.64d + 12$ **b** $T = 10 + 35p$

c $B = 560 + 30m + 10n$

3 $\frac{ab}{b} = b$

4 $P = \sqrt[3]{\frac{k}{m}}$

5 $\frac{8(D + k)}{ab} = c$, $8(D + k) = abc$, $D + k = \frac{1}{8}abc$, $D = \frac{1}{8}abc - k$

6 False, third formula is different

7 a i $11.6\,\text{cm}^2$ **ii** $29.4\,\text{cm}^2$ **iii** $9.9\,\text{cm}^2$ **iv** $63.7\,\text{cm}^2$

b $c = s - \frac{A^2}{s(s - a)(s - b)}$

8 a $126\pi^2\,\text{cm}^3$ **b** $r = 7\,\text{mm}$, $R = 15\,\text{mm}$

6.2S

1 a $3x$ **b** $x - 2$ **c** $2x + 1$ **d** x^2

e $\frac{1}{x}$ **f** x^3

2 a 5 **b** 11 **c** -5 **d** 1

3 a 2 **b** 7 **c** -1 **d** 0

4 (Top to bottom) $-12, -2, 3, 4.5$

5 a 0.25 **b** 0.5 **c** 0.4 **d** Undefined

6 a $x - 4$ **b** $x + 3$ **c** $2x$ **d** $\frac{x}{5}$

e $\frac{x + 3}{2}$ **f** $\frac{x - 2}{7}$ **g** $3(x + 4)$ **h** $2x - 5$

7 a $\frac{1}{x}$ **b** $2 - x$ The functions are self-inverse

8 a 17 **b** 80 **c** 27 **d** 36

e 43 **f** 44 **g** $8x + 3$ **h** $8x + 12$

9 a $(3x + 1)^2$ **b** $3x^2 + 1$ **c** $x = -1, 0$

10 a 1 **b** 10 **c** 1 **d** 3

e $2x^2 + 3$ **f** $4x^2 + 4x + 2$ **g** $x = 1$

11 $ff^{-1}(x) = 4\left(\frac{x - 5}{4}\right) + 5 = x$

12 The inverse function is found by swapping each y-coordinate with its corresponding x-coordinate.

13 $f^{-1}(x) = \sqrt{x}$ which is not a real value if $x < 0$.

6.2A

1 (mapping diagram values are given from top to bottom)

a i $3, 6, 0, 9, 4$ **ii** $y = 3x$ **iii** $f(x) = 3x$

iv $(1, 3), (2, 6), (0, 0), (3, 9), (4, 12)$ plotted.

b i $2, 4, -3, 5, 3$ **ii** $y = x + 2$ **iii** $f(x) = x + 2$

iv $(0, 2), (2, 4), (-3, -1), (5, 7), (1, 3)$ plotted.

c i $3, 2, 1, -2, 4$ **ii** $y = \frac{x}{2} + 1$ **iii** $f(x) = \frac{x}{2} + 1$

iv $(4, 3), (2, 2), (2, 1.5), (-2, 0), (6, 4)$ plotted.

d i $1, 7, 1.5, 5, 1$ **ii** $y = 2x + 1$ **iii** $f(x) = 2x + 1$

iv $(0, 1), (3, 7), (1.5, 4), (2, 5), (1, 3)$ plotted.

2 a i $f^{-1}(x) = \frac{x}{3}$

ii, iii

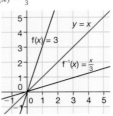

The two functions are reflections in $y = x$. The inverse is found by swapping the x- and y- coordinates.

b i $f^{-1}(x) = x - 2$

ii, iii

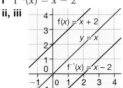

The two functions are reflections in $y = x$. The inverse is found by swapping the x- and y- coordinates.

c i $f^{-1}(x) = 2(x - 1)$

ii, iii

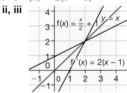

The two functions are reflections in $y = x$. The inverse is found by swapping the x- and y- coordinates.

d i $f^{-1}(x) = \frac{x - 1}{2}$

ii, iii

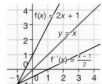

The two functions are reflections in $y = x$. The inverse is found by swapping the x- and y- coordinates.

3 a

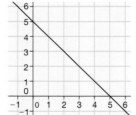

b

c

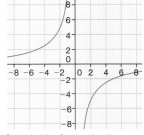

d

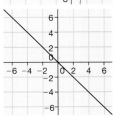

Each function is identical to its inverse.

4 a Yes   **b** No, it does not map one input to one output

5 a

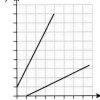

   b $f(x) = (f^{-1})^{-1}(x)$

6

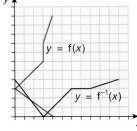

$y = f(x)$ is not a function as it maps one input to more than one output.

7 a $f(x) = c - x$ or $f(x) = x$

   b $f(x) = x$, or $f(x) = c$ where c is a number

8 a $f_n f_m(x) = (x^m)^n = x^{mn}$

   b i $f_m f_n(x) = (x^n)^m = x^{nm} = x^{mn}$ **ii** $f^{-1}(x) = \sqrt[n]{x} = x^{\frac{1}{n}} = f_{\frac{1}{n}}(x)$

9 a $g(x) = 3x + 2$ **b** $g(x) = \frac{x}{2} - 2$

   c $g(x) = 3x - 4$ **d** $g(x) = 3x - 5$

6.3S

1 a $7(x + 4) \equiv 7x + 28$ **b** $3(x - 2) \equiv 3x - 6$

   c $2(3 + x) \equiv 6 + 2x$ **d** $5(2 - x) \equiv 10 - 5x$

2 a $4(2m + 1)$ **b** $3(4n - 3)$ **c** $5(3p + 11)$ **d** $q(q + 2)$

   e $4(4r - 7)$ **f** $2q(2p - 5)$

3 a Identity **b** Function **c** Formula **d** Equation

   e Expression **f** Formula **g** Formula **h** Function

4 a Formula **b** Identity **c** Equation **d** Identity

   e Equation **f** Formula **g** Formula **h** Identity

   i Equation

5 Students' examples.

6 a False, $4a + 8$ **b** True

   c True   **d** False, $y^2 + 3y$

   e True

7 a $5 \times a + 5 \times 2$ **b** $3 \times x + 3 \times 4$

   c $5 \times y - 5 \times 3$ **d** $y \times y + y \times 3$

   e $x \times x - x \times 2 + 2 \times x - 2 \times 2$

8 a $4 \times a + 4 \times 2 + 2 \times a + 2 \times 1$

   b $3 \times x + 3 \times 2 + 4 \times x - 4 \times 1$

   c $5 \times y - 5 \times 2 + 3 \times y - 3 \times 3$

   d $y \times y + y \times 3 + 2 \times y + 2 \times 3$

   e $x \times x - x \times 4 + x \times x + x \times 2$

9 a $a = 7, b = 9$ **b** $a = 7, b = 6$

   c Many answers possible, e.g. $a = -2, b = 3$

   d $a = 1, b = 2$ **e** $a = 3, b = 10$

   f $a = 6, b = -22$

10 a Formula **b** Identity **c** Equation **d** Formula

   e Formula **f** Identity **g** Identity **h** Formula

   i Formula **j** Equation **k** Identity

6.3A

1 a i $5(a + 2) \equiv 5a + 10$ **ii** $5(a + 2) = 80$

   b i $7(b + 5) \equiv 7b + 35$ **ii** $7(b + 5) = 105$

   c i $2(c + 1) \equiv 2c + 2$ **ii** $2(c + 1) = 24$

   d i $12(b + 10) \equiv 12b + 120$ **ii** $12(b + 10) = 240$

2 a i $x < 10$ **ii** $10 \times 12 + 7 \times (10 - x) \equiv 10 \times 19 - 7x$

   iii e.g. 183   **iv** $190 - 7x = 183$

   b i $x < 11$ **ii** $8 \times 11 + 4 \times (11 - x) \equiv 11 \times 12 - 4x$

   iii e.g. 120   **iv** $132 - 4x = 120$

3 a $2 \times 7 + 2 \times g \equiv 2(7 + g)$ **b** $2m + 2n \equiv 2(m + n)$

   c $ab + ac \equiv a(b + c)$ **d** $3f + bf + df \equiv f(3 + b + d)$

   e $4a^2 \equiv (2a)^2$ **f** $(a + b)^2 \equiv a^2 + 2ab + b^2$

4 a Sometimes true   **b** True   **c** False

   d Sometimes true   **e** False   **f** False

   g Sometimes true   **h** True

5 a Students' answers

   b $5n + 10 = 5(n + 2)$ has a factor of 5 for all n

6 a $7 + 7 = 14$ **b** $3 + 5 = 8$

   c $0.5^2 < 0.5$ **d** $3 \times 4 = 12$

7 a $(2n)^2 = 4n^2$ has a factor of 4 for all n

   b $n(n + 1) - n = n^2 + n - n = n^2$

   c $2n \times (2m + 1) = 2(2mn + n)$ has a factor of 2 for all n, m

   d $2n + 1 + 2m + 1 = 2(n + m + 1)$ has a factor of 2 for all n, m

6.4S

1 a $x^2 + 5x + 6$ **b** $p^2 + 11p + 30$

   c $w^2 + 5w + 4$ **d** $c^2 + 10c + 25$

   e $x^2 + 2x - 8$ **f** $y^2 + 5y - 14$

   g $t^2 + 4t - 12$ **h** $x^2 - 7x + 10$

   i $y^2 - 14y + 40$ **j** $w^2 - 3w + 2$

   k $p^2 - 10p + 25$ **l** $q^2 - 24q + 144$

2 a $6x^2 + 17x + 7$ **b** $10p^2 + 19p + 6$

   c $6y^2 + 11y + 4$ **d** $4y^2 + 24y + 36$

   e $10t^2 + 12t - 16$ **f** $15w^2 + 42w - 9$

   g $6x^2 - 6y^2$ **h** $9m^2 - 24m + 16$

   i $6p^2 - pq - 40q^2$ **j** $4m^2 - 12nm + 9n^2$

3 a $(x + 2)(x + 5)$ **b** $(x + 3)(x + 5)$

   c $(x + 2)(x + 6)$ **d** $(x + 5)(x + 7)$

   e $(x - 5)(x + 2)$ **f** $(x - 7)(x + 5)$

   g $(x - 3)(x - 5)$ **h** $(x - 5)(x + 4)$

   i $(x - 20)(x + 12)$ **j** $(x - 9)(x + 12)$

   k $(x - 5)(x + 5)$ **l** $(x - 3)(x + 2)$

4 a $x^2 - 1$ **b** $25x^2 - 1$ **c** $4x^2 - 1$ **d** $x^2 - y^2$

5 a $(x - 10)(x + 10)$ **b** $(y - 4)(y + 4)$

   c $(m - 12)(m + 12)$ **d** $(p - 8)(p + 8)$

   e $(x - \frac{1}{2})(x + \frac{1}{2})$ **f** $(k - \frac{5}{6})(k + \frac{5}{6})$

   g $(w - 50)(w + 50)$ **h** $(7 - b)(7 + b)$

   i $(2x - 5)(2x + 5)$ **j** $(3y - 11)(3y + 11)$

   k $(4m - \frac{1}{2})(4m + \frac{1}{2})$ **l** $(20p - 13)(20p + 13)$

   m $(x - y)(x + y)$ **n** $(2a + 5b)(2a - 5b)$

   o $(3w - 10v)(3w + 10v)$ **p** $(5c - \frac{1}{2}d)(5c + \frac{1}{2}d)$

   q $x(x - 4)(x + 4)$ **r** $2y(5 - y)(5 + y)$

   s $\left(\frac{4x}{7} - \frac{8y}{9}\right)\left(\frac{4x}{7} + \frac{8y}{9}\right)$

6 a $3(2x - 3y)(x - y)$ b $(4a - 3b)(4a + 3b)$
c $(x - 7)(x - 4)$ d $(2x - 3)(x + 7)$
e $x(x - 3)(x + 6)$ f $5ab(1 + 2ab)$
g $(2 - x)(5 + x)$ h $10(1 - x)(1 + x)$
i $(x - 7)(x + 9)$ j $2x(x^2 - 66)$
k $(3x - 2)(2x - 3)$ l $(x^2 - y^2)(x^2 + y^2)$

7 a $(2x + 3)(x + 1)$ b $(3x + 2)(x + 2)$
c $(2x + 5)(x + 1)$ d $(2x + 3)(x + 4)$
e $(3x + 1)(x + 2)$ f $(2x + 1)(x + 3)$
g $(2x + 7)(x - 3)$ h $(3x + 1)(x - 2)$
i $(4x - 3)(x - 5)$ j $(6x - 1)(x - 3)$
k $(3x + 2)(4x + 5)$ l $(2x - 3)(4x + 1)$
m $3(2x - 5)(x - 2)$ n $(2x - 3)(2x + 3)$

8 a $(3 + 2y)(1 - 3y)$ b $(1 + 4p)(3 - 2p)$
c $(y - 2)(3y - 5)$ d $3(m - 2)(5 - 2m)$
e $y(2 - x)(3x + 1)$

9 a $x + 2$ b $x - 7$ c $\frac{1}{x + 7}$ d $\frac{x - 2}{3x + 4}$
e $2y - 5$ f $\frac{1}{x + 9}$ g $2x - 5$ h $\frac{3x + 4}{x - 2}$

10 a $3(x + 7)$ b $\frac{2(x - 2)(x - 9)}{x^2 + 17x + 18}$ c $\frac{2(x - 5)}{x^2(x - 6)}$

6.4A

1 a $(x + 6)(x - 3)$ b $(2m - 3)^2$
2 a $x^2 + 9x - 22$ b $(x + 11)(x - 2)$
c Two side lengths would be negative.
3 a $(a + b)^2 = a^2 + 2ab + b^2$ b 16
c Students' answers
4 $(2.3 + 1.7)^2 = 4^2 = 16$
5 a $3, 18$ b $2, 2$ c $4, 2, 14$ d $2, 2, 5$
6 $2(x^2 + 17)$
7 $12x^2 + 32x - 12, 6x - 2, 2x + 6, 3x - 1, 2x + 3$
8 $\frac{1}{4}(6x^2 + 10x + 23 + 2x^2 + 20x - 11 + 4x^2 + 13x + 21 + 9x - 4x^2 + 27) = 2x^2 + 13x + 15 = (2x + 3)(x + 5)$
9 $x = 20\sqrt{3}$
10 $97^2 - 57^2 = (97 + 57)(97 - 57) = 154 \times 40 = 6160$
11 $y = -3$

Review 6

1 a -11 b 20 c 9 d 5
2 a $X = \frac{A - 3}{2}$ b $X = \frac{3C + B}{A}$
c $X = \frac{20Z - Y}{3}$ d $X = 4Y^2 - 4$
e $X = \pm\sqrt{L^2 + 2K}$
3 a i 35 ii -2 b i $\frac{x}{5}$ ii $x - 3$
c i $5x + 3$ ii $5(x + 3)$
4 a $v^2 = u^2 + 2as$ b $3(x + 2) \equiv 3x + 6$
c $5x^2 + 3$ d $7a + 5 = 19$
5 a $x > -2$ b $y \leqslant 0$
6 $5(2x + 3) + 2(4 - 5x) \equiv 10x + 15 + 8 - 10x \equiv 23$
7 Let n be an integer.
$n + (n + 1) + (n + 2) \equiv 3n + 3 \equiv 3(n + 1)$
8 a $x^2 + 11x + 18$ b $x^3 - 13x^2 + 42x$
c $3x^2 + 31x - 22$ d $12x^2 + 2x - 2$
9 a $(x + 9)(x - 9)$ b $(4x + 7)(4x - 7)$
c $x(x - 7)$ d $7x(3x + 4)$
10 a $(x - 4)(x + 1)$ b $(x - 2)(x - 5)$
c $(x - 1)(x + 9)$ d $(2x + 1)(x + 3)$
e $(2x + 3)(3x - 1)$ f $2(2x - 3)(2x - 1)$
11 a $x + 2$ b $\frac{x - 4}{x + 4}$ c $\frac{x - 1}{x + 1}$

Assessment 6

1 a Stina, $s = \frac{1}{2}(-4 + 12)8 = 32$.
b Stina, $v^2 = 0^2 + 2 \times 2 \times 16$, $v = \sqrt{64} = 8$.
2 a $\frac{a}{n^2} = -x + \frac{h}{n^3}$, $x = \frac{h - a}{n^2}$, $n^2 = \frac{h - a}{x}$, $n = \sqrt{\frac{h - a}{x}}$
b $\frac{t}{x} = \sqrt{\frac{1 - e}{1 + e}}$, $\frac{t^2}{x^2} = \frac{1 - e}{1 + e}$, $t^2(1 + e) = x^2(1 - e)$,
$t^2 + et^2 = x^2 - ex^2$, $e(x^2 + t^2) = x^2 - t^2$, $e = \frac{x^2 - t^2}{x^2 + t^2}$

c $\frac{1}{S} = \frac{r}{Rr} + \frac{R}{Rr} = \frac{(r + R)}{(Rr)}$
3 a $C = 15p + 12q + 14t$ b £5.08 c £13.55
4 a 58.8 m (1 dp) b 44.7 m c 101 km/h
5 a 10.4 km b 5.73 m c Yes
6 9
7 a i No, $p^2 - 11p + 28$. ii No, $26v^2 - 38v - 47$.
b i Yes ii No, $(v + 10)(v - 10)$.
iii No, $(6x - y)(5x + 3y)$.
8 a i $(2.4 - 3.6)(2.4 - 3.6)$ ii 1.44
b i $= (89 - 11)(89 + 11) = 78 \times 100 = 7800$
ii $= (6.89 - 3.11)(6.89 + 3.11) = 3.78 \times 10 = 378$
9 $(x^{-4})^2 + 14$
10 a F $5 \times 4 = 20$ b F 5 has 2 factors, 1 and 5.
c T Let the two numbers be $2x$ and $2y$. $2x + 2y = 2(x + y)$ so the sum is even.
d F $0^2 = 0$ e F $2 \times 7 = 14$
f T Let the numbers be $2x$, $2x + 2$ and $2x + 4$.
$2x + 2x + 2 + 2x + 4 = 6x + 6 = 6(x + 1)$, is divisible by 6.
g F x^2 has 3 factors, 1, x and x^2.
11 a $y = 4x + 1$ b $y = \frac{x}{3} - 2$
12 a $g^{-1}(x) = \frac{x - 1}{2}$
b i 17 ii 107

Revision 1

1 Jenni would save £17.50 − £15 = £2.50
2 a £4928.40 b 30 096 miles
3 a i $18x + 9$ ii $9x - 18$
iii $2x^2 + 51x - 20$ iv $18x^2 - 27x - 18$
b 76.6 cm³ (3 sf)
4 a i QRS $= 28°$ (**Isosceles** $\triangle$) so QSR $= 180 - 90 - 28 = 62°$ (Angles in a $\triangle$); PSQ $= 62°$ (Angles in a $\triangle$) so RST $= 180 - 2 \times 62 = 56°$ (Angles on a straight line); RSP $= 2 \times 62 = 124°$
ii SP = SR, $\angle$PSQ = $\angle$RSQ and QS is common to both triangles so the triangles are congruent by SAS, PQ and QR are corresponding sides. (Other reasons possible)
b i Isosceles ii $\angle$ABD $= 72°$, $\angle$DBC $= 36°$
iii No. It has different angles.
5 a i Correct. HBD is an isosceles right-angled triangle.
ii Incorrect. HBDF is a square.
iii Incorrect. HBGC is an isosceles trapezium.
b i One of the following: OPF, OBQ, ODQ
ii One of the following: HOF, HOB, BOD, FOD, BHF, BDF, HBD, HFD, ACO, CEO, EGO, GAO.
c i $\angle$ABC $= 135°$ ii $\angle$POF $= 45°$
iii $\angle$HOC $= 135°$ iv $\angle$ABH $= 22.5°$

6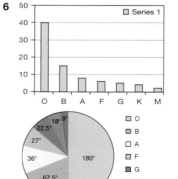

7 a i 6.5 ii 7 nights.
b i 11 ii 5 nights
8 a 22 450 b 31.2%
c Maximum = 35 271, Minimum = 35 216

Answers

9 a 55.7 **b** 69
10 a $\frac{1}{5}$
  b i 16 ml **ii** 4 ml
  c i Acid: 64 ml, Water: 16 ml
    ii Acid: 16 ml, Water: 84 ml
11 She is wrong. For example, if $x = -1$, $10x = -10 < -1 = x$
12 a 248.5 ft **b** 88.2 ft (3 sf)
  c 68.9 mph (3 sf)
  d Yes. He will pass the train in 2326 ft < 5280 ft.

Chapter 7

Check in 7

1 a i 16 cm **ii** 160 mm **iii** 12 cm²
  b i 18 cm **ii** 180 mm **iii** 20.25 cm²
  c i 19.1 cm **ii** 191 mm **iii** 14 cm²
2 A(3, 2), B(5, 5), C(4, −1), D(−2, −3), E(1, 6)
3 a $x = 3$ **b** $x = 5$ **c** $y = 2$
  d $y = -1$ **e** $y = x$

7.1S

1 a 1800 mm **b** 4.5 cm **c** 3.5 m **d** 2 km
  e 3.5 km **f** 4.5 m
2 a $AB = 5$ cm, $CD = 7$ cm, $EF = 3.5$ cm, $GH = 9.5$ cm,
    $IJ = 2.5$ cm, $KL = 7.3$ cm, $MN = 2.7$ cm, $OP = 2.8$ cm,
    $QR = 4.9$ cm, $ST = 4.4$ cm
  b $AB = 50$ mm, $CD = 70$ mm, $EF = 35$ mm, $GH = 95$ mm,
    $IJ = 25$ mm, $KL = 73$ mm, $MN = 27$ mm, $OP = 28$ mm,
    $QR = 49$ mm, $ST = 44$ mm
3 a Line AB exactly 9 cm. **b** Midpoint at 4.5 cm
4 a Acute **b** Students' estimates **c** 30°
5 a Acute **b** Students' estimates **c** 45°
6 a Obtuse **b** Students' estimates **c** 120°
7 a Reflex **b** Students' estimates **c** 330°
8 a acute **b** obtuse **c** right angle
  d acute **e** obtuse **f** acute
  g obtuse **h** acute **i** obtuse
  j acute **k** reflex **l** reflex
  m reflex **n** reflex **o** reflex
9 a 50° **b** 319°
10 Check students' bearings.

7.1A

1 a 056° **b** 170° **c** 238°
  d 275° **e** 349°
2 Kim is wrong.
3 a St. Mawes positioned at the intersection of the bearings.
  b 2.2 km **c** 13.8 km
4 a Triangle HSY with H most northern point where SH = 5 cm,
    HY = 3.8 cm, SY = 4 cm (horizontal, S left of Y).
  b 142°
5 a i and **b i**

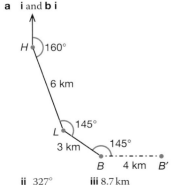

  ii 327° **iii** 8.7 km
  b ii 354° **iii** 7.4 km

7.2S

1 a 15 cm² **b** 27 m² **c** 200 cm² **d** 81 cm²
2 a 16 cm² **b** 100 m² **c** 9 cm² **d** 64 m²
3 a 40 cm² **b** 16 m² **c** 18 cm² **d** 36 mm²
4 a 6 square units **b** 12 square units
5 a 6 square units **b** 6 square units
6 a 80 cm² **b** 800 m² **c** 120 mm² **d** 384 cm²
7 a 50 cm² **b** 375 mm² **c** 28 m² **d** 160 cm²
8 a 150 cm² **b** 150 cm²
9 a 6 m **b** 14 cm **c** 8 mm **d** 8 cm

7.2A

1 a perimeter = 42 cm **b** perimeter = 40 cm
    area = 74 cm²   area = 72 cm²
  c perimeter = 33 cm **d** perimeter = 36 cm
    area = 32 cm²   area = 44 cm²
2 area of shape = 121 cm²
  area remaining = 479 cm²
3 205.5 cm²
4 a

  b 19 cm² **c** 36 cm² − 17 cm² = 19 cm²
5 54 cm
6 a 20 cm
  b Many possible solutions. Check the perimeter of students'
    shapes is 28 cm.
7 a Many answers possible check base × height = 36
  b Many answers possible check base × height = 48
  c Many answers possible check height × sum of parallel
    sides = 40
8 No. Smallest diameter possible is a square with sides 12 m and
  the perimeter of this would be 48 m, which would cost £480.
9 11.5 cm²

7.3S

1 a $\begin{bmatrix} 3 \\ 1 \end{bmatrix}$ **b** $\begin{bmatrix} 5 \\ 3 \end{bmatrix}$ **c** $\begin{bmatrix} -3 \\ 1 \end{bmatrix}$ **d** $\begin{bmatrix} 5 \\ -3 \end{bmatrix}$
  e $\begin{bmatrix} 2 \\ -4 \end{bmatrix}$ **f** $\begin{bmatrix} -4 \\ 2 \end{bmatrix}$
2 a Vertices at (−4, 3), (−3, 4), (−2, 3) and (−3, 1)
  b Vertices at (1, 1), (2, −1), (1, −2) and (0, −1)
3 a Vertices at (1, −3), (1, −5), (3, −7) and (4, −6)
  b Vertices at (3, 1), (5, 1), (6, 4) and (7, 3)
4 a Vertices at (1, 1), (2, −1), (0, −1) and (−2, 1)
  b Vertices at (1, 2), (1, −1), (−1, −2) and (−1, 0)
5 a Vertices at (2, −1), (4, 0) and (5, −3)
  b Vertices at (−3, 5), (−2, 2) and (0, 3)
6 a Vertices at (6, 6), (6, 9) and (8, 6)
  b Vertices at (8, 0), (8, 3) and (10, 0)
7 a Vertices at (−4, 6), (−3, 7), (−2, 6) and (−3, 4)
  b Vertices at (2, −1), (3, 0), (4, −1) and (3, −3)

7.3A

1 a $x = 2$ **b** $y = 3$
2 a Rotate 90° clockwise around (0, 0)
  b Rotate 180° around (0, 0)
  c Rotate 90° anti-clockwise around (0, 0)
3 Rotate 135° clockwise
4 a Vertices at (2, −1), (4, −1) and (4, −4)
  b Vertices at (−2, 1), (−4, 1) and (−4, 4)
  c reflection in $x = 0$

5 a Vertices at $(1, -2)$, $(1, 0)$, $(2, 1)$ and $(4, 1)$

b Vertices at $(0, 0)$, $(0, 2)$, $(1, 3)$ and $(3, 3)$

c Translate $\begin{bmatrix} -4 \\ 1 \end{bmatrix}$

6 No, A is mapped onto B by a rotation $90°$ clockwise about $(0,4)$.

7 Rotation $180°$ around $(3, 3)$, reflection in $(x = 3)$, translation $\begin{bmatrix} 4 \\ 0 \end{bmatrix}$

7.4S

1 a **b** **c**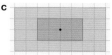

2 a Ensure axes extended to 16 and drawn to scale.

b Vertices at $(4, 4)$, $(4, 10)$ and $(8, 4)$

c Vertices at $(6, 6)$, $(6, 15)$ and $(12, 6)$

3 a Ensure axes extended and drawn to scale.

b Vertices at $(1, -1)$, $(3, 3)$, $(1, 5)$ and $(-1, 3)$

c Vertices at $(1, -5)$, $(5, 3)$, $(1, 7)$ and $(-3, 3)$

4 b Vertices of U: $(1, 2)$, $(5, 2)$ and $(2, 3)$

c Vertices of V: $(2, 3)$, $(6, 3)$ and $(3, 4)$

5 a Ensure axes extended and drawn to scale. $(1,3)$, $(1, 6)$, $(3, 9)$, $(4, 6)$ plotted and joined to form a quadrilateral.

b Vertices of R: $(3, 3)$, $(4, 4)$, $(3, 4)$ and $(3\frac{2}{3}, 5)$

c Vertices of S: $(1, 2)$, $(0.5, 3.5)$, $(-0.5, 2)$ and $(-0.5, 0.5)$

6 b Vertices of E: $(3, -3)$, $(-6, -6)$, $(-12, 0)$

c Vertices of F: $(0.5, -0.5)$, $(-1, -1)$, $(-2, 0)$

d 3

7 b Vertices of K: $(5.5, -3.5)$, $(5.5, -8)$, $(-3.5, -3.5)$

c Vertices of L: $(2, 0)$, $(2, -1)$, $(0, 0)$

d 3

7.4A

1 a

scale factor 2

b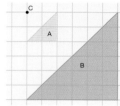

scale factor 3

c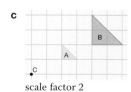

scale factor 2

d

scale factor 2

2 a scale factor 3, centre of enlargement $(0, 0)$

b scale factor $\frac{1}{3}$, centre of enlargement $(0, 0)$

3 a scale factor 2, centre of enlargement $(-3, 0)$

b scale factor $\frac{1}{2}$ centre of enlargement $(-3, 0)$

4 a i scale factor $\frac{1}{2}$, centre of enlargement $(0, 0)$

ii scale factor 2, centre of enlargement $(0, 0)$

iii scale factor 4, centre of enlargement $(0, 0)$

b perimeter of C = 4 × perimeter of A

5 b Vertices of Y: $(2, -2)$, $(0, -6)$, $(-6, 4)$

c Vertices of Z: $(-3, 3)$, $(0, 9)$, $(9, 6)$

d scale factor 3, centre of enlargement $(0, 0)$

e scale factor $\frac{1}{3}$, centre of enlargement $(0, 0)$

f Same centre, scale factors which are reciprocals of one another.

6 a The value of each coordinate is tripled.

b Enlargement of scale factor 3 with centre $(0, 0)$.

Answers

Review 7

1 a $063°$ **b** $243°$ **c** $17.5\,\text{km}$

2 a $60\,\text{m}^2$ **b** $30\,\text{cm}^2$ **c** $2.25\,\text{mm}^2$ **d** $49.5\,\text{cm}^2$

e $32.5\,\text{m}^2$

3 a, b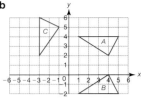

c Reflection in the line $y = -x$

4 a 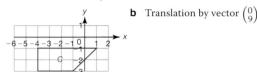   **b** Translation by vector $\begin{pmatrix} 0 \\ 9 \end{pmatrix}$

5 a, b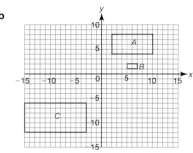

c Enlargement of scale factor 0.5 centre of enlargement $(-10, 6)$.

6 a The centre of rotation.

b The points on the mirror line.

c None.

d The centre of enlargement, assuming scale factor $\neq 1$.

Assessment 7

1 a Marta   **b** Karl   **c** Marta

2 a Check student's drawings.

b i $6.5\,\text{cm}\ (\pm 3\,\text{mm})$ **ii** $60°\ (\pm 3°)$

iii $135°\ (\pm 3°)$

c Kite   **d** Rectangle

3 a Yes   **b** No, right angle.

c Yes   **d** Yes

e Yes   **f** No, acute.

g No, right angle.   **h** No, acute.

i Yes   **j** No, acute.

k No, reflex.

4 a Check students' drawings.   **b** $15.9\,\text{cm}^2\ (3\,\text{sf})$

5 a i $195°$ **ii** $110°$ **iii** $015°$ **iv** $290°$

v $245°$

b The bearings differ by $180°$.

6 a 49

b i No, $2 \times 2 = 4$ and 4 is not a factor of 49.

ii No, $3 \times 3 = 9$ and 9 is not a factor of 49.

c No. Each side has length 7 m so must include two 2×2 slabs and a 3×3 slab. There are exactly three ways of arranging slabs around the edges, all of which include gaps or overlap.

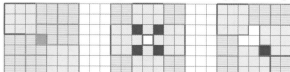

(Other lines of reasoning possible.)

7 a $74.25\pi\,\text{mm}^2$ **b** $233\,\text{mm}^2$

8 a i $264.5\pi\,\text{mm}^2$ **ii** $831\,\text{mm}^2\,(3\,\text{sf})$

  b $16.3\,\text{mm}$

9 Check students' drawings.

10 a, c

  b $\begin{pmatrix}25\\45\end{pmatrix}$ **d** $\begin{pmatrix}15\\20\end{pmatrix}$

  e $\begin{pmatrix}40\\65\end{pmatrix}$ **f** $\begin{pmatrix}25\\45\end{pmatrix}+\begin{pmatrix}15\\20\end{pmatrix}=\begin{pmatrix}40\\65\end{pmatrix}$

11 a Rotation $180°$ about the point $(2, 2)$.

  b, c

  d Reflection in the line $y = 0$.

Chapter 8

Check in 8

1 a $\frac{3}{5}$ **b** $\frac{3}{7}$ **c** $\frac{5}{8}$ **d** $\frac{7}{11}$

  e $\frac{5}{6}$ **f** $\frac{9}{20}$ **g** $\frac{23}{24}$ **h** $\frac{5}{8}$

2 a 40   **b** 40   **c** 21   **d** 56

3 a 0.7   **b** 0.75   **c** 0.375   **d** 0.4

  e $0.\dot{3}$ **f** 0.0625   **g** $0.\dot{1}$ **e** 0.625

8.1S

1 a i Even chance, relative frequency suggests a probability of 0.5.

    ii Unlikely, he is expected to win 2 or 3.

  b i Likely, this reflects the average for the time of year.

    ii Impossible, 1913 was not a leap year so there was no February 29th.

2 a i $\frac{1}{5}$ **ii** 0.2   **iii** 20%

  b i $\frac{7}{40}$ **ii** 0.175   **iii** 17.5%

  c i $\frac{13}{40}$ **ii** 0.325   **iii** 32.5%

  d i $\frac{3}{10}$ **ii** 0.3   **iii** 30%

3 a 15 times.   **b** 5 times.

4 a Bag A. In bag A the probability of getting a red is $\frac{1}{3}$ compared with $\frac{1}{5}$ in bag B.

  b Bag B. Bag A only contains 1 red ball so it is impossible to take two reds without replacement from this bag.

5 a 50

  b i $\frac{9}{50}$ **ii** $\frac{7}{25}$ **iii** $\frac{27}{50}$

  c 2 red, 3 green, 5 blue.

  d Increase the number of trials.

6 9 days.

7 14 shares.

8 a Students' frequency tables.

  b There are infinite possible outcomes with decreasing probabilities.

8.1A

1 No. The results should start to reflect these proportions after a higher number of trials.

2 a Xavier's statement is consistent with his experiment but his sample size is too small to be reliable.

  b No. This is an estimate, not a fact.

  c i This estimate combines all the results (115 reds) and divides by the total number of trials (125). A higher number of trials gives a more reliable probability.

    ii Not necessarily, this is an estimate only.

3 a

Colour	Red	White	Blue
Frequency	10	5	10
Relative Frequency	0.4	0.2	0.4

  The relative frequencies sum to 1.

  b A or C. B could not be Alik's as his table shows that white was an observed result and this spinner does not include white.

  c Multiple answers possible provided they include the colours red, white and blue.

4 As Keith's friends were born in different years their birthday this year is not a good indicator of variation. Keith may need a larger and more diverse sample before a difference in numbers becomes apparent. The difference between weekend and weekday births may be too small to be observable in this sample.

5 a 10

  b More often; white balls make up a higher proportion of the total in the blue bag than they do in the red bag. From the blue bag, one would estimate 15 white balls.

6 a 6 times   **b** 25p per go

8.2S

1 a $\frac{1}{6}$ **b** $\frac{2}{6}=\frac{1}{3}$ **c** $\frac{3}{6}=\frac{1}{2}$ **d** $\frac{2}{6}=\frac{1}{3}$

2 a Yes, each outcome has probability $\frac{1}{6}$.

  b No. It is more likely that the pin will land point down due to the instability of the object.

  c Yes, each outcome has probability $\frac{1}{7}$.

  d No. Getting 999 tails in a row is less likely than getting 4 tails in a row.

  e No, some letters are more common than others. For example, 'e' and 't' are more common than 'x' and 'q'.

3 6 black counters.

4 a Highest relative frequency $\frac{9}{20}=0.45$.

    Lowest relative frequency $\frac{3}{20}=0.15$

  b Total 6's = 161, total throws = $24 \times 20 = 480$; relative frequency = $161 \div 480 = 0.34$

5 a $\frac{1}{6}$ **b** $\frac{4}{9}$ **c** 1

6 a 0.5, 0.45, 0.47, 0.43, 0.38, 0.38, 0.41, 0.4, 0.39, 0.38, 0.39, 0.39, 0.38, 0.38

  b 0.38

7 Dominika's method is best as she has used the biggest sample available.

8.2A

1 a $\frac{14}{30}=\frac{7}{15}$ **b** $\frac{9}{30}=\frac{3}{10}$

2 1, 2, 3, 4 and 6 are factors of 12, 5 is not a factor of 12. The theoretical probability of getting a 5 is $\frac{1}{6}=0.17$ and the experimental probability of this is $(100-71)\div100=0.29$. The experimental probability is a lot higher than the theoretical probability so there is reason to believe the dice is biased towards 5.

3 $\frac{1}{8}$

4 0.785 (3 sf)

5 a $\frac{3}{5}$ **b** $\frac{13}{32}$

6 $\frac{1}{6}$

7 The experimental probability suggests the following:

  $(17 \div 80) \times 19 = 4.0$ 4 white sectors

  $(32 \div 80) \times 19 = 7.6$ 7.6 black sectors

  $(31 \div 80) \times 19 = 7.4$ 7.4 red sectors

Given that there are equal numbers of black sectors and red sectors, these results suggest two possibilities. There could be 7 black sectors, 7 red sectors and $19 - 2 \times 7 = 5$ white sectors, in which case the experiment results suggest the spinner is biased **against** white. There could be 8 black sectors, 8 red sectors and $19 - 2 \times 8 = 3$ white sectors, in which case the experiment results suggest the spinner is biased **in favour** of white.

8.3S

1 a Not mutually exclusive.   **b** Mutually exclusive.
   c Not mutually exclusive.   **d** Not mutually exclusive.

2 a Not mutually exclusive   **b** Mutually exclusive
   c Mutually exclusive   **d** Mutually exclusive

3 a 0   **b** $1 - \frac{1}{3} = \frac{2}{3}$

4 a $\frac{1}{6}$ **b** $\frac{1}{2}$

5 a $\frac{1}{3}$ **b** $\frac{1}{2}$

6 a $\frac{1}{7}$ **b** $\frac{3}{7}$ **c** $\frac{11}{21}$

7 a 0.2   **b** 0.6   **c** 0.5

8 $P(\text{green}) = \frac{1}{7}$ $P(\text{blue}) = \frac{2}{7}$ $P(\text{white}) = \frac{4}{7}$
Green, white and blue are mutually exclusive options so
$P(\text{green or white or blue}) = \frac{1}{7} + \frac{2}{7} + \frac{4}{7} = 1$
Green, white and blue are exhaustive, there can be no other
colours in the bag.

9 a $P(\text{green}) = 0.1$ **b** $P(\text{white}) = 0.6$

8.3A

1 a

	1	2	3	4	5	6
1	0	1	2	3	4	5
2	1	0	1	2	3	4
3	2	1	0	1	2	3
4	3	2	1	0	1	2
5	4	3	2	1	0	1
6	5	4	3	2	1	0

   b i $\frac{10}{36} = \frac{5}{18}$ **ii** $\frac{6}{36} = \frac{1}{6}$ **iii** $1 - \frac{6}{36} - \frac{10}{36} = \frac{20}{36} = \frac{5}{9}$

2 No, taking a green, taking a blue, and taking a black ball are
mutually exclusive events $P(\text{green or blue or black}) = \frac{1}{4} + \frac{1}{3} + \frac{1}{2}$
$= \frac{3}{12} + \frac{4}{12} + \frac{6}{12} = \frac{13}{12} > 1$
A probability cannot be greater than 1 so Serena must be wrong.

3 a $\frac{1}{6} + \frac{1}{3} + \frac{1}{8} + \frac{3}{8} = 1$
   Arinda **may** be correct as her probabilities do not exceed 1.
   b If Arinda is correct there are no other colours. Her
   probabilities indicate $P(\text{green or blue or white or black}) = 1$
   so the probability of another colour is 0.

4 a $\frac{1}{2}$ **b** $\frac{3}{4}$ **c** $\frac{5}{12}$

5 a C and D   **b** C and D
   c A and C are not mutually exclusive because the number 2 is
   both even and prime.

6 a True, a student cannot be both male and female.
   b True, a student must be either male or female.
   c False, there are 9 students in the class who are both male
   and dark-haired.
   d True, there are no students in the class that are both female
   and have red hair.

7 a

	1	2	3	4	5	6
1	0	−1	−2	−3	−4	−5
2	1	0	−1	−2	−3	−4
3	2	1	0	−1	−2	−3
4	3	2	1	0	−1	−2
5	4	3	2	1	0	−1
6	5	4	3	2	1	0

   b i $\frac{4}{36} = \frac{1}{9}$ **ii** $\frac{1}{36}$ **iii** $\frac{35}{36}$

8 a 0.1   **b** 0.3   **c** 0.5

9 a 0.001   **b** 0.500

Review 8

1 a 0.15   **b** 10

2 a 0.15   **b** Red 0.15; Green 0.15; Yellow 0.45
   c 60

3 a $\frac{6}{19}$ **b** 0   **c** 1

4 a $\frac{1}{9}$ **b** $\frac{1}{18}$

5 $\frac{9}{20}$

6 a $\frac{7}{25}$ **b** $\frac{6}{25}$ **c** $\frac{12}{25}$

7 a $0.2\dot{3}$ **b** $\frac{1}{6}$
   c it would be close to $0.1\dot{6} = \frac{1}{6}$

Assessment 8

1 No, the probability of this event is greater than zero.

2 $P(H) = 0.5$ because each throw is an independent event.

3 $\frac{51}{100}$

4 a 200   **b** 2   **c** 4

5 a If $x = 0.2$ then the total probability $= 1.02$, $x = 0.19$
   b 52   **c** 128   **d** 118

6 a i 0.18   **ii** 0.17
   b Ben's, because he sampled more packets.

7 a $\frac{1}{5}$ **b** $\frac{1}{2}$

8 9 packets

9 a Yes, $P(\text{Late}) = 0.51 > 0.49$ **b** 70

10 a i $\frac{3}{25}$ **ii** $\frac{8}{25}$ **iii** $\frac{7}{25}$ **iv** $\frac{11}{25}$
   b $P(\text{Blue 4}) = 0$

11 a $\frac{2}{5}$ **b** $\frac{1}{5}$ **c** $\frac{1}{2}$ **d** $\frac{3}{5}$

12 a i No, 102 is a multiple of 3.   **ii** No, 5 is prime.
   iii Yes, no white numbers are multiples of 7
   b i No, 81 is a multiple of 3.   **ii** No, 29 is prime.
   iii No, 84 is a multiple of 7.

13 No, sunny and snowing are not mutually exclusive events.

14 a $P(\text{blue}) = \frac{72}{360} = \frac{1}{5}$
   b You have assumed that all angles are equally likely to be
   chosen.

Chapter 9

Check in 9

1 a i 38.5   **ii** 39   **b i** 16.1   **ii** 16
   c i 103.9   **ii** 100   **d i** 0.1   **ii** 0.082
   e i 0.4   **ii** 0.38

2 a 954   **b** 337.415
   c 48.99 (2 dp)   **d** 22.37 (2 dp)
   e 105.59 (2 dp)   **f** −45.19 (2 dp)

9.1S

1

	(i)	(ii)	(iii)
a	8.37	8.4	8
b	18.8	19	20
c	35.8	36	40
d	279	280	300
e	1.39	1.4	1
f	3890	3900	4000
g	0.00837	0.0084	0.008
h	2400	2400	2000
i	8.99	9.0	9
j	14.0	14	10
k	1400	1400	1000
l	140000	140000	100000

2 a 8   **b** 560   **c** 0.008   **d** 10
   e 3.2   **f** 50   **g** 16   **h** 400
   i 2   **j** 5

3 a 30   **b** 40   **c** 1.8   **d** 25
   e 50   **f** 12100

4 a The denominator would become zero, and you can't divide
   by zero
   b The answer would not be a good estimate since $1^8 = 1$
   c The answer would not be a good estimate since $1^2 = 1$

5 a 1.4   **b** 2.8   **c** 3.2   **d** 3.9
   e 4.5   **f** 5.1   **g** 5.7   **h** 6.7
   i 8.4   **j** 9.2

6 a 100   **b** 1.5   **c** 900   **d** 60
   e 6.3   **f** 7.1

7 a 6000 **b** 3500 **c** 3 **d** 44
e 250 **f** 12.8
8 a 2.5 **b** 10 **c** 100 **d** 4
e 5 **f** 15
9 a 10 **b** 10^5 **c** 0.8 **d** 1×10^4
e 3.2×10^4 **f** 340

9.1A

1

	a	b	c
(i)	$260 + 360 = 620$	exact	620
(ii)	$60 \div 30 = 2$	underestimate	2.37
(iii)	$60 \times 200 = 12000$	underestimate	13279.86
(iv)	$100 - 65 = 35$	overestimate	31.9

2 30 slabs
3 a 83 **b** 4 photos
4 a £2000 **b** £500 **c** £3500
5 a 5 hours **b** 35 litres **c** £40 **d** Less
6 a Accept 40 000–52 000 grains
b Yes, accept 850 g – 1200 g as estimate for weight.
7 32 years
8 50 kg oxygen = 1.9×10^{27} 15 kg carbon = 0.8×10^{27}
7.5 kg hydrogen = 4.5×10^{27} Total $\sim 7 \times 10^{27}$
9 Students' responses

9.2S

1 a 21, 20.1 **b** 19, 18.6 **c** 13, 12.5 **d** 9, 9.8
e 10, 11.1 **f** 3, 3.1
2 a 71.1 **b** 29.624 **c** 2.07885304659
d 186.408 **e** 0.1508856039 **f** 19.05
3 a 5.8 **b** 1.3 **c** 1.7 **d** 2.2
4 a 5.6 **b** 73 **c** 0.085 **d** −35
e 110 **f** 38
5 a 464.5923967 **b** 0.4536084142
6 a 178.4123835 **b** 0.1967089505
c 3.210178253 **d** 3.350190476
e 1.157007415 **f** 0.1356045007
7 a 6790.558376 **b** 3796.885246
c 2.921455939 **d** 38.342
e 248.636363636 **f** 10.64011765
8 a 2.361087696 **b** 10.10395528
c 106.2121833 **d** 3.716757935
e 5.408 **f** 17.41017481
9 Students' responses

9.2A

1 a $2.4 \times (4.2 + 3.7) = 19.2$ **b** $6.8 \times (3.75 - 2.64) = 7.548$
c $(3.7 + 2.9) \div 1.2 = 5.5$ **d** $(2.3 + 3.4^2) \times 2.7 = 37.422$
e $5.3 + 3.9 \times (3.2 + 1.6) = 24.02$
f $3.2 + 6.4 \times (4.3 + 2.5) = 46.72$
2 a $3.4 \times (2.3 + 1.6) = 13.26$ **b** $3.5 \times (2.3 - 1.04) = 4.41$
c $(2.6 + 6.5) \div 1.3 = 7$ **d** $(1.4^2 - 1.2) \times 2.3 = 1.748$
e $(2.4^2 \div 1.8) \times (3.2 + 1.6) = 15.36$
f $(3.2 + 5.3) \times (2.4 - 1.2) = 10.2$
3 a i 15.23 m² **ii** £103.41 **b** £66.67
4 a 30 hours 21 mins **b** 1821 mins
5 a £21.13 **b** £16.37
c Yes – in both cases it would save money to use the new offer.
6 a £187 **b** £28.20
7 a i 1087.2 feet per second **ii** 1122.4 feet per second
b i 741.3 mph **ii** 765.3 mph

9.3S

1 a centimeters (cm) **b** milliliters (ml)
c kilograms (kg) **d** centimeters (cm)
e kilograms (kg) **f** kilometers (km)
g milliliters (ml) **h** liters (l)
i tonne (t) **j** gram (g)

2 a 2 cm **b** 4 m **c** 4.5 m **d** 4 km
e 5 mm **f** 4500 g **g** 6 kg **h** 6.5 kg
i 2.5 tonnes **j** 3000 ml
3 2.5 mph
4 37.5 miles
5 2.5 hours (1 hr 30 mins)
6 a 70 km/hr **b** 14 km/litre
7 a 333.3 m/min **b** 5.56 m/s
8 a 5 g/cm³ **b** 87.88 g
9 a 9.4575 kg **b** 6.15 l
10 a 5.75 – 5.85 m **b** 16.45 – 16.55 l
c 0.85 – 0.95 kg **d** 6.25 – 6.35 N
e 10.05 – 10.15 s **f** 104.65 – 104.75 cm
g 15.95 – 16.05 km **h** 9.25 – 9.35 m/s
11 a 6.65 – 6.75 m **b** 7.735 – 7.745 l
c 0.8125 – 0.8135 kg **d** 5.5 – 6.5 N
e 0.0005 – 0.0015 s **f** 2.535 – 2.545 cm
g 1.1615 – 1.1625 km **h** 14.5 – 15.5 m/s
12 a Max: 174 kg, min: 162 kg **b** Max: 28.4 kg, min: 27.6 kg
13 a 32.5 – 37.5 mm **b** 37.5 – 42.5 mm
c 107.5 – 112.5 mm **d** 42.5 - 47.5 mm

9.3A

1 a 141 kg **b** 8.4 m² **c** £13.63 **d** 0.8 mins
2 Max: 526.75 cm², min: 481.75 cm²
3 Max: 65 g, min: 63 g
4 21.7 cm
5 49 crates
6 14 boxes
7 7 crates
8 Group does not exceed maximum weight
9 a 85.5 – 86.5 cm
b If path is maximum, and stride is minimum, 34 strides would be required.
10 Maximum speed: 1.55 m/s   Minimum speed: 1.45 m/s
11 Yes, the minimum density is 7497 kg/m³ and the maximum is 8226 kgm³. Steel is the only metal with the density in that range.

Review 9

1 a i 93021.00 **ii** 93 000 **b i** 27.94 **ii** 28
c i 0.01 **ii** 0.0063 **d i** 0.90 **ii** 0.90
2 a i 64 **ii** 106 **iii** −1.1 **iv** 2
b i 61.13 **ii** 108.26 **iii** −3.67 **iv** 2.22
3 a e.g. centimetres, millimetres **b** e.g. grams, kilograms
c cm³, m³ **d** millilitres, litres
4 a 6.72 litres **b** 0.205 litres **c** 3.5 litres
5 20:15
6 16:25
7 300 cm³
8 a i 12.55 cm **ii** 12.45 cm **b i** 11.55 kg **ii** 11.45 kg
c i 1.005 m **ii** 0.995 m **d i** 0.0255 km **ii** 0.0245 km
9 $84.95 \leqslant x < 85.05$
10 a 11.55, 5.95 **b** 0.37 (2 dp), 0.23
c 225, 90.25 **d** 4.31 (2 dp), 3.60 (2 dp)
11 a 20.625 cm² **b** 27.625 cm²

Assessment 9

1 a 3.23 (3 sf) **b** 29 (2 sf)
c 0.2 (1 sf) **d** 310 (nearest 10)
e 5700 (nearest 100) **f** 256000 (nearest 1000)
2 a Bart **b** Bart **c** Christian **d** Ahmed
e Christian
3 $6 \times 60 \times 25 = 9000$, Mia used 25 as an estimate for the number of hours in a day.
4 14
5 a $2 \times 2^2 = 8$, 2
b 0.1911090054, the difference is larger than his estimate.
6 40

7 68%

8 **a** $4.6 + (4.1 + 1.2) \times 2.6 = 18.38$
   b $14.9 - 6.8 \div (3.7 - 1.2) - 12.18$
   c $(3.4 \times 1.6) + (5.9 - 2.8) = 8.54$
   d $2.6 + 7.56 \div (1.8 - 0.72) = 9.6$
   e $(12.3 - 5.2 \times 1.6 + 3.4) \times 2 = 14.76$

9 **a** ≈ 100 − 1000 litres   **b** ≈ 20 kilograms
   c ≈ 2 grams   **d** Students' answers in metres
   e ≈ 150 metres   **f** ≈ 50 − 300 tonnes
   g ≈ 30 millilitres   **h** ≈ 15 millimetres
   i ≈ 75 litres   **j** ≈ 40 kilometres
   k ≈ 15 centimetres   **l** ≈ 300 millilitres

10 **a** 66 km/h   **b** 41.7 km/h
   c 481.25 km   **d** 3410 miles
   e 17 min 52 s

11 **a** Yes, 0.81 g/cm³ < 1.   **b** 2.28 cm³ (3 sf)
   c 5400 tonnes

12 **a** LB 221455, UB 221465   **b** LB 85 cm, UB 95 cm
   c LB 452.5 g, UB 457.5 g
   d LB 3 min 28.75 s, UB 3 min 28.85 s
   e LB 238 bags, UB 242 bags
   f LB 27.5 tonnes, UB 28.5 tonnes
   g LB 585.5 mm, UB 586.5 mm

13 **a** 390.1625 cm³, 261.4375 cm³
   b 0.293 g/cm³ (3 sf), 0.194 g/cm³ (3 sf)

Chapter 10

Check in 10

1 **a** $\frac{2}{5}$ **b** $\frac{19}{45}$ **c** $\frac{1}{2}$ **d** $5\frac{17}{30}$
   e $\frac{7}{12}$ **f** $3\frac{1}{8}$

2 **a** 9   **b** 4   **c** 30   **d** 10

3 **a** $x + 16y$ **b** $9x^2 + 3x$ **c** $7p - 9$ **d** $24x - 27$
   e $5y + 11$ **f** $3x^2 - 2xy$ **g** $x^2 + 2x - 63$
   h $6w^2 - 32w + 32$ **i** $p^2 + q^2 - 2pq$

4 **a** $(x + 2)(x + 3)$ **b** $(x - 6)(x + 4)$
   c $(x - 3)^2$ **d** $(x - 10)(x + 10)$

10.1S

1 **a** 8   **b** 41   **c** 9   **d** 24
2 **a** 10   **b** 9   **c** 15   **d** 70
   e 30   **f** 60   **g** −2   **h** $\frac{25}{3}$
3 **a** −6   **b** −1   **c** 2   **d** −2
   e 1.5   **f** $\frac{2}{3}$ **g** 0.75   **h** −0.5
4 **a** 4.2   **b** $\frac{29}{6}$ **c** 9   **d** −0.25
   e $\frac{-13}{3}$ **f** −1.2
5 **a** 3.8   **b** $\frac{16}{7}$ **c** 1.4   **d** 5
   e − 3.4   **f** $-\frac{1}{3}$ **g** $-\frac{1}{3}$ **h** $\frac{19}{16}$
   i 2.8   **j** 2.375   **k** $\frac{7}{9}$
6 **a** −1   **b** 2   **c** 3   **d** $-\frac{13}{16}$
7 **a** 6   **b** 8   **c** 2   **d** 2
8 **a** 1.1   **b** $-\frac{27}{22}$ **c** 1.75   **d** 5.125
   e 4.3   **f** 5   **g** −13   **h** 1
9 **a** $\frac{1}{3}$ **b** $\frac{1}{3}$ **c** $\frac{4}{3}$ **d** 0.875
   e −5   **f** $\frac{5}{11}$ **g** $\frac{-7}{3}$ **h** $\frac{1}{3}$
10 **a** 5   **b** 10   **c** 28   **d** −1
11 **a** −8   **b** 0.75   **c** $\frac{2}{11}$ **d** −1.25
12 **a** −7   **b** 2   **c** −4   **d** 2
13 **a** 47   **b** $\frac{30}{17}$ **c** 3   **d** $\frac{28}{3}$
   e $\frac{135}{14}$ **f** $\frac{4}{23}$

10.1A

1 Lengths: 1 mm, 14 mm
2 Angles 40°, 70°, 70°

3 **a** Angles 40°, 60°, 80°
   b Square 8 units, rectangle 10 units by 6 units
4 **a** $(x + 4) \div 12 = 7, 80$ **b** $(x - 3) \div 11 = x \div 8, -8$
5 5 units
6 $x = 36$, Set 1 Values: 71, 110, 184, 252, 212, −61
   Set 2 Values: 101, 188, −23, 218, 156
7 **a** 32   **b** 7.2 units
8 $\frac{15}{11}$
9 35 years
10 Lengths: 3 cm, 4 cm, 1 cm, 4 cm
11 $x = 11$; 4, 7 and 16
12 $x = 9$; 5, 10 and 15

10.2S

1 **a** 0, 3   **b** 0, −8   **c** 0, 4.5   **d** 0, 3
   e 0, 5   **f** 0, 7   **g** 0, 6   **h** 0, 3
2 **a** −3, −4   **b** −7, 2   **c** 5, −1   **d** 3, 2
   e −0.5, −3   **f** $-2, -\frac{1}{3}$ **g** −0.5, −2   **h** $-0.5, \frac{-2}{3}$
   i 6, 2   **j** 1.5, −5   **k** 3, 2   **l** −7, −3
3 **a** ±4   **b** ±8   **c** ±5   **d** $\pm\frac{2}{3}$
   e ±0.5   **f** ±13   **g** ±2.5   **h** ±2
4 **a** 0, $\frac{1}{3}$ **b** 5, −3   **c** $\frac{2}{3}$, 3   **d** $\pm\frac{4}{3}$
   e ± 1.2   **f** 0, 1   **g** 0.6, −0.25   **h** 2, 6
5 **a** 10, 2   **b** 3, −5   **c** 5, −1   **d** −1
   e 7, −9   **f** 7   **g** 8, 0
   h $-6 + \sqrt{37}, -6 + \sqrt{37}$ **i** $\frac{3}{2} + \sqrt{4.25}, \frac{3}{2} - \sqrt{4.25}$
6 **a** −0.785, −2.55   **b** 1, 0.2
   c 5, −1.5   **d** −0.268, −3.73
   e 0.158, −3.16   **f** 2.21, −0.377
   g 3, $\frac{1}{3}$ **h** 0.158, −3.16
   i $\frac{5}{3}$, −4
7 **a** 1.61, −5.61   **b** 4.46, −2.46   **c** 1.17, −0.284
   d 2.87, −4.87   **e** 2.32, −0.323
8 **a** −5, −2   **b** −6, 2   **c** +7, −7   **d** 0, 8
   e 2, −6   **f** $\frac{1}{3}$, 2
9 **a** −0.838, −7.16   **b** −0.227, −0.631
   c 2.41, −0.414   **d** 0.657, −0.457
   e 0.425, −1.18   **f** 2.65, 0.849
10 **a** 0.5, $\frac{-5}{3}$ **b** 3   **c** 0.6, −0.5   **d** 1
   e ±3, ±2
11 **a** 0.219, 2.28   **b** 9.47, 0.528
   c −7, −3   **d** 0.3, −1   **e** 0.762, −2.36

10.2A

1 **a** $w(w + 7) = 60$
   b $w^2 + 7w = 60$, $w^2 + 7w - 60 = 0$
   c 5 cm, 12 cm
2 $p = -3$ – graph has just one solution so touches axis but doesn't go below.
3 No solution since root of a negative number cannot be calculated.
4 **a** 9, −13   **b** 17, −21   **c** 6.16, −0.162
   d 0.6, −0.5   **e** 8, −15   **f** 10
5 8 cm, 15 cm
6 **a** 34 cm   **b** 25 cm
7 −13, 46, 39 or 11, 22, 39
8 Alexa: mistake on line 4. She did not take the square root of 4.
   $x = 1$ or 5
   Briana: mistake on line 2. She did not complete the square correctly.
   $x = 1 + \sqrt{2}$ or $1 - \sqrt{2}$
9 $a\left(x^2 + \frac{b}{a}x + \frac{c}{a}\right) = 0$ (divide both sides by a)
   $\left(x + \frac{b}{2a}\right)^2 - \frac{b^2}{4a^2} + \frac{c}{a} = 0$
   $\left(x + \frac{b}{2a}\right)^2 = \frac{b^2}{4a^2} - \frac{4ac}{4a^2}$
   $x = \frac{-b}{2a} \pm \frac{\sqrt{b^2 - 4ac}}{2a}$

1 a, b

2 a i e.g. $(3, 5), (5, 1), (7, -3)$ **ii** e.g. $(3, 9), (4, 3), (5, -3)$
  b $x = 4, y = 3$

3 a $x = 4, y = 3$ **b** $x = 2, y = -3$
  c $x = -1, y = 4$ **d** $x = 3, y = 0.5$
  e $n = 2, m = 5$ **f** $x = -2, y = 1$

4 a $x = 7, y = -1$ **b** $x = 3, y = 0.5$
  c $a = 3, b = -1$ **d** $x = 2, y = 5$
  e $x = 3.25, y = \frac{-2}{7}$ **f** $p = 2, q = 7$

5 a $x = 5, y = -2$ **b** $x = 2, y = 3$
  c $a = -2, b = 3$ **d** $v = 4, w = 2$
  e $p = 2, q = 2$ **f** $x = 5, y = 2$

6 a $x = 12, y = -16$ **b** $x = 7, y = -1$
  c $a = 3, b = -2$ **d** $v = 3, w = 2$
  e $q = 3, p = 0$ **f** $x = 5, y = 2$

7 a $x = -1, y = -5$ **b** $x = \frac{2}{3}, y = -3$
  c $x = -0.8, y = \frac{1}{3}$ **d** $x = \frac{1}{6}, y = \frac{-3}{7}$

8 a $x = 6, y = 2$ **b** $a = 8, b = -1$
  c $q = -1, p = 8$ **d** $x = 12, y = -8$

9 a $x = 7, y = 4$ **b** $x = -4, y = -5$
  c $c = 2, d = -1$ **d** $x = 2, y = 0.5$
  e $a = \frac{-2}{3}, b = 0.75$ **f** $p = 20.5, q = 51$

10 a ± 7 **b** ± 5 **c** ± 10 **d** ± 3

11 a $x = 4, y = 8$ or $x = -1, y = 3$ **b** $x = 1, y = 0$
  c $x = 0.5, y = 0.5$ or $x = 32, y = 4$ **d** $x = 1, y = \pm\sqrt{5}$

12 a $x = 5, y = 17$ or $x = 3, y = 7$
  b $p = -2, q = -1$ or $p = 2, q = 1$

1 a 17p **b** 6.4 cm

2 a 17, 24 **b** 17, 23
  c 4 large, 1 small **d** $105°, 37.5°, 37.5°$
  e 4, −2

3 Sun = 29, moon = 17

4 a 2, 0 or −0.4, 1.6 **b** James is 3, Isla is 1

5 The lines are parallel so they never intersect.

6 (1, 7)

7 $x = -3, y = 2$ or $x = 3, y = -2$

8 $p = -1, q = 6$

9 a i 2.56, −1.56 **ii** 1.62, −0.62 **iii** 3, −1
  b The x-coordinates or the intersection of the red curve and the x-axis: 2, −1
  c $y = 2 - x$

1 a $x^3 = 200 + x^2, x = \sqrt[3]{200 + x^2}$
  b 6.201 **c** 199.9909376. Solution is accurate to 2 dp.

2 a i $p^2 + p = 100, p^2 = 100 - p, p = \sqrt{(100 - p)}$
   ii $p^2 + p = 100, p = 100 - p^2$
  b 9.51
  c Sequence diverges – does not reach a solution.

3 a −2.627
  b $0.00573 > 0$. This solution is not exact but is very close to the true value.

4 a −2.1044754, −2.1117932, −2.1139788, −2.1146306, −2.1148250
  b These values appear to converge to −2.115.
  c 1.8608

5 a $x^3 - 6x + 3 = 0, x^3 = 6x - 3,$
  $x = \sqrt[3]{6x - 3}, x_{n+1} = \sqrt[3]{6x_n - 3}$
  b 2.1451

1 a $12\,000(1 + A)^3 = 6000(3 + 3A + A^2)$
  $2(1 + 3A + 3A^2 + A^3) = 3 + 3A + A^2$
  $2A^3 + 5A^2 + 3A - 1 = 0$

  b No, solution does not converge.
  c 24.3%

2 a $1000(1+0.04n)=1000(1.03)^n, 1+0.04^n=1.03^n, 1=1.03^n-0.04n$
  b If one result is below 1 and the other is above 1, choose a new value, x, halfway between the two values chosen in the first step. Otherwise choose new starting values. Work out the value of $y = 1.03^x - 0.04x$. Choose one of your starting values so that out of x and the chosen starting value one gives a result for the formula greater than 1 and the other gives a result less than 1. Repeat the process.
  c 19.5 years

3 a 1.414215686
  b 2.00001

4 a 1.75
  b The sequence oscillates between 1 and 3.

5 a 240, 278, 312, 340, 360
  b Rapid growth over the first few years, then the rate of growth decreases.
  c 400

1 a $x \leqslant 7$ Solid dot at 7, arrow to left.
  b $x > 11$ Hollow dot at 11, arrow to right.
  c $p \leqslant -16$ Solid dot at −16, arrow to left.
  d $x > -3$ Hollow dot at −3, arrow to right.
  e $x \leqslant \frac{2}{3}$ Solid dot at $\frac{2}{3}$, arrow to left.
  f $y < -3$ Hollow dot at −3, arrow to left.
  g $x \leqslant 2$ Solid dot at 2, arrow to left.
  h $x > -5$ Hollow dot at −5, arrow to right.
  i $x \leqslant 10$ Solid dot at 10, arrow to left.
  j $p \leqslant -3$ Solid dot at −3, arrow to left.
  k $x > -18$ Hollow dot at −18, arrow to right.
  l $x \geqslant 0.5$ Solid dot at 0.5, arrow to the right.

2 a $x > 7\frac{2}{3}$ Hollow dot at $7\frac{2}{3}$, arrow to right.
  b $p \leqslant -32$ Solid dot at −32, arrow to left.
  c $x \leqslant -2\frac{3}{4}$ Solid dot at $-2\frac{3}{4}$, arrow to left.
  d $y < -4$ Hollow dot at −4, arrow to left.
  e $q \leqslant 10$ Solid dot at 10, arrow to left.
  f $z \leqslant -3$ Solid dot at −3, arrow to left.
  g $y > -10\frac{4}{5}$ Hollow dot at $-10\frac{4}{5}$, arrow to right.

3 a $-1 < x < 4$ **b** $-5 < x < 0$

4 a $-1 \leqslant y \leqslant 6$ **b** $-2 < z < 6$
  c $\frac{1}{3} < p < 5$

5 a Region to the right of $x = 4$ (solid line).
  b Region below $y = 3$ (solid line).
  c Region to the right of $x = -2$ (dotted line).
  d Region below $y = -3$ (dotted line).
  e Region between $x = -1$ (solid line), and $x = 6$ (solid line).
  f Region between $y = 1$ (dotted line), and $y = 8$ (dotted line).
  g Region to the left of $x = -2$ (solid line) and to the right of $x = 8$ (dotted line).
  h Region below $y = 2$ (dotted line) and above $y = 9$ (dotted line).
  i Region to the right of $x = 2.5$ (solid line).
  j Region below $y = 8$ (dotted line).
  k Region inside the rectangle formed by the lines $x = -2.5$ (dotted line), $x = 7$ (dotted line), $y = 0$ (solid line) and $y = 5.25$ (solid line).

6 a $-2 < y \leqslant 5$ **b** $1 \leqslant x < 7.5$

7 $(-1, 0), (0, 0), (1, 0), (-1, 1), (0, 1), (1, 1)$

8 a $-8 < x < 8$ **b** $x < -1, x > 1$
  c $x < -2, x > 0$ **d** $0 \leqslant x \leqslant 6$
  e $-4 < x < -2$ **f** $-4 < x < 3$
  g $-0.5 \leqslant x \leqslant 3$ **h** No solutions

9 Several solutions possible, for example $x^2 \leqslant 1$

10 No – the inequality simplifies to $x \geqslant 7$ which is an incomplete solution. The correct answer is $x \geqslant 7$ and $x \leqslant -7$

10.5A

1 a Solid line $y = x + 5$, origin in shaded region.
   b Dotted line $y = 2x + 1$, origin not in shaded region.
   c Dotted line $y = 1 - 3x$, origin not in shaded region.
   d Dotted line $y = 0.25x - 2$, origin not in shaded region.
   e Solid line $y = 1.5x + 2$, origin not in shaded region.
   f Dotted line $y = 3x - \frac{7}{3}$, origin not in shaded region.
   g Dotted line $y = 9 - x$, origin in shaded region.
   h Dotted line $y = \frac{7}{4} - \frac{1}{2}x$, origin not in shaded region.

2 a $y < 2x + 4$ b $y \geq 2\frac{2}{3}x - 1$
3 a $y \leq 9, y \geq x^2$ b $y < 12, y \geq (x-3)(x+2)$
4 a Solid vertical line through (-1, 0). Dashed horizontal line through (0, 2). Solid line through (0, -1) and (2, 5). Enclosed triangle shaded. $(-1, 1), (-1, 0), (-1, -1), (-1, -2), (-1, -3), (-1, -4), (0, 1), (0, 0), (0, -1)$
   b Dashed line through (2, 2) and (5, 0). Dashed vertical line through (2, 0). Dashed horizontal line through (0, -0.5). Enclosed triangle shaded. (3, 1), (3, 0), (4, 0)
5 Region enclosed by the lines $y = x^2 - 4$ and $y = x - 1$.
6 a $x \geq 0$ and $y \geq 0$ because the number of each type of coach cannot be negative.
      The total number of seats is $20x + 48y$, which must be at least enough for 316 people, so $20x + 48y \geq 316$ or $5x + 12y \geq 79$. To have at least 2 adults per coach, there can be no more than 8 coaches, so $x + y \leq 8$.
   b Solid line through (0, 8) and (8, 0). Solid line through (11, 2) and $(-1, 7)$. Indicated region to right of y- axis, above $5x + 12y = 79$ and below $x + y = 8$.
7 a e.g. $x > 0, y > 0, x + y < 1$
   b e.g. $y > 1$ or $y < 0, y > x$ or $y > 3 - x$

Review 10

1 a 14   b 100   c -3
   d 6   e 12   f 3.5
2 a $2x + 0.6 = 7$
   b Sam = 3 hr, 12 min; Andy = 3 hr, 48 min
3 a $(x + 3)(x - 4) = 0, x = -3, 4$
   b $(x - 5)(x - 7) = 0, x = 5, 7$
   c $(x + 6)^2 = 0, x = -6$
   d $(4x + 11)(4x - 11) = 0, x = -\frac{11}{4}, \frac{11}{4}$
   e $3x(2x - 3) = 0, x = 0, \frac{3}{2}$
   f $(3x + 4)(x + 1) = 0, x = -\frac{4}{3}, -1$
   g $(2x - 3)(5x - 2) = 0, x = \frac{3}{2}, \frac{2}{5}$
4 $(2x + 3)(x - 5) = 0, x = -\frac{3}{2}, 5$
5 a i $(x - 2)^2 - 5$ ii $(x + \frac{3}{2})^2 - \frac{1}{4}$
   b i $x = 2, 8$ ii $x = -6, 2$
6 a $x = -1.14, 6.14$ b $x = -2.26, -0.736$
7 a $x = 3, y = -1$ b $v = 0.5, w = 1.5$
   c $(-1, 1)$ and $(2, 4)$ d $(-5, 0)$ and $(1, 6)$
8 a $2(L + W) = 17$ and $LW = 15$
   b $L = 6$ cm, $W = 2.5$ cm or vice versa
9 a $2.89, -1.44$ b 8.08
10 a $x > 5$ b $x \leq -3$
   c $-7 < x < -1$
   d $x \leq -6, x \geq 3$

11 a
   b

c

Assessment 10

1 a 5.5 kg   b Albert 3, Oliver 9   c 154 g
   d Brendan €740, Arsene €370, José €890
   e Fizz 4, Milo 3   f 135 miles
2 a $5k - 2(30 - k) = 115, k = 25$ b 9 km
3 a $1 + 3\sqrt{11}$ b 53, 59   c 100
4 a 1 s, the ball is going up or 3 s, the ball is coming down.
   b 2 s, the ball has reached its greatest height.
   c $25 = 20t - 5t^2, t = 2 \pm \sqrt{-1}, \sqrt{-1}$ is not real and 25 m > 20 m.
   d $t = 4. t = 0$ is the start, $t = 4$ the ball hits the ground.
5 12
6 No, 0.47 or -8.53.
7 a $x = 4, y = 6$ and $x = -6, y = -4$
   b $x = 4, y = -3$ and $x = -3, y = 4$
8 a i £6.50   ii £4.50   b 50 g
   c i 72p   ii 63p   d i £4.99   ii £11.95
   e i £250   ii £500   f i 4p   ii 3p
9

10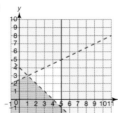

11 a 22 g ≤ 345, g ≤ 15.7 (3 sf)   b 100, 101 or 102
   c i $f = 3, p = 5; f = 4, p = 6; f = 5, p = 7$
      ii $f = 5, p = 7$ iii $f = 3, p = 5$

Lifeskills 2

1 a 10000 : 1
   b, c

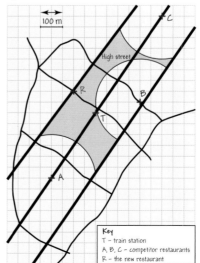

2 a 7.24 cm² **b** 45.25 cm² **c** 640

3 a $x + y \leqslant 32$

$x \leqslant 8y$

b

c 28

4 a Style B. **b** Style A 1.54 m², Style B 1.62 m².

c Multiple answers possible.

5 a 11

b Upper bound 97.0025 m², lower bound 95.0025 m².

c No. If the lower bound is accurate the tables will be less than 1 m apart which does not meet the requirements.

6 a Juliet

b 20

c Multiple answers possible e.g. they call at similar times to the previous week, they don't repeat calls to people who are rarely in.

Chapter 11

Check in 11

1 $a = 43°, b = 107°, c = 233°, d = 139°$

2 $e = 102°, f = 29°, g = 123°, h = 73°$

3 a 17.5 cm² **b** 15 m² **c** 50 m²

11.1S

1 a 31.4 cm **b** 25.1 m **c** 37.7 cm **d** 62.8 m

2 a 75.4 mm **b** 144.5 cm **c** 329.9 mm **d** 3.8 cm

e 22.6 cm **f** 392.7 cm

3 a 78.5 cm² **b** 50.3 m² **c** 113.1 cm² **d** 314.2 m²

4 a 452.4 mm² **b** 1661.9 cm² **c** 8659 mm² **d** 1.1 cm²

e 40.7 cm² **f** 12271.8 cm²

5 a 6.0 cm **b** 5.0 m **c** 9.0 cm **d** 15.0 m

6 a 39.3 cm² **b** 81.4 cm² **c** 904.8 mm² **d** 127.2 cm²

e 402.1 mm² **f** 372.5 cm²

7 a 226.2 cm² **b** 8.3 cm² **c** 65.3 m² **d** 195.3 cm²

e 141.8 mm² **f** 4.5 cm²

8 a 25.7 cm **b** 37.3 cm **c** 123.4 mm **d** 46.3 cm

e 82.3 m **f** 79.2 cm

9 a 61.7 cm **b** 11.8 cm **c** 33.2 m **d** 57.3 cm

e 48.8 mm **f** 8.7 cm

10 a 13.85 cm² **b** 3.79 cm²

11 112.6 m

12 2513.3 cm²

13 a 4.90 cm **b** 4.47 m **c** 6.00 cm **d** 7.74 m

e 19.99 cm

14 7.0 cm

15 50 cm

11.1A

1 No, £52 300 (3 sf) is needed.

2 a 20 plants **b** 0.28 m²

3 See diagram in question (diameter of end pieces = 63.7 m).

4 19.2 cm²

5 a Yes, perimeter = 11.82 m < 12 m.

b No, area = 9.78 m², 35 × 9.78 × 6 = 2054 g > 2 kg.

6 a $\frac{1}{2}\pi D + \frac{1}{2}\pi d + \frac{1}{2}\pi(D - d) = \pi D$

b $\frac{1}{2}\pi\left(\frac{d}{2}\right)^2 + \frac{1}{2}\pi\left(\frac{D}{2}\right)^2 - \frac{1}{2}\pi\left(\frac{D-d}{2}\right)^2 = \frac{1}{4}\pi Dd$

7 No. For a semi-circle of radius 2r, the factor of 2 is also squared giving and extra factor of 2 in the calculation.

8 $\frac{3}{\pi}$

9 a i 600

ii Yes, 21% is wasted (< 25%).

b Maximum diameter = 4.05 cm. At this value, part **a i** decreases to 551 and part **a ii** changes to no as 26% is wasted (> 25%). Minimum diameter = 3.95 cm. At this value the answers to parts **a i** and **a ii** remain the same.

10 $h < 13.04$ in (2 dp)

11.2S

1 a 12.6 cm **b** 50.3 cm²

2 a i $\frac{5}{12}$ **ii** $\frac{5}{9}$ **iii** $\frac{2}{5}$

b i 8π **ii** 10π **iii** 8π **iv** 54π

3 a 15.1 cm² **b** 15.7 cm² **c** 101.8 cm² **d** 15.7 cm²

e 6.0 cm² **f** 12.3 cm²

4 a 7.5 cm **b** 10.5 cm **c** 33.9 cm **d** 6.3 cm

e 5.0 cm **f** 7.0 cm

5 a 12.6 mm, 113.1 mm² **b** 3.77 m, 3.0 m²

c 15.4 cm, 26.9 cm² **d** 2.50 km, 0.576 km²

6 a 45.4 mm **b** 14.8 cm **c** 29.5 cm

7 143.2°

8 229.2°

9 123.6 cm²

10 23.6 cm²

11.2A

1 a 11.05 m²

b 12.9 bricks (assuming inner radius of brick is 30 cm long)

2 50 cm²

3 $96 + 24\pi$

4 10.9 cm² (3 sf)

5 a $2\left(\frac{1}{4} \times 2 \times \pi \times 20\right) = 20\pi$ cm

b $2\left(\frac{1}{4} \times \pi \times 20^2\right) - 20^2 = 200(\pi - 2)$ cm²

6 Difference = $\pi \times 16^2 - (64\pi - 128) - (64\pi - 128) = 128(\pi + 2)$ cm²

7 a Area = $\frac{30}{360}\pi R^2 - \frac{30}{360}\pi r^2 = \frac{1}{12}\pi(R^2 - r^2) = \frac{1}{12}\pi(R + r)(R - r)$

b $\frac{3\pi}{2}$

8 $\frac{45\pi}{2}$

9 a 19.1 m

b Yes. Total area = area of sector + area of rectangle − area of triangle that overlaps

= 24.8 m² > 24 m²

10 $1600 - 400\pi$

11.3S1

1 a 66° – Circle Theorem 1 **b** 43.5° – Circle Theorem 1

c 126° – Circle Theorem 1 **d** 72° – Circle Theorem 1

e 107° – Circle Theorem 1 **f** 133° – Circle Theorem 1

g 180° – Circle Theorem 1 **h** 45° – Circle Theorem 1

2 a $a = 97°$ – Circle Theorem 2

b $b = c = d = 49°$ – Circle Theorem 2

c $e = 34°, f = 67°$ – Circle Theorem 2

d $g = 72°, h = 28°$ – Circle Theorem 2

e $i = 90°$ – Circle Theorem 2

f $j = 63°, k = 30°$ – Circle Theorem 2

g $l = 40°$ – Circle Theorem 2, $m = 75°$ – the sum of angles inside a triangle.

h $n = 35°, o = 53°$ – Circle Theorem 2

3 a 90° – Circle Theorem 3 **b** 63° – Circle Theorem 3

c 32° – Circle Theorem 3 **d** 30° – Circle Theorem 3

e 15° – Circle Theorem 3

f $f = 48°, g = 132°$ – Circle Theorem 3

g 36° – Circle Theorem 3 and sum of angles inside a triangle.

h 40° – Circle Theorem 3 and sum of angles inside a triangle.

4 a $a = 128°$ – Circle Theorem 4
  b $b = 59°$ – Circle Theorem 4
  c $c = 152°$, $d = 110°$ – Circle Theorem 4
  d $e = 134°$, $f = 67°$ – Circle Theorem 4
  e $g = 117°$ – Circle Theorem 4, $h = 126°$ – Circle Theorem 1
  f $i = j = 98°$ – Circle Theorem 4
  g $k = 86°$ – Angles on a straight line, $l = 94°$ – Circle Theorem 4
  h $m = 122°$ – Angles on a straight line and Circle Theorem 4

11.3S2

1 a $90°$ – Circle Theorem 5
  b $b = 44°$, $c = 46°$ – Circle Theorem 5, 8
  c $d = 28°$, $e = 28°$ – Circle Theorem 5, 8
  d $f = 20°$, $g = 20°$ – Circle Theorem 5, 8
  e $h = 48°$ – Circle Theorem 8
  f $k = 69°$ (internal angles in a triangle sum to 180°), $l = 111°$ (Theorem 5, and the internal angles in a quadrilateral must sum to 360°)
  g $28°$ – Circle Theorem 5 and angles in isosceles triangle
  h $m = 12°$ (Circle Theorem 5, and the internal angles in a quadrilateral must sum to 360°), $n = 84°$ (Circle Theorem 1)
  i $r = 73°$, $s = 53.5°$ – Circle Theorem 1, 5
  j $v = 65°$ – Circle Theorem 5 and internal angles in a quadrilateral must sum to 360°. $w = 32.5°$ – angles on a straight line sum to 180° and base angles of an isosceles triangle are equal.
  k $x = 90°$ (Theorem 5), $y = 66°$ (internal angles in a triangle sum to 180°)
  l $53°$ – internal angles in a triangle sum to 180°

2 a $43°$ – Circle Theorem 8
  b $72°$ – Circle Theorem 8
  c $57°$ – Circle Theorem 8
  d $19°$ – Circle Theorem 8
  e $e = 50°$, $f = 40°$ – Circle Theorem 5, 8
  f $g = 35°$, $h = 55°$ – Circle Theorem 5, 8
  g $26°$ – Circle Theorem 5
  h $42°$ – Circle Theorem 8
  i $k = 36°$, $l = 36°$ – Circle Theorem 8
  j $m = 59°$, $n = 59°$ – Circle Theorem 8
  k $s = 59°$ (Theorem 8, and the internal angles in a triangle sum to 180°), $t = 56°$ – angles on a straight line sum to 180°, base angles in an isosceles triangle are equal, angles in a triangle sum to 180°.
  l $u = 64°$, $v = 128°$, $w = 26°$ (Theorem 1, 8, and the internal angles in a triangle sum to 180°)
  m $x = 58°$, $y = 32°$ – Circle Theorem 8, and the internal angles in a triangle sum to 180°
  n $17°$ – Circle Theorem 8, and the internal angles in a triangle sum to 180°

11.3A

1 Opposite angles inside a cyclic quadrilateral must sum to 180° $> 176°$.
2 Square. $\angle ABC = \angle CDA = 90°$, $\angle DAB = \angle BCD = 90°$ (opposite angles of a parallelogram are equal, opposite angles of a cyclic quadrilateral sum to 180°).
3 If DF is the diameter, angle E must equal 90°, so would be able to use Pythagoras Theorem to calculate the hypotenuse, which should equal the diameter.
$DF^2 = (120^2 + 64^2) \rightarrow DF = 136\,mm =$ diameter of circle
4 If C is the centre of the circle, the angle at the centre of the circle must equal twice the angle at the circumference from the same arc.
$2 \times 62 \neq 122$ so C is not the centre
5 $\angle PQR = \angle QPS$ (base angles of an isosceles triangle are equal, angles on a straight line sum to 180°).

$\angle QRS = 180° - \angle QPS = 180° - \angle PQR = \angle PSR$ (opposite angles in a cyclic quadrilateral sum to 180°).
$\angle PQR + \angle QRS = 180°$ so QP is parallel to RS.
6 The internal angles inside a triangle must sum to 180°
$OY = OX$ so angle Y = angle X = 38°
The angle between a tangent and the radius at the point where the tangent touches the circle is a right angle but $38° + 62° = 100°$. Therefore, XP is not a tangent.
7 The angle between a tangent and the radius at the point where the tangent touches the circle is a right angle. Therefore, angle B equals angle A, OA=OB so AOBP is a kite and so BP must equal AP.
8 Triangle OXT is congruent to triangle OXS (SSS) so $\angle OXT = \angle OXS$.
9 Let the line segment AB be a chord on a circle with centre O. The perpendicular line from O to AB intersects AB at X.
$\angle OAX = \angle OBX$ (base angles of an isosceles triangle are equal), $\angle AXO = \angle BXO = 90°$ so $\angle AOX = \angle BOX$ (angle sum of a triangle). Triangles OAX and OBX are congruent by ASA so $AX = BX$.
10 Let line segment XY be a tangent to the circle at A such that $\angle XAB < 90°$.
$\angle BCA = \angle XAB$ and $\angle CBA = \angle YAC$- Circle Theorem 8, $\angle CBA = \angle BCA$ (base angles of an isosceles triangle are equal), so $\angle XAB = \angle CBA$ these are alternate angles so XY is parallel to BC.
11 Let line segment XY be the tangent to the circle at P such that $\angle XPR < 90°$ and let line segment YZ be the tangent to the circle at Q such that $\angle ZQR < 90°$.
$\angle RPQ = \angle PQY = \angle PRQ$ (alternate angles and Circle Theorem 8). $\angle RQP = \angle QPY = \angle PRQ$ (alternate angles and Circle Theorem 8). So $\angle RPQ = \angle PQY = \angle PRQ$ (alternate angles and Circle Theorem 8). So $\angle RPQ = \angle PRQ = \angle RQP$.

11.4S

1 a, b, c, d Check the angle between lines is 90°.
2 a, b, c, d Check the angle between lines is 90°.
3 a Two angles of 28°.   **b** Two angles of 41°.
  c Two angles of 51°.
4 a Two angles of 35°.   **b** Two angles of 55°.
  c Two angles of 45°.   **d** Two angles of 65°.
  e Two angles of 25°.   **f** Two angles of 42.5°.
5 a Triangle with side lengths 4 cm, 7 cm, 8 cm.
  b Triangle with side lengths 2.1 cm, 3 cm, 4 cm.
  c Triangle with side lengths 6 cm, 7.5 cm, 10 cm.
  d Triangle with side lengths 2 cm, 7.7 cm, 8 cm.
  e Triangle with side lengths 5 cm, 6 cm, 9 cm.
  f Triangle with side lengths 2.8 cm, 2.8 cm, 4 cm.
6 a Isosceles triangle with base angles 69° and base 5 cm.
  b Isosceles triangle with base angles 46° and base 7 cm.
7 a Triangle with three sides of length 5 cm, three angles of 60°.
  b Six angles of 30°.
  c The angle bisectors cross at the centre of the triangle and bisect the sides.
8 a Triangle with three angles of 60°.
  b Two angles of 30°.
9 a Check lines meet at 90°, line divided into two 3 cm lengths.
  b Check lines meet at 90°, line divided into two 4.5 cm lengths.
  c Check lines meet at 90°, line divided into two 2.8 cm lengths.
  d Check lines meet at 90°, line divided into two 5 cm lengths.
  e Check lines meet at 90°, line divided into two 5.6 cm lengths.
10 a Triangle with three sides of length 5 cm, three angles of 60°.
  b Six line segments of length 2.5 cm.
  c The diagrams are the same.

Answers

11.4A

1 a, b

2 a, b

3 a, b

4 a, b

a, c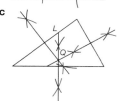

d In triangle A the angle bisectors meet at point P. A circle can be drawn with centre P such that each side of triangle A is a tangent. In triangle L the side bisectors meet at point Q. A circle can be drawn with centre Q such that the vertices of triangle L lie on the circle.

5 a, b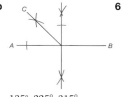

6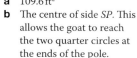

c 135°, 225°, 315°.

7

8 a 109.6 ft²

b The centre of side SP. This allows the goat to reach the two quarter circles at the ends of the pole.

9 a Yes. Let the original angle be $\angle AXB$ such that the initial arcs are drawn at A and B, let C be the point at which the third and fourth arcs cross. Triangles $\angle ACX$ and $\angle BCX$ are congruent by SSS so $\angle AXC = \angle BXC$.

b i The construction creates a triangle with three equal sides so the triangle has three equal angles, $180° \div 3 = 60°$.

ii The construction creates a rhombus and the diagonals of a rhombus bisect each other at right angles.

Review 11

1 a 133 cm² **b** 40.8 cm

2 14.3 cm

3 a 124 m² **b** 48.0 m

4 a 45.8 cm² **b** 11.4 cm **c** 27.4 cm

5 $a = 65°$, angles in same segment are equal
$b = 33°$, angles in same segment are equal
$c = 90°$, angles in semi-circle are 90°
$d = 36°$, angles in triangle add up to 180°

6 $a = 70.5°$, angles in an isosceles triangle
$b = 39°$, alternate segment theorem

7 a, b check students' drawing.

8 Check circle of radius 4 cm.

9 Check students' drawing.

Assessment 11

1 a i A, $61.58 \div 2\pi = 9.80$ cm (3 sf) $= x$
ii C, $314.16 \div 2\pi = 50.0$ cm (3 sf)
iii B, $2.01 \div 2\pi = 0.320$ cm (3 sf)
b i B, $\sqrt{\frac{176.7}{\pi}} = 7.50$ m (3 sf)   **ii** A, $\sqrt{\frac{81.1}{\pi}} = 5.08$ m (3 sf)
iii C, $\sqrt{\frac{0.407}{\pi}} = 0.360$ m (3 sf) $= y$

2 63.7 times

3 a 233 mm²
b Outer = 59.7 mm, inner = 25.1 mm.

4 36 589 times

5 24 minutes

6 a 44.0 cm (3 sf) **b** 117 cm² (3 sf)

7 $r = \frac{\theta}{360} \times 2 \times \pi \times r$, $\theta = \frac{360}{2\pi} = 57.3°$

8 $p = 55.5°$ (isosceles triangle), $q = 90 - 55.5 = 34.5°$ (angle in a semicircle is 90°), $r = q = 34.5°$ (isosceles triangle).

9 $\angle BOC = 90 - 34 = 56°$
$\angle BCO = \frac{1}{2}(180 - 56)$ 62° (Isosceles triangle)
$\angle ABC = 90°$ (Angle in a semicircle)
$\angle CAB = 180 - 62 - 90 = 28°$ (Angle sum of a triangle)

10 $x = 90°$ (angle in semicircle), $w = 49°$ (Alternate Segment Theorem), $v = 41°$ (angle sum of a triangle).

11 $\angle ABO = 90 - x$ (tangent perpendicular to radius), $\angle OAB = 90 - x$ (isosceles triangle), $\angle AOB = 180 - 2(90 - x) = 2x$ (angle sum of a triangle), $y = x$ (angle at centre) $= 2 \times$ angle at circumference

12 a 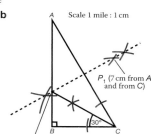 **b**

c 7.5 miles (±0.5 miles). **d** 3 miles/second

13 a Angles in triangle: $\angle LPF = 107°$, $\angle PFL = 29.5$, $\angle FLP = 43.5$. LP = 21.3 cm, PF = 29.7 cm (different scales possible)
b 413 (± 10) miles on a bearing of 285° (± 10°).

Chapter 12

Check in 12

1 a 80 **b** 21 **c** 19.6 **d** 168
2 a 60% **b** $\frac{7}{20}$ **c** $\frac{13}{20}$ **d** 0.35
3 a 1 : 3 **b** 1 : 6 **c** 2 : 3 **d** 5 : 2

12.1S

1 a 75 g **b** 4.5 g **c** 100 g **d** 125 g
2 a 90 cm³ **b** 228 cm³ **c** 1.2 m³ **d** 168 cm²
3 a 60 g **b** 36 mm **c** 380 g **d** 31.2 km
4 a $\frac{13}{50}$, 0.26 **b** $\frac{71}{100}$, 0.71 **c** $\frac{1}{50}$, 0.02 **d** $\frac{51}{50}$, 1.02
5 a $\frac{19}{50}$, 0.38 **b** $\frac{23}{50}$, 0.46 **c** $\frac{4}{5}$, 0.80 **d** $\frac{7}{100}$, 0.07
6 a $\frac{1}{3}$ **b** $\frac{1}{2}$ **c** $\frac{3}{10}$ **d** $\frac{1}{4}$
e $\frac{2}{5}$ **f** 2 **g** $\frac{1}{8}$ **h** $\frac{1}{16}$
i $\frac{1}{12}$ **j** $\frac{1}{4}$
7 a 33.3% **b** 50% **c** 30% **d** 25%
e 40% **f** 200% **g** 12.5% **h** 6.3%
i 8.3% **j** 25%
8 a i $\frac{3}{10}$ **ii** 30% **b i** $\frac{5}{8}$ **ii** 62.5%
c i $\frac{4}{9}$ **ii** 44.4% **d i** $\frac{5}{12}$ **ii** 41.7%

9 a i $\frac{7}{20}$ **ii** 35% **b i** $\frac{2}{25}$ **ii** 8%

c i $\frac{3}{10}$ **ii** 30% **d i** $\frac{3}{40}$ **ii** 7.5%

e i $\frac{9}{40}$ **ii** 22.5% **f i** $\frac{3}{20}$ **ii** 15%

g i $6\frac{2}{3}$ **ii** 666.7% **h i** $2\frac{1}{2}$ **ii** 250%

i i $4\frac{3}{25}$ **ii** 412% **j i** $6\frac{2}{5}$ **ii** 640%

10 a 135 kg **b** 32.085 m **c** £105 **d** 22.5 cm³
  e 510.15 g **f** £16.50
11 A 30%, B 25.9%, C 23.3%, D 20.8%

12.1A

1 a $\frac{3}{25}$ **b** 12%
2 a £2100 **b** $\frac{8}{15}$, or 53.3%
3 a 64 cheeses **b** 32%
4 a i Maths **ii** English
  b Overall score = $(48 + 39 + 55) \div (60 + 50 + 70) = \frac{71}{90}$
5 50 cars
6 Yes, $\frac{(45 \times 1 + 60 \times 2)}{220} = \frac{3}{4}$
7 Yes. Tropical ~ 0.124 g/ml > Pineapple ~ 0.115 g/ml, Blackcurrant ~ 0.117 g/ml.
8 a 40, 50 **b** 50 employees.
9 No, the shapes are both half shaded.
10 $\frac{5}{16}$

12.2S

1 a 1:3 **b** 3:1 **c** 1:3 **d** 1:7
  e 1:10 **f** 5:1 **g** 3:1 **h** 1:15
2 a 2:1 **b** 3:1 **c** 9:7 **d** 5:3
3 a 2:1 **b** 1:3 **c** 2:3 **d** 9:8
  e 3:4 **f** 4:3 **g** 10:9 **h** 4:5
4 a £27:£63 **b** 287 kg:82 kg
  c 64.5 tonnes:38.7 tonnes
  d 19.5 litres:15.6 litres
  e £6:£12:£18
5 a £40, £35 **b** £350, £650 **c** 260 days, 104 days
  d 142.86 g, 357.14 g **e** 214.29 m, 385.71 m
6 a 2:5 **b** 11:16 **c** 5:2 **d** 5:3
  e 5:3 **f** 8:5
7 a £100:£250:£150 **b** 180°:45°:135°
  c 900 m:400 m:700m
8 a 1:3 **b** 1:4 **c** 1:2 **d** 1:5
  e 1:2 **f** 1:3 **g** 1:5 **h** 1:3
  i 1:2 **j** 1:5 **k** 1:2 **l** 1:6
9 a 1000 m **b** 4000 m **c** 5000 m
  d 250 m **e** 7250 m
10 a 325 m **b** 0.6 cm
11 a 50 m **b** 4 cm
12 a 200 m **b** 12 cm
13 a 116 m **b** 180 cm

12.2A

1 a 66.7% **b** 360 cm
2 a 120% **b** 102 kg
3 a 72 g **b** 115 g Copper, 69 g Aluminium
4 $\frac{11}{50}$
5 a 300 m **b** 120 m
6 75 cm by 200 cm
7 a 4 m, 6 m, 12 m **b** 72 m²
8 0.7
9 a £280 **b** £80, £200
10 a 3:6:2 **b** $20 ($60 compared to $40)
11 8 sides
12 P = 55°, Q = 125°

12.3S

1 a 0.5 **b** 0.6 **c** 0.25 **d** 0.085
  e 0.0015 **f** 0.0001

2 a £40 **b** 260 cm **c** 3.2 kg **d** 20 m
  e 190 p **f** £35 **g** 3 kg **h** £6.20
3 a 325.35 kg **b** $120 **c** 10.35 kg **d** 5.88 kg
  e £21.70 **f** 78.2 m
4 a £224 **b** £385.20 **c** €1458 **d** £77
  e €13.35 **f** £465.83
5 10% of £350 (£35 > £30)
6 8% of £28 (£2 < £2.24)
7 a £549.60 **b** £2519 **c** £842.72 **d** £1167.90
8 a £790 **b** £1109.25 **c** £54.60 **d** £132.43
9 a 1.2 **b** 1.3 **c** 1.45
10 a 0.6 **b** 0.4 **c** 0.65
11 a £495 **b** 672 kg **c** £756 **d** 392 km
  e £658 **f** 256 m
12 a £275 **b** £2264 **c** £18 060 **d** £2520
  e £4.23 **f** £2000
13 a £385 **b** 70.3 kg **c** £550.20 **d** 491.4 km
  e 1128 kg **f** £216
14 a £397.80 **b** 524.9 kg **c** £1758.96
  d 599.56 km **e** $3423.55 **f** 2154.75 m

12.3A

1 a 60% **b** 48% **c** 60%
2 a 51.35% **b** 28.57% **c** 3.83%
3 a 20% **b** 20%
4 a 12.5% **b** 20%
5 10.8% per year
6 £56 is 80% of the original price, the original price is not 120% of £56. The original price is £70.
7 £5
8 a 50 **b** 40 **c** 80 **d** 110
9 a £6 **b** £80
10 a Francesca £13.13, Frank £12.80. Francesca has the bigger pay increase.
  b £78.03
11 a £50 **b** £30.94
12 £8450
13 13 600
14 a £321.63 **b** £1749.60

Review 12

1 a $\frac{16}{25}$ **b** 36% **c** 45
2 a $\frac{1}{3}$ **b** 1:2
3 a 2:3 **b** 5:3 **c** 1:250
4 a £35, £7 **b** £22, £44, £33
5 a 2:1 **b** $\frac{1}{2}$
6 a 400 g milk, 600 g flour **b** 300 g **c** $\frac{2}{5}$
7 a 35 m **b** 962 m²
8 a 8 km **b** 3.125 cm
9 a 80.4 **b** 80.64
10 £6114
11 a 6.6% **b** £464 000 **c** 232%
12 a £1.24 **b** £9060

Assessment 12

1 a $N = 1$, $W = 4$
  b Brazil nuts $\frac{3}{5}$, walnuts $\frac{1}{4}$, chestnuts $\frac{3}{20}$.
  c Brazil nuts 60%, walnuts 25%, chestnuts 15%.
  d Chestnuts **e** Brazil nuts
  f 75% **g** $\frac{3}{4}$
  h $\frac{1}{4}$
2 a i 0.52 = 52% **ii** 117 000 rand
  b £13 110
3 Chocolate 27, plain 36.
4 a 8:2:3 **b** Butter 45 g, cheese 67.5 g
5 a Small 150 ml, large 600 ml.
  b Small 250 ml, medium 625 ml.

6 a 6:5 **b** $\frac{5}{3}:1$

7 a Girls 32, boys 56. **b** Girls 44, boys 77, $44:77 = 4:7$

8 a 3:4 **b** 9:16

9 a 9.1 s **b** 21.1 s

10 a 40.5 miles **b** 25.6 in **c** 236 miles

11 a 200 m **b** 1.17 km **c** 4.8 cm

  d Distance $= 0.2 \times 20.75 = 4.15$ km. Time $= (4.15 \div 80)$ hours $= 187$ seconds (3 sf) < 200 seconds.

12 a Gavin

  b Multiply by $(100\% - 45.4\%)$. $550 \times (1 - 0.454) = 300.3$ ml

13 a 42.22% **b** 30%

14 a £2 767 292.50 **b** 90.8%

  c Yes, $(35 - 24) \div 24 = 45.83\% > 45\%$.

16 a £4236.31 **b** £2480 **c** £237.11

Revision 2

1 9

2 72 cm²

3 a  **b** 90° **c** 30 km²

4 a 39 609 m² **b** 3.96 Ha (3 sf)

5 a Enlargement, scale factor 3, centre (0, 0)

  b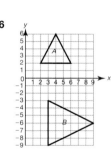

6 Rotation 90° clockwise about (0, 0) and enlargement of scale factor 1.5, centre (0, 0) (either order).

7 a 120 cm **b** 40 000 g or 40 kg

  c Length $= 110.2$ cm, mass $= 41\,800$ g or 41.8 kg.

  d Percentage error for length $= 8.89\%$, percentage error for mass $= -4.31\%$

8 a LB 76.637 m³, UB 75.880 m³

  b Upper bound $= 214\,583.6$ kg $= 215$ tonnes (3 sf)
    Lower bound $= 212\,464$ kg $= 212$ tonnes (3 sf)

9 a, b **c** 1 Lupin, 2 Stocks.

10 4 cm $\times$ 25 cm

11 26 10p coins and 15 50p coins.

12 $r < 100$

13 a 21.5% **b i** 9 **ii** 11

14 a $\angle BD = 120°$ (ASL), $AB = AD$ (TEL), thus $\angle BAD = \angle BDA = 30°$ (Isosceles $\triangle$). Thus $\angle DEA = 30°$ (AST), $\angle EDA = 90°$ (ASC) so $\angle EDF = 180 - 90 - 30 = 60°$ (ASL).

  b $\angle DOA = 2 \times \angle DEA = 60°$ (ACC). Thus $\angle OAD = ODA = 60°$ (Isosceles triangle), so $\triangle ODA$ is equilateral and $AD = AO = OD = r$.

  c Arc $AD = \frac{\pi r}{3}$, area of sector $AOD = \frac{\pi r^2}{6}$

15 a, b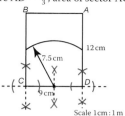

  c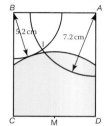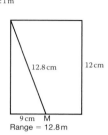

    Range = 12.8 m

  No. The point of intersection of the two arcs is not in the shaded region.

16 a 56 men and 48 women.

  b The numbers become 70 men and 60 women. $70:60 = 7:6$ so the ratio does not change.

  c 70 men and 60 women.

  d 52 men and 52 women.

17 7.3 inches

18 a 2.97% (3 sf) **b** £10 669 (nearest £)

19 a 1 can pair with any of the five remaining cards, 3 can pair with any of the four remaining cards (as the pair 1 and 3 has already been counted), and so on. Number of pairings $= 5 + 4 + 3 + 2 + 1 = 15$

  b i $\frac{2}{15}$ **ii** $\frac{4}{15}$

Chapter 13

Check in 13

1 a 1, 2, 3, 4, 6, 12

  b 1, 2, 3, 5, 6, 10, 15, 30

  c 1, 2, 3, 4, 5, 6, 8, 10, 12, 15, 20, 24, 30, 40, 60, 120

  d 1, 2, 3, 4, 5, 6, 8, 9, 10, 12, 15, 18, 20, 24, 30, 36, 40, 45, 60, 72, 90, 120, 180, 360

2 2, 3, 5, 7, 11, 13, 17, 19, 23, 29, 31, 37, 41, 43, 47, 53, 59, 61, 67, 71, 73, 79, 83, 89, 97

3 a 2^7 **b** 3^3 **c** 5^7 **d** 6^2

  e 7^3 **f** 4^4

4 a 1 **b** 1 **c** 5 **d** 0.5

13.1S

1 a 12 **b** 8 **c** 4 **d** 6

2 a 7×11 **b** 3×17 **c** 5×13 **d** 7×13

  e 7×17 **f** 13×17

3 9, 3, 3; $18 = 2 \times 3 \times 3$

4 a 225 **b** 40 **c** 63 **d** 360

  e 441 **f** 300

5 a $2^2 \times 3^2$ **b** $2^3 \times 3 \times 5$ **c** 2×17 **d** 5^2

  e $2^4 \times 3$ **f** $2 \times 3^2 \times 5$ **g** 3^3 **h** $2^2 \times 3 \times 5$

  i $3 \times 5 \times 7$ **j** $3^2 \times 11$ **k** 37 **l** 7×13

6 a $2^2 \times 263$ **b** $2^9 \times 5$

  c $2 \times 3^2 \times 5 \times 7$ **d** $3 \times 5^2 \times 11$

  e $5 \times 11 \times 13$ **f** $7 \times 11 \times 13$

  g 3×73 **h** 17^2

  i $2^3 \times 5 \times 71$ **j** $5 \times 7^2 \times 11$

k $7 \times 13 \times 19$ **l** $2 \times 3^2 \times 11 \times 17$

m $2^2 \times 11 \times 13 \times 17$ **n** $2 \times 5 \times 7 \times 13^2$

o $2^2 \times 23 \times 31$

7 a 12, 24, 36, 48, 60, 72; 9, 18, 27, 36, 45, 54

  b 36

8 a 20   **b** 36   **c** 30   **d** 60

  e 70   **f** 40

9 a 5   **b** 16   **c** 3   **d** 5

  e 14   **f** 15

10 a 48   **b** 800   **c** 66   **d** 416

  e 280   **f** 5040

11 a 1260, 60   **b** 8085, 7   **c** 1680, 48   **d** 9216, 2

  e 314706, 2   **f** 82944, 16

12 a Yes, the last digit is 5 so the number is divisible by 5 and the digits sum to a multiple of 3 so the number is divisible by 3.

  b Yes, the 6-digit number repeats a 3-digit sequence: $262(1000 + 1) = 262\,000 + 262$.

  c Yes, the digits sum to a multiple of 9 so it is divisible by 9 and the last digit is 5 so it is divisible by 5.

13.1A

1 9:40 am

2 120 seconds or 2 minutes

3 18 cm × 18 cm

4 a $30 = 2 \times 3 \times 5$, $105 = 3 \times 5 \times 7$, $45 = 3 \times 3 \times 5$ (many more solutions)

  b $210 = 2 \times 3 \times 5 \times 7$, $36 = 2 \times 2 \times 3 \times 3$, $81 = 3 \times 3 \times 3 \times 3$, $910 = 2 \times 5 \times 7 \times 13$, $46189 = 11 \times 13 \times 17 \times 19$ (many more solutions)

  c $108 = 2^2 \times 3^3$, $162 = 2 \times 3^4$, $243 = 3^5$, $200 = 2^3 \times 5^2$ (more solutions)

  d $64 = 2^6$ or $96 = 2^5 \times 3$

5 a $1815 = 3 \times 5 \times 11^2$

  b 15 cm × 11 cm × 11 cm; 55 cm × 3 cm × 11 cm; 33 cm × 5 cm × 11 cm; 121 cm × 3 cm × 5 cm

6 Let n be prime. Then n^3 is divisible by 1, n, n^2 and n^3 only.

7 a 9, 25 (any product of two odd numbers)

  b 2

  c i e.g. 5, 11, 17   **ii** e.g. 35

8 30 and 900, 60 and 450 or 90 and 300 are other possibilities

9 45

10 a It is true

  b The HCF is the product of the numbers in the intersection of the circles, the LCM is the product of the numbers in the union of the circles, and to find the product of the two numbers we multiply all of the numbers in one circle with all of the numbers in the other circle.

11 a 6, 1890   **b** 2, 34650   **c** 1, 510510

12 a $1 + 2 + 14 + 4 + 7 = 28$

  b $1 + 2 + 60 + 4 + 30 + 8 + 15 + 24 + 5 + 40 + 3 + 6 + 20 + 10 + 12 = 240 > 120$

  c $1 + 2 + 248 + 4 + 124 + 8 + 62 + 16 + 31 = 496$

13.2S

1 a 15.21   **b** 4.41   **c** 0.49   **d** 175.5625

2 a ±36.67   **b** ±6.21   **c** ±84.22   **d** ±15.32

3 a −157.46   **b** 970.30   **c** −0.001   **d** 4784.09

4 a 23   **b** −6   **c** −4.12   **d** 0.25

5 a $81 = 9^2$ **b** $125 = 5^3$ **c** $128 = 2^7$ **d** $100\,000 = 10^5$

  e $81 = 3^4$ **f** $343 = 7^3$

6 a 2   **b** 3   **c** 2   **d** 3

  e 5   **f** 6

7 a 6^5 **b** 4^9 **c** 2^{13} **d** 11^7

  e 1   **f** 7^{12}

8 a 7^2 **b** 8^4 **c** 3   **d** 9^3

  e 4^6 **f** 1

9 a 8^5 **b** 5^5 **c** 2^6 **d** 9^2

  e 8^8 **f** 7^7 **g** 4^{10} **h** 11

10 a 3   **b** 5^{10} **c** 4^2 **d** 7^2

11 a 4^2 **b** 6^2 **c** 9^2 **d** 8

  e 5^3 **f** 6^5

12 a $3^3 \times 4^4$ **b** $6^4 \times 5^6$ **c** $2^3 \times 5^2$ **d** $5^3 \times 8^5$

  e $2^2 \times 9^5$

13 a $5^2 \times 8^3$ **b** $6^2 \times 7^2$ **c** $5^2 \times 6^2$ **d** $5^3 \times 7^6$

  e $8^7 \div 3^2$ **f** $4^5 \times 5^2$ **g** $6^4 \times 7^2$ **h** $4^4 \times 7^5$

13.2A

1 a 6.999998354, 2.645751 is a rounded number and not exactly equal to $\sqrt{7}$

  b 56 and 57

2 15

3 a 27

  b Yes, using clues 1 and 2 or 2 and 3, but we need to know how old a teacher might reasonably be (e.g. not 1, 8 or 729)

4 a $6^4 = 2^6 = (2^3)^2 = 8^2$ and $64 = 2^6 = (2^2)^3 = 4^3$

  b $3^6 = (3^2)^3 = 9^3$ and $3^6 = (3^3)^2 = 27^2$

  c $4^6 = (4^2)^3 = 16^3$ and $4^6 = (4^3)^2 = 64^2$

5

		$\sqrt[3]{27}$	$\sqrt{20}$ $\sqrt[3]{125}$ $\sqrt[3]{210}$	$\sqrt{50}$	$\sqrt{75}$ $\sqrt[3]{999}$ $\sqrt{81}$
0	2		4	6	8 10

6 $a = 2$, $b = 3$ (many solutions)

7 $p = 3$, $q = 6$ (many solutions)

8 Kira needs to evaluate each term before adding them together. The correct answer is 27.

9 a $36 \times 216 = 6^2 \times 6^3 = 6^5 = 7776$

  b 1296

10 a $10^3, 10^6, 10^9, 10^{12}$ **b** 1024, 1048576

  c 24   **d** 48576   **e** 4.6%

  f Yes, since $(2^{10})^2 = 2^{10 \times 2} = 2^{20}$

11 a 16   **b** 2   **c** $-\frac{2}{3}$ **d** 2

12 $2 \times 4^{x+3} \div 2^{2x+4} = 2 \times 4^{x+3} \div 4^{x+2} = 2 \times 4^1 = 8$

13.3S

1 a Rational (an exact fraction).

  b Irrational (π has no repeating decimal patterns).

  c Rational (an exact fraction).

  d Irrational (3 is not a square number).

  e Rational (25 is a square number).

  f Irrational (10 is not a cube number).

2 a $\sqrt{6}$ **b** $\sqrt{15}$ **c** $\sqrt{231}$

3 a $\sqrt{2}\sqrt{7}$ **b** $\sqrt{3}\sqrt{11}$ **c** $\sqrt{3}\sqrt{7}$ **d** $\sqrt{5}\sqrt{7}$

4 a $2\sqrt{5}$ **b** $3\sqrt{3}$ **c** $7\sqrt{2}$ **d** 5

5 a $\sqrt{48}$ **b** $\sqrt{50}$ **c** $\sqrt{80}$ **d** $\sqrt{200}$

6 a 6   **b** 12   **c** 66

7 a 30   **b** $7\sqrt{7}$ **c** $8\sqrt{3}$

8 a 30; 18.520; 13.856   **b** 30; 18.520; 13.856

  c The answers are the same.

9 a $2\sqrt{3}$ **b** $2\sqrt{5}$ **c** 5   **d** $3\sqrt{7}$

10 a 2   **b** 5   **c** $3 + 3\sqrt{3}$

  d $2\sqrt{3} + 8$ **e** $3\sqrt{15} + 5\sqrt{3}$ **f** $(8 - 2\sqrt{5})\pi$

  g $7 + 3\sqrt{5}$ **h** $38 - 14\sqrt{7}$

11 a 5.66   **b** 3.24   **c** 4.24   **d** 106.10

12 a 20.9   **b** 44.7   **c** 12.3   **d** 32.2

13 a $4 + \sqrt{3}$ **b** $3 + \sqrt{2}$

  c $7 + 2\sqrt{7}$ **d** $14 + \sqrt{17}$

14 a $\frac{\sqrt{2}}{2}$ **b** $\frac{\sqrt{3}}{3}$ **c** $\frac{\sqrt{7}}{7}$ **d** $\frac{\sqrt{6}}{6}$

  e $\frac{\sqrt{5}}{5}$ **f** $\frac{\sqrt{2}}{2}$ **g** $\frac{\sqrt{10}}{5}$ **h** $\frac{\sqrt{3}}{2}$

i $\frac{\sqrt{30}}{6}$ **j** $\frac{2\sqrt{10}}{5}$

15 a 3   **b** $\frac{7}{16}$ **c** $\frac{\sqrt{2}}{5}$ **d** $\frac{4 + 3\sqrt{2}}{10}$

13.3A

1 **a** $\sqrt{2} + \sqrt{3} \approx 1.414 + 1.732 = 3.146$
   b $\sqrt{10} = \sqrt{2}\sqrt{5} \approx 1.414 \times 2.236 = 3.162$
   c $\sqrt{125} = (\sqrt{5})^3 \approx (2.236)^3 = 11.18$
   d $\sqrt{120} = \sqrt{2^3} \times \sqrt{3} \times \sqrt{5} \approx 1.414^3 \times 1.732 \times 2.236$
     $= 10.95$

2 Students' suggestions.

3 **a** $2 + 4\sqrt{5}, 5 + \sqrt{5}$ **b** $4, 2\sqrt{3} - 3$

4 **a** $7 + 3\sqrt{5}$ **b** $1 + \sqrt{3}$ **c** $38 - 14\sqrt{7}$
   d 23   **e** 16   **f** -26

5 **a** $\sqrt{64}$ and $\sqrt{100}$ **b** $2 - \sqrt{5}$ and $2 + \sqrt{5}$
   c $\sqrt{5}$ and $\sqrt{20}$ **d** $\sqrt{10}$ and $\sqrt{100}$
   e $3 + \sqrt{5}$ and $3 - \sqrt{5}$ **f** $8\sqrt{2}$ and $3\sqrt{2}$
   g $8\sqrt{2}$ **h** $2 - \sqrt{5}$

6 **a** $\frac{-1 - 2\sqrt{3}}{11}$ **b** $\frac{5 - \sqrt{5}}{4}$ **c** $2\sqrt{2} - 3$ **d** $16 + 11\sqrt{2}$

7 **a** $\frac{6 + 3\sqrt{3}}{4}$ **b** $\frac{5 - 3\sqrt{7}}{3}$ **c** $\frac{30 + 10\sqrt{2}}{9}$ **d** $\frac{20 + 11\sqrt{5}}{9}$

8 $2(10 + 11\sqrt{3})$

9 $5 - 2\sqrt{2}$

10 **a** $\frac{4\sqrt{3} - 6}{3} = -2 + \frac{4}{3}\sqrt{3}$
   b $\sqrt{5} - 5 < 0$ and the square of a real number cannot be negative.

Review 13

1 **a** 2, 3, 37, 101   **b** 1, 3, 15, 105   **c** 63, 105
2 **a** $3 \times 5 \times 7$ **b** 37   **c** $2^2 \times 3 \times 5^2$ **d** $2 \times 3^2 \times 7$
3 **a** **i** 35   **ii** 1   **b** **i** 39   **ii** 13
   c **i** 180   **ii** 12   **d** **i** 540   **ii** 6
4 **a** 5.48   **b** 3.56
5 **a** 4   **b** 5   **c** 64   **d** 81
6 **a** 7^4 **b** 3^9 **c** 3^3 **d** 7^{21}
   e 5
7 **a** $6\sqrt{3}$ **b** $2\sqrt{3} + 2\sqrt{2}$ **c** 10   **d** $2\sqrt{3}$
   e 6   **f** $\frac{\sqrt{2}}{2}$ **g** $13 - 7\sqrt{3}$ **h** 3
8 **a** $\frac{\sqrt{7}}{7}$ **b** $\frac{\sqrt{6}}{2}$ **c** $\frac{\sqrt{5}}{2}$ **d** $\frac{2\sqrt{6} + 3}{3}$
   e $\frac{6 - 5\sqrt{2}}{8}$ **f** $\frac{12 + 5\sqrt{6}}{27}$

Assessment 13

1 Isa, $2^3 \times 3^2 \times 5^2 \times 11$.
2 **a** $16 = 3 + 13 = 5 + 11$
   b $64 = 3 + 61 = 5 + 59 = 11 + 53 = 17 + 47 = 23 + 41$
   c Odd square number = even + odd, 2 is the only even prime, there is only one answer.
3 23
4 **a** $41 - 0 + 0^2 = 41, 41 - 3 + 3^2 = 47, 41 - 6 + 6^2 = 71$ are all prime.
   b $41, 41 - 41 + 41^2 = 41^2$.
5 **a** No, HCF $= 2 \times 3^3 = 54$.
   b $96 = 2^5 \times 3, 270 = 2 \times 3^3 \times 5$, LCM $= 2^5 \times 3^3 \times 5 = 4320$.
6 6
7 **a** HCF $= 6$ **b** HCF $= 84$
   c Yes. Students' answers, for example HCF (84, 228, 504) $= 12$
8 9 pm
9 **a** Saquib, 3^5.   **b** Gino, 14^6.
10 **a** $p = 5$ **b** $q = 2$
11 **a** **i** 45   **ii** $15\sqrt{3} + 6\sqrt{21}$
   iii $25\sqrt{3} + 10\sqrt{21}$ **2**
   b $90 + 80\sqrt{3} + 32\sqrt{21}$ **c** $10 + 26\sqrt{3} + 4\sqrt{7}$
   d $45(5 + 2\sqrt{7})$
12 **a** $2\sqrt{3}, \sqrt{6} = \sqrt{2}\sqrt{3}$.   **b** $120, \sqrt{36} = 6$.
   c $8\sqrt{3} + 19$, there are two $4\sqrt{3}$ terms.

d $10 - 4\sqrt{6}$, there are two $-2\sqrt{6}$ terms.
e $18 + 4\sqrt{3}$, the $\sqrt{3}$ terms don't cancel.
f $2, \sqrt{5} \times \sqrt{-5} = -5$
g $-14 - 28\sqrt{22}, 12\sqrt{64} = 12 \times 8 = 96$, and the $\sqrt{11}$ terms don't cancel.

13 a $\frac{4\sqrt{5}}{5}$ **b** $\sqrt{8}$ **c** $\frac{\sqrt{2} - 2}{6}$ **d** $\frac{4\sqrt{5} - 3\sqrt{10}}{5}$
   e $\frac{15\sqrt{7} - 7\sqrt{3}}{35}$ **f** $\frac{1}{2}$

14 a $h^2 = (4 + \sqrt{2})^2 + (4 - \sqrt{2})^2 = 16 + 8\sqrt{2} + 2 + 16 - 8\sqrt{2} + 2 = 36, h = 6.$
   b $P = 6 + 4 + \sqrt{2} + 4 - \sqrt{2} = 14, A = \frac{1}{2}(4 + \sqrt{2})(4 - \sqrt{2})$
     $= \frac{1}{2}(16 - 4\sqrt{2} + 4\sqrt{2} - 2) = 7$

Chapter 14

Check in 14

1 **a** **i** 9   **ii** 4   **b** **i** 27   **ii** -8
   c **i** 18   **ii** 8   **d** **i** 30   **ii** -10
   e **i** 24   **ii** 14   **f** **i** 60   **ii** -20
   g **i** 63   **ii** 8   **h** **i** 15   **ii** -30
2 **a** Straight line passing through (3, 1) and (3, 3).
   b Straight line passing through (1, −2) and (3, −2).
   c Straight line passing through (1, 3) and (3, 7).
   d Straight line passing through (1, 4) and (3, 2).
3 **a** $n = 2$ **b** $m = 1$ **c** $p = \frac{1}{2}$
4 **a** **i** 3   **ii** 4   **iii** Positive slant
   b **i** -4 **ii** 10   **iii** Negative slant
   c **i** 4   **ii** 5   **iii** Positive slant
   d **i** 2   **ii** 7.5   **iii** Positive slant
   e **i** 0   **ii** 7   **iii** Horizontal
   f **i** 0.5   **ii** 2   **iii** Positive slope

14.1S

1 **a** **i** 1; 2   **ii** $y = x + 2$
   b **i** 2; 3   **ii** $y = 2x + 3$
   c **i** 3; 0   **ii** $y = 3x$
   d **i** -2; 3   **ii** $y = -2x + 3$
2 $y = 3x + 5; y = 5x - 2; y = -2x + 7; y = \frac{1}{2}x + 9; y = -\frac{1}{4}x - 3;$
   $y = 4; y = x$
3 **a** $y = -x + 5$ **b** $y = x + 3$
   c $y = -x - 2$ **d** $y = x - 3$
   e $y = -2x + 6$ **f** $y = -5x + 9$
   g $y = -3x - 2$ **h** $y = 2x + 5$
   i $y = -\frac{1}{2}x + 2$ **j** $y = \frac{1}{2}x + 4$
   k $y = -\frac{1}{2}x + 4$ **l** $y = -3x + 4$
4 **a** $y = 6x + 2$ **b** $y = 5 - 2x$
   c $y = \frac{1}{2} - x$ **d** $y = -3x - 4$
5 **a** $y = 2x + 7$ (any solution of the form $y = 2x + c$)
   b $y = -5x + 1$ (any solution of the form $y = -5x + c$)
   c $y = -\frac{1}{4}x + 4$ (any solution of the form $y = -\frac{1}{4}x + c$)
   d $y = -4x + 2$ (any solution of the form $y = -4x + c$)
   e $y = \frac{3}{4}x + 7$ (any solution of the form $y = \frac{3}{4}x + c$)
   f $y = \frac{9}{2}x + 4$ (any solution of the form $y = \frac{9}{2}x + c$)
6 **a** $y = -4x - 2$ **b** $y = \frac{3}{2}x - 2$
7 **a** $y = -\frac{1}{2}x + 2$ (any solution of the form $y = -\frac{1}{2}x + c$)
   b $y = \frac{1}{5}x + 2$ (any solution of the form $y = \frac{1}{5}x + c$)
   c $y = 4x + 2$ (any solution of the form $y = 4x + c$)
   d $y = \frac{1}{4}x + 2$ (any solution of the form $y = \frac{1}{4}x + c$)
   e $y = -\frac{4}{3}x + 2$ (any solution of the form $y = -\frac{4}{3}x + c$)
   f $y = -\frac{2}{9}x + 2$ (any solution of the form $y = -\frac{2}{9}x + c$)
8 C
9 **a** $y = 19 - x$ **b** $y = -\frac{1}{2}x + \frac{17}{2}$
   c $y = 2x + 4$ **d** $y = \frac{3}{2}x + 5$
10 **a** 4   **b** 2   **c** $\frac{1}{3}$

11 a $y = 4x - 7$ **b** $y = 2x - 4$ **c** $y = \frac{1}{3}x + \frac{1}{3}$

14.1A

1 a A, B, E   **b** C, F   **c** D   **d** A
e C, F

2 a $y = 6x + 50$. The gradient, m, tells you that for every year older you get, you grow 6 cm, and the intercept, c, tells you that at birth you are 50 cm long.

b $y = \frac{1}{2}x + 10$. The gradient, m, tells you that for every year older you get you have to have another half a driving lesson, and it is not sensible to interpret the intercept, c, because it says that a newborn baby would need 10 lessons to pass the driving test!

3 The gradient tells you that for every mark more scored on Paper 1 another mark was scored on Paper 2. It does not make sense to interpret the intercept because it says that if you scored nothing in Paper 1 you would score -10 in Paper 2, and this is impossible.

4 a $y = -2x + 21$ **b** $y = \frac{3}{2}x + \frac{55}{4}$
c $y = \frac{6}{17}x - \frac{63}{34}$

5 $\frac{t}{3}$

6 False

7 $y = -\frac{1}{2}x + 4$

14.2S

1 a Straight line passing through (5, 0) and (5, 2).
b Straight line passing through (0, 2) and (2, 2).
c Straight line passing through (1.6, 0) and (1.6, 2).
d Straight line passing through (0, −3) and (2, −3).
e Straight line passing through (0, 1) and (2, 1).
f Straight line passing through (−1.25, 0) and (−1.25, 2).

2 a Straight line passing through (0, −2) and (1, 1).
b Straight line passing through (0, 4) and (2, 0).
c Straight line passing through (0, 3) and (−6, 0).
d Straight line passing through (0, 5) and (5, 0).

3 a Straight line passing through (0, 4) and (4, 0).
b Straight line passing through (0, −3) and (1.5, 0).
c Straight line passing through (0, 4) and (2, 0).
d Straight line passing through (0, 1.5) and (3, 0).

4 a Straight line passing through (0, −3) and (1, 2).
b Straight line passing through (0, 0.5) and (2, 1).
c Straight line passing through (0, 3) and (1, −1).
d Straight line passing through (0, −2) and (3, 0).

5 a No. If $x = 3$ then $y = (2 \times 3) + 1 = 7 \neq 8$.
b (3, 7) (many solutions possible)

6 a Straight line passing through (0, −1) and (1, 2), straight line passing through (0, 2) and (1, 5).
b The lines are parallel because they have the same gradient, so never intersect.
c $y = 3x + 7$ (any line of the form $y = 3x + c$)

7

Straight line	Parabola
$y = 3x - 2$	$y = x^2 - 2$
$y = x$	$y = 10 + x^2$
$3x + 2y = 8$	$y = x^2 + 2x + 1$
Any line of the form $y = mx + c$ for example, $y = 3$.	Any line of the form $y = ax^2 + bc + c$ for example, $y = x^2$.

8 a, c ∪-shaped parabola with minimum at (0, −2) passing through (−2, 2) and (2, 2).
b x^2: 9, 4, 1, 0, 1, 4, (9), 16.
$y = x^2 - 2$: 7, 2, −1, −2, −1, 2, (7), 14.

9 a i

x	−4	−3	−2	−1	0	1	2	3	4
x^2	16	9	4	1	0	1	4	9	16
$y = x^2 + 3$	19	12	7	4	3	4	7	12	19

ii and **iii** ∪-shaped parabola passing through (−1, 4) and (1, 4).
iv (0, 3)

b i

x	−4	−3	−2	−1	0	1	2	3	4
x^2	16	9	4	1	0	1	4	9	16
$y = 2x^2$	32	18	8	2	0	2	8	18	32

ii and **iii** ∪-shaped parabola passing through (−1, 2) and (1, 2).
iv (0, 0)

c i

x	−4	−3	−2	−1	0	1	2	3	4
x^2	16	9	4	1	0	1	4	9	16
$3x^2$	48	27	12	3	0	3	12	27	48
$y = 3x^2 - 1$	47	26	11	2	−1	2	11	26	47

ii and **iii** ∪-shaped parabola passing through (−1, 2) and (1, 2).
iv (0, −1)

d i

x	−4	−3	−2	−1	0	1	2	3	4
x^2	16	9	4	1	0	1	4	9	16
$y = x^2 + x$	12	6	2	0	0	2	6	12	20

ii and **iii** ∪-shaped parabola passing through (−1, 0) and (0, 0).
iv (−0.5, −0.25)

10 a x^2: 4, 1, 0, 1, 4, 9; $y = 3$, 1, 1, 3, 7, 13
b ∪-shaped parabola passing through (1, 3) and (−2, 3).
c $y = \frac{3}{4}, x = -\frac{1}{2}$

11 B, C and D

14.2A

1 a ∩-shaped parabola passing through (−2, 0), (0, 10) and (5, 0).
b ∪-shaped parabola passing through (0, 9), (3, 0) and (6, 9).
*c ∪-shaped parabola passing through (−1, 0), (0, −5) and (2.5, 0).

2 a $20x$: 0, 20, 40, 60, 80, 100. $4x^2$: 0, 4, 16, 36, 64, 100.
y: 0, 16, 24, 24, 16, 0.
b ∩-shaped parabola passing through (0, 0), (2.5, 25) and (5, 0).
c i 25 m, 2.5 s   **ii** 0.70 s, 4.3 s   **iii** $0.92 < t < 4.1$

3 a 100, 480, 660, 640, 420, 0, −620. ∩-shaped parabola passing through (0, 100) and (50, 0).
b 676 cm   **c** 50 m

4 a −1.6, 2.6   **b** −0.6, 1.6   **c** −1, 3

5 a −1.2, 3.2   **b** 0, 2   **c** 1   **d** −0.3, 3.3
e −1.6, 2.6   **f** −1, 3

*6 a −2.6, 2.6   **b** −1.3   **c** 0, 0.5   **d** 1.7

14.3S

1 a $y = x^2$: ∪-shaped parabola passing through (−2, 4), (0, 0) and (2, 4).
$y = x^2 - 2$: ∪-shaped parabola passing through (−2, 2), (0, −2) and (2, 2).
$y = -x^2$: ∩-shaped parabola passing through (−2, −4), (0, 0) and (2, −4).
b $y = -x^2$ **c** $y = x^2 - 2$

2 a ∩-shaped parabola passing through (−2, 0), (3, 5) and (8, 0).
b Maximum

3 a ∩-shaped parabola passing through (−3, 0), (0, 6) and (1, 0).
b Maximum

4 a Minimum   **b** One root

5 a ∪-shaped parabola passing through (0, 1) and (4, −3).
b $x \approx 0.5$ or 7.5

6 a $x = 0$ or −7   **b** $x = 0$ or 8
c $x = 0$ or −5   **d** $x = 0$ or −2
e $x = 0$ or −4   **f** $x = 0$ or $-\frac{5}{2}$
g $x = 0$ or −3   **h** $x = 0$ or $-\frac{3}{5}$

7 a $x = -1$ or −6   **b** $x = 1$ or 2
c $x = -2$ or −3   **d** $x = 3$ or −4
e $x = 3$ or −2   **f** $x = 1$ or 4
g $x = -3$ or 5   **h** $x = -2$ or −7

8 a $x = -2$ **b** $x = -4$ or 4
c $x = -2$ or 2   **d** $x = -5$

9 a $y = (x + 3)^2 - 6$ **b** (0, 3)
c (−3, −6)

d ∪-shaped parabola passing through (−1, −2), (−3, −6) and (−5, −2).

e Two roots

10 a i $y = (x + 1)^2 - 4$ **ii** (−1, −4)

   iii ∪-shaped parabola passing through (−3, 0), (−1, −4), and (1, 0).

   iv Two roots

 b i $y = (x + 1)^2 + 2$ **ii** (−1, 2)

   iii ∪-shaped parabola passing through (−3, 6), (−1, 2), and (1, 6).

   iv No roots

 c i $y = (x + 4)^2 - 4$ **ii** (−4, −4)

   iii ∪-shaped parabola passing through (−6, 0), (−4, −4), and (−2, 0).

   iv Two roots

 d i $y = (x - 3)^2 + 3$ **ii** (3, 3)

   iii ∪-shaped parabola passing through (1, 7), (3, 3), and (5, 7).

   iv No roots

 e i $y = (x + \frac{5}{2})^2 - \frac{37}{4}$ **ii** $(-\frac{5}{2}, -\frac{37}{4})$

   iii ∪-shaped parabola passing through (−5, −3), (−2.5, −9.25), and (0, −3).

   iv Two roots

 f i $y = (x - 3)^2$ **ii** (3, 0)

   iii ∪-shaped parabola passing through (0, 9), (3, 0), and (6, 9).

   iv One root

 g i $y = -(x - 5)^2 + 24$ **ii** (5, 24)

   iii ∩-shaped parabola passing through (0, −1), (5, 24), and (10, −1).

   iv Two roots

 h i $y = -(x - \frac{3}{2})^2 + \frac{37}{4}$ **ii** $(\frac{3}{2}, \frac{37}{4})$

   iii ∩-shaped parabola passing through (−1, 3), (1.5, 9.25), and (4, 3).

   iv Two roots

*11 a $y = 3(x + 1)^2 - 5$ **b** (0, −2)   **c** (−1, −5)

 d ∪-shaped parabola passing through (−2, −2), (−1, −5) and (0, −2).

 e Two roots

12 −20; the function has no roots since $\sqrt{-20}$ has no real solutions

13 $\sqrt{100} = \pm 10$ so the function has two roots

14 17 m

14.3A

1 a

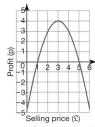

 b £1 and £5   **c** £3, 4 pence

2 a £225, £1250   **b** £200, £250

3 a Sketch the graph of the function

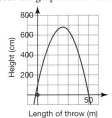

 b 676 cm   **c** 50 m

4 a 50 cm   **b** 62.5 cm

5 a

 b 44.1 metres   **c** 3 seconds

6 a

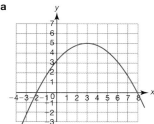

 b Maximum

7 a

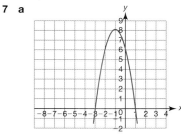

 b Maximum

8 a Minimum   **b** 1

9 a

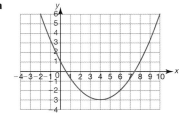

 b $x \approx 0.5, x \approx 7.5$

*10 a $\frac{1}{20}(10)^2 + 5 = 5 + 5 = 10$

 b Multiple solutions possible, e.g. (20, 25), (30, 50)

 c, d Students' research.

14.4S

1 a 13.3 km/h (2 dp)   **b** 40 km/h

2 a 6.1 km/h (1 dp)   **b** 10.5 km/h (1 dp)

3 a 2:00 pm and 2:30 pm because the gradient of the line is steepest here

 b 46.7 km/h (1dp)   **c** 35 km/h

 d The velocity is negative (though the speed is not).

4 a 6 m/s   **b** 1 and 12 seconds

 c 10 m/s²   **d** 115 m

5 a 2 m/s²   **b** 18 m   **c** 143.5 m

*6 Between 0 and 2 seconds there is non-uniform acceleration to 2 m/s, then between 2 and 5 seconds there is acceleration of 0.6 m/s² to 4 m/s, and between 5 and 15 seconds there is deceleration of 0.4 m/s² to rest.

7 a Line 1: straight line segment passing through (11:00, 0) and (12:00, 65). Line 2: straight line segment passing through (11:30, 0) and (12:00, 75).

 b 11:53 am

14.4A

1 In the following distance-time graphs distance (m) is on the y-axis and time (s) is on the x-axis.

In the following acceleration-time graphs acceleration (m/s²) is on the y-axis and time (s) is on the x-axis.

a i Between (0, 0) and (6, 30) curve with increasing gradient. Between (6, 30) and (16, 130) straight line segment. Between (16, 130) and (20, 150) curve with decreasing gradient.

ii Between (0, $1\frac{2}{3}$) and (6, $1\frac{2}{3}$) straight line segment. Between (6, 0) and (16, 0) straight line segment. Between (16, −2.5) and (20, −2.5) straight line segment.

b i Between (0, 0) and (6, 36) curve with increasing gradient. Between (6, 36) and (12, 72) curve with decreasing gradient.

ii Between (0, 2) and (6, 2) straight line segment. Between (6, −2) and (12, −2) straight line segment.

c i Between (0, 0) and (4, 48) straight line segment. Between (4, 48) and (6, 66) curve with decreasing gradient. Between (6, 66) and (9, 84) straight line segment. Between (9, 84) and (18, 111) curve with decreasing gradient.

ii Between (0, 0) and (4, 0) straight line segment. Between (4, −3) and (6, −3) straight line segment. Between (6, 0) and (9, 0) straight line segment. Between (9, $\frac{-2}{3}$) and (18, $\frac{-2}{3}$) straight line segment.

2 a Both cyclists are the same distance from a point.

b Joanna's average speed is 8 km/h and Chris's average speed is 15 km/h

c Joanna's average velocity is 8 km/h and Chris's average velocity is −15 km/h

3 a 111 m (3sf)

b A: Between (0 s, 111.1 m) and (10 s, 0 m) curve segment with increasing gradient.
B: Between (0 s, 111.1 m) and (10 s, 0 m) curve segment with decreasing gradient.

c A: A straight line through the points (0 s, 2.22 m/s²) and (10 s, 2.22 m/s²)
B: A straight line through the points (0 s, −2.22 m/s²) and (10 s, −2.22 m/s²)

4 a The second and fourth sections of the graph are impossible: they show the object going back in time

b The first section of the graph is impossible: it shows negative distance travelled

c The second section of the graph is impossible: it shows an instantaneous change in velocity

d The second and fourth sections of the graph are impossible: they show the object going back in time.

5 Initially Sara runs faster than Becky (8 m/s against 5 m/s). After 20 seconds Becky stops for a 5 second rest. 25 seconds in to the race Becky starts running again at 13.3 m/s and Sara slows her pace to 5 m/s. 37 seconds into the race Becky overtakes Sara. 40 seconds in to the race Becky slows her pace to 6.7 m/s and runs for a further 15 seconds. 45 seconds into the race Sara increases her pace to 6.7 m/s and runs for a further 15 seconds. Becky wins the race by 5 seconds.

Review 14

1 a 2.5   **b** $y = 2.5x + 2$

2 $y = -3x + 5$

3 a Gradient −2, y-intercept 7   **b** Gradient 1, y-intercept 9
   c Gradient −1, y-intercept 3   **d** Gradient $-\frac{2}{3}$, y-intercept $\frac{5}{3}$

4 a $y = 5x + 6$ **b** $y = -\frac{1}{5}x + 1$

5

6 a

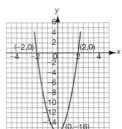

b

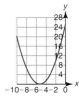

c

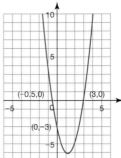

d

7 a (0, −6)   **b** (−2, 0), (1.5, 0)
c $x = -2, 1.5$ **d** −0.25

8 a i $(x + 1)^2 - 6$ **ii** $2(x - 3)^2 - 13$
   iii $-\left(x - \frac{5}{2}\right)^2 + \frac{25}{4}$

b i (−1, −6), min   **ii** (3, −13), min
   iii (2.5, 6.25), max

9 a 50 m   **b** 3.33 m/s   **c** 21.2 km/h
   d Increasing/accelerating.   **e** 2 m/s²

Assessment 14

1 a A: 2   B: $\frac{3}{2}$ C: $-\frac{5}{3}$ D: $-\frac{2}{3}$
   b i C   **ii** B   **iii** A   **iv** D
   c i $\frac{20}{3}$ **ii** $\frac{9}{2}$ **iii** 0   **iv** 0

2 a 2, 7   **b** 4, 9   **c** 6, −11   **d** −4, 12
   e $\frac{4}{7}$, −2   **f** $-\frac{15}{14}, \frac{35}{14}$

3 a $y = -5x - 61$ **b** $y = 3x - \frac{1}{2}$
   c $y = \frac{1}{3}x - 2$ **d** $y = \frac{1}{4}x - \frac{3}{4}$

4 a 1   **b** −8   **c** −1   **d** $-\frac{1}{2}$

5 a $y = 7x + 5$ **b** $y = -3x - \frac{29}{4}$
   c $y = -\frac{1}{3}x + \frac{8}{3}$

6 a A, H, J   **b** I and B or G. D and A, H, or J
   c I   **d** B, G   **e** D, H   **f** C, E

7 a (3, 4), $-\frac{1}{4}$ **b** $y = 4x - 8$

8 a $7y = 780 - 25x$ **b** 'decreases', '3.57'
   c Yes, it lies far from the line of best fit.

9 a

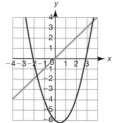

b (−1.6, −1.6) and (3.6, 3.6)
c −2, 3
d −2, 3. The estimate was accurate.

Answers

10a 8.7, 9.3, 10, 10.7, 11.3, 12, 12.7, 13.3, 14, 14.7

b

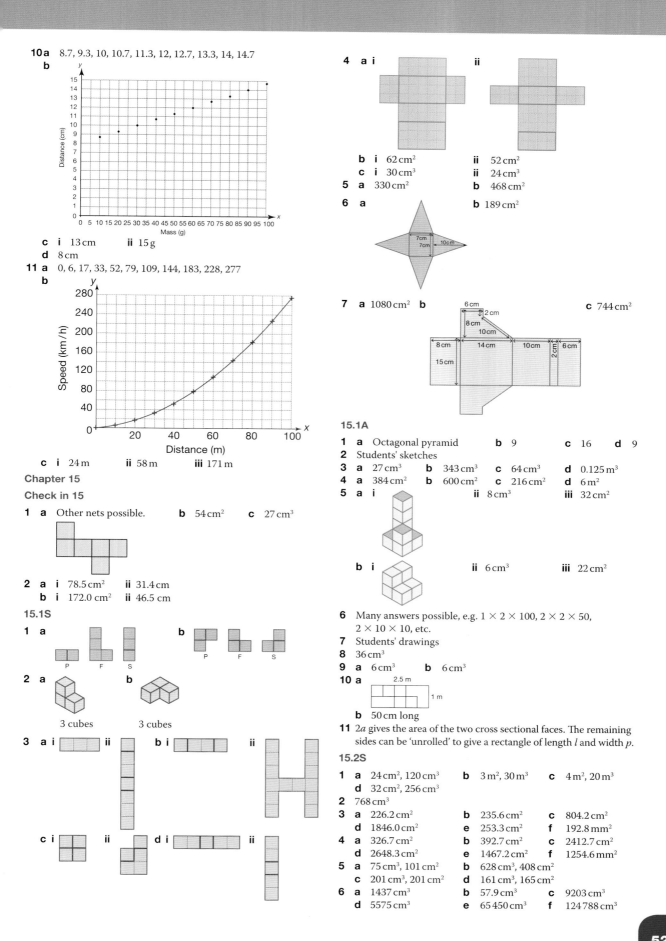

c i 13 cm   **ii** 15 g

d 8 cm

11 a 0, 6, 17, 33, 52, 79, 109, 144, 183, 228, 277

b

c i 24 m   **ii** 58 m   **iii** 171 m

Chapter 15

Check in 15

1 a Other nets possible.   **b** 54 cm²   **c** 27 cm³

2 a i 78.5 cm²   **ii** 31.4 cm
   b i 172.0 cm²   **ii** 46.5 cm

15.1S

1 a   **b**

2 a   **b**

   3 cubes   3 cubes

3 a i   **ii**   **b i**   **ii**

   c i   **ii**   **d i**   **ii**

---

4 a i   **ii**

b i 62 cm²   **ii** 52 cm²
c i 30 cm³   **ii** 24 cm³

5 a 330 cm²   **b** 468 cm²

6 a   **b** 189 cm²

   7cm 7cm 10cm

7 a 1080 cm²   **b**   **c** 744 cm²

   6 cm 2 cm 8 cm 10 cm
   8 cm 14 cm 10 cm 2 cm 6 cm
   15 cm

15.1A

1 a Octagonal pyramid   **b** 9   **c** 16   **d** 9
2 Students' sketches
3 a 27 cm³   **b** 343 cm³   **c** 64 cm³   **d** 0.125 m³
4 a 384 cm²   **b** 600 cm²   **c** 216 cm²   **d** 6 m²
5 a i   **ii** 8 cm³   **iii** 32 cm²

   b i   **ii** 6 cm³   **iii** 22 cm²

6 Many answers possible, e.g. 1 × 2 × 100, 2 × 2 × 50,
   2 × 10 × 10, etc.
7 Students' drawings
8 36 cm³
9 a 6 cm³   **b** 6 cm³
10 a

   2.5 m
   1 m

   b 50 cm long
11 2*a* gives the area of the two cross sectional faces. The remaining
   sides can be 'unrolled' to give a rectangle of length *l* and width *p*.

15.2S

1 a 24 cm², 120 cm³   **b** 3 m², 30 m³   **c** 4 m², 20 m³
   d 32 cm², 256 cm³
2 768 cm³
3 a 226.2 cm²   **b** 235.6 cm²   **c** 804.2 cm²
   d 1846.0 cm²   **e** 253.3 cm²   **f** 192.8 mm²
4 a 326.7 cm²   **b** 392.7 cm²   **c** 2412.7 cm²
   d 2648.3 cm²   **e** 1467.2 cm²   **f** 1254.6 mm²
5 a 75 cm³, 101 cm²   **b** 628 cm³, 408 cm²
   c 201 cm³, 201 cm²   **d** 161 cm³, 165 cm²
6 a 1437 cm³   **b** 57.9 cm³   **c** 9203 cm³
   d 5575 cm³   **e** 65 450 cm³   **f** 124 788 cm³

g 21688 cm³   **h** 2226 cm³   **i** 9855 mm³
j 8.4 m³
7 a 615.8 cm²   **b** 72.4 cm²   **c** 2123.7 cm²
d 1520.5 cm²   **e** 7854.0 cm²   **f** 12076.3 cm²
g 3761.0 cm²   **h** 824.5 cm²   **i** 2222.9 mm²
j 20.0 m²

15.2A

1 a 5.96 g/cm³   **b** 66
2 1.36 km
3 160, the lengths given are accurate or rounded down.
4 Yes, 6 cm³ extra space.
5 45.3 cm
6 7 cm
7 785.4 cm³
8 a Cube   **b** Cube
9 a i 6 cm   **ii** 12 cm
   b 452.4 cm²   **c** 452.4 cm²
   d i $4\pi r^2$ **ii** $4\pi r^2$ **e** $4\pi r^2 = 2\pi r \times 2r$
10 a 1.64 kg
   b i Students' answers e.g. 75 cm × 16 cm × 4 cm
     ii 78.5%

15.3S1

1 a 48 cm³   **b** 75 m³   **c** 56 cm³   **d** 48 mm³
2 a 42 cm³   **b** 1.77 cm³   **c** 69.2 cm³   **d** 32 cm³
3 a 443.4 mm²   **b** 249.4 cm²   **c** 46.8 cm²   **d** 106 cm²
   e 138.5 cm²   **f** 197.2 cm²   **g** 85.1 cm²   **h** 94.5 cm²
   i 100 cm²
4 618.7 cm³, 422.4 cm²
5 8 cm
6 15 mm

15.3S2

1 a 50 cm³   **b** 88 m³   **c** 263.9 cm³   **d** 87.1 cm³
2 a 128.8 cm²   **b** 56.5 cm²   **c** 42.4 cm²   **d** 23.6 cm²
   e 221.0 cm²   **f** 784.6 cm²
3 a 207.3 cm²   **b** 84.8 cm²   **c** 106.0 cm²   **d** 43.2 cm²
   e 362.0 cm²   **f** 1357.2 cm²
4 a 264 cm²   **b** 126 cm²
5 a 188.5 cm²   **b** 82.4 cm²   **c** 316.5 cm²
6 a 15.7 cm²   **b** 15.7 cm²   **c** 68 cm²   **d** 184.4 cm²
   e 101.8 cm²   **f** 40.2 m²

15.3A

1 a Y   **b** 134 cm³
2 3 cm
3 a 93.5 mm²   **b** 4699 mm²
4 a 69.2 cm³   **b** 95 cm²
5 a i 490 cm³   **ii** 659.5 cm²   **b i** 1186 cm³   **ii** 706.9 cm²
   c i 1057.3 cm³   **ii** 704.5 cm²   **d i** 1982.1 cm³   **ii** 1151.6 cm²
   e i 638.4 cm³   **ii** 467.3 cm²   **f i** 159.2 cm³   **ii** 252.2 cm²
   g i 650.1 cm³   **ii** 510 cm²   **h i** 1985 cm³   **ii** 1020 cm²
6 a $\frac{7}{3}\pi r^2 h$ **b** $5\pi r^2 + 3\pi r\sqrt{h^2 + r^2}$

Review 15

1 a 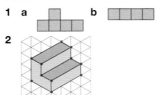 **b**

2

3 a 245 m³   **b** 273 m²   **c** 367.5 kg
4 a 9557 cm³   **c** 2532 cm²
5 a 1767 cm³   **b** 66.7 cm³
6 Volume = 9.42 cm³, Surface area = 102 cm²

---

7 a 6 cm   **b** 10 cm²

Assessment 15

1 a Triangle-based pyramid
   b

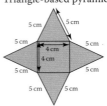

2 a **A**, the width of the front elevation is split into two values with ratio 3:1.
   b **B**, the height of the front elevation is split into two values with ratio 1:1.
3 a 512   **b** 64   **c** 8
   d i 18   **ii** 26
4 a 432
   b No, ratio of volumes = 432,
     ratio of surface areas = 51.2 (3 sf).
5 a Yes. 13.5 × 0.3 × 0.3 = 6 × 0.45 × 0.45 = 1.215 cm³
   b Fat chips, 11.205 cm² < 16.38 cm²
6 a i 21.2 cm³ (3 sf)   **ii** 70.4 cm³ (3 sf)
   b 168 cm² (3 sf)
7 a i 50.4 in²   **ii** 13.5 in³
   b 9.93 cm³
   c i 47 666.7 cm³ (1 dp)   **ii** 91
   d i 4.71 cm³ (3 sf)   **ii** 12.3 g (3 sf)
8 a 94.2 cm³ (3 sf)   **b** 104 cm³ (3 sf)
9 a 41.6 mm² (3 sf)   **b** 2700 mm³ (3 sf)
   c 2630 mm³
10 a 10 900 in³ (3 sf)   **b** 1020 in³ (3 sf)
   c 4520 in² (3 sf)
11 a 654 ml   **b** 3 hr 48 min   **c** 3 cm   **d** 141 ml
   e 513 ml   **f** 02:51
   g As the cross section of the cone decreases, the water level decreases at an increasingly faster rate. A cylinder.

Lifeskills 3

1 40
2 a

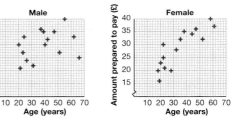

   b Males – no correlation. There is no observable connection between age and amount prepared to pay. Females – positive correlation. Older females are prepared to pay more than younger females.
   c No. Several reasons possible e.g. data may be unreliable due to a small sample size, sample may not give an indication of the relative populations of different age groups.

3 a

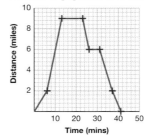

b 1:41 pm

c 41 minutes.

4 a

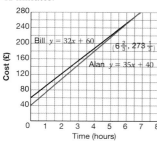

b $x > 6$ hours 40 minutes

c Multiple answers possible. If the quote is accurate Alan is cheaper, if it takes more than 10 minutes extra Bill is cheaper.

5 a 17241 cm³

b $30 \times 17240.7312 = 517\,220$
$70 \times 115 \times 180 = 1\,449\,000$
$\frac{517\,220}{1\,449\,000} \times 100 = 35.7\%$

c $1449000\,\text{cm}^3 = 1.449\,\text{m}^3$
$1\,\text{m}^3 = 1000\,000\,\text{cm}^3$

d 396.5 cm²

e 3.72 cm

f 18

g They offer the same cost per volume but the cartons allow more efficient use of space in the restaurant.

6 a £854

b

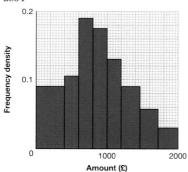

c

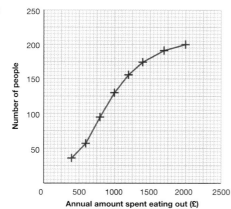

d Median ≈ 830, IQR ≈ 600.

e

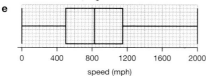

f The median is close to the estimate for the mean so supports this estimate.

g £213.50

Chapter 16

Check in 16

1 a 62   **b** 60   **c** 18.5   **d** 15
   e 7   **f** 30   **g** 108   **h** 150

2 a £50   **b** £70   **c** 2 days

16.1S

1 a 4, 9, 16, 11, 10   **b** 50

2 a 8, 6, 7, 10, 3, 6   **b** 40

3 a 8, 8, 13, 6, 5

b

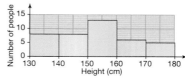

4

5

6 a i 2   **ii** 1

b 85–90 m   **c** 70–75 m   **d** 10 athletes

16.1A

1 a Class width: 2, 1, 1, 1, 3;
Frequency density: 6, 17, 19, 11, 6

b

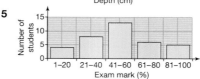

2 a Class width: 5, 5, 10, 20, 20, 40;
Frequency density: 1.2, 2, 2.3, 1.45, 1.2, 0.2

b

3 a Class width: 10, 20, 20, 30, 30;
Frequency 8, 16, 28, 39, 9
Frequency density: 0.8, 0.8, 1.4, 1.3, 0.3

b

4 a 10   **b** 4, 18, 53, 20, 10   **c** 105

5 a 5, 8, 11

b
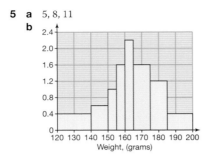

16.2S

1 a 1, 6, 8, 3, 4, 3   **b** 70 to 74   **c** 70 to 74
2 a 3   **b** 13, 18, 23, 28
   c 13, 108, 46, 28. Mean = 19.5 mph
3 a i $10 < t \leqslant 15$ **ii** $15 < t \leqslant 20$ **iii** 16.1
   b i $10 < t \leqslant 20$ **ii** $20 < t \leqslant 30$ **iii** 23.5
   c i $5 < t \leqslant 10$ **ii** $10 < t \leqslant 15$ **iii** 14.7
   d i $15 < t \leqslant 25$ **ii** $25 < t \leqslant 35$ **iii** 30
4 a $165 \leqslant h < 170$ **b** 164.3 cm   **c** $165 \leqslant h < 170$

16.2A

1 a December = 50, January = 59.7
   b December: modal $40 < m \leqslant 60$, median $40 < m \leqslant 60$.
       January: modal $40 < m \leqslant 60$, median $40 < m \leqslant 60$
   c Less variation in miles travelled in January, less short
       journeys.
       The most common journey length does not change.
       The mean is greater for January than December.
2 a Teachers = 23.1 miles, office workers = 35.3 miles
   b Teachers: $20 < t \leqslant 30$, office workers: $30 < t \leqslant 40$
   c On average, office workers take longer travelling home.
3 Range is 40 g, close to 39 g, but mean is 53.125 g, much higher
    than 45 g. Unlikely to be Granny Smith apples.
4 Yes. Modal and median class is $24.5 < w \leqslant 25.5$. Estimated
    mean = 25.28 g which is close to 25 g. However as data is
    grouped the weight of the packets could be less than 25 g.
5 Machine A estimated mean = 0.3007 mm, machine B
    estimated mean = 0.2907. The estimated mean indicates that
    machine A produces paper closest to the required thickness.

16.3S

1
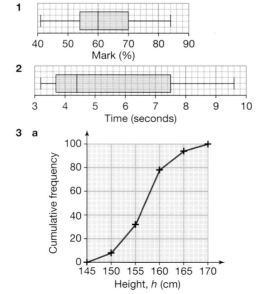

b
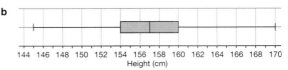

c i 17 girls   **ii** 12 girls

4 a
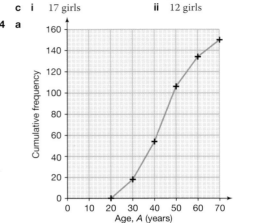

b

c i 37   **ii** 30

5 a
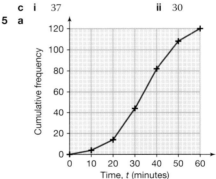

b
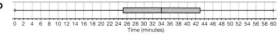

c i 30   **ii** 26

6 a
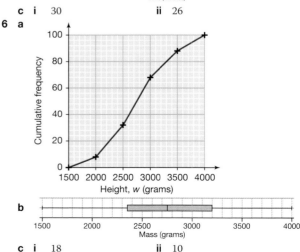

b

c i 18   **ii** 10

7 a

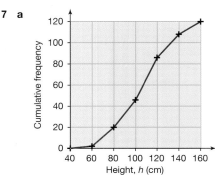

b

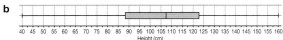

c i 98 **ii** 87

8 a

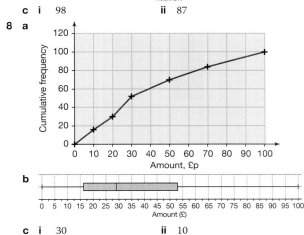

b

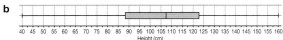

c i 30 **ii** 10

16.3A

1 The boys' results are higher than the girls', on average. The middle half of the girls' results is less varied than that of the boys. The range is the same for the girls and the boys.

2 Farmer Jenkins' sunflowers are shorter than Farmer Giles', on average. The middle half of Farmer Jenkins' sunflowers vary in height more than those of Farmer Giles. The range of heights of Farmer Jenkins' sunflowers is greater than that of Farmer Giles'.

3 Boys have higher mobile phone bills, on average. The middle half of the mobile phone bills varies the same for girls and boys. The range of mobile phone bills is the same for girls and boys.

4 On average, waiting times are higher at the dentist. The range of waiting times is greater at the doctor. The middle half of waiting times varies more at the doctor than at the dentist. The waiting times at the doctor are symmetrical, but those at the dentist are negatively skewed.

5 The average reaction time is the same for boys and girls. The range of reaction times is greater for girls. The middle half of reaction times varies less for girls than for boys. Reaction times for girls are symmetrical, but for boys they are negatively skewed.

6 On average, results are the same in the English and French tests. The ranges of results are the same. The middle half of results varies more in the English test. The English test results are negatively skewed, but the French test results are symmetrical.

7 On average, 17-year-old girls make longer phone calls than 13-year-olds. The range of the length of calls made is greater for 17-year-olds. The middle half of the calls made varies more for 13-year-olds than for 17-year-olds. The lengths of calls made by 13-year-olds is negatively skewed, but those for 17-year-olds are symmetrical.

1 a Negative correlation **b** No correlation
  c Positive correlation

2 a A: Poor exam mark, lots of revision;
    B: Very good exam mark, lots of revision;
    C: Very good exam mark, little revision;
    D: Poor exam mark, little revision;
    E: Average exam mark, average amount of revision

  b A: Not much pocket money, equal eldest;
    B: Lots of pocket money, equal eldest;
    C: Lots of pocket money, equal youngest;
    D: Not much pocket money, equal youngest;
    E: Average pocket money, middle age

  c A: Low fitness level, lots of hours in gym,
    B: Good fitness level, lots of hours in gym,
    C: Good fitness level, few hours in gym,
    D: Low fitness level, few hours in gym,
    E: Medium fitness level, medium hours in gym

3 a No correlation **b** Positive correlation
  c Negative correlation

4 a

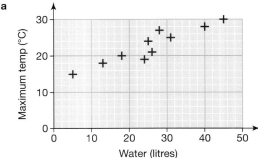

  b Positive correlation
  c i Increases **ii** Decreases

5 a

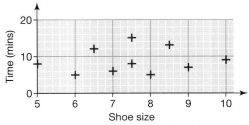

  b No correlation **c** No relationship

16.4A

1 a Positive correlation.
  b Paper 1 = 20, paper 2 = 80. Student performed poorly in paper 1 and well in paper 2.
  c 30
  d The line predicts that the student scores 110% which is impossible.

2 a 16 **b** 24
  c One student scored 20 marks on paper 1, but only 5 marks on paper 2

3 a, c Check students' diagrams.
  b Positive correlation.
  d 20 years
  e, f 1 year old gives a negative diameter. Predictions outside the range of data values can be unreliable.

4 a, c Check students' diagrams.
  b Positive correlation **d** 8°C
  e Predictions outside the range of data values can be unreliable.

16.5S

1

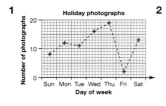

Holiday photographs

2

DVD rentals.

3

Peter's mass

4

Monthly sunshine

5

Television viewing figures

16.5A

1 a

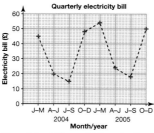

Monthly phone bill

b Typical phone bills are about £15. These fall during the summer months and peak in December

2 a

Quarterly electricity bill

b Electricity bills are highest in the winter months and lowest in the summer months. This annual pattern repeats itself; there is a slight trend for bills to rise from year-to-year.

3 a

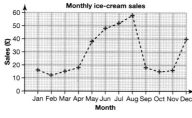

Monthly ice-cream sales

b Ice cream sales grow steadily during spring and summer but drop sharply in the autumn. Sales are low during autumn and winter except for a peak in December.

4 a

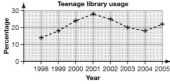

Teenage library usage

b The percentage of students using the library grow steadily from about 15% in 1998 to 28% in 2001. It then falls back to around 20%.

5 The average suggests Sell-a-lot. (116 is a valid range for both data sets.)

Review 16

1

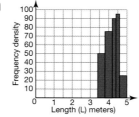

2 a $30 < s \leq 40$ **b** $30 < s \leq 40$
   c £35 750

3 a

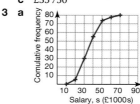

b i £26 000 **ii** £42 000 **iii** £33 000
   iv £16 000 **v** 80%

c

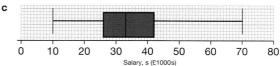

4 a, c

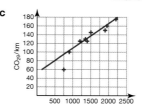

b Positive correlation, as engine capacity increases CO_2 emissions increase.

d Approximately 165 g/km
e This would be extrapolation, unreliable.

5 a

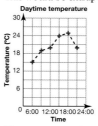

Daytime temperature

b The temperature increases to a peak at 18:00 and then begins to decrease.

Answers

Assessment 16

1 a

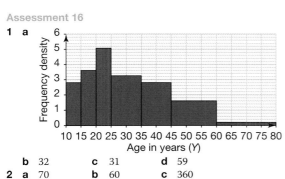

b 32 **c** 31 **d** 59

2 a 70 **b** 60 **c** 360

3 a 114

b

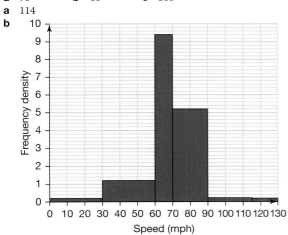

4 a i $25 \leqslant Y < 35$ **ii** $20 < \text{Age} \leqslant 30$ **iii** $70 < V \leqslant 90$
b i $25 \leqslant Y < 35$ **ii** $20 < \text{Age} \leqslant 30$ **iii** $60 < V \leqslant 70$
c i 32.9 years **ii** 30.3 years (3 sf) **iii** 68.8 mph (3 sf)

5 a, b

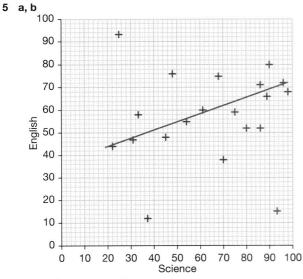

c Weak positive correlation
d (93, 15) shows a student who is very good at Science but poor at English, (25, 93) shows a student who is very good at English but poor at Science.
e i Approximately 44 **ii** Approximately 57

6 a 23 **b** 18 **c** 24
d The graph just shows a trend and there are no actual figures recorded at this time.
e 41 **f** Students' answers.

7 a

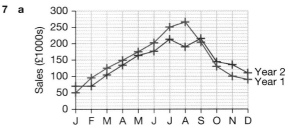

b Winter sales in Year 1 were lower than in Year 2. Summer sales in Year 1 were higher than in Year 2.

8 a

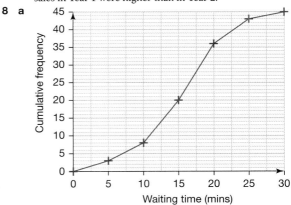

b i 8 **ii** 25
c i 16 **ii** 8
d

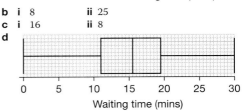

Chapter 17

Check in 17

1 a 7π **b** 13π **c** 36π **d** 4
 e $3\sqrt{2}$ **f** $\sqrt{3}$
2 a i 1, 2, 3, 6 **ii** 1, 2, 3, 4, 6, 12
  iii 1, 2, 4, 7, 14, 28 **iv** 1, 2, 3, 4, 6, 9, 12, 18, 36
 b 2, 3, 5, 7, 11, 13, 17, 19, 23, 29, 31, 37, 41, 43, 47
3 a 3^2 **b** 4^5 **c** 6^3 **d** 5^4
4 a 16 **b** 27 **c** 16 **d** 125
 e 128 **f** 7

17.1S

1 a 10 **b** 4 **c** 7 **d** 2
 e 11 **f** 12 **g** 2 **h** 3
 i 10 **j** 9 **k** 3 **l** 5
 m 0 **n** 10 **o** 4
2 a 2^{-1} **b** 5^{-1} **c** -7^{-1} **d** 2^{-1}
 e 10^{-1} **f** 3^{-1}
3 a 7^{-2} **b** 9^{-2} **c** 2^{-2} **d** 2^{-5}
 e 3^{-4} **f** 6^{-4}
4 a $\frac{1}{8^2}$ **b** $\frac{1}{7^3}$ **c** $\frac{1}{5^2}$ **d** $\frac{1}{9^4}$
 e $\frac{1}{3^2}$ **f** $\frac{1}{9^3}$
5 a i $\frac{1}{9}$ **ii** $0.\dot{1}$ **b i** $\frac{1}{8}$ **ii** 0.125
 c i $\frac{1}{100\,000}$ **ii** 0.00001 **d i** 1 **ii** 1
 e i $\frac{1}{8}$ **ii** 0.125 **f i** $\frac{1}{16}$ **ii** 0.0625

a 0.0625 **b** 0.25 **c** 1 **d** 2
e 4 **f** 16 **g** 64 **h** 0.5
i 0.015625
7 a $3^{-\frac{1}{2}}$ **b** $5^{-\frac{1}{2}}$ **c** $11^{-\frac{1}{2}}$
8 a 5 **b** 0.2 **c** 1 **d** 125
e 0.04 **f** 0.008 **g** 64 **h** 9
i 0.125 **j** 0.3 **k** 0.04 **l** 0.000 000 1
m 8 **n** 4 **o** 243 **p** 0.1
q 0.015625 **r** 100 **s** 0.05 **t** 2197
u 0.125
9 a $2^{\frac{1}{2}}$ **b** $2^{\frac{1}{6}}$ **c** $2^{\frac{9}{2}}$ **d** $2^{-\frac{2}{5}}$
e $2^{\frac{5}{2}}$ **f** 1
10 a 3 **b** 3^{-6} **c** 3^{-2} **d** 3^{-6}
e 3^{-1} **f** 3^{4}
11 a 5 **b** $5^{\frac{2}{3}}$ **c** $5^{\frac{2}{3}}$ **d** $5^{\frac{1}{6}}$
e 5^{-4} **f** 5^{-6} **g** 5^{3} **h** $5^{\frac{5}{2}}$
i $5^{-\frac{3}{2}}$ **j** $5^{-\frac{3}{5}}$ **k** $5^{-\frac{2}{5}}$ **l** $5^{-\frac{5}{6}}$
12 a $2^{-\frac{5}{4}}$ **b** $2^{\frac{1}{6}}$ **c** $2^{\frac{5}{2}}$ **d** $2^{-\frac{3}{4}}$
e $2^{-\frac{3}{8}}$ **f** $2^{-\frac{9}{2}}$
13 a i 4^{-1} **ii** $16^{-\frac{1}{2}}$
b i 4^{2} **ii** 16^{1}
c i 4^{-2} **ii** 16^{-1}
d i $4^{-\frac{1}{2}}$ **ii** $16^{-\frac{1}{4}}$
14 a 10^{-1} **b** $10^{\frac{1}{2}}$ **c** $10^{\frac{3}{2}}$ **d** $10^{-\frac{5}{2}}$
15 a 5^{-2} **b** 5^{-3} **c** $5^{-\frac{1}{3}}$ **d** $5^{\frac{2}{3}}$
e $5^{-\frac{3}{2}}$ **f** $5^{-\frac{4}{3}}$
16 a i $\frac{3}{2}$ **ii** $\frac{2}{3}$ **b i** $\frac{5}{2}$ **ii** $\frac{2}{5}$
c i $\frac{4}{3}$ **ii** $\frac{3}{4}$ **d i** $-\frac{3}{4}$ **ii** $-\frac{4}{3}$
17 $\frac{2}{3}$

17.1A

1 a 4^{-1} **b** 4^{2} **c** 4^{-2} **d** $4^{-\frac{1}{2}}$
2 a $16^{-\frac{1}{2}}$ **b** 16^{1} **c** 16^{-1} **d** $16^{-\frac{1}{4}}$
3 a 10^{-1} **b** $10^{\frac{1}{2}}$ **c** $10^{\frac{3}{2}}$ **d** $10^{-\frac{5}{2}}$
4 a 5^{-2} **b** 5^{-3} **c** $5^{-\frac{1}{3}}$ **d** $5^{\frac{2}{3}}$
e $5^{-\frac{3}{2}}$ **f** $5^{-\frac{4}{3}}$
5 a $6x^{\frac{2}{3}}$ **b** $27\,\text{cm}^3$ **c** No, $6x^2 = x^3 = 216$
6 a $\frac{1}{9}, \frac{1}{3}, 1, 3, 9$
b Points $(-1, \frac{1}{9})$, $(-\frac{1}{2}, \frac{1}{3})$, $(0, 1)$, $(\frac{1}{2}, 3)$, $(1, 9)$ joined by a smooth curve.
c Points $(-1, \frac{1}{4})$, $(-\frac{1}{2}, \frac{1}{2})$, $(0, 1)$, $(\frac{1}{2}, 2)$, $(1, 4)$ joined by a smooth curve. The gradient of $y = 9^x$ increases more rapidly than the gradient of $y = 4^x$, both curves pass through $(0, 1)$.
7 a i $25^{\frac{3}{2}}$ **ii** $125^{\frac{2}{3}}$ **b i** $4^{\frac{3}{2}}$ **ii** $32^{\frac{2}{5}}$
c i $27^{\frac{4}{3}}$ **ii** $81^{\frac{3}{4}}$ **d i** $16^{-\frac{3}{4}}$ **ii** $0.125^{-\frac{4}{3}}$
8 a 729 **b** 32 **c** 0.125 **d** 25
9 $\frac{2}{3}$
10 a -1 **b** 3 **c** -1 **d** $\frac{1}{6}$
e $-\frac{5}{2}$
11 a $1, 2^{\frac{2}{3}}, 2^{-\frac{2}{3}}, 2^{\frac{1}{2}}, 2^{\frac{1}{6}}, 2^{-\frac{5}{6}}$ **b** $2^{-\frac{3}{2}}, 2^{-\frac{5}{4}}, 2^{-\frac{1}{4}}, 2^{-\frac{1}{2}}, 2^{-\frac{3}{4}}, 2^{\frac{1}{2}}$
c Students' examples

17.2S

1 a $\frac{8}{15}$ **b** $\frac{29}{35}$ **c** $\frac{5}{56}$ **d** $\frac{44}{45}$
e $4\frac{7}{8} = 4.785$ **f** $\frac{59}{60}$ **g** $\frac{19}{40} = 0.475$ **h** $9\frac{17}{36}$
2 a $\frac{1}{2} = 0.5$ **b** $\frac{5}{3}$ **c** $\frac{25}{16} = 1.5625$ **d** $\frac{8}{15}$
e $\frac{14}{9}$ **f** $14\frac{3}{10} = 14.3$ **g** $2\frac{37}{55}$ **h** $26\frac{4}{9}$
3 No exact truncated decimal form exists if the denominator has factors other than 2 or 5
4 a $\frac{17}{30}$ **b** $1\frac{67}{87}$ **c** $\frac{1}{3}$ **d** $\frac{207}{1715}$
e $1\frac{181}{324}$ **f** $-\frac{8}{21}$
5 No. **a** $\frac{5}{4}$ **b** $\frac{5}{4}$ **c** $\frac{5}{4}$
d $\frac{5}{4}$ **e** $\frac{5}{4}$ **f** $\frac{7}{4}$

6 a 2π **b** 3π **c** $2 + 2\pi$ **d** 10π
e 18π **f** 16π
7 a 32π **b** $7 + 2\pi$ **c** $28 + 4\sqrt{2}$ **d** 62π
8 a 5.66 **b** 3.24 **c** 4.24 **d** 106.10
9 a $5\sqrt{5}$ **b** $3 - \sqrt{3}$ **c** $\frac{3}{2} + \frac{3\sqrt{3}}{4}$ **d** $\frac{5}{3} - \sqrt{7}$
e $\frac{30 + 10\sqrt{2}}{9}$ **f** $\frac{20 + 7\sqrt{5}}{9}$
10 a $5 + \sqrt{5}$ **b** $-3 + 5\sqrt{3}$ **c** $9 + 5\sqrt{3}$ **d** $11 - 6\sqrt{2}$
e 18 **f** $9 + 14\sqrt{5}$
11 a $\frac{\sqrt{15} - \sqrt{5} + \sqrt{3} - 1}{2}$ **b** $4 - 2\sqrt{3} + 2\sqrt{7} - \sqrt{21}$
c $\sqrt{11} - 3$ **d** $8 + 5\sqrt{3}$ **e** $\frac{-27 - 16\sqrt{3}}{3}$ **f** $\frac{18 + \sqrt{5}}{22}$

17.2A

1 $40(\pi + 2)\,\text{cm}, 800\pi\,\text{cm}^2$
2 $4(4 - \pi)\,\text{cm}^2$
3 a $x = \sqrt{5} + \pi, y = 2\sqrt{5} - \pi$
b $x = 2\sqrt{5} + \pi, y = 3\sqrt{5} - 2\pi$
c $x = 2\pi - \frac{12\sqrt{5}}{5}, y = \frac{13\sqrt{5}}{5} - 2\pi$
4 $10\sqrt{\pi}\,\text{cm}$
5 $\frac{5\sqrt{5}}{\pi}\,\text{cm}$
6 $\frac{25}{3} + 40\sqrt{3}\,\text{cm}^2$
7 $1\,\text{cm}$
8 $\frac{80\pi\sqrt{5}}{3}\,\text{mm}^3$
9 $4\sqrt{3}\,\text{cm}$
10 a height $= \sqrt{2^2 - 1^2} = \sqrt{3}$ **b** $3\,\text{cm}$ **c** $2\sqrt{3}\,\text{cm}$
11 $250\pi\,\text{cm}^3$
12 Laura, the equation implies $2n^2 = m^2$, m^2 is even so m is even. m can be written $m = 2k$ so by substitution $2n^2 = 4k^2$, $n^2 = 2k^2$ so n^2 is even, n is even.

17.3S

1 a 1 000 **b** 1 000 000 **c** 100 000
2 a 1 **b** 0.01 **c** 0.000 01
3 a 9×10^3 **b** 6.5×10^2 **c** 6.5×10^3 **d** 9.52×10^2
e 2.358×10^1 **f** 2.5585×10^2
4 a 3.4×10^{-4} **b** 1.067×10^{-1}
c 9.1×10^{-6} **d** 3.15×10^{-1}
e 5.05×10^{-5} **f** 1.82×10^{-2}
g 8.45×10^{-3} **h** 3.06×10^{-10}
5 $9 \times 10^3, 1.08 \times 10^4, 3.898 \times 10^4, 4.05 \times 10^4, 4.55 \times 10^4, 5 \times 10^4$
6 a 63 500 **b** 910 000 000 000 000 000
c 111 **d** 299 800 000
7 a 0.004 5 **b** 0.000 031 7
c 0.000 001 09 **d** 0.000 000 979
8 a 6×10^2 **b** 4.5×10^4 **c** 6.5×10^0 **d** 5×10^6
e 2.15×10^4 **f** 7×10^{13} **g** 1.22516×10^{20}
h 1.5×10^7 **i** 2.8×10^{-1} **j** 4×10^{-2} **k** 1.35×10^{-3}
l 1.2×10^{-7}
9 a 2.5×10^8 **b** 2.4×10^{13} **c** 5×10^{-1} **d** 9.2×10^{-8}
10 a 5×10^1 **b** 7.5×10^2 **c** 2×10^{-2} **d** 4.2×10^{-8}
11 a 6×10^7 **b** 4×10^{12} **c** 1.5×10^9 **d** 7×10^0
e 1.5×10^{10} **f** 1.2×10^{11}
12 a 2×10^{-2} **b** 1×10^{-1} **c** 1.5×10^{-8} **d** 2×10^3
e 2.5×10^{-6} **f** 3.1×10^4
13 a 5.2×10^{-1} **b** 4.6×10^{-2} **c** 2.09×10^{-2} **d** 1.3×10^4
14 a 7.74×10^{-3} **b** 9.63×10^5 **c** 4.38×10^{-5} **d** 2.55×10^2
e 3.4×10^5 **f** 4.47×10^{-2}

17.3A

1 3.33×10^5
2 10^{45}
3 500
4 1.79×10^6
5 107.7
6 $7.83 \times 10^7\,\text{km}, 3.775 \times 10^8\,\text{km}$
7 a $3.3 \times 10^{-9}\,\text{s}$ **b** $9.47 \times 10^{15}\,\text{m}$

8 $1.641 \times 10^{-27}\,\text{kg}$

9 a Jupiter, $2.668\,612 \times 10^{27}\,\text{kg}$ **b** 0.22%

10 Yes, ratio is roughly 1.93

11 a Venus' year $= \frac{2 \times \pi \times 1.08 \times 10^8}{1.26 \times 10^5}$ hours $= \frac{12\,000\pi}{7}$ hours $=$ 224 days $<$ 243 days

b If Venus follows a circular orbit then the distance it travels in one year is πr^2, eccentric orbits might be longer.

Review 17

1 a 9^{-1} **b** 5^{-7} **c** $12^{\frac{1}{3}}$ **d** $8^{-\frac{1}{2}}$

2 a $10\,000\,000$ **b** 125 **c** 1 **d** 729

e 10 **f** 8 **g** $\frac{1}{5}$ **h** $\frac{1}{36}$

i 8 **j** 13 **k** 9 **l** $\frac{1}{11}$

3 a $2^{\frac{5}{7}}$ **b** $2^{\frac{1}{7}}$ **c** $2^{\frac{6}{7}}$ **d** 2^{-7}

e 2 **f** 2^{12} **g** 1 **h** $2^{-\frac{5}{3}}$

4 a 3 **b** $\frac{1}{2}$ **c** $\frac{2}{3}$ **d** -3

5 a $2^{\frac{6}{7}}$ **b** $\frac{1}{12}$ **c** $1\frac{2}{15}$ **d** $\frac{27}{56}$

e $3\frac{13}{15}$ **f** 5 **g** $\frac{2}{33}$ **h** $\frac{1}{6}$

i $\frac{27}{40}$

6 a $4 - 3\pi$ **b** $1 - \sqrt{3}$ **c** 33π **d** $\frac{34 + 19\sqrt{3}}{6}$

e $-5 - 7\sqrt{7}$ **f** $\frac{\sqrt{5} - 1}{3}$

7 a $\frac{\sqrt{6}}{2}$ **b** $2\sqrt{10}$

8 a i $2\sqrt{2}$ **ii** 12

b i $2\sqrt{2}\pi$ **ii** $6\pi + 12$

9 a 6.3×10^7 **b** $1.496 \times 10^8\,\text{km}$

c $2.2 \times 10^{-5}\,\text{mm}$ **d** $3 \times 10^{-2}\,\text{kg}$

10 a $2\,180\,000\,000$ **b** $310\,000$

c $0.000\,000\,5$ **d** $0.000\,099\,2$

11 a 8.4×10^{11} **b** 4.8×10^{10}

c 4.53×10^7 **d** -1.916×10^{-4}

Assessment 17

1 a Soraya **b** Peter **c** Soraya

2 a No, 15^{12} **b** Yes, 3^{20}

c i No **ii** Yes **iii** No

d No, 7^3

3 a Yes **b** No, 36 **c** No, $\frac{1}{7}$ **d** No, 2744

e No, $\frac{1}{243}$ **f** No, 1

4 a $\frac{1}{2}$ **b** $\frac{1}{5}$ **c** 1 **d** $\frac{1}{4}$

5 a i 3^1 **ii** $9^{\frac{1}{2}}$ **b i** 3^2 **ii** 9^1

c i 3^4 **ii** 9^2 **d i** 3^{-1} **ii** $9^{-\frac{1}{2}}$

e i 3^{-2} **ii** 9^{-1} **f i** 3^{-4} **ii** 9^{-2}

g i 3^3 **ii** $9^{\frac{3}{2}}$ **h i** 3^{-3} **ii** $9^{-\frac{3}{2}}$

i i $3^{\frac{1}{2}}$ **ii** $9^{\frac{1}{4}}$

6 a $V = \frac{\pi}{3}r^2h, r = h$ so $V = \frac{\pi}{3}x^3, k = \frac{\pi}{3}$

b $p = \pi\sqrt{2}, q = 2$ **c** $\frac{64\pi}{3}\,\text{m}^3$

7 a $\frac{9\sqrt{3}}{2}\,\text{in}^2$ **b** $9\pi\sqrt{3}\,\text{in}^3$ **c** $0.04\,\pi^2\,\text{in}^3$

d $(9\pi\sqrt{3} - 0.04\,\pi^2)\,\text{in}^3$

8 a English Channel $29\,000\,\text{mi}^2$, Baltic Sea $146\,000\,\text{mi}^2$, Bering Sea $876\,000\,\text{mi}^2$, Caribbean Sea $1\,060\,000\,\text{mi}^2$, Malay Sea $3\,140\,000\,\text{mi}^2$, Indian Ocean $28\,400\,000\,\text{mi}^2$

b 1×10^9 by 1

c i 4 **ii** 2 **iii** 6 **iv** -3

d i $10\,763\,\text{km}$ **ii** 0.0008 joules

e i $4 \times 10^{-6}\,\text{km}$ **ii** $4\,\text{mm}$

9 a 8706 **b** 199 times **c** $\frac{1}{751\,000}$

d $226\,\text{km}\,(3\,\text{sf}) = 2.26 \times 10^2\,\text{km}$

10 $1.9848 \times 10^4\,\text{m} = 19\,848\,\text{m}$

11 a i $1.46092 \times 10^7\,\text{mi}^2 = 14\,609\,200\,\text{mi}^2$ **ii** $6.6768 \times 10^6\,\text{mi}^2$

b i $1 : 3.94\,(3\,\text{sf})$ **ii** $1.39 \times 10^8\,\text{mi}^2$ **iii** 70.8%

Chapter 18

Check in 18

1 a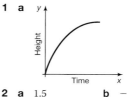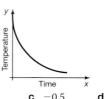

2 a 1.5 **b** -1 **c** -0.5 **d** $\frac{1}{3}$

e 0.6

3 a $x = 3$ or 4 **b** $y = 0$ or 8

c $x = 3.56$ or -0.56 **d** $x = -3$ or $-\frac{1}{2}$

e $y = 8, 3$ **f** $x = -\frac{1}{3}$ or 1

4 a 40 **b** 6 **c** -80

18.1S

1 b

x	-3	-2	-1	0	1	2	3
y	-26	-7	0	1	2	9	28

c Cubic passing through $(-3, -26), (-2, -7), (-1, 0), (0, 1),$ $(1, 2), (2, 9)$ and $(3, 28)$

d i y is approximately 4.4
ii y is approximately 2.4

2 a

x	-2	-1	0	1	2	3
x^3	-8	-1	0	1	8	27
$x^3 - 4$	-12	-5	-4	-3	4	23

b

x	-2	-1	1	2	3
$\frac{1}{x}$	$-\frac{1}{2}$	-1	1	$\frac{1}{2}$	$\frac{1}{3}$
y	-1	-2	2	1	$\frac{2}{3}$

c

x	-2	-1	0	1	2	3
x^3	-8	-1	0	1	8	27
$x + 1$	-1	0	1	2	3	4
y	-9	-1	1	3	11	31

d

x	-2	-1	1	2	3
$\frac{3}{x}$	$-\frac{3}{2}$	-3	3	$\frac{3}{2}$	1
y	$-\frac{1}{2}$	-2	4	$\frac{5}{2}$	2

3 a Cubic passing through $(-3, -36), (-2, -14), (-1, -4),$ $(0, 0), (1, 4), (2, 14)$ and $(3, 36)$

b Cubic passing through $(-3, -32), (-2, -12), (-1, -4),$ $(0, -2), (1, 0), (2, 8)$ and $(3, 28)$

c Cubic passing through $(-3, -34), (-2, -14), (-1, -6),$ $(0, -4), (1, -2), (2, 6)$ and $(3, 26)$

d Cubic passing through $(-3, -45), (-2, -18), (-1, -5),$ $(0, 0), (1, 3), (2, 10)$ and $(3, 27)$

4 a Cubic passing through $(-3, -44), (-2, -14), (-1, 0), (0, 4),$ $(1, 4), (2, 6), (3, 16)$

b i x is approximately -2.3. **ii** y is approximately 4.8.

5 a Reciprocal curve, with asymptotes $x = 2$ and $y = 0$, passing through $(-2, -3), (-1, -4), (0, -6), (1, -12), (3, 12), (4, 6),$ $(5, 4), (6, 3)$

b Reciprocal curve, with asymptotes $x = -4$ and $y = 0$, passing through $(-3, -3), (-2, -1), (-1, -\frac{1}{3}), (0, 0), (1, \frac{1}{5}),$ $(2, \frac{1}{3}), (3, \frac{3}{7})$ and $(4, \frac{1}{2})$

c Reciprocal curve, with asymptotes $x = 0$ and $y = -2$, passing through $(-3, -3), (-2, -3.5), (-1, -5), (1, 1),$ $(2, -0.5)$ and $(3, -1)$

6 a Blue **b** Orange **c** Green **d** Pink

18.1A

1 a $-3.375, -1, -0.125, 0, 0.125, 1, 3.375, 8; -4.5, -2, -0.5, 0,$ $-0.5, -2, -4.5, -8; +3, +3, +3, +3, +3, +3, +3, +3; -4.875,$ $0, 2.375, 3, 2.625, 2, 1.875, 3$

b Smooth curve through points $(-1.5, -4.875), (-1, 0),$ $(-0.5, 2.375), (0, 3), (0.5, 2.625), (1, 2), (1.5, 1.875), (2, 3)$

c The curve and the line $y = 2 - x$ intersect once.

d -0.47

2 a $-10.125, -3, -0.375, 0, 0.375, 3, 10.125; 6, 4, 2, 0, -2, -4, -6;$ $+3, +3, +3, +3, +3, +3, +3; -1.125, 4, 4.625, 3, 1.375, 2, 7.125$

b Smooth curve passing through $(-1.5, -1.125)$, $(-1, 4)$, $(-0.5, 4.625)$, $(0, 3)$, $(0.5, 1.375)$, $(1, 2)$, $(1.5, 7.125)$.

c Smooth curve passing through $(0.5, -4)$, $(1, -2)$, $(2, -1)$ not touching either axis. Smooth curve passing through $(-2, 1)$, $(-1, 2)$, $(-0.5, 4)$ not touching either axis.

d The two functions have two points of intersection.

e $x = -1.31$ and $x = -0.44$.

3 a, b Smooth curve through the points $(-0.5, -3.75)$, $(-0.25, -2.03)$, $(0, -1)$, $(0.25, -0.47)$, $(0.5, -0.25)$, $(0.75, -0.25)$, $(1, 0)$, $(1.25, 0.41)$, $(1.5, 1.25)$.

c $x = 1$. The curve intersects the line $y = 2 - 2x$ once only.

4 a Smooth curve passing through $(-1.5, 0.5)$, $(-1, -3)$, $(-0.5, -3)$, $(0, -1)$, $(0.5, 1.5)$, $(1, 3)$, $(1.5, 2)$, $(2, -3)$.

b Smooth curve passing through $(0.5, 2)$, $(1, 1)$, $(2, 0.5)$ not touching either axis. Smooth curve passing through $(-0.5, -2)$, $(-1, -1)$, $(-2, -0.5)$ not touching either axis.

c $\frac{1}{x} = -2x^3 + x^2 + 5x - 1$ or $1 = -2x^4 + x^3 + 5x^2 - x$

d $x = -1.38, x = -0.38, x = 0.56, x = 1.69$

5 $y = x^3 + 7x^2 + 14x + 11$ blue, this is an increasing function for large and small values of x.
$y = -x^3 + 6x^2 - 8x + 1$ pink, this is a decreasing function for large and small values of x.
$y = x^3$ orange, this is an increasing function.

6

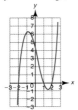

7

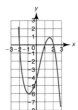

*8 $y = x^2(x - 3)$ $y = x(x - 3)^2$

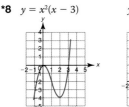 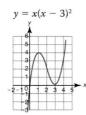

9 a A smooth curve passing through $(-2, -7)$, $(-1, -9)$, $(0, -9)$, $(1, -7)$, $(2, -3)$. A smooth curve passing through $(1, 9)$, $(3, 3)$, $(6, 1.5)$, $(9, 1)$, $(12, 0.75)$ not touching either axis. A smooth curve passing through $(-1, -9)$, $(-3, -3)$, $(-6, -1.5)$, $(-9, -1)$, $(-12, -0.75)$ not touching either axis.

b $x^2 + x - 9 = \frac{9}{x}$ or $9 = x^3 + x^2 - 9x$

c $x = -3, x = -1, x = 3$

d Smooth curve passing through $(-4, -21)$, $(-3, 0)$, $(-2, 5)$, $(-1, 0)$, $(0, -9)$, $(1, -16)$, $(2, -15)$, $(3, 0)$, $(4, 35)$.

e $x^2 + x - 9 = \frac{9}{x}$
Multiplying both sides by x and subtracting 9 from both sides results in
$x^3 + x^2 - 9x - 9 = 0$ which is the cubic function with $y = 0$.

1 a iv   **b** i   **c** ii   **d** iii

2 a, b Exponential graph passing through $(-4, 0.125)$, $(-3, 0.25)$, $(-2, 0.5)$, $(-1, 1)$, $(0, 2)$, $(1, 4)$, $(2, 8)$, $(3, 16)$ and $(4, 32)$.

c 3.7

3 a Exponential graph passing through $(-2, \frac{1}{256})$, $(-1, \frac{1}{64})$, $(0, \frac{1}{16})$, $(1, \frac{1}{4})$, $(2, 1)$, $(3, 4)$, $(4, 16)$, $(5, 64)$ and $(6, 256)$

b Exponential graph passing through $(-4, 80)$, $(-3, 26)$, $(-2, 8)$, $(-1, 2)$, $(0, 0)$, $(1, -\frac{2}{3})$, $(2, -\frac{8}{9})$, $(3, -\frac{26}{27})$ and $(4, \frac{80}{81})$.

4 a $x = 44°$ or $136°$ **b** $x = 53°$ or $127°$
c $x = 233°$ or $307°$ **d** $x = 210°$ or $330°$

5 a 0.5   **b** 0.5   **c** 0.71   **d** -0.71

6 a $x = 46°$ or $314°$ **b** $x = 37°$ or $323°$
c $x = 143°$ or $217°$ **d** $x = 120°$ or $240°$

7 a 0.5   **b** -0.77 **c** 0.71   **d** -0.71

8 a $x = 27°$ or $207°$ **b** $x = 103°$ or $283°$
c $x = 88°$ or $268°$ **d** $x = 172°$ or $354°$

9

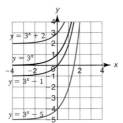

$f(x) + a$ is a vertical **translation** of $f(x)$ by a units.

10

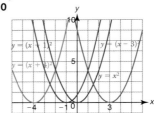

$f(x + a)$ is a **horizontal translation** of $f(x)$ by $-a$ units

11 a

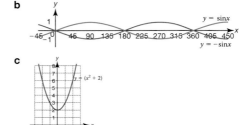

b

c

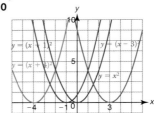

$-f(x)$ is a **reflection** of $f(x)$ in the **x-axis**

12 a

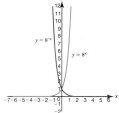

b

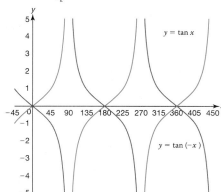

c

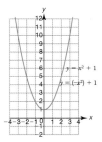

$f(-x)$ is a **reflection** of $f(x)$ in the **y-axis**

18.2A

1 a, b

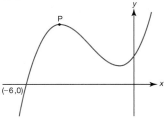

c $(-4, 4)$

2 a $(2, 2)$ **b** $(3, 0)$ **c** $(3, -2)$ **d** $(-3, 2)$

3 a–d

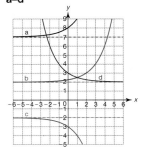

4 a True. The graph is symmetric about the y-axis.
b False. The period of sine is $360° > 180°$ so a horizontal translation of $180°$ will not map it onto itself.
c True. The period of tangent is $180°$ so a horizontal translation of $180°$ will map it onto itself.
d True. If the sine curve is translated $90°$ to the left it maps onto the cosine curve.

e True. If the cosine curve is reflected in the x-axis and translated $90°$ to the left the result will be $y = \sin(x)$.
f True. If the tangent curve is reflected in the x-axis and the y-axis it maps onto itself.

5 a $(135, 1)$ **b** $p = 45$ (or $405°$, $765°$, ..)
c $q = 135$ (or $495°$, $855°$, $1215°$, ...)

6 a 1
b Any two of: $45°$, $225°$, $405°$, $585°$, $765°$,..
c Keep adding (or subtracting) $180°$.
d Use $30 + 360m$ and $150 + 360n$ for integer values of m and n.
e Use $60 + 360m$ and $300 + 360n$ for integer values of m and n.

7 a An increase in the size of a corresponds with an increase in amplitude.
b An increase in the size of a corresponds with a decrease in period.

18.3S

1 a 200 **b** 225 **c** 25
d Points plotted at (O, 400), (N, 325) and (D, 400) joined with straight-lines.
e The sales dropped in November and picked up in December, possibly due to sales season over Christmas holiday.
f From January to September, the sales is quite flat and at low level; from October to December the sales increases a lot and peaks in October and December.

2 a $0.7\,\text{m}$ **b** $0.5\,\text{m}$ **c** $1.9\,\text{m}$ **d** $0.1\,\text{m}$
e Unlikely. Because its rate of growth is slowing down.

3 A-R, B-P, C-S, D-Q.

18.3A

1 a Quadratic function
2 a, b **A** Quadratic **B** Cubic (not exact)

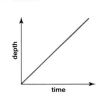
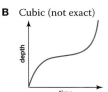
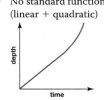

C Linear **D** No standard function (linear + quadratic)

3 a, b

Winners	1	2	3	4	5	6
Prize (£)	1000	500	333	250	200	167

c Reciprocal
4 a Check coordinates. **b** Reciprocal
c Yes, because as the temperature goes down, the pulse size goes up. Moreover, there is a lower limit on pulse size, which is the feature of reciprocal functions.
*5 a ∩-shaped quadratic through the points $(0, 0)$, $(2, 8)$, $(4, 12)$, $(6, 12)$, $(8, 8)$, $(10, 0)$.
b, c $y = -0.5(x - 5)^2 + 12.5$ **d** $(10,0)$
e Students' answers

18.4S

1 a 2 **b** $-\frac{2}{3}$ **c** 2 **d** -1
e 6 **f** -4 **g** $-\frac{1}{3}$ **h** $\frac{1}{2}$
i 1 **j** $-\frac{3}{8}$

2 a

b i 2
ii 6
iii −2
iv −4
v 0
vi −6

3 For the function x^2, the gradient is double the x coordinate. In the example the area found is close to $3^3/3$.

4 a

b i 2
ii −3
iii −2
iv 0
v 1
vi 3

c The gradient will be 5 since it is always equal to the value of x in this case.

5 a 10.5

b An overestimate since the errors in the smaller trapezium cancel each other out.

*6 a 20 **b** 19.625

7 31 m

8 a i 1 **ii** 4 **iii** 9 **iv** 16

b

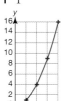

The curve is a section of $y = x^2$.

9 a i 1.5 **ii** 9 **iii** 28.5 **iv** 66

b

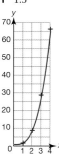

This curve is an approximation to $y = x^3$.

18.4A

1 a 3 m/s² **b** 0 m/s² **c** 4.5 m/s² **d** 20 metres

2 a £5560 **b** 153 **c** £153 pounds per year

d i £247 pounds per year **ii** £399 pounds per year

3 a £15,081.69

b £23,000 pounds per year (2 sf)

4 −6.25 °C per minute

5 $T = 6$ seconds

*6 a $9\frac{1}{3}$ **b** 6.7 **c** 12.95

18.5S

1 a $x^2 + y^2 = 25$ **b** $x^2 + y^2 = 36$
c $x^2 + y^2 = 121$ **d** $x^2 + y^2 = 196$
e $x^2 + y^2 = 6.25$ **f** $x^2 + y^2 = 20.25$

2 a i (0, 0) **ii** 1 **b i** (0, 0) **ii** 9
c i (0, 0) **ii** 10 **d i** (0, 0) **ii** $2\sqrt{15}$

3 a–c

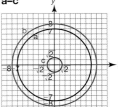

d–f

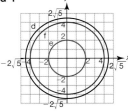

4 a $x^2 + y^2 = 9$ **b** $x^2 + y^2 = 64$
c $x^2 + y^2 = 144$ **d** $x^2 + y^2 = 81$

5 a (0, −5) and (0, 5) **b** (−3, 4) and (3, 4)
c (−4, −3) and (4, −3) **d** (5, 0)
e ($\sqrt{21}$, 2) and ($-\sqrt{21}$, 2)
f (−1, $2\sqrt{6}$) and (−1, $-2\sqrt{6}$)
g ($\frac{1}{2}$, $\sqrt{\frac{99}{4}}$) and ($\frac{1}{2}$, $-\sqrt{\frac{99}{4}}$)
h ($\sqrt{\frac{221}{9}}$, $-\frac{2}{3}$) and ($-\sqrt{\frac{221}{9}}$, $-\frac{2}{3}$)

6 a $\frac{4}{3}$ **b** $y = \frac{4}{3}x$ **c** $-\frac{3}{4}$ **d** $\frac{25}{4}$
e $y = -\frac{3}{4}x + \frac{25}{4}$

7 a $y = -\frac{3}{4}x + \frac{25}{2}$ **b** $y = -\frac{4}{3}x + \frac{50}{3}$
c $x = 10$

8 a $y = \frac{1}{4}x + \frac{17}{4}$ **b** $y = -4x + 17$ **c** $y = -\frac{1}{4}x - \frac{17}{4}$

9 a $x = 3, y = 4$ and $x = -4, y = -3$
b $x = 4, y = 3$ and $x = 0, y = -5$
c $x = 8, y = -6$ and $x = -8, y = 6$
d $x = 5, y = 12$ and $x = -\frac{16}{5}, y = -\frac{63}{5}$
*e $x = 1-\sqrt{17}, y = -1-\sqrt{17}$ and $x = 1+\sqrt{17}, y = -1+\sqrt{17}$
*f $x = -\frac{2}{5} - \frac{\sqrt{19}}{5}, y = \frac{1}{5} - \frac{2\sqrt{19}}{5}$ and $x = -\frac{2}{5} + \frac{\sqrt{19}}{5}, y = \frac{1}{5} + \frac{2\sqrt{19}}{5}$

18.5A

1 $x^2 + y^2 = 53$

2 $y = 2x + 10, y = -2x - 10$

3 a $x^2 + y^2 = 37$ **b** $x^2 + y^2 = 32$
c $x^2 + y^2 = 673$ **d** $x^2 + y^2 = 58$
e $x^2 + y^2 = 40$ **f** $x^2 + y^2 = 116$
g $x^2 + y^2 = 169$ **h** $x^2 + y^2 = 10$

4 a $y = 20 - 3x$ and $y = 3x - 20$
b $y = \frac{34}{5} - \frac{3}{5}x$ and $y = \frac{3}{5}x + \frac{34}{5}$
c $y = -\frac{3}{2}x - \frac{13}{2}$ and $y = \frac{3}{2}x - \frac{13}{2}$
*d $y = \frac{2}{\sqrt{6}}x + \frac{5\sqrt{6}}{3}, y = -\frac{2}{\sqrt{6}}x - \frac{5\sqrt{6}}{3}$
*e $y = \frac{\sqrt{8}}{2}x + 6, y = -\frac{\sqrt{8}}{2}x + 6$

5 $x^2 + y^2 = 25.8$ (3 sf)

6 $x^2 + y^2 = 45.0$ (3 sf)

7 a

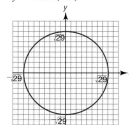

b $y = -2.5x - 14.5$ or $y = -2.5x + 14.5$

8 $x^2 + y^2 = 45$

*9 A circle with centre (2, 4) and radius 3. A circle with equation $(x - a)^2 + (y - b)^2 = r^2$ has centre (a, b) and radius r.

10 $\frac{1}{2}$, the green chord is parallel to the tangent.

11 The line $y = \frac{1}{2}x + 5$ is parallel to the tangent to the circle at the point (−2, 2) and lies above this tangent, therefore it does not intersect the circle.

12 Multiple answers that give one solution e.g. $a = 0, b = 2$.

13 a i $\dfrac{y}{x+r}$ **ii** $\dfrac{-y}{r-x}$

   b $\dfrac{y}{x+r} = \dfrac{(r-x)}{y}$

   c $x^2 + y^2 = r^2$. This is the equation of a circle with centre $(0, 0)$ and radius r, so P must lie on this circle.

Review 18

1 a C   **b** A

2 a   **b**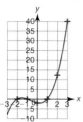

3 a 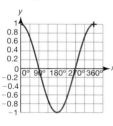   **b**

4 a i $x = 30°, 150°$ **ii** $x = 120°, 240°$
   b i -1 **ii** 0.5

5 a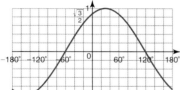

   b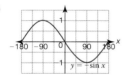

6 a A: translated up two units.
      B: translated left two units.
      C: reflected in x-axis.
   b A: $y = x^2 + 2$, B: $y = (x+2)^2$, C: $y = -x^2$

7 a 320   **b**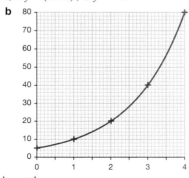

   c Exponential growth.
   d $y = 5(2^x)$ **e** 13.9 (3 sf)

8 a $2.5\,\text{m/s}^2$ **b** $-12\,\text{m/s}^2$ **c** $81\,\text{m}$

9 $x^2 + y^2 = 81$

10 a i $(0, 0)$ **ii** $2\sqrt{2}$ **b** $y = 4 - x$ and $y = x - 4$

1 i D   **ii** C   **iii** B   **iv** A
2 i E   **ii** C   **iii** B   **iv** A
  v D

3 a $y = \dfrac{36}{x}$ **b**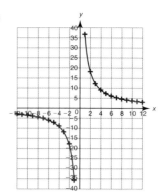

   c 4 and 9. The two points of intersection between $y = 13 - x$ and $y = \dfrac{36}{x}$ are $(4, 9)$ and $(9, 4)$ which give the same pair of numbers.

4 a, b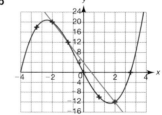

   c $x = \dfrac{1}{2}$ **d** $x = -4, 0, 3$
   e $(-2, 20), (-1, 12), (2, -12)$ **f** $x^3 + x^2 - 4x - 4 = 0$

5 a, d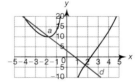

   b i $y = 11.05$ **ii** $y = 1.45$ **iii** $y = 13$
   c i $x = -2.8, -1.1, 3.9$ **ii** $x = -4.1, -0.6, 4.7$
     iii $x = 1.1$ **iv** $x = 0.8$ **v** $x = 2.3$
   e $x = -4, -1.7, 1.7$
   f The point of intersection are the points where $3 - 4x = x^2 - \dfrac{12}{x}$. This rearranges to give $x^3 + 4x^2 - 3x - 12 = 0$.

6 a $0, 1.5, 2.12, 2.6, 3, 2.12, 0, -3, 0$
   b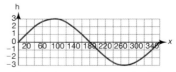

   c $-3 \leqslant h \leqslant 3$ **d** $42°$ and $138°$

7 a 169π **b** $5^2 + 12^2 = 169 = 13^2$ **c** $\dfrac{12}{5}$
   d Gradient of line segment $(0, 0)$ to $(5, 12)$ is $\dfrac{12}{5}$ so gradient of tangent is the negative reciprocal $-\dfrac{5}{12}$. Tangent passes through $(5, 12)$ so $y - 12 = -\dfrac{5}{12}(x - 5)$, $y = -\dfrac{5}{12}x + \dfrac{169}{12}$.

8 a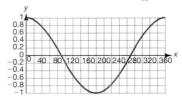

   b Twice. $y = \cos(x)$ and $y = \tan(x)$ cross exactly twice in $0 \leqslant x \leqslant 360$.

9 a 280, 314, 351, 393

b Each year the population increases by another 12%

c

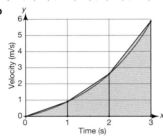

d i 6.1 years   **ii** 9.7 years

e 50 fish per year   **f** 691 fish

10 a A(−2, 0) B(0, 5) C(2, 4) D(3, 5)

b A(−1, −4) B(1, 1) C(3, 0) D(4, 1)

c A(2, −4) B(0, 1) C(−2, 0) D(−3, 1)

d A(−2, 4) B(0, −1) C(2, 0) D(3, −1)

11 a, b

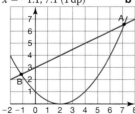

b Area under curve ≈ (0.5 × 1 × 1) + 0.5(1 + 2.5) + 0.5(2.5 + 6) = 6.5

Revision 3

1 a HCF = 30 , LCM = 485 100

b $3300 \times 33 = (2^2 \times 3 \times 5^2 \times 11) \times 3 \times 11$
$= 2^2 \times 3^2 \times 5^2 \times 11^2$;
$4410 \times 10 = (2 \times 3^2 \times 5 \times 7^2) \times 2 \times 10$
$= 2^2 \times 3^2 \times 5^2 \times 7^2$

2 a $p = 4, q = 13$ **b** $a = \frac{9 - 7\sqrt{3}}{2}, b = \frac{9}{2}$

3 a iv, vii, iii, viii, ii, vi, i, v   **b** iv and vii   **c** iii and viii

4 6 and 8

5 a No. $a = 2$ **b** No. $c = 3$ **c** No. $c = 2$ or 4

6 a $y = -x + 7$ **b** $y = x + 3$

c (2, 5)   **d** Check centre (2, 5), radius = 4

7 a $x = -1.1, 7.1$ (1 dp)   **b** $x = 3 \pm \sqrt{17}$

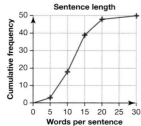

8 a i 1.0827×10^{12} km³   **ii** 5.95×10^{24} kg

b 11 074 631 years (nearest year)

9 a $\frac{(3 \times 3 + 15 \times 7.5 + 21 \times 12.5 + 9 \times 17.5 + 2 \times 22.5)}{50} = \frac{586.5}{50} = 11.73$

b Median = 11.5, IQR = 6

Sentence length

(cumulative frequency graph)

10 a

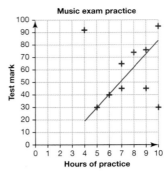

Music exam practice

b A, H, students' answers.   **c**   36

11 a

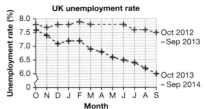

UK unemployment rate

b The trend over the two years is one of consistently falling unemployment rates. Year 2 is lower overall than Year 1

12 a i 216   **ii** 100 000   **iii** 5   **iv** 8

v 0   **vi** −7

b i No. 12^9 **ii** No. 6^6 **iii** No. 210^1 **iv** Yes

v No. 9^{16}

c i 5   **ii** 10   **iii** 4   **iv** 3

13 a 564 000 units   **b** 1.95×10^{22}

14 a

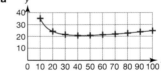

b i 26 l/h   **ii** 22 l/h

c i 13 mph   **ii** 24 mph, 80 mph

d 44 mph as this is where fuel consumption is lowest.

Chapter 19

Check in 19

1 a 49   **b** 52   **c** 34

d 48   **e** 45   **f** 8

2 a $x = 6y$ **b** $x = 5y$ **c** $x = 10y$

d $x = \frac{2}{y}$ **e** $x = \frac{5}{y}$ **f** $x = \frac{8}{y}$

19.1S

1 a 20 cm²   **b** 8 cm²   **c** 4 cm²   **d** 15 cm²

2 a 5 cm   **b** 17 cm   **c** 13 cm   **d** 10.3 cm

e 10.8 cm   **f** 9.9 cm   **g** 12.5 cm   **h** 7.3 cm

3 (3, 4, 5), (8, 15, 17), (12, 13, 5)

4 a 8 cm   **b** 19.8 cm   **c** 10 cm   **d** 9 cm

e 7.5 cm   **f** 5.4 cm   **g** 5.5 cm   **h** 7.7 cm

5 a (6, 8, 10), (10, 24, 26), (9, 12, 15)

b Different sets of triplets

c Some sets are multiples of others.

6 a 5   **b** 8.6   **c** 10.0   **d** 10.6

7 8.6 cm

8 9.6 cm

9 11.3 cm

19.1A

1 5.7 cm

2 a 12 cm   **b** 60 cm²

3 1.2 m

4 29.4 m further

5 a 5.9 **b** 6.6 **c** 9.7 **d** 4.9

6 15.3 cm

7 16.7 cm

8 13.6 cm

9 a $a = 8$ cm, $b = 14.4$ cm **b** $c = 9.6$ cm, $d = 16.3$ cm

10 $20^2 + 21^2 = 29^2$, Pythagoras' theorem applies for right-angled triangles

11 a $6^2 + 13^2 < 15^2$, Pythagoras's theorem does not apply to this triangle therefore it is not right-angled.

b obtuse

12 (3, 4, 5), (8, 15, 17), (5, 12, 13), (21, 20, 29), (7, 24, 25), (35, 12, 37), (9, 40, 41)

19.2S

1 a 3.63 cm **b** 11.3 cm **c** 10.8 cm **d** 6.38 cm
e 3.02 cm **f** 3.79 cm

2 a 6.89 cm **b** 8.60 cm **c** 7.88 cm **d** 4.62 cm
e 4.77 cm **f** 39.7 cm **g** 10.6 cm **h** 9.56 cm

3 a 26.4° **b** 25.7° **c** 56.2° **d** 67.4°
e 17.6° **f** 23.6° **g** 69.2° **h** 55.1°
i 19.8° **j** 35.7° **k** 27.7° **l** 40.1°
m 73.0° **n** 41.0°

19.2A

1 a 28.9° **b** 35.5°

2 a 8.51 cm **b** 58.4°

3 a 15.2 cm **b** 6.91 cm

4 49.4 cm²

5 63.8 cm²

6 7.5 km

7 a 130.11° (or 49.89°) **b** 20.8 km (or 5.4 km)

8 Not both correct, $20\tan 52° \neq 28\tan 44°$

9 a 14.2 cm **b** 13.66 m

10 $\cos 30 = \frac{\sqrt{3}}{2}$, $\sin 30 = \frac{1}{2}$, $(\cos 60)^2 + (\sin 60)^2 = (\frac{1}{2})^2 + (\frac{\sqrt{3}}{2})^2 = \frac{1}{4} + \frac{3}{4} = 1$

11 $\cos 45 = \frac{\sqrt{2}}{2}$ $\sin 45 = \frac{\sqrt{2}}{2}$; it equals 1.

12 a 1, 0
b hyp, increases rapidly in size
opp, decreases until it is zero
adj, increases rapidly in size
c 0, 1
d hyp, increases rapidly in size
opp, increases rapidly in size
adj, decreases until it is zero

19.3S

1 a 6.53 cm **b** 14.3 cm **c** 5.70 cm **d** 10.3 cm
e 9.47 cm **f** 18.1 cm

2 a 33.4° **b** 26.1° **c** 13.0° **d** 44.8°
e 18.3° **f** 59.8°

3 a 6.19 cm **b** 13.9 cm **c** 12.0 cm **d** 6.59 cm
e 14.0 cm **f** 11.3 cm

4 a 33.0° **b** 34.8° **c** 85.3° **d** 55.8°
e 19.9°

19.3A

1 $B = 41°$, $C = 59°$, $x = 10.4$ cm

2 11 km, 191°

3 18 km, 262°

4 47.7 cm

5 41.3 cm

6 47.3 cm

7 a 13.4 cm **b** 10.5 cm

8 43.7 cm

9 12.7 cm

10 11.8 cm

11 a In triangle ADC $\sin A = \frac{h}{b}$, $h = b \sin A$
In triangle BDC $\sin B = \frac{h}{a}$, $h = a \sin B$
b $a \sin B = b \sin A$
Dividing by $\sin A \sin B$ gives $\frac{a}{\sin A} = \frac{b}{\sin B}$
c Area $= \frac{1}{2} \times AB \times h = \frac{1}{2} \times c \times b \sin A$ or $\frac{1}{2} \times c \times a \sin B$
Area $= \frac{1}{2}bc \sin A$ or $\frac{1}{2}ca \sin B$

*12 a $b^2 = h^2 + x^2$ (Pythagoras' theorem)
$a^2 = h^2 + (c - x)^2 = h^2 + c^2 - 2cx + x^2$ (Pythagoras' theorem)
Subtracting gives $a^2 - b^2 = c^2 - 2cx$
so $a^2 = b^2 + c^2 - 2cx$ $\qquad$ (1)
In triangle ADC $\cos A = \frac{x}{b}$ so $x = b \cos A$
Substituting in (1) gives $a^2 = b^2 + c^2 - 2bc \cos A$
b $2bc \cos A = b^2 + c^2 - a^2$
Dividing by $2bc$ gives
$\cos A = \frac{b^2 + c^2 - a^2}{2b}$
c $a^2 = h^2 + (x + c)^2 = h^2 + x^2 + 2cx + c^2$ (Pythagoras' theorem)
$b^2 = h^2 + x^2$
Subtracting gives $a^2 - b^2 = c^2 + 2cx$
so $a^2 = b^2 + c^2 + 2cx$ $\qquad$ (1)
In triangle ADC $\cos (180° - \angle A) = \frac{x}{b}$
so $x = b \cos (180° - \angle A)$
Substituting in (1) gives
$a^2 = b^2 + c^2 + 2bc \cos (180° - \angle A)$
But $\cos (180° - \angle A) = -\cos A$
Therefore $a^2 = b^2 + c^2 - 2bc \cos A$

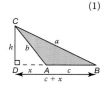

19.4S

1 a 20 cm **b** 25 cm **c** 36.9° **d** 43.2°

2 a 15 cm
b i 28.1°
ii $\cos \angle QPR = \frac{17^2 + 17^2 - 16^2}{2 \times 17 \times 17} = 0.5570...$
$\angle QPR = \cos^{-1} 0.5570... = 56.144...°$
$\angle SPR = 56.144...° \div 2 = 28°$ (nearest °)

3 a 7.18 cm **b** 63.8° **c** 70.9°

4 a i $20\sqrt{2}$ cm **ii** $20\sqrt{3}$ cm
b i 45° **ii** 35.3°

5 a 40 cm **b** 78 cm **c** 87.7 cm **d** 27.1°
e 53.1°

6 a, b i 60.8 cm **ii** 25.3° **iii** 909.3 cm²
c i 209 cm **ii** 8.1°

7 a 40.89 mm, 54.33 mm **b** 130.1°

8 a 22.2° **b** 3.5 cm² **c** 25.3° **d** 3.1 cm²

*9 a 47.1 cm **b** 63.2° **c** 70.3°

10 a 1.44 m **b** 56.3°

19.4A

1 Yes, over 50 m

2 160.5°

3 a 22.2 km, 64.8° **b** Yes, 9.44 km < 10 km

4 a i $\frac{3}{5}$ **ii** $\frac{4}{5}$
b i $\frac{5}{13}$ **ii** $\frac{12}{5}$
c i 0.6 **ii** 0.75

5
$(\sin \theta)^2 + (\cos \theta)^2 = \left(\frac{O}{H}\right)^2 + \left(\frac{A}{H}\right)^2$
$= \frac{O^2 + A^2}{H^2} = \frac{H^2}{H^2} = 1$

6 Angle subtended at centre of circle by one side $= \frac{360°}{n}$
Let r be the radius of the circle.
Area of each triangle $= \frac{1}{2}r^2 \sin \left(\frac{360°}{n}\right)$
Area of polygon, $A = \frac{1}{2}nr^2 \sin \left(\frac{360°}{n}\right)$

7 a $33.7°$ **b** £7632.23

8 $1.56\,m^2$

9 $19.02°$

10 $136\pi\,cm^2$

*11 $26\pi\,kg$

*12 42 seconds (nearest second)

*13 Using Pythagoras

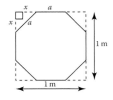

$a^2 = 2x^2$, $a = \sqrt{2}x$

$a + 2x = 1$, so $a + \frac{2}{\sqrt{2}}a = 1$

$a(1 + \sqrt{2}) = 1$

$a = \frac{1}{\sqrt{2} + 1} = \sqrt{2} - 1$

Perimeter $= 8a = 8(\sqrt{2} - 1)$ metres

19.5S

1 a

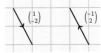

b Vectors are parallel and the same size

2 a DX, EF, CB, XA **b** FX, XC, ED, AB

  c FE, AX, XD, BC **d** BA, CX, XF, DE

3 a OJ, NK **b** OM, QK **c** OP, LK **d** JO, KN

  e MO, KQ **f** PO, KL

4 Diagrams

5 a $\underline{r}$ **b** $\underline{p} + \underline{r}$ **c** $-\underline{p} - \underline{r}$ **d** $\underline{p} - \underline{r}$

6 a $3\underline{p}$ **b** $5\underline{p}$ **c** $5\underline{p} + 2\underline{q}$ **d** $5\underline{p} + 6\underline{q}$

  e $4\underline{q}$ **f** $\underline{p} + 3\underline{q}$ **g** $\underline{p} + 6\underline{q}$ **h** $3\underline{p} + 3\underline{q}$

  i $\underline{p} + 2\underline{q}$ **j** $\underline{p} + 6\underline{q}$ **k** $\underline{p} + \underline{q}$ **l** $-\underline{p} + \underline{q}$

  m $-4\underline{q}$ **n** $-\underline{p} - 4\underline{q}$ **o** $-3\underline{p} - 4\underline{q}$ **p** $-\underline{p} + 3\underline{q}$

  q $\underline{p} - 3\underline{q}$ **r** $-\underline{p} - 3\underline{q}$ **s** $-4\underline{p} - 4\underline{q}$ **t** $-\underline{p} - 4\underline{q}$

  u $-3\underline{p} + 4\underline{q}$

7

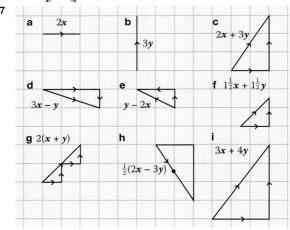

a $2x$	**b**	**c** $2x + 3y$
d $3x - y$	**e** $y - 2x$	**f** $1\frac{1}{2}x + 1\frac{1}{2}y$
g $2(x + y)$	**h** $\frac{1}{2}(2x - 3y)$	**i** $3x + 4y$

19.5A

1 a $\underline{z}$ **b** $\underline{z} + \underline{w}$ **c** $\frac{1}{2}(\underline{z} + \underline{w})$ **d** $\frac{1}{2}(\underline{z} + \underline{w})$

  e $\frac{1}{2}(-\underline{z} + \underline{w})$ **f** $\frac{1}{2}(\underline{z} - \underline{w})$

2 a $\underline{t}$ **b** $\underline{t} + \underline{r}$ **c** $\frac{3}{4}(\underline{t} + \underline{r})$ **d** $\frac{1}{4}(-\underline{r} - \underline{t})$

  e $\frac{3}{4}\underline{r} - \frac{1}{4}\underline{t}$

3 a $-\underline{a} + \underline{b}$ **b** $\frac{4}{5}\underline{a} + \frac{1}{5}\underline{b}$

4 AB multiple of CD therefore parallel

5 a i $2\underline{r} + 2\underline{s}$ **ii** $\underline{r}$ **iii** $\underline{s}$ **iv** $\underline{r} + \underline{s}$

  b MN multiple of RT therefore parallel

6 a i $3\underline{p}$ **ii** $\underline{r} + 3\underline{p}$ **iii** $-\underline{r} + \underline{p}$ **iv** $\underline{r} + 2\underline{p}$

  b XQ multiple of OX and both pass through same point, therefore straight line

7 a i $\underline{j} + \underline{k}$ **ii** $\underline{k}$ **iii** $\underline{k} - \underline{j}$

  b XM multiple of JX and both pass through same point, therefore straight line

*8 Let $\overrightarrow{AB} = \underline{x}$, $\overrightarrow{BC} = \underline{y}$ and $\overrightarrow{CD} = \underline{z}$

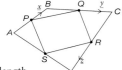

$\overrightarrow{PQ} = \frac{1}{2}\underline{x} + \frac{1}{2}\underline{y}$

$\overrightarrow{AD} = \underline{x} + \underline{y} + \underline{z}$

$\overrightarrow{SR} = \overrightarrow{SD} + \overrightarrow{DR} = \frac{1}{2}\underline{x} + \frac{1}{2}\underline{y}$

PQ and SR are parallel and equal in length.

This is sufficient to prove that $PQRS$ is a parallelogram.

*9 The statement can be proved by showing that the mid-points of the diagonals coincide.

In parallelogram $PQRS$, let $\overrightarrow{PQ} = \underline{a}$ and $\overrightarrow{PS} = \underline{b}$

$\overrightarrow{PR} = \underline{a} + \underline{b}$ and its mid-point M is such that $\overrightarrow{PM} = \frac{1}{2}(\underline{a} + \underline{b})$

$\overrightarrow{QS} = -\underline{a} + \underline{b}$ and its mid-point N is such that $\overrightarrow{QN} = \frac{1}{2}(-\underline{a} + \underline{b})$

Therefore $\overrightarrow{PN} = \overrightarrow{PQ} + \overrightarrow{QN} = \underline{a} + \frac{1}{2}(-\underline{a} + \underline{b}) = \frac{1}{2}(\underline{a} + \underline{b})$

Therefore $\overrightarrow{PN} = \overrightarrow{PM}$ and the mid-points of the diagonals are the same point.

Review 19

1 $a = 20.2\,mm$, $b = 12.8\,cm$

2 $a = 8.56\,cm$, $b = 1.21\,cm$, $c = 6.57\,cm$

3 $a = 45.3°$, $b = 19.8°$

4 a i $15.3\,cm$ **ii** $15.8\,cm$ **b i** $1.3°$ **ii** $75.3°$

5 a $142\,km$ **b** $212°$

6 a 1 **b** $\frac{\sqrt{3}}{2}$

7 $x = 7.12\,cm$, $\theta = 103°$

8 $25.2\,cm^2$

9 $\underline{u} = \begin{pmatrix} 6 \\ -5 \end{pmatrix}$ $\underline{v} = \begin{pmatrix} -3 \\ 10 \end{pmatrix}$

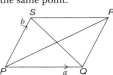

10 a i $\underline{u} + \underline{v}$ **ii** $1.5\underline{v}$ **iii** $0.5\underline{v} - \underline{u}$

  b $0.25v - u$

Assessment 19

1 a Hannah needs to calculate the square root of 181.

  $a = 13.45$ (2 dp)

  b Pawel calculated the square root of $24^2 + 25^2$ instead of the square root of $25^2 - 24^2$. Side $= 7$ in.

2 a $2^2 + 3^2 = 13$, not 13^2 as in a Pythagorean triple.

  b $20^2 + 99^2 = 10201 \neq 100^2$. 60, 80, 100 or 28, 96, 100.

3 9.85 (3 sf)

4 a Yes, $5^2 + 12^2 = 169 = 13^2$ **b** Yes, $9^2 + 12^2 = 225 = 15^2$

  c No, $9^2 + 14^2 = 277 \neq 17^2$

  d Yes, $1.6^2 + 3.0^2 = 11.56 = 3.4^2$

  e No, $11^2 + 19^2 = 482 \neq 22^2$

  f Yes, $3.6^2 + 7.7^2 = 72.25 = 8.5^2$

5 a Right angled. **b** $10.0\,cm$ (3 sf) **c** $59.6°$

6 a $\angle HBS = 62°$ (Alternate), $\angle SBL = 180 - 152 = 28°$, $\angle HBL = 62 + 28 = 90°$.

  b $4.38\,mi$ (3 sf) **c** $10.5\,mi$ (3 sf)

  d $55°$ **e** $117°$

7 a $320\,m$ (3 sf) **b** $111\,m$ (3 sf)

8 a $14°$ **b** $38.7°$

9 a $8.09\,m$ (3 sf) **b** $7.89\,m$ (3 sf)

10 a $7\,cm$. **b** $28.1°$

11 a Check students' drawings

b **i** $\binom{2}{13}$ **ii** $\binom{-6}{1}$ **iii** $\binom{6}{-1}$ **iv** $\binom{8}{12}$

v $\binom{4}{-14}$ **vi** $\binom{14}{-9}$

Check students' drawings

c **i** $\sqrt{173}$ **ii** $\sqrt{37}$ **iii** $\sqrt{37}$ **iv** $4\sqrt{13}$

v $2\sqrt{53}$ **vi** $\sqrt{277}$

d The vectors are parallel, and the same length.

12 a Correct **b** Incorrect, $-2\mathbf{x} + 7\mathbf{y}$

c Incorrect, $-6\mathbf{y}$ **d** Correct

e Incorrect, $2\mathbf{x} + 3\mathbf{y}$ **f** Correct

13 a $\binom{12}{3}$ **b** $\binom{35}{-10}$ **c** $\binom{16}{-10}$ **d** $\binom{-1}{3}$

e $\binom{3}{-3}$ **f** $\binom{3}{4}$ **g** $\binom{-40}{18}$ **h** $\binom{56}{-15}$

i $\binom{-35}{16}$

Chapter 20

Check in 20

1 a 0.55 **b** 0.04 **c** 0.72 **d** 0.625
e 0.6 **f** 0.34 **g** 0.9 **h** 0.18
i 0.17 **j** 0.192 **k** 0.235 **l** 0.92

2 a $\frac{1}{6}$ **b** $\frac{4}{5}$ **c** $\frac{2}{9}$ **d** $\frac{13}{15}$
e $\frac{11}{12}$ **f** $\frac{5}{9}$ **g** $\frac{8}{45}$ **h** $\frac{2}{5}$

20.1S

1 a 1, 4, 9, 16, 25, 36, 49, 64, 81, 100
b Canada, United States of America, Mexico
c 2, 3, 5, 7, 11, 13, 17, 19, 23, 29
d 1, 2, 3, 4, 6, 9, 12, 18, 36

2 a {1, 4, 9, 36} **b** {2, 3} **c** ∅
d {1, 2, 3, 4, 5, 7, 9, 11, 13, 16, 17, 19, 23, 25, 29, 36, 49, 64, 81, 100}

3 a Factors of 10 **b** Multiples of 2
c Vowels **d** Outcomes of flipping a coin twice
e Coins of the pound sterling
f First ten multiples of 3

4 a 1, 2, 3, 4, 5, 6, 7, 8, 9, 10 **b** 1, 3, 5, 7, 9
c 2, 3, 5, 7 **d** 1, 4, 9

5 a {3,5,7} **b** {1, 9}
c {1, 3, 4, 5, 7, 9} **d** {1, 2, 3, 4, 5, 7, 9}

6 Single digit numbers are less than ten and are integers.

7 a Yes, 10 and 20. **b** Yes, 2.
c Yes, 4, 9, and 36. **d** Yes, 9 and 36.
e No **f** No
g No **h** Yes, 3.
i No

8 a **i** 8 **ii** 13 **iii** 9 **iv** 21
b Students who play **neither** hockey **nor** tennis.
c $\frac{13}{25}$

9 a **i** $\frac{29}{50}$ **ii** $\frac{19}{50}$ **iii** $\frac{17}{50}$ **iv** $\frac{43}{50}$
b 14

20.1A

1 a 58 **b** 110 **c** 5
d A pupil in this set is an only child who walks to school and takes Biology.

2 a (right-angled isosceles triangle)
b An equilateral triangle has three 60° angles and cannot contain a right angle.

3 a $x = 8$ **b** A and B are disjoint.

4 3

5 Maximum 0.15, Minimum 0

6 5

*7 a 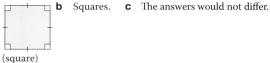 **b** Squares. **c** The answers would not differ.

(square)

*8 0.06

20.2S

1 a 20 **b** 48

2 a $\frac{1}{12}$ **b** $\frac{1}{6}$ **c** $\frac{7}{36}$ **d** $\frac{6}{36} = \frac{1}{6}$
e {2, 3, 4, 5, 6, 7, 8, 9, 10, 11, 12}

3 a $\frac{2}{36} = \frac{1}{18}$ **b** $\frac{19}{36}$ **c** $\frac{8}{36} = \frac{2}{9}$ **d** $\frac{8}{36} = \frac{2}{9}$
e {1, 2, 3, 4, 5, 6, 8, 9, 10, 12, 15, 16, 18, 20, 24, 25, 30, 36}

4 a

	1	2	3	4	5	6
1	0	1	2	3	4	5
2	1	0	1	2	3	4
3	2	1	0	1	2	3
4	3	2	1	0	1	2
5	4	3	2	1	0	1
6	5	4	3	2	1	0

b {0, 1, 2, 3, 4, 5}
c **i** $\frac{1}{6}$ **ii** $\frac{1}{6}$ **iii** 0 **iv** $\frac{4}{9}$

5 a {HHH, HHT, HTH, THH, TTH, THT, HTT, TTT}
b 3 **c** 2

6 a

Spinner 1 \ Spinner 2	1	3	5
1	2	4	6
2	3	5	7
4	5	7	9

b {2, 3, 4, 5, 6, 7, 9}
c **i** $\frac{1}{9}$ **ii** $\frac{1}{9}$ **iii** $\frac{1}{3}$

7 a

Spinner 1 \ Spinner 2	1	3	5
1	1	3	5
2	2	6	10
4	4	12	20

b {1, 2, 3, 4, 5, 6, 10, 12, 20}
c **i** $\frac{1}{9}$ **ii** $\frac{1}{9}$ **iii** $\frac{2}{3}$

8 The product.

9 a $\frac{1}{12} \times 100 = 8$ (1 sf)
b $\frac{1}{18} \times 100 = 6$ (1 sf)
c You should expect to see a 6 more often in the list of sums than in the list of products.
d You should expect to see a 3 about the same number of times in the list of sums as in the list of products.
e No, it is not certain that 7 will appear in the list of sums. He should expect to see 7 more often in the list of sums than in the list of products for the reason stated.

10 a

	1	2	2	3	3	3
1	2	3	3	4	4	4
2	3	4	4	5	5	5
3	4	5	5	6	6	6
4	5	6	6	7	7	7
5	6	7	7	8	8	8
6	7	8	8	9	9	9

b **i** $P(6) = \frac{1}{6}$ **ii** $P(7) = \frac{1}{6}$ **iii** $P(9) = \frac{1}{12}$ **iv** $P(3) = \frac{1}{12}$
c 4, 5 and 7.
d 6 can be obtained from any of the outcomes on the unusual dice. This is not so for 2, 3, 8, and 9 which all have lower probabilities than 6.

20.2A

1 a $\frac{1}{6}$ **b** $\frac{2}{3}$ **c** $\frac{5}{6}$

2 $\frac{1}{4}$

3 a

	B1	R2	Y2	G3	B2	R4
1	1	2	0	3	2	4
2	2	4	0	6	4	8
3	3	6	0	9	6	12
4	4	8	0	12	8	16
5	5	10	0	15	10	20
6	6	12	0	18	12	24

b i $\frac{1}{9}$ **ii** $\frac{5}{12}$

4 a $\frac{9}{100}$ **b** $\frac{9}{10}$ **c** $\frac{19}{100}$

5 a i $\frac{2}{15}$ **ii** $\frac{3}{5}$ **b** $\frac{2}{3}$

*6 a $\frac{5}{12}$ **b** $\frac{1}{2}$

*7 a i $\frac{1}{18}$ **ii** $\frac{1}{6}$ **iii** $\frac{1}{9}$
  b One of the following pairs: A and C, A and E, B and D, B and E.
  c Four pairs

20.3S

1 a

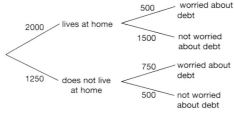

b 0.38

2 a

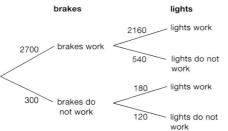

b 840
c Tommy is wrong because there are other features that can cause a car to fail an MOT which are not discussed here.

3 a, e

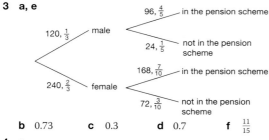

b 0.73 **c** 0.3 **d** 0.7 **f** $\frac{11}{15}$

4 a

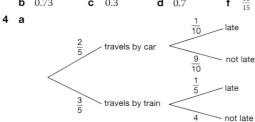

b

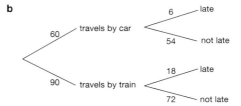

c 24 days **d** 0.16

5 a

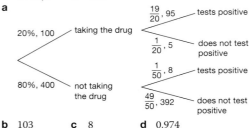

b 103 **c** 8 **d** 0.974

20.3A

1 a 0.84
  b P(Draw then win) = P(Draw) × P(Win),
    P(Lose then win) = P(Lose) × P(Win)
  c Students' own responses e.g. No, losing the first match results in a drop in confidence, reducing the chance of winning the next match.

2 a

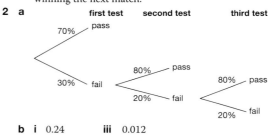

b i 0.24 **iii** 0.012

3 a

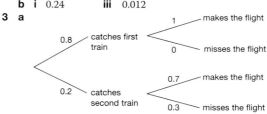

b 0.94

4 a Yes, (X ∩ Y) contains all outcomes where both balls are the same colour **and** that colour is blue.
  b $P(X) = \frac{1}{2}$, $P(Y) = \frac{3}{7}$
    $P(X \cap Y) = P(Z) = \frac{1}{2} \times \frac{3}{7} = P(X) \times P(Y)$

5

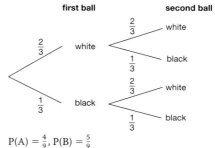

$P(A) = \frac{4}{9}$, $P(B) = \frac{5}{9}$
$P(A \cap B) = \frac{4}{9} > P(A) \times P(B) = \frac{20}{81}$
A and B are **not** independent.

*6 a $\frac{1}{9}$

b

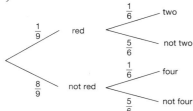

c $\frac{1}{6}$

d Using the spinner does not affect Sara's likelihood of getting a four. If the spinner lands on red, the possible scores are 2, 4, 6, 8, 10, and 12 and if the spinner does not land on red the possible scores are 1, 2, 3, 4, 5, and 6. Either way the possible scores are equally likely and $P(4) = \frac{1}{6}$.

e i $P(\text{finish}) = P(\text{not red and } 5) = \frac{4}{27} < \frac{1}{6}$.

ii $P(\text{finish}) = P(\text{red and } 4) = \frac{1}{54} < \frac{1}{6}$.

20.4S

1 a 0.9 **b** 0.5 **c** 0.8

d $P(S \cap R) = 0.4$

$P(S) \times P(R) = 0.5 \times 0.8 = 0.4$ S and R are independent.

2

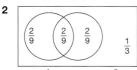

b i $\frac{4}{9}$ **ii** $\frac{2}{9}$ **iii** $\frac{1}{2}$

c P and F are not independent because $P(P \mid F) \neq P(P)$.

3 a $\frac{43}{179}$ **b** $\frac{98}{181}$

c No. P(Walks to school | in year 11) $= \frac{98}{181}$ is higher than P(Walks to school) $= \frac{141}{360}$.

4 a The following tree diagram describes drivers who are involved in a serious accident.

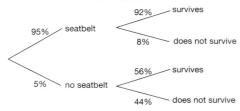

b P(No seatbelt and survives) $= 0.05 \times 0.56 = 0.028$

5 a

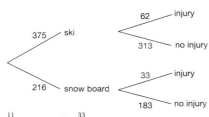

b $\frac{11}{72}$ **c** $\frac{33}{95}$

6 a

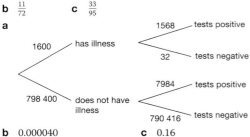

b 0.000040 **c** 0.16

20.4A

1 a

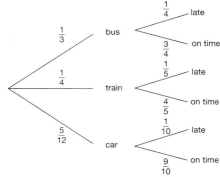

$P(\text{late}) = \frac{7}{40}$

b $\frac{10}{21}$

2 a 0.036 **b** 0.056 **c** 0.64

3 a $\frac{23}{38}$ **b** $\frac{1}{2}$ **c** $\frac{6}{23}$

4 a

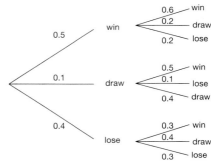

b 0.67 **c** 0.22

5 a 0.29 **b** 0.31 **c** 0.72

Review 20

1

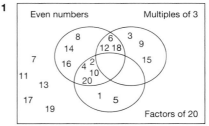

2 a 0.08 **ii** 0.1025

3 a

	Egg	Cheese	Tuna
White	W, E	W, C	W, T0
Brown	B, E	B, C	B, T
Granary	G, E	G, C	G, T

b i $\frac{1}{3}$ **ii** $\frac{1}{9}$ **iii** $\frac{4}{9}$

4 a

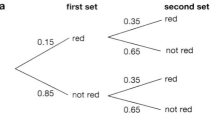

b i 0.5525 ii 0.4475 c 0.65
5 a i $\frac{3}{8}$ ii $\frac{1}{8}$ iii $\frac{1}{2}$
b $\frac{1}{4}$ c $\frac{2}{7}$
6 a

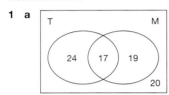

b i $\frac{171}{200}$ ii $\frac{21}{200}$

Assessment 20

1 a

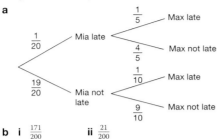

b i $\frac{41}{80}$ ii $\frac{11}{20}$ iii $\frac{3}{4}$ iv $\frac{17}{80}$

2 a b $n = 16$

3 a

+	1	2	3	4	5	6	7	8
1	2	3	4	5	6	7	8	9
2	3	4	5	6	7	8	9	10
3	4	5	6	7	8	9	10	11
4	5	6	7	8	9	10	11	12
5	6	7	8	9	10	11	12	13
6	7	8	9	10	11	12	13	14
7	8	9	10	11	12	13	14	15
8	9	10	11	12	13	14	15	16

b i $\frac{1}{32}$ ii $\frac{7}{64}$ iii $\frac{5}{64}$
c 9
d All outcomes were assumed to be equally likely.
4 Tea 17, Coffee 15; Tea-Chocolate 12, Tea-Plain 5; Coffee-Chocolate 10, Coffee-Plain 5
5 a B $\frac{1}{4}$, A $\frac{3}{4}$; BB $\frac{1}{16}$, BA $\frac{3}{16}$; AB $\frac{3}{16}$, AA $\frac{9}{16}$ b $\frac{1}{16}$ c $\frac{7}{16}$
6 a 0.1
b

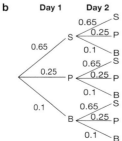

c i 0.025 ii 0.4225 iii 0.165 iv 0.19
7 No, $P(A) \times P(R) = 0.07 \neq P(A \text{ and } R) = 0.15$.

8 a $P(H) + P(T) = 1$, so $P(H) = 1 - P(T) = 1 - p$
b $p(1 - p) = \frac{6}{25}, p = \frac{2}{5}$
c $P(T) < P(H)$, so p must be the smaller solution.
9 a

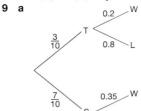

b $P(W|T) = \frac{2}{10} \neq P(W) = \frac{61}{200}$
c $P(T|W) = \frac{12}{61}$ d $P(S|L) = \frac{49}{55}$

Lifeskills 4

1 118
2 a 15.4 m b 33.9° c 244 m³ d 158 m²
e The marquee comprised of a cylinder and cone is cheaper.
3 a $\overrightarrow{TB} = \begin{pmatrix} 200 \\ 50 \end{pmatrix}$ $\overrightarrow{TA} = \begin{pmatrix} -200 \\ -300 \end{pmatrix}$ $\overrightarrow{TR} = \begin{pmatrix} -100 \\ 100 \end{pmatrix}$
b $TB = 206$ m, $TA = 361$ m, $TR = 141$ m c 137.5°
4 0.412
5 1257 customers, 11 142 customers, 23rd month.

Chapter 21

Check in 21

1 a 10, 12 b 70, 64 c 16, 22
d −2, −5 e 48, 96 f $\frac{1}{6}, \frac{1}{7}$
2 $4(n - 1), 15 - n, 2n + 7, 2n^2, \frac{9}{n} + 15$
3 Square numbers. They form a pattern of squares when drawn.

21.1S

1 a 2, 4, 6, 8, 10 b 17, 19, 21, 23, 25
c 4, 8, 12, 16, 20 d 24, 30, 36, 42, 48
e 7, 12, 17, 22, 27 f 1, 4, 9, 16, 25
g 2, 5, 10, 17, 26 h 2, 4, 8, 16, 32
2 a 4, 7, 11, 14, 17 b 25, 19, 13, 7, 1
c 4, 8, 12, 16, 20 d 30, 22, 14, 6, −2
e −16, −10, −4, 2, 8
3 a 29, 34 b 65, 58
c 16, 22 d 999 999, 9 999 999
e 13, 21 f 3.375, 1.6875
4 a 7, 13 b 8, 16 c 93, 91, 89 d 8, 125
5 a 6, 11, 16; 51 b 11, 14, 17; 38
c 4, 12, 20; 76 d −2, 4, 10; 52
e 22, 20, 18; 4 f 10, 5, 0; −35
g −13, −6, 1; 50 h −2, 2, 6; 34
6 a $6n + 5$ b $9n - 8$ c $7n + 8$ d $4n - 14$
e $-3n + 23$ f $-4n + 19$ g $-8n + 24$ h $-8n + 39$
7 a $4n + 3$ b $4n - 10$ c $-9n + 41$ d $-6n + 21$
8 a $5n - 1$ b $2n - 1$ c $2n + 8$ d $0.5n + 0.5$
e $2n - 6$ f n g $13n$ h $10n - 6$
i $12 - 2n$ j $105 - 5n$ k $50.25 - 0.25n$
l $79 - 4n$
9 a $6 - 5n$ b $0.5 + 1.5n$ c $9.5 - 1.5n$ d $0.8 + 0.6n$
e $1.5 + 0.5n$ f $3.5 - 0.5n$

21.1A

1 a $W = 2B + 6$
b i $B = 2W + 2$ ii $L = 3W + 1$
2 a $E = 5H + 1$ b $M = 2L(L + 1)$
3 a False b False c False

4 a $8n - 20$ **b** $3n + 2$ **c** $2n - 46$

5 a $\frac{2n + 1}{3n + 4}$ **b** $\frac{2n + 8}{33 - 3n}$ **c** $\frac{n + 6}{n^2}$ **d** $\frac{n^3}{13 - 2n}$

6 No integer solution to $5n - 3 = 75$, sequence is ... 72, 77 ...

7 36

*8 a First sequence is $-7n + 140$ and the second is $5n - 64$. When $n = 17$, both become 21.

   b $126, 91, 56, 21, -14, -49$

*9 The two sequences might contain the same term in different positions.

21.2S

1 a 2, 3, 6, 11 **b** 5, 7, 11, 17 **c** 10, 11, 13, 16
   d 5, 8, 14, 23 **e** 10, 9, 6, 1 **f** 8, 6, 2, −4

2 a 31, 43 **b** 24, 34 **c** 23, 31
   d 25, 36 **e** 10, 15 **f** 24, 18
   g −32, −55

3 a 36 **b** 2 **c** 5 **d** 11, 29
   e 50, 76 **f** 15 **g** −1

4 a i 4, 7, 12 **ii** 3, 8, 15
   iii 0, 4, 10 **iv** 2, 8, 18
   v 6, 12, 22
   b i 103 **ii** 120 **iii** 108
   iv 200 **v** 204

5 a i 2, 5, 10, 17 **ii** −2, 1, 6, 13
   iii 4, 9, 16, 25 **iv** −2, 2, 8, 16
   v 3, 10, 21, 36 **vi** 2.5, 5, 8.5, 13
   vii 6, 14, 26, 42 **viii** −1, 0, 3, 8
   b i 2 **ii** 2 **iii** 2 **iv** 2
   v 4 **vi** 1 **vii** 4 **viii** 2

6 a $n^2 + 3$ **b** $n^2 - 4$ **c** $2n^2$ **d** $n^2 + n$
   e $n^2 + 4n$ **f** $n^2 + 2n$ **g** $2n^2 + 2n + 2$
   *h $4n - n^2$

7 a $10 - n^2$ **b** $20 - 2n^2$
   c $n - n^2$ **d** $n + 10 - n^2$

8 a Estimates between 30 and 40 **b** $n^2 + n$
   c If $n^2 + n \geqslant 1000$ then $n^2 + n - 1000 \geqslant 0$
   Smallest value of $n = 32$.

9 No – if $n^2 + 3 = 150$ then $n^2 = 147$. 147 is not a square number so 150 is not in the sequence.

10 Yes: $3n^2 - n = n(3n - 1)$
   If n is even: $3n - 1$ is odd so and even × odd = even
   If n is odd: $3n - 1$ is even and odd × even = even

11 a i 3, 5, 7, 9 **ii** 2, 7, 12, 17
   b i $(2n + 1)^2 = 4n^2 + 4n + 1$
   ii $(2n + 1)^2 + 1 = 4n^2 + 4n + 2$
   iii $(5n - 3)^2 = 25n^2 - 30n + 9$
   iv $(5n - 3)^2 + 5 = 25n^2 - 30n + 14$
   c $(3n + 4)^2$

21.2A

1 No – the 10th term is not usually double the 5th term (or the 10th term = 103)

2 a No – as the second difference = 2, the nth term will start with 'n^2'
   b $n^2 - n + 5$

3 $T(n) = n^2 - n$: 0, 2, 6, 12; $T(n) = n^2 - 1$: 0, 3, 8, 15;
   $T(n) = n^2 + 2$: 3, 6, 11, 18; $T(n) = n(n + 1)$: 2, 6, 12, 20;
   $T(n) = 4 - n^2$: 3, 0, −5, −12

4 Many answers possible including $n^2 + n + 1$, $2n^2 - 2n + 3$, $\frac{1}{2}n^2 + 2\frac{1}{2}n$, $3n^2 - 5n + 5$, $4n^2 - 8n + 7$

5 a

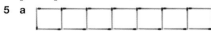

   b $m = 4, 7, 13, 22$
   c $m = \frac{3}{2}n^2 - \frac{3}{2}n + 4$ **d** 10th pattern
   e 3679 matches

6 e.g.

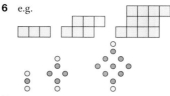

7 316

8 a 24 moves **b** 15 moves, 8 moves.
   c

Number of pairs	1	2	3	4
Number of moves	3	8	15	24

   d Number of moves $= n^2 + 2n$ where n = number of pairs. number of moves for 50 pairs = 2600

9 a 1 3 6 10 15
   1st Diff 2 3 4 5
   2nd Diff 1 1 1
   The second difference is 1 so the first part of the nth term is $\frac{1}{2}n^2$

$\frac{1}{2}n^2$	$\frac{1}{2}$	2	$4\frac{1}{2}$	8
Sequence	1	3	6	10
Difference	$\frac{1}{2}$	1	$1\frac{1}{2}$	2

   The nth term for the difference is $\frac{1}{2}n$
   So the overall nth term is: $\frac{1}{2}n^2 + \frac{1}{2}n = \frac{1}{2}n(n + 1)$
   b Consider each term of the triangular sequence being doubled and rearranged as rectangle:

   x   x xx   xxxx   xxxxx
              xx x   xxxx   xxxxx
                            xxxx   xxxxx
                                          xxxxx

   The nth term of the sequence can be arranged as a rectangle with dimensions $n \times (n + 1)$.
   But this is twice the value of the triangular number and so: nth triangular number $= \frac{1}{2}n(n + 1)$

*10 a n^3 **b** $2n^3$
   c $n^3 - n$ ****d** $n^3 - 2n^2 + 3n - 4$

21.3S

1 a 1, 3, 6, 10, 15, 21, 28, 36, 45, 55
   b 1, 4, 9, 16, 25, 36, 49, 64, 81, 100
   c 1, 8, 27, 64, 125, 216, 343, 512, 729, 1000

2 a $21 + 3 + 6$ **b** $28 + 3$ **c** $28 + 3 + 1$

3 a arithmetic **b** arithmetic
   c Fibonacci-type **d** quadratic
   e geometric **f** arithmetic
   g Fibonacci-type **h** geometric
   i quadratic **j** arithmetic
   k geometric **l** geometric
   m geometric **n** geometric
   o arithmetic **p** geometric

4 Yes, as they have a constant second difference of 1. They can be represented by the quadratic expression $\frac{1}{2}n^2 + \frac{1}{2}n$.

5 a 6, 8, 10 **b** 8, 16, 32 **c** 6, 10, 16
   d *Multiple answers possible, for example* 7, 11, 16

6 a 3, 6, 12, 24 **b** 10, 50, 250, 1250
   c 3, 1.5, 0.75, 0.375 **d** 2, −6, 18, −54
   e $\frac{1}{2}, \frac{1}{4}, \frac{1}{8}, \frac{1}{16}$ **f** −3, 6, −12, 24
   g $4, 4\sqrt{3}, 12, 12\sqrt{3}$ **h** $\sqrt{3}, 3, 3\sqrt{3}, 9$
   i $2\sqrt{5}, 10, 10\sqrt{5}, 50$

7 a 15, arithmetic sequence with difference 5.
   b 20, geometric sequence with common ratio 2.
   c 15 Fibonacci-type sequence.
   d 17, quadratic sequence with second difference 2.

8 $\frac{a}{b} = \frac{b}{c}$ so by multiplying both sides of the equation by bc, $ac = b^2$ and thus $\sqrt{ac} = b$.

9 Several answers possible including: $1^2 = 1^3 = 1$, $4^3 = 8^2 = 64$, $9^3 = 27^2 = 729$.

10 a 3, 4, 7, 11, 18   **b** 0, −1, −1, 0, 1
    c 3, 5, 11, 21, 43   **d** 1, 1, 1, 1, 1
    e 2, 3, 4, 5, 6

11 Yes – using this rule, T(1) = 1, T(2) = 3, T(3) = 6,... which is the triangular number sequence.

12 Both are correct, the square numbers form a quadratic sequence.

13 a **i** $\frac{1}{2}, \frac{2}{3}, \frac{3}{4}, \frac{4}{5}, \frac{5}{6}$ **ii** $\frac{1}{2}, \frac{2}{5}, \frac{3}{10}, \frac{4}{17}, \frac{5}{26}$ **iii** $1, \frac{1}{4}, \frac{1}{9}, \frac{1}{16}, \frac{1}{25}$
    iv $\frac{1}{2}, \frac{4}{3}, \frac{9}{4}, \frac{16}{5}, \frac{25}{6}$ **v** 0.9, 0.81, 0.729, 0.6561, 0.59049
    vi 1.1, 1.21, 1.331, 1.4641, 1.61051

   b **i** The sequence approaches 1.   **ii** The terms get smaller.
      iii The terms get smaller.   **iv** The terms get larger.
      v The terms get smaller.   **vi** The terms get larger.

21.3A

1 Many answers possible including:
   a 4, 7, 10, 13, 16, 19   **b** 4, 8, 16, 32, 64, 128
   c 4, 5, 9, 14, 23, 37   **d** 4, 6, 10, 16, 24, 34

2 a 32   **b** 7   **c** −19.5   **d** −1, 7
   e 768   **f** 1   **g** $9\sqrt{3}$ **h** 16

3 a Yes. This method creates the pattern shown below which represents the triangle numbers.

     x x   x
        xx   xx
            xxx

   b Pentagonal numbers: 1, 5, 12, 22 …

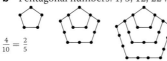

$\frac{4}{10} = \frac{2}{5}$

     Hexagonal numbers: 1, 6, 15, 28 …

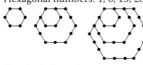

     Tetrahedral numbers: 1, 4, 10, 20, …

4 a Students own presentation
   b **i** The diagonals of Pascal's Triangle sum to Fibonacci numbers.
      ii The ratios of consecutive terms converge to the golden ratio.

5 a Option 3: By the end of the February: £2.7 million
   b Option 3: By the end of the September: £10.7 million

6 a This geometric sequence has first term $\frac{1}{2}$ and common ratio $\frac{1}{2}$.
   b $\frac{1}{2}, \frac{3}{4}, \frac{7}{8}, \frac{15}{16}, \frac{31}{32}$
   c The square in the diagram has area 1. S(n) covers a proportion of the square and the remainder is uncovered. S($n+1$) will cover the space of S(n) plus half the remaining space. As S(n) gets large the amount of uncovered space gets smaller but never disappears, so the terms of S(n) approach 1 but never reach 1.

7 a In the second step multiply the number by 7 instead.
   b In the second step multiply the number by 5 instead.
   c Students own responses.

8 a $ab, ab^3, ab^5, ab^7, ab^9$
   b $c^2d^7, c^4d^5, c^6d^3, c^8d, c^{10}d^{-1}$
   c $3x − 6, 6x^2 − 12x, 12x^3 − 24x^2, 24x^4 − 48x^3, 48x^5 − 96x^4$

---

9 In all cases assuming $a \neq 0$. If $r > 1$ or $r < −1$ the terms will diverge. If $r = 1$ the terms will remain constant. If $r = −1$ the terms will oscillate between two values. If $−1 < r < 1$ the terms will converge.

Review 21

1 a **i** 44, 53, 62   **ii** 19, 6, −7   **iii** 9.2, 10.8, 12.4
   b **i** +9   **ii** −13   **iii** +1.6

2 a 73   **b** 141

3 a $6n − 5$ **b** $7n + 8$ **c** $63 − 12n$ **d** $−1.5n − 5$

4 a 175   **b** 340

5 a $x^2 + 3$ **b** $3x^2 − x$ **c** $2x^2 + 3x − 1$

6 a Geometric   **b** Quadratic   **c** Fibonacci-type
   d Arithmetic   **e** Quadratic   **f** Geometric

7 a **i** 15, 21   **ii** $\frac{1}{25}, \frac{1}{36}$ **b** **i** $\frac{1}{2}n(n + 1)$ **ii** $\frac{1}{n^2}$

8 a 32, 64, 128   **b** 2^n

9 a $2\sqrt{2}n − \sqrt{2}; 19\sqrt{2}$ **b** $\frac{n}{2n + 1}; \frac{10}{21}$

Assessment 21

1 a v   **b** iv   **c** viii   **d** vi
   e vii   **f** i   **g** iii   **h** ix
   i ii

2 a +10   **b** $10n − 9$

3 a **i** $2n + 8$ **ii** $2n + 2$ **b** 54   **c** 42

4 a Each term is the sum of the two previous terms.
   b 55, 89   **c** $144 = 12^2$ **d** Fibonacci

5 a Correct. $2 \times 10 + 7 = 27$
   b Incorrect. $6 \times 1 − 5 = 1, 6 \times 2 − 5 = 7, 6 \times 3 − 5 = 13$
   c Incorrect. $13 − 3 \times 100 = −287$
   d Incorrect. $10^2 − 10 = 100 − 10 = 90$
   e Correct. $15 − 3 \times 100^2 = 15 − 30000 = −29985$

6 No. The ratio is decreasing, $1 : 2, 1 : 4, 1 : 6$

7 a Dawn   **b** $D = \frac{t(t+1)}{2}$

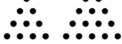

   c **i** 55 dots   **ii** 1275 dots   **iii** 5050 dots

8 a 9, 14, 20   **b** $D = \frac{p(p-3)}{2}$
   c **i** 35   **ii** 1175   **iii** 4850

9 a False. −1   **b** False. $15 \times 16 \div 2 = 120$.
   c False 210   **d** True. All terms would be the same.

10 a −2   **b** $\frac{20}{9}$ **c** $a + d, a + 2d, a + 9d$

Chapter 22

Check in 22

1 a **i** £49.50   **ii** 66 mm
      iii 52.8 km   **iv** 4 hours 24 minutes
   b **i** 4.5 miles   **ii** 43.5 minutes
      iii 10.5 kg   **iv** £46.50

2 a 40 mph   **b** 42.5 mph   **c** 54 mph   **d** 32 mph

3 a 600 cm³   **b** 351.9 cm³   **c** 45.92 cm³

4 a 500 cm²   **b** 276.5 cm²   **c** 107 cm²

22.1S

1 a 7.7 m/s   **b** 7.1 m/s   **c** 6.8 m/s   **d** 5.1 m/s

2 a 160 km   **b** 161 miles   **c** 54 m   **d** 288 miles

3 a 3 hours   **b** 4 hours   **c** 20 minutes **d** 15 minutes

4 1.425 g/cm³

5 a 9.46 kg   **b** 6.15 litres

6 a 5704 kg   **b** 3400 kg   **c** 61 824 kg   **d** 125 cm³
   e 24.7 cm³   **f** 2000 cm³

7 a 1.29 N/m²   **b** 19.3 N/cm² **c** 0.46875 m²
   d 7424 N   **e** 12.64 cm²   **f** 68.88 N

8 £2.10/m

9 £11.95/hour

10 a 2.5 litres/s **b** 1.6 litres/s

11 1200 litres

12 a 2.4 units/hour **b** 57.6 units

13 a 500 km **b** 20 litres **c** 12.5 km/litre

14 a $1.5 per £ **b** $187.5 **c** £80

22.1A

1 89.25 mph

2 300 kg

3 8.5 g/cm³

4 32 mph

5 525 km, 2 hours 12 minutes 30 seconds, 37.3 kmph, 215 km, 37.14 kmph.

6 a 5 g/cm³ **b** 87.88 g

7 a 5.96 g/cm³ **b** 105 blocks

8 a 10 500 kg/m³ **b** 4 500 kg/m³ **c** 2 700 kg/m³

9 a £89.25 **b** 15 hours

10 £42.24

11 4 hours 15 seconds

12 a 550 N/cm² **b** 17.3 N/cm²

22.2S

1 a 50 mm **b** 80 mm **c** 150 mm **d** 67 mm
 e 193 mm **f** 45 mm **g** 43 mm **h** 106 mm
 i 800 mm **j** 1000 mm

2 a 6 cm **b** 8.5 cm **c** 24 cm **d** 6.3 cm
 e 0.4 cm **f** 400 cm **g** 1000 cm **h** 350 cm
 i 160 cm **j** 163 cm

3 a 4 m **b** 4.5 m **c** 4.75 m **d** 4.7 m
 e 0.5 m **f** 1000 m **g** 4000 m **h** 500 m
 i 3500 m **j** 18 000 m

4 a 8 m² **b** 80 000 cm²

5 a 24 m² **b** 240 000 cm²

6 a 400 mm² **b** 730 mm² **c** 1090 mm² **d** 250 mm²
 e 40 000 mm²

7 a 6 cm² **b** 12 cm² **c** 8.5 cm² **d** 65 cm²
 e 100 cm²

8 a 4 m² **b** 8.5 m² **c** 100 m² **d** 12.5 m²
 e 0.5 m²

9 a 50 000 cm² **b** 100 000 cm² **c** 65 000 cm²
 d 77 500 cm² **e** 6000 cm²

10 a 232.5 cm² **b** 193.75 cm³

22.2A

1 a 2268 cm² **b** 240.57 litres

2 244 cm²

3 125 cm³

4 a Their sides are all the same length, so any scale factor will be constant on all dimensions.
 b The scale factor on one dimension could be different from that on another.
 c i 1:49 **ii** 1:343

5 a 2:3 **b** 34 cm² **c** 81 cm³

6 Deal A

7 Volume of original cone : volume of removed cone = 3³ : 2³ so volume removed cone : volume frustum = 27−8 : 8.

8 No, $\left(\frac{260.3}{1017.9}\right)^{\frac{1}{2}} \neq \left(\frac{188.5}{1352.2}\right)^{\frac{1}{3}}$.

22.3S

1 a false **b** true **c** false **d** true
 e true

2 a  **b**
 c **d**

Two variables that are in direct proportion have a straight-line graph passing through (0, 0).

3 a $w = kl$ **b** 2.48 kg/m **c** 7.192 kg

4 a doubled **b** halved
 c multiplied by 6 **d** divided by 10
 d multiplied by 0.7

5 a halved **b** doubled
 c divided by 6 **d** multiplied by 10
 d divided by 0.7

6 a 10 people **b** 25 hours **c** 40 hours

7 a 2 **b** 8 **c** 0.4
 d 16 **e** 0.16

8 $y = \frac{500}{w}$

9 800

10 4

11 a $d = kt$ **b** 48 **c** 120 miles
 d 3 hours 20 minutes

22.3A

1 Pack A: 1.2p per pin > Pack B: 1.15p per pin. Pack B is better value.

2 Regular, 0.7 pence/sheet < 0.77 pence/sheet.

3 a R = 100s² **b** 64 **c** 1.41

4 A = πr^2 so the area is directly proportional to the square of the radius, with constant of proportionality π.
 Students' answers e.g. C = $2\pi r$, the circumference of a circle is directly proportional to the radius, with constant of proportionality 2π.

5 615 cm² (3 sf)

6 a F = $\frac{100}{d^2}$; 11.i N, 6.25 N
 b

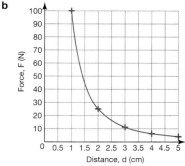

7 a ii **b** iii **c** i **d** iv

8 1 hour 2 minutes 30 seconds

9 a $y = \frac{10}{\sqrt{x}}$ **b i** 1 **ii** $\frac{25}{36}$

10 a $P = \frac{32}{t^3}$ **b i** $1\frac{5}{27}$ **ii** 4

22.4S

1 a 13
 b Students' answers with $1 < x_{\min} < 2$ and $2 < x_{\max} < 3$, e.g. $\frac{2.5^3 - 1.5^3}{2.5 - 1.5} = 12.25$
 c $y = 12x - 16$

2 a 4

b Students' answers with $1 < x_{min} < 2$ and $2 < x_{max} < 3$,

e.g. $\frac{(2.5^3 - 2.5 \times 9) - (1.5^3 - 1.5 \times 9)}{2.5 - 1.5} = 3.25$

c $y = 3x - 16$

3 a, b

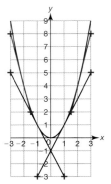

c $-4, -2, 0, 2, 4$

d Gradient $= 2p$

4 a i 1   **ii** 0

b $\frac{2}{3}$

c The section with gradient 1 is twice as long as the section with gradient 0.

5 a, d

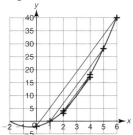

b i 7   **ii** 7   **iii** 7

c 7   **d** All the chords and the tangent are parallel.

6 The gradient at P is $2p$ and the equation of the tangent is $y = 2px - p^2$. This crosses the y-axis at $(0, -p^2)$, as required.

*7 a $\frac{(4 + h)^2 - 10 - (4 - h)^2 + 10}{2h} = \frac{16h}{2h} = 8$

b The chords which tend closer and closer to the tangent at $(4, 6)$ all have gradient precisely 8. The tangent itself therefore has gradient precisely 8.

22.4A

1 a $-10\,$m/s   **b** The ball is falling 10 m each second.

2 a No profit is made if the price is zero or if the price is too high for potential customers.

b At this point the profit is maximised.

c 0.6 pounds (i.e. 60p) gives the greatest profit. At this point the profit is maximised.

3 a $-4, -1, -0.5$ **b** °C per minute.

c Initially the temperature drops at approximately 4°C per minute. The temperature continues to fall at a gradually reducing rate until it reaches approximately 10°C.

4 a

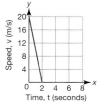

b Acceleration   **c** $-10\,$m/s^2

5 a Approximately 375 m/s   **b** Approximately 750 m/s

22.5S

1 a 1.25   **b** 0.75   **c** 1.025   **d** 0.975

2 a

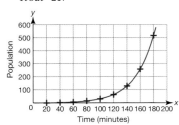

b i Trout $= 800 \times 0.85^n$

   ii e.g. $n = 2$, Trout $= 578$; $n = 5$, Trout $= 354$; $n = 8$, Trout 217

3 a

b i 181   **ii** 362

c The population has doubled.

4 a A number is increased by 50% by multiplying by 1.5, this increase occurs n times in n hours from a starting value of 200.

b 400×1.35^n **c** 7 hours

5 a $16\,000 \times 0.85^n$ **b** £8352.10   **c** 9 years

6 a 56859   **b** 10 years

7 a

End of year	Amount in the account (£)
2	2163.20
3	2249.73
4	2339.72
5	2433.31
6	2530.64
7	2631.86
8	2737.14
9	2846.62
10	2960.49

b 48%

c i Multiplier for percentage increase $= 1 + \frac{r}{100}$. Number of times increase occurs $= n$. Initial amount $= P$.

   ii $A = 2000(1 + \frac{4}{100})^{10} = 2960.49$

8 7

9 a, b i

   ii 1.25%

22.5A

1 a £66.33   **b** £235.27   **c** £723.40

2 She is incorrect. The yearly interest on the half-yearly saver account is 4.04% > 4%.

3 **a** The number of vehicles per day when the road opens is 2400. The number of vehicles increases by 8% each year.

b

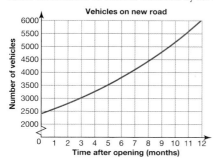

c This implies the annual increase in vehicles continues to rise, in reality it is likely to level off at a certain value.

4 **a** Each year the number of trees is 70% of the previous number plus the 60 new trees.

b 208 trees

c Smooth curve passing through (0, 250), (1, 235), (2, 224), (3, 217), (4, 212) and (5, 208).

5 **a** 6 months **b** 7.07%

6 £8934.34

7 **a** Liam calculated the simple interest. He should multiply by 1.045^6 to work out the compound interest.

b Interest = $\pounds P(1.045^n - 1)$ **c** £6400

8 3.085%

9 **a** **i**

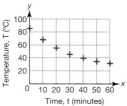

ii 0.8°C/minute

b **i** Yes, the formula gives values close to Tanya's data.

* ii 20 °C is room temperature, the values are approaching this limit. 65 °C is the difference between the initial temperature of the coffee and room temperature. 0.97^t refers to the fact that the temperature decreases by 3% each minute.

*10 **a**

Number of years	Amount left (kg)
$30 = 30 \times 1$	$\frac{1}{2}$
$60 = 30 \times 2$	$\frac{1^2}{2}$
$90 = 30 \times 3$	$\frac{1^3}{2}$
$n = 30 \times \frac{n}{30}$	$\frac{1^{\frac{n}{30}}}{2}$

b

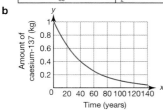

c The y-values would be multiplied by 1000.

Review 22

1 1.25 Pa

2 £0.21 or 21 p per 100 g

3 **a** **i** 364.5 kcal **ii** 14.2 g **iii** 1.2 g
b 190 g

4 **a** 6700 mm **b** 44 min, 11 s **c** 0.45 litres
d 25.2 km/h **e** 0.067253 m²

5 **a** 526.5 cm³ **b** 40 cm²

6 **a** $y = 1.2x$ **b** **i** 9.48 **ii** 9.67 (3 sf)

7 **a** $y = 2x^2$ **b** **i** 242 **ii** 5

8 **a** $Y = \frac{66}{X}$ **b** **i** 3 **ii** 0.5

9 **a**

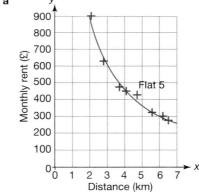

b

10 **a** 13 m/h (allow 12.5 − 13.5) **b** 11 m/h

11 **a** **i** £1537.50 **ii** £1697.11 **b** $V = 1500 \times 1.025^t$

Assessment 22

1 8 min 20 s

2 **a** 39.37 in **b** £9.75 **c** 14 min 56 s

3 The wombat, Bolt's speed = 36.73 km/h.

4 **a** 7.28 g/cm³ **b** 23.81 cm³ **c** 103 g

5 **a** 45 tins **b** Yes, $100 \div \left(\frac{3}{2}\right) = 67$ days.
c No, $30 \div \left(\frac{9}{2}\right) = 7$ days.

6 **a** 157.5 litres **b** 379 cm

7 **a** 69.12 kg **b** $x = 24$ cm, $y = 10$ cm

8 0.21 km²

9 **a** £10 **b** 150 000 cm³

10 **a** $T = 4e$ **b** 4.25 cm **c** 18 N

11 **a** $A = 3.142x^2$ **b** Yes, $3.142 \approx \pi$ and x = radius.
c $A = 314.2$ **d** $x = 3$

12 **a**

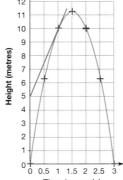

b Flat 5, it lies furthest from the curve. **c** 3.5 km

13 **a** $P = \frac{10}{\sqrt{t}}$ **b** 1

14 **a**

Height of a ball

b 5 m/s

15 **a** £27 000, 5% **b** £9546

Revision 4

1 7.14 km

2 161 cm

3 $8.63°$

4 $BC = 5.28\,m$ (3 sf)

5 a 17.0 **b** $17.9°$

6 a $75\,m$ **b** $36.9°$
  c $BO = 180\,m$ (3 sf), $PB = 195\,m$ (3 sf) **d** $22.6°$

7 $RQ = \mathbf{p} - 2\mathbf{q}$, $NM = \frac{\mathbf{p}}{2} - \mathbf{q}$, $NM = \frac{1}{2}RQ$

8 a Yes. $1^2 + 1 = 2$, $2^2 + 2 = 6$, $3^2 + 3 = 12$, $4^2 + 4 = 20$
  b 2550, $10\,100$ **c** $2n + 2$ **d** 102, 202

9 a

3	4	5
6	10	15
3	6	10
9	16	25

  b i 1275 **ii** 5050
  c r^2 **d** $\frac{r(r-1)}{2}$
  e $\frac{2(2-1)}{2} = 1$ **f** $\frac{60(60-1)}{2} = 1770$

10 a $1300\,ml$ **b** $0.076\,kg$

11 a 6 **b** Liddi, Liddi 27 p $<$ Addle 27.5 p

12 a $40\,mph$ **b** 1 hr 26 min
  c 6 minutes **d** 10:48 am

13 a $15.75\,g/cm^3$ **b** $5.48\,kg$ (3 sf)
  c $2.86\,cm^3$ (3 sf)

14 a i $m \propto \frac{1}{d^3}$ **ii** $T \propto \sqrt{l}$ **iii** $S \propto r^2 - 3$
  b i 75 **ii** 4.84

15 $1.125 \times 10^5\,Nm^{-2}$

16 a $P = 1 - 0.28 - 0.35 = 0.37$
  b

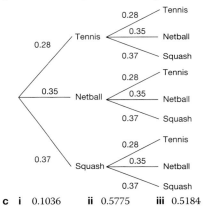

  c i 0.1036 **ii** 0.5775 **iii** 0.5184

A

acceleration 292, 294, 296, 382
accuracy 180–3, 184
acute angle 44, 60, 132
addition 8–11
  of algebraic fractions 36
  of algebraic terms 22, 24
  brackets and 120
  of decimals 8, 10
  of denominators 34
  of fractions 92
  of indices 26, 28
  of negative number 8
  of numerators 34
  of powers 26, 28
  of probabilities 430, 432
  in standard form 358
  of vectors 412, 414
  using compensation 8
  using mental methods 8, 10
  using partitioning 8
  using written method 8, 10
adjacent side 400, 416
algebraic conventions 22
algebraic fractions 34–7
alternate angle 44, 46, 60
alternate segment 228
angle 44–7, 148
  acute 44, 60, 132
  alternate 44, 46, 60
  corresponding 44, 46, 60
  exterior 47, 56, 58, 60
  interior 44, 56, 58, 60
  measurement of 132–5
  obtuse 44, 60, 132
  of pie chart 74, 76
  at a point 44, 46, 132
  in polygons 56–9
  in quadrilateral 48, 50
  reflex 44, 60, 132
  right 44, 60
  on a straight line 44, 46
  supplementary 44
  in a triangle 48, 50
  vertically opposite 44, 46
angle bisector 232, 234
angle of rotation 140, 142
annulus 223
anticlockwise 140
approximate solutions 202–5
approximation 172–5, 184, 270, 354, 362
arc 222, 230, 236
arc length 222, 224
area 148
  of circle 218, 220, 356
  of parallelogram 136, 138
  of rectangle 136, 138
  of sector 222, 224
  of semi-circle 218, 220
  of trapezium 136, 138
  of triangle 136, 138, 220, 408, 410
  of 2D shape 136–9
  under graph 380–3
area ratio 468, 470
arithmetic sequence 454, 456, 458
ascending order 94
asymptote 370, 372, 388
averages 78–81, 328–31

B

bar chart 74, 76, 82, 324, 326
bar-line chart 74, 82
base 26, 28, 38, 136, 362
bearing, (three-figure) 46, 60, 132, 134, 406, 410
bias 66, 68, 82, 160, 166
BIDMAS 12, 14, 16, 22, 24, 172
bisection 232, 236
box plot 332–5, 344
bracket key 176

brackets 26, 30
  double, expansion of 122
  double, factorisation into 122
  expansion of 120–3
  multiplication and 30, 32
  surds within 272

C

calculation
  exact 354–7, 362
  with roots and indices 350–3
calculator methods 176–9
cancelling 34, 36, 88, 92, 100, 120
capacity 180, 182, 184
category 74
centimetres 132
centre of enlargement 144, 146, 148
centre of rotation 140, 142, 148
certainty 154, 156, 166
chord 228, 236, 386
circle 218–25, 236
  area of 356
  equation of 384–7
  theorems 226–31
circumference 218, 220, 230, 236, 312
class interval 324, 328, 330
clockwise 140
coefficient 26, 38, 120
collinear points 414, 416
common factor 30, 43, 88, 100, 120
compensation 8, 16
complement of set 422, 438
complementary event 164
completing the square 194, 196, 210
composite function 112, 114, 124
compound interest 254, 480, 482
compound measures 180, 182
compound units 464–7
conditional probability 434–7, 438
cone 316
  volume of 312, 356
congruence 52–5, 60, 140, 142
congruent trapeziums 136
congruent triangle 52, 54
consecutive numbers 118
constant 122
constant of proportionality 472, 474
construct 232, 236
construction 232–5, 236
continuous data 180, 324, 326, 340, 344
conversion between units 468–71
coordinates 338
correlation 336–9, 344
  of variables 336, 338
corresponding angle 44, 46, 60
cosine graph 372
cosine ratio (cos) 400, 402, 416
cosine rule 404, 406, 408, 410
counter-example 118, 124
cross-multiplication 190, 192
cross-section 316
cube, volume of 302
cube numbers 454, 456
cube root 266, 274
cubic function 368–71, 388
cumulative frequency graph 332–5, 344
curved surface area 312
cyclic quadrilateral 226, 230
cylinder 306, 316
  surface area of 306, 308
  volume of 306, 356

D

data 66, 68
  continuous 180, 324, 340, 344
  discrete 324, 326, 340, 344
  grouping 324, 326, 330
  ordered 70
  organisation of 70–3

data collection sheet 70
data representation 74–7
decagon 56
decay 480–3
decimal 100
  conversion to fraction 96, 98
  equivalent 88, 90
  proportion and 242, 244
  recurring 88, 96, 98, 100, 354
  terminating 96, 98, 100, 354
decimal approximation 270, 354
decimal places (dp) 4, 16
decreasing trend 376
degrees 44
denominator 34, 96, 100, 120, 190
  rationalisation of 270, 272
density 180, 182, 184, 464, 466, 484
derivation of formula 108
descending order 94
diagonal line 284
diameter 218, 236
difference of two squares (DOTS) 120
direct proportion 472–5
directed number 16
discrete data 324, 326, 340, 344
disjointed sets 438
distance–time graph 292, 294
division 12–15
  of algebraic terms 22, 24
  by fractions 92, 94
  of inequality 206
  by negative number 12
  by power of 10 4, 6
  of powers 26, 28, 266, 268
  of scale 246, 248
  in standard form 358
domain 112, 114, 124

E

edge (solid) 302, 316
element of set 422, 438
elevation 302, 304, 316
elimination of variables 198, 200, 210
empty set 422, 424, 438
enlargement 52, 134, 144, 146, 148
equal likelihood 166
equal to 6
equation 22, 116, 118, 124, 190
  of circle 384–7
  of straight line 280–3, 296
equidistance 140, 232, 234
equilateral triangle 48, 54
equivalences in algebra 116–19
equivalent decimal 88, 90
equivalent fraction 34, 88, 90, 92
error interval 184
estimated mean 328, 330
estimates of averages 328
estimation 8, 10, 12, 14, 16, 172–5, 184
even chance 156
event 154, 158, 166
exact calculations 354–7, 362
exhaustive set 162, 164
expansion 30–3, 38, 120–1, 124
  of brackets 120, 190, 192
expected frequency 154, 156, 166
experimental probability 166
exponential decay 480
exponential function 372–5, 378, 388
exponential growth 480
expression 22, 38, 116, 118
  simplification of 22–5, 28
exterior angle 47, 56, 58, 60
extrapolation 338

F

face 302, 316
factor 34, 262–5, 274
factor tree 262

factorisation 30–3, 38, 108, 120–3, 124, 194, 196
Fibonacci-type sequence 454, 456, 458
FOIL acronym 122
force 464
formula 22, 108–22, 124, 446
formula triangle 464, 466
fractional index 350, 352, 362
fractions 88–91, 100
    addition of 92, 94
    algebraic 34–5
    calculations with 92–5
    cancelling 92
    conversion of decimal to 96, 98
    conversion of percentage to 96, 98
    conversion to equivalent 34
    division by 92, 94
    equations involving 190
    equivalent 88, 90, 92
    improper 92, 94, 100
    multiplication of 88, 90, 92, 94
    probability as 430
    proportion and 242, 244
    reciprocal of 92, 94
    subtraction of 92, 94
frequency
    expected 154, 156, 166
    relative 154, 156, 160, 166
frequency density 326, 344
frequency diagram 324–7
frequency table 70, 72
frequency tree 430, 438
frustum 314, 316
function 112–15, 116, 124

G

general term *see n*th term
geometric sequence 454, 456, 458
gradient 280, 282, 292, 294, 296, 378, 380–3
    at a point 476, 478
    between two points 476
    of line segment 476
graph
    area under 380–3
    cosine 372, 374
    cubic 368
    cumulative frequency 332–5, 348
    distance–time 292
    exponential 372, 374, 378
    inequalities on 206, 208
    kinematic 292–5
    line 340, 342
    of linear equation 284
    periodic 372
    quadratic 284, 286, 288, 290
    real-life 376–9
    reciprocal 368, 370, 472, 474
    S-shaped 368
    scatter 336–9, 344
    sine 372, 374
    speed–time 380, 382
    tangent 372, 374
    time series 340, 342, 344
    velocity–time 292
greater than 4, 6, 206, 208
greater than or equal to 4, 6, 206, 208
grouped data 324, 326, 330
grouped frequency table 324, 328
grouping data 324
growth 480–3

H

heptagon 56
hexagon 56, 58
    regular 56
highest common factor 30, 262, 264, 274
histogram 324, 326, 344
horizontal line 284
hypotenuse 52, 396, 398, 400, 416

I

identity 116, 118, 124
image 148
implied accuracy 180, 184
impossibility 154, 156, 166
improper fraction 92, 94, 100
increasing trend 376
independent event 432, 438
index (indices) 26–9, 38, 362, 372
    calculations with 350–3
    negative 26
    positive 26
    see also power
index laws 26, 28, 38, 350
index notation 266
indices *see* index (indices)
inequality 118, 206–9, 210
input 112, 114
integer 206
interest 250, 254
interior angle 44, 56, 58, 60
interpolation 338
interquartile range (IQR) 78, 80, 82, 334
intersection 44, 200, 264, 422, 424, 438
invariance 148
inverse function 112, 114, 404
inverse operations 110
inverse proportion 472–5, 484
irrational numbers 270, 272, 354, 356
isosceles trapezium 48
isosceles triangle 48, 54
iteration 202, 204, 210, 482

K

kilometres 132
kinematic graph 292–5
kinematics 296
kite 48

L

least common multiple (LCM) 262, 264, 274
length 148, 180, 182, 184
    measurement of 132–5
length ratio 468, 470
less than 4, 6, 206, 208
less than or equal to 4, 6, 206, 208
life skills
    business plan 104–5
    getting ready 320–1
    launch party 442–3
    starting the business 214–15
like terms 22, 38, 108, 122
likelihood 154
line 44–7
    degrees on 44, 46, 48
    equation of straight 280–3, 284, 286, 296
    horizontal 284
    parallel 44, 46, 280, 282, 414
    perpendicular 44, 280, 282
    of symmetry 288, 290
    vertical 284
line graph 340, 342
line of best fit 336, 338, 344
line of symmetry 56
linear equation 190–3, 198, 284
linear function 284–7
linear relationship 336
linear sequence 36, 446–7, 458
locus (loci) 232–5, 236, 384
lower bound 180, 182, 184, 332
lower quartile 78, 80, 82
lowest common denominator 92

M

map scale 246, 248
mapping diagrams 112, 114
mass 180, 182, 184
maximum 288, 296
mean 78, 80, 82, 330
    estimated 328, 330

measurement
    of angles 132–3
    of lengths 132–3
measures 180–3
median 78, 80, 82, 330, 334
member of set 424, 438
mental methods
    of addition 8
    of percentages 88
    of subtraction 8
meters 132
metric units 180, 468
millimetres 132
minimum 288, 296
mirror line 140, 142, 148
mixed number 10, 92, 94
modal class 328, 330
mode 78, 80, 82, 330
multiple 262–5, 274, 416
multiplication 12–15
    of algebraic terms 22, 24
    brackets and 30, 32, 120
    double brackets 122
    expansion of 30, 32
    of fractions 88, 90, 92
    of indices 26, 28, 266, 268
    of inequality 206
    by negative number 12
    by power of 10 4, 6
    of powers 26, 28, 266, 268
    of probabilities 430, 432
    of scale 246
    in standard form 358
    of vectors 412
multiplication pyramid 36
multiplier 172, 480
mutually exclusive events 162–5

N

negative index 350, 352, 362
negative number 16
    addition 8, 10
    division 12
    multiplication 12
    subtraction 8, 10
net 302, 304, 316
nonagon 56
not equal to 6
*n*th term 446, 448, 450, 452, 458
number value 4–5
numerator 34, 100, 120

O

object 148
obtuse angle 44, 60, 132
octagon 56
operation 12, 16
opposite side 400, 416
order of operations 12, 14, 16, 176, 178
order of rotational symmetry 56
ordered data 70
outcome 158, 166
outlier 78, 336, 338
output 112, 114

P

parabola 288, 296
    line of symmetry 288, 290
parallel lines 44, 280, 282, 414
parallelogram 48
    area of 136, 138
partitioning 8, 16
pentagon 56
    regular 56
percentage 88–91, 100, 254
    conversion to fraction 96, 98
    proportion and 242, 244
    reverse 252, 254
percentage change 250–3
percentage decrease 250, 252, 254

percentage increase 250, 252, 254
perimeter 28, 136, 138, 148
    of semicircle 218, 220
periodic function 388
periodic graph 372
perpendicular bisector 232, 234, 236, 282
perpendicular height 136
perpendicular line 44, 228, 280, 282
pi 218, 356
pie chart 74, 76, 82
place value 4–7, 12, 14, 16
plan 302, 304, 316
polygon 56, 60
    angles in 56–9
    regular 56, 58
population 66
position 458
position-to-term 450, 452
possibility spaces 426–9
power 26, 28, 38, 110, 266–9, 362
    estimating 172
power of a power 26, 28, 266, 268
pressure 464, 466, 484
prime factor 262, 264
prime factor decomposition 262, 264, 274
prime number 262, 274
prism 316
    triangular 304, 306
    volume of 302, 304, 306, 308–9
probability
    conditional 434–7, 438
    experimental 154–7, 158, 166
    as fraction 430
    theoretical 158–61, 166
    Venn diagrams and 422
probability experiments 154–7
product of factors 262
product of variables 472
proof 124
proportion 74, 242–5, 254, 484
protractor 132, 134, 232
pyramid 316
    square-based 310
    volume of 310, 312, 314
Pythagoras' theorem 384, 396–9, 416
    trigonometric problems 408–11
    vectors 412

Q

quadratic equation 194–7, 198
quadratic expression 122, 124
quadratic formula 194, 196, 210
quadratic function 284–7, 296, 388
    properties 288–91
quadratic sequence 450–3, 454, 456, 458
quadrilateral 48–51, 56, 60
    angles in 48, 50
    cyclic 226, 230
quadrilaterals 48–9
questionnaire 66, 68

R

radius (radii) 218, 226, 228, 236, 312, 384, 386
random sample 66, 68
range 78, 80, 82, 112, 114, 124, 330, 334
rate 464, 466, 484
rate of change 382, 476–9
rate of flow 464
rate of pay 464
ratio 246–9, 254, 468, 470
    simplified 246, 248
rational number 270, 272, 354
rationalisation 270, 274
real-life graph 376–9
rearranging formula 108, 110, 124
reciprocal 92, 100, 350, 352, 362
reciprocal function 368–71, 388
reciprocal graph 472, 474

rectangle 48, 50
    area of 136, 138
recurrence formula 202
recurring decimal 96, 98, 100, 354
recurring fraction 96
recursive process 210
reduction 134
reflection 140, 142, 148
reflex angle 44, 60, 132
regular polygon 56
regular shape 56
regular tetrahedron 310
relative frequency 154, 156, 160, 166, 430
reliability 160
resultant 412, 416
reverse percentage 252, 254
rhombus 48, 50
right angle 44, 60
right-angled triangle 48, 52, 396, 398, 400,
    402, 408
root(s) 266–9, 296, 362
    calculations with 350–3
root of function 288, 290
rotation 140, 142, 148
rounding 4–7, 16, 172, 218
ruler 132, 134, 232

S

S-shaped graph 368
sample 66
    random 66, 68
    stratified 68
sample space 426
sampling 66–9
scalar 416
scale 46, 246, 248, 254, 378
scale diagram/drawing 134, 246, 248, 254
scale factor 52, 60, 134, 144, 146, 148, 246
scalene triangle 48
scatter graph 336–9, 344
sector 76, 222, 224, 236
segment 228, 236
    alternate 228
self-similar shape 54
semicircle
    area of 218, 220
    perimeter of 218, 220
sequence 458
set 438, 422–5
significant figures (sf) 4, 16
similar shapes 468, 484
similarity 52–5, 60
simple interest 250, 254, 480
simplest form 246
simplification
    of algebraic fractions 36
    of expressions 22–5
    with powers 26
    of ratio 254
simultaneous equations 198–201, 210
sine graph 372
sine ratio (sin) 400, 402, 416
sine rule 404, 406, 408, 410
solution
    approximate 202–5
    of equations 190, 192
speed 180, 182, 184, 296, 464, 466, 484
speed–time graph 380, 382
sphere 316
    surface area 306, 308
    volume of 306, 308, 356
spread 78–81, 328–31
square 48, 50
square-based pyramid 310
square numbers 454, 456
square root 266, 270, 272, 274
standard form 172, 358–61, 362
stem-and-leaf diagram 70, 72
stratified sample 68

subject of formula 108, 110, 124
substitution 22, 24, 38, 210
    in simultaneous equations 198, 200
subtraction 8–11
    of algebraic fractions 36
    of algebraic terms 22, 24
    of decimals 8, 10
    of fractions 92
    of indices 26, 28, 266, 268
    of negative number 8
    of powers 26, 28
    of vectors 412, 414
    in standard form 358
    using compensation 8
    using mental methods 8, 10
    using partitioning 8
    using written method 8, 10
supplementary angle 44
surd 270–3, 274, 354, 356, 362
surd form 270, 272, 354, 362
surface area 302, 304, 310–15, 316
    curved, of cone 312
    of cone 312, 314
    of cylinder 306, 308
    of sphere 306, 308
survey 66, 68

T

tangent 228, 236, 380, 382, 386
tangent graph 372
tangent ratio (tan) 400, 402, 416
tangent to a curve 388, 476, 478
term 38, 116, 458
terminating decimal 96, 98, 100, 354
term-to-term rule 450, 452, 458
tetrahedron, regular 310
theoretical outcome 154
theoretical probability 158–61, 166
theta 222
three-dimensional shapes 302–5
three-figure bearing 60
time series graph 340–3, 344
transformation 140–7, 148
translation 140, 142, 148, 414
trapezium 48, 136
    area of 136, 138
    isosceles 48
tree diagram 430–3, 438
trend (time series) 340, 342, 344, 376
trial 158, 166
triangle 48–51, 56
    angles in 48, 50
    area of 136, 138, 220, 408, 410
    ASA 52, 232
    congruent 52, 54
    construction of 232
    equilateral 48, 54
    isosceles 48, 54
    RHS 52, 232
    right-angled 48, 52, 396, 398, 400, 402,
        408, 410
    SAS 52, 232
    scalene 48
    SSS 52, 232
triangular number 454, 456, 458
triangular prism 304, 306
trigonometric function 372–5, 388
trigonometry 400–7
    problems 408–11
turn 140, 142
turning point 288, 290, 296
two-dimensional shape, area of 136–9
two-way table 70, 72, 82

U

union of sets 422, 424, 438
units, converting between 468–71
universal set 422, 438
unknown 24, 118, 190, 192

unlikelihood 154
upper bound 180, 182, 184, 332
upper quartile 78, 80, 82

V
variable 38, 198, 336, 338
vector 412–15, 416, 442
velocity 292, 294
velocity–time graph 292, 294
Venn diagram 262, 264, 422, 424, 434, 436, 438
vertex (vertices) 56, 302, 316

vertical line 284
vertically opposite angles 44, 46
vinculum 98
volume 184, 310–15, 316
    metric units of 468
    of cone 312, 356
    of cube 302
    of cylinder 306, 356
    of frustum 314
    of prism 302, 304, 306, 308–9

of pyramid 310, 312, 314
of sphere 306, 308, 356
volume ratio 468, 470

X
x-coordinate 284
x-intercept 288

Y
y-coordinate 284
y-intercept 280, 282, 286, 296